Mathematica by Example

Mathematica by Example

Martha L. Abell
Department of Mathematics and Computer Science
Georgia Southern University
Statesboro, Georgia

James P. Braselton
Department of Mathematics and Computer Science
Georgia Southern University
Statesboro, Georgia

ACADEMIC PRESS, INC.
Harcourt Brace Jovanovich, Publishers
Boston San Diego New York
London Sydney Tokyo Toronto

This book is printed on acid-free paper. ♾

ACADEMIC PRESS, INC.
1250 Sixth Avenue, San Diego, CA 92101

United Kingdom Edition published by
ACADEMIC PRESS LIMITED
24–28 Oval Road, London NW1 7DX

LCCCN: 91-58715
ISBN: 0-12-041540-2

Printed in the United States of America
92 93 94 95 9 8 7 6 5 4 3 2 1

CONTENTS

Sections that discuss features of Version 1.2 will begin with symbols like ■ ▨ □;
unless otherwise noted, these commands are supported under Version 2.0.

Sections that discuss the features of Version 2.0 will begin with symbols like ● ◍ ○.
These sections are NOT pertinent to Version 1.2.

PREFACE

Mathematica by Example is intended to bridge the gap which has existed between the very elementary handbooks available on *Mathematica* and those reference books written for the more advanced *Mathematica* users. This book is an extension of a manuscript which was developed to quickly introduce enough *Mathematica* commands to a group of students at Georgia Southern University that they could apply *Mathematica* towards the solution of nonlinear ordinary differential equations. In addition to these most basic commands, these students were exposed to the vast uses of lists in *Mathematica*. Having worked through this material, these students were successfully able to take advantage of the capabilities of *Mathematica* in solving problems of interest to our class.

Mathematica by Example is an appropriate reference book for readers of all levels of *Mathematica* experience. It introduces the very basic commands and includes examples of applications of these commands. It also includes commands useful in more advanced areas such as ordinary and partial differential equations. In all cases, however, examples follow the introduction of new commands. Of particular note are the sections covering *Mathematica* Packages (Chapters 7, 8, and 9), because the commands covered in these chapters are absent from most *Mathematica* reference books. The material covered in this book applies to all versions of *Mathematica* as well with special notes concerning those commands available only in Version 2.0. Other differences in the various versions of *Mathematica* are also noted.

Of course, appreciation must be expressed to those who assisted in this project. We would like to thank our department head Arthur Sparks for his encouragement and moral support and for being the instigator of the Computer Calculus Project which initiated the idea of writing a book like *Mathematica* by Example. We would also like to thank Prof. William F. Ames for suggesting that we publish our work and for helping us contact the appropriate people at Academic Press. We would like to express appreciation to our editor, Charles B. Glaser, and our production manager, Simone Payment, for providing a pleasant environment in which to work. We would also like to thank our colleagues for taking the time to review our manuscript as it was being prepared for publication. We appreciated their helpful comments. Finally, we would like to thank those close to us for enduring with us the pressures of meeting a deadline and for graciously accepting our demanding work schedules. We certainly could not have completed this task without your care and understanding.

M. L. Abell

J. P. Braselton

Chapter 1
Getting Started

■ *Mathematica*, first released in 1988 by Wolfram Research, Inc., is a system for doing mathematics on a computer. It combines symbolic manipulation, numerical mathematics, outstanding graphics, and a sophisticated programming language. Because of its versatility, *Mathematica* has established itself as the computer algebra system of choice for many computer users. Overall, *Mathematica* is the most powerful and most widely used program of this type. Among the over 100,000 users of *Mathematica*, 28% are engineers, 21% are computer scientists, 20% are physical scientists, 12% are mathematical scientists, and 12% are business, social, and life scientists. Two-thirds of the users are in industry and government with a small (8%) but growing number of student usrs. However, due to its special nature and sophistication, beginning users need to be aware of the special syntax required to make *Mathematica* perform in the way intended.

■ The purpose of this text is to serve as a guide to beginning users of *Mathematica* and users who do not intend to take advantage of the more specialized applications of *Mathematica.* The reader will find that calculations and sequences of calculations most frequently used by beginning users are discussed in detail along with many typical examples. We hope that *Mathematica* by Example will serve as a valuable tool to the beginning user of *Mathematica.*

■ A Note Regarding Different Versions of Mathematica

For the most part, *Mathematica* by Example was created with Version 1.2 of *Mathematica.* With the release of Version 2.0 of *Mathematica*, several commands from earlier versions of *Mathematica* have been made obsolete. In addition, Version 2.0 incorporates many features not available in Version 1.2. *Mathematica* by Example adopts the following conventions:

Sections that discuss features of Version 1.2 will begin with symbols like ■ ▩ □;
unless otherwise noted, these commands are supported under Version 2.0.

Sections that discuss the features of Version 2.0 will begin with symbols like ● ◍ ○.
These sections are NOT pertinent to Version 1.2.

▩ 1.1 Macintosh Basics

Since *Mathematica* by Example was created using Macintosh computers, we will quickly review several of the fundamental Macintosh operations common to all application programs for the Macintosh, in particular to *Mathematica*. However, this book is not meant to be an introduction to the Macintosh and the beginning user completely unfamiliar with the Macintosh operating system should familiarize himself with the Macintosh by completing the **Macintosh Tour** and consulting the **Macintosh Reference**. The material that appears in *Mathematica* by Example should be useful to anyone who uses *Mathematica* in a **windows** environment. Non-Macintosh users may either want to quickly read **Chapter 1** or proceed directly to **Chapter 2**, provided they are familiar with their computer.

After the *Mathematica* program has been properly installed, a user can access *Mathematica* by first clicking twice on the hard disk icon located in the upper right hand corner of the computer screen. The following window will appear:

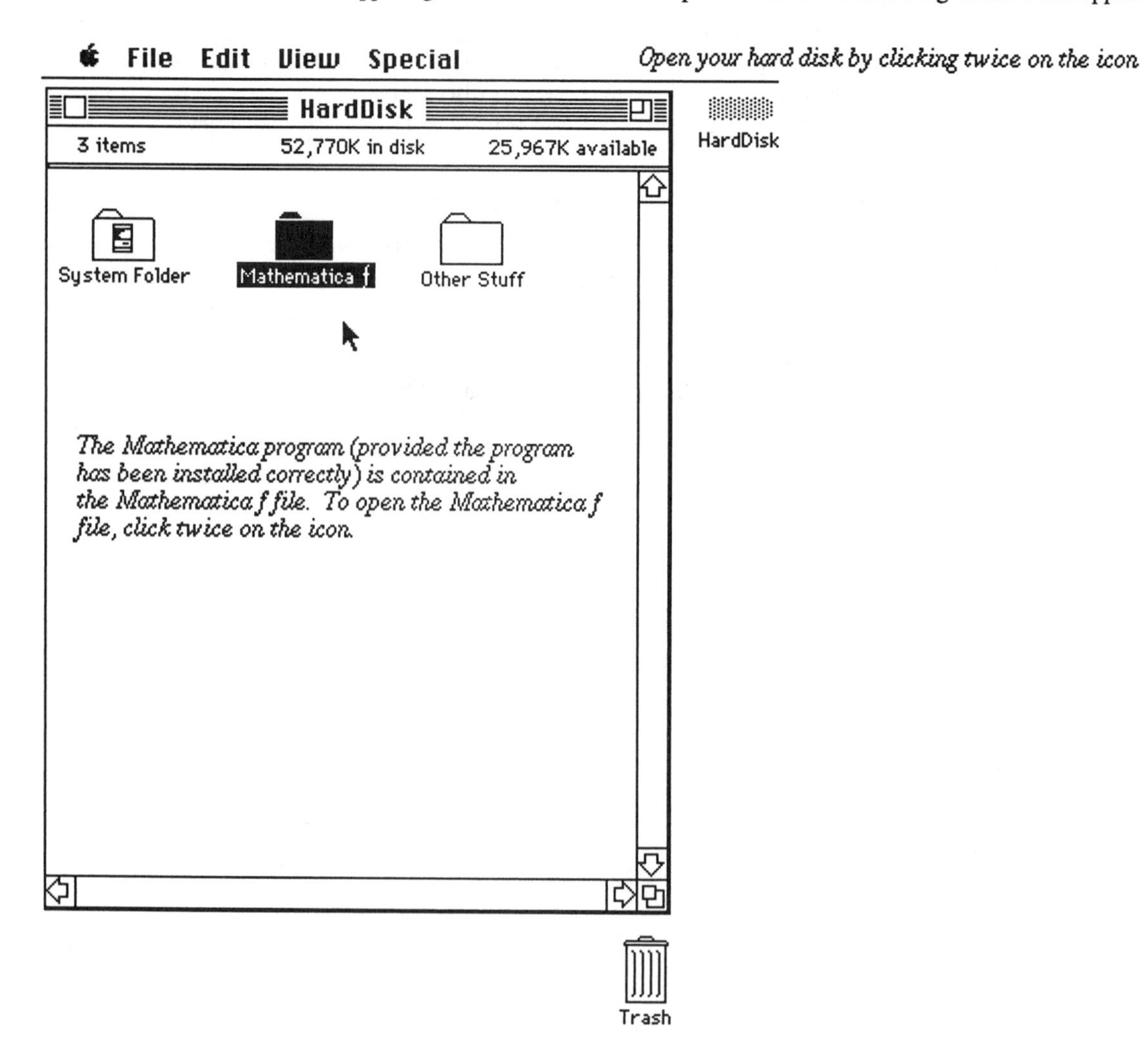

The *Mathematica* f folder can be opened by clicking twice on its icon. After opening the *Mathematica* f folder, start *Mathematica* by double clicking on the icon labeled *Mathematica*. These steps are illustrated below:

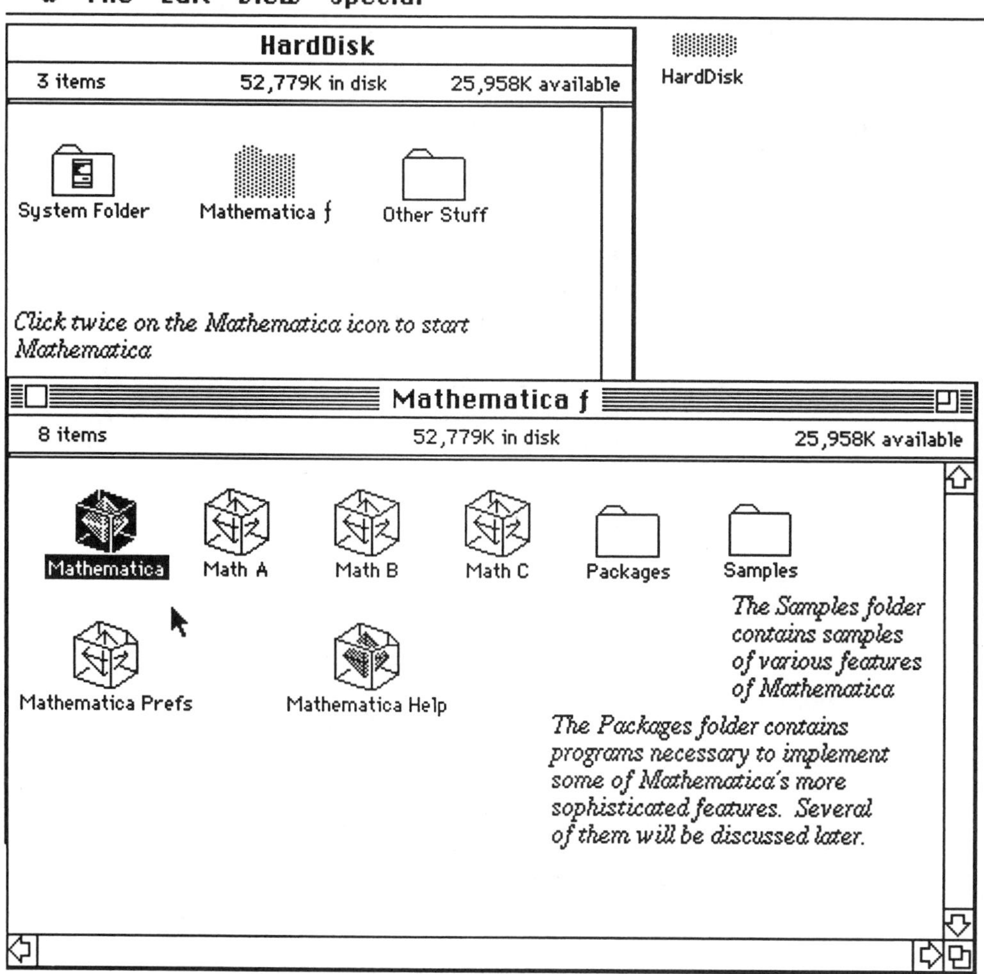

After double-clicking on the *Mathematica* icon, an empty *Mathematica* document appears; the *Mathematica* session can be initiated by typing anything. When you begin typing, *Mathematica* automatically creates an **input cell** for you. If an input cell contains a *Mathematica* command, the command is evaluated by pressing **ENTER** or **Shift-Return**.

In general, the **ENTER** key and **RETURN** key are not the same. The **ENTER** key is used to evaluate *Mathematica* commands; the **RETURN** key gives a new line.

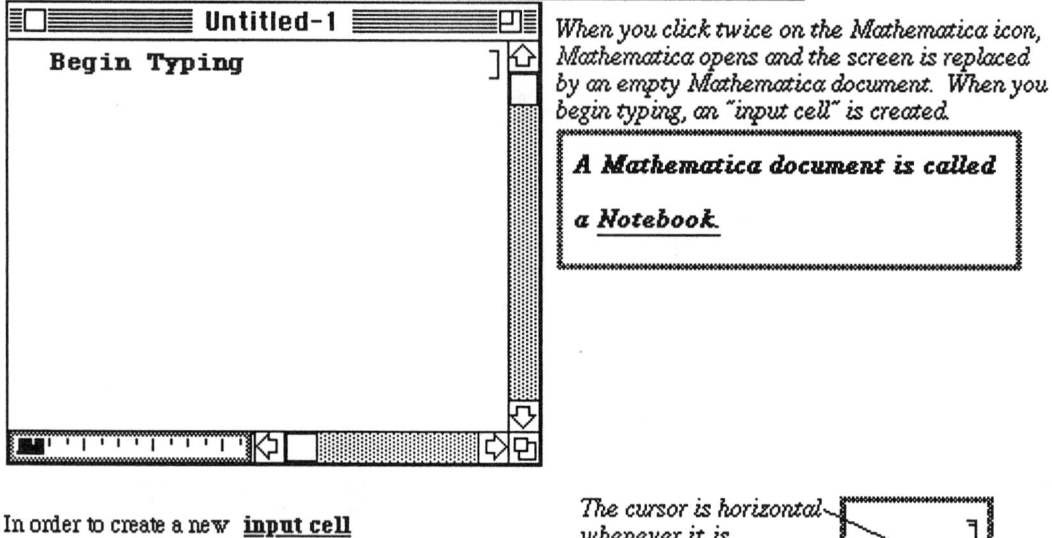

In order to create a new **input cell** move the cursor below the original cell so that the cursor is horizontal. When the cursor is horizontal, click the mouse once:

The cursor is horizontal whenever it is between two cells:

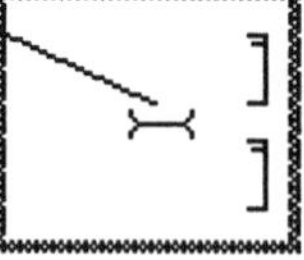

When the cursor is horizontal and the mouse is clicked once, a black line appears across the document window:

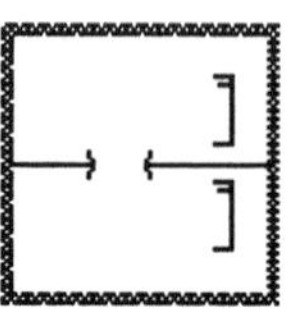

A horizontal black line appears after clicking the horizontal cursor once. Additional typing causes *Mathematica* to replace this line with a new **input cell** containing the most recently typed information.

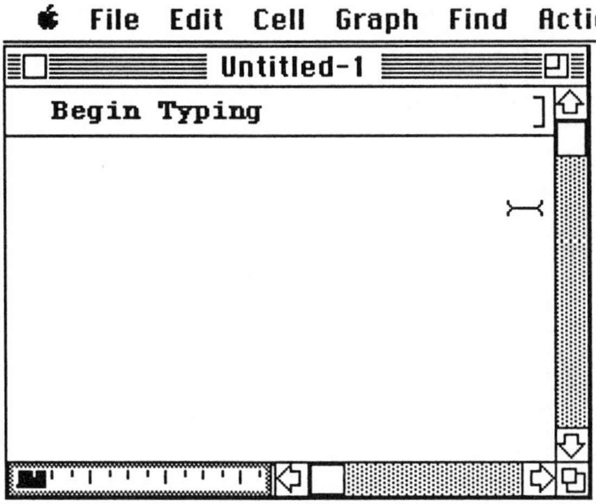

 File Edit Cell Graph Find Action Style Window

Untitled-1

Begin Typing

To create a new "input cell", move the cursor below the existing cell, click once. Notice that a horizontal black line appears.
When you begin typing, Mathematica replaces the black line with a cell to hold your text.

■ 1.2 Introduction to the Basic Types of Cells, Cursor Shapes, and Evaluating Commands

In the following example, **2+3** is a *Mathematica* command. The input cell containing **2+3** can be evaluated by pressing **ENTER** after the command has been typed.

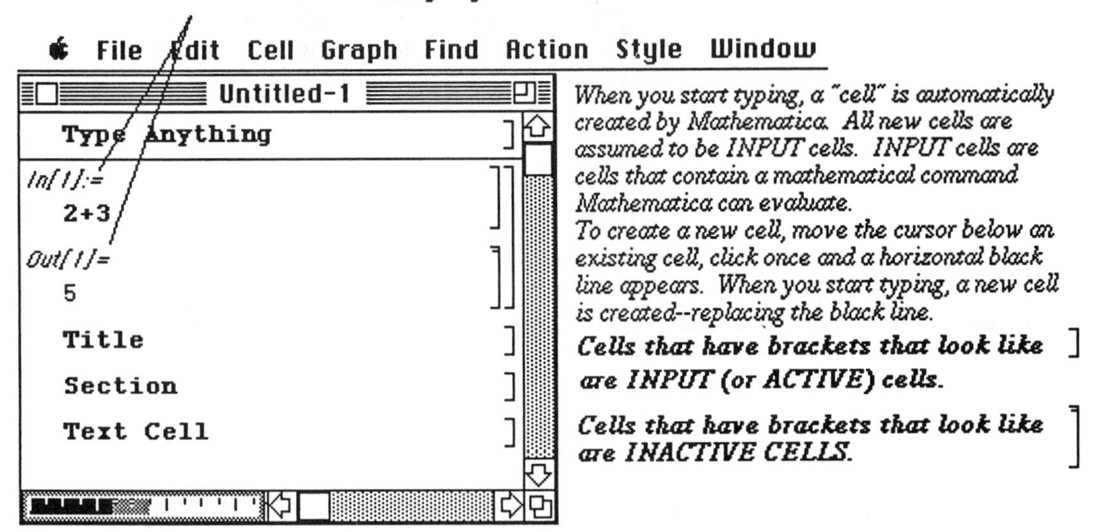

__Inactive cells__ are cells that cannot be evaluated by *Mathematica*. Inactive cells include output cells, graphics cells, and text cells. __Output cells__ are cells that contain the results of calculations performed by *Mathematica*; __graphics cells__ are cells that contain two- or three-dimensional graphics produced by *Mathematica*; and __text cells__ are cells that contain explanations or other written material that cannot be evaluated by *Mathematica.*

To verify that you are able to evaluate input cells correctly, carefully type and **ENTER** each of the following commands:

Notice that **every** *Mathematica* command begins with capital letters and the argument is enclosed by square brackets "[]".

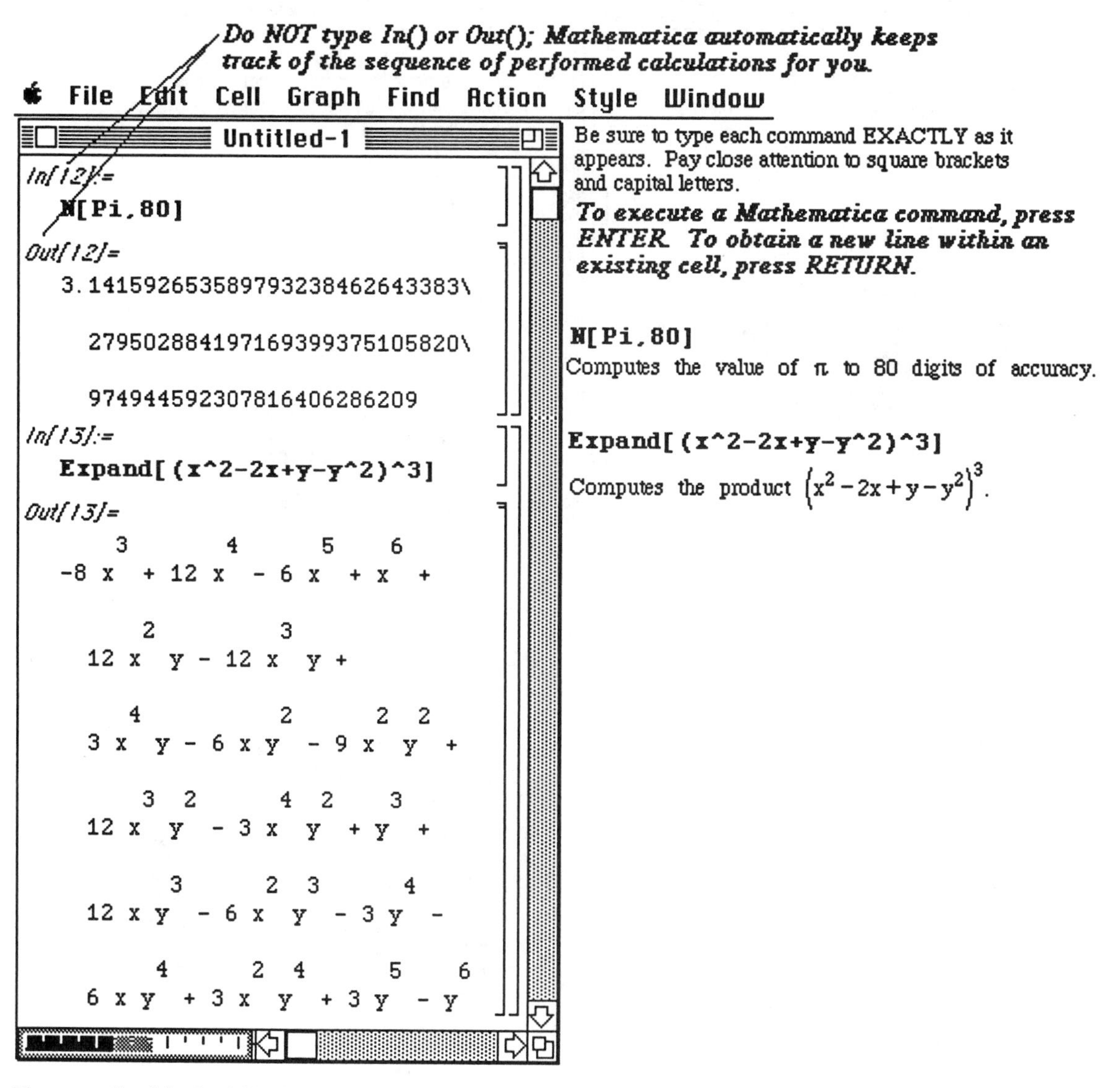

The arrow "->" in the following example is obtained by typing the minus key "-" followed by the greater than key ">".

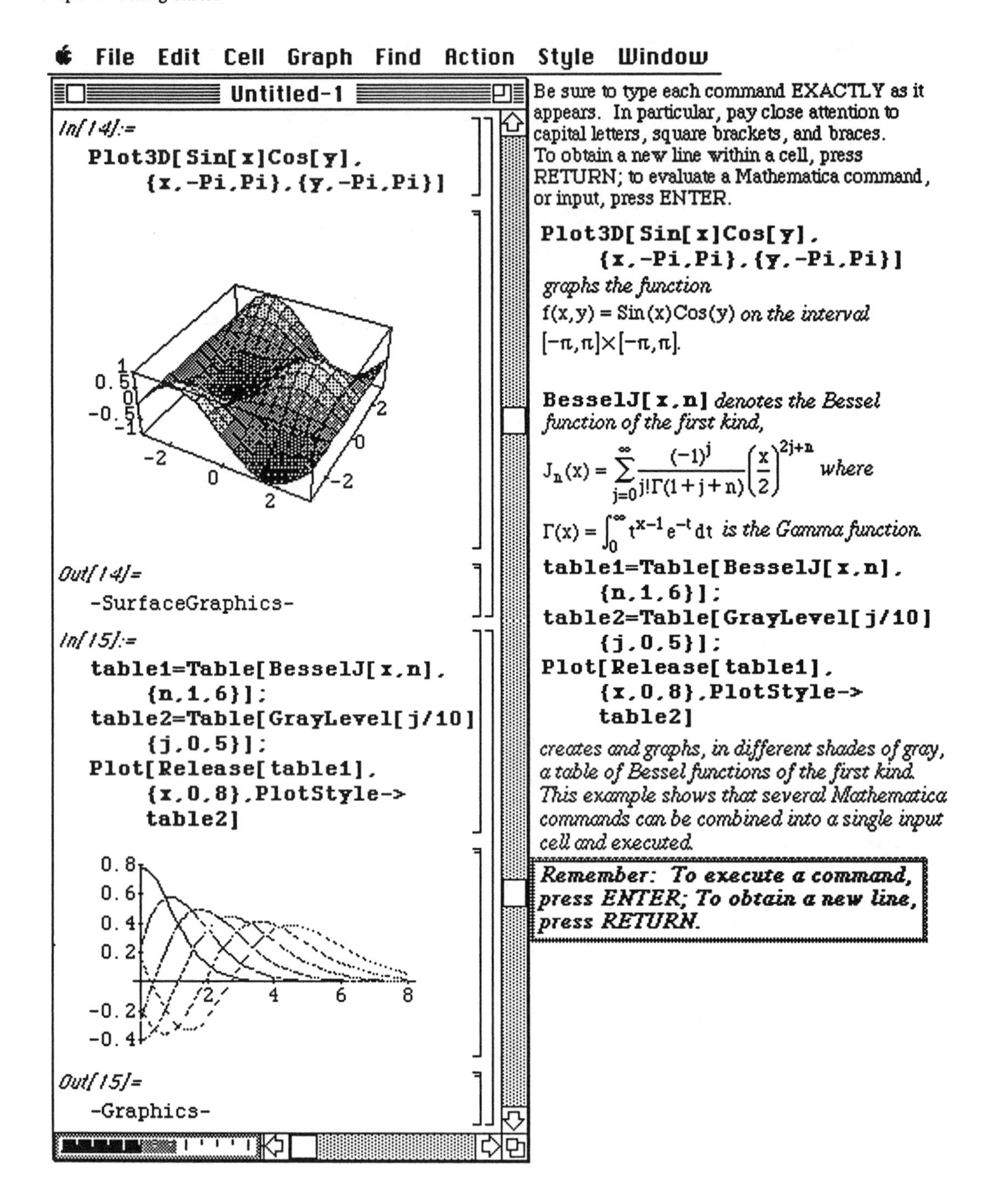

Be sure to type each command EXACTLY as it appears. In particular, pay close attention to capital letters, square brackets, and braces. To obtain a new line within a cell, press RETURN; to evaluate a Mathematica command, or input, press ENTER.

```
Plot3D[Sin[x]Cos[y],
    {x,-Pi,Pi},{y,-Pi,Pi}]
```

graphs the function $f(x,y) = \text{Sin}(x)\text{Cos}(y)$ *on the interval* $[-\pi,\pi]\times[-\pi,\pi]$.

BesselJ[x,n] *denotes the Bessel function of the first kind,*

$$J_n(x) = \sum_{j=0}^{\infty} \frac{(-1)^j}{j!\Gamma(1+j+n)}\left(\frac{x}{2}\right)^{2j+n} \quad \textit{where}$$

$\Gamma(x) = \int_0^{\infty} t^{x-1} e^{-t}\, dt$ *is the Gamma function.*

```
table1=Table[BesselJ[x,n],
    {n,1,6}];
table2=Table[GrayLevel[j/10]
    {j,0,5}];
Plot[Release[table1],
    {x,0,8},PlotStyle->
    table2]
```

creates and graphs, in different shades of gray, a table of Bessel functions of the first kind. This example shows that several Mathematica commands can be combined into a single input cell and executed.

Remember: To execute a command, press ENTER; To obtain a new line, press RETURN.

Often when using a notebook, users need to convert **active** cells to **inactive** cells. This may be accomplished as follows:

□ To convert Active Cells to Inactive Cells:

1) Use the mouse to click on the cell bracket of the cell to be modified. The cell bracket will become highlighted.

2) Go to **Style** and select **Cell Style**.

3) Use the mouse and cursor to choose the desired cell style.

Notice how the cells from the first example have been modified; the Title Cell is highlighted.

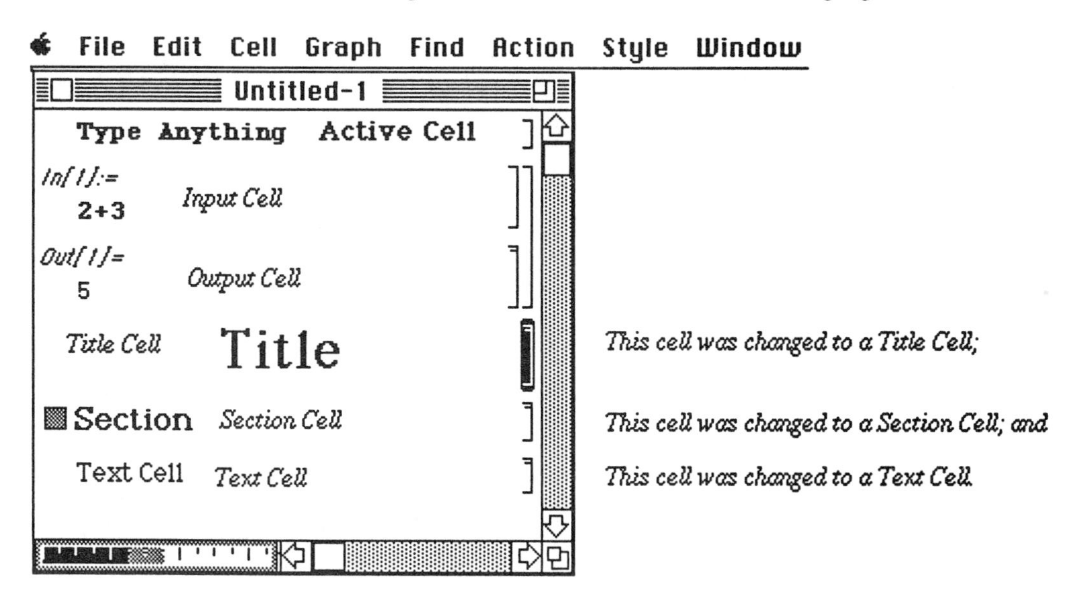

As the cursor is moved within a *Mathematica* notebook, the cursor changes shape. The shape depends on whether (a) the cursor is within an active or inactive cell or (b) the cursor is between two cells.

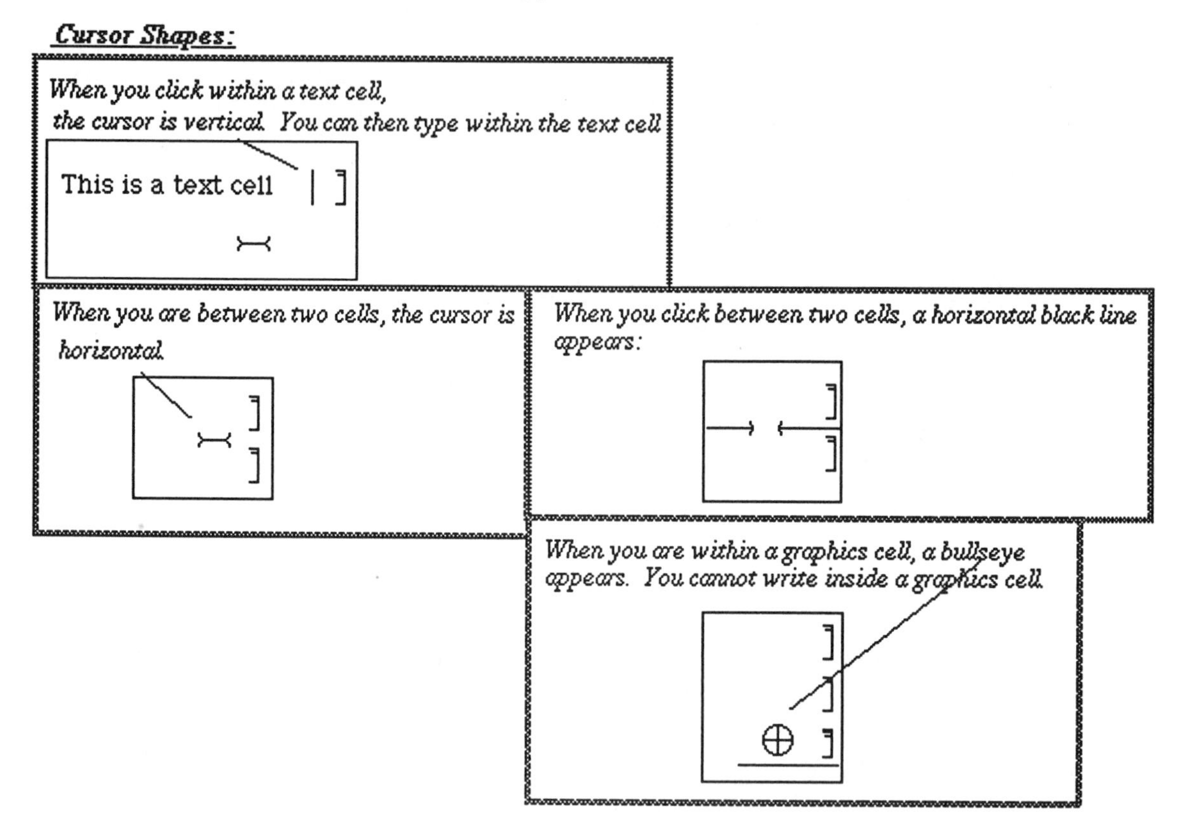

1.3 Introduction to the *Mathematica* Menu

After *Mathematica* has started, the *Mathematica* Menu appears at the top of the screen. The purpose of this section is to introduce the most frequently used operations from the Menu. The Menu will be described in more detail in **Chapter 10.**

- The Menu discussed here is as it appears in Version 1.2. The Version 2.0 Menu is somewhat different from the Version 1.2 Menu. For a discussion of the Version 2.0 Menu, see **Chapter 10.**

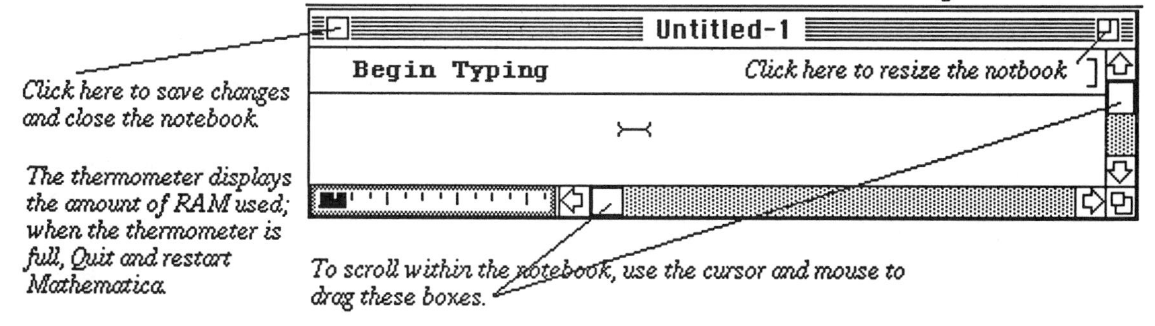

To use the Menu, use the mouse to move the cursor to either **File**, **Edit**, **Cell**, **Graph**, **Find**, **Action**, **Style**, or **Window**. We briefly describe several of the features available under **File**, **Edit**, **Style**, and **Window**.

Use the mouse to move the cursor to FILE in order to create a new Mathematica notebook, Open an existing Mathematica notebook, Save changes to a notebook, Print a notebook, or Quit Mathematica.

File

- **New** *Creates a new Mathematica Notebook*
- **Open...** *Opens an existing Mathematica Notebook*
- **Save** *Saves (but does not close) the open Mathematica Notebook*
- **Save As...**
- **Save As Other...**
- **Show Page Breaks** *Marks where page breaks will occur*
- **Show Keywords**
- **Show Names**
- **Page Setup...** *Use to specify type of printer and paper used*
- **Printing Options...** *Use to modify margins and page numbering*
- **Print...** *Prints the open Mathematica Notebook*
- **Print Selection...** *Prints highlighted cells*
- **Quit** *Saves changes to Mathematica Notebooks then quits Mathematica*

To take advantage of the standard Macintosh editing commands (Cut, Copy, Paste) select EDIT. One can also divide a cell into two cells or merge two (or more) cells of the same type into a single cell. The various Mathematica settings will be discussed later.

Edit

- Undo/Can't Undo
- Cut
- Copy
- Paste
- Clear
- Paste and Discard
- Convert Clipboard
- Select All Cells — *Highlights all cells*
- Nesting
- Divide Cell — *Divides a single cell into two cells*
- Merge Cells — *Merges highlighted cells of the same type into a single cell*
- Settings — *Contains various startup and display settings for Mathematica*

To modify highlighted text or cells, use the mouse to move the cursor to STYLE. Fonts, faces, sizes, color and cell style can be modified.

Style

- Font — *Use to change highlighted text to different fonts*
- Face — *Convert highlighted text to italics, bold, or underline*
- Size — *Change size of highlighted text*
- Color — *Change color of highlighted text*
- Format
- Cell Style — *Change cell style of highlighted cells*
- Uniform Style
- Default Styles
- All Default Styles

WINDOW lists all open notebooks, several options for viewing several open notebooks simultaneously, and contains lists of the various Mathematica defaults and styles which will be discussed in detail later.

Window

- Stack Windows — *Various ways of viewing several open notebooks simultaneously.*
- Tile Windows Wide
- Tile Windows Tall
- Network Window
- Defaults
- Styles
- Clipboard
- (Open Files) — *Mathematica displays a list of the open notebooks*

▩ Preview:

In order for the *Mathematica* user to take full advantage of the capabilities of this software, an understanding of its syntax is imperative. The goal of *Mathematica* by Example is to introduce the reader to the *Mathematica* commands and sequences of commands most frequently used by beginning users. Although all of the rules of *Mathematica* syntax are far too numerous to list here, knowledge of the following five rules equips the beginner with the necessary tools to start using the *Mathematica* program with little trouble.

▩ Remember these Five Basic Rules of *Mathematica* Syntax

■ 1. The ARGUMENTS of functions are given in *square brackets*.

■ 2. The NAMES of built-in functions have their first letters capitalized.

■ 3. Multiplication is represented by a space.

■ 4. Powers are denoted by a ^.

■ 5. If you get no response or an incorrect response, you have entered or executed the command incorrectly.

Chapter 2
Mathematical Operations on Numbers, Expressions and Functions in *Mathematica*

■ **Chapter 2** introduces the essential commands of *Mathematica.* Basic operations on numbers, expressions, and functions are introduced and discussed.

■ Commands introduced and discussed in this chapter from **Version 1.2** are:

Operations:
- **+**
- **-**
- *****
- **^**
- **/**

Constants:
- **E**
- **I**
- **Pi**

Built-In Functions:
- **N[number]**
- **number // N**
- **N[number,digits]**
- **Abs[number]**
- **Sqrt[number]**
- **Exp[number]**
- **Sin[number]**
- **Cos[number]**
- **Tan[number]**
- **ArcCos[number]**
- **ArcSin[number]**
- **ArcTan[number]**
- **Log[b]**
- **Log[a,b]**
- **Mod[a,b]**
- **Prime[n]**

Operations on Equations:
- **Solve**
- **NRoots**
- **FindRoot**

Operations on Expressions and Functions:
- **Simplify[expression]**
- **Factor[expression]**
- **Expand[expression]**
  - **Modulus->p**
- **Together[expression]**
- **Apart[expression]**
- **Numerator[fraction]**
- **Denominator[fraction]**
- **Cancel[expression]**
- **Clear[functions]**
- **Compose[function1, function2, . . .,x]**
- **Nest[function,n,x]**

Evaluation:
- **expression /. variable->number**
- **Out[n]**
- **%**
- **@**

Graphics:

Plot[f[x],{x,a,b},options] or **Plot[{f[x], g[x], . . .},{x,a,b},options]**

Options:
- **PlotStyle**
- **DisplayFunction**
- **AspectRatio**
- **Framed**
- **Ticks**
- **AxesLabel**
- **PlotLabel**
- **GrayLevel[number]**
- **RGBColor[number(1),number(2),number(3)]**
- **PlotRange->{a,b}**
- **Axes->{a,b}**

Show[graphics,options]

■ Commands introduced and discussed in this chapter from **Version 2.0** are:

<u>Operations on Expressions and Functions:</u>

```
Composition[function1,function2,...,functionm][x]
ComplexExpand[expression]
PolynomialMod[poly,p]
```

<u>Graphics:</u>

```
GraphicsArray[{{graph1.1,graph1.2,...,graph1.n},
               {graph2.1,...,graph2.n},...,
               {graphn.1,...graphn.n}}]
Rectangle[{xmin,ymin},{xmax,ymax},graphics]
```

<u>Options:</u>

```
Background
GridLines
Frame
DefaultFont
PlotLabel->FontForm
```

■ <u>Application:</u> Locating intersection points of graphs of functions

2.1 Numerical Calculations and Built-In Functions

Numerical Calculations and Built-In Constants

The basic arithmetic operations (addition, subtraction, multiplication, and division) are performed in the natural way with *Mathematica* . Whenever possible, *Mathematica* gives an exact answer and reduces fractions:

"a plus b" is entered as **a+b**;
"a minus b" is entered as **a-b**;
"a times b" is entered as either **a*b** or **a b** (note the space between a and b); and
"a divided by b" is entered as **a/b**. Executing the command **a/b** results in a reduced fraction.

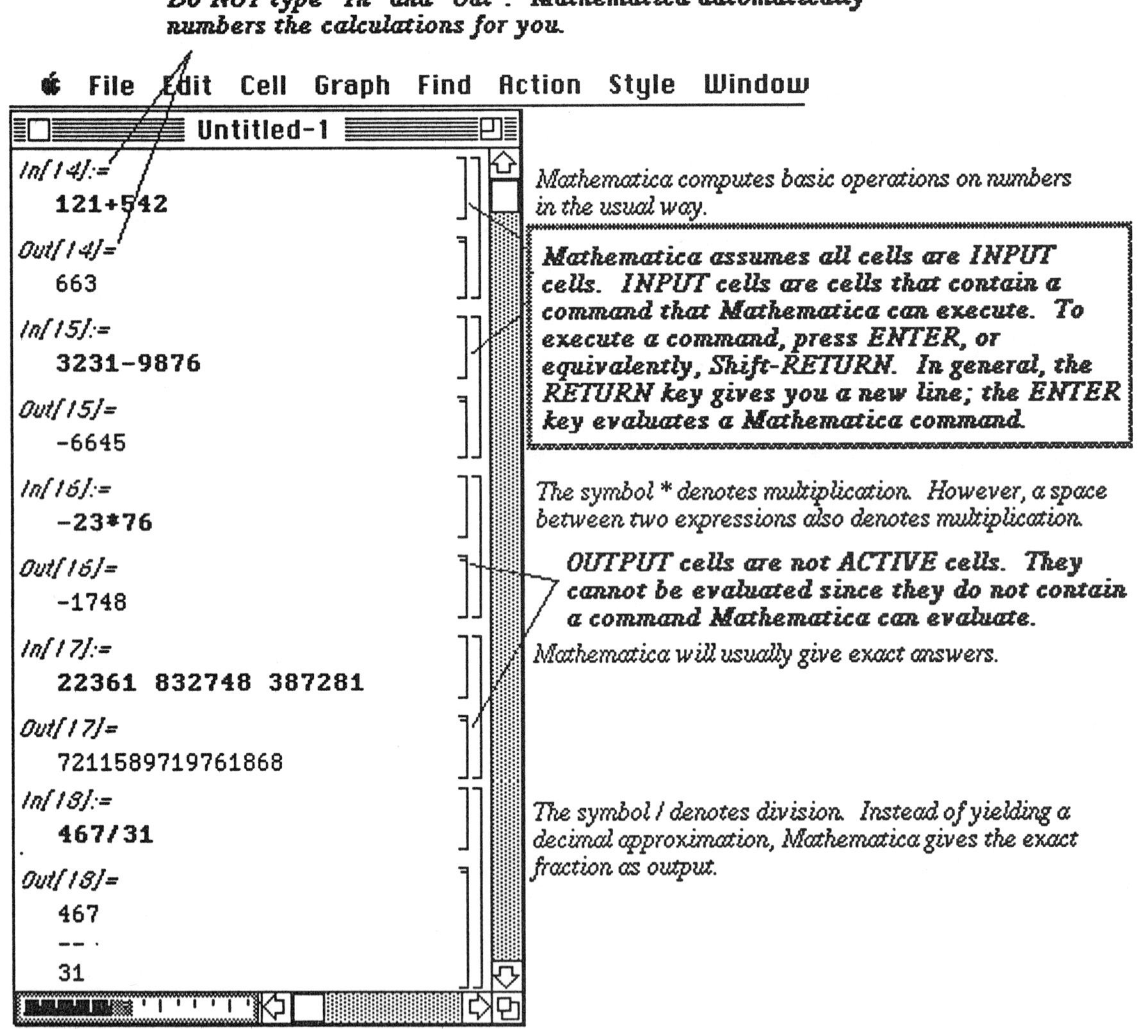

a^b, "a raised to the bth power", is entered as **a^b**.

$\sqrt{a} = a^{1/2}$ can be evaluated as either **a^(1/2)** or **Sqrt [a]**; $\sqrt[3]{a} = a^{1/3}$ can be evaluated by **a^(1/3)**.

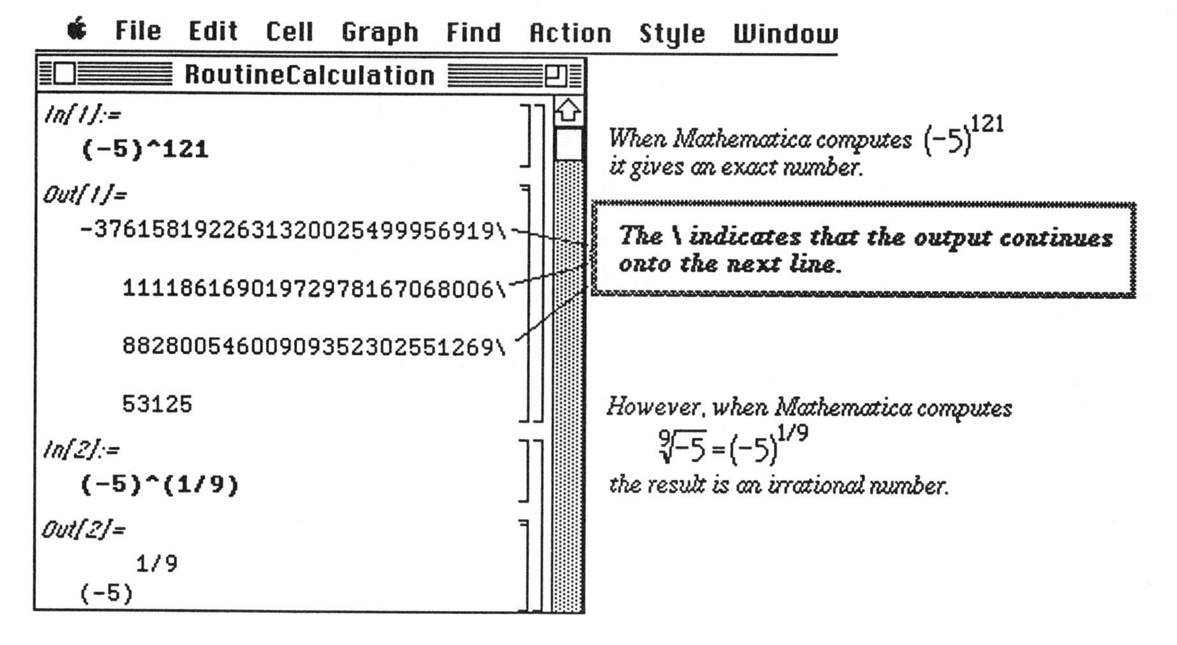

Notice that *Mathematica* gives an exact answer whenever possible. For a variety of reasons, however, numerical approximations of results are either more meaningful or more desirable. The command used to obtain a numerical approximation of the number **a**, is `N[a]` or equivalently
`a // N`. The command to obtain a numerical approximation of **a** to **n** digits of precision is `N[a,n]`.

The exact values computed in the previous window are approximated numerically below:

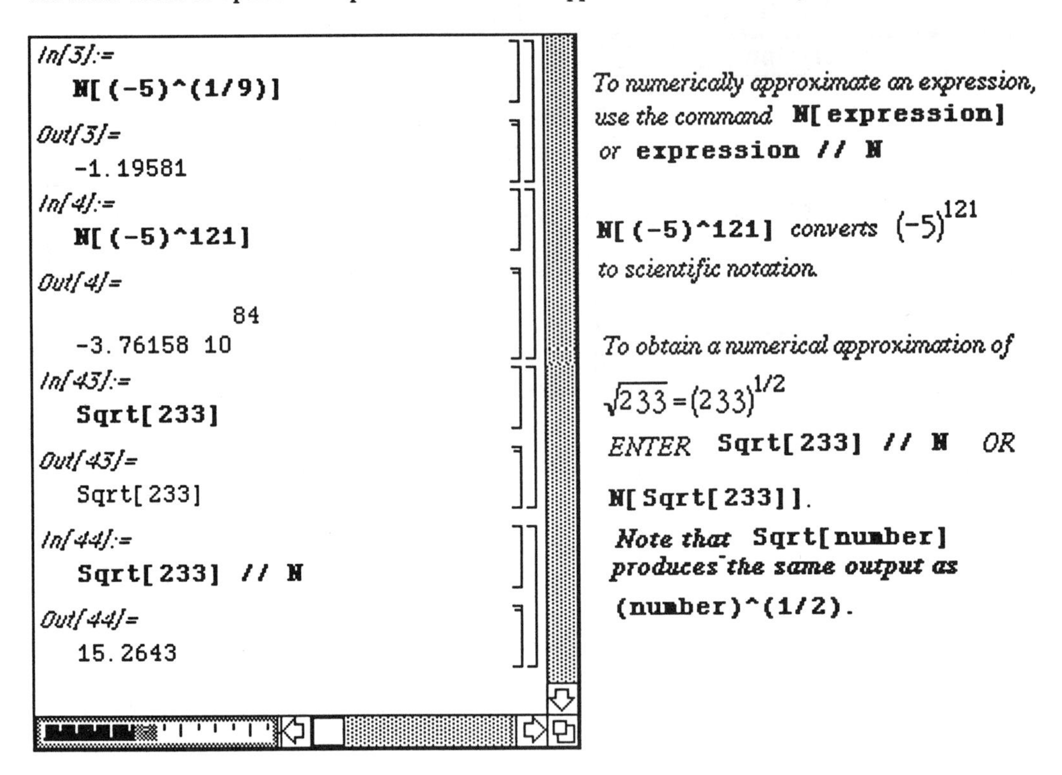

Mathematica has built-in definitions of many commonly used constants. In particular, e is denoted by **E**; π is denoted by **Pi**; and $i=\sqrt{-1}$ is denoted by **I**.

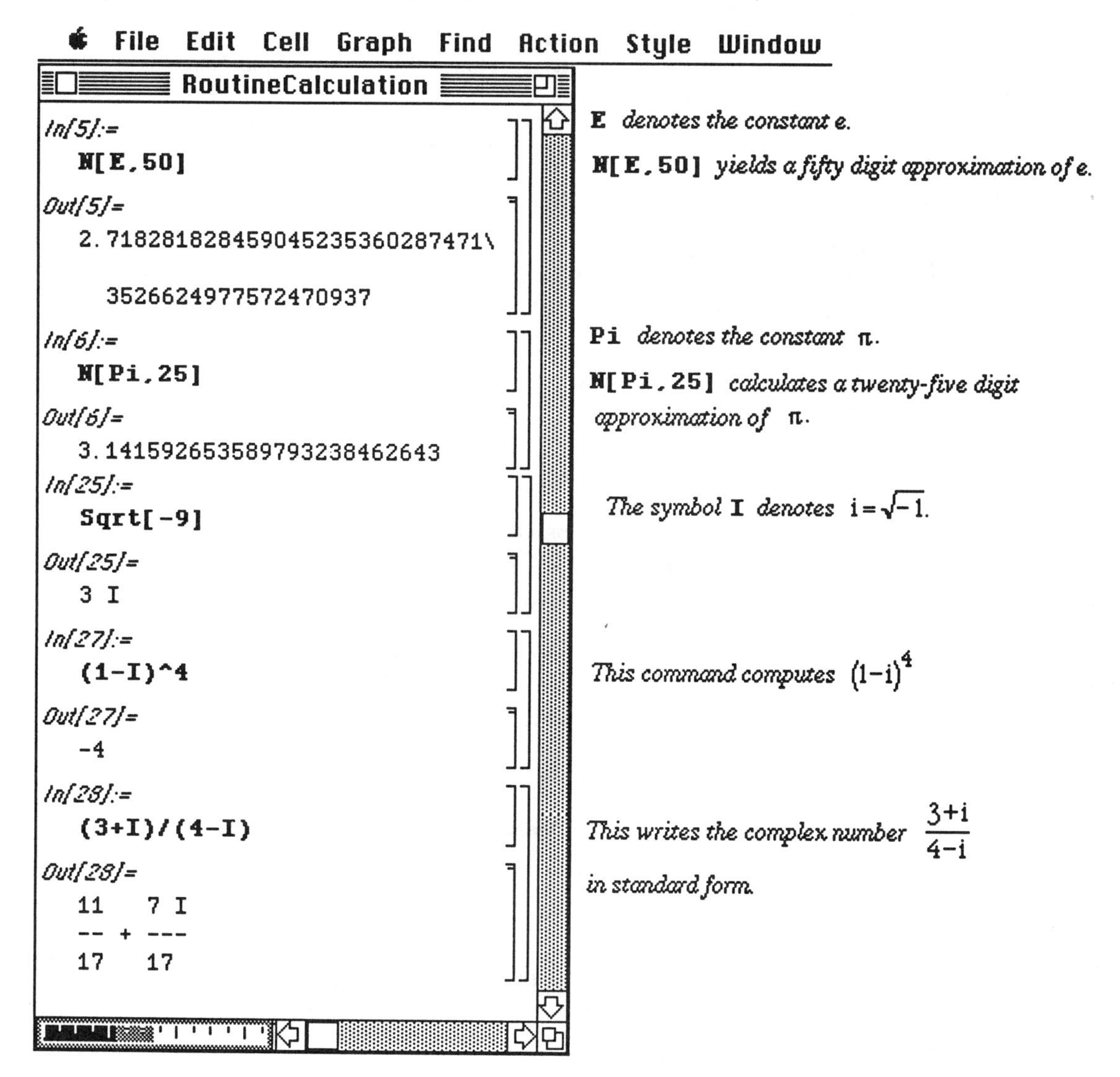

■ Built-In Functions

Mathematica recognizes numerous built-in functions. These include the exponential function, **Exp[x]**; the absolute value function, **Abs[x]**; the trigonometric functions **Sin[x]**, **Cos[x]**, **Tan[x]**, **Sec[x]**, **Csc[x]**, and **Cot[x]**; and the inverse trigonometric functions **ArcCos[x]**, **ArcSin[x]**, **ArcTan[x]**, **ArcSec[x]**, **ArcCsc[x]**, and **ArcCot[x]**. Notice that each of these functions is capitalized and uses square brackets.
(Note that the inverse trigonometric functions include two capital letters!) If both of these requirements are not met, then *Mathematica* will not recognize the built-in function and undesirable results will be obtained.

□ The Absolute Value, Exponential and Logarithmic Functions

Calculations involving the functions **Abs[x]**, **Exp[x]**, and **Log[x]** appear in the following windows. Notice that in order to obtain a numerical value of **Exp[x]**, a numerical approximation must be requested by either the command **N[Exp[x]]** or **Exp[x]//N**. Otherwise, the exact value is given which, in many cases, is not as useful as the numerical approximation.

File Edit Cell Graph Find Action Style Window

RoutineCalculation

```
In[7]:=
  Exp[-5]
Out[7]=
  -5
 E
In[8]:=
  Exp[-5] // N
Out[8]=
  0.00673795
In[2]:=
  Abs[-5]
Out[2]=
  5
In[3]:=
  Abs[14]
Out[3]=
  14
```

To compute $\frac{1}{e^5}=e^{-5}$ *ENTER either* **Exp[-5]** *or equivalently* **E^(-5)**.

Exp[-5] // N *numerically approximates the irrational number* $\frac{1}{e^5}=e^{-5}$; *the identical result would be produced by the commands* **N[Exp[-5]]** *OR* **N[E^(-5)]**.

Abs[-5] *computes* $|-5|$.

Abs[14] *computes* $|14|$.

In addition to real numbers, the function **Abs[x]** can be used to find the absolute value of the complex number **a+bI**, where **Abs[a+bI] = Sqrt[a^2+b^2]**.

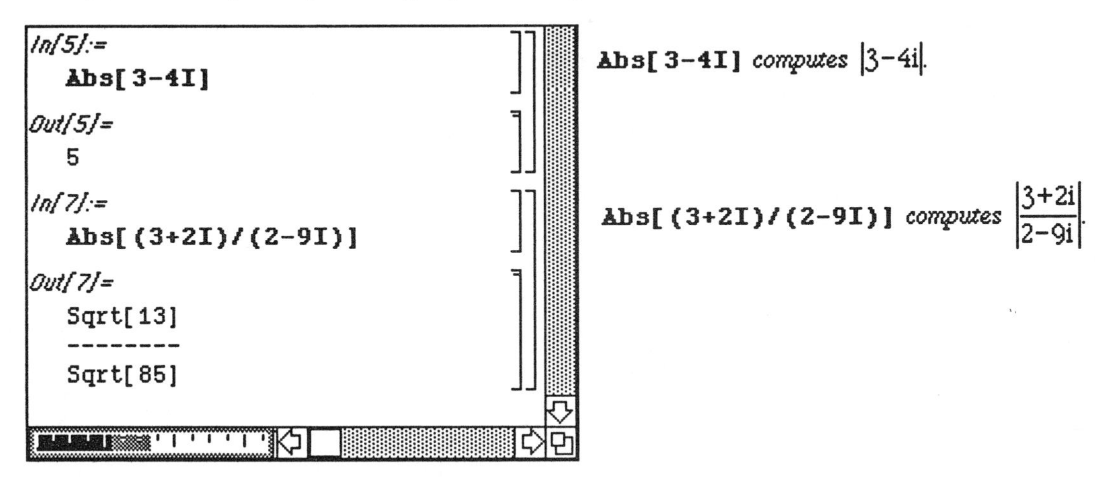

```
In[5]:=
  Abs[3-4I]
Out[5]=
  5
In[7]:=
  Abs[(3+2I)/(2-9I)]
Out[7]=
  Sqrt[13]
  --------
  Sqrt[85]
```

Abs[3-4I] *computes* $|3-4i|$.

Abs[(3+2I)/(2-9I)] *computes* $\left|\frac{3+2i}{2-9i}\right|$.

Log[x] computes the natural logarithm of **x** which is usually denoted as either $\text{Ln}(x)$ or $\text{Log}_e(x)$:

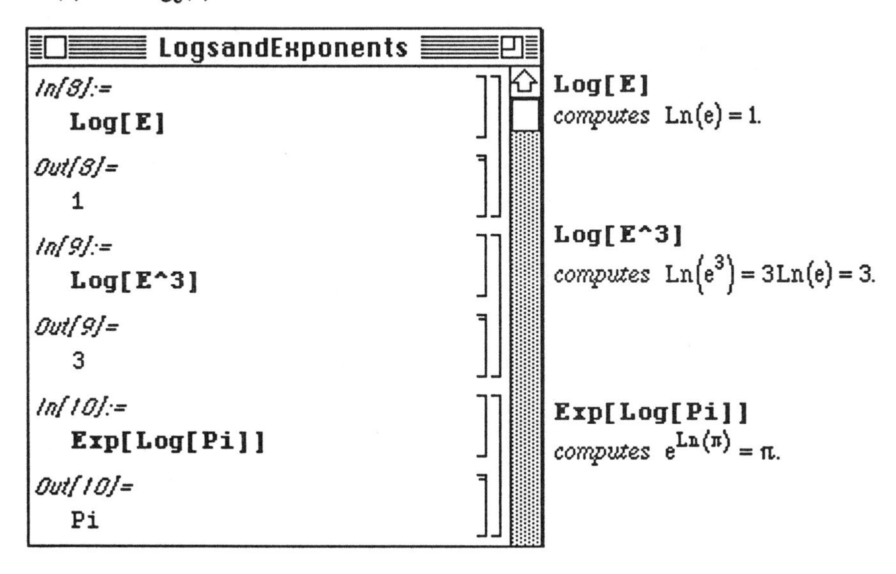

```
LogsandExponents
In[8]:=
  Log[E]
Out[8]=
  1
In[9]:=
  Log[E^3]
Out[9]=
  3
In[10]:=
  Exp[Log[Pi]]
Out[10]=
  Pi
```

Log[E] *computes* $\text{Ln}(e) = 1$.

Log[E^3] *computes* $\text{Ln}(e^3) = 3\text{Ln}(e) = 3$.

Exp[Log[Pi]] *computes* $e^{\text{Ln}(\pi)} = \pi$.

Log [a,b] computes $\text{Log}_b(a) = \frac{\text{Ln}(a)}{\text{Ln}(b)}$:

```
In[11]:=
  Log[3,9]
Out[11]=
  2
In[12]:=
  Log[2,10]
Out[12]=
  Log[10]
  -------
  Log[2]
In[13]:=
  N[Log[2,10],10]
Out[13]=
  3.321928095
```

Log[3,9]
computes $\text{Log}_3(9) = 2$.

Log[2,10]
computes $\text{Log}_2(10) = \frac{\text{Ln}(10)}{\text{Ln}(2)}$.

N[Log[2,10],10]
computes the numerical value of $\text{Log}_2(10) = \frac{\text{Ln}(10)}{\text{Ln}(2)}$ *to ten decimal places.*

◻ Trigonometric Functions

Examples of typical operations involving the trigonometric functions **Sin[x]**, **Cos[x]**, and **Tan[x]** are given below. (Although not illustrated in the following examples, the functions **Sec[x]**, **Csc[x]**, and **Cot[x]** are used similarly.) Notice that *Mathematica* yields the exact value for trigonometric functions of some angles, while a numerical approximation must be requested for others.

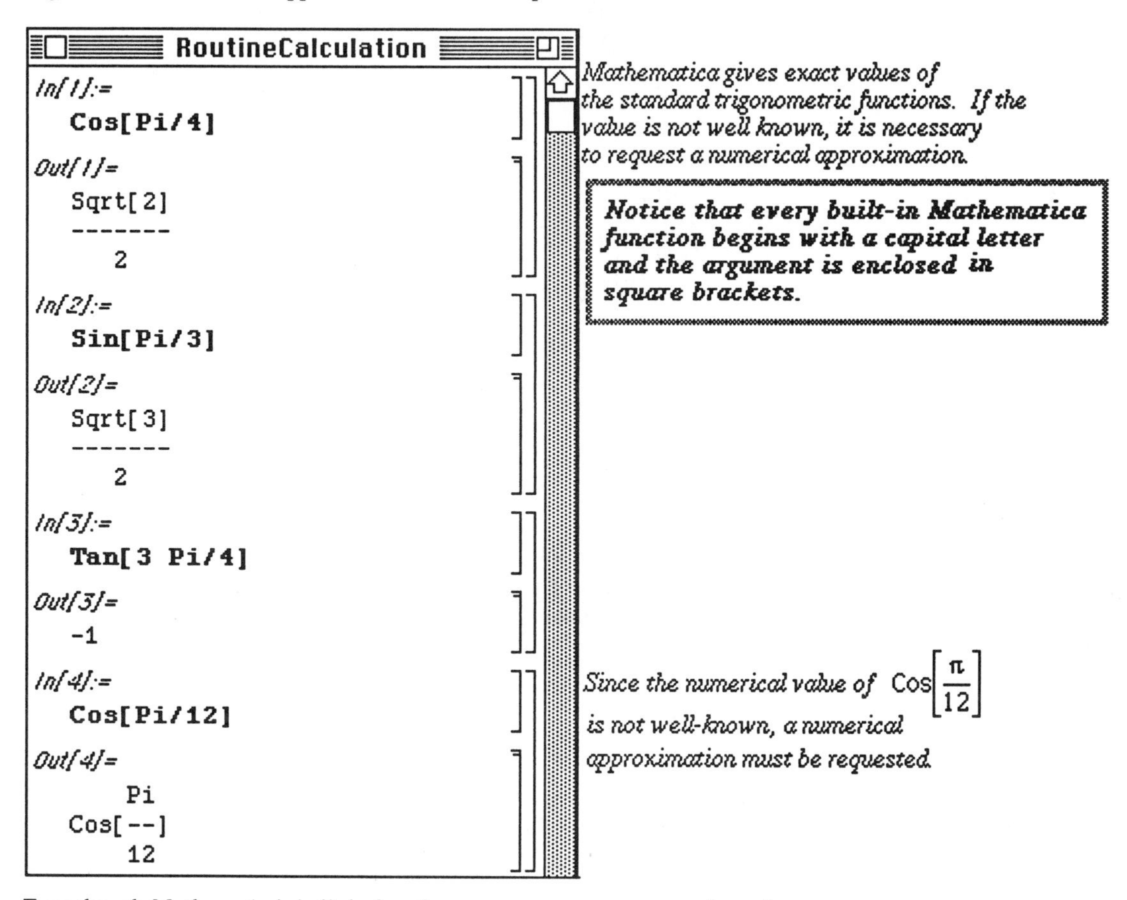

Even though *Mathematica*'s built-in functions cannot compute exact values of

$\text{Cos}\left(\frac{\pi}{12}\right)$ and $\text{Sin}\left(\frac{-9\pi}{8}\right)$, numerical approximations can be obtained by entering

N[Cos[Pi/12]] or **Sin[-9Pi/8]//N**.

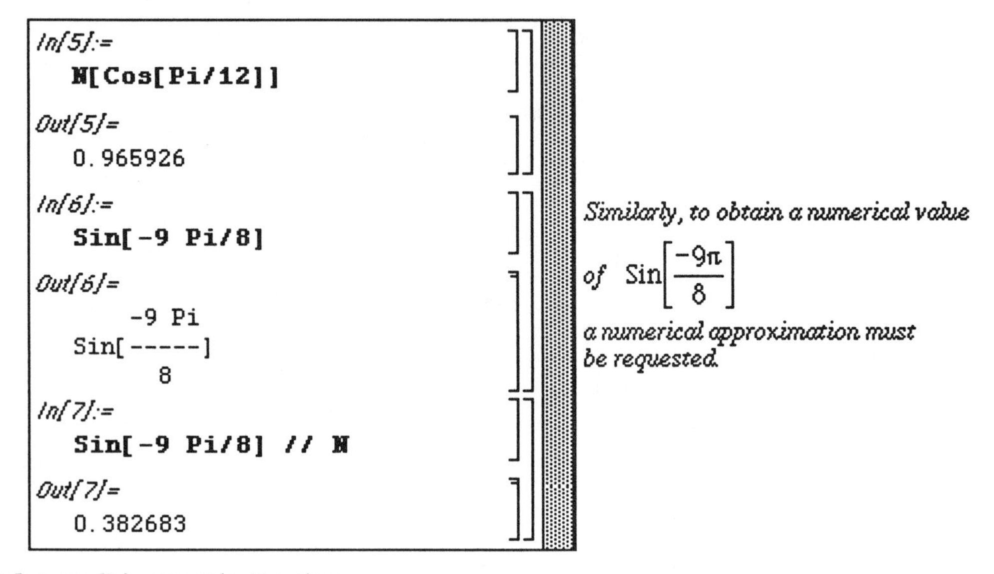
```
In[5]:=
  N[Cos[Pi/12]]
Out[5]=
  0.965926
In[6]:=
  Sin[-9 Pi/8]
Out[6]=
        -9 Pi
  Sin[-----]
          8
In[7]:=
  Sin[-9 Pi/8] // N
Out[7]=
  0.382683
```

Similarly, to obtain a numerical value of $\text{Sin}\left[\frac{-9\pi}{8}\right]$ *a numerical approximation must be requested.*

☐ Inverse Trigonometric Functions

Commands involving the inverse trigonometric functions are similar to those demonstrated in the earlier section on trigonometric functions. Again, note the two capital letters in each of the inverse trigonometric functions. The (built-in) inverse trigonometric functions are:

(i) **ArcCos[x]**; (ii) **ArcCoth[x]**; (iii) **ArcSec[x]**; (iv) **ArcSinh[x]**; (v) **ArcCosh[x]**; (vi) **ArcCsc[x]**; (vii) **ArcSech[x]**; (viii) **ArcTan[x]**; (ix) **ArcCot[x]**; (x) **ArcCsch[x]**; (xi) **ArcSin[x]**; and (xii) **ArcTanh[x]**.

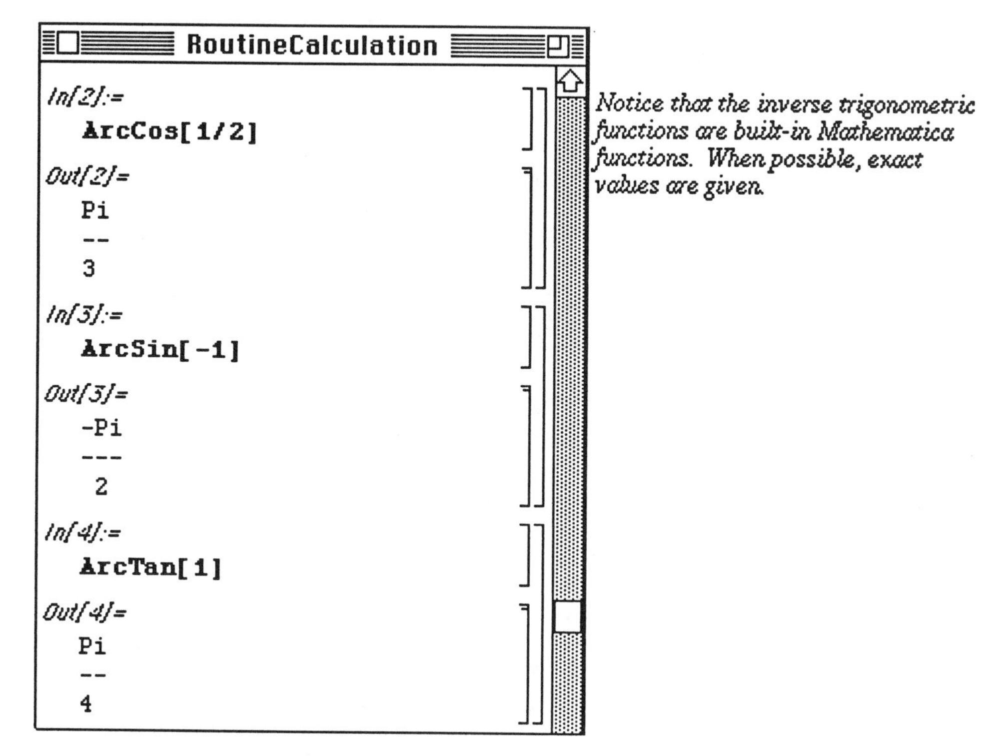

In most instances, a numerical approximation must be requested:

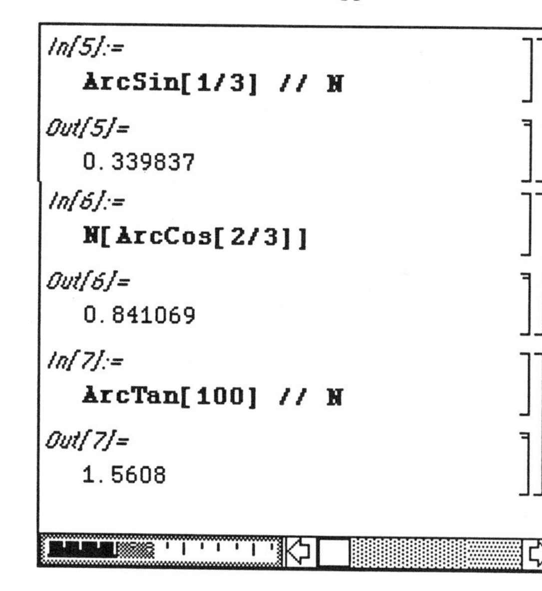

Since **ArcSin[1/3]** *is not well known, a numerical approximation is obtained.*

Notice that **N[ArcCos[2/3]]** *gives the same numerical approximation to* **ArcCos[2/3]** *as* **ArcCos[2/3] // N** *if it were evaluated.*

2.2 Expressions and Functions

Basic Algebraic Operations on Expressions

Mathematica performs standard algebraic operations on mathematical expressions. For example, the command **Factor[expression]** factors **expression**; **Expand[expression]** multiplies **expression**; **Together[expression]** writes **expression** as a single fraction.

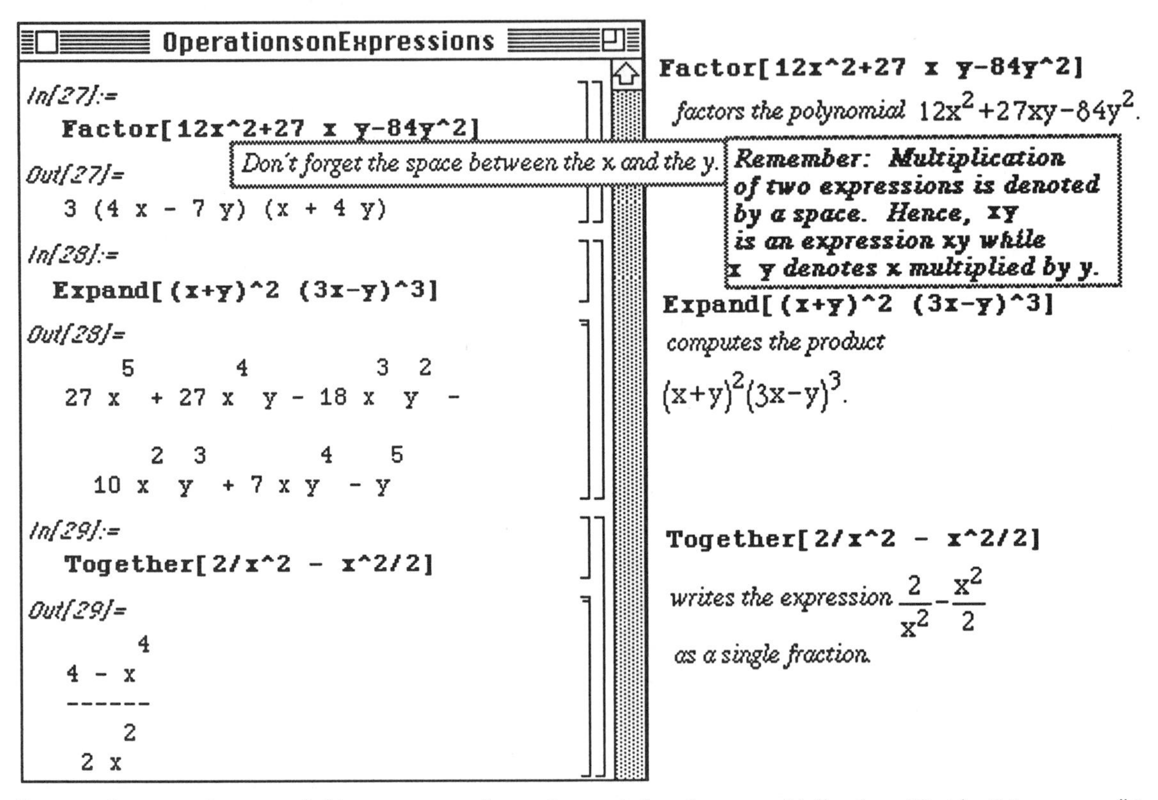

In general, a space is not needed between a number and a symbol to denote multiplication. That is, **3dog** means "**3** times variable **dog**"; *Mathematica* interprets **3 dog** the same way. However, when denoting multiplication of two variables, either include a space or *****: **cat dog** means "variable **cat** times variable **dog**", **cat*dog** means the same thing but **catdog** is interpreted as a variable **catdog**.

The command **Apart[expression]** computes the partial fraction decomposition of **expression**; **Cancel[expression]** factors the numerator and denominator of **expression** then reduces **expression** to lowest terms.

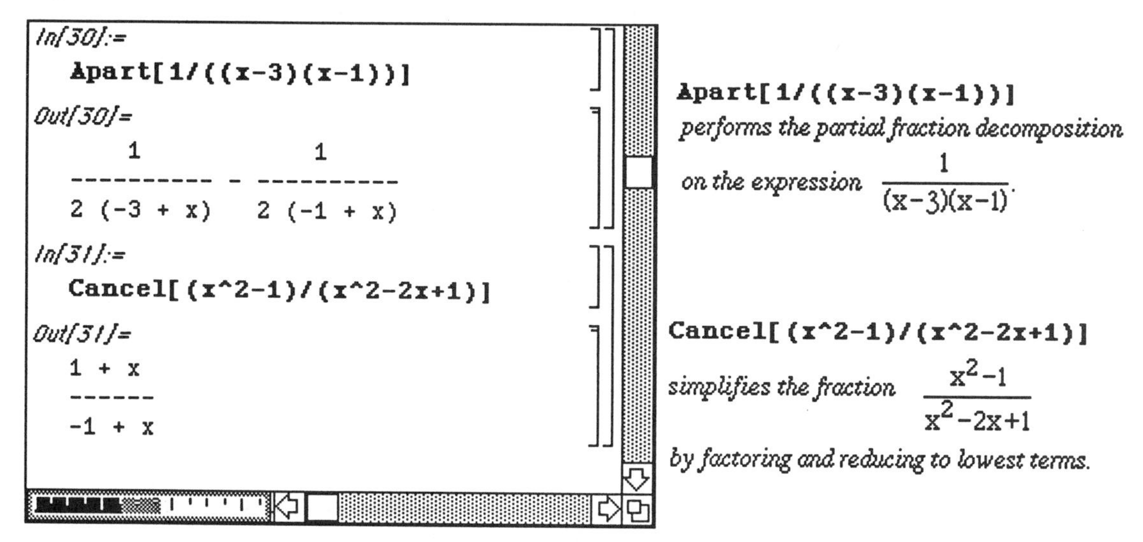

Apart[1/((x-3)(x-1))] *performs the partial fraction decomposition on the expression* $\frac{1}{(x-3)(x-1)}$.

Cancel[(x^2-1)/(x^2-2x+1)] *simplifies the fraction* $\frac{x^2-1}{x^2-2x+1}$ *by factoring and reducing to lowest terms.*

■ Naming and Evaluating Expressions

In *Mathematica,* mathematical objects can be named. Naming objects is convenient: we can avoid typing the same mathematical expression repeatedly and named expressions can be referenced throughout a notebook. Since every built-in *Mathematica* function begins with a capital letter, we will adopt the convention that every mathematical object we name will begin with a lower-case letter. Consequently, we will be certain to avoid any possible ambiguity with a built-in *Mathematica* object. An expression is named by using a single equals sign (=).

Expressions can be evaluated easily. To evaluate an expression we introduce the command **/.** . The command /. means "replace by". For example, the command **x^2 /. x-> 3** means evaluate the expression

x^2 when $x = 3$.

The following example illustrates how to name an expression. In addition, *Mathematica* has several built-in functions for manipulating fractions:

1) **Numerator[fraction]** yields the numerator of a fraction; and

2) **Denominator[fraction]** yields the denominator of a fraction.

The naming of expressions makes the numerator and denominator easier to use in the following examples:

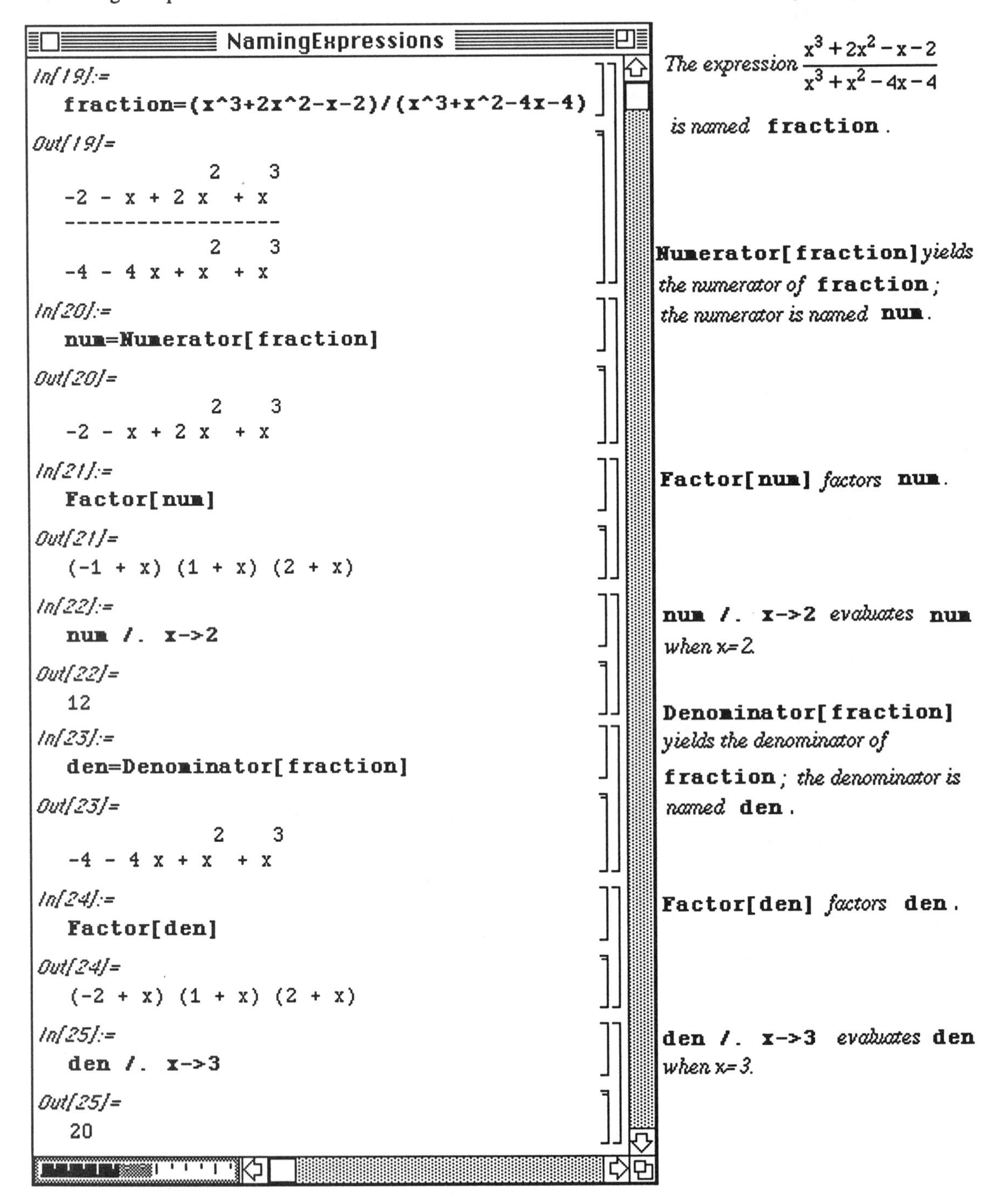

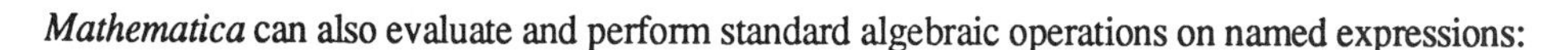

Mathematica can also evaluate and perform standard algebraic operations on named expressions:

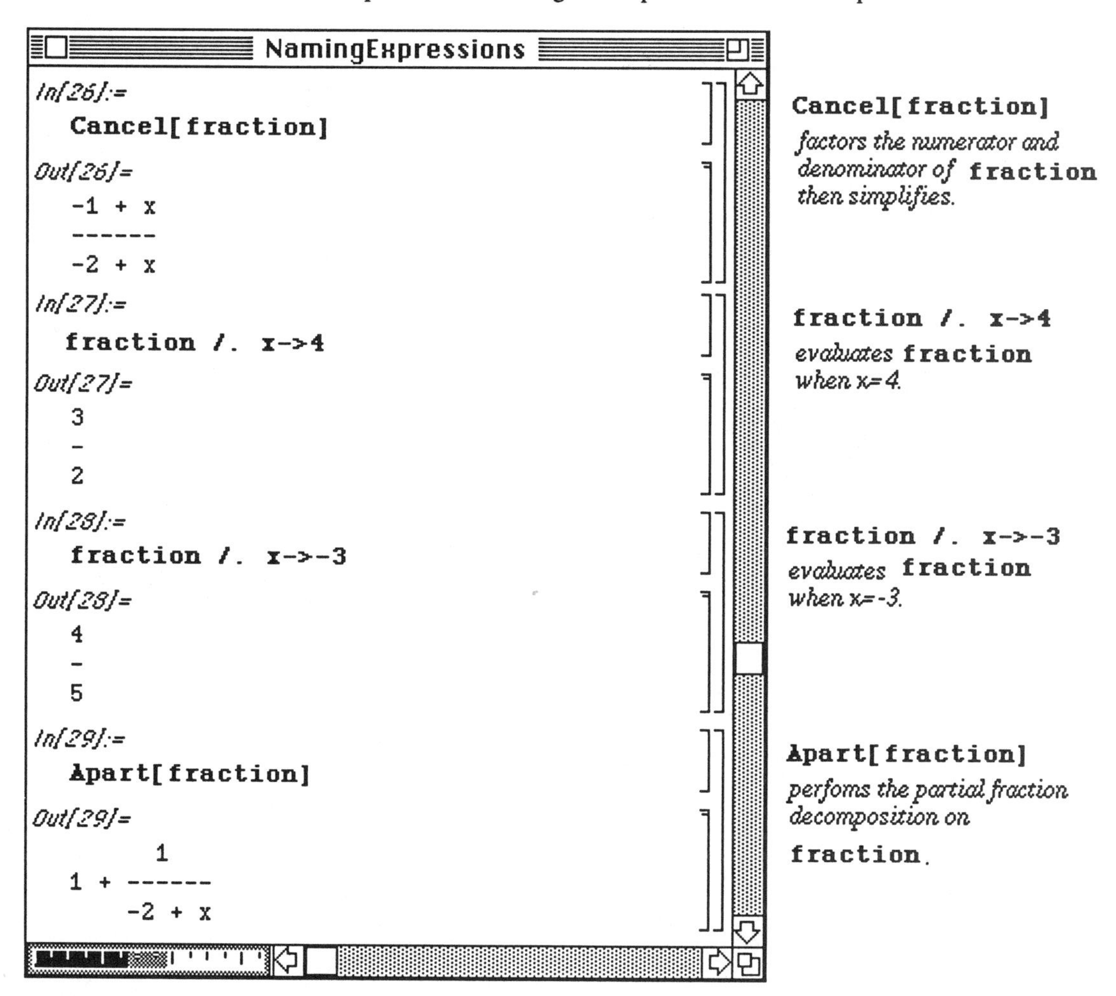

Every *Mathematica* object can be named; even graphics and functions can be named with *Mathematica*.

■ Defining and Evaluating Functions

It is important to remember that functions, expressions, and graphics can be named anything that is not the name of a built-in *Mathematica* function or command. Since every built-in *Mathematica* function begins with a capital letter, every user-defined function or expression in this text will be defined using lower case letters. This way, the possibility of conflicting with a built-in *Mathematica* command or function is completely eliminated. Also, since definitions of functions are frequently modified, we introduce the command **Clear**. **Clear[expression]** clears all definitions of **expression**. Consequently, we are certain to avoid any ambiguity when we create a new definition of a function. When you first define a function, you must always enclose the argument in square brackets and place an underline after the argument on the left-hand side of the equals sign in the definition of the function.

□ **Example:**

Use *Mathematica* to define $f(x)=x^2$, $g(x)=\sqrt{x}$, and $h(x)=x+\mathrm{Sin}(x)$.

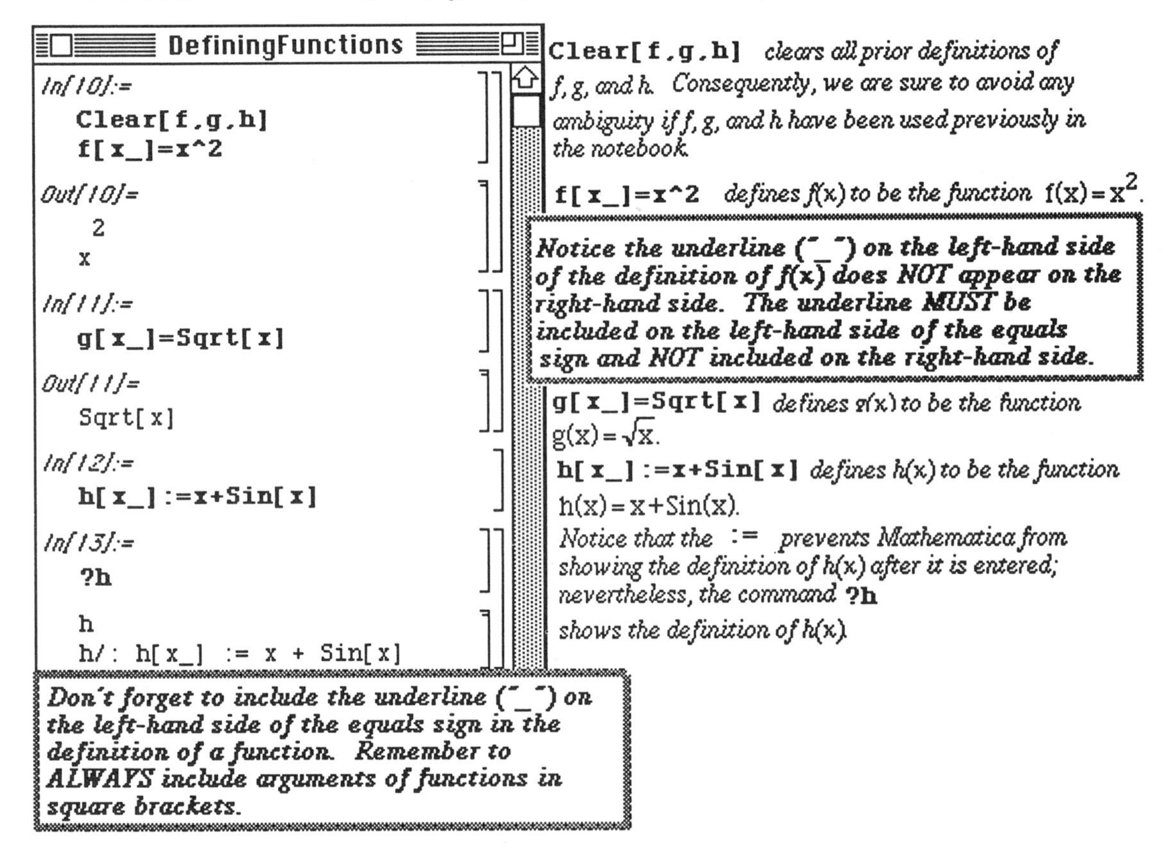

When you evaluate a function, type **`functionname[point]`** **ENTER**. Notice that functions can be evaluated for any real number (in the function's domain):

□ **Example:**

Using the definitions of f, g, and h from above, compute f(2), g(4) and h(π/2).

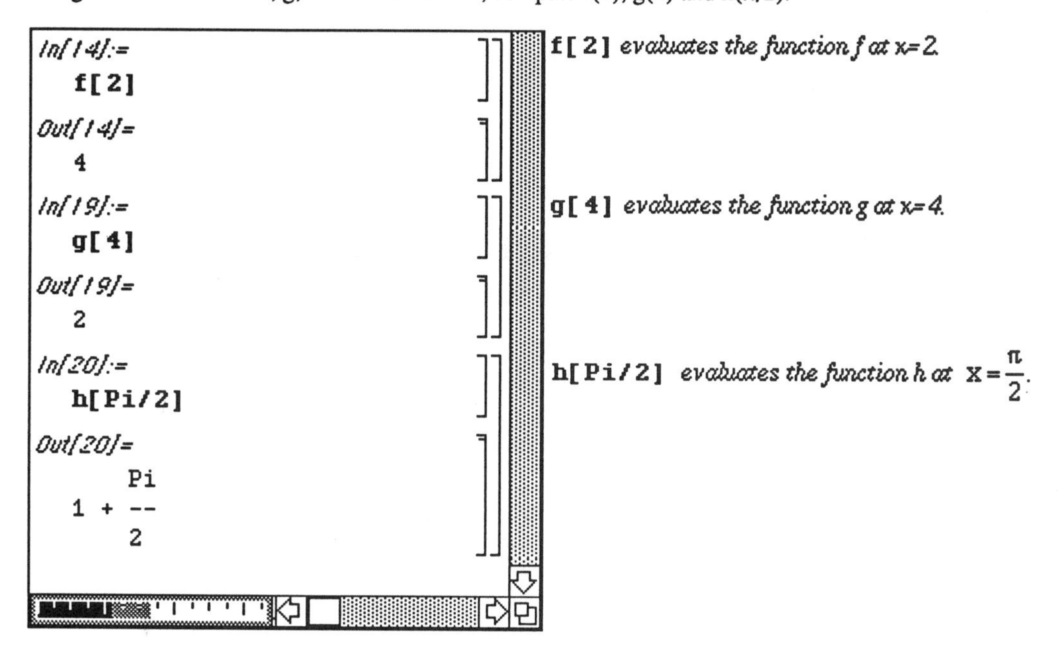

f[2] *evaluates the function f at x=2.*

g[4] *evaluates the function g at x=4.*

h[Pi/2] *evaluates the function h at* $x = \frac{\pi}{2}$.

Moreover, *Mathematica* can symbolically evaluate and manipulate functions.

□ Example:

Several examples follow which involve the function $f(x) = x^2$ defined above.

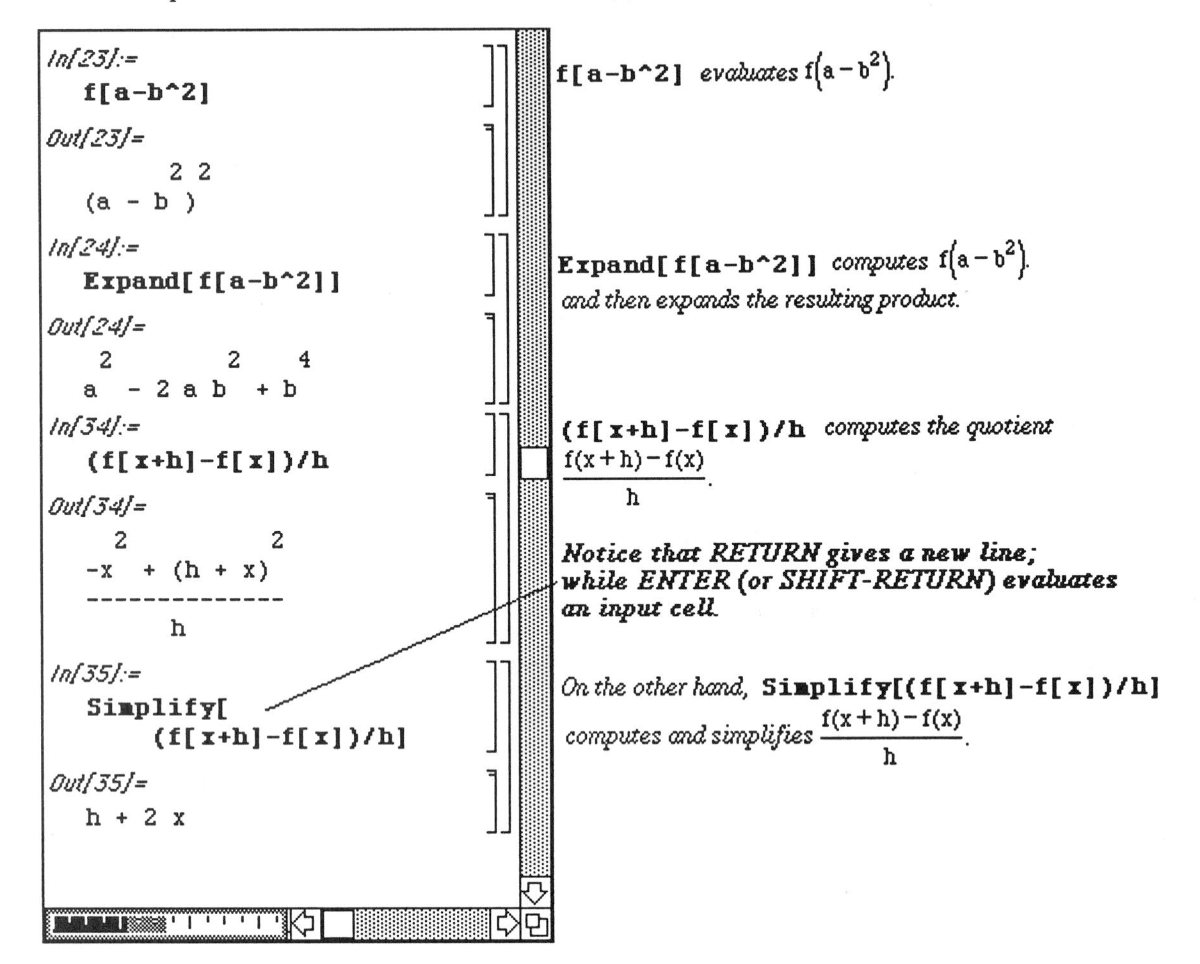

Many different types of functions can be defined using *Mathematica*. An example of a function f of two variables is illustrated below.

Additional ways of defining functions will be discussed in later parts of this text.

□ Example:

Define $f(x,y)=1-\mathrm{Sin}\left(x^2+y^2\right)$. Compute $f(1,2)$, $f\left(2\sqrt{\pi},\frac{3}{2}\sqrt{\pi}\right)$, $f(0,a)$, and $f\left(a^2-b^2,b^2-a^2\right)$.

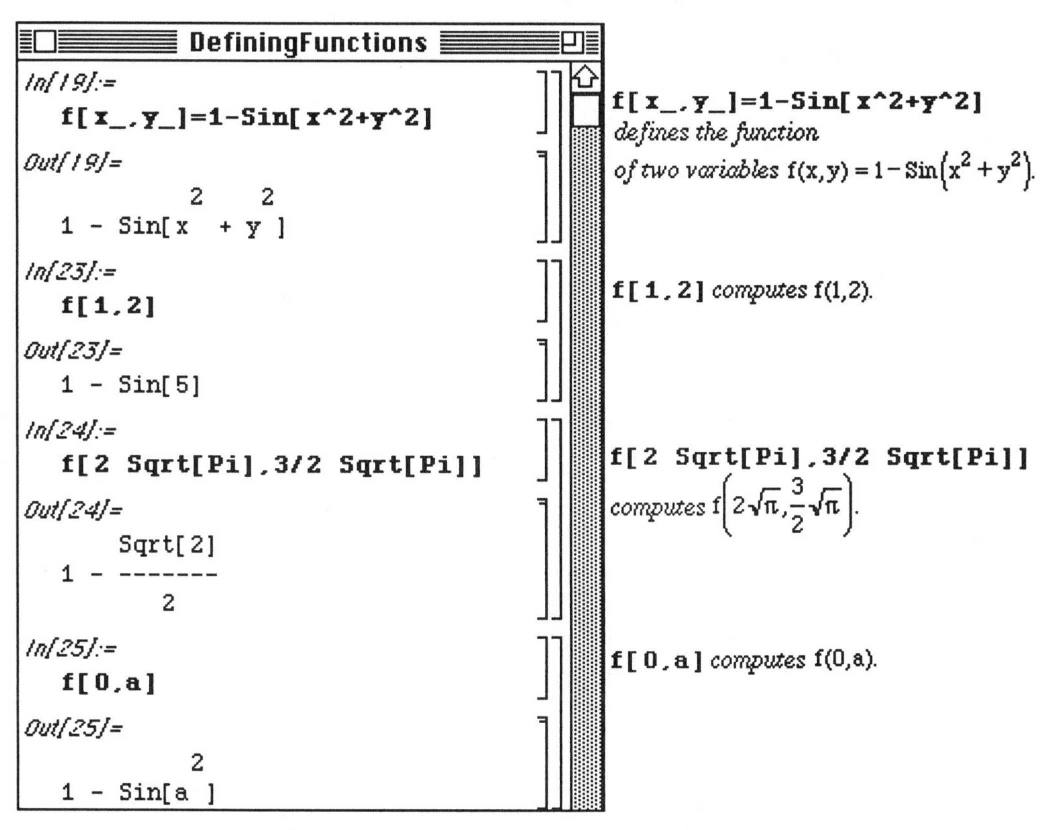

Evaluating $f\left(a^2-b^2,b^2-a^2\right)$ is done the same way as in the previous examples:

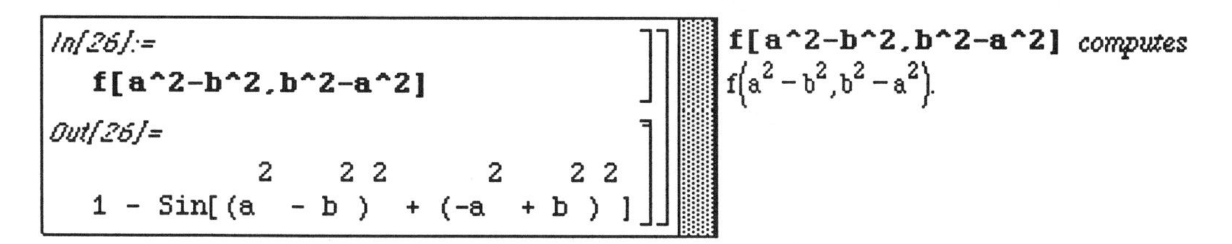

Vector-valued functions, such as g below, can also be defined:

□ Example:

Define the vector-valued function $g(x)=\{x^2,1-x^2\}$; compute g(1) and g(Sin(b)).

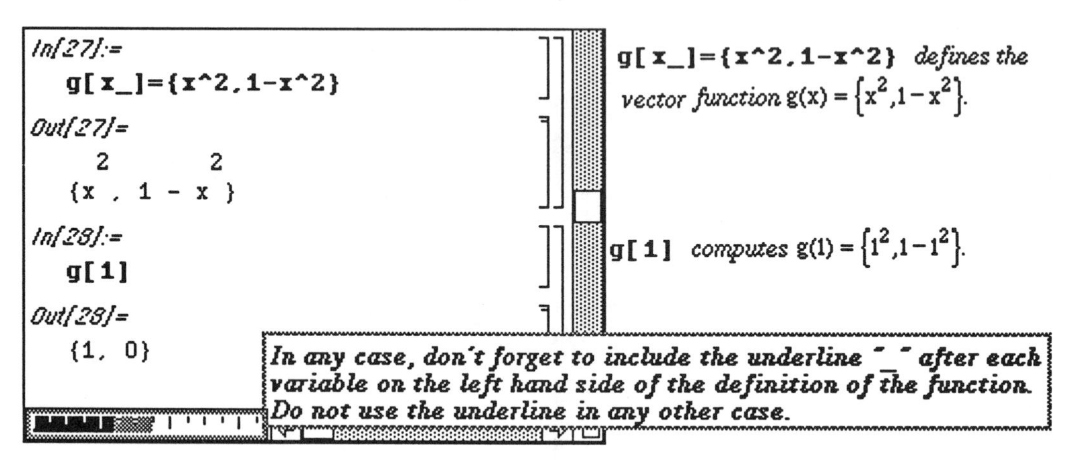

g[x_]={x^2,1-x^2} *defines the vector function* $g(x)=\{x^2,1-x^2\}$.

g[1] *computes* $g(1)=\{1^2,1-1^2\}$.

g(Sin(b)) is computed the same way:

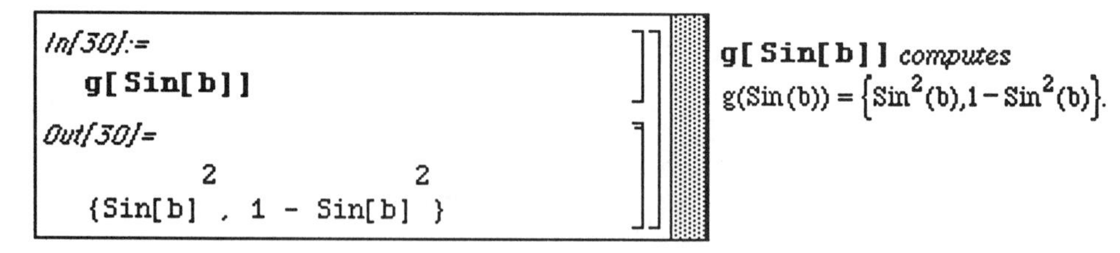

g[Sin[b]] *computes* $g(\text{Sin}(b))=\{\text{Sin}^2(b),1-\text{Sin}^2(b)\}$.

□ Example:

Define the vector-valued function of two variables $h(x,y)=\{\text{Cos}(x^2-y^2), \text{Sin}(y^2-x^2)\}$.

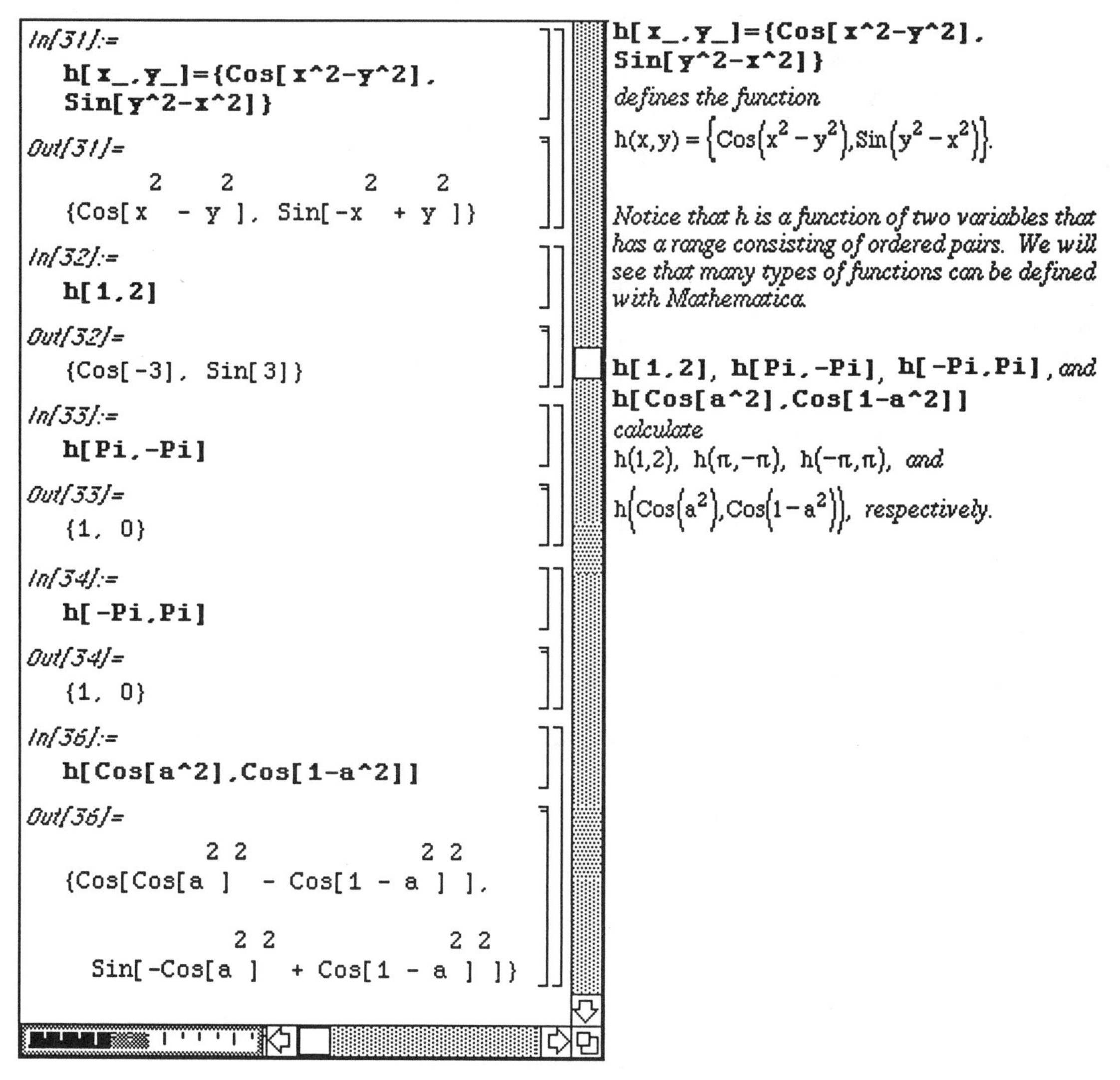

h[x_,y_]={Cos[x^2-y^2], Sin[y^2-x^2]} *defines the function* $h(x,y)=\{\text{Cos}(x^2-y^2),\text{Sin}(y^2-x^2)\}$.

Notice that h is a function of two variables that has a range consisting of ordered pairs. We will see that many types of functions can be defined with Mathematica.

h[1,2], **h[Pi,-Pi]**, **h[-Pi,Pi]**, *and* **h[Cos[a^2],Cos[1-a^2]]** *calculate* $h(1,2)$, $h(\pi,-\pi)$, $h(-\pi,\pi)$, *and* $h(\text{Cos}(a^2),\text{Cos}(1-a^2))$, *respectively.*

■ Additional Ways to Evaluate Functions and Expressions

Not only can a function **f[x]** be evaluated by computing **f[a]** where **a** is either a real number in the domain of f or an expression, functions and expressions can be evaluated using the command **/.** . In general, to evaluate the function f[x] when x is replaced by **expression**, the following two commands are equivalent and yield the same output:

1) **f[expression]** replaces each variable in **f** by **expression**; and

2) **f[x] /. x-> expression** replaces each variable **x** in **f[x]** by **expression**.

□ **Example:**

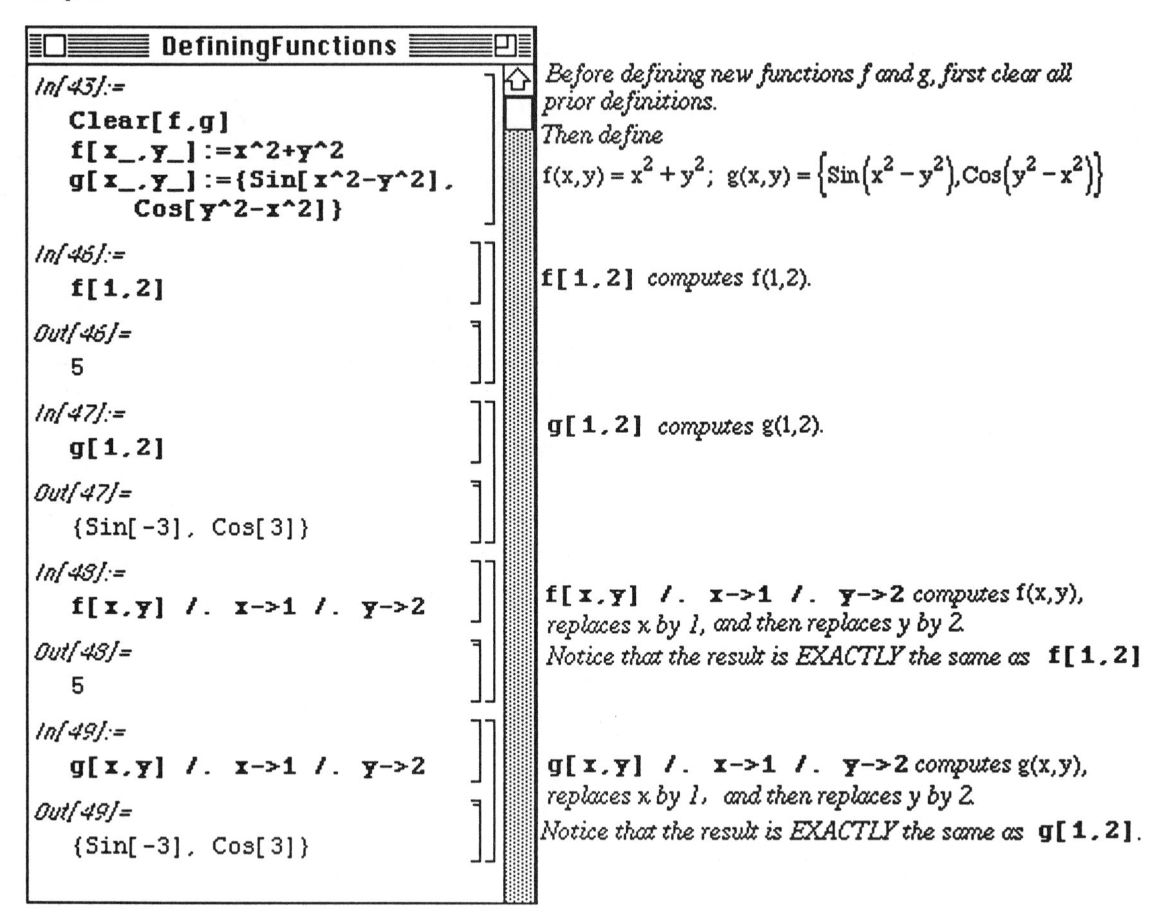
DefiningFunctions
In[43]:=
Clear[f,g]
f[x_,y_]:=x^2+y^2
g[x_,y_]:={Sin[x^2-y^2],
Cos[y^2-x^2]}
In[46]:=
f[1,2]
Out[46]=
5
In[47]:=
g[1,2]
Out[47]=
{Sin[-3], Cos[3]}
In[48]:=
f[x,y] /. x->1 /. y->2
Out[48]=
5
In[49]:=
g[x,y] /. x->1 /. y->2
Out[49]=
{Sin[-3], Cos[3]}
Before defining new functions f and g, first clear all prior definitions.
Then define
f(x,y) = x² + y²; g(x,y) = {Sin(x² − y²), Cos(y² − x²)}
f[1,2] computes f(1,2).
g[1,2] computes g(1,2).
f[x,y] /. x->1 /. y->2 computes f(x,y), replaces x by 1, and then replaces y by 2.
Notice that the result is EXACTLY the same as f[1,2]
g[x,y] /. x->1 /. y->2 computes g(x,y), replaces x by 1, and then replaces y by 2.
Notice that the result is EXACTLY the same as g[1,2].

There are several other methods available for evaluating functions. However, depending on the situation, one method may prove to be more appropriate than others. Some of these methods are discussed here in order to make the reader aware of alternate approaches to function evaluation. In the example which follows, a function **f** is defined which maps a list of two elements, **{a,b}**, to the real number, a Modulo b using the built-in function **Mod**. If **a** and **b** are real numbers, **Mod[a,b]** returns **a** modulo **b**. The typical approach to evaluating **f** at **{a,b}** is to directly substitute **{a,b}** into **f** with **f[{a,b}]**. However, two another approaches which yield the same result are **f@{a,b}** and **{a,b}//f**. These are demonstrated below with **{5,3}**.

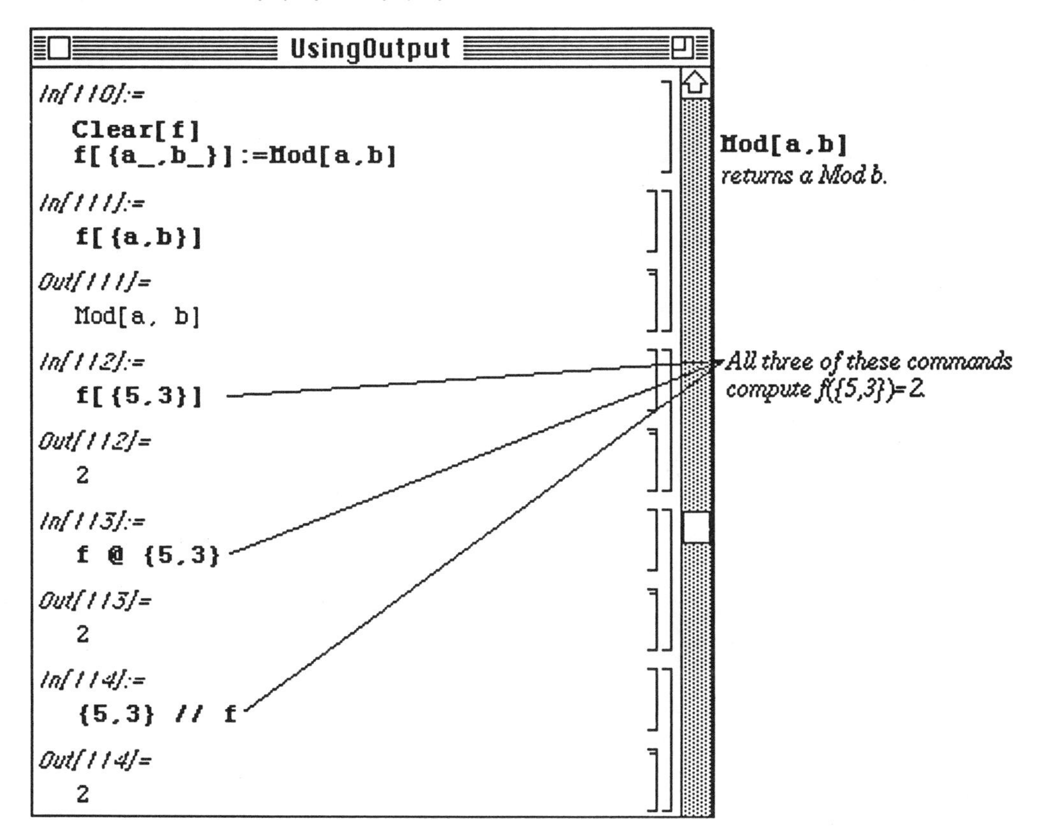

■ Retrieving Unnamed Output

Although naming *Mathematica* objects is convenient, occasionally, one may want to use previous results in subsequent calculations even though these objects were not necessarily named. Fortunately, *Mathematica* provides two convenient ways to refer to previously generated output. First, the symbol **%** refers to the most recent output; **%%** refers to the second most recent output; **%%%** refers to the third most recent output and, in general **%%%...%** (k-times) refers to the kth most recent output. Second, **Out[n]**, where **n** is a positive integer, refers to the **n**th output.

Several examples are given below which illustrate these ideas. First, functions **f**, **g**, and **h** are defined. Then, these functions are evaluated using several different methods. The commands **f[%]** and **f[Out[30]]** given below yield the same output since both evaluate the function **f** at x = .077.

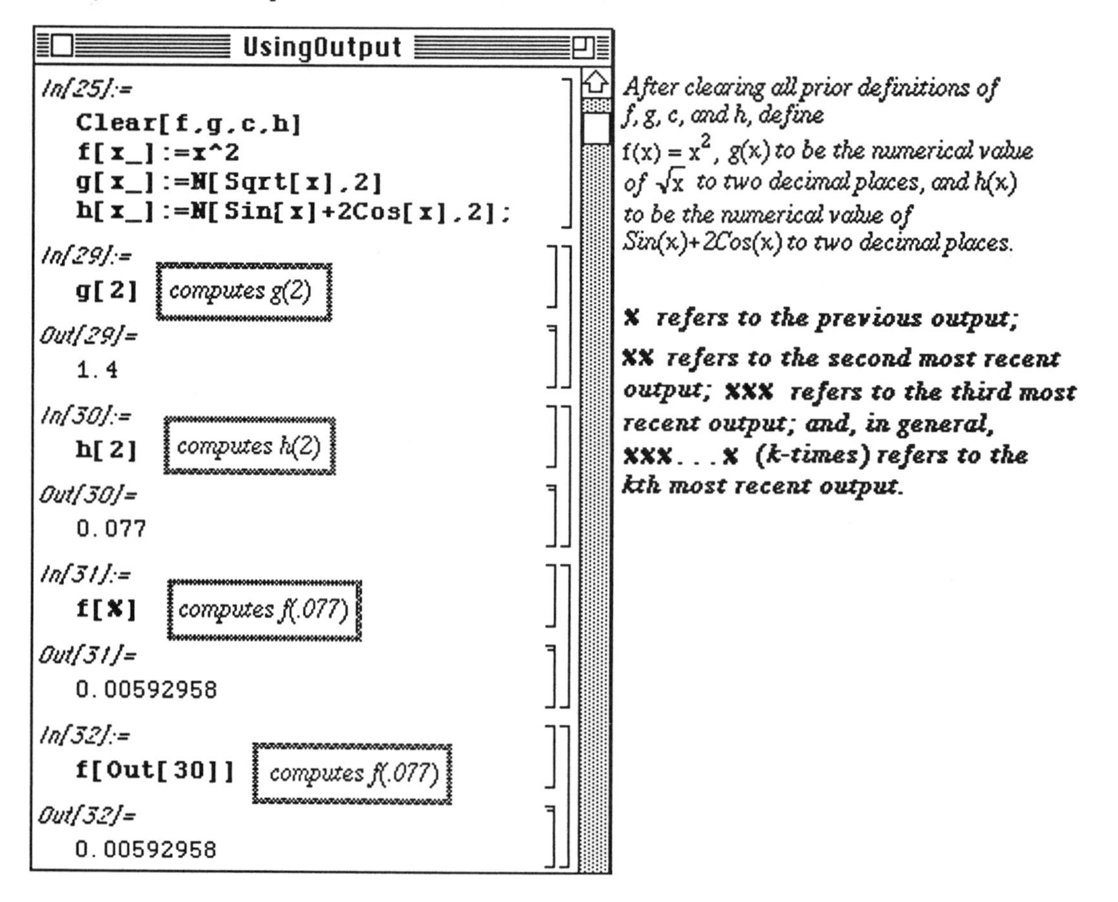

These methods of retrieving output are useful as input is altered. For example, a new variable **c** is defined below in terms of **a** and **b**. The function **f** can then be evaluated at **c** in several ways which are demonstrated below. **g[%]** computes **g** at the previous output, **Out[[35]]**. Hence, **g[%%%%]** computes **g** at the fourth previous output, **Out[[32]]**. In the last example below, **h** is evaluated at the second previous output, **Out[[35]]**.

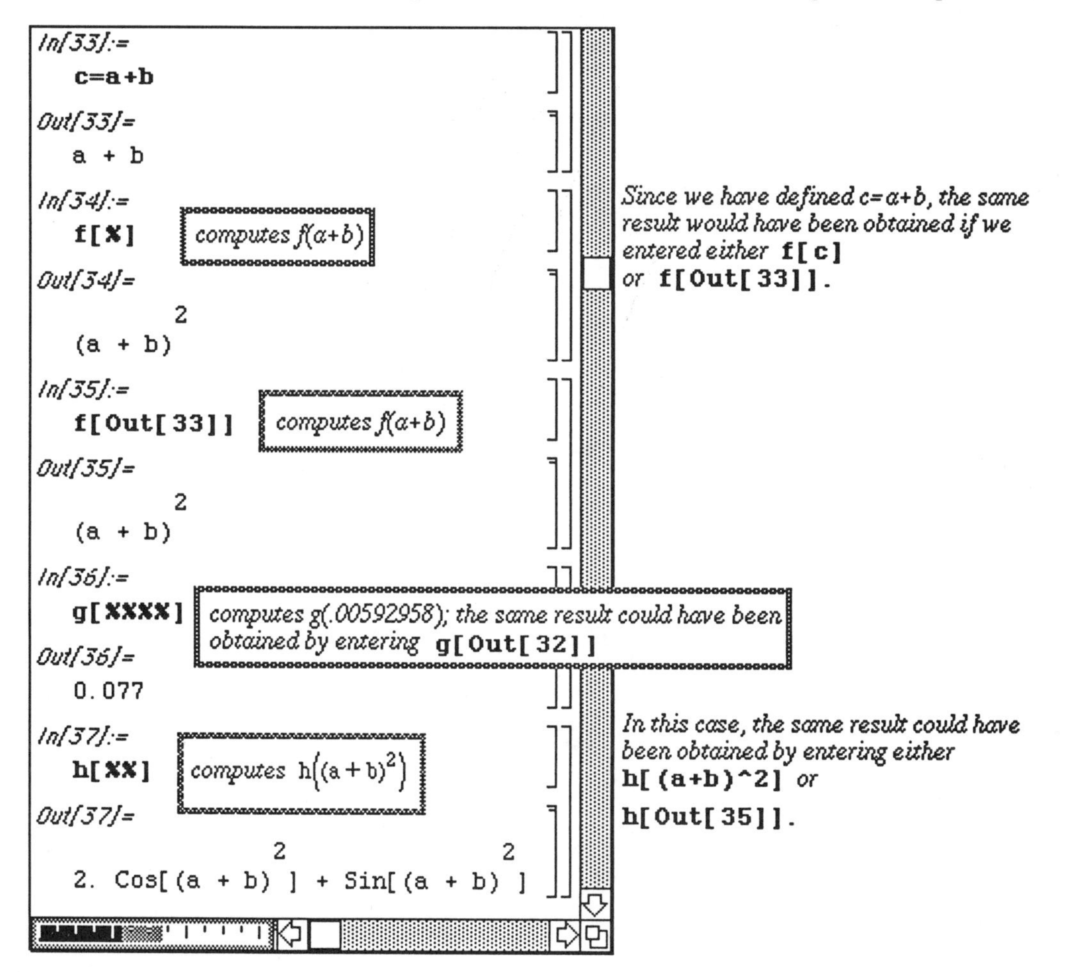

■ Composition of Functions

Mathematica can easily perform the calculation f[g[x]]. However, when composing several different functions or repeatedly composing a function with itself, two additional commands are provided:

1) **Compose[f1, f2, f3, . . . ,fn,x]** computes the composition

$f1 \circ f2 \circ f3 \circ \ldots \circ fn(x)$ where f1, f2, f3, ..., and fn are functions and x is an expression.

o In Version 2.0, the function **Compose** is replaced by the function **Composition**. In Version 2.0, **Composition[f1,f2,...,fn][x]** computes the composition

$f1 \circ f2 \circ f3 \circ \ldots \circ fn(x)$ where f1, f2, f3, ..., and fn are functions and x is an expression.

2) **Nest[f, x, n]** computes the composition

$$f \circ f \circ f \circ \ldots \circ f(x)$$

(f composed with itself n times)

where f is a function, n is a positive integer, and x is an expression.

□ **Example:**

In the following example $f(x) = x^2$ and $h(x) = x + \mathrm{Sin}(x)$.

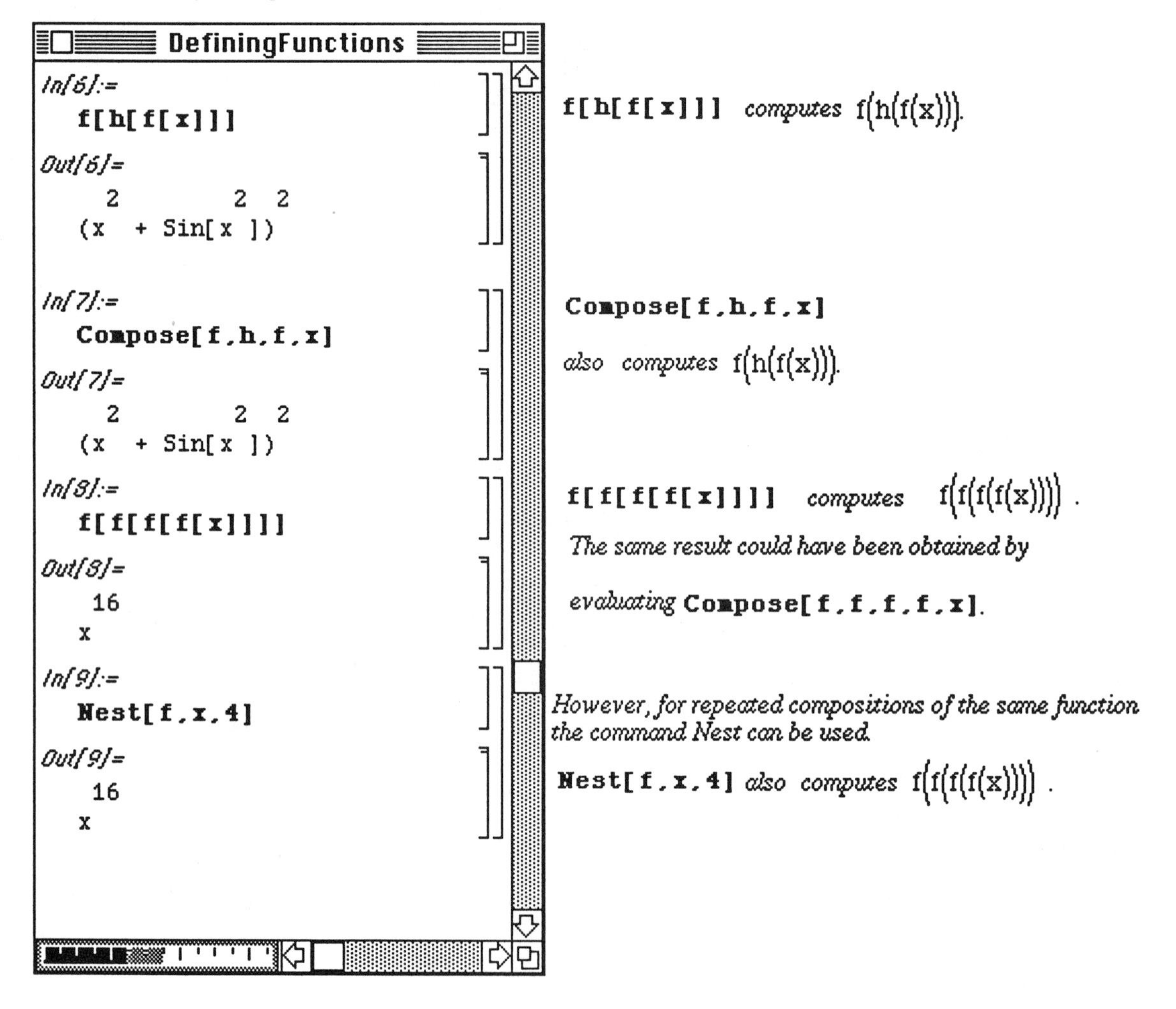

o In Version 2.0 *Mathematica* displays output for **EACH** command as it is generated unless a semi-colon is included at the end of the command. Hence, in the following example, output is displayed for all except the last command:

o **Example:**

Let $f(x) = \text{Log}\left(\frac{2x+1}{x-1/2}\right)$, $g(x) = \text{Sin}(3x) - \text{Cos}(4x)$, $h(x) = x^2$ and $k(x) = h(g(x))$. Compute and simplify k, compute $\text{Exp}[f(x)] = e^{f(x)}$, and write $e^{f(x+i\,y)} = \text{Exp}[f(x+i\,y)]$ in terms of its real and imaginary parts, assuming x and y are real.

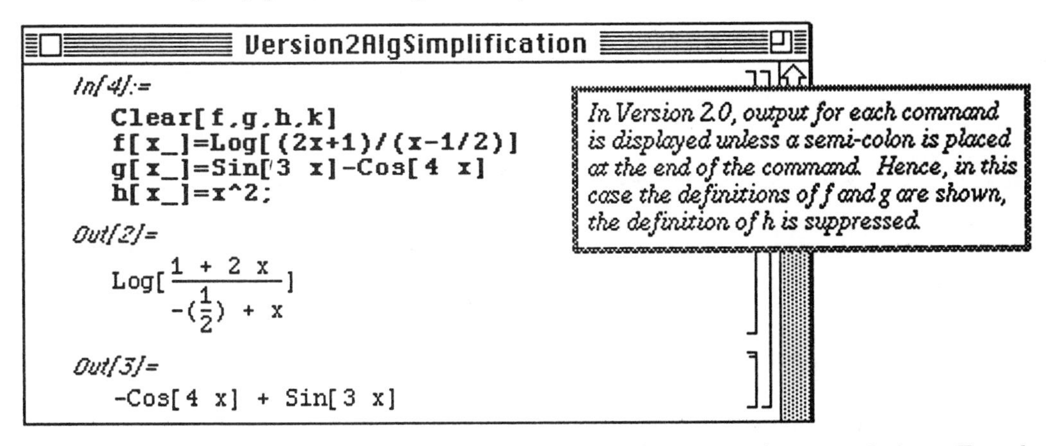

o In Version 2.0, the command **`Compose`** has been replaced by the command **`Composition`**. Even though entering the command **`Compose[f,g,x]`** yields **`f[g[x]]`**, *Mathematica* issues a warning that **`Compose`** is an obsolete function, replaced by **`Composition`**.

- Also notice that the option **`Trig->True`** has been added to the command **`Expand`**. The effect of the option **`Trig->True`** is to eliminate powers of Sines and Cosines in trigonometric expressions:

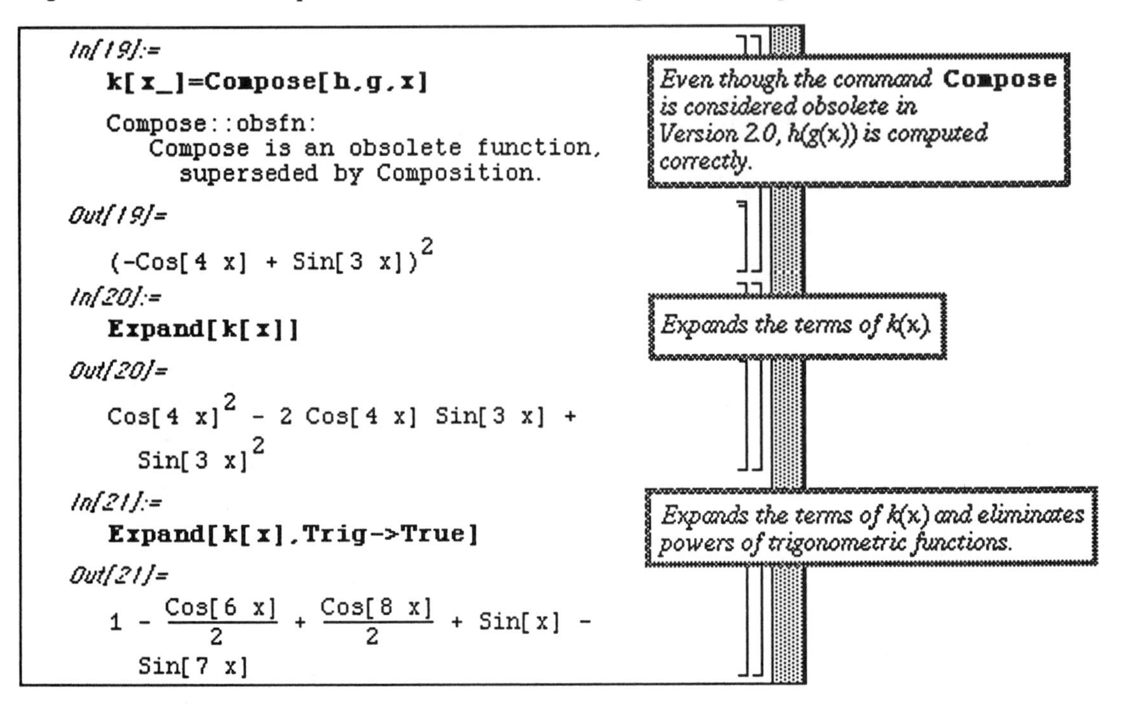
```
In[19]:=
  k[x_]=Compose[h,g,x]

  Compose::obsfn:
     Compose is an obsolete function,
       superseded by Composition.
Out[19]=
                         2
  (-Cos[4 x] + Sin[3 x])
In[20]:=
  Expand[k[x]]
Out[20]=
           2
  Cos[4 x]  - 2 Cos[4 x] Sin[3 x] +
             2
    Sin[3 x]
In[21]:=
  Expand[k[x],Trig->True]
Out[21]=
      Cos[6 x]   Cos[8 x]
  1 - -------- + -------- + Sin[x] -
         2          2
    Sin[7 x]
```

o Version 2.0 also includes the new command **ComplexExpand**. If **expression** is a *Mathematica* expression in terms of **x+I y**, the command **ComplexExpand[expression]** rewrites **expression** in terms of its real and imaginary components, assuming that **x** and **y** are both real.

In order to compute h(g(x)) in Version 2.0, enter **Composition[h,g][x]**:

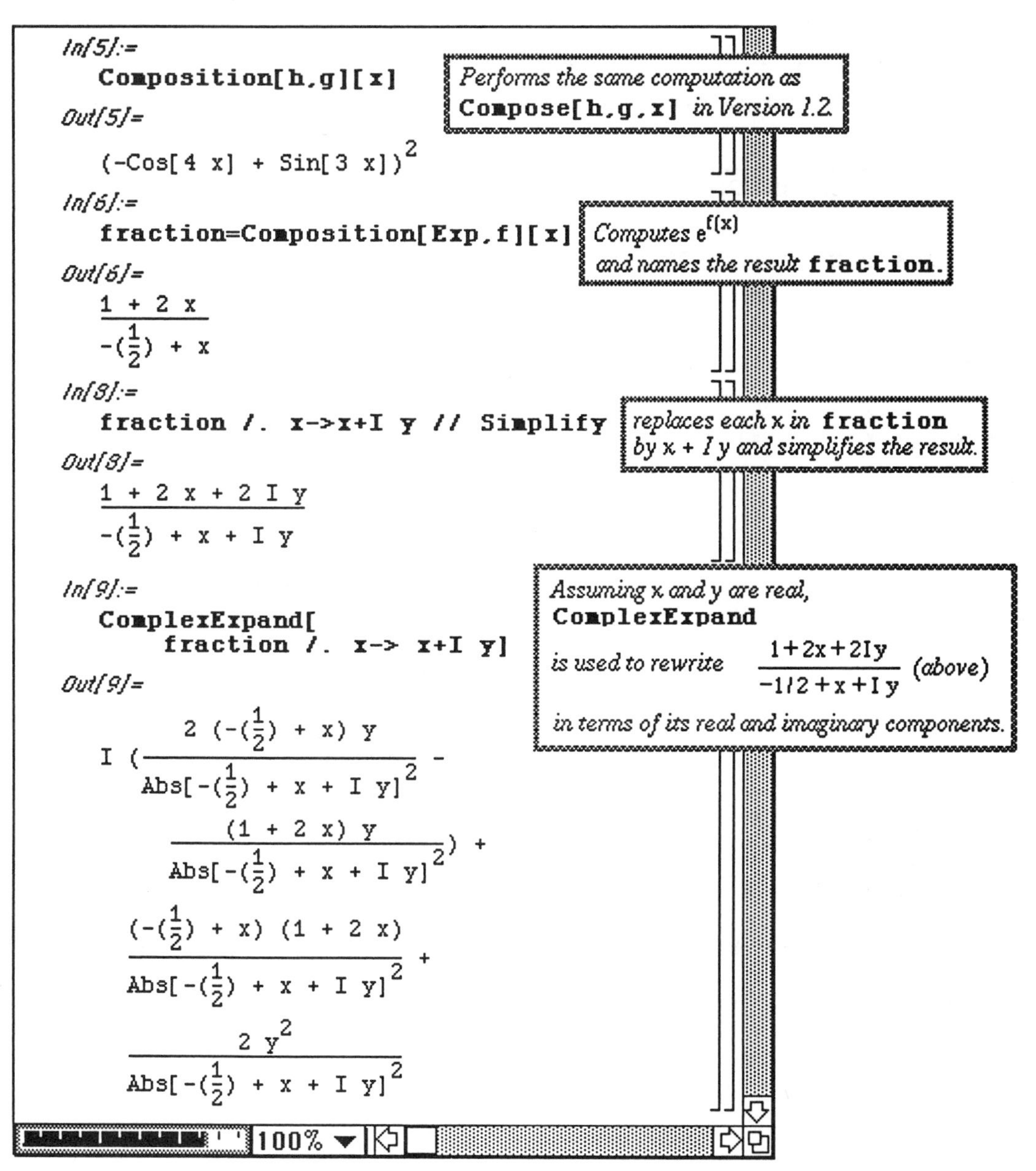

2.3 Mod Math

The command **Mod[a,b]** reduces the number **a** modulo **b**. If **p** is a polynomial, the command **Mod[p,b]** reduces the coefficients of **p** modulo **b**.

Example:

In the following example, the factors of $x^4+x^3+x^2+x+1$ modulo 5 are found and verified. A function **modexpand[poly,p]** which expands and factors the polynomial **poly** modulo **p** is then defined for later use.

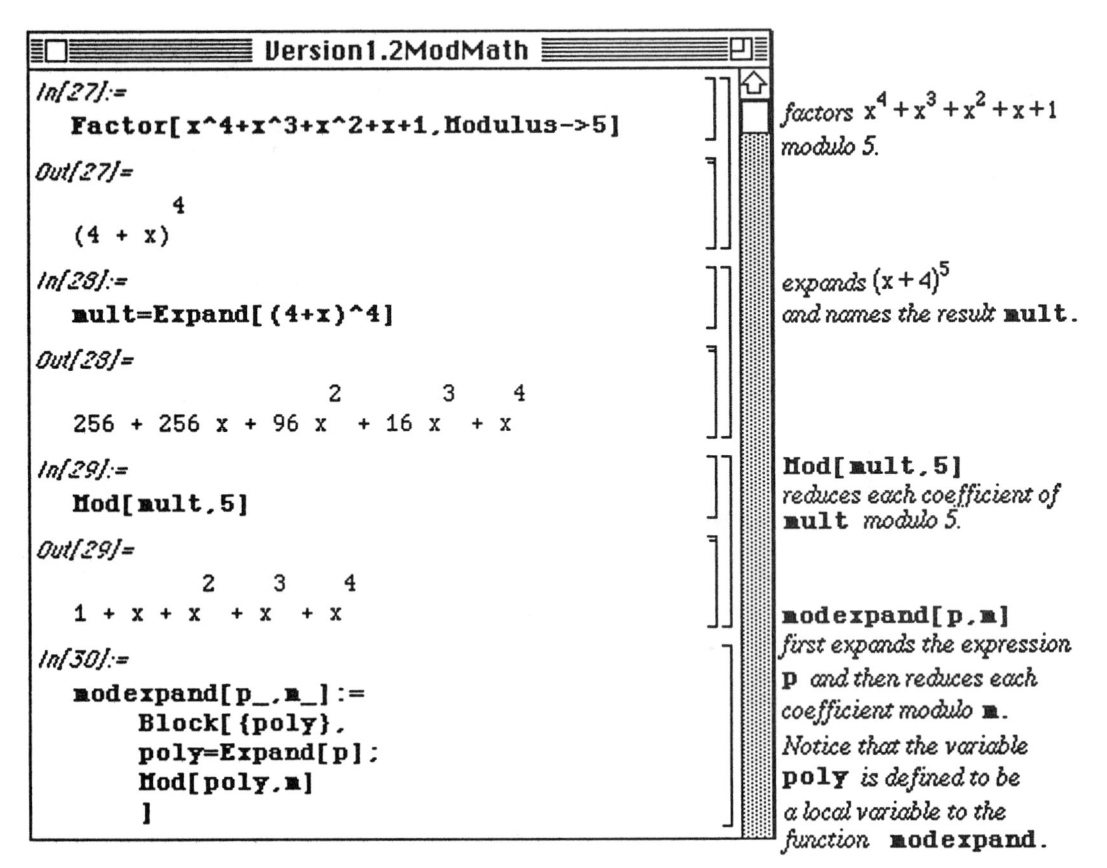

◻ Example:

It is well known that if F is a field of characteristic p, p a prime number,

and a, b $\in$ F, then $(a+b)^{p^m} = a^{p^m} + b^{p^m}$. Illustrate this fact when $m=1$ for the first five prime numbers.

We proceed by using the user-defined command **modexpand** from above and the built -in command **Prime**. **Prime[i]** returns the **i**th prime number.

The command **Table** is discussed in more detail in **Chapters 4** and **5**.

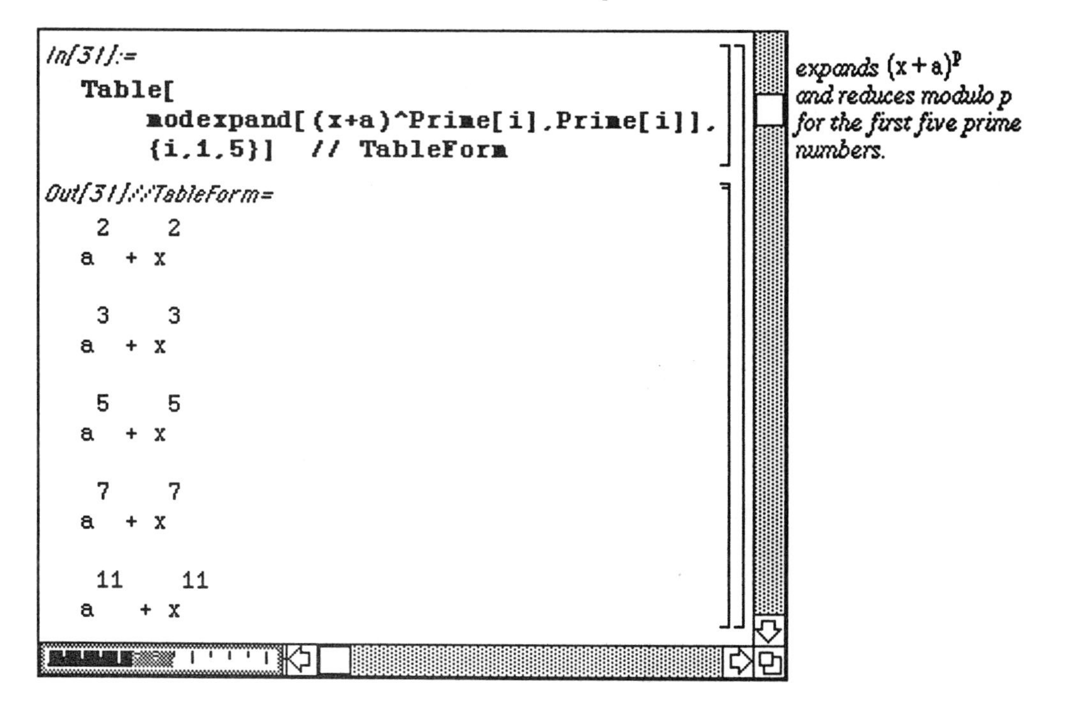

- In Version 2.0, **`Mod[a,b]`** reduces the number **`a`** modulo **`b`**. Notice that unlike prior versions of *Mathematica*, **`a`** must be a number. To reduce the coefficients of a polynomial p modulo b, use the command **`PolynomialMod[p,b]`**:

Example:

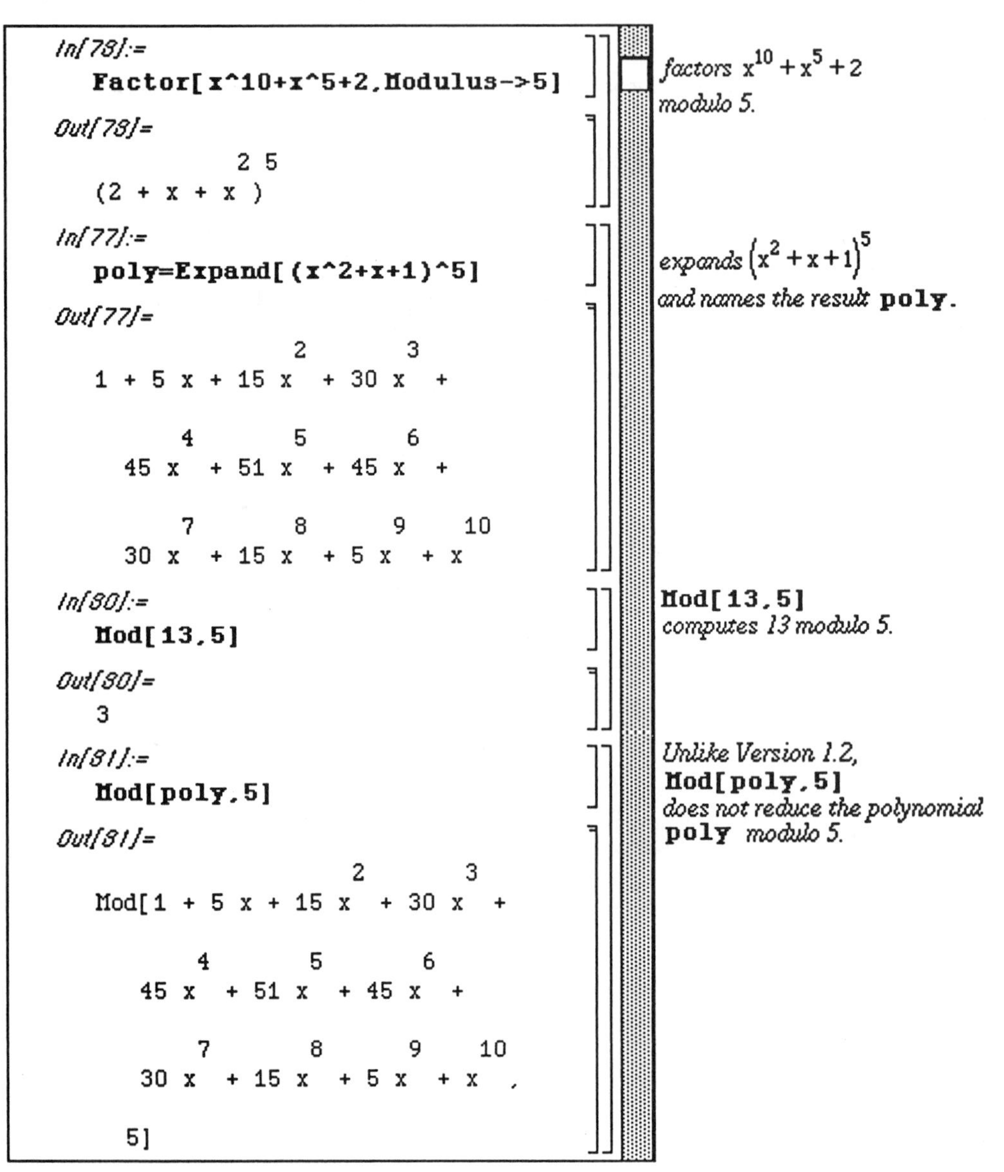

```
In[78]:=
  Factor[x^10+x^5+2,Modulus->5]
Out[78]=
            2 5
  (2 + x + x )
In[77]:=
  poly=Expand[(x^2+x+1)^5]
Out[77]=
                    2       3
  1 + 5 x + 15 x  + 30 x  +

       4        5        6
    45 x  + 51 x  + 45 x  +

       7        8       9    10
    30 x  + 15 x  + 5 x  + x
In[80]:=
  Mod[13,5]
Out[80]=
  3
In[81]:=
  Mod[poly,5]
Out[81]=
                         2       3
  Mod[1 + 5 x + 15 x  + 30 x  +

       4        5        6
    45 x  + 51 x  + 45 x  +

       7        8       9    10
    30 x  + 15 x  + 5 x  + x  ,

    5]
```

factors $x^{10}+x^5+2$ *modulo 5.*

expands $(x^2+x+1)^5$ *and names the result* **`poly`**.

`Mod[13,5]` *computes 13 modulo 5.*

Unlike Version 1.2, **`Mod[poly,5]`** *does not reduce the polynomial* **`poly`** *modulo 5.*

However, coefficients of a polynomial **p** can be reduced modulo **n** with the command **PolynomialMod[p,n]**:

```
In[82]:=
  PolynomialMod[poly,5]
Out[82]=
          5    10
  1 + x   + x
```

PolynomialMod[poly,5] *reduces* **poly** *modulo 5.*

Hence, the previous definition of modexpand must be altered to include PolynomialMod In Version 2.0. This command is then illustrated by creating a table similar to that in the previous example for the prime numbers 13, 17, 19, 23, and 29. Note that this table also includes the prime number as well as the reduced polynomial.

```
In[87]:=
  modexpand[p_,m_]:=
       PolynomialMod[Expand[p],m]
In[89]:=
  modexpand[(x^2+x+1)^11,11]
Out[89]=
          11    22
  1 + x    + x
In[90]:=
  Table[
       {Prime[i],
       modexpand[(x^2+x+b)^Prime[i],Prime[i]]},
       {i,6,10}]  // TableForm
Out[90]//TableForm=
        13     13     26
  13   b   + x    + x

        17     17     34
  17   b   + x    + x

        19     19     38
  19   b   + x    + x

        23     23     46
  23   b   + x    + x

        29     29     58
  29   b   + x    + x
```

2.4 Graphing Functions and Expressions

One of the best features of *Mathematica* is its graphics capabilities. In this section, we discuss methods of graphing functions and several of the options available to help graph functions. The command used to plot real-valued functions of a single variable is **Plot**. The form of the command to graph the function **f[x]** on the domain **[a,b]** is **Plot[f[x],{x,a,b}]**. To plot the graph of **f[x]** in various shades of gray or colors, the command is **Plot[f[x],{x,a,b},PlotStyle->GrayLevel[w]]** where **w** is a number between 0 and 1. **PlotStyle->GrayLevel[0]** represents black; **PlotStyle->GrayLevel[1]** represents a white graph.

If a color monitor is being used, the command is **Plot[f[x],{x,a,b},PlotStyle->RGBColor[r,g,b]]** where **r**, **g**, and **b** are numbers between 0 and 1. **RGBColor[1,0,0]** represents red, **RGBColor[0,1,0]** represents green, and **RGBColor[0,0,1]** represents blue.

□ Example:

Use *Mathematica* to define and graph $f(x) = \mathrm{Sin}(x)$ on the interval $[-2\pi, 2\pi]$ and $g(x) = e^{-x^2}$ on the interval $[-1,1]$.

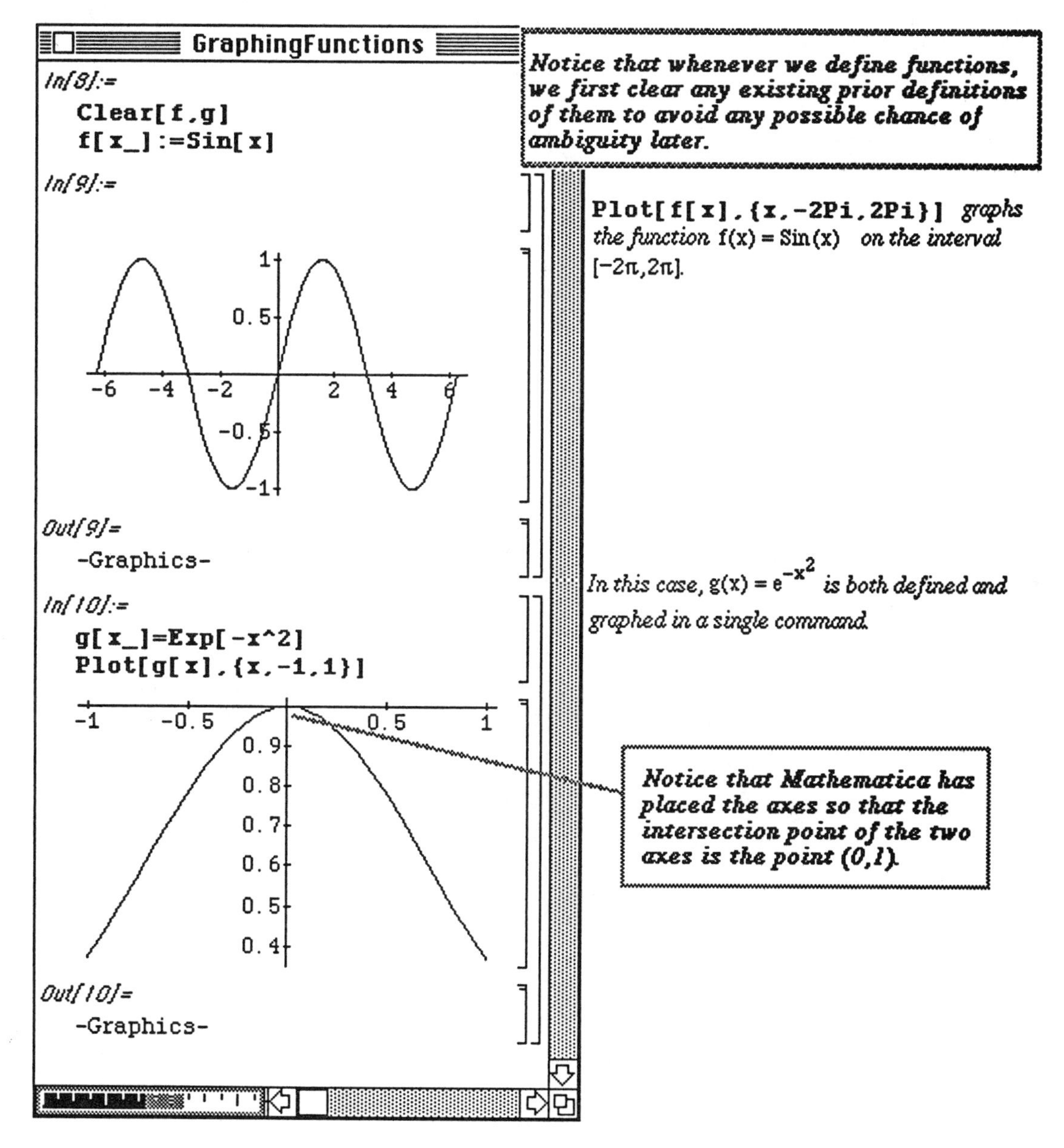

Graphs of functions, like expressions, can be named. This is particularly useful when one needs to refer to the graph of particular functions repeatedly or to display several graphs on the same axes.

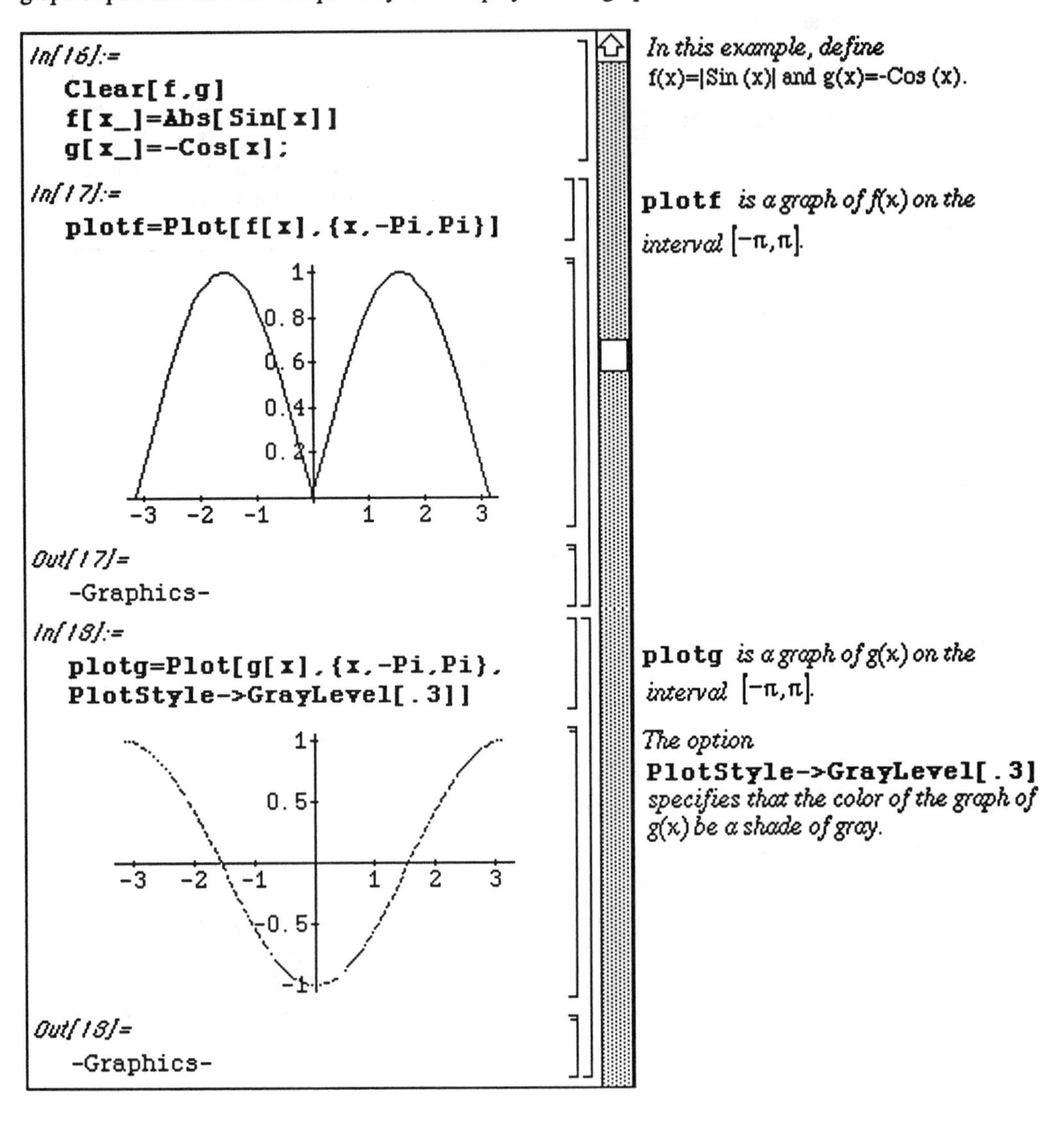

The command used to display several graphs on the same axes is **Show**. To show two graphs named **graph1** and **graph2**, the command entered is **Show[graph1, graph2]**. This command is shown below using **plotf** and **plotg** from above:

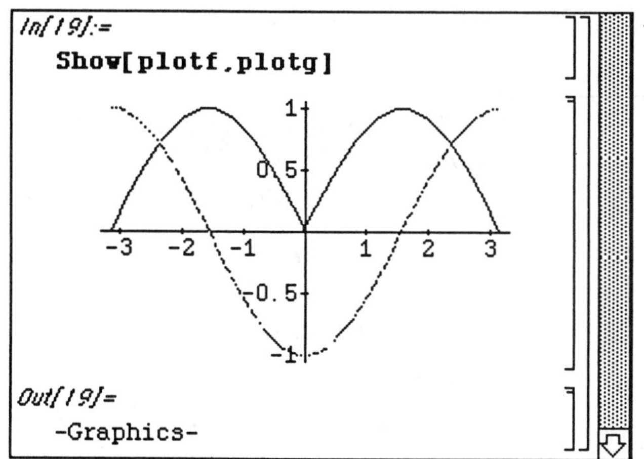

The command **Show[plotf,plotg]** *shows* **plotf** *and* **plotg** *simultaneously.*

More generally, the commands **Plot** and **Show** have many options. To implement the various options, the form of the command **Plot** is **Plot[f[x],{x,a,b},options]**; the form of the command **Show** is **Show[graphs, options]**. The option **DisplayFunction->Identity** prevents the graph from being shown; the option **DisplayFunction->$DisplayFunction** causes the display of a graph which previously was suppressed. For example, one can create several graphs without displaying any of them, and then display all of them simultaneously:

□ Example:

Let $f(x) = e^{-x^2}$, $g(x) = e^{-x^2} + 1 = f(x) + 1$, and $h(x) = e^{-(x-1)^2} = f(x-1)$.

Graph f, g, and h on the intervals [-1,1], [-1,1], and [-2,1], respectively. Show the graphs of all three functions simultaneously.

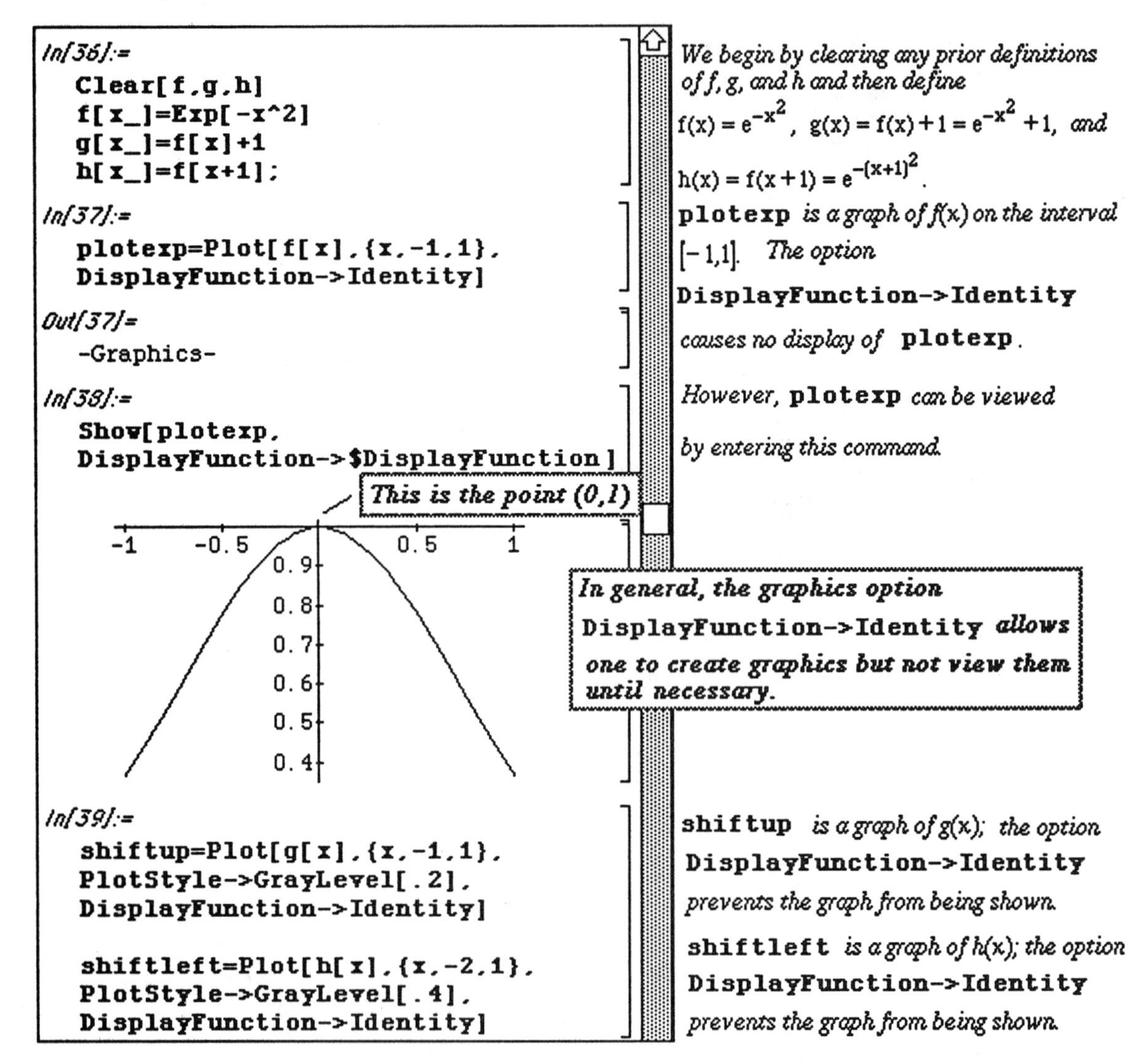

```
In[36]:=
  Clear[f,g,h]
  f[x_]=Exp[-x^2]
  g[x_]=f[x]+1
  h[x_]=f[x+1];
```

We begin by clearing any prior definitions of f, g, and h and then define $f(x) = e^{-x^2}$, $g(x) = f(x)+1 = e^{-x^2}+1$, *and* $h(x) = f(x+1) = e^{-(x+1)^2}$.

```
In[37]:=
  plotexp=Plot[f[x],{x,-1,1},
  DisplayFunction->Identity]
Out[37]=
  -Graphics-
```

plotexp *is a graph of f(x) on the interval* $[-1,1]$. *The option* **DisplayFunction->Identity** *causes no display of* **plotexp**.

```
In[38]:=
  Show[plotexp,
  DisplayFunction->$DisplayFunction]
```

However, **plotexp** *can be viewed by entering this command.*

In general, the graphics option **DisplayFunction->Identity** ***allows one to create graphics but not view them until necessary.***

```
In[39]:=
  shiftup=Plot[g[x],{x,-1,1},
  PlotStyle->GrayLevel[.2],
  DisplayFunction->Identity]

  shiftleft=Plot[h[x],{x,-2,1},
  PlotStyle->GrayLevel[.4],
  DisplayFunction->Identity]
```

shiftup *is a graph of g(x); the option* **DisplayFunction->Identity** *prevents the graph from being shown.*

shiftleft *is a graph of h(x); the option* **DisplayFunction->Identity** *prevents the graph from being shown.*

Even though **shiftup** and **shiftleft** are not shown, they may be viewed along with **plotexp**, using the **Show** command together with the option **DisplayFunction->$DisplayFunction**.

Note that no graphs would be displayed if the **DisplayFunction->$DisplayFunction** option were omitted from the following **Show** command:

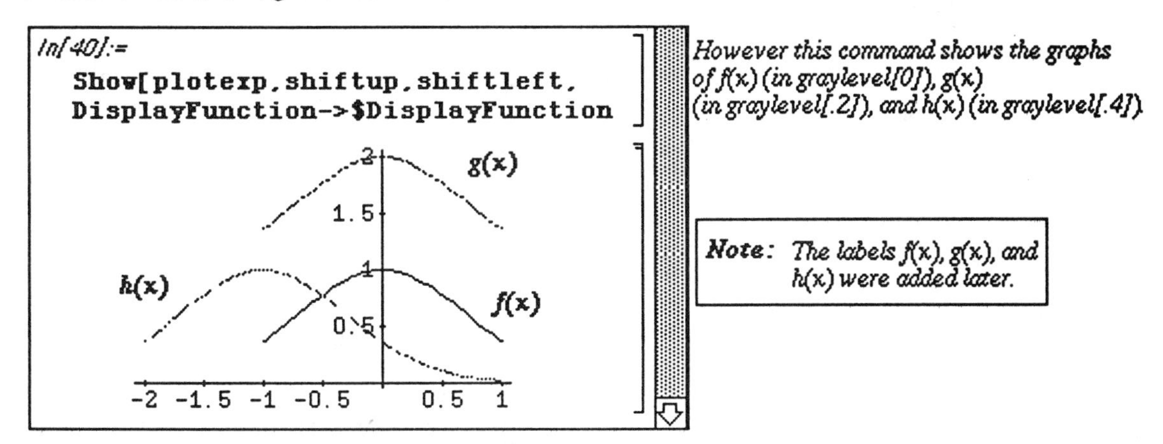

The **Plot** command can also be used to **Plot** several functions simultaneously. To display the graphs of the functions **f[x]**, **g[x]**, and **h[x]** on the domain [a,b] on the same axes, enter commands of the form **Plot[{f[x],g[x],h[x]},{x,a,b},options]**. This command can be generalized to include more than three functions.

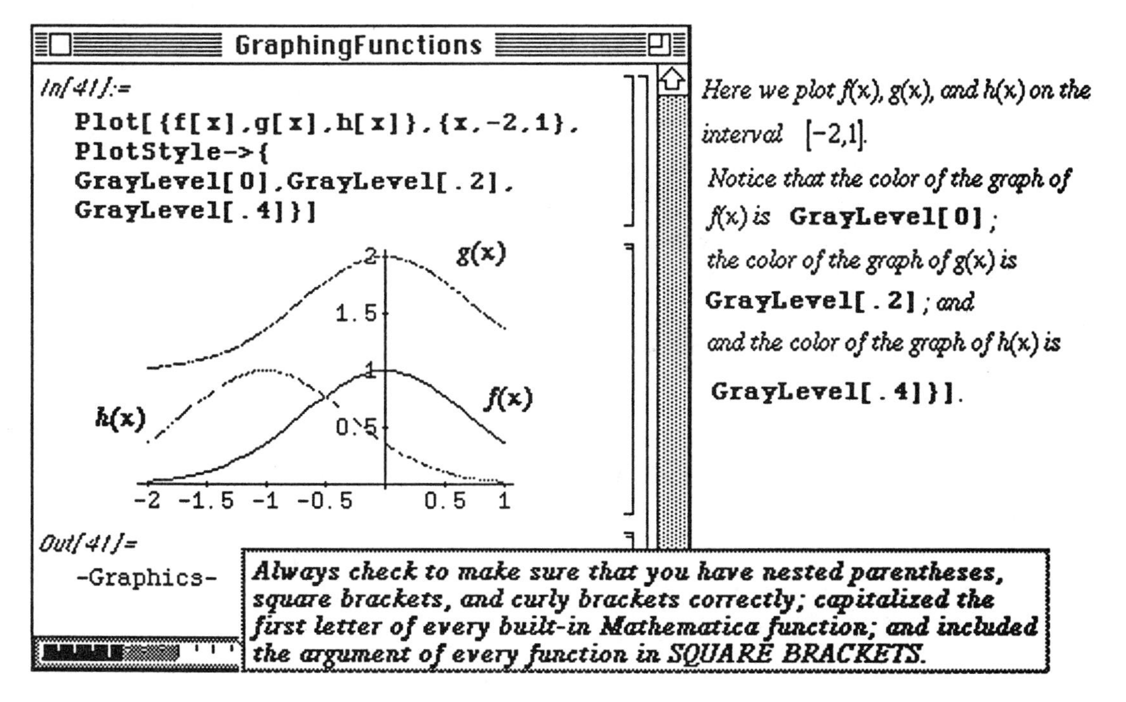

■ Other Available Options

Additional **Plot** options include:

1) **AspectRatio->number**
This makes the ratio of the length of the x-axis to the y-axis **number**. The default value is **1/GoldenRatio**. **GoldenRatio** is a built-in *Mathematica* constant (like **E** and **Pi**)

with value $\frac{1+\sqrt{5}}{2}$ (approximately 1.61803).

2) **Framed->True**
This draws a frame around the graph; the default value is **False**--no frame is drawn.

o In Version 2.0, the option **Framed** is replaced by the option **Frame**. Hence, if you are using Version 2.0, including **Frame->True** instructs *Mathematica* to place a frame around the graph.

3) **Ticks->None** or **Ticks->{{x-axis ticks},{y-axis ticks}}**
This specifies that either no tick marks be placed on either axis <u>OR</u> tick marks be placed on the x-axis at **x-axis ticks** and on the y-axis at **y-axis ticks**.

4) **AxesLabel->{"x-axis label","y-axis label"}**
This labels the x-axis **x-axis label** and the y-axis **y-axis label**. For example, the command **Plot[f[x],{x,xmin,xmax,AxesLabel->{"jane","mary"}]** graphs the function **f[x]** on the interval **[xmin,xmax]**; and labels the x-axis **jane** and the y-axis **mary**. The default for the option is that no labels are shown.

5) **PlotLabel->{"name"}**
This centers **name** above the graph. The default for the option is that the graph is not labeled.

6) **Axes->{x-coordinate,y-coordinate}**
This option specifies that the x-axis and y-axis intersect at the point {**xcoordinate,ycoordinate**}.

o In Version 2.0, **Axes** has been redefined. The option **Axes->False** specifies that the graph is to be drawn without axes; the option **AxesOrigin->{x-coordinate,y-coordinate}** places the axes so they intersect at the point **{x-coordinate,y-coordinate}**.

7) **PlotRange->{y-minimumn,y-maximum}**
specifies the range displayed on the final graph to be the interval **[y-minimum,y-maximum]**; **PlotRange->All** attempts to show the entire graph.

□ Example:

These graphing options are illustrated below:

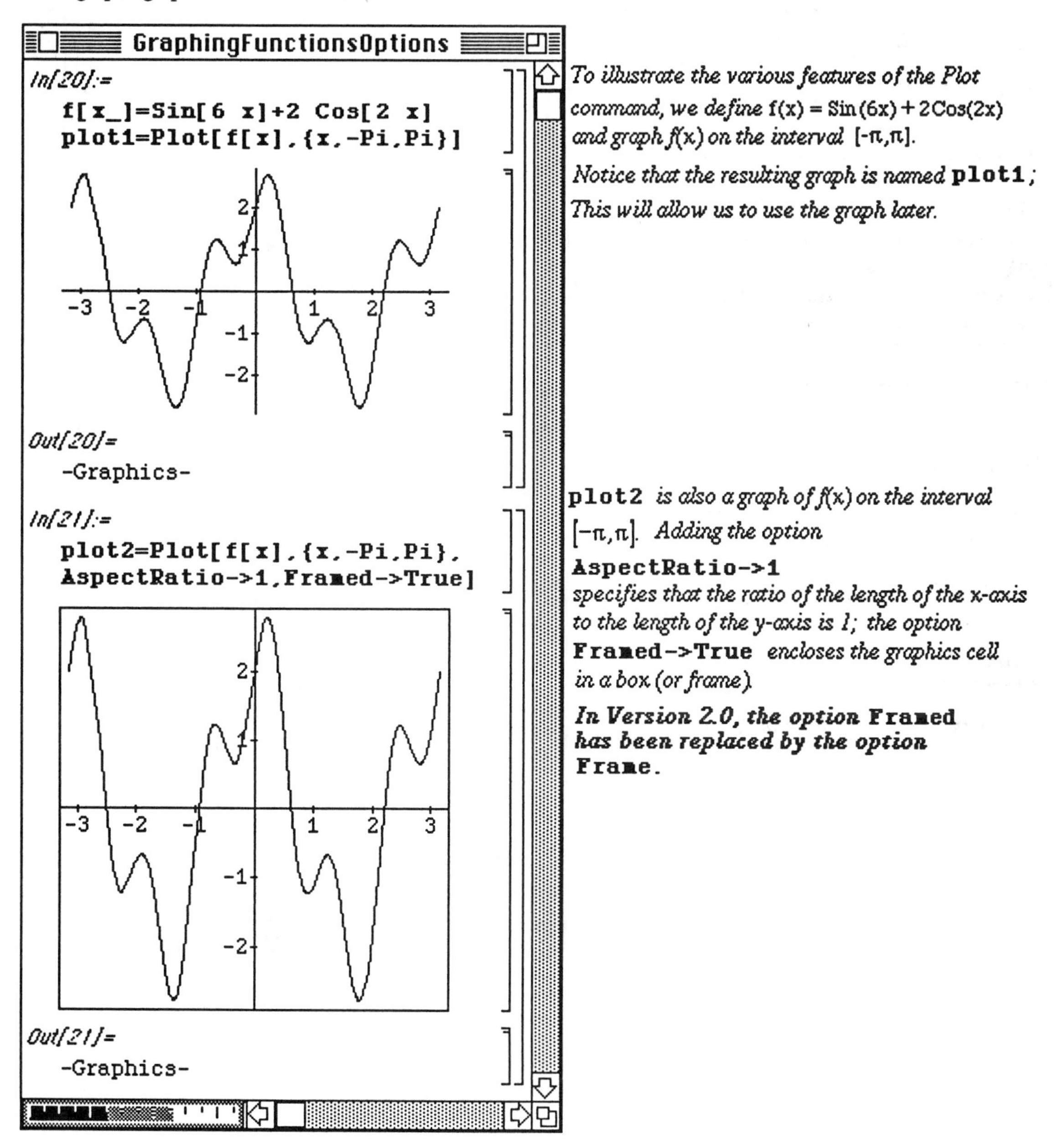

To illustrate the various features of the Plot command, we define f(x) = Sin(6x) + 2Cos(2x) *and graph f(x) on the interval* [-π,π].

Notice that the resulting graph is named **plot1**; *This will allow us to use the graph later.*

plot2 *is also a graph of f(x) on the interval* [-π,π]. *Adding the option* **AspectRatio->1** *specifies that the ratio of the length of the x-axis to the length of the y-axis is 1; the option* **Framed->True** *encloses the graphics cell in a box (or frame).*

***In Version 2.0, the option* Framed *has been replaced by the option* Frame.**

plot3 *is a graph of f(x). The option* **Ticks->None** *specifies that no tick marks are placed on either the x-axis or y-axis; the option* **AxesLabel-> {"x-axis","y-axis"}** *specifies that the x-axis is marked* **x-axis** *and the y-axis is marked* **y-axis**.

> ***When working with the Plot command, be sure to begin with a CAPITAL letter and enclose the entire command in square brackets.***

For **plot4** *the option* **Ticks->{{-Pi,0,Pi},{-2,0,2}}** *specifies that tick marks be placed at* x=-π and x=π *on the x-axis and* y=-2 and y=2 *on the y-axis; the option* **PlotLabel-> "f[x]=Sin[6 x]+2 Cos[2 x]"** *spcifies that the top of the graph is marked* f[x]=Sin[6 x]+2 Cos[2 x].

GraphingFunctionsOptions

```
plot3=Plot[f[x],{x,-Pi,Pi},
Ticks->None,
AxesLabel->
{"x-axis","y-axis"}]
```

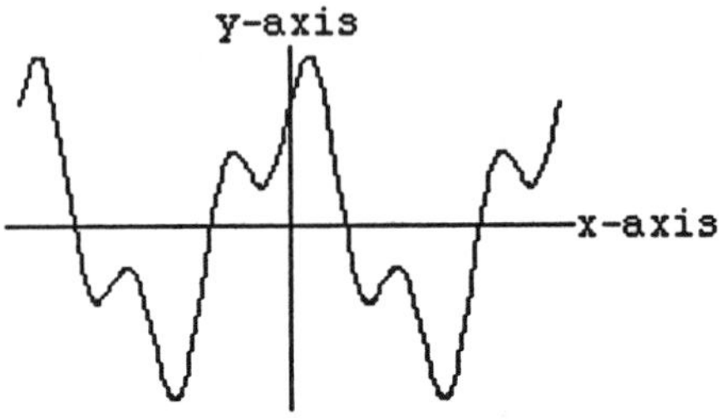

y-axis
x-axis

Out[23]=
-Graphics-

```
plot4=Plot[f[x],{x,-Pi,Pi},
Ticks->{{-Pi,0,Pi},{-2,0,2}},
PlotLabel->
"f[x]=Sin[6 x]+2 Cos[2 x]"]
```

f[x]=Sin[6 x]+2 Cos[2 x]
2
-Pi
Pi
-2

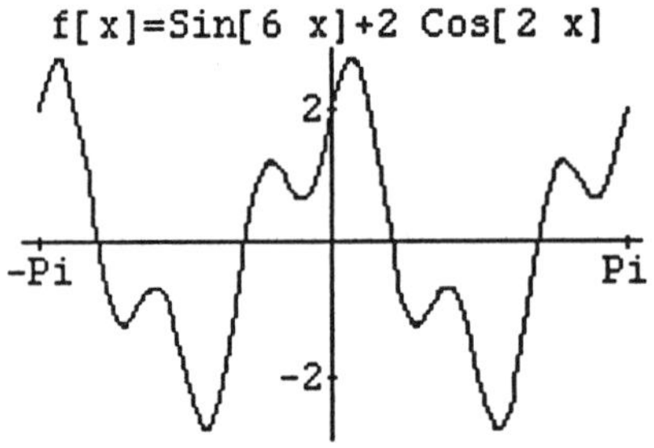

Out[25]=
-Graphics-

• Graphing Features and Options of Version 2.0

Version 2.0 of *Mathematica* offers several plotting options which are not available or differ from those in Version 1.2. In the first example below, the fact that a semi-colon must follow a command in Version 2.0 in order that it be suppressed is illustrated. (In Version 1.2, only the output of the last command in a single input cell is given even if semi-colons are not used.) After defining the function **f**, the graph of **f** is plotted and called **plotf**. Since a semi-colon follows the definition of **f**, the formula for **f** is not given in the output. Also shown below is the **GridLines** option in the **Plot** command. Notice in **feature1**, **GridLines->Automatic** causes horizontal and vertical gridlines to be shown on the graph.

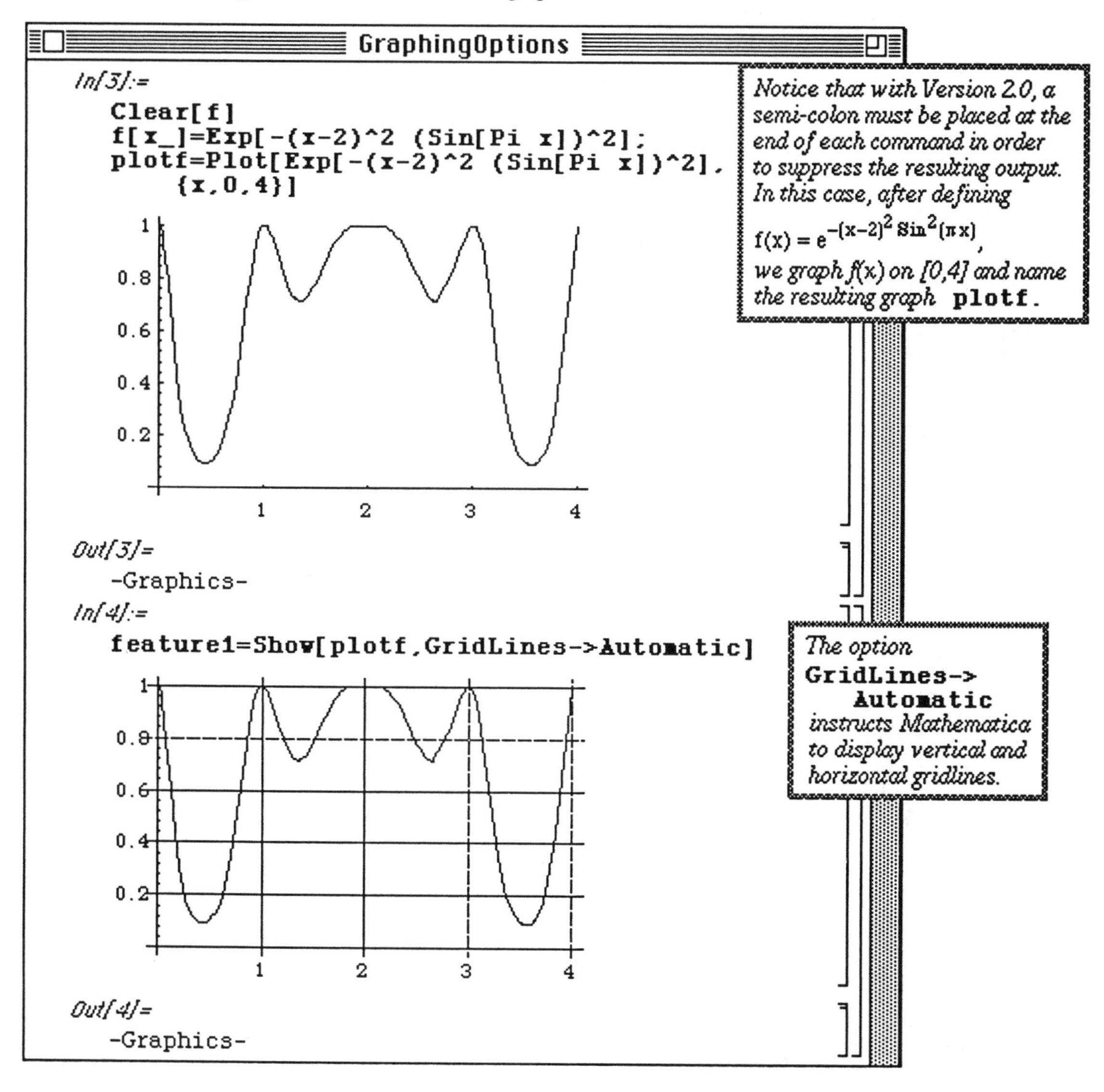

The **GridLines** option can be altered slightly. The following examples illustrate how one type of gridline is requested. In **feature2, GridLines->{None, Automatic}** specifies that only horizontal gridlines be displayed while in **feature3, GridLines->{{1,2,3},None}** gives vertical gridlines at x = 1, 2, and 3. Also in **feature3, Ticks->{Automatic,None}** causes tick marks to be placed on the x-axis but none on the y-axis. Finally in **feature3**, the x and y axes are labeled with the option **AxesLabel->{"x-axis", "y-axis"}**.

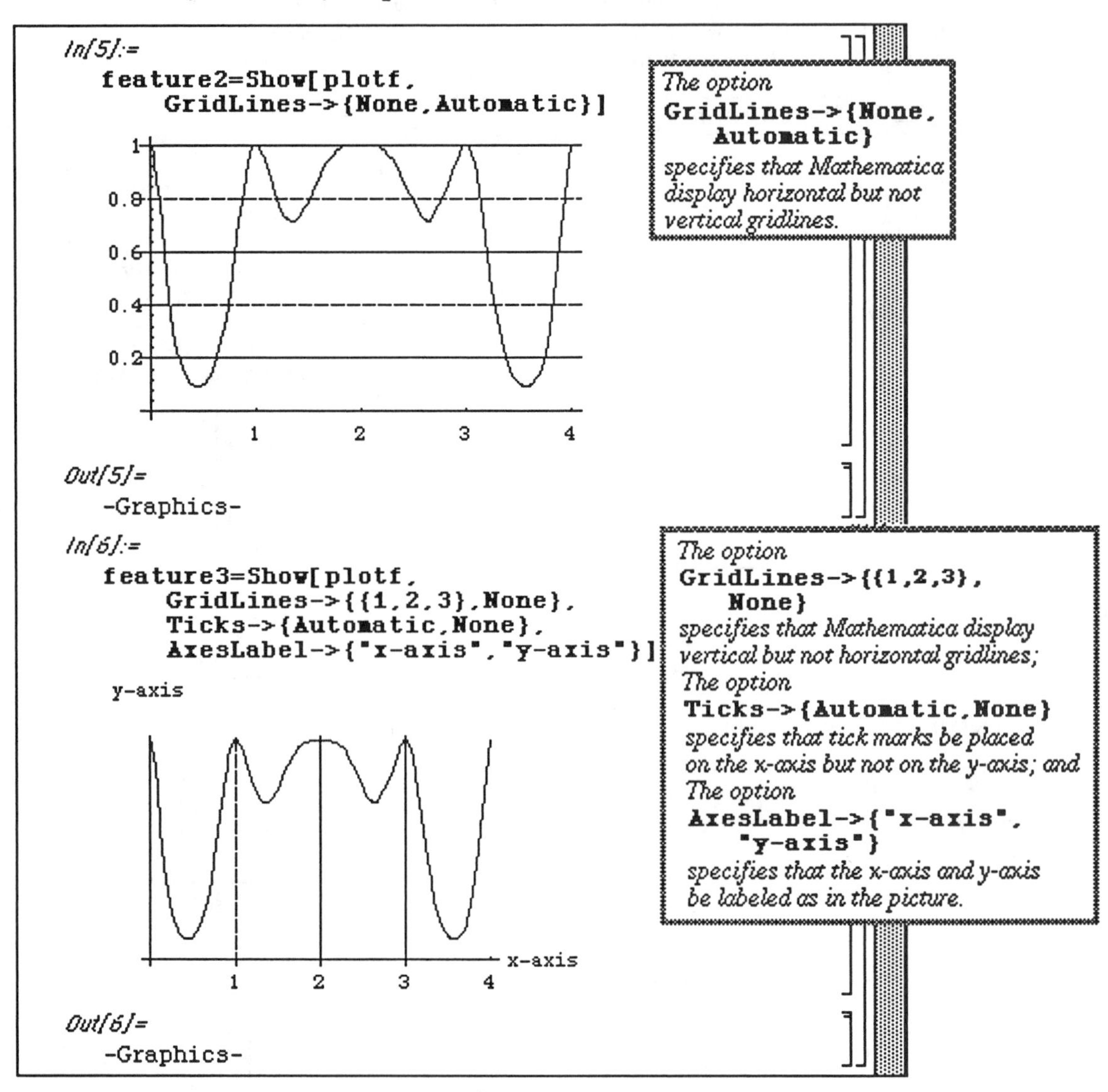

Several other **Plot** options are shown in the examples below. In **feature4, AxesOrigin->{x0,y0}** is illustrated. This causes the major axes to be drawn in such a way that they meet at the point **{x0,y0}**. Another option is **Frame** which is demonstrated in both examples. **Frame->True** encloses the graph in a frame.

o In Version 2.0, **AxesOrigin** replaces **Axes** from Version 1.2 and **Frame** replaces **Framed**.

Note the tick marks which accompany the frame in **feature4**. In **feature5**, however, the **FrameTicks->None** option prohibits the marking of ticks on the frame. Also notice the **PlotLabel** option which appears in each **Plot** command. In **feature4**, the label is given in quotation marks. This causes the function within the quotations to be printed exactly as it appears in the **PlotLabel** option. Since the label does not appear in quotations in **feature5**, the label is given in mathematical notation.

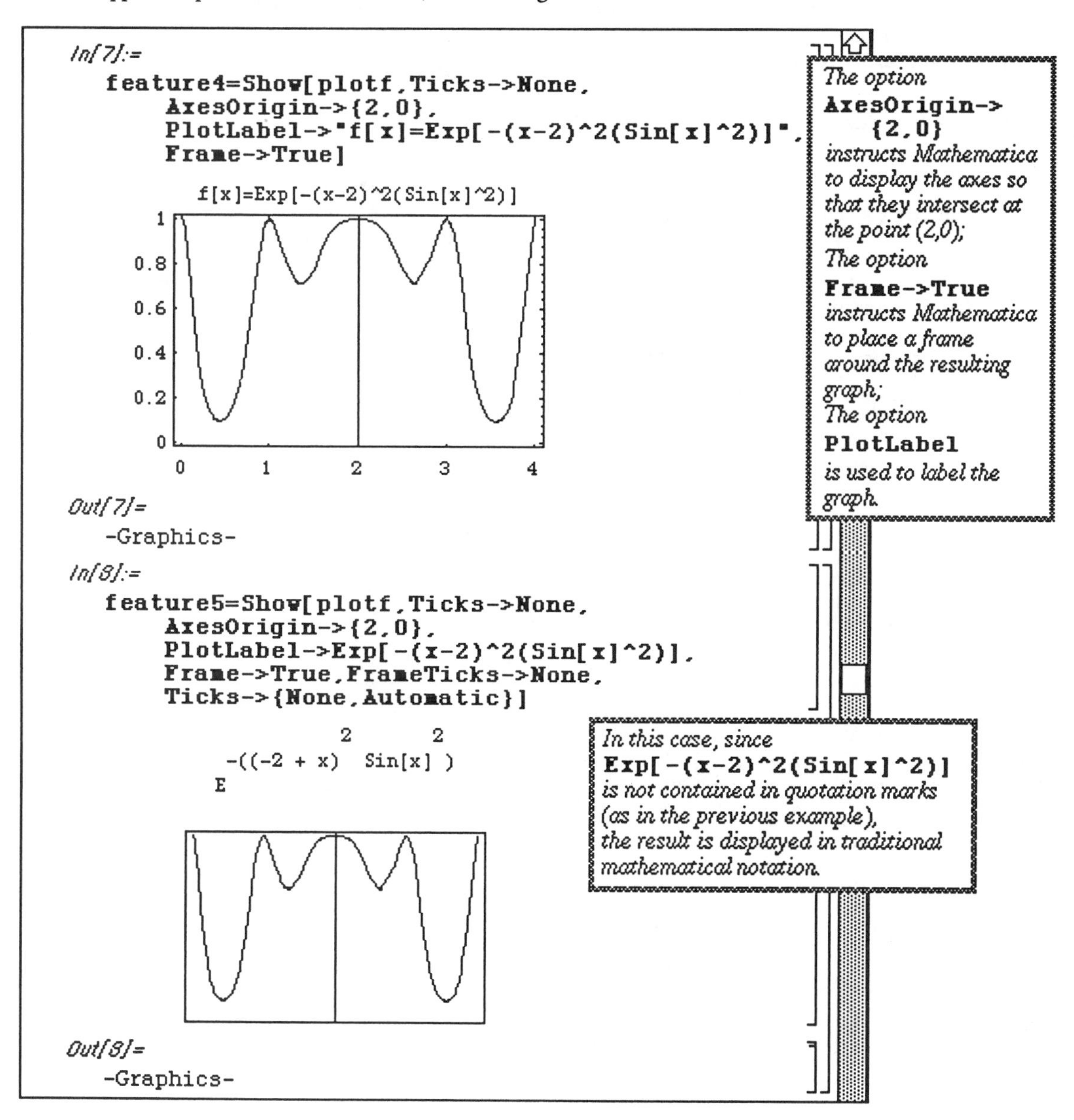

• Displaying Several Graphs with Version 2.0

The plots given in **feature1**, **feature2**, **feature3**, and **feature4** are viewed below in a single graphics cell with the **GraphicsArray** option. **GraphicsArray[{feature1,feature2},{feature3,feature4}}]** produces an array of graphics objects called **features** which is viewed in pairs with **Show[features]**, where **features=GraphicsArray[{{feature1,feature2},{feature3,feature4}}]**. In general, **GraphicsArray** can be used to visualize any m x n array of graphics objects.

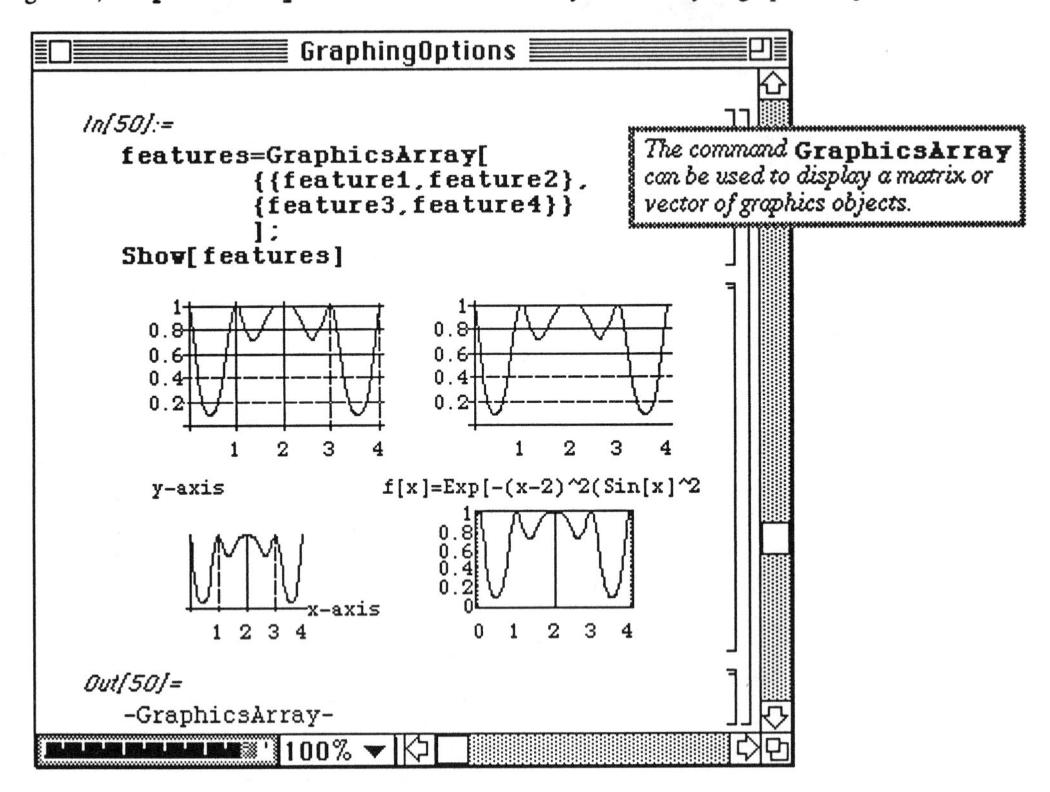

The command **Rectangle[{x0,y0},{x1,y1}]** creates the graphics primitive for a filled rectangle with sides along the lines **x** = **x0**, **y** = **y0**, **x** = **x1**, and **y** = **y1**. Hence, other *Mathematica* commands must be used to visualize the rectangles represented by **Rectangle[{x0,y0},{x1,y1}]**. Visualization is accomplished with **Show** and **Graphics** as illustrated below.

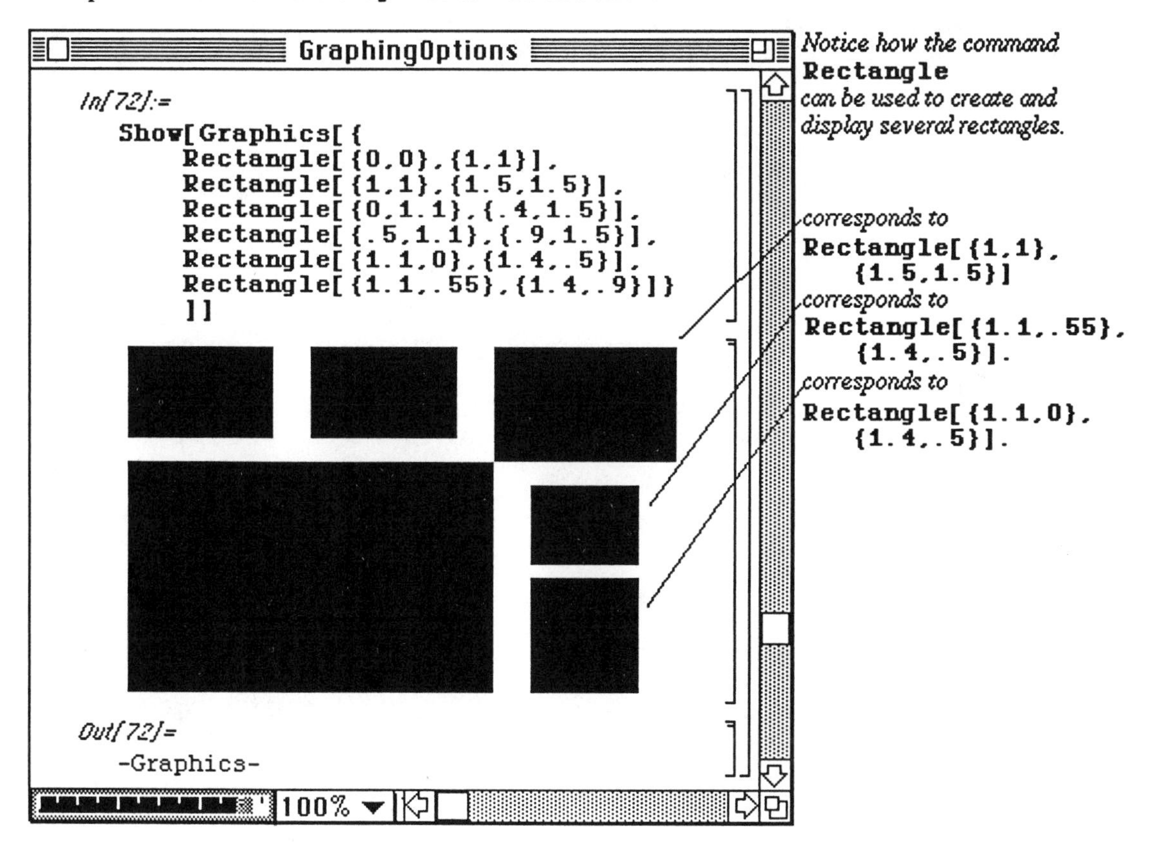

Rectangle[{x0,y0},{x1,y1}] can be used in conjunction with other graphics cells to produce graphics of a particular size. The command **Show[Rectangle[{x0,y0},{x1,y1},plot]]** displays **plot** within the rectangle determined with **Rectangle[{x0,y0},{x1,y1}]**. This is illustrated below with rectangles from the previous example as well as earlier plots.

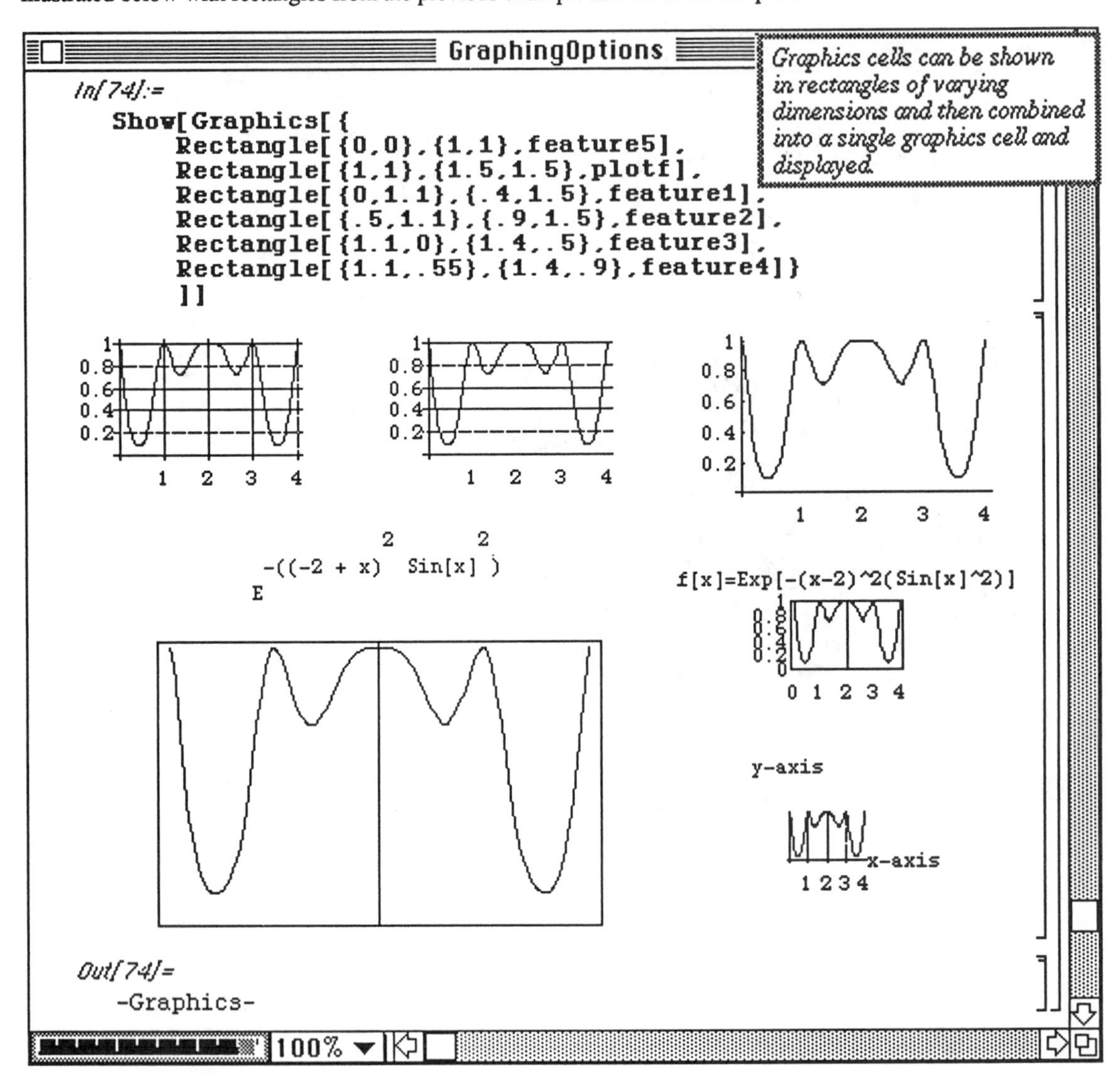

o Labeling Graphs in Version 2.0

In addition to the above features, graphs created with Version 2.0 can be labeled in a variety of ways. For example, in the following example the options **`DefaultFont->{"font",size}`** and **`PlotLabel->FontForm["label",{"font",size}]`**, where **`font`** and **`size`** is a font available on your computer and **`label`** is the desired graphics label, are used to create several trigonometric graphs.

The numbering of the tick marks of **`plotsin`** are in size 12 Times font; the graph is labeled in size 14 Times font:

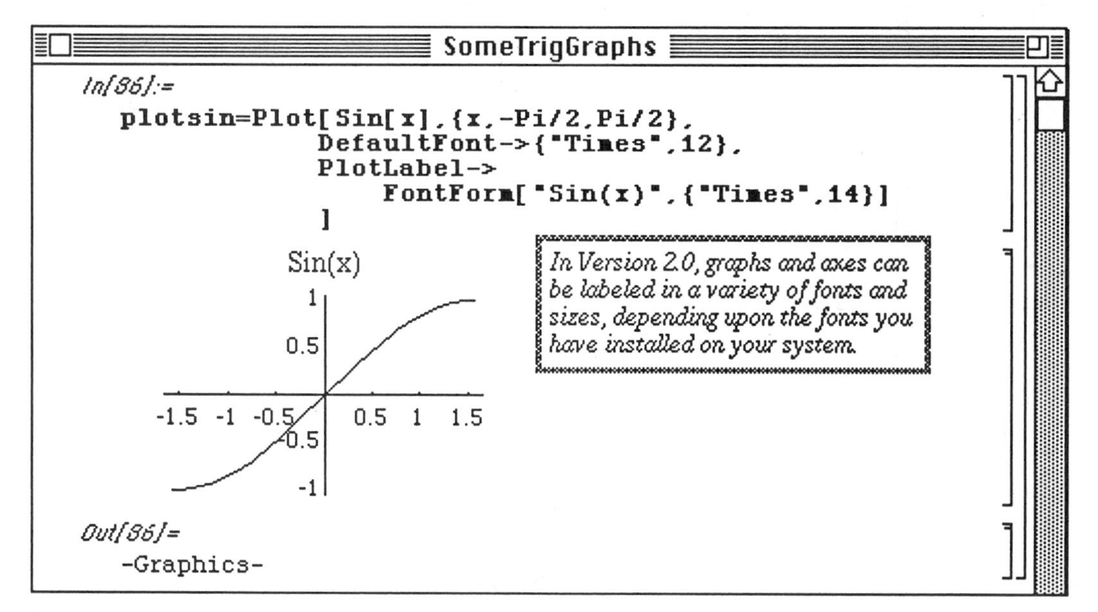

Similarly, the axes can also be labeled in different fonts and sizes using the option

```
AxesLabel->{
  FontForm["x-axis  label",{"font",size}],
  FontForm["y-axis  label",{"font",size}]
          }
```

In the following example, the function $\text{ArcSin}(x) = \text{Sin}^{-1}(x)$ is graphed on the interval $[-1,1]$. The axes are labeled "x-axis" and "y-axis" in size 12 Times font. The graph is labeled "ArcSin (x)" in size 12 Venice font since the **DefaultFont** is chosen to be size 12 Venice font:

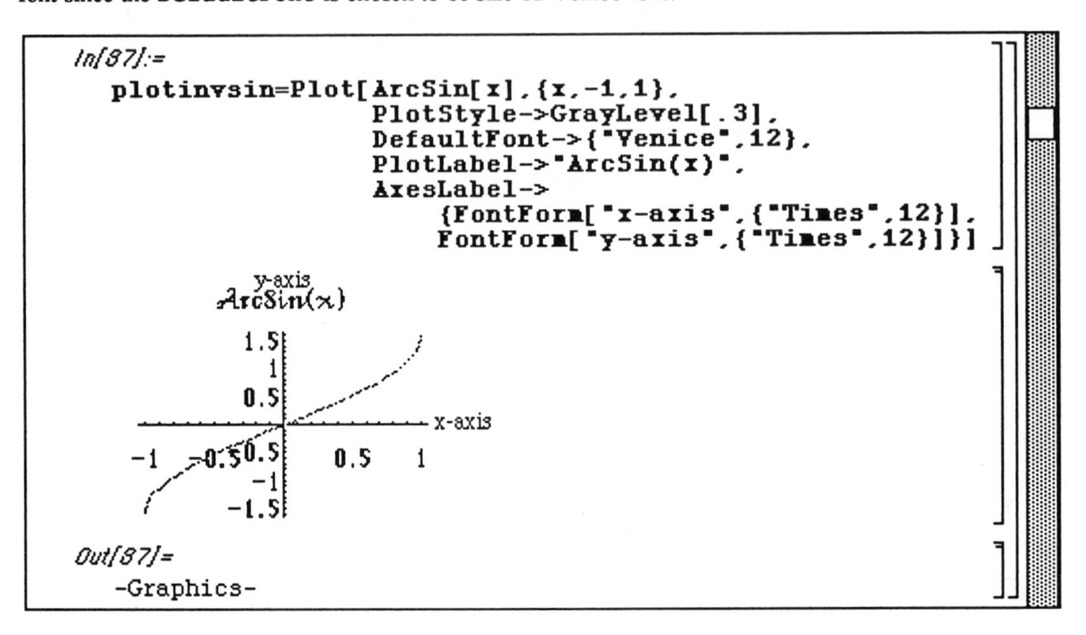

Naturally, many different options can be combined together. In the following window, the previous two graphs are displayed. The option **`Ticks->None`** specifies that no tick marks are to be drawn on either axis; the graph is labeled "Sin(x) and ArcSin(x)" in size 14 London font:

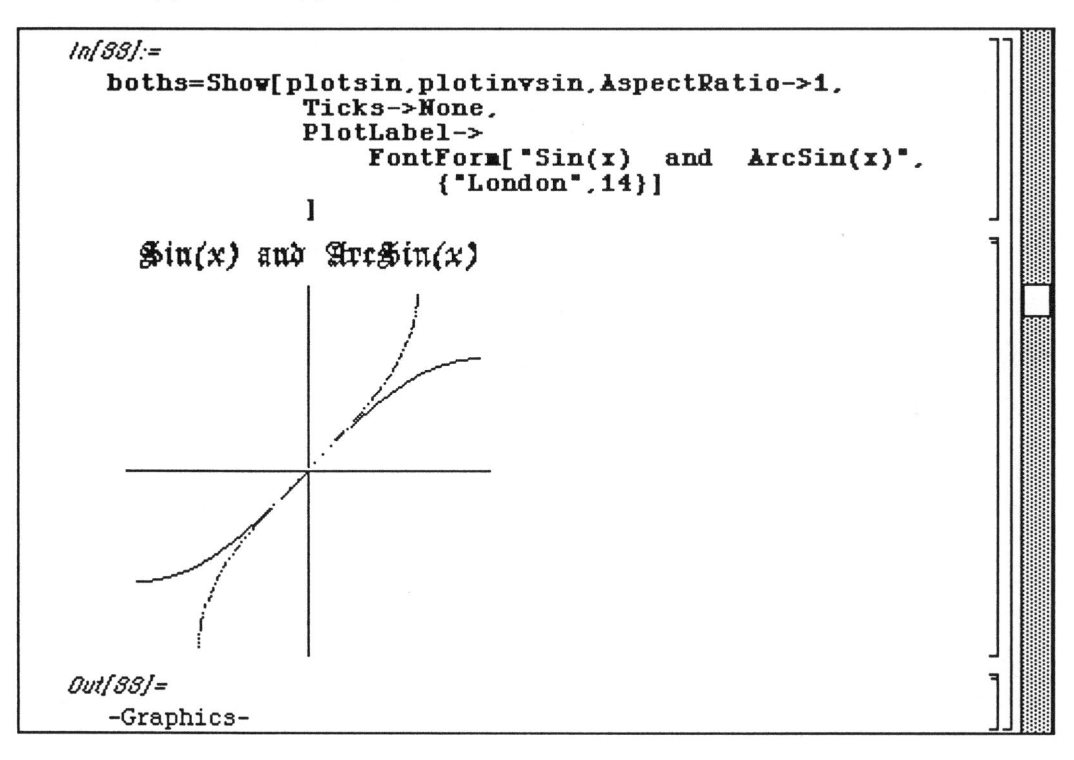

In the next example we graph Cos(x) and ArcCos(x). The option **DisplayFunction->Identity** is used so the graphs are not immediately displayed. Instead, these three graphs are shown simultaneously with the three previous graphs, **plotsin**, **plotinvsin**, and **boths**, as a graphics array:

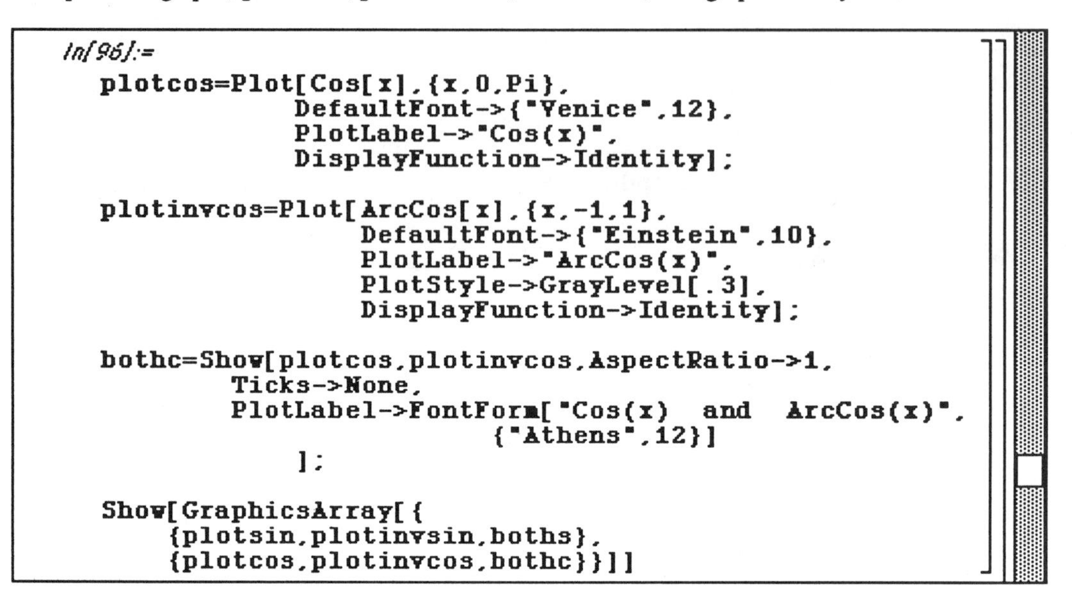

```
In[96]:=
plotcos=Plot[Cos[x],{x,0,Pi},
             DefaultFont->{"Venice",12},
             PlotLabel->"Cos(x)",
             DisplayFunction->Identity];

plotinvcos=Plot[ArcCos[x],{x,-1,1},
               DefaultFont->{"Einstein",10},
               PlotLabel->"ArcCos(x)",
               PlotStyle->GrayLevel[.3],
               DisplayFunction->Identity];

bothc=Show[plotcos,plotinvcos,AspectRatio->1,
        Ticks->None,
        PlotLabel->FontForm["Cos(x)  and  ArcCos(x)",
                        {"Athens",12}]
           ];

Show[GraphicsArray[{
     {plotsin,plotinvsin,boths},
     {plotcos,plotinvcos,bothc}}]]
```

All six graphs are then displayed as a graphics array, illustrating the various options we have used:

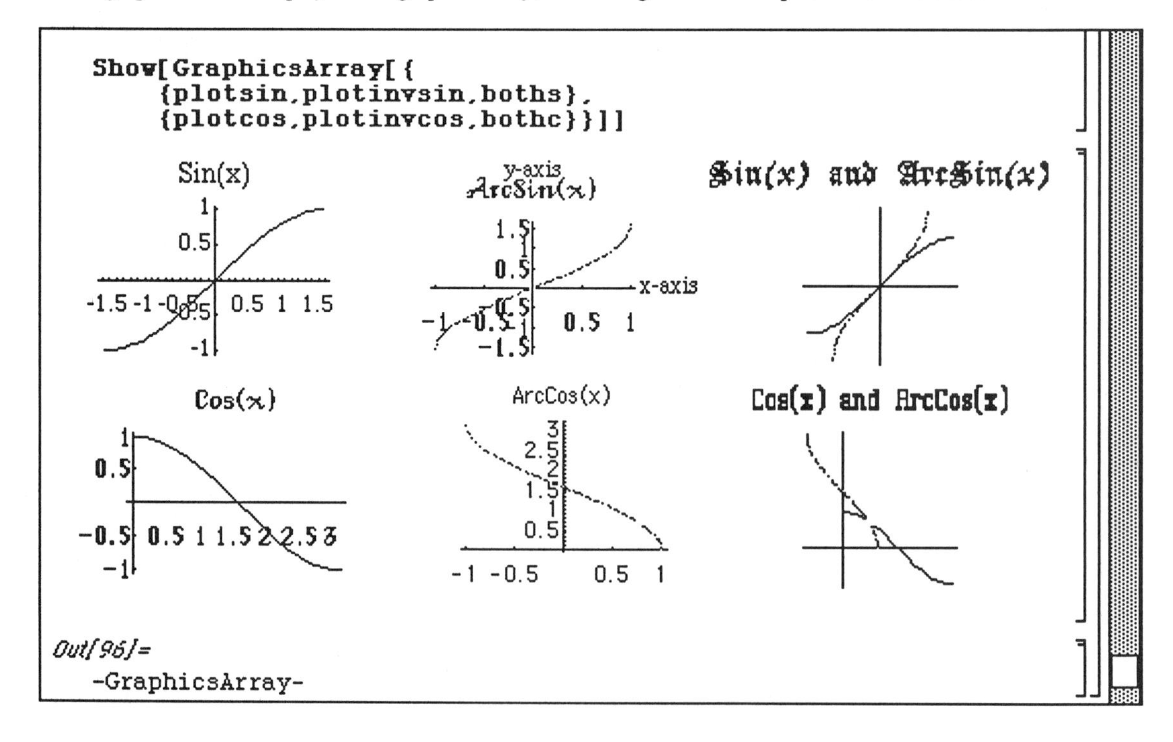

```
Show[GraphicsArray[{
     {plotsin,plotinvsin,boths},
     {plotcos,plotinvcos,bothc}}]]
```



```
Out[96]=
-GraphicsArray-
```

■ Piecewise Defined Functions

Piecewise defined functions may also be defined and graphed with *Mathematica*. In the following example, h(x) is defined in three "pieces". Notice that **/;** designates the definition of h(x) for different domain values.

□ Example:

Use *Mathematica* to graph h(x) on the interval [−3, 3] if $h(x)=\begin{cases} 6+2x & \text{for } x \le -2 \\ x^2 & \text{for } -2 < x \le 2. \\ 11-3x & \text{for } x > 2 \end{cases}$

Not that **<=** represents a less than or equal to symbol.

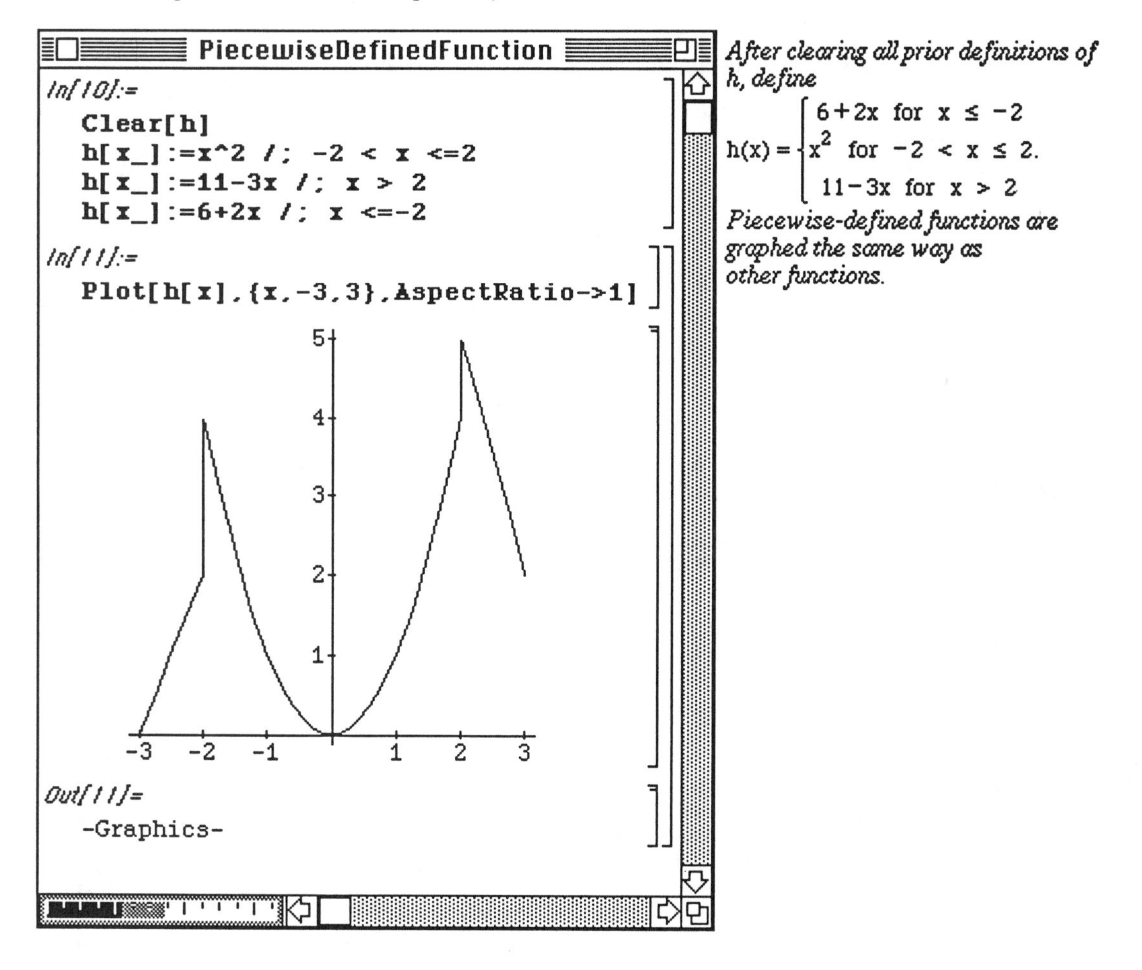

After clearing all prior definitions of h, define

$h(x)=\begin{cases} 6+2x & \text{for } x \le -2 \\ x^2 & \text{for } -2 < x \le 2. \\ 11-3x & \text{for } x > 2 \end{cases}$

Piecewise-defined functions are graphed the same way as other functions.

Functions can be defined recursively. For example, if the function **f[x]** is defined on the interval [**a**,**b**], then f can be defined for x > b with **f[x_]:=f[x-(b-a)] /; x>b**. Two examples are given below. Functions of this type are useful in the study of Fourier series.

o Example:

(A) If $-1 \le x \le 1$, let $f(x)=\begin{cases} 1 \text{ for } 0 \le x \le 1; \\ -1 \text{ for } -1 \le x < 0 \end{cases}$. If $x > 1$, define $f(x)$ recursively by $f(x)=f(x-2)$. Use *Mathematica* to define f and graph f on the interval [0,6].

(B) If $-1 \le x \le 3$, let $h(x)=\begin{cases} \frac{1}{2}x \text{ for } -1 \le x \le 2; \\ 1 \text{ for } 2 \le x < 3 \end{cases}$. If $x > 3$, define $h(x)$ recursively by $h(x)=h(x-4)$. Use *Mathematica* to define h and graph h on the interval [0,12].

o Version 2.0 was used in the following solution to illustrate the Version 2.0 graphics option **Background->GrayLevel[n]**, where n is between 0 and 1. If using Version 2.0, the functions are defined the exact same way; however, the option **Background->GrayLevel[n]** is not available in Version 1.2:

```
In[84]:=
   Clear[f]
   f[x_]:=1 /; 0<=x<=1
   f[x_]:=-1 /; -1<=x<0
   f[x_]:=f[x-2] /; x > 1
```

After clearing all prior definitions of f, f is defined by $f(x)=\begin{cases} 1 \text{ for } 0 \le x \le 1 \\ -1 \text{ for } -1 \le x < 0 \end{cases}$. *Then f(x)=f(x-2) for x > 1 defines f recursively for x greater than one.*

```
In[86]:=
   graphf=Plot[f[x],{x,0,6},DisplayFunction->Identity];
   graphftwo=Plot[f[x],{x,0,6},
        Background->GrayLevel[.2],
        PlotStyle->{{Thickness[.01],GrayLevel[1]}},
        DisplayFunction->Identity];
```

graphf graphftwo *are different graphs of f illustrating Plot options available in Version 2.0.*

```
In[90]:=
   Clear[h]
   h[x_]:=1/2x /; -1<=x<=2
   h[x_]:=1 /; 2<=x<=3
   h[x_]:=h[x-4] /; 3<x
In[92]:=
   graphh=Plot[{h[x],1/2 h[x-1]},{x,0,12},
        PlotStyle->{GrayLevel[0],GrayLevel[.3]},
        DisplayFunction->Identity];
   graphhtwo=Plot[{h[x],1/2 h[x-1]},{x,0,12},
        Background->GrayLevel[0],
        PlotStyle->{{GrayLevel[1]},{GrayLevel[.8]}},
            DisplayFunction->Identity];
```

In Version 2.0, arrays of graphics cells can be visualized with the command **GraphicsArray**. Since **graphf**, **graphftwo**, **graphh**, and **graphhtwo** are graphics cells, **{{graphf,graphftwo},{graphh,graphhtwo}}** is an array of graphics cells. All four can be viewed in a single graphics cell using the command **GraphicsArray**:

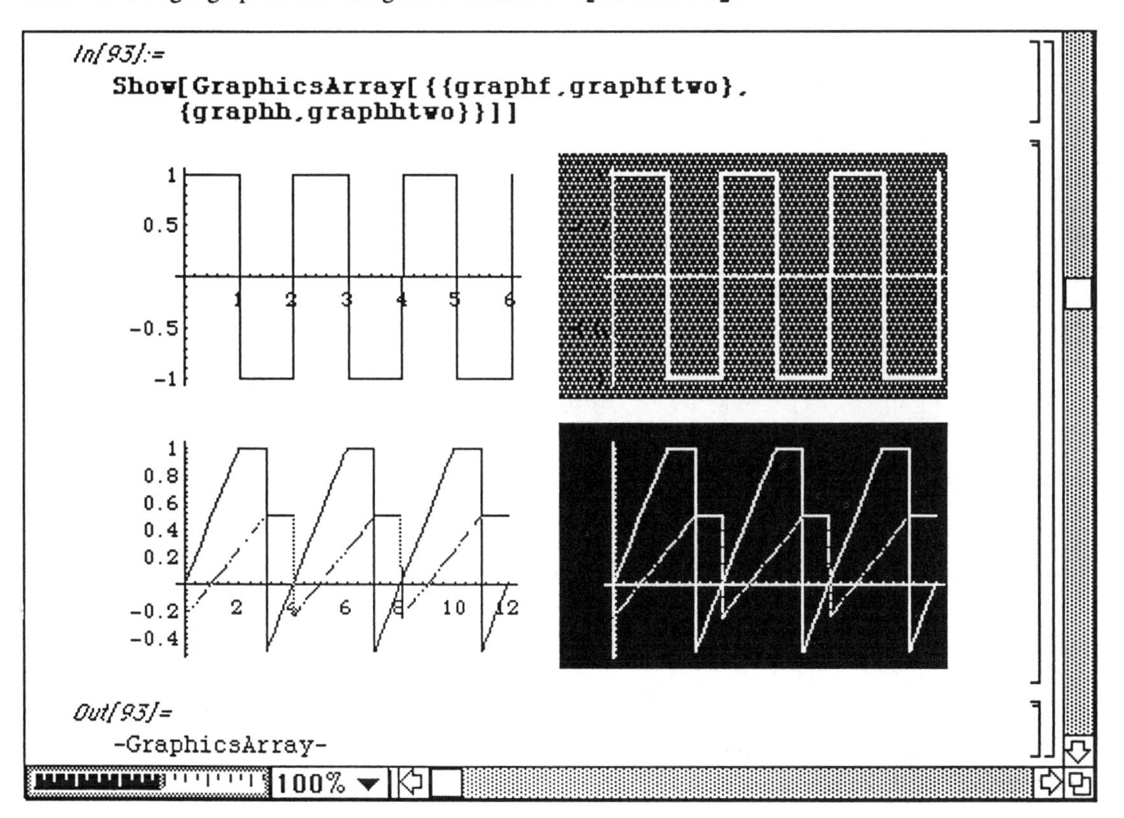

Sometimes it is useful to have *Mathematica* remember functional values it computes. For example, this is particularly useful when defining functions recursively as in the previous examples. In general, to define a function **f** to remember the values it computes enter the definition in the form
f[x_]:=f[x]=*mathematical expression*.

□ Example:

Use *Mathematica* to define $k(x) = \mathrm{Exp}\left[-(x-2)^2\left|\mathrm{Cos}(\pi(x-2))\right|\right]$. For $0 \le x \le 4$, define $g(x) = k(x)$ and for $x > 4$, define g recursively by $g(x) = g(x-4)$. Graph g on the interval [0,16].

In the following example, notice that k is defined so that *Mathematica* remembers the values of k(x) it computes. Since g is defined recursively in terms of k, evaluation of g for values of x greater than four proceeds quickly since the corresponding k-values have already been computed and remembered:

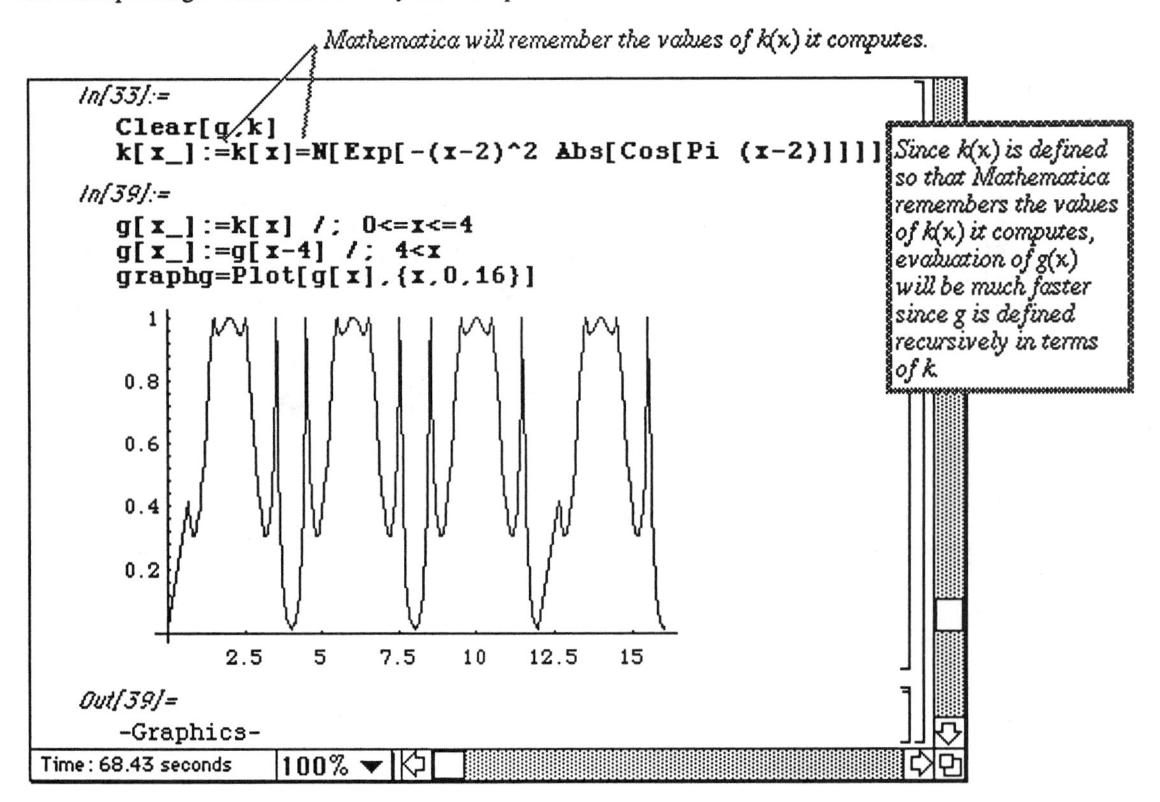

2.5 Exact and Approximate Solutions of Equations

Exact Solutions of Equations

Mathematica can solve many equations exactly. For example, *Mathematica* can find exact solutions to systems of equations and exact solutions to polynomial equations of degree four or less. Since a single equals sign (=) is used to name objects in *Mathematica*, equations in *Mathematica* are of the form **`left-hand side==right-hand side`**. The "double-equals" sign (==) between the left hand side and right hand side specifies that the object is an equation. For example, to represent the equation 3x+7=4 in *Mathematica*, type **`3x+7==4`**. The command **`Solve[lhs==rhs,x]`** solves the equation lhs=rhs for x. If the **<u>only</u>** unknown in the equation lhs=rhs is x and *Mathematica* does not need to use inverse functions to solve for x, then the command **`Solve[lhs==rhs]`** solves the equation lhs=rhs for x. Hence, to solve the equation 3x+7=4, both the command **`Solve[3x+7==4]`** and **`Solve[3x+7==4,x]`** produce the same result.

<u>Notice</u> that the representation of equations in *Mathematica* involves replacing the traditional single equals sign by a double equals sign:

Example:

Use *Mathematica* to find exact solutions of the equations $3x+7=4$, $\frac{x^2-1}{x-1}=0$, and $x^3+x^2+x+1=0$.

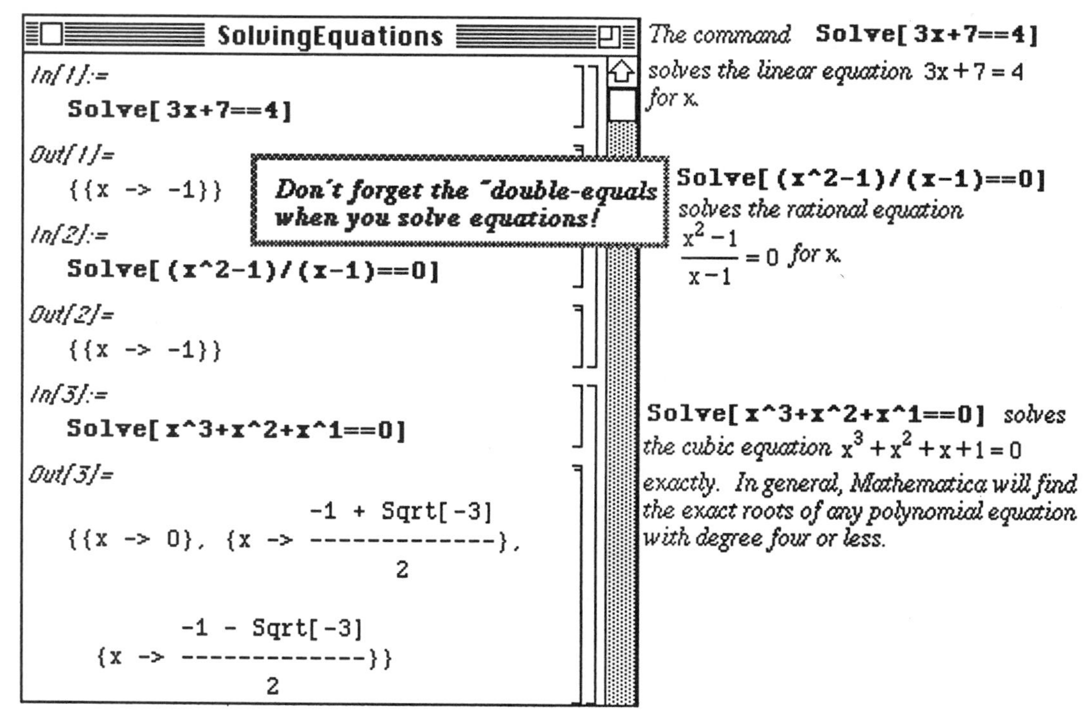

As stated above, the exception to the above rule is when using the command **Solve** to find solutions of equations where inverse functions must be used:

□ Example:

Solve $\mathrm{Sin}^2(x) - 2\mathrm{Sin}(x) - 3 = 0$.

When the command **Solve[Sin[x]^2-2Sin[x]-3==0]** is entered, *Mathematica* solves the equation for **Sin[x]**. However, when the command **Solve[Sin[x]^2-2Sin[x]-3==0,x]** is entered, *Mathematica* attempts to solve the equation for x. In this case, *Mathematica* succeeds in finding one solution:

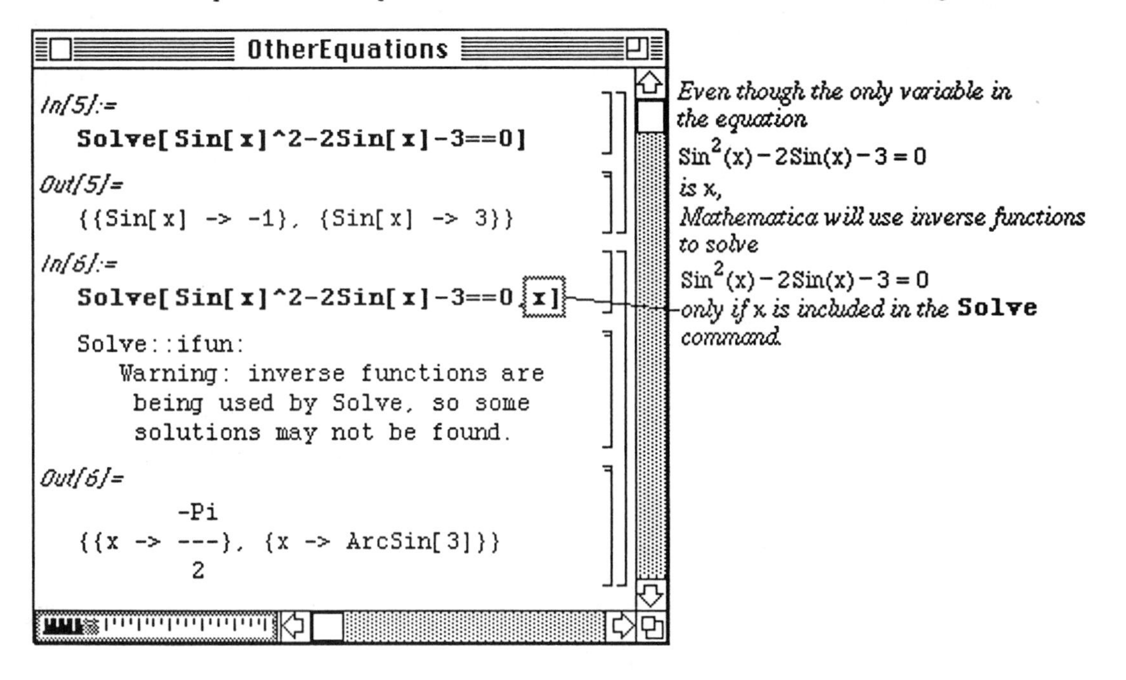

`Solve[{lhs1=rhs1,lhs2==rhs2},{x,y}]` solves a system of two equations for x and y. **`Solve[{lhs1==rhs1,lhs2==rhs2}]`** attempts to solve the system of equations for all unknowns. In general, **`Solve`** can find the solutions to a system of linear equations. In fact, if the systems to be solved are inconsistent or dependent, *Mathematica* 's output will tell you so.

□ **Example:**

Use *Mathematica* to solve each of the systems of equations (i) $\begin{cases} 3x - y = 4 \\ x + y = 2 \end{cases}$; and

(iii) $\begin{cases} 2x - 3y + 4z = 2 \\ 3x - 2y + z = 0. \\ x + y - z = 1 \end{cases}$

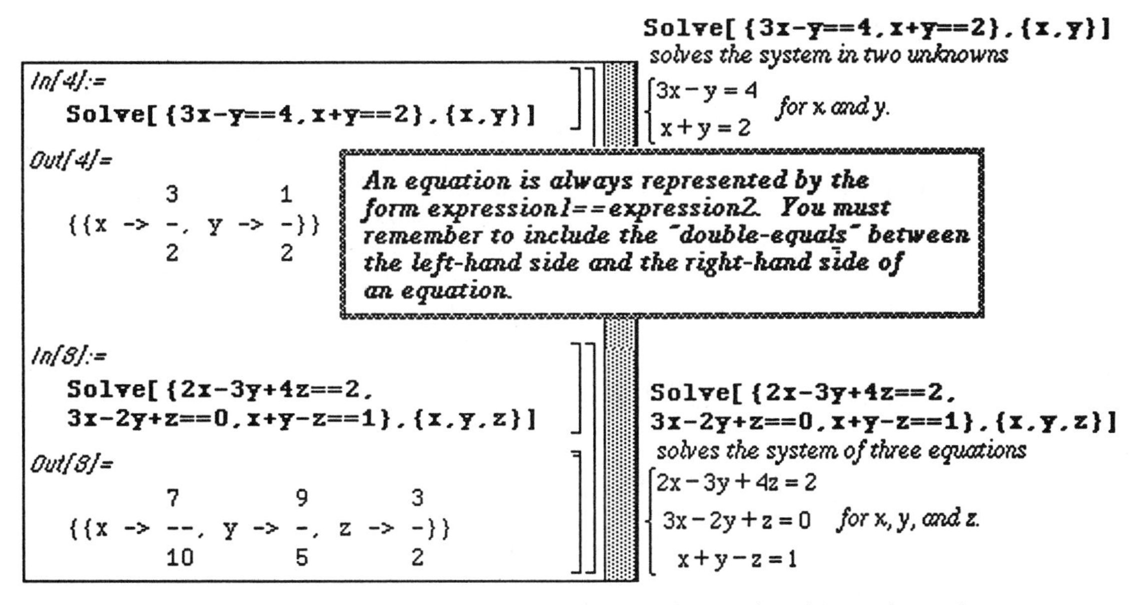

Although *Mathematica* can find the exact solution to every polynomial equation of degree four or less, exact solutions to some equations that *Mathematica* can solve may not be meaningful. In those cases, *Mathematica* can provide approximations of the exact solutions using either the **`N[expression]`** or the **`expression // N`** command:

Remember that *Mathematica* denotes $\sqrt{-1}$ by `I`.

□ **Example:**

Approximate the values of x that satisfy the equation (i) $x^4 - 2x^2 = 1 - x$; and (ii) $1 - x^2 = x^3$.

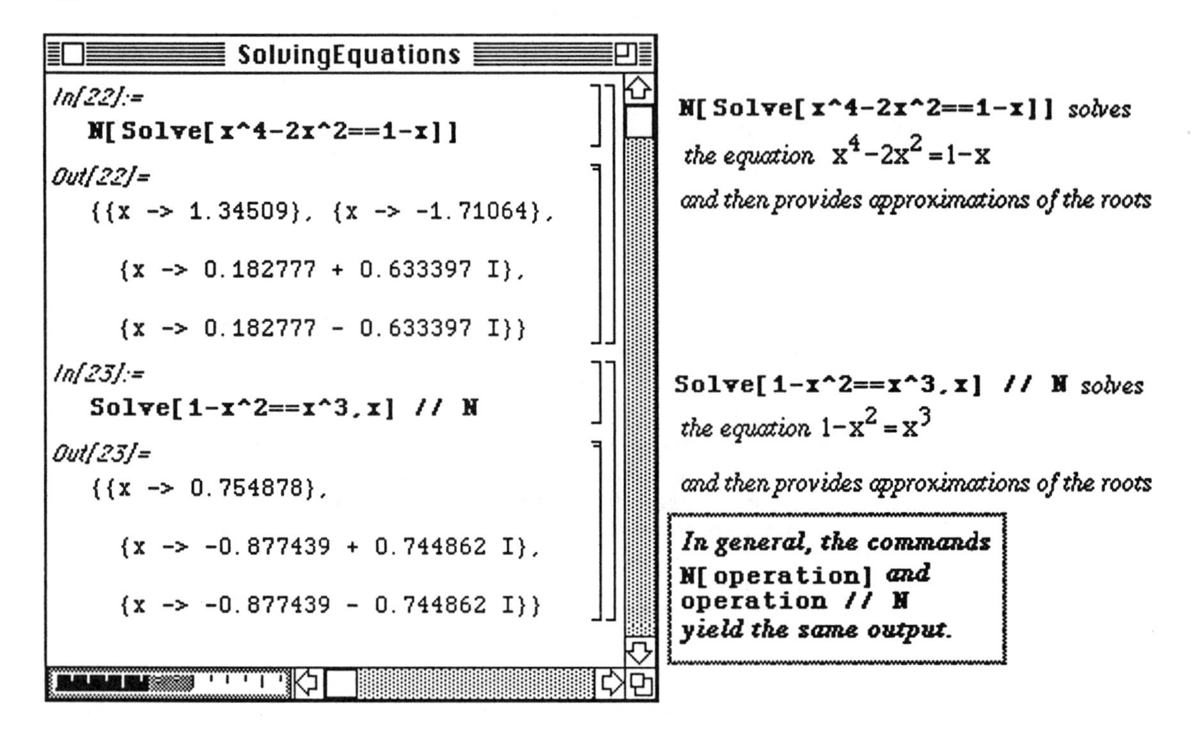

Mathematica can also solve equations involving more than one variable for one of the other variables in terms of other unknowns.

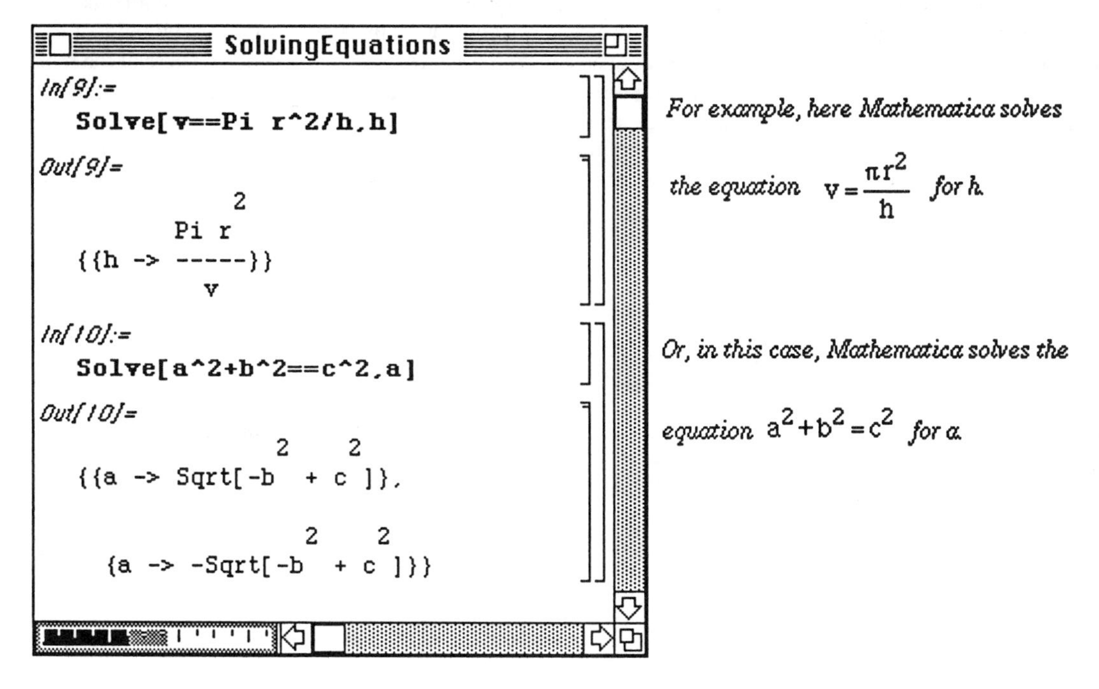

■ Numerical Approximation of Solutions of Equations

When solving an equation is either impractical or impossible, *Mathematica* provides two functions to approximate roots of equations: **FindRoot** and **NRoots**. **NRoots** numerically approximates the roots of any polynomial equation. **FindRoot** attempts to approximate a root to an equation provided that a "reasonable" guess of the root is given. **FindRoot[lhs==rhs,{x,firstguess}]** searches for a numerical solution to the equation **lhs==rhs**, starting with x=**firstguess**. (**firstguess** can be obtained by using the **Plot** command.) Thus, **FindRoot** works on functions other than polynomials. Moreover, to locate more than one root, **FindRoot** must be used several times. **NRoots** is easier to use when trying to approximate the roots of a polynomial.

□ Example:

Approximate the solutions of $x^5+x^4-4x^3+2x^2-3x-7=0$.

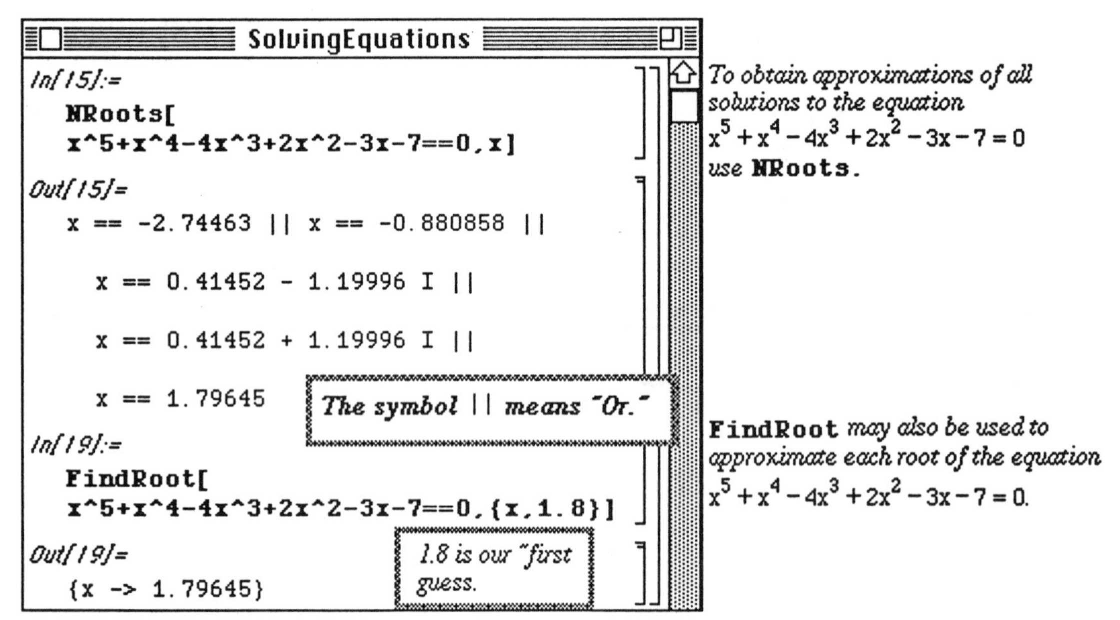

□ Example:

In order to approximate the roots of the equation Cos(x)-x=0, **FindRoot** must be used since Cos(x)-x=0 is not a polynomial equation.

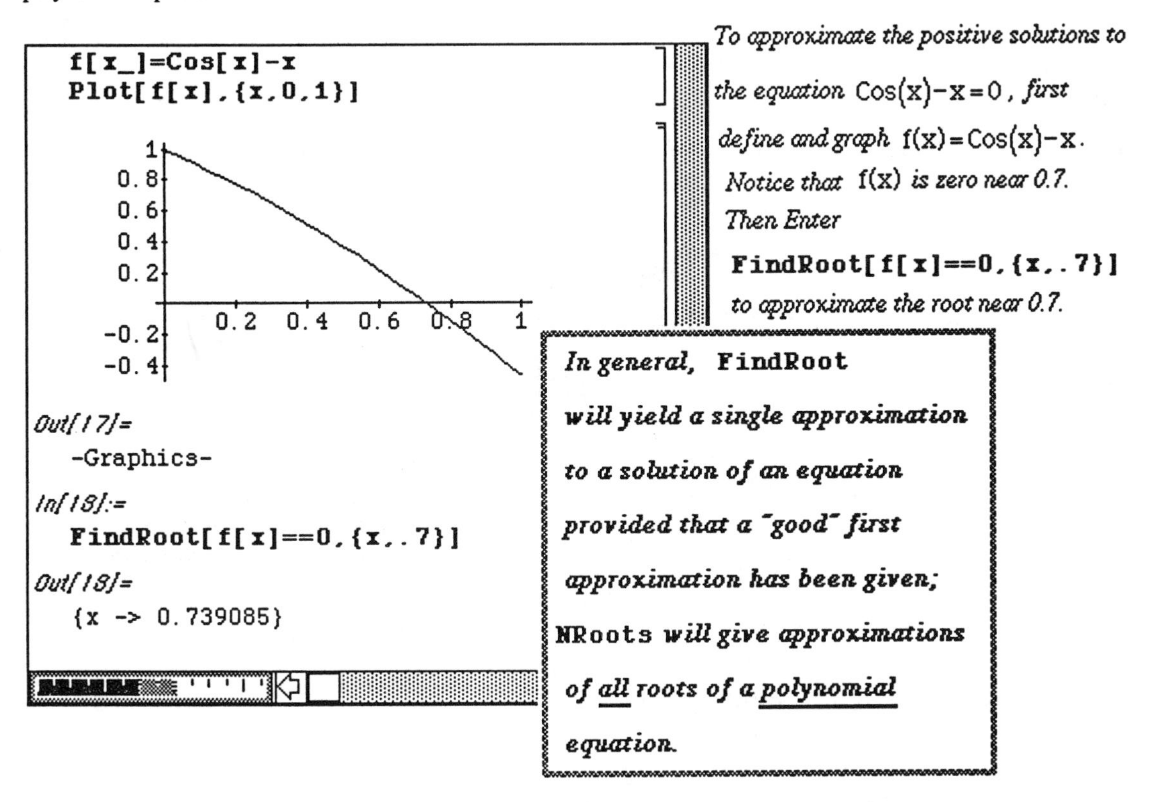

• Approximating Solutions of Equation in Version 2.0

In addition to the commands **FindRoot** and **NRoots**, Version 2.0 contains the command **NSolve** which can also be used to approximate roots of some equations.

o Example:

If $h(x)=x^3-8x^2+19x-12$ and $k(x)=\frac{1}{2}x^2-x-\frac{1}{8}$, use *Mathematica* Version 2.0 to compute approximations of the solution of $h(x)=k(x)$ using (i) **NRoots**; and (ii) **NSolve** .

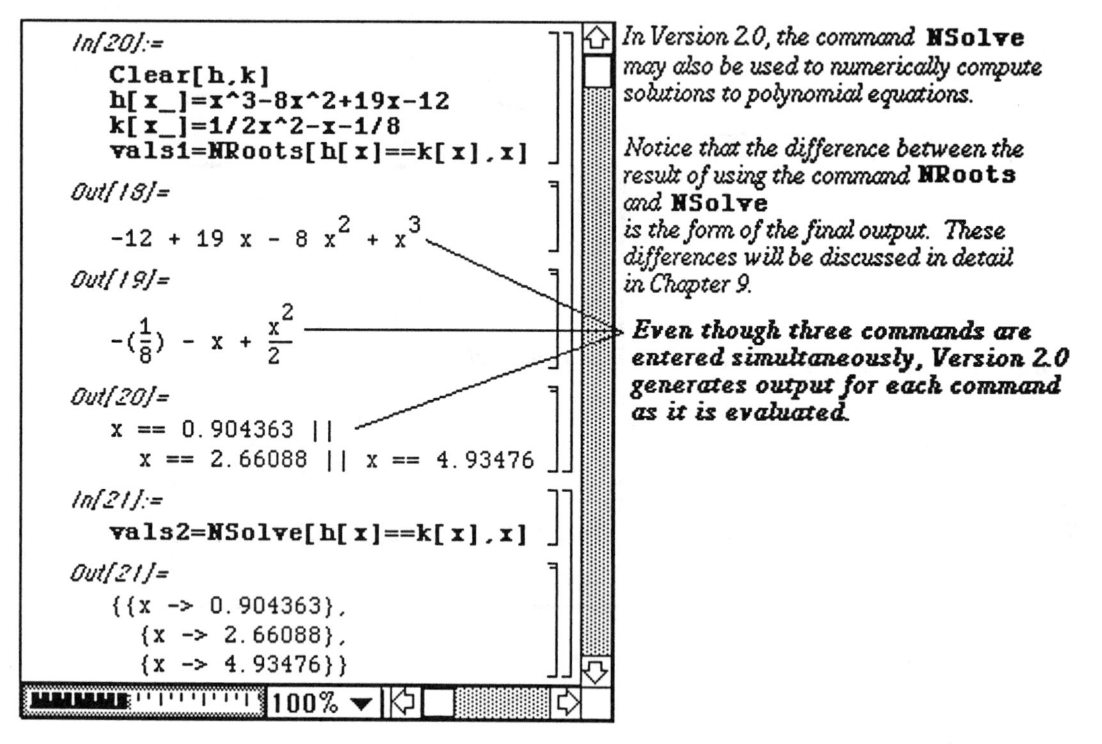

■ Application: Intersection Points of Graphs of Functions

■ (A)

Locate the points where the graphs of $h(x)=x^3-8x^2+19x-12$ and $k(x)=\frac{1}{2}x^2-x-\frac{1}{8}$ intersect.

Notice that the x-coordinates of the intersection points satisfy the equation h(x)=k(x). Consequently, to locate the intersection points, it is sufficient to solve the equation h(x)=k(x). Although this step is not necessary to solve the problem, we first graph h and k and notice that h and k intersect three times.

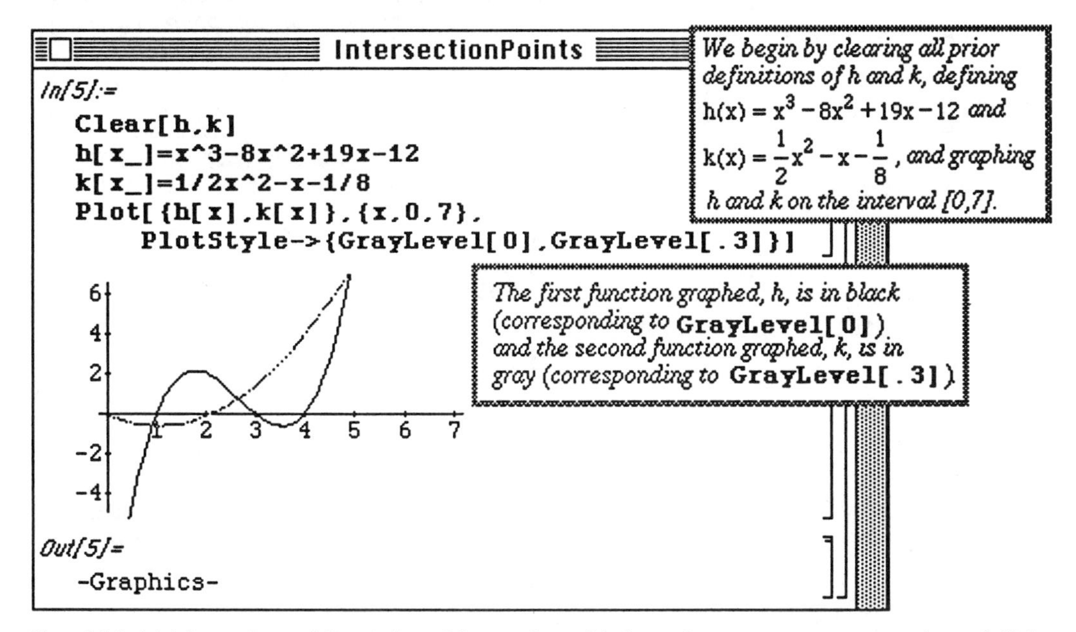

Since h(x)=k(x) is a polynomial equation of degree three, *Mathematica* can compute exact values of all three roots. However, the roots are complicated so we approximate the solutions. Moreover, since h(x)=k(x) is a polynomial equation we use the command `NRoots[h[x]==k[x],x]`:

o Notice that in Version 2.0, `NSolve[h[x]==k[x],x]` produces the same result.

In the following example, the exact solutions of the equation h(x)=k(x) are computed with the command **Solve[h[x]==k[x]]**. Notice that the resulting solution is expressed as a <u>list</u>. Lists are discussed in detail in **Chapters 4** and **5**. Nevertheless, the results of the command **Solve[lhs==rhs]** attempts to solve the equation lhs=rhs for all variables that appear in the equation. The solutions, if any, are displayed as a list. In general, the command **Solve[lhs==rhs][[1]]** yields the first element of the list of solutions, **Solve[lhs==rhs][[2]]** yields the second element of the list of solutions, and **Solve[lhs==rhs][[j]]** yields the jth element of the list of solutions.

```
In[6]:=
   NRoots[h[x]==k[x],x]
Out[6]=
   x == 0.904363 || x == 2.66088 || x == 4.93476
In[7]:=
   Solve[h[x]==k[x]]
Out[7]=
            17                    49
   {{x -> -- + ----------------------------+
            6         151    Sqrt[-16585] 1/3
                  36 (--- + ------------)
                      432     48 Sqrt[3]

         151    Sqrt[-16585] 1/3
        (--- + ------------)    },
         432     48 Sqrt[3]

          17
     {x -> -- + (Sqrt[-3]
```

Computes numerical approximations of all solutions to the polynomial equation h(x)=k(x). Be sure to use "double-equals" signs when working with equations.

The command **Solve[h[x]==k[x]]** *computes the exact solutions of the polynomial equation h(x)=k(x). The command* **Solve[h[x]==k[x]][[1]]** *yields the exact value of the first solution; the command* **Solve[h[x]==k[x]][[2]]** *yields the exact value of the second solution; and the command* **Solve[h[x]==k[x]][[3]]** *yields the exact value of the third solution. Since the degree of the polynomial equation h(x)=k(x) is 3, there are at most three distinct solutions.*

■ (B)

Locate the points where the graphs of $f(x) = e^{-(x/4)^2}\mathrm{Cos}\left(\frac{x}{\pi}\right)$ and $g(x) = \mathrm{Sin}\left(x^{3/2}\right) + \frac{5}{4}$ intersect.

Notice that the x-coordinates of the intersection points satisfy the equation f(x)=g(x). Consequently, to locate the intersection points, it is sufficient to solve the equation f(x)=g(x). Since this problem does not involve polynomials, we must first graph f and g and notice that they intersect twice.

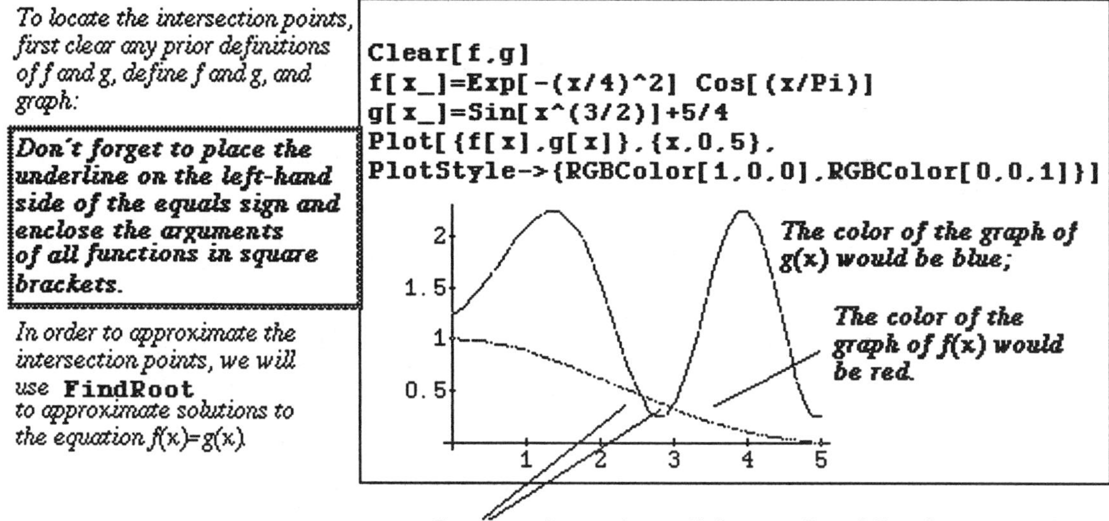

Mathematica cannot solve f(x)=g(x) exactly. Since f(x)=g(x) is NOT a polynomial equation, the command **NRoots** cannot be used to numerically approximate the roots. However, we can use the command **FindRoot** to approximate each root provided we have a "good" initial approximation of the root. To obtain a "good" initial approximation of each root:

1) Move the cursor within the graphics cell and click once. **Notice that a box appears around the graph:**
2) Press and hold down the **Open-Apple** key; as you move the cursor within the graphics cell, notice that the thermometer at the bottom of the screen has changed to ordered pairs approximating the location of the cursor within the graphics cell:

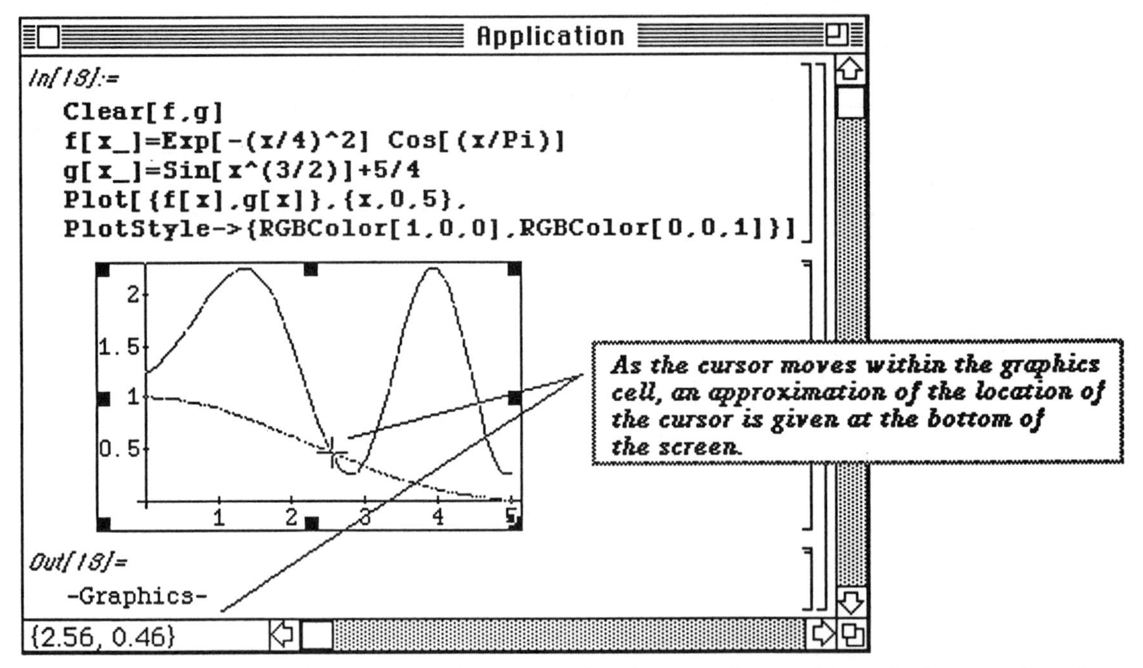

When the cursor is near the intersection point, we see that the x-coordinate of the point is approximately 2.56. Hence, we will use 2.56 as our initial approximation in the **FindRoot** *command.*
An approximation of the second intersection point is similarly obtained.

We then use **FindRoot** twice to compute an approximation of each solution:

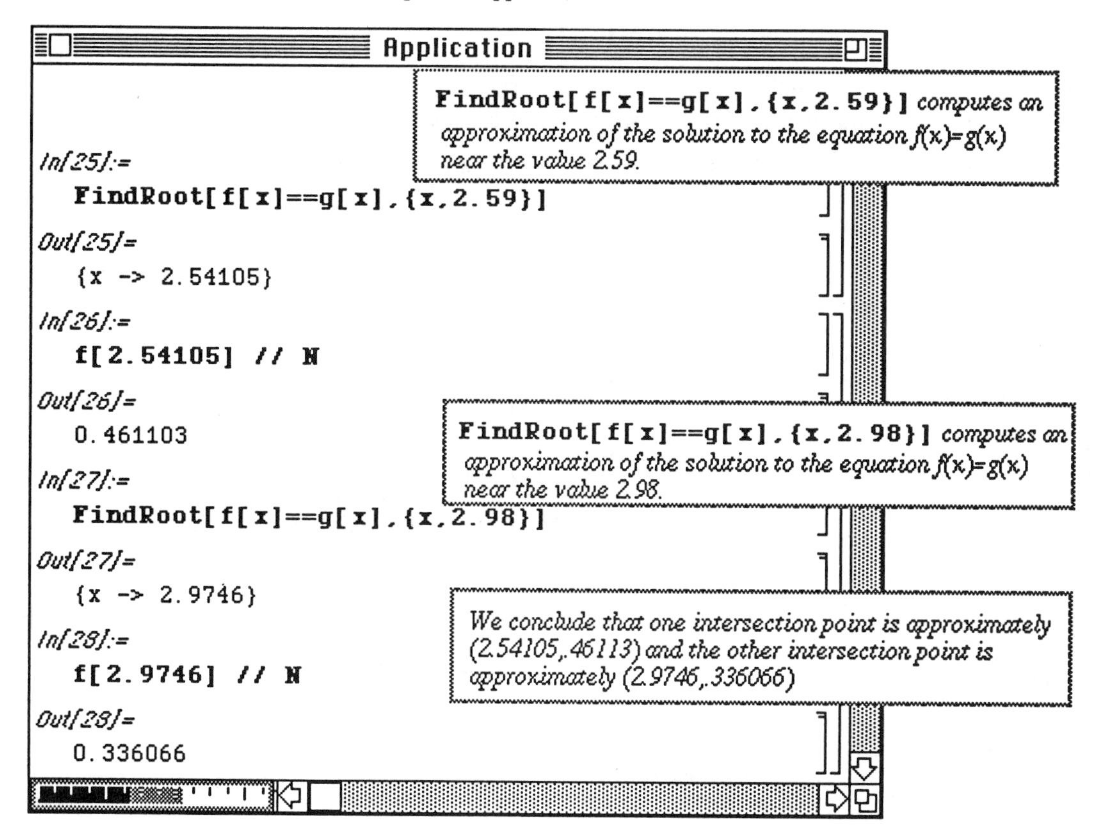

Chapter 3
Calculus

■ **Chapter 3** introduces *Mathematica*'s built-in calculus commands. The examples used to illustrate the various commands are similar to examples routinely done in first-year calculus courses.

■ Commands introduced and discussed in this chapter from **Version 1.2** are:

The command **Chop[smallnumber]** produces zero when |**smallnumber**|$\leq 10^{-10}$

Other commands include:

Limits

```
Limit[expression, x->a]
```

Differential Calculus

```
f'[x]
f''[x]
f'''[x]
D[f[x],x]
D[f[x],{x,n}]
Dt[equation,x]
```

Integral Calculus

```
Integrate[expression,x]
Integrate[expression, {x,a,b}]
NIntegrate[expression, {x,a,b}]
```

Multi-Variable Calculus

```
D[f[x,y],x]
D[f[x,y],y]
D[f[x,y],{x,n}]
D[f[x,y],{y,n}]
D[f[x,y],x,y]
Derivative[n,m][f][x,y]
Integrate[f[x,y],{x,a,b},{y,c,d}]
NIntegrate[f[x,y],{x,a,b},{y,c,d}]
```

Series Calculus

```
Series[f[x],{x,a,n}]
Normal[Series[f[x],{x,a,n}]]
LogicalExpand[series1==series2]
```

Graphics

```
ContourPlot[f[x,y],{x,a,b},{y,c,d}]
Plot3D[f[x,y],{x,a,b},{y,c,d},options]
```

Options

```
PlotPoints->n
Shading->False
Boxed->False
PlotLabel->"text"
AxesLabel->{"x-axis text","y-axis text","z-axis text"}
Ticks->None
Ticks->{{x-axis tick marks},{y-axis tick marks},{z-axis tick marks}}
Axes->None
BoxRatios->{a,b,c}
Mesh->False
DisplayFunction->Identity
```

■ Commands and options discussed in this chapter from **Version 2.0** include:

Graphics
ImplicitPlot[equation,{x,xmin,xmax},{y,ymin,ymax}]
Version 2.0 Graphics Options Include:

ContourSmoothing, **ContourShading**, **Contours** } **ContourPlot**

Show, **GraphicsArray** } **Plot3D**

Other Options:
Direction->±1} **Limit**

■ Applications in this chapter include:

□ Differential Calculus
Locating Horizontal Tangent Lines
Graphing Functions and Tangent Lines
Maxima and Minima

□ Integral Calculus
Area between Curves
Volumes of Solids of Revolution
Arc Length

□ Series
Approximating the Remainder
Computing Series Solutions to Differential Equations

□ Multi-Variable Calculus
Classifying Critical Points
Tangent Planes
Volume

■ 3.1 Computing Limits

One of the first topics discussed in calculus is that of limits. *Mathematica* uses the command **`Limit[expression,x->a]`** to find the limit of **`expression`** as **`x`** approaches the value **`a`**, where **`a`** can be a finite number, positive infinity (**`Infinity`**), or negative infinity (**`-Infinity`**). The "**->**" is obtained by typing a minus sign "**-**" followed by a greater that sign "**>**".

□ **Example:**

Use *Mathematica* to compute (i) $\lim_{x\to -3} \frac{3x^2+4x-15}{13x^2+32x-21}$; (ii) $\lim_{x\to 0} \frac{\text{Sin}(x)}{x}$; (iii) $\lim_{x\to\infty} \frac{50-17x^2}{200x+3x^2}$;

and (iv) $\lim_{x\to\infty} \frac{3-x^2}{4-1000x}$.

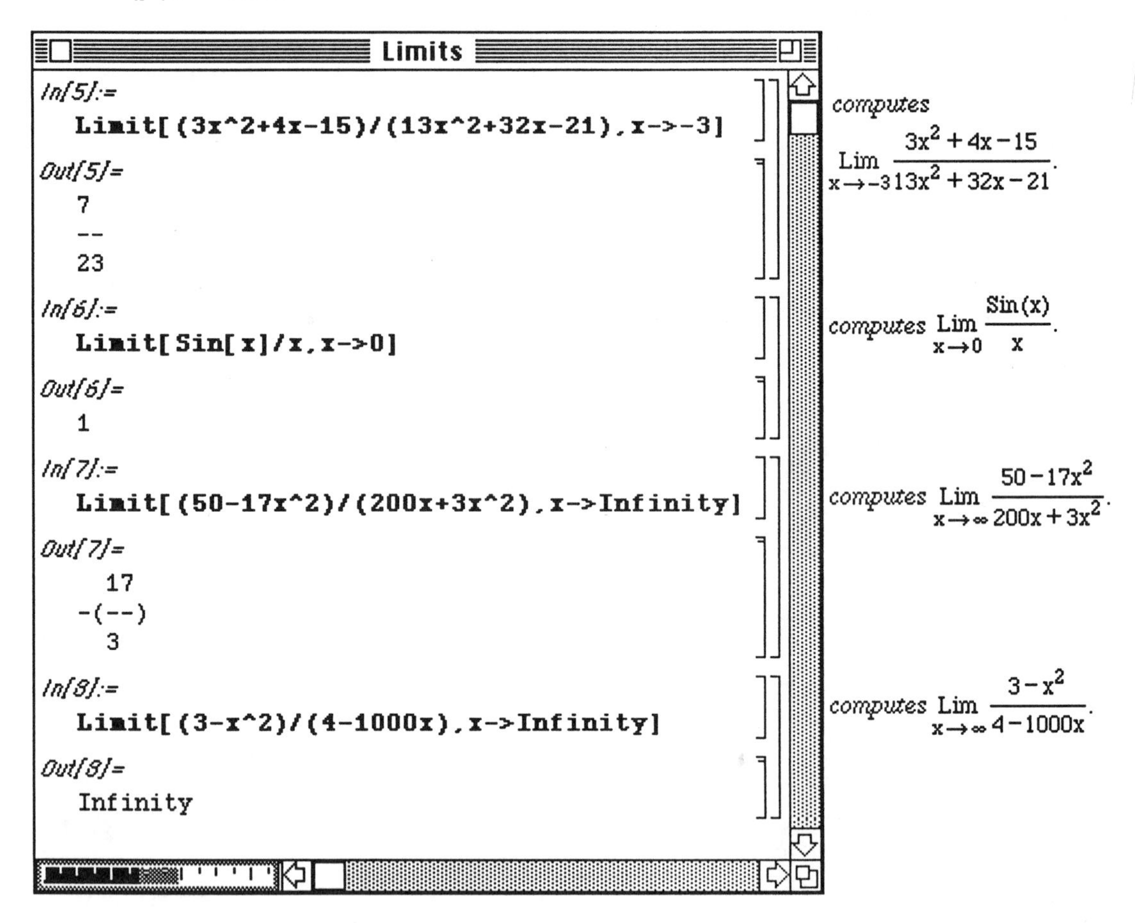

The **Limit** command can also be used along with **Simplify** to assist in determining the derivative of a function by using the definition of the derivative. This is illustrated in the following example. (This example also shows that an expression can be assigned any name, as long as that name is not a built-in *Mathematica* function or constant. ***Remember:*** Since every built-in *Mathematica* object begins with a capital letter, we have adopted the convention that all user-defined objects will be named using lower-case letters.)

□ **Example:**

Let $g(x) = x^3 - x^2 + x + 1$. Use *Mathematica* to compute and simplify (i) $\frac{g(x+h)-g(x)}{h}$; and (ii) $\lim_{h\to 0} \frac{g(x+h)-g(x)}{h}$.

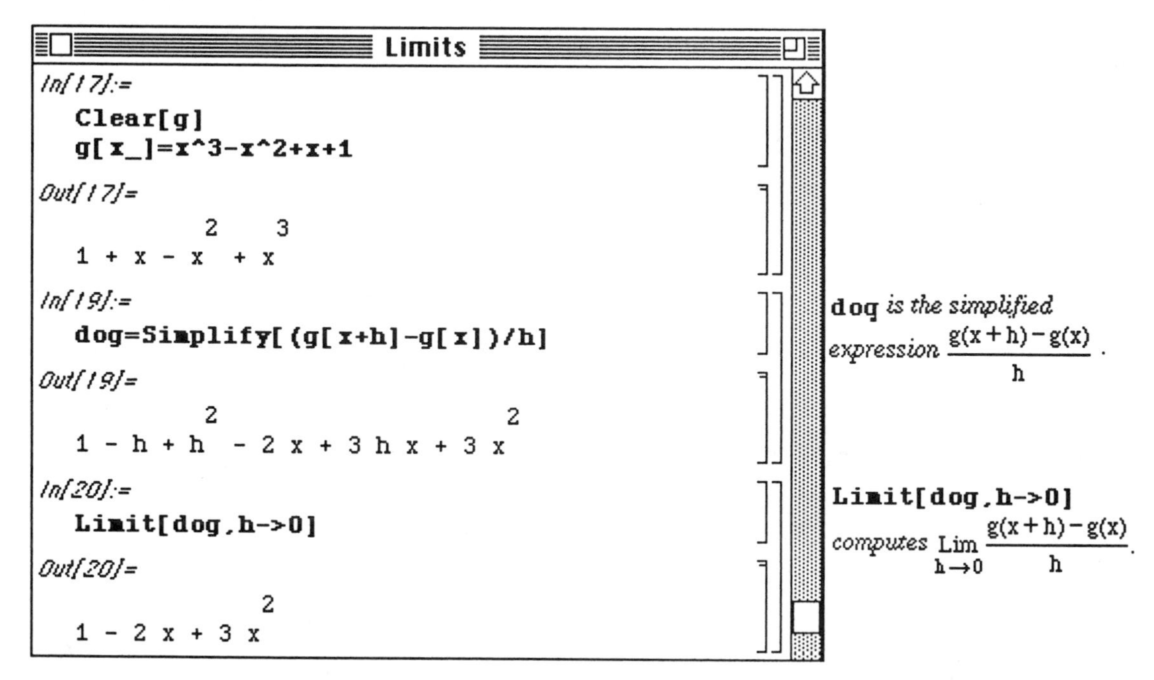

The next example illustrates how several *Mathematica* commands can be combined in a single statement to obtain the desired result.

□ Example:

Let $f(x) = \frac{1}{\sqrt{x}} + \sqrt{x}$. Use *Mathematica* to compute and simplify (i) $\frac{f(x+h)-f(x)}{h}$; and (ii) $\lim_{h\to 0} \frac{f(x+h)-f(x)}{h}$.

Remember that *Mathematica* denotes $\sqrt{x}$ by **Sqrt [x]**. Hence, f is defined by **f[x_]=1/Sqrt[x]+Sqrt[x]**. Entering **f[x_]=x^(-1/2)+x^(1/2)** would yield the same result.

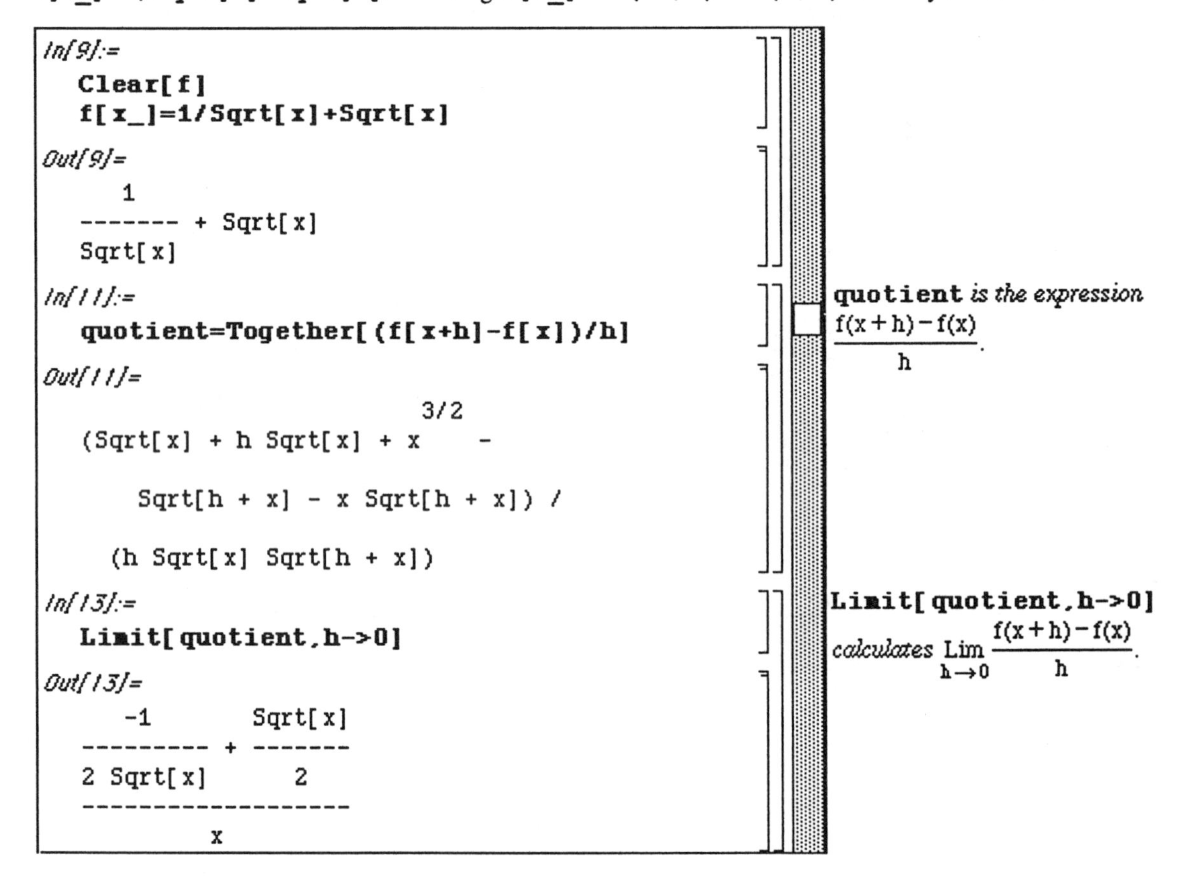

```
In[9]:=
  Clear[f]
  f[x_]=1/Sqrt[x]+Sqrt[x]
Out[9]=
     1
  ------- + Sqrt[x]
  Sqrt[x]
In[11]:=
  quotient=Together[(f[x+h]-f[x])/h]
Out[11]=
                              3/2
  (Sqrt[x] + h Sqrt[x] + x        -
      Sqrt[h + x] - x Sqrt[h + x]) /
    (h Sqrt[x] Sqrt[h + x])
In[13]:=
  Limit[quotient,h->0]
Out[13]=
     -1        Sqrt[x]
  --------- + -------
  2 Sqrt[x]      2
  -------------------
          x
```

quotient *is the expression* $\frac{f(x+h)-f(x)}{h}$.

Limit[quotient,h->0] *calculates* $\lim_{h\to 0} \frac{f(x+h)-f(x)}{h}$.

Note that the square brackets must be properly nested in order to correctly perform the combined operations. Also note that **Simplify** can be used to express the result in a more reasonable form:

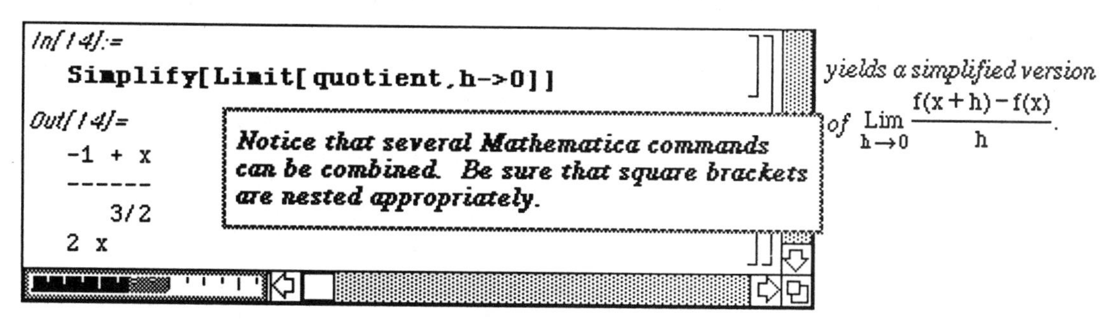

Sometimes Version 1.2 yields surprising results. For example, if **f** is an unknown function, **Limit[f[x],x->a]** yields **f[a]**.

In other cases, the command **Limit** returns results that do not make sense:

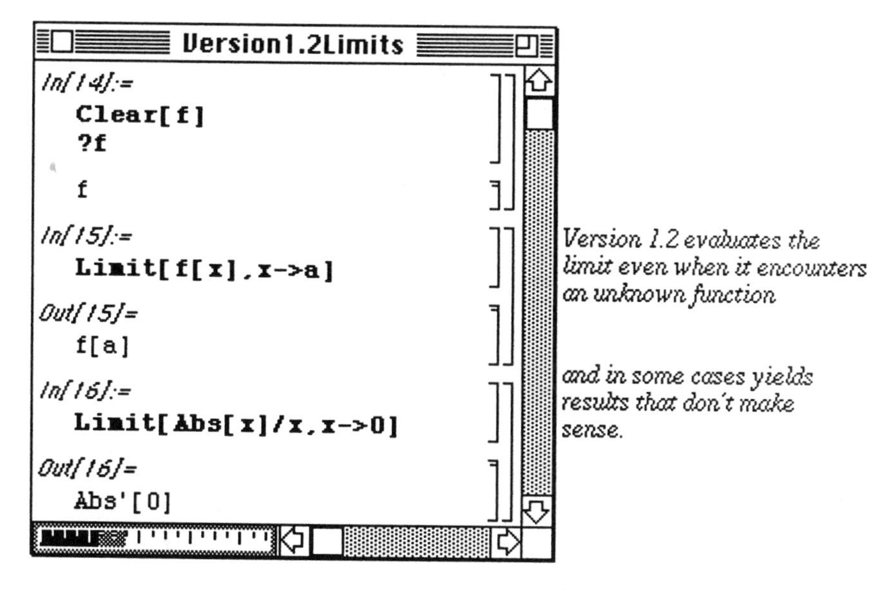

• Computing Limits with Version 2.0

In Version 2.0, the command **Limit** does not evaluate when it encounters an unknown function unless the option **Analytic->True** is included.

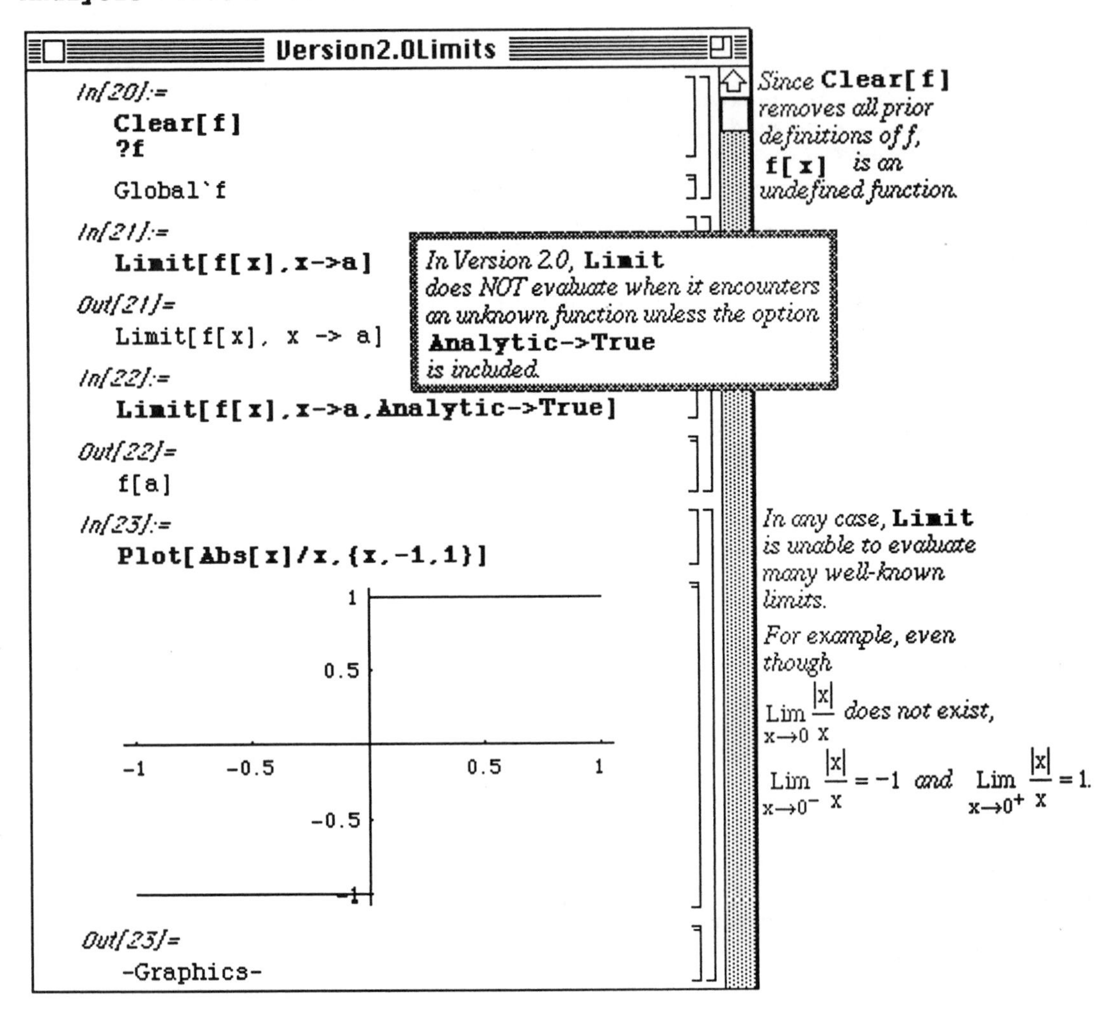

$\lim_{x \to 0} \frac{|x|}{x}$ *does not exist,*

$\lim_{x \to 0^-} \frac{|x|}{x} = -1$ *and* $\lim_{x \to 0^+} \frac{|x|}{x} = 1.$

In addition, in some cases, the options **Direction->1** or **Direction->-1** may help in computing limits. The command **Limit[f[x],x->a,Direction->-1]** computes

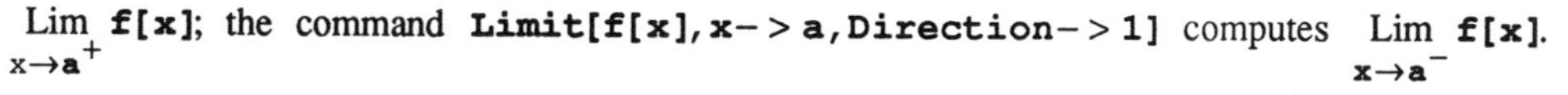

$\lim_{x\to a^+}$ **f[x]**; the command **Limit[f[x],x->a,Direction->1]** computes $\lim_{x\to a^-}$ **f[x]**.

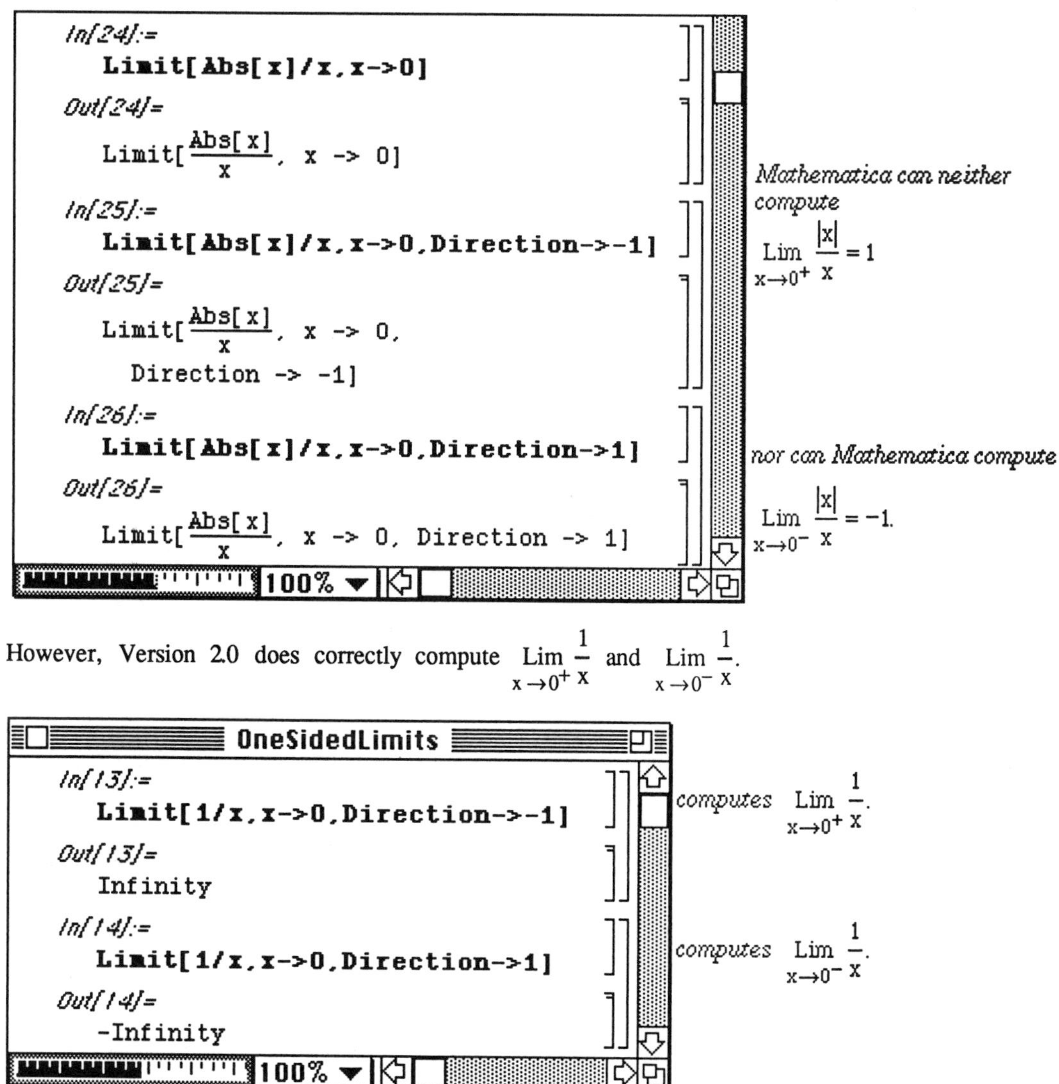

However, Version 2.0 does correctly compute $\lim_{x\to 0^+} \frac{1}{x}$ and $\lim_{x\to 0^-} \frac{1}{x}$.

3.2 Differential Calculus

Calculating Derivatives of Functions and Expressions

If we are given a differentiable function f(x), *Mathematica* can compute the derivative of f(x) in at least two ways once f(x) has been properly defined using *Mathematica* :

1) The command `f'[x]` computes the derivative of `f[x]` with respect to `x`.

2) The command `D[f[x],x]` computes the derivative of `f[x]` with respect to `x`.

3) The command `D[f[x],{x,n}]` computes the `n`th derivative of `f[x]` with respect to `x`.

4) The command `D[expression, variable]` computes the derivative of `expression` with respect to `variable`.

Other ways *Mathematica* can compute derivatives of functions and expressions are discussed in **Section 3.6.**

□ Example:

For example, in order to compute the derivative of $(7x-3)^3(5-4x^2)^2$, we may either directly compute the derivative of $(7x-3)^3(5-4x^2)^2$ or we may define $h(x)=(7x-3)^3(5-4x^2)^2$ and compute $h'(x)$.

Derivative

```
In[1]:=
 D[(7x-3)^3(5-4x^2)^2,x]
Out[1]=
                    3
 -16 x (-3 + 7 x)

              2
    (5 - 4 x ) +

                   2          2 2
   21 (-3 + 7 x)   (5 - 4 x )
In[2]:=
 Clear[h]
 h[x_]=(7x-3)^3(5-4x^2)^2;
In[3]:=
 h'[x]
Out[3]=
                    3
 -16 x (-3 + 7 x)

              2
    (5 - 4 x ) +

                   2          2 2
   21 (-3 + 7 x)   (5 - 4 x )
```

`D[(7x-3)^3(5-4x^2)^2,x]` *computes the derivative (with respect to x) of the expression* $(7x-3)^3(5-4x^2)^2$.

A semi-colon placed at the end of a command suppresses the output. Remember that RETURN gives a new line; ENTER evaluates Mathematica input.

`Clear[h]`
`h[x_]=(7x-3)^3(5-4x^2)^2;`
first clears all previous definitions of h and then defines $h(x)=(7x-3)^3(5-4x^2)^2$.

`h'[x]` *computes the derivative of h.*

Notice that both **h'[x]** and **D[h[x],x]** produce the same result.

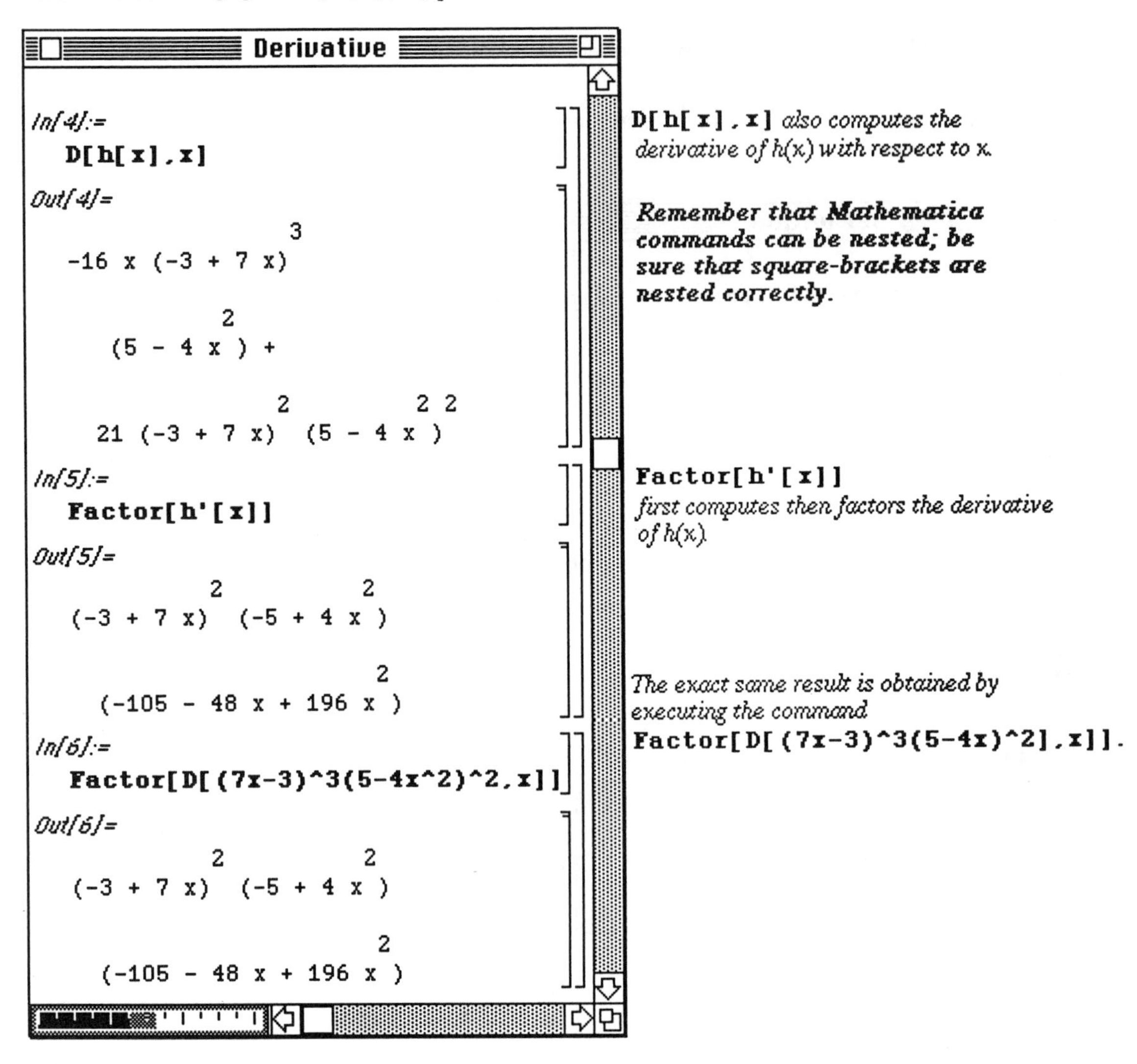

D[h[x],x] *also computes the derivative of h(x) with respect to x.*

Remember that Mathematica commands can be nested; be sure that square-brackets are nested correctly.

Factor[h'[x]] *first computes then factors the derivative of h(x).*

The exact same result is obtained by executing the command **Factor[D[(7x-3)^3(5-4x)^2],x]]**.

■ Graphing Functions and Derivatives

Moreover, since **f'[x]** is a function of **x**, **f'[x]** can be graphed. The following example shows how to compute the derivative of a function and then plot the original function and its derivative simultaneously :

□ Example:

Let $f(x) = x\mathrm{Sin}^2(x)$. Compute f' and graph both f and f' on the same axes.

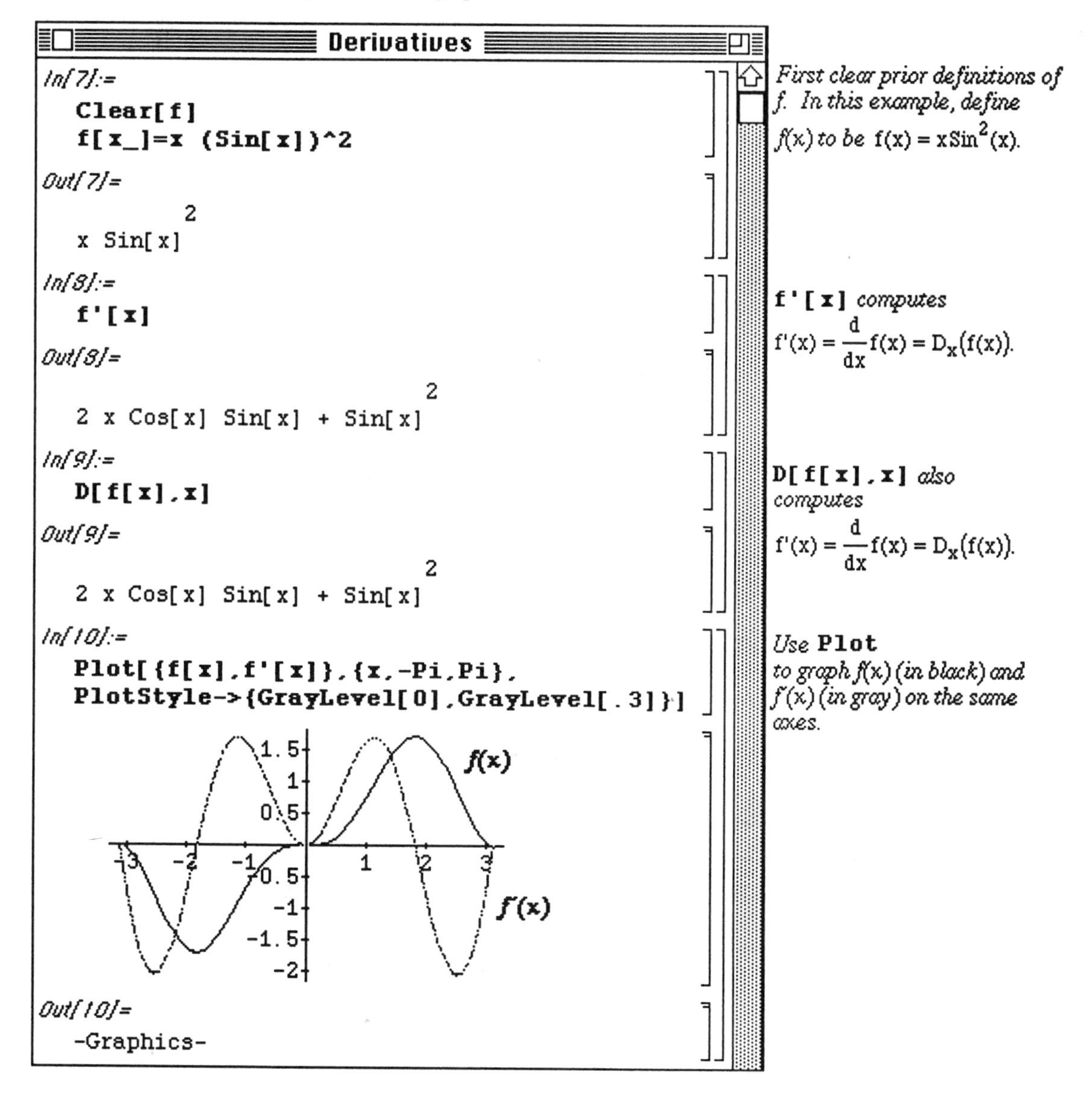

First clear prior definitions of f. In this example, define f(x) to be $f(x) = x\mathrm{Sin}^2(x)$.

f'[x] *computes* $f'(x) = \frac{d}{dx}f(x) = D_x(f(x))$.

D[f[x],x] *also computes* $f'(x) = \frac{d}{dx}f(x) = D_x(f(x))$.

Use **Plot** *to graph f(x) (in black) and f'(x) (in gray) on the same axes.*

■ Computing Higher Order Derivatives

The command **D[f[x],{x,n}]** computes the **n**th derivative of **f[x]** with respect to **x**:

$$f^{(n)}(x) = \frac{d^n f(x)}{dx^n}.$$

Using the same definition of f as above, the following calculations compute the second, third, and fourth derivatives of f:

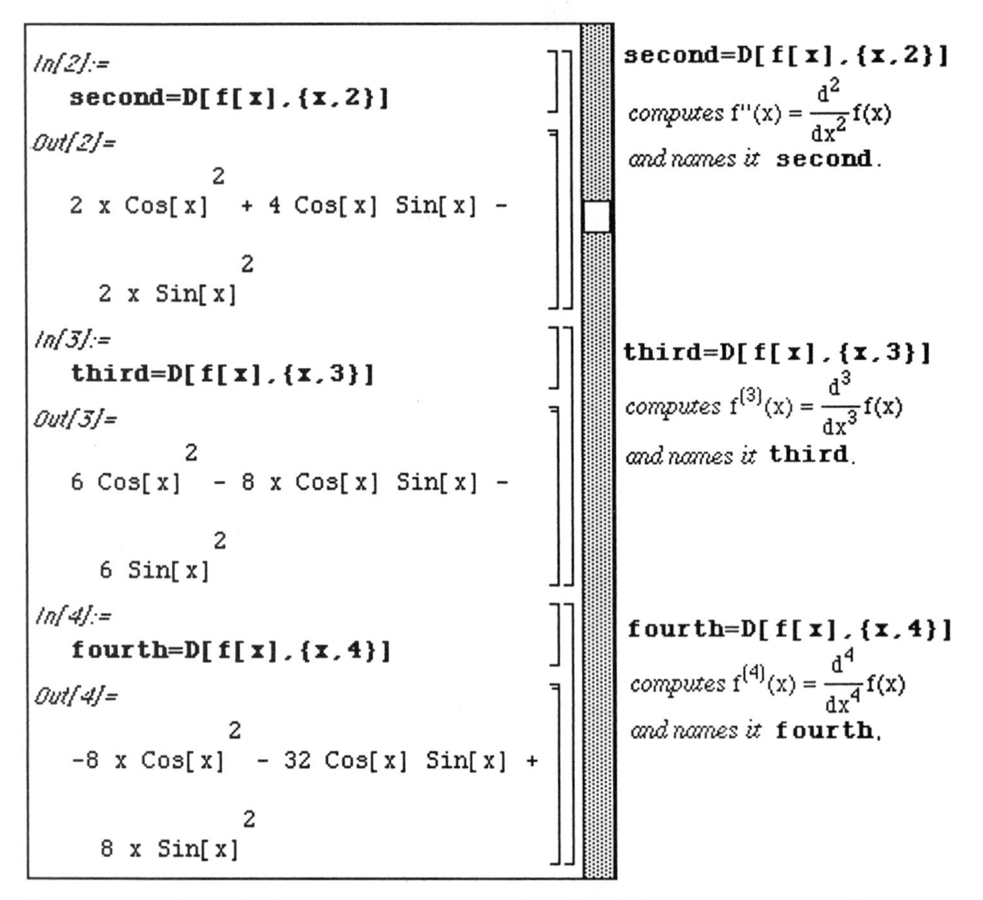

These higher order derivatives can be graphed together. Recall that any expression in *Mathematica* can be assigned a name. The names **plotsecond**, **plotthird**, and **plotfourth** are given to the graphs of the second, third, and fourth derivatives of the function f(x), respectively. By naming these graphs individually, the **Show** command can be used to plot the three graphs at one time. Notice the relationship of the behavior of the graphs of these three derivatives.

Remember, $f^{(3)}(x)$ is the derivative of $f^{(2)}(x)$, so $f^{(3)}(x) > 0$ when $f^{(2)}(x)$ is increasing and $f^{(3)}(x) < 0$ when $f^{(2)}(x)$ is decreasing.

Also, $f^{(4)}(x)$ is the second derivative of $f^{(2)}(x)$, so $f^{(4)}(x) > 0$ when $f^{(2)}(x)$ is concave up and $f^{(4)}(x) < 0$ when $f^{(2)}(x)$ is concave down.

Notice how **GrayLevel** is used to distinguish between the three curves; the option **DisplayFunction->Identity** suppresses the resulting graph; and the option **DisplayFunction->$DisplayFunction** is used in the **Show** command to display the graphs which were suppressed initially with the option **DisplayFunction->Identity**:

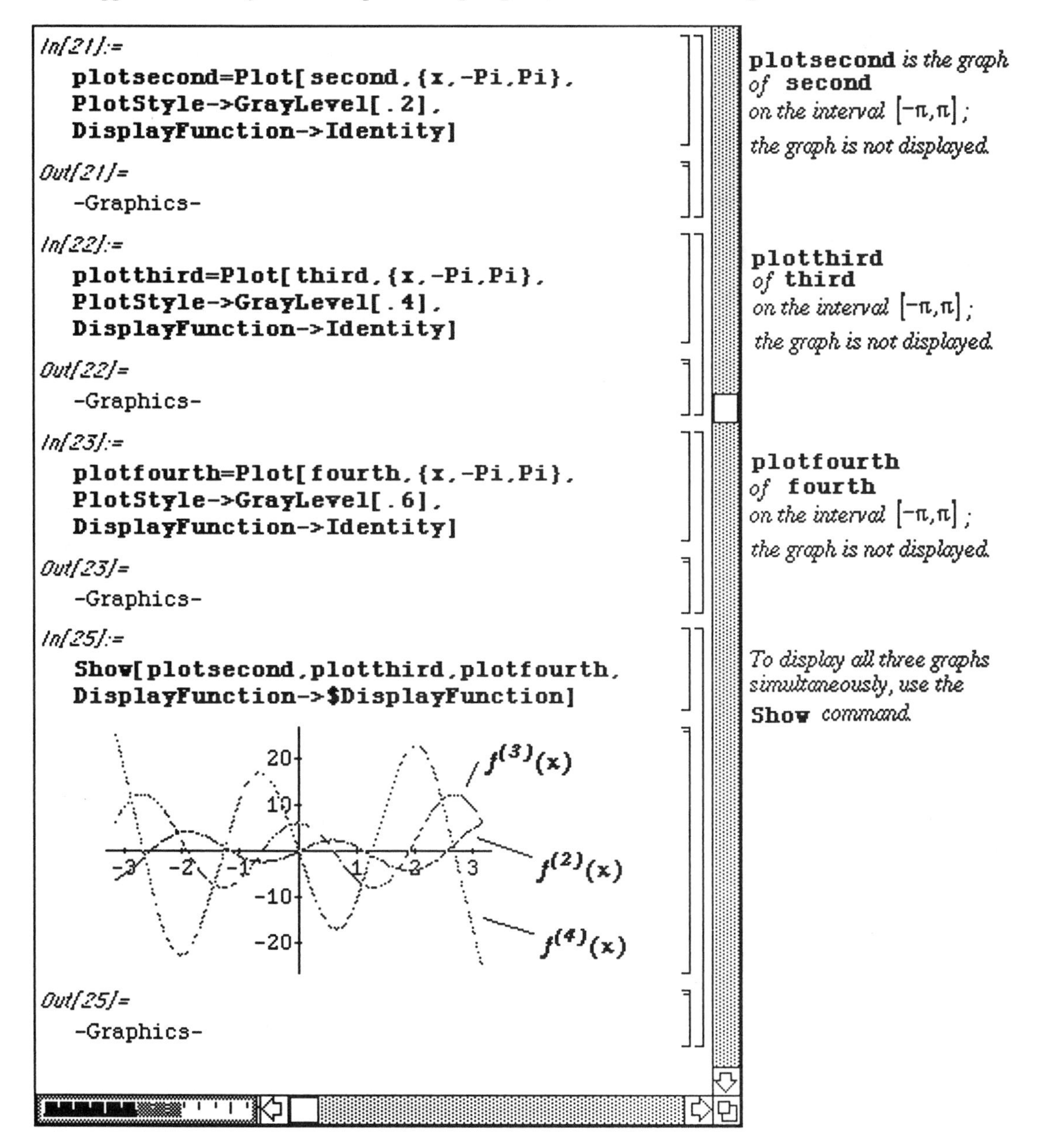

■ Locating Critical Points and Inflection Points

Since derivatives of functions are expressions, algebraic procedures can be performed on them. Hence, in addition to finding the roots of a function, g(x), *Mathematica* can also be used to locate the critical points and inflection points of g(x).

In order to observe the location of these points, the **Plot** command is used to graph g(x), g'(x), and g''(x) simultaneously.

□ Example:

Let $g(x) = 2x^3 - 9x^2 + 12x$. Graph g, g', and g'' on the interval $[-1, 4]$. Locate all critical points and inflection points.

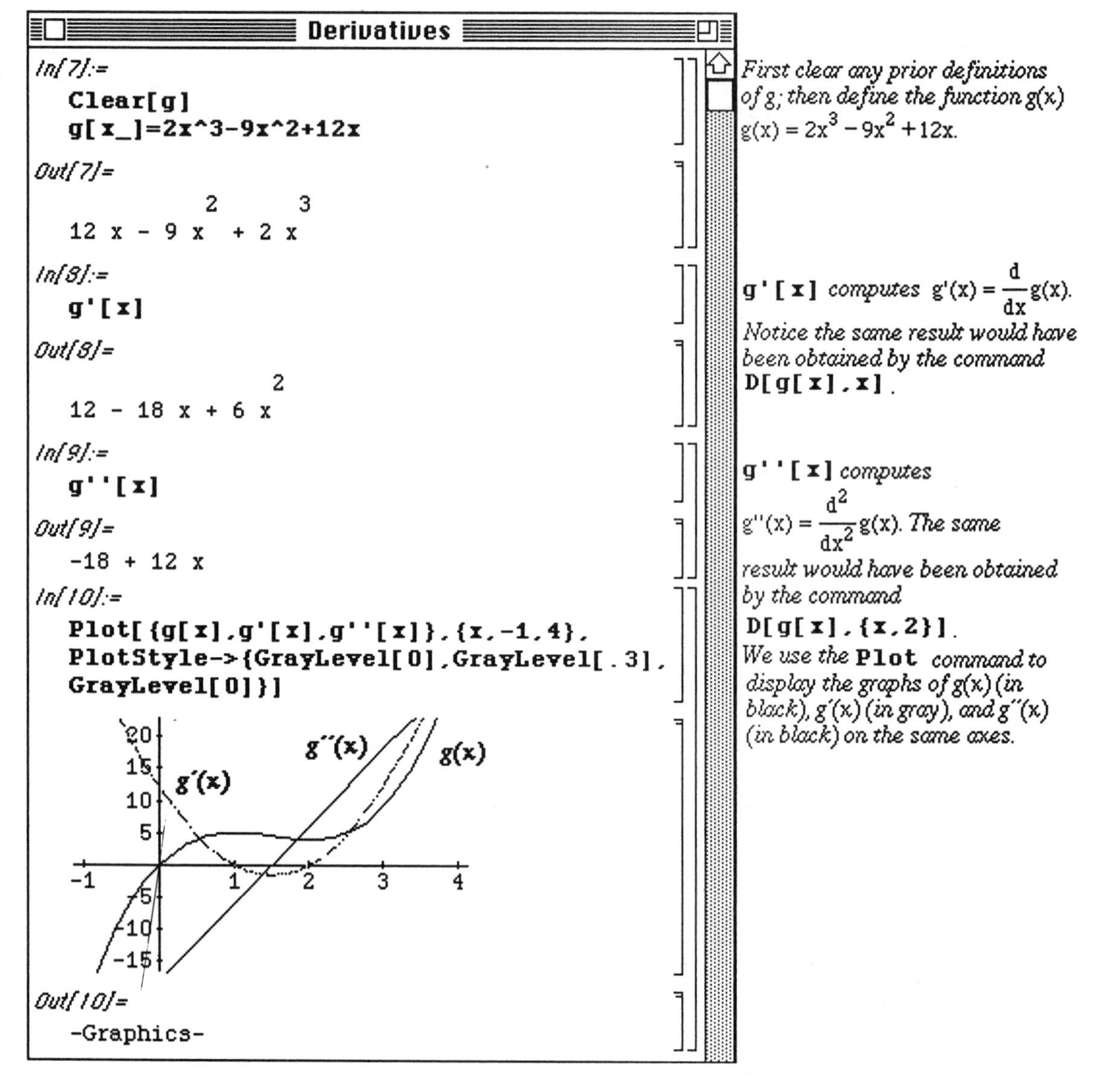

Solving each of the equations g(x)=0, g'(x)=0, and g''(x)=0 locates the roots of g, the x-coordinates of the critical points of g, and the x-coordinates of the inflection points of g. Since g is a polynomial with degree less than five, these three equations can be solved with the **`Solve`** command by entering **`Solve[g[x]==0]`**, **`Solve[g'[x]==0]`** and **`Solve[g''[x]==0]`**. Be sure to include the double-equals sign between the left- and right-hand sides of equations when using the **`Solve`** command:

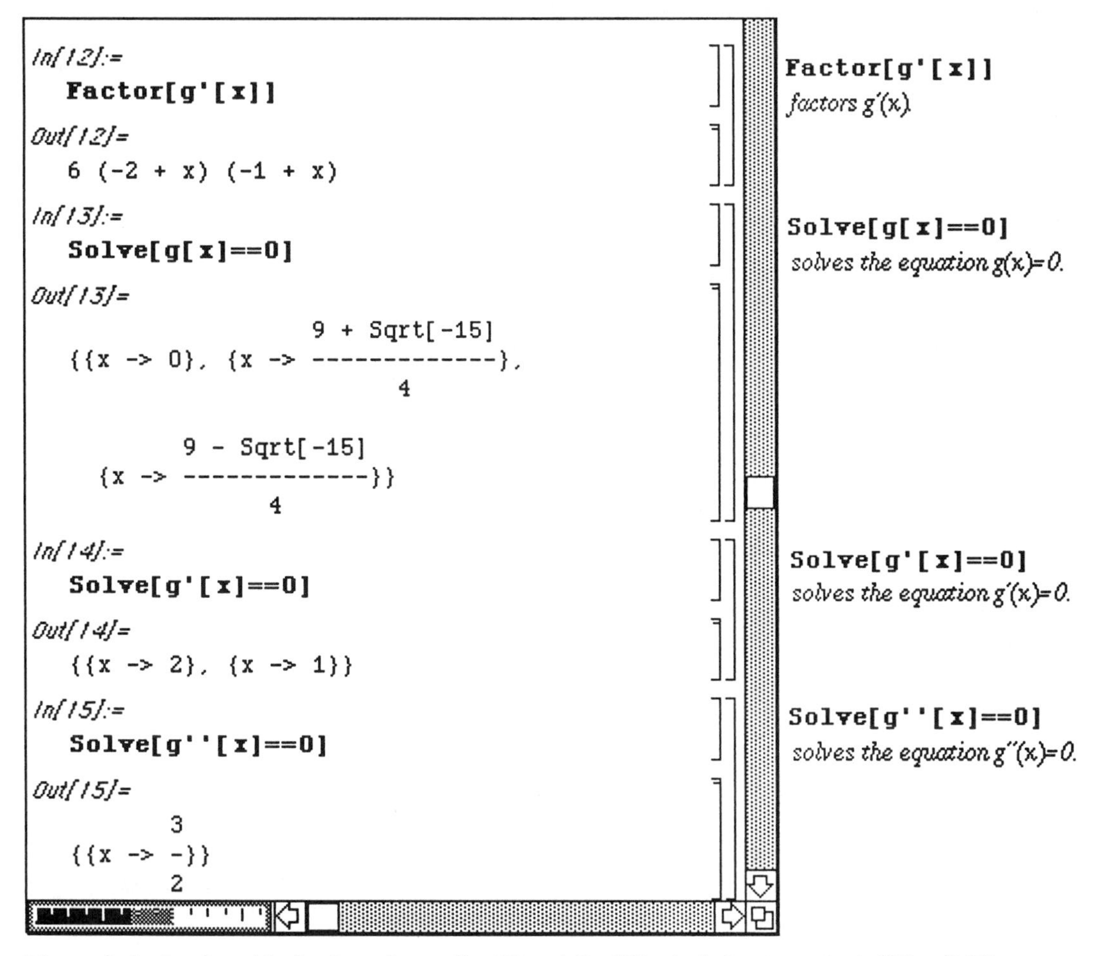

We conclude that the critical points of g are (1,g(1)) and (2,g(2)); the inflection point is (3/2, g(3/2)).

A similar type of problem which can be solved using *Mathematica* is as follows :

□ **Example:**

Locate the values of x for which the line tangent to the graph

$p(x) = \frac{1}{2}x^6 - 2x^5 - \frac{25}{2}x^4 + 60x^3 - 150x^2 - 180x - 25$ is horizontal.

Notice that the function p(x) is a polynomial of degree 6, so p'(x) is a polynomial of degree 5. Therefore, when determining the values of x such that p'(x) = 0, the command **NRoots** must be used instead of **Solve**. (Recall that **Solve** finds exact solutions of polynomial equations of degree four or less.) Some of the roots of p'(x) are complex numbers. These values are ignored since we are only concerned with real numbers.

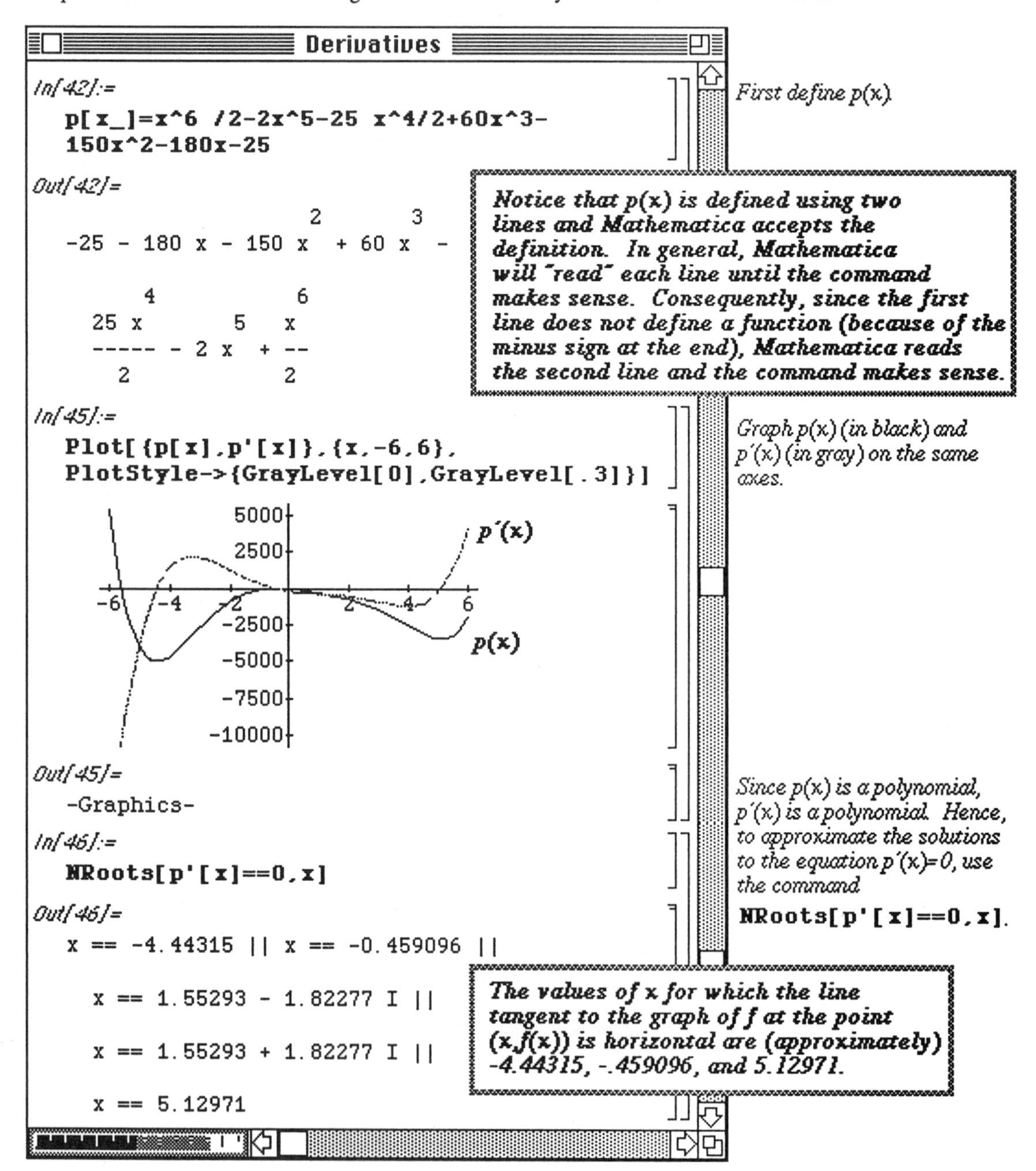

Even though **p''[x]** is a polynomial of degree 4, the inflection points are found using the command **NRoots.** (These points can also be determined with the **Solve** command.)

o If you are using Version 2.0, **NSolve** can also be used to approximate the solutions of the polynomial equation p"(x)=0 by entering **NSolve[p''[x]==0,x]**.

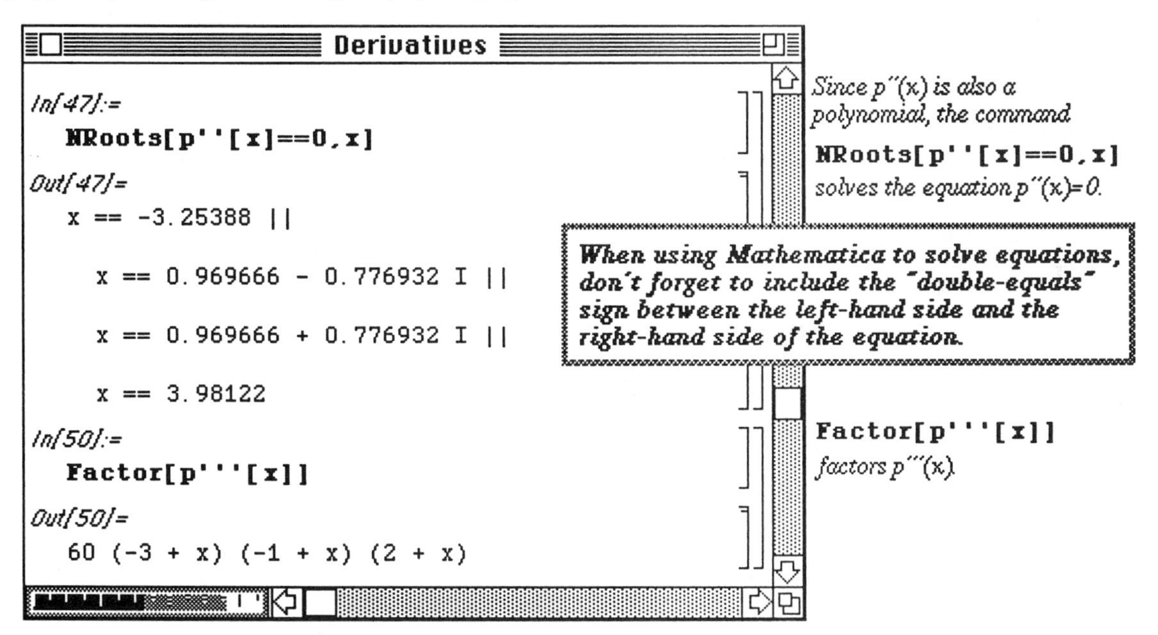

Up to this point in our example problems, we have only considered polynomial functions. This next example involves a function which is not a polynomial. Hence, the **FindRoot** command, which depends on an initial guess, must be employed.

□ **Example:**

Let $w(x) = 2\mathrm{Sin}^2(2x) + \frac{5}{2} x \mathrm{Cos}^2\left(\frac{x}{2}\right)$ on $(0,\pi)$. Locate the values of x for which the line tangent to the graph of w at the point (w,w(x)) is horizontal.

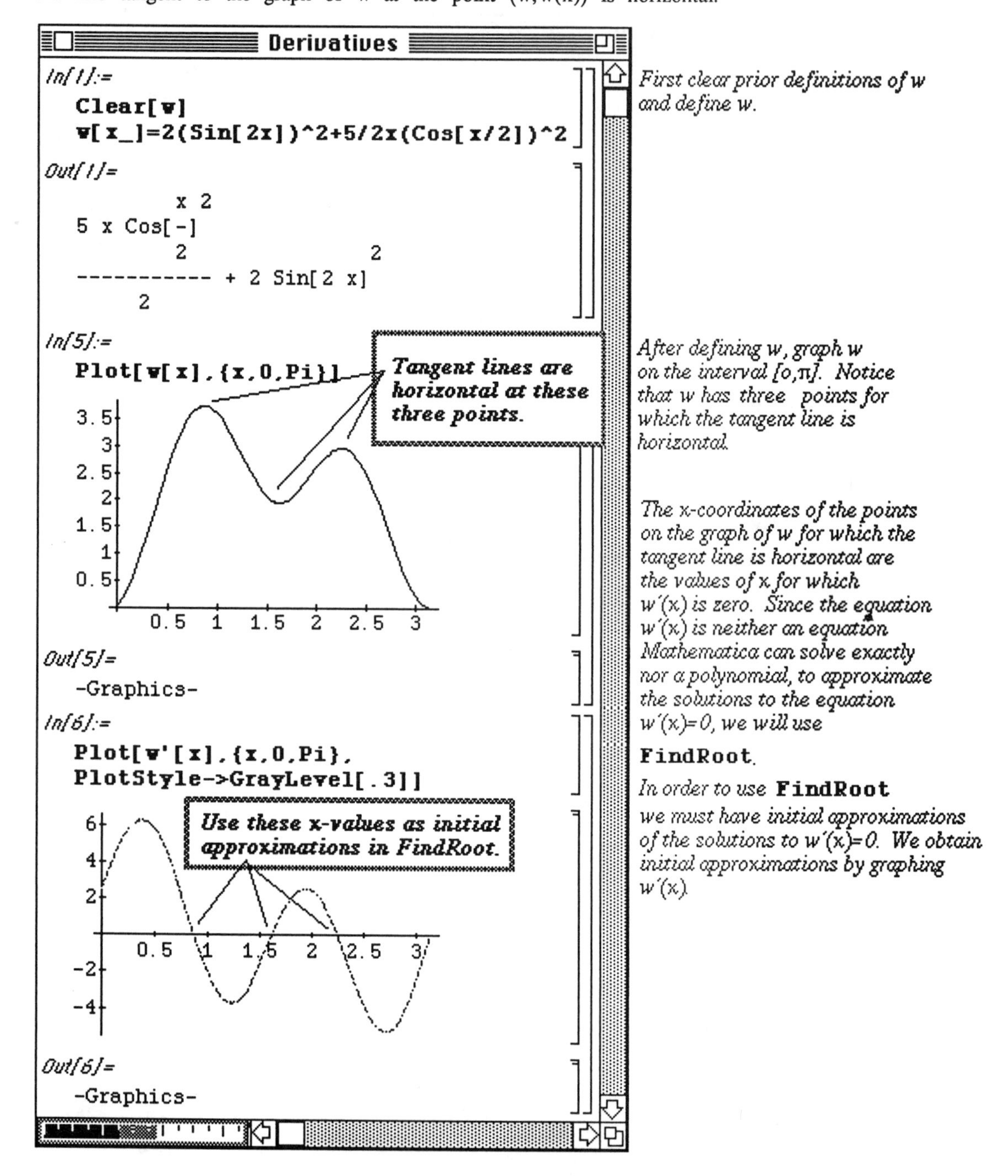

First clear prior definitions of w and define w.

After defining w, graph w on the interval [o,π]. Notice that w has three points for which the tangent line is horizontal.

The x-coordinates of the points on the graph of w for which the tangent line is horizontal are the values of x for which w′(x) is zero. Since the equation w′(x) is neither an equation Mathematica can solve exactly nor a polynomial, to approximate the solutions to the equation w′(x)=0, we will use **FindRoot**.

In order to use **FindRoot** *we must have initial approximations of the solutions to w′(x)=0. We obtain initial approximations by graphing w′(x).*

After using the graph of w'(x) to find the initial guesses, the x-values such that w'(x) = 0 can be approximated using **FindRoot**. These three calculations are given below using initial guesses x=.863, x=1.63, x=2.25, the values where w'(x) appears to cross the x-axis.

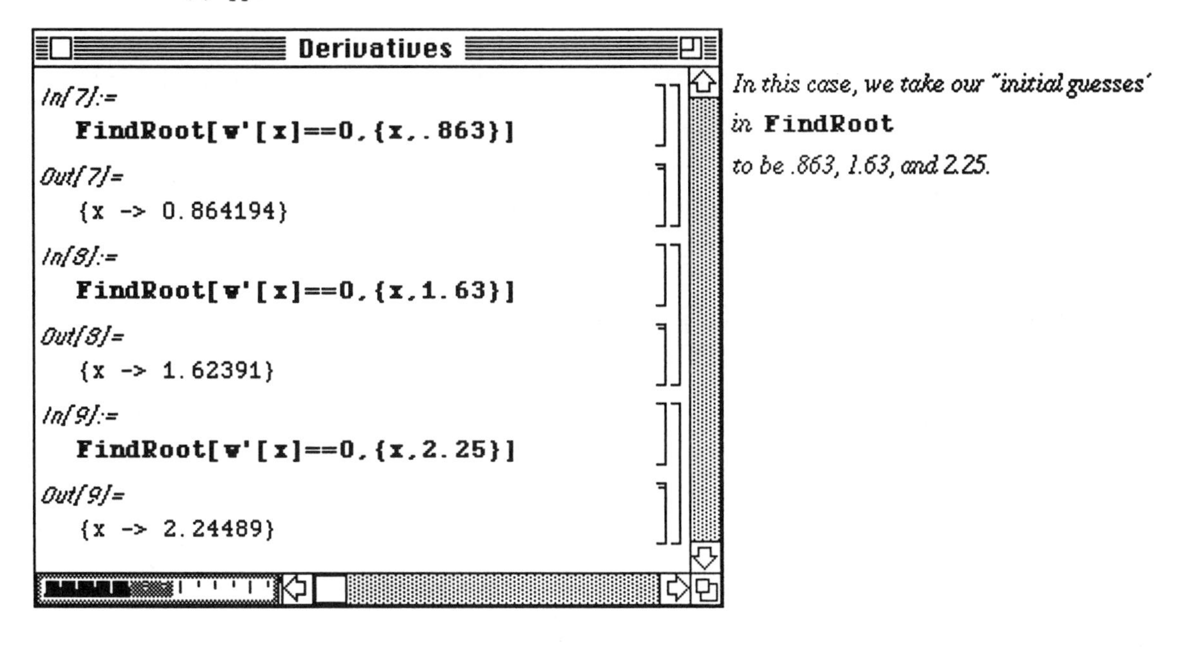

■ Application: Graphing Functions and Tangent Lines

An equation of the line with slope m that passes through the point (a,b) is given by the relationship y-b=m(x-a). To write a function that describes the line with slope m that passes through the point note that y-b=m(x-a) is equivalent to the statement y=m(x-a)+b.

The following example illustrates how the tangent line to the graph of a function can be determined and plotted simultaneously with the function.

□ Example:

Let $h(x) = \text{Sin}(6x) + 2\text{Cos}(2x)$. Graph h, the line tangent to the graph of h when $x = \frac{\pi}{3}$, and the line tangent to the graph of h when $x = \frac{2\pi}{3}$ on the interval $[0, \pi]$.

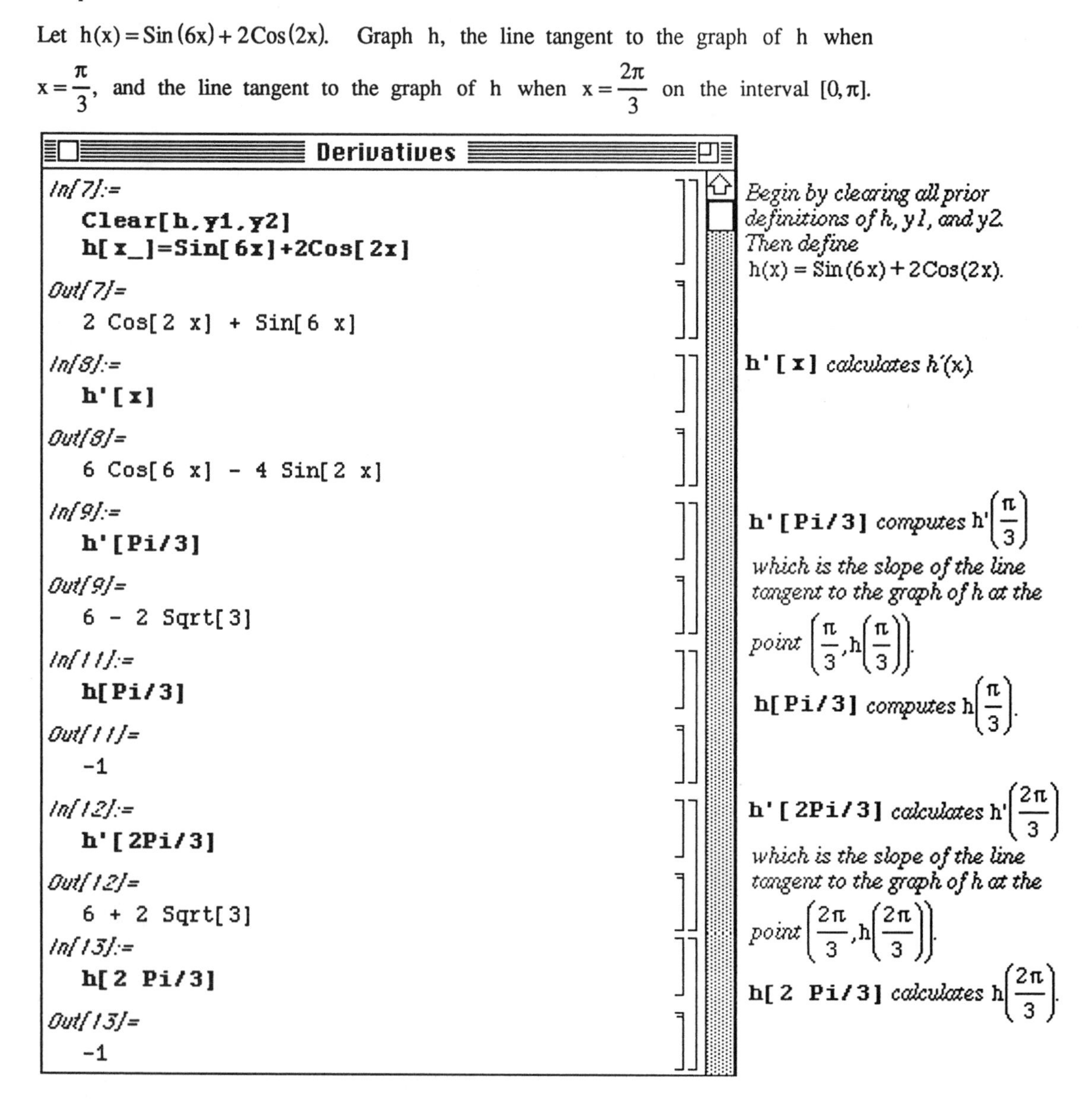

Derivatives

```
In[7]:=
  Clear[h,y1,y2]
  h[x_]=Sin[6x]+2Cos[2x]
Out[7]=
  2 Cos[2 x] + Sin[6 x]
In[8]:=
  h'[x]
Out[8]=
  6 Cos[6 x] - 4 Sin[2 x]
In[9]:=
  h'[Pi/3]
Out[9]=
  6 - 2 Sqrt[3]
In[11]:=
  h[Pi/3]
Out[11]=
  -1
In[12]:=
  h'[2Pi/3]
Out[12]=
  6 + 2 Sqrt[3]
In[13]:=
  h[2 Pi/3]
Out[13]=
  -1
```

Begin by clearing all prior definitions of h, y1, and y2. Then define $h(x) = \text{Sin}(6x) + 2\text{Cos}(2x)$.

h'[x] *calculates* $h'(x)$.

h'[Pi/3] *computes* $h'\left(\frac{\pi}{3}\right)$ *which is the slope of the line tangent to the graph of h at the point* $\left(\frac{\pi}{3}, h\left(\frac{\pi}{3}\right)\right)$.

h[Pi/3] *computes* $h\left(\frac{\pi}{3}\right)$.

h'[2Pi/3] *calculates* $h'\left(\frac{2\pi}{3}\right)$ *which is the slope of the line tangent to the graph of h at the point* $\left(\frac{2\pi}{3}, h\left(\frac{2\pi}{3}\right)\right)$.

h[2 Pi/3] *calculates* $h\left(\frac{2\pi}{3}\right)$.

Consequently, a linear function tangent to the graph of h at the point $\left(\frac{\pi}{3},-1\right)$ is given by $y_1(x)=\left(6-2\sqrt{3}\right)\left(x-\frac{\pi}{3}\right)-1$ and a linear function tangent to the graph of h at the point $\left(\frac{2\pi}{3},-1\right)$ is given by $y_2(x)=\left(6+2\sqrt{3}\right)\left(x-\frac{\pi}{3}\right)-1.$

These are defined with *Mathematica* as follows:

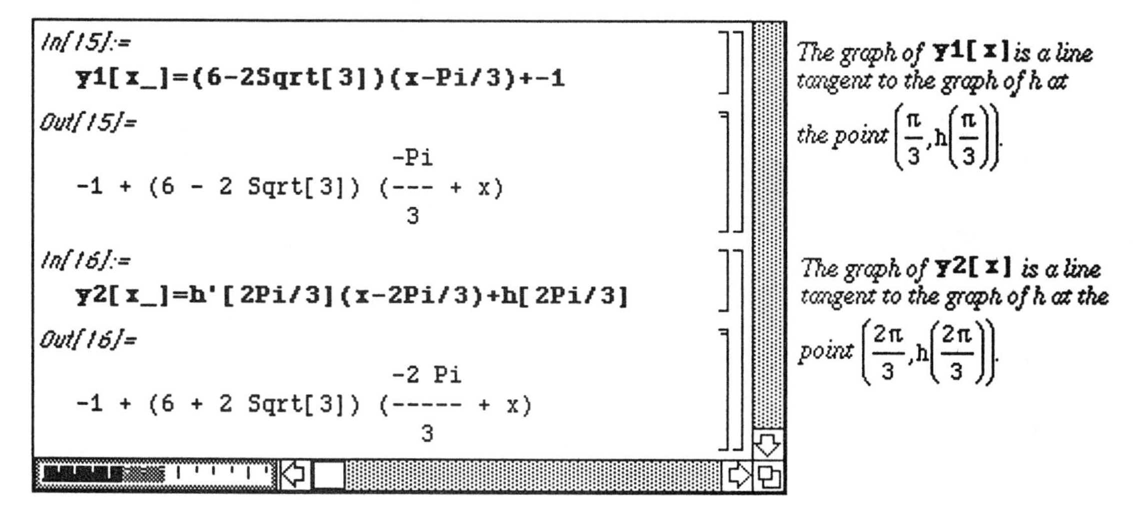

The two lines **y1[x]** and **y2[x]** are plotted with the function **h[x]** with the single command below. Note that the option **AspectRatio -> 1** is used in the **Plot** command. If this option is not used with this particular function, then the graph is difficult to read (and the lineas do not appear tangent to the curve).

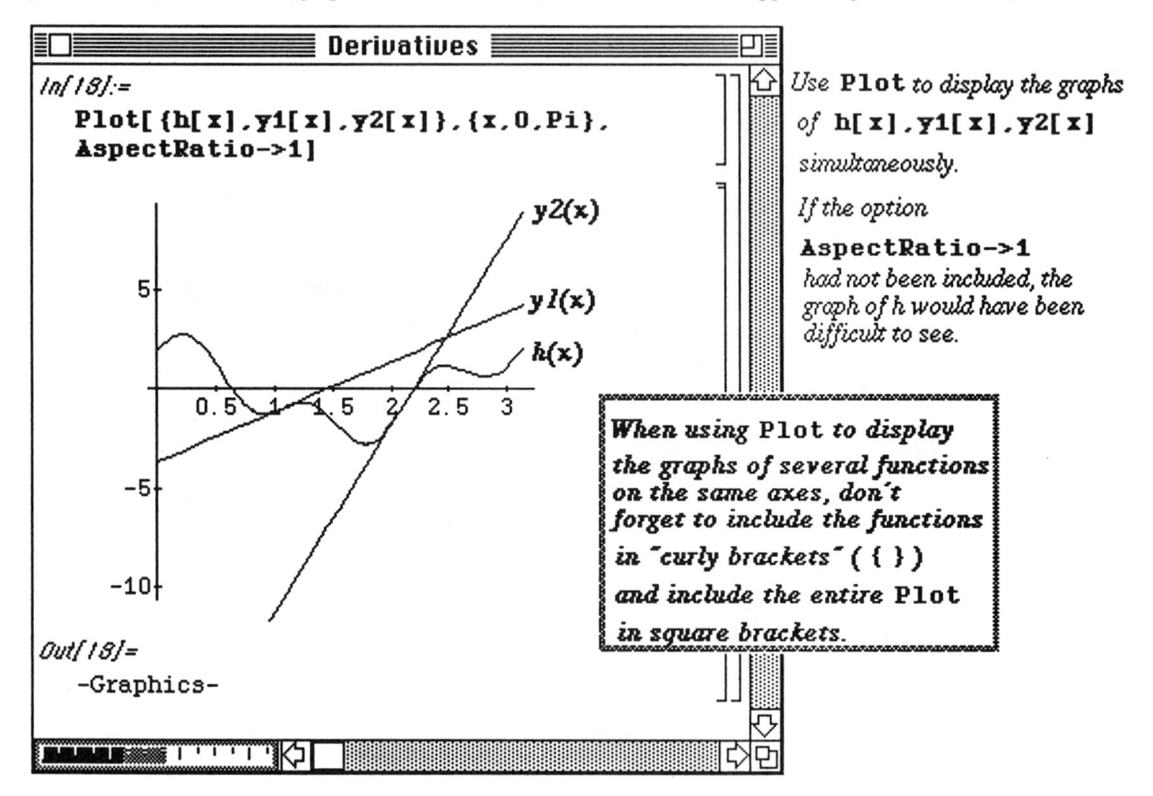

■ Application: Maxima and Minima

Mathematica can be used to solve maximization/minimization problems. An example of this type of problem is as follows :

□ Example:

A farmer has 100 feet of fencing to construct five dog kennels by first constructing a fence around a rectangular region, and then dividing that region into five smaller regions by placing four fences parallel to one of the sides. What dimensions will maximize the total area?

First, let y denote the length across the top and bottom of the rectangular region and let x denote the vertical length. Then, since 100 feet of fencing are used, a relationship between x and y is given by the equation: $2y + 6x = 100$. Solving this equation for y, we obtain $y = 50 - 3x$
which is shown in the diagram below:

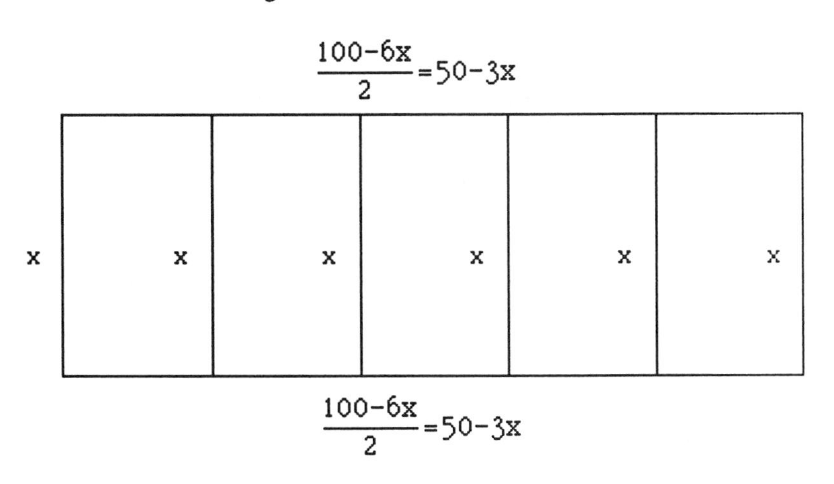

Since the area of a rectangle is A = x y, the function to be maximized is defined by entering **area[x_]=x(50-3x)**. The value of x which maximizes the area is found by finding the critical value and observing the graph of **area[x]**.

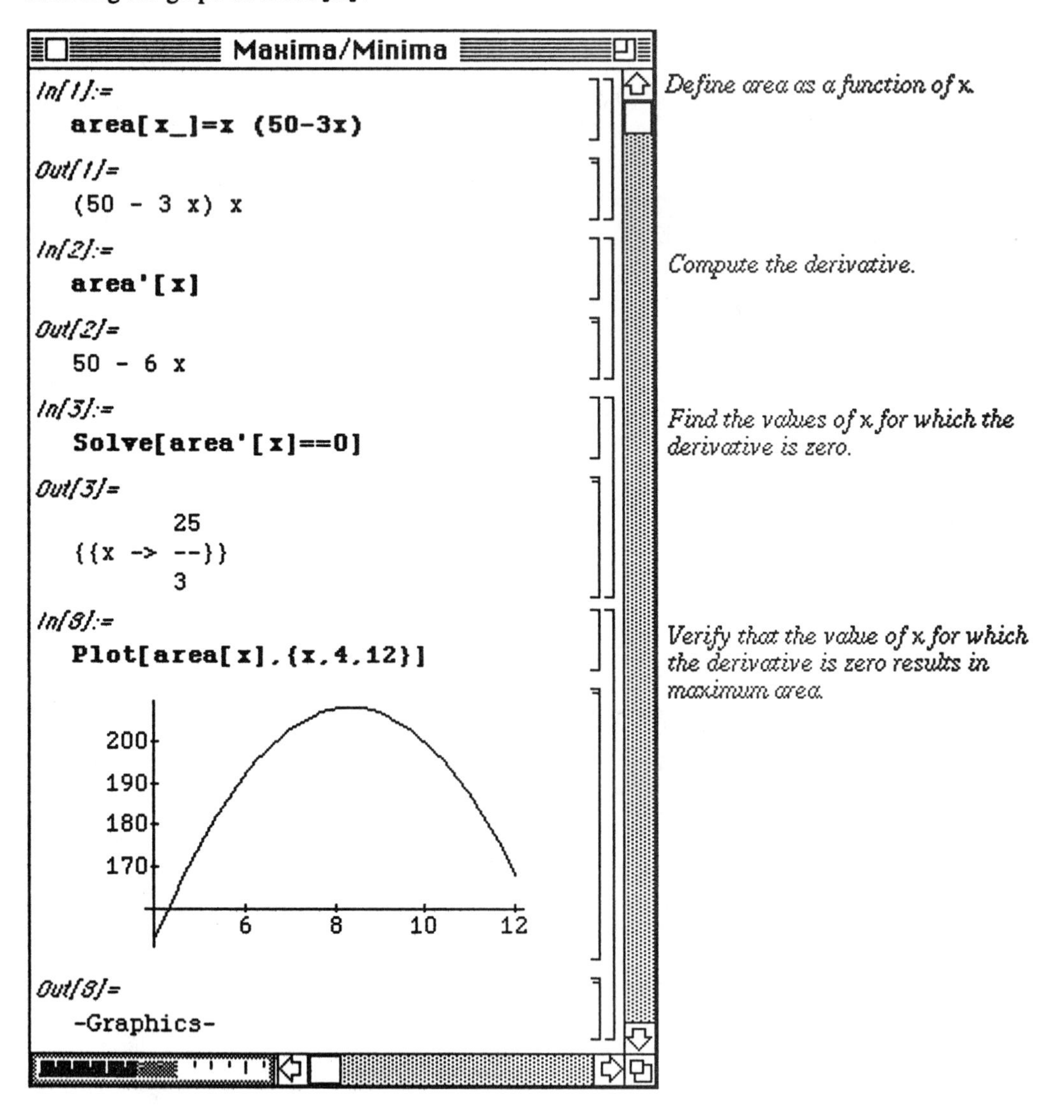

The next problem is slightly different.

□ Example:

A farmer wants to construct five dog kennels of total area 500 square feet by first constructing a fence around a rectangular region, and then dividing that region into five smaller regions by placing four fences parallel to one of the sides. What dimensions will minimize the fencing used?

In this case, the total amount of fencing needed to construct the kennels is to be minimized using the constraint that the total area is 500 square feet. (In the first problem, we maximized area using a constraint on the perimeter.) Again, let y = length across the top and bottom of the rectangular region. Using the fact that area = 500, we have x y = 500 or y = 500/x:

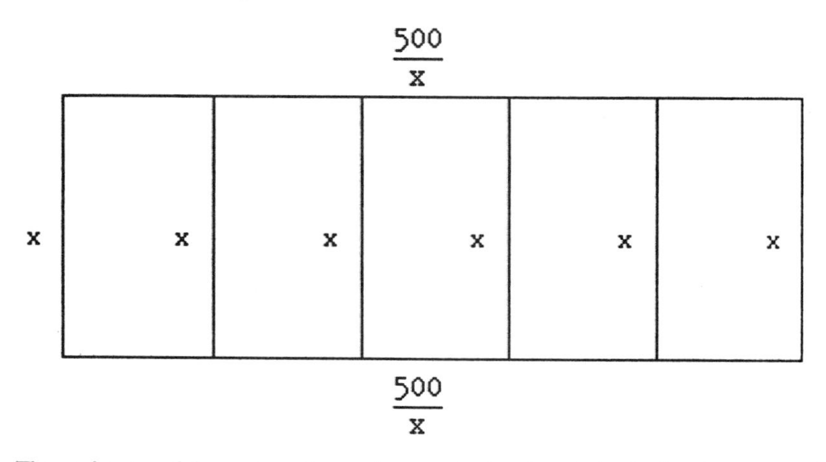

The perimeter of the rectangular region equals 2 x+2 y. Substituting for y, we define the function **perimeter** by entering **perimeter[x_]=6x+2(500/x)** which is to be minimized. The steps involved in solving this problem are shown below. (Note that only positive values of x are considered since x represents length.)

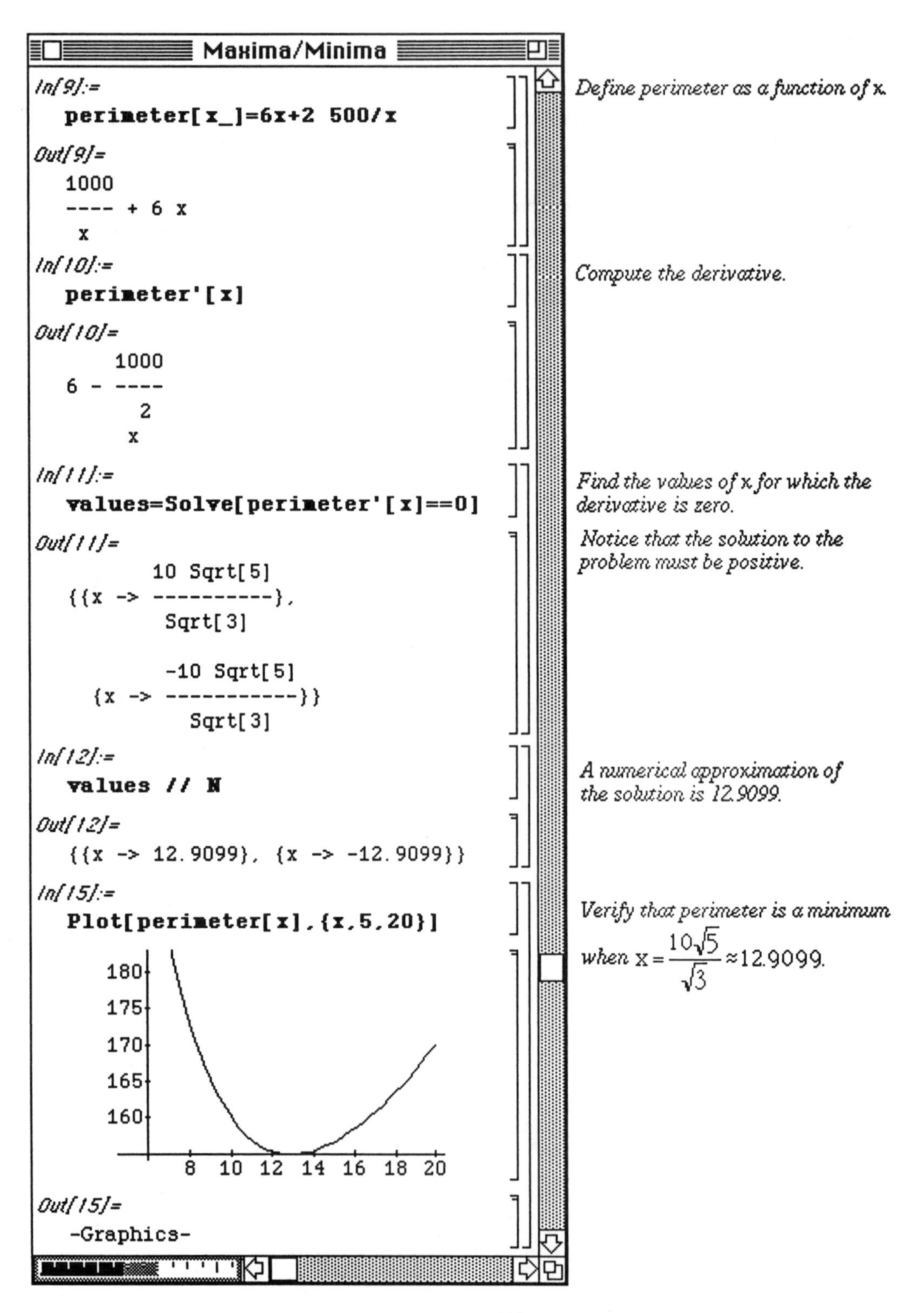

Maxima/Minima

```
In[9]:=
  perimeter[x_]=6x+2 500/x

Out[9]=
  1000
  ---- + 6 x
   x

In[10]:=
  perimeter'[x]

Out[10]=
      1000
  6 - ----
        2
       x

In[11]:=
  values=Solve[perimeter'[x]==0]

Out[11]=
          10 Sqrt[5]
  {{x -> ----------},
           Sqrt[3]

          -10 Sqrt[5]
    {x -> -----------}}
             Sqrt[3]

In[12]:=
  values // N

Out[12]=
  {{x -> 12.9099}, {x -> -12.9099}}

In[15]:=
  Plot[perimeter[x],{x,5,20}]
```



```
Out[15]=
  -Graphics-
```

Define perimeter as a function of x.

Compute the derivative.

Find the values of x for which the derivative is zero.

Notice that the solution to the problem must be positive.

A numerical approximation of the solution is 12.9099.

Verify that perimeter is a minimum when $x = \dfrac{10\sqrt{5}}{\sqrt{3}} \approx 12.9099.$

Our final example illustrates *Mathematica*'s ability to symbolically manipulate algebraic expressions.

□ **Example:**

Let f(x)=mx+b and (x0,y0) be any point. Find the value of x for which the distance from (x0,y0) to (x,f(x)) is a minimum.

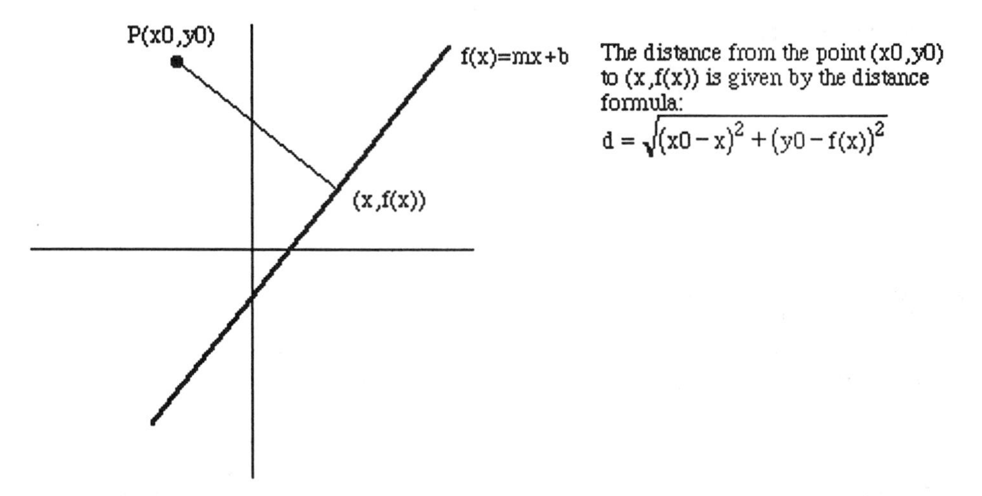

In order to determine the value of x which minimizes the distance between (x0,y0) and (x, f(x)), a function which determines this distance must first be defined. This is accomplished by defining the function **`distance`** by entering **`distance[{a_,b_},{c_,d_}]`** which gives the distance between any two points (a,b) and (c,d). Then the particular distance function for this problem is obtained by substituting the appropriate points (x0,y0) and (x,f(x)) into **`distance`**. The value of x that minimizes this function is obtained in the usual manner. (Notice how naming the distance function **`expression`** simplifies the solution of the problem.)

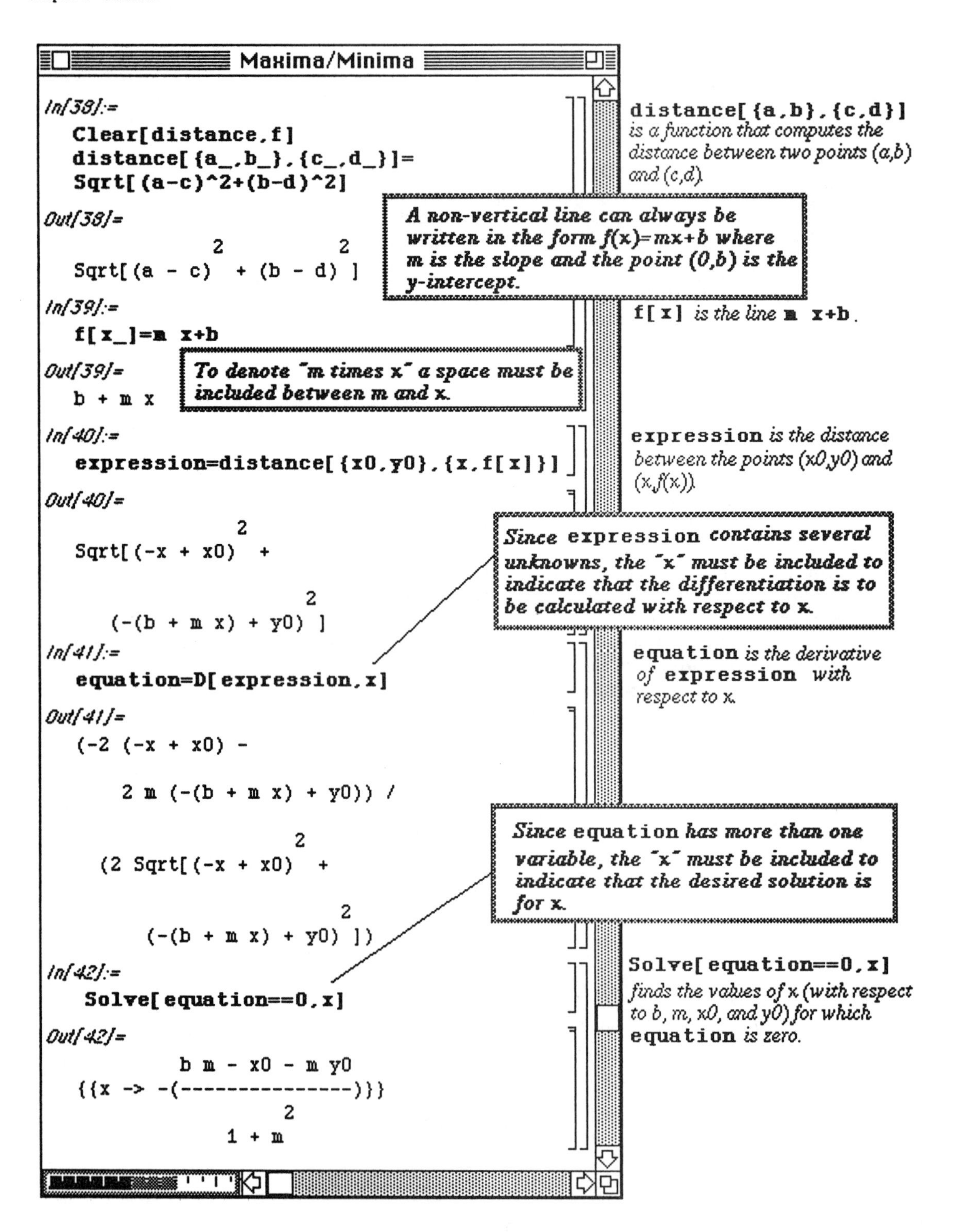
Maxima/Minima
In[38]:=
Clear[distance,f]
distance[{a_,b_},{c_,d_}]=
Sqrt[(a-c)^2+(b-d)^2]
Out[38]=
Sqrt[(a - c)^2 + (b - d)^2]
distance[{a,b},{c,d}] is a function that computes the distance between two points (a,b) and (c,d).
A non-vertical line can always be written in the form f(x)=mx+b where m is the slope and the point (0,b) is the y-intercept.
In[39]:=
f[x_]=m x+b
f[x] is the line m x+b.
Out[39]=
b + m x
To denote "m times x" a space must be included between m and x.
In[40]:=
expression=distance[{x0,y0},{x,f[x]}]
expression is the distance between the points (x0,y0) and (x,f(x)).
Out[40]=
Sqrt[(-x + x0)^2 + (-(b + m x) + y0)^2]
Since expression contains several unknowns, the "x" must be included to indicate that the differentiation is to be calculated with respect to x.
In[41]:=
equation=D[expression,x]
equation is the derivative of expression with respect to x.
Out[41]=
(-2 (-x + x0) - 2 m (-(b + m x) + y0)) / (2 Sqrt[(-x + x0)^2 + (-(b + m x) + y0)^2])
Since equation has more than one variable, the "x" must be included to indicate that the desired solution is for x.
In[42]:=
Solve[equation==0,x]
Solve[equation==0,x] finds the values of x (with respect to b, m, x0, and y0) for which equation is zero.
Out[42]=
{{x -> -((b m - x0 - m y0)/(1 + m^2))}}

■ Application

As was previously stated, *Mathematica* can define many different types of functions. The following example is a function **plotderiv** which is a function that depends on a function and an interval. When given a function, f(x), and an interval, {a,b}, **plotderiv** simultaneously plots f(x) with **GrayLevel[0]** and f'(x) with **GrayLevel[.3]**. (Remember, **GrayLevel[0]** indicates the darker curve.) Note that the labels f, f', f, and g' on the plots below were added later.

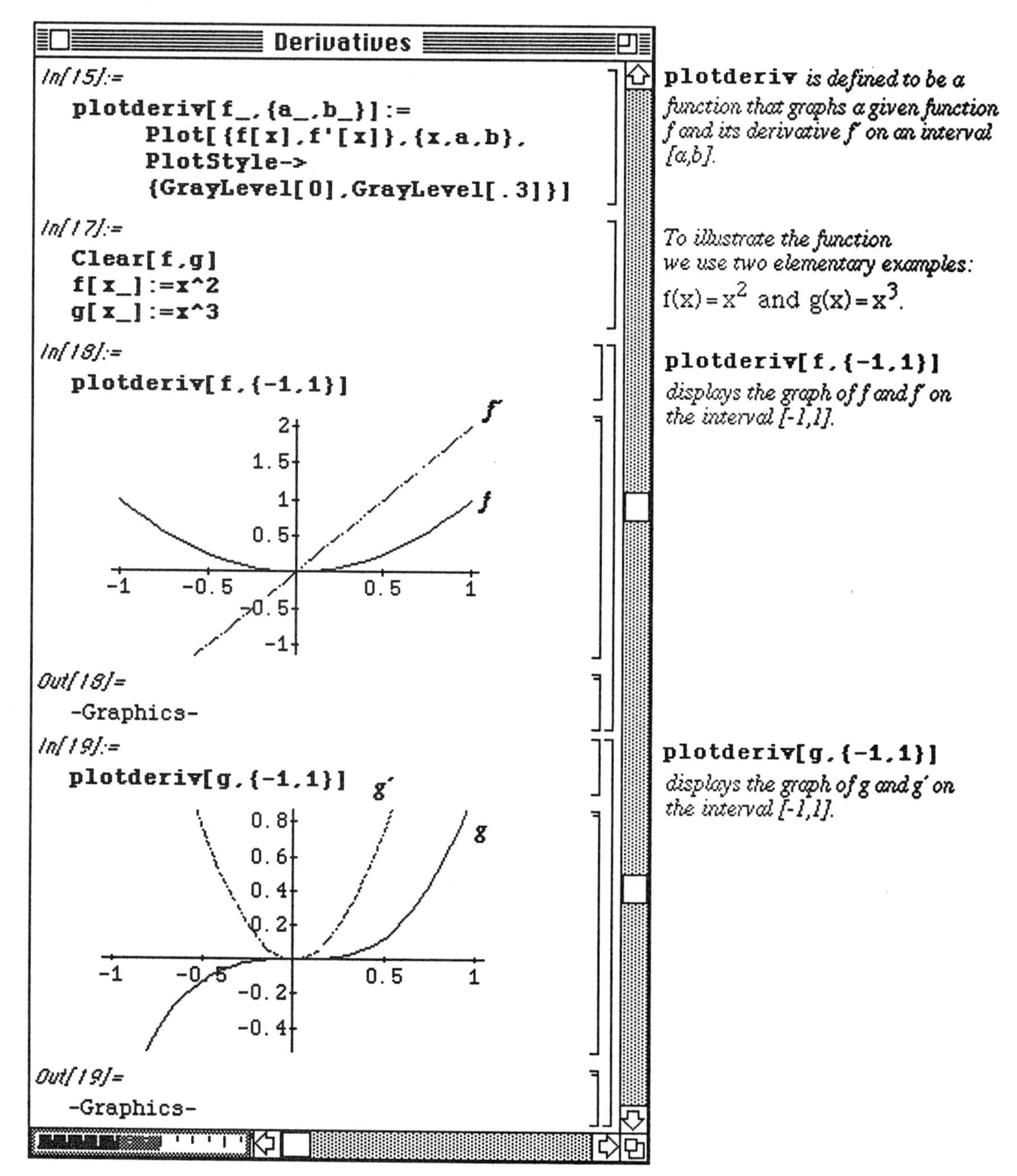

3.3 Implicit Differentiation

Computing Derivatives of Implicit Functions

If **equation** is an equation with variables **x** and **y**, *Mathematica* computes the implicit derivative of **equation** with the command **Dt[equation,x]** where **equation** is differentiated with respect to the variable **x.**

The expression **Dt[y,x]** encountered when using implicit differentiation represents the derivative of y with respect to x, dy/dx . (Hence, **Dt[x,y]** represents dx/dy.)

The built-in command **Dt** is versatile. Although here **Dt** is used to perform implicit differentiation,

Dt[expression,variable] computes the total derivative: $\frac{d(\mathtt{expression})}{d\,\mathtt{variable}}$; and

Dt[expression] computes the total differential $d(\mathtt{expression})$.

The following examples demonstrate the use of the implicit differentiation command, **Dt[equation,x]** and show how this command can be used with **Solve** to obtain the desired derivative in a single command.

□ Example:

Find $\frac{dy}{dx}$ for: (A) $x^3+y^3=1$; (B) $\mathrm{Cos}\left(x^2-y^2\right)=y\,\mathrm{Cos}(x)$; and (C) $3y^4+4x-x^2\mathrm{Sin}(y)-4=0$.

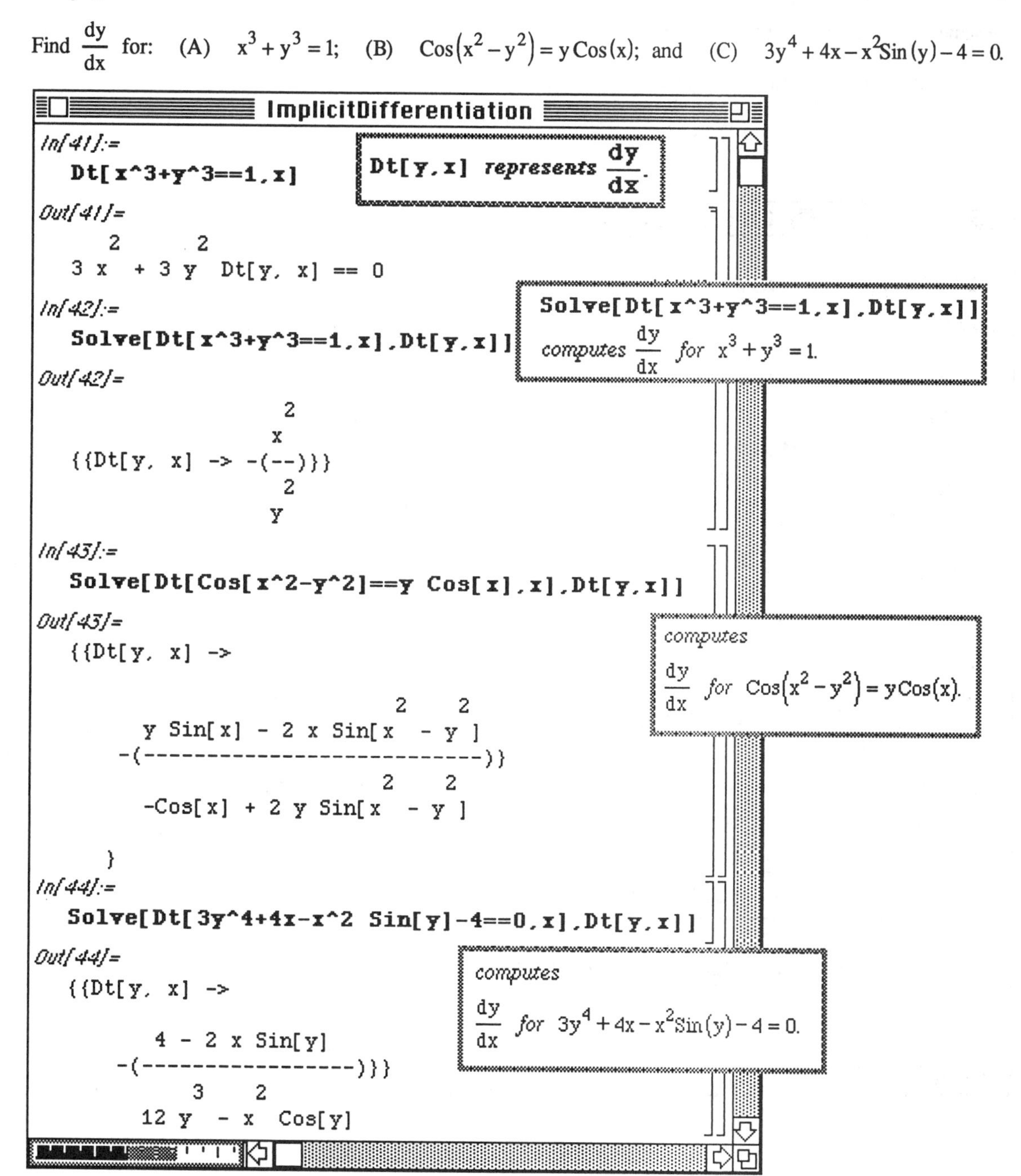

■ Other Methods to Compute Derivatives of Implicit Functions

The same results as above can be obtained if y is declared to be a function of x. Hence instead of entering the equation $x^3+y^3=1$ as **x^3 + y^3 == 1**, enter it as **x^3 +** $\underbrace{\texttt{y[x]}}_{\text{y is a function of x.}}$ **^3 == 1.**

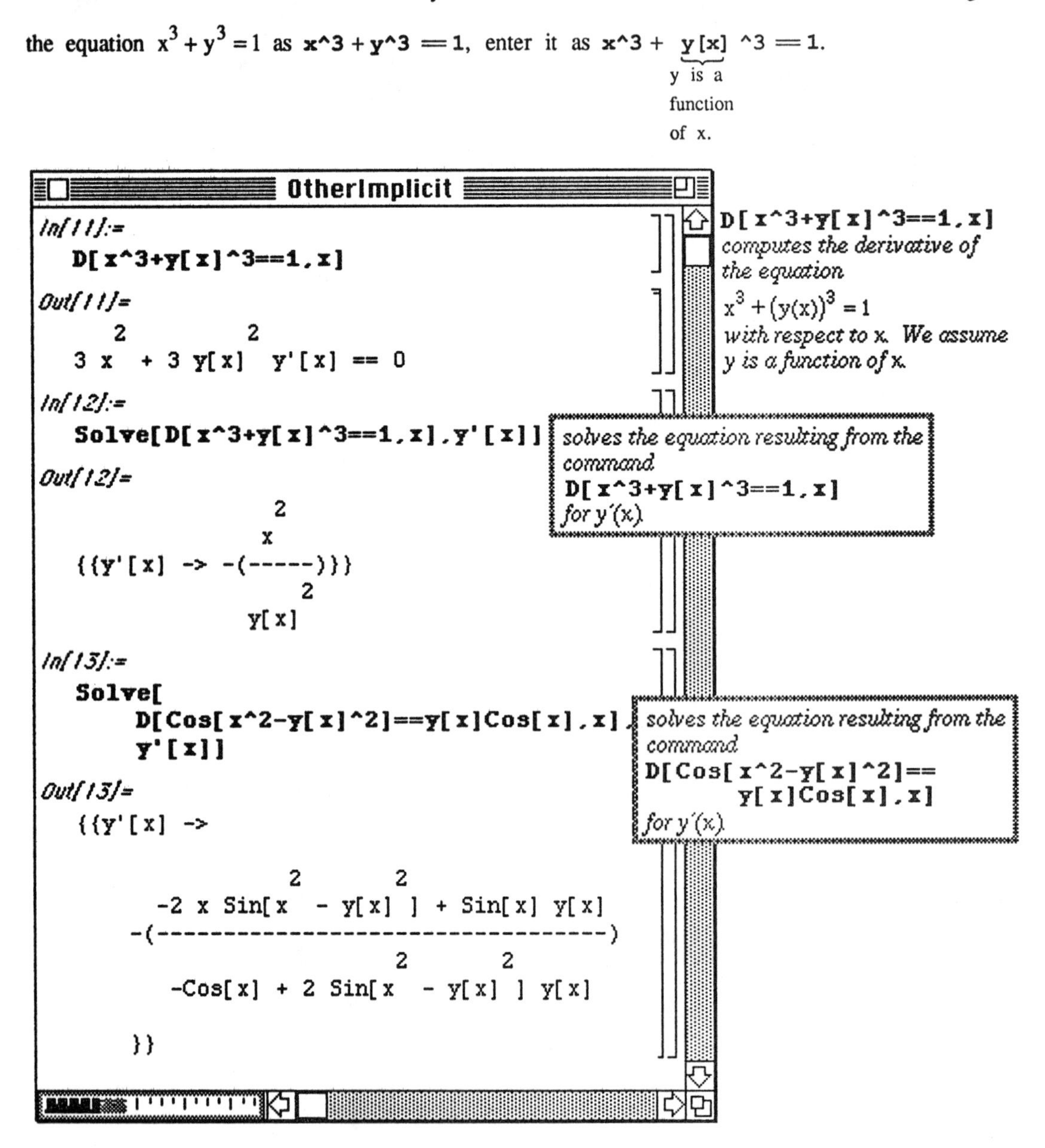

• Graphing Implicit Functions with Version 2.0

The Version 2.0 package **ImplicitPlot.m** contains the command **ImplicitPlot** which can be used to graph some equations; **ImplicitPlot.m** is not included in earlier versions of *Mathematica*. **ImplicitPlot** is discussed in more detail in **Chapter 9**.

The most basic form of the syntax for the command **ImplicitPlot** is **ImplicitPlot[equation,{x,xmin,xmax}]**. The set of y-values displayed can also be specified by entering the command in the form **ImplcitPlot[equation,{x,xmin,xmax},{y,ymin,ymax}]**. Be sure to always include the double-equals sign between the right- and left-hand side of equations.

o Example:

Use **ImplicitPlot** to graph the equation $x^3+y^3=1$.

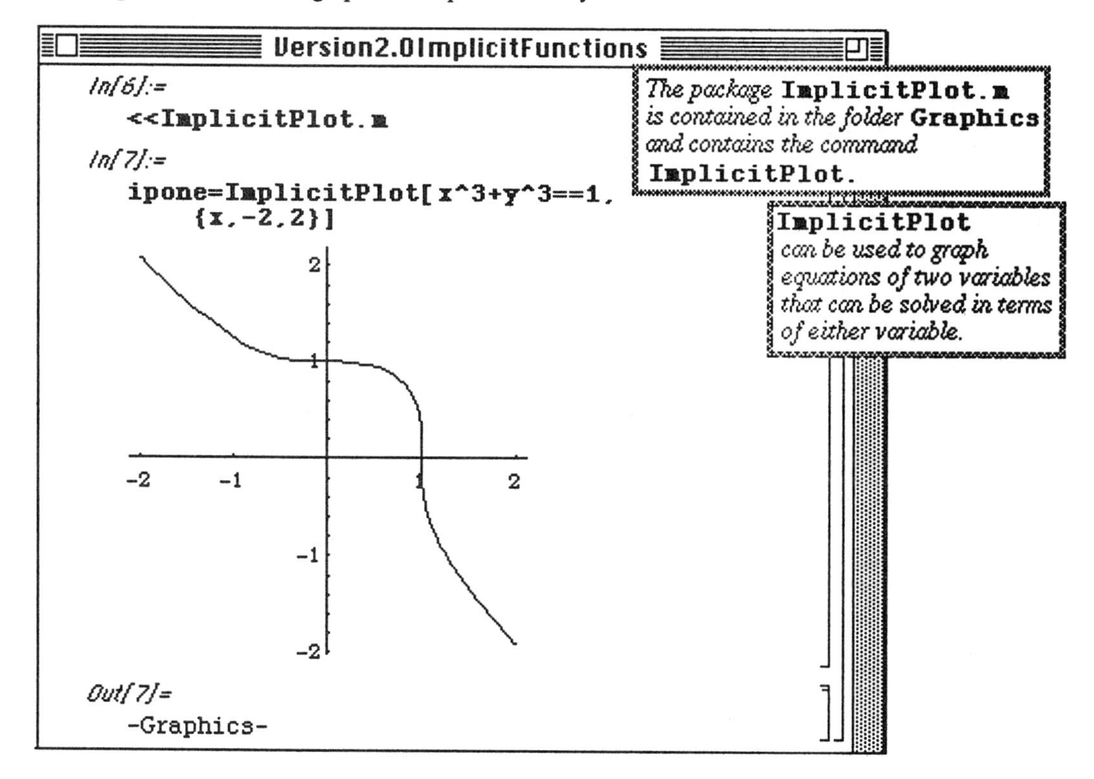

The command **ImplicitPlot** works best with equations that are (easily) solvable. Notice that **ImplicitPlot** cannot be used to graph the equation $\mathrm{Cos}\left(x^2-y^2\right)=y\,\mathrm{Cos}(x)$:

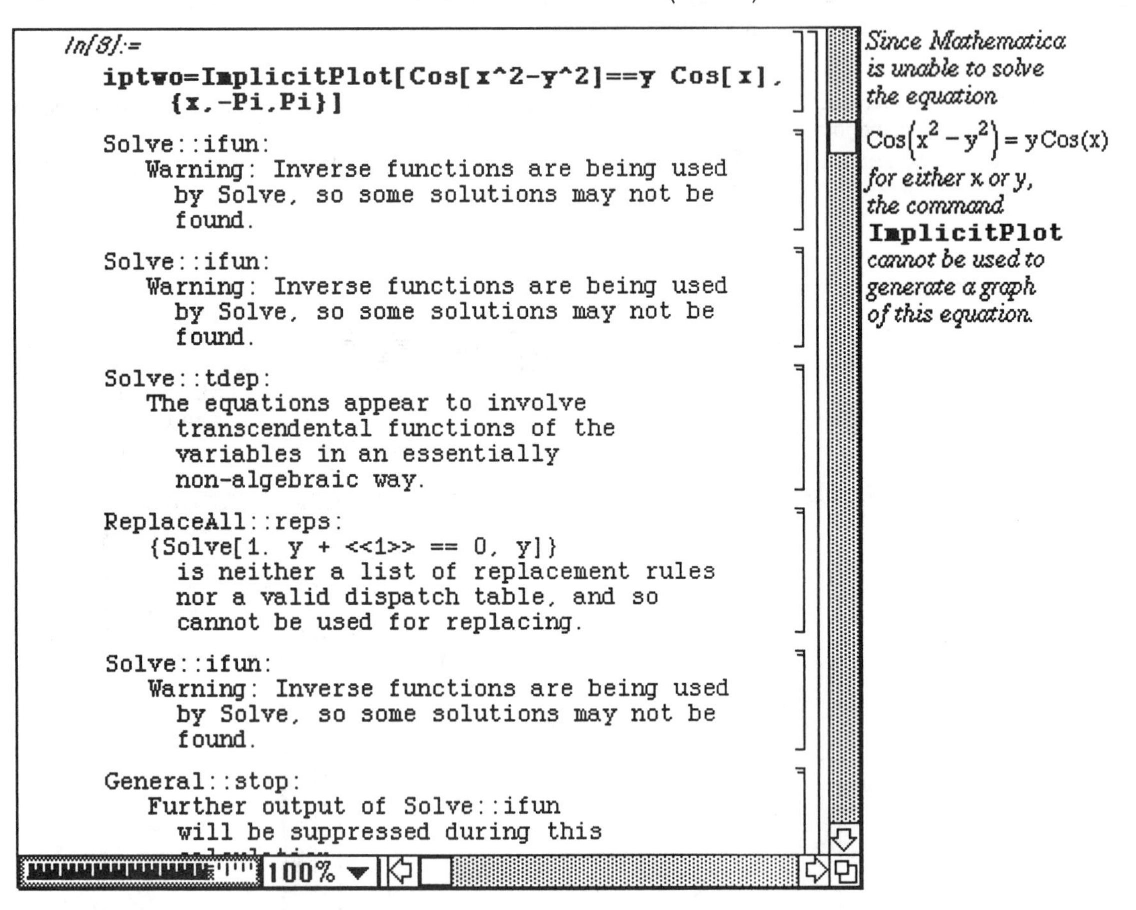

```
In[8]:=
   iptwo=ImplicitPlot[Cos[x^2-y^2]==y Cos[x],
       {x,-Pi,Pi}]

   Solve::ifun:
      Warning: Inverse functions are being used
        by Solve, so some solutions may not be
        found.

   Solve::ifun:
      Warning: Inverse functions are being used
        by Solve, so some solutions may not be
        found.

   Solve::tdep:
      The equations appear to involve
        transcendental functions of the
        variables in an essentially
        non-algebraic way.

   ReplaceAll::reps:
      {Solve[1. y + <<1>> == 0, y]}
        is neither a list of replacement rules
        nor a valid dispatch table, and so
        cannot be used for replacing.

   Solve::ifun:
      Warning: Inverse functions are being used
        by Solve, so some solutions may not be
        found.

   General::stop:
      Further output of Solve::ifun
        will be suppressed during this
```

Since Mathematica is unable to solve the equation $\mathrm{Cos}\left(x^2-y^2\right)=y\,\mathrm{Cos}(x)$ *for either* x *or* y, *the command* **ImplicitPlot** *cannot be used to generate a graph of this equation.*

Instead a different approach is used taking advantage of the built-in function **ContourPlot**. The contour graphs shown here were created with the **ContourPlot** command as it is in Version 2.0. The Version 2.0 **ContourPlot** command is substantially different from earlier versions of **ContourPlot** which are discussed later.

To use **ContourtPlot** to graph the equation $\mathrm{Cos}(x^2-y^2)=y\,\mathrm{Cos}(x)$, we begin by noticing that graphing the equation $\mathrm{Cos}(x^2-y^2)=y\,\mathrm{Cos}(x)$ is equivalent to defining $f(x,y)=\mathrm{Cos}(x^2-y^2)-y\,\mathrm{Cos}(x)$ and graphing $f(x,y)=0$.

ContourPlot[f[x,y],{x,xmin,xmax},{y,ymin,ymax}] graphs a set of level curves of the function **f[x,y]** on the rectangle [**xmin,max**] x [**ymin,ymax**]:

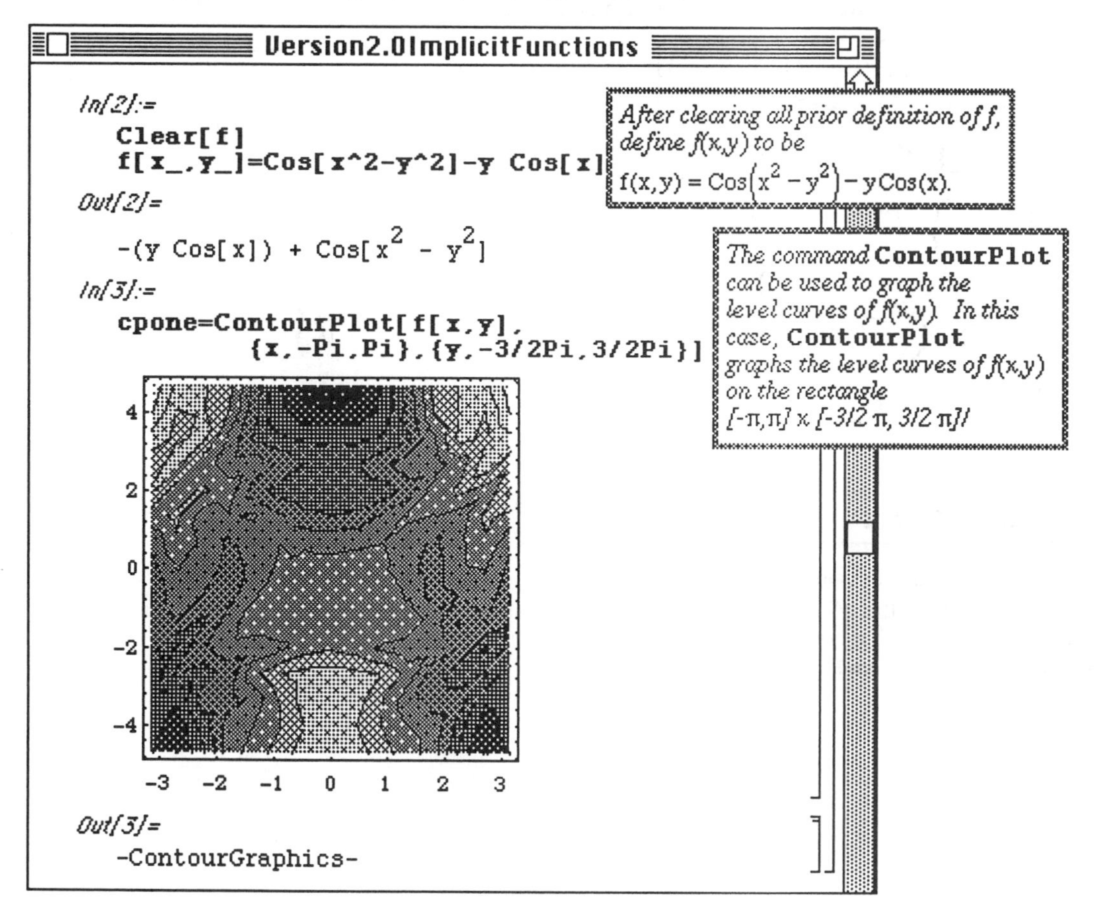

ContourPlot has many available options. For example, *Mathematica* can apply a smoothing algorithm to each contour which results in a smoother graph with the option **ContourSmoothing->Automatic**; the option **ContourShading->False** specifies that *Mathematica* not shade the resulting graph:

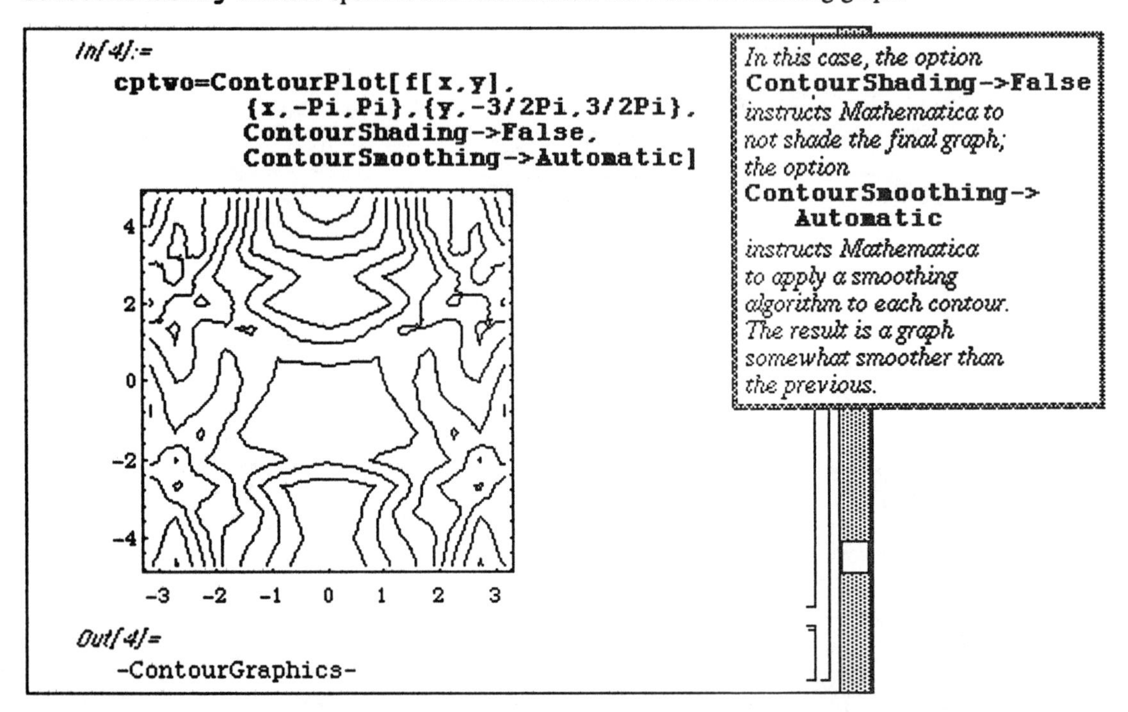

In addition, the actual contour values or the number of contours to be graphed can be specified. The option **Contours->n** specifies that *Mathematica* draw **n** evenly-spaced level curves (the default value is 10). The option **Contours->{val1,val2,...,valn}** specifies that *Mathematica* graph level curves corresponding to **val1, val2,** ... , **valn**.

Since the graph of $f(x,y)=0$ corresponds to a level curve of $f(x,y)$ for the value 0, the desired graph is obtained as follows:

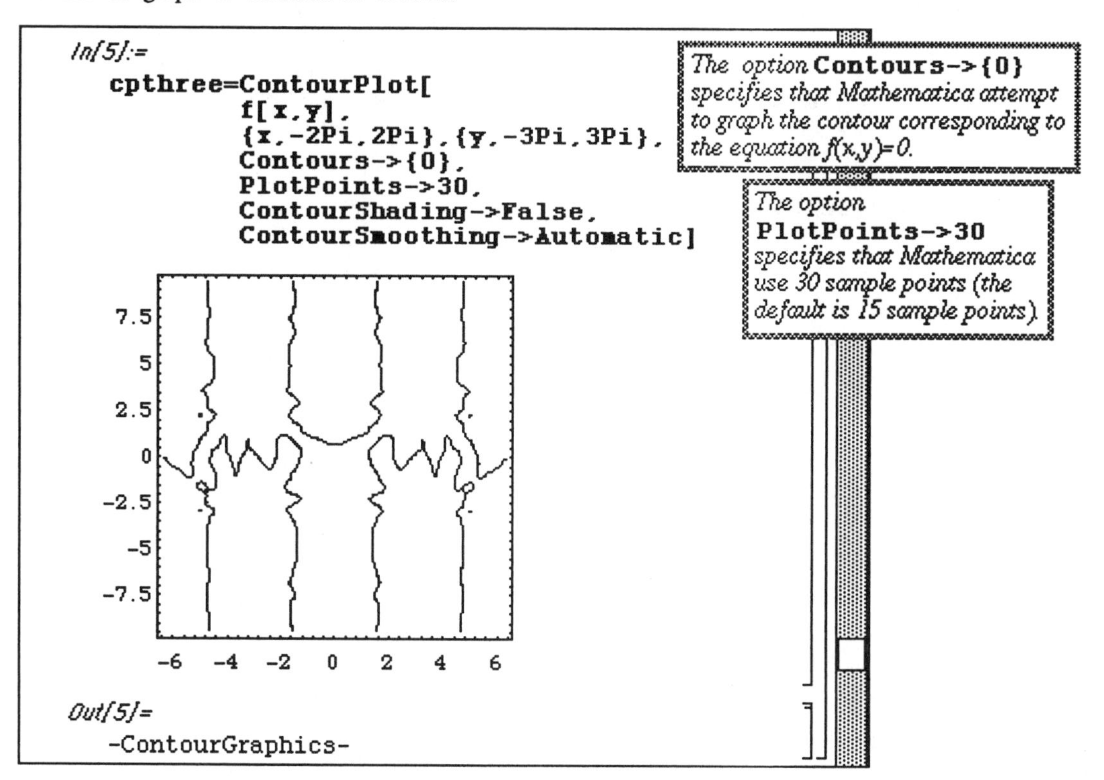

In the same manner as above, the graph of the equation $3y^4 + 4x - x^2\text{Sin}(y) - 4 = 0$ corresponds to a contour graph of $3y^4 + 4x - x^2\text{Sin}(y) - 4$ for the contour with value 0:

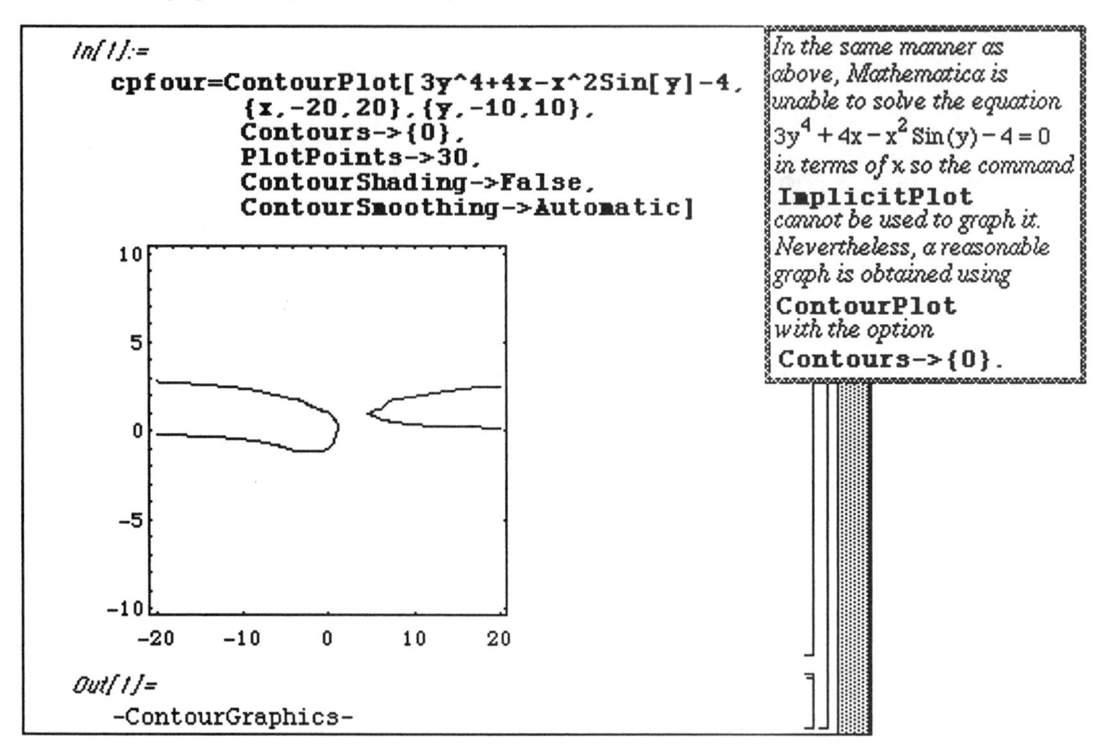

3.4 Integral Calculus

Computing Definite and Indefinite Integrals

- In order to compute definite integrals, Version 1.2 (or earlier) of *Mathematica* must load the package **`IntegralTables.m`**. The package **`IntegralTables.m`** is contained in the folder **StartUp** which is contained in the folder **Packages** in the ***Mathematica*** **f** folder. The easiest way to load the package **`IntegralTables.m`** is to **Enter** the command **`<<IntegralTables.m`**. If you are using Version 2.0, this procedure is not applicable since Version 2.0 automatically loads the package **`IntegralTables.m`**.

- Version 2.0 automatically loads the package **`IntegralTables.m`**; hence, the above procedure is **not** pertinent to Version 2.0 users.

Note Regarding Frequent Computations of Definite Integrals

This note is only applicable if you are NOT using Version 2.0.

If you are going to be computing definite integrals frequently you will want to have *Mathematica* automatically load up **`IntegralTables.m`** when the *Mathematica* kernel is started. To do this, proceed as follows.

1) Go to **Edit** and select **Settings**;

2) Select **StartUp**;

3)

Click inside the box beside Integration Rules. Boxes with x's in them indicate which packages are automatically loaded when the kernel is started.

Click in this box

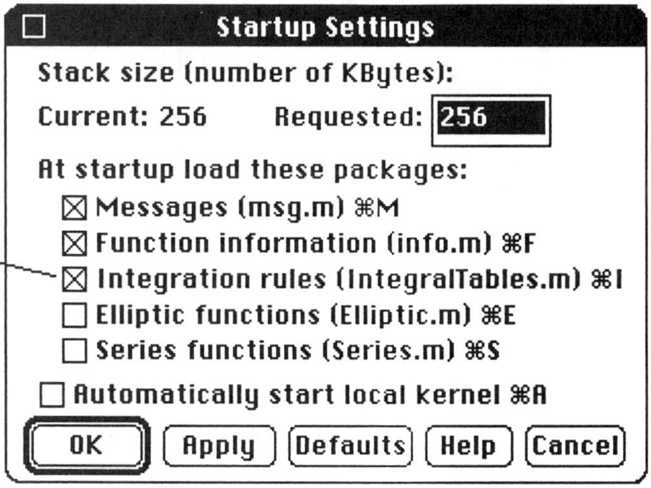

Startup Settings
Stack size (number of KBytes):
Current: 256 Requested: 256
At startup load these packages:
☒ Messages (msg.m) ⌘M
☒ Function information (info.m) ⌘F
☒ Integration rules (IntegralTables.m) ⌘I
☐ Elliptic functions (Elliptic.m) ⌘E
☐ Series functions (Series.m) ⌘S
☐ Automatically start local kernel ⌘A
OK | Apply | Defaults | Help | Cancel

Each of the following examples illustrate typical commands used to compute indefinite integrals.

The *Mathematica* command to compute $\int f(x)dx$ is **`Integrate [f[x],x]`**.

The command **`Integrate[expression,variable]`** instructs *Mathematica* to integrate **`expression`** with respect to **`variable`**.

□ **Example:**

Use *Mathematica* to compute $\int x^2\left(1-x^3\right)^4 dx$, $\int e^x \mathrm{Cos}(x)dx$, $\int \frac{\mathrm{Ln}(x)}{x^3}dx$,

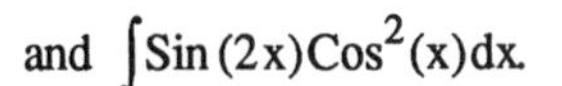

and $\int \mathrm{Sin}(2x)\mathrm{Cos}^2(x)dx$.

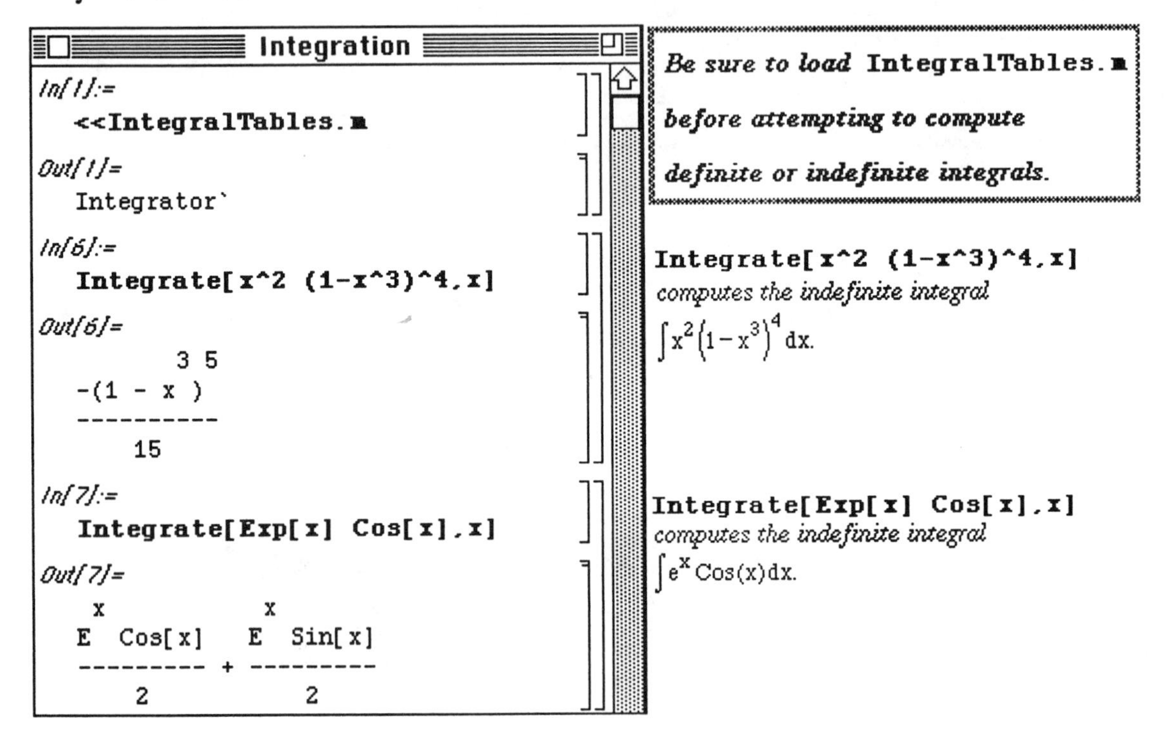

To compute $\int \frac{\text{Ln}(x)}{x^3}dx$, remember that the *Mathematica* function **Log [x]** denotes the natural logarithm function Ln(x):

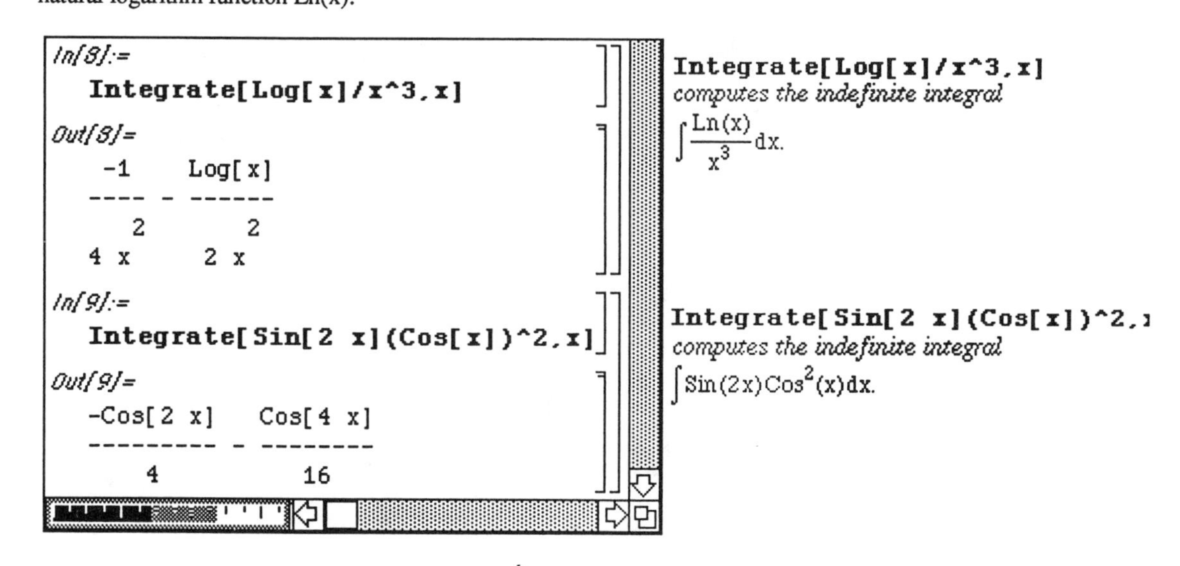

Mathematica computes the definite integral $\int_a^b f(x)dx$ with the command **Integrate [f[x],{x,a,b}]**.

In general, the command
Integrate[expression,{variable,lower limit,upper limit}]
integrates **expression** with respect to **variable** and evaluates from **lower limit** to **upper limit**.

□ Example:

Use *Mathematica* to compute $\int_0^{\pi}\mathrm{Sin}(x)dx$, $\int_0^2\sqrt{4-x^2}\,dx$, $\int_1^2 x^3 e^{4x}\,dx$, and $\int_{-\pi}^{2\pi} e^{2x}\,\mathrm{Sin}^2(2x)dx$.

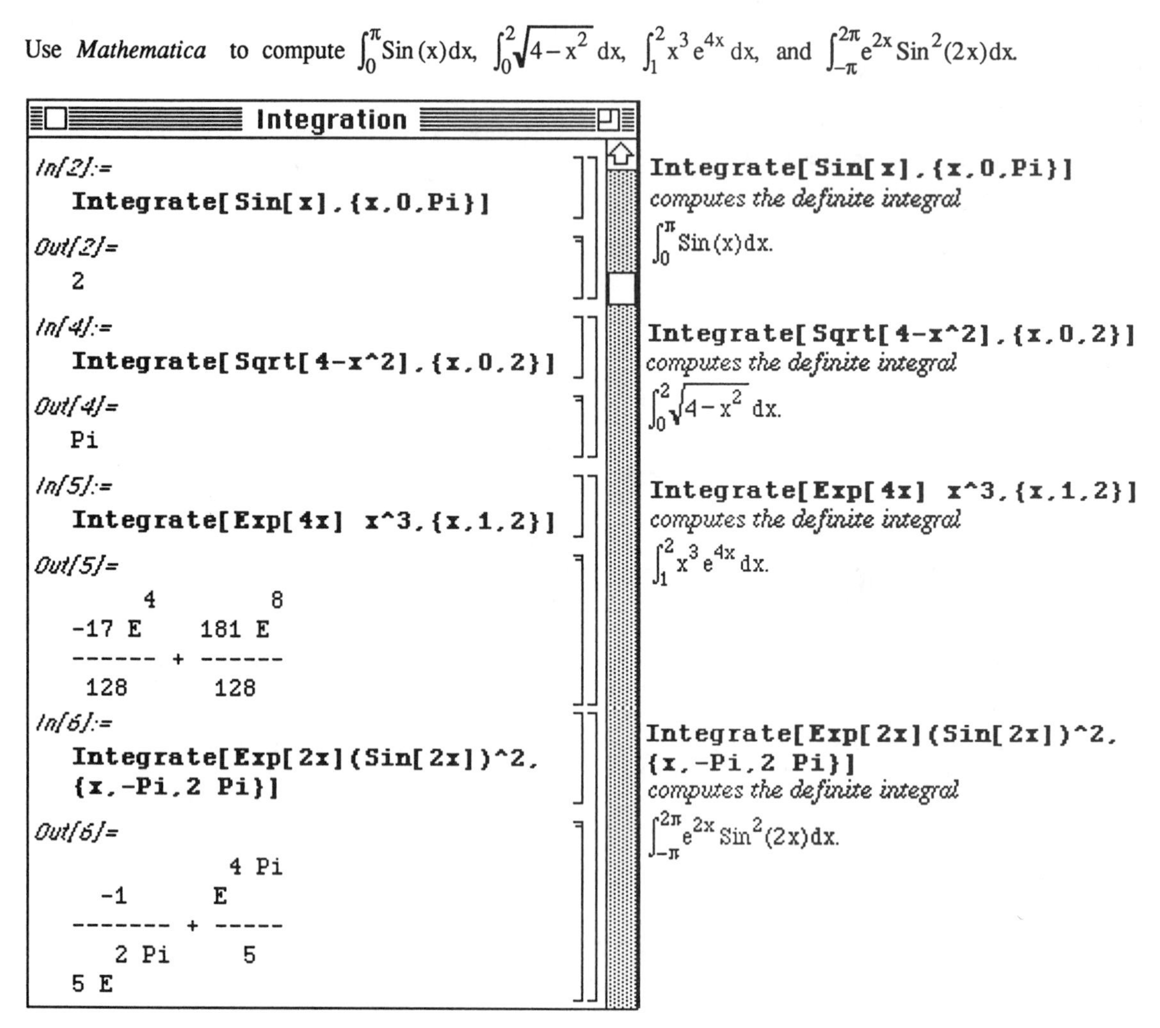

When the command **`Integrate[f[x],{x,xmin,xmax}]`** is entered, *Mathematica* computes an anti-derivative **`F`** of **`f`** and computes **`F[xmax]-F[xmin]`**. Nevertheless, *Mathematica* does not apply the Fundamental Theorem of Calculus since *Mathematica* does not verify that **`f`** is continuous on the interval [**`xmin,xmax`**]. In cases when **`f`** is not continuous on [**`xmin,xmax`**], errors often occur:

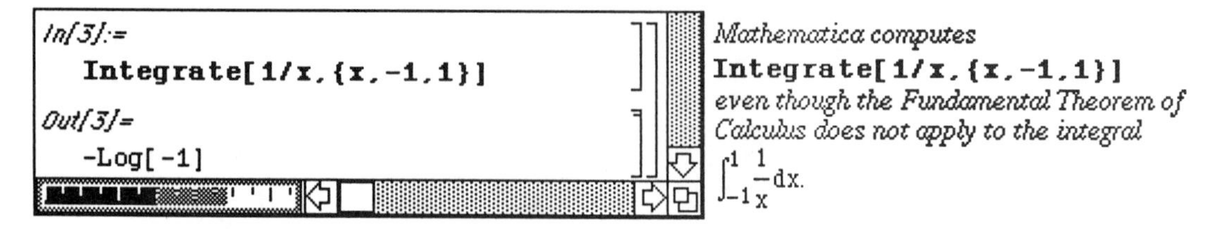

■ Numerically Computing Definite Integrals

Mathematica can also numerically integrate definite integrals of the form

$\int_{\mathbf{lowerlimit}}^{\mathbf{upperlimit}} \mathbf{expression}\, d\, \mathbf{variable}$ with the command

NIntegrate[expression,{variable,lowerlimit,upperlimit}]. NIntegrate is a built-in command and is NOT contained in the package **IntegralTables.m**. Consequently, Version 1.2 users should be aware that it is **not** necessary to load the package **IntegralTables.m** to use the command **NIntegrate**.

The command **NIntegrate** is useful when an anti-derivative of **expression** cannot be (easily) found and **expression** is fairly smooth on the interval **[lower limit, upper limit]** Also, in those cases in which an anti-derivative can be determined, the value of the definite integral can usually be computed more quickly by an approximation with **NIntegrate** rather than **Integrate**.

o In Version 2.0, the package **GaussianQuadrature.m** contained in the **Numerical Math** folder can also be used to numerically compute integrals. The package **GaussianQuadrature.m** is discussed in **Chapter 9**.

Integrate[f[x],{x,a,b}] applies the Fundamental Theorem of Calculus, if applicable: it finds an anti-derivative of **f[x]**, evaluates the anti-derivative at the upper limit of integration, and subtracts the value of the anti-derivative evaluated at the lower limit of integration. As noted above, if **f** is not continuous on **[a,b]**, error often occurs.

□ Examples:

(A) Approximate $\int_0^{\pi^{1/3}} e^{-x^2}\text{Cos}(x^3)dx$; and

(B) Compute both exact and approximate values of $\int_0^{\pi/8} e^{3x}\,\text{Cos}(4x)\,dx$.

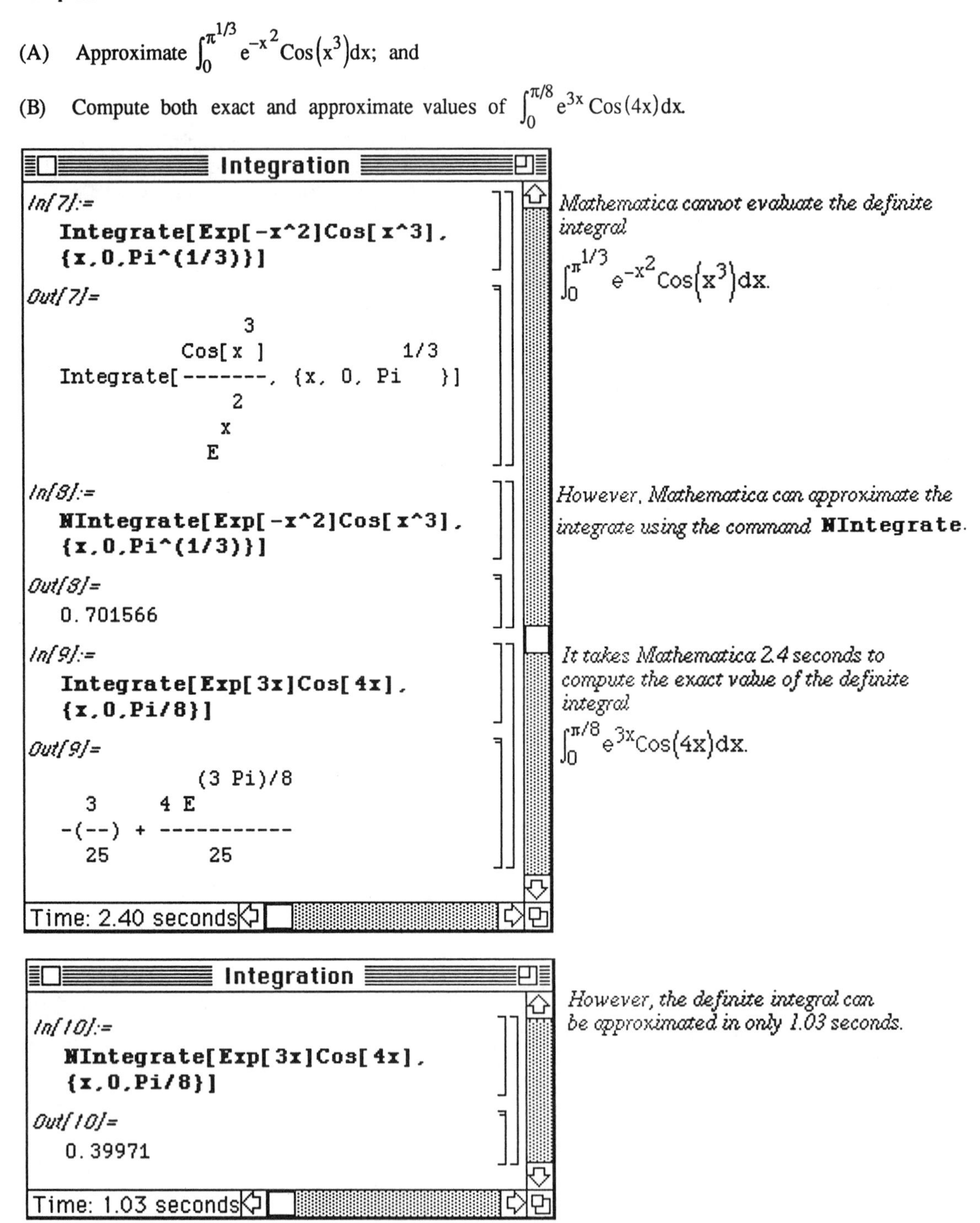

• Definite Integration with Version 2.0

The point to be made concerning integration in Version 2.0 is that **IntegralTables.m** does not have be loaded before evaluating definite integrals. Several examples are given below. In each case, the same results are obtained with Version 1.2:

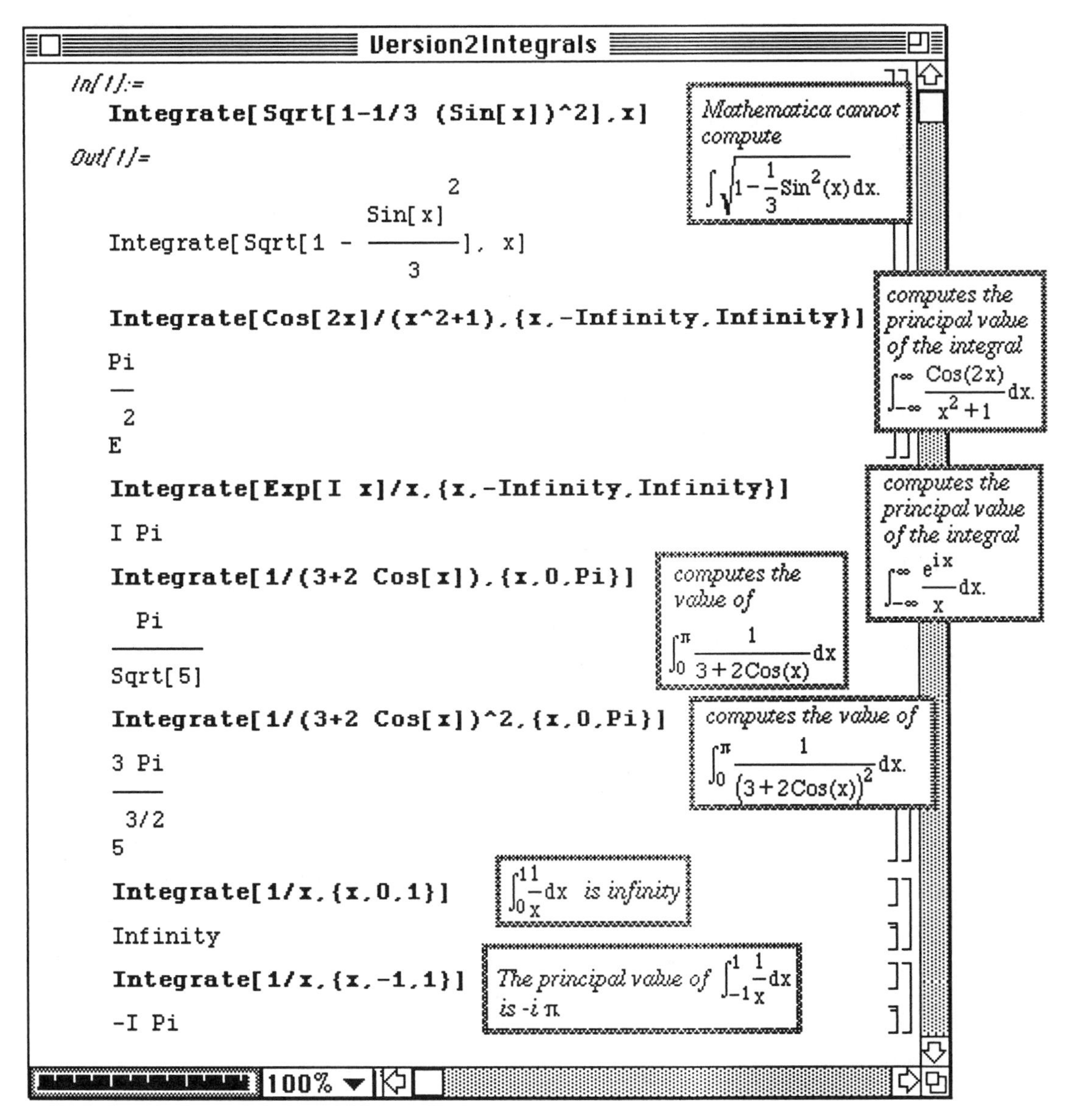

■ Application: Area Between Curves

A type of problem which incorporates the commands **Integrate** and **NIntegrate** is that of finding the area between curves. These problems also use several other *Mathematica* commands (**Plot**, **NRoots**, **FindRoot**, ...) which were introduced earlier in the text.

□ Example:

Let $p(x) = \frac{3}{10}x^5 - 3x^4 + 11x^3 - 18x^2 + 12x + 1$ and $q(x) = -4x^3 + 28x^2 - 56x + 32$

Approximate the area of the region bounded by the graphs of p and q.

Mathematica is quite helpful in problems of this type. We can observe the region whose area we are seeking using the **Plot** command, and we can locate the points of intersection with one of the commands used in solving equations (**NRoots**, **FindRoot**, **Solve**, or **NSolve** (Version 2.0 only)) These steps are carried out below:

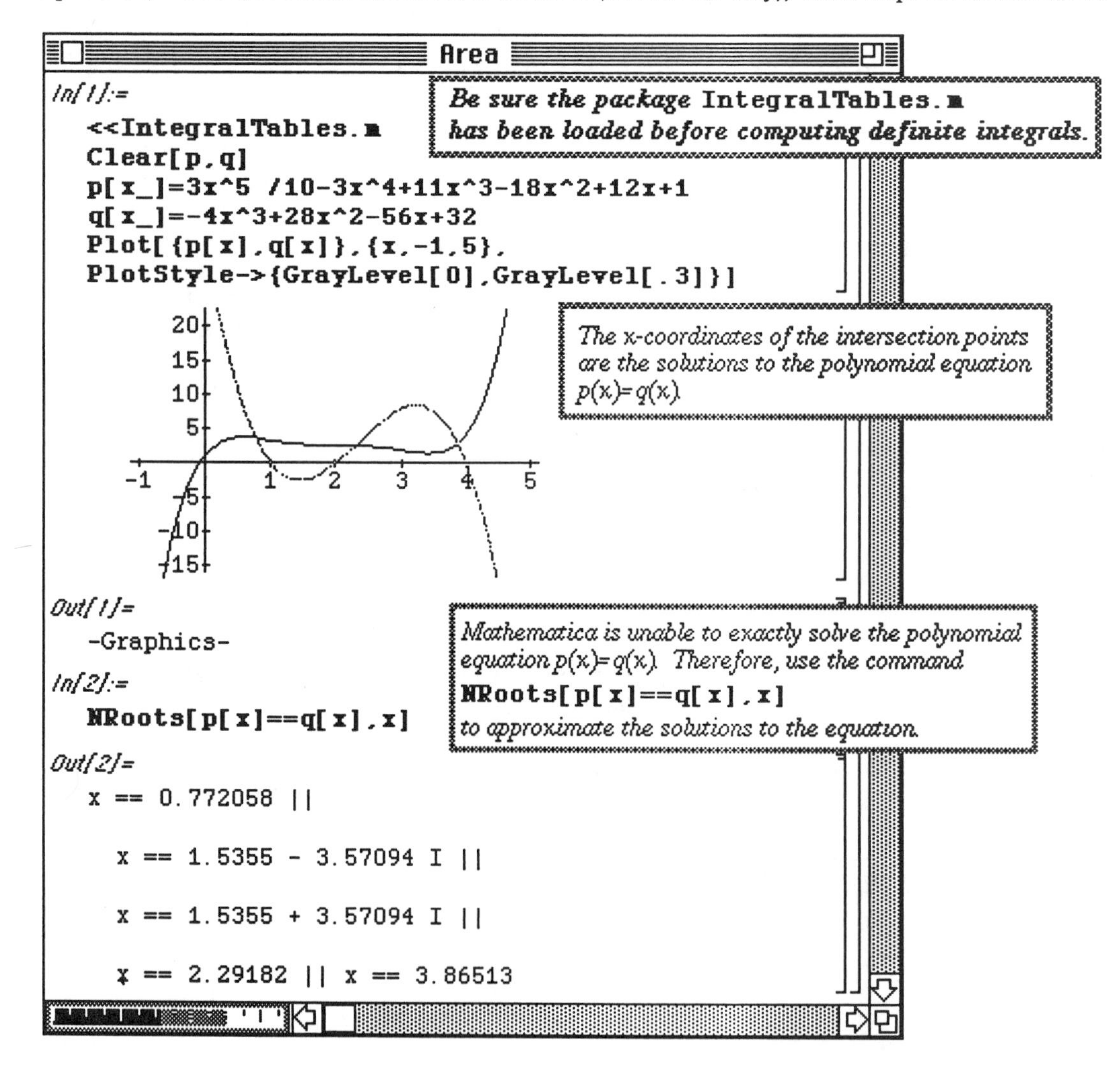

Using the roots to the equation p(x) = q(x) found above, the graph clearly shows that
p(x) > q(x) between x=0.772058 and x=2.29182 ; and
q(x) > p(x) between x=2.29182 and x=3.86513.
Hence, an approximation of the area bounded by p(x) and q(x) given by the integral

$\int_{.772058}^{2.29182}(p(x)-q(x))\,dx+\int_{2.29182}^{3.86513}(q(x)-p(x))\,dx$ is computed with either of the following commands

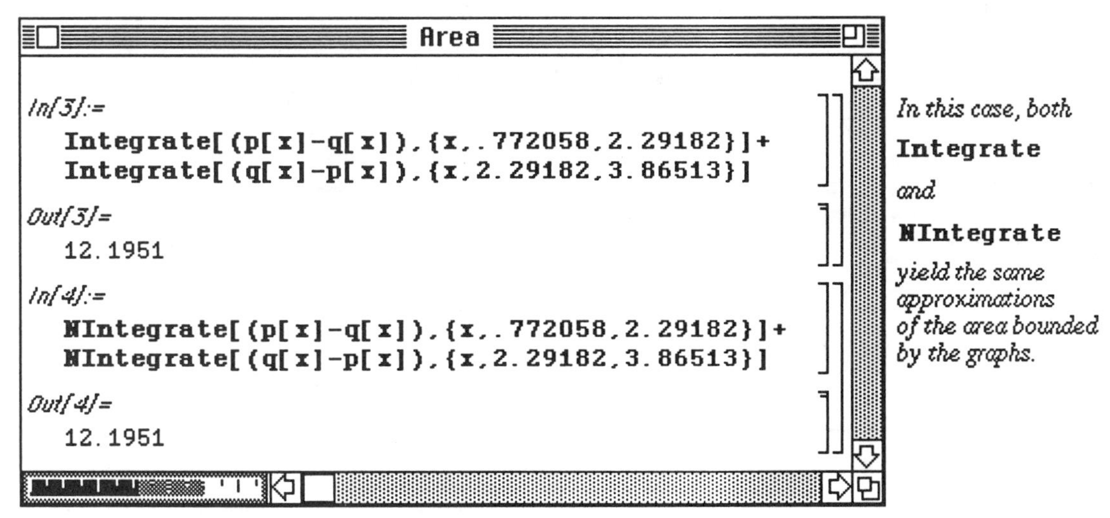

In this case, both **Integrate** *and* **NIntegrate** *yield the same approximations of the area bounded by the graphs.*

Next, consider a problem which involves functions which are <u>not</u> polynomials.

□ Example:

Let $f(x) = e^{-(x-2)^2 \mathrm{Cos}[\pi x]}$ and $g(x) = 4\mathrm{Cos}(x-2)$ on the interval $[0,4]$.

Approximate the area of the region bounded by the graphs of f and g. Since these functions are not polynomials, **FindRoot** must be used to determine the points of intersection. Recall that **FindRoot** depends on an initial guess of the root. Therefore, the first step towards solving this problem is to graph the functions f and g. Then the cursor is used to locate the initial guesses. (This topic was discussed earlier in the text.)

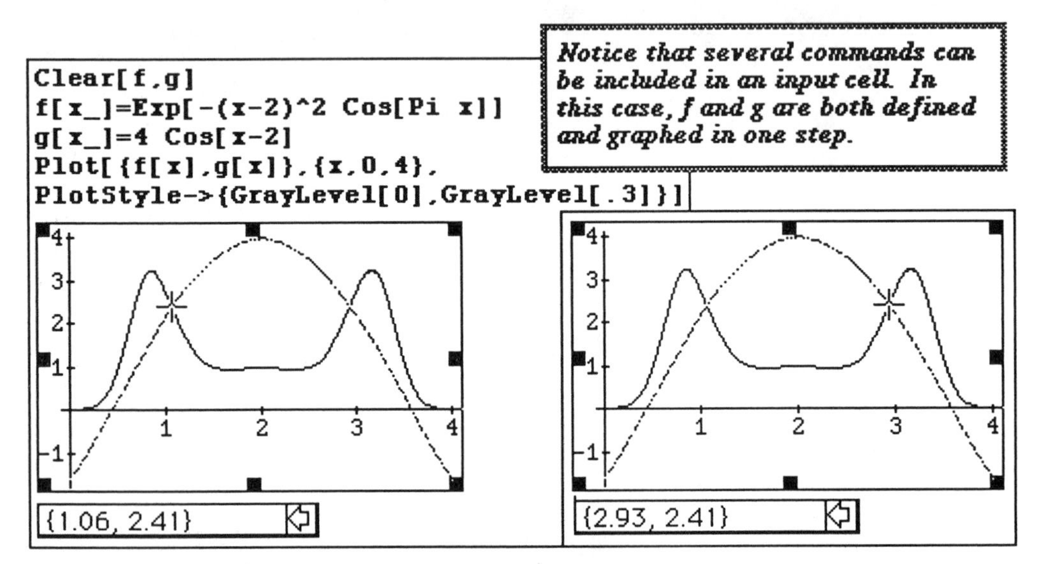

Once the initial guesses have been determined with the cursor, **FindRoot** is used to approximate the solutions to the equation f(x) = g(x), and the area is approximated with **NIntegrate**.

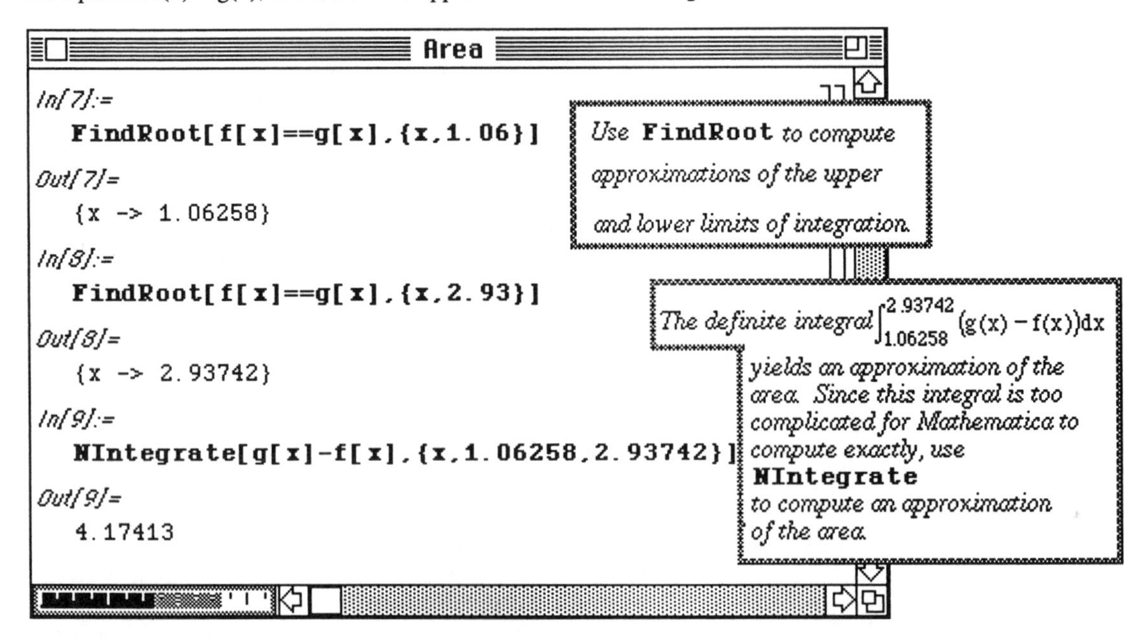

■ Application: Arc Length

□ Example:

Approximate the arc length of the graph of $f(x) = \text{Sin}\left(\pi \text{Sin}(x-2)^2\right)$ on the interval [4,5].

Recall the formula for the length of the smooth curve g(x) from the point (a,g(a)) to (b,g(b))

is given by: $\text{Length} = \int_a^b \sqrt{1 + (g'(x))^2}\, dx$.

The resulting definite integrals used for determining arc length are usually difficult to compute since they involve a radical. Since the built-in command `NIntegrate[f[x],{x,a,b}]` numerically approximates integrals, *Mathematica* is very helpful with approximating solutions to these types of problems!

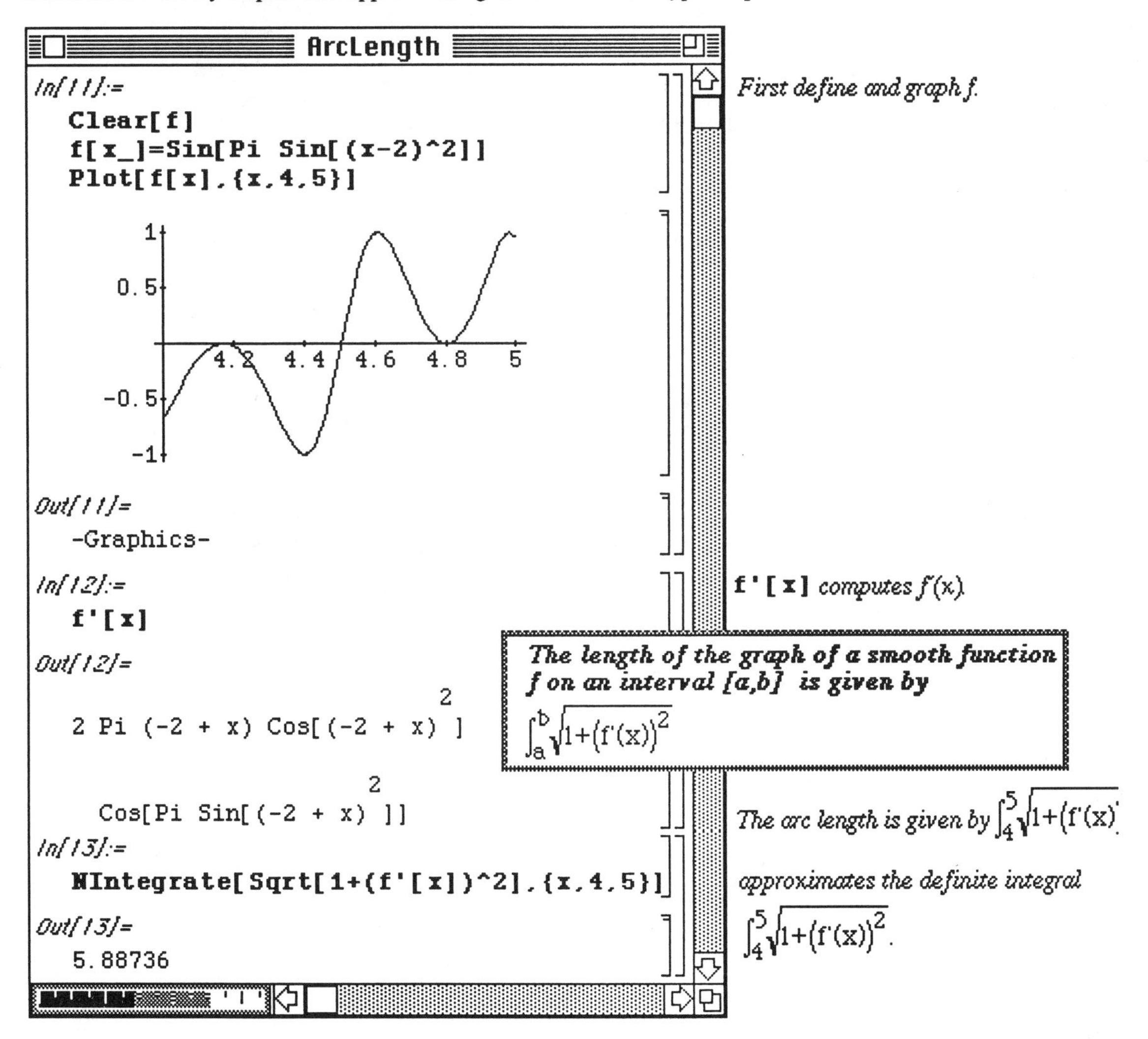

■ Application: Volumes of Solids of Revolution

□ Example:

Find the volume of the solid generated by revolving the region bounded by the graphs of

$g(x) = x^2 \mathrm{Sin}(x)$, $x = 0$, $x = \pi$, and $y = 0$ about the $y-$axis.

Before solving the problem, we remark that solids generated by revolving the graph of a function about the x- or y-axis can be visualized with *Mathematica*. The commands used to generate the following graphics are discussed in the **Appendix**.

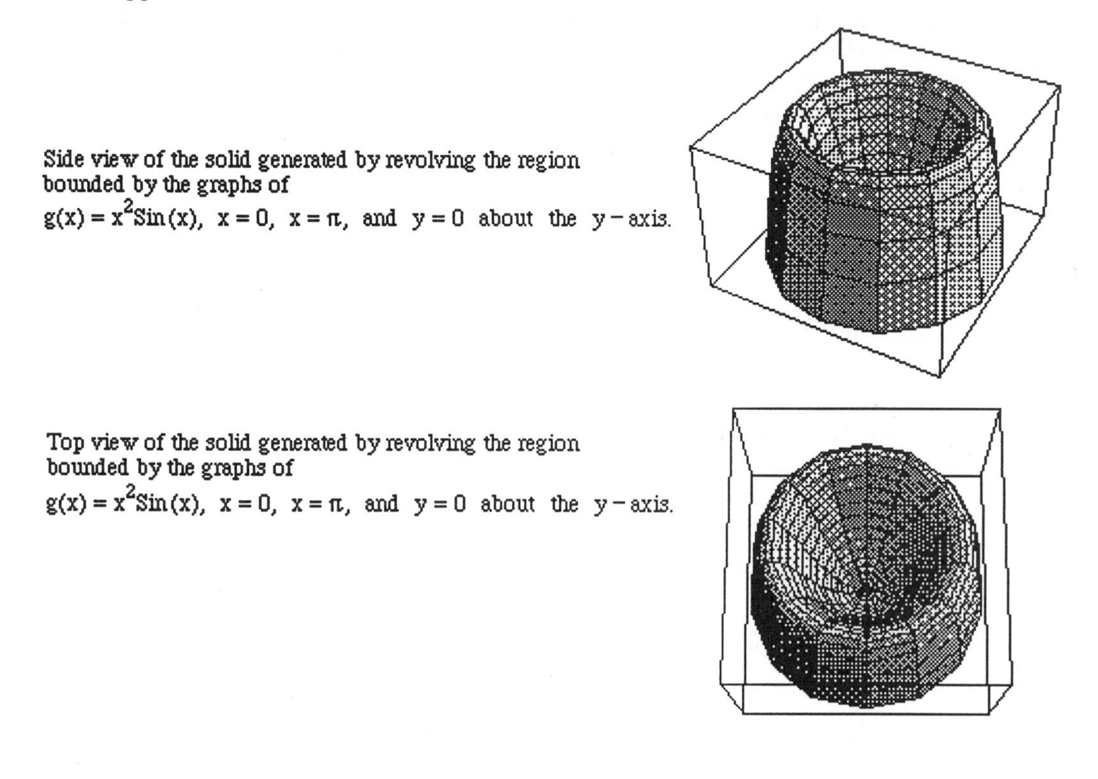

Side view of the solid generated by revolving the region bounded by the graphs of $g(x) = x^2 \mathrm{Sin}(x)$, $x = 0$, $x = \pi$, and $y = 0$ about the $y-$axis.

Top view of the solid generated by revolving the region bounded by the graphs of $g(x) = x^2 \mathrm{Sin}(x)$, $x = 0$, $x = \pi$, and $y = 0$ about the $y-$axis.

The method of cylindrical shells is used to compute the volume of this solid.

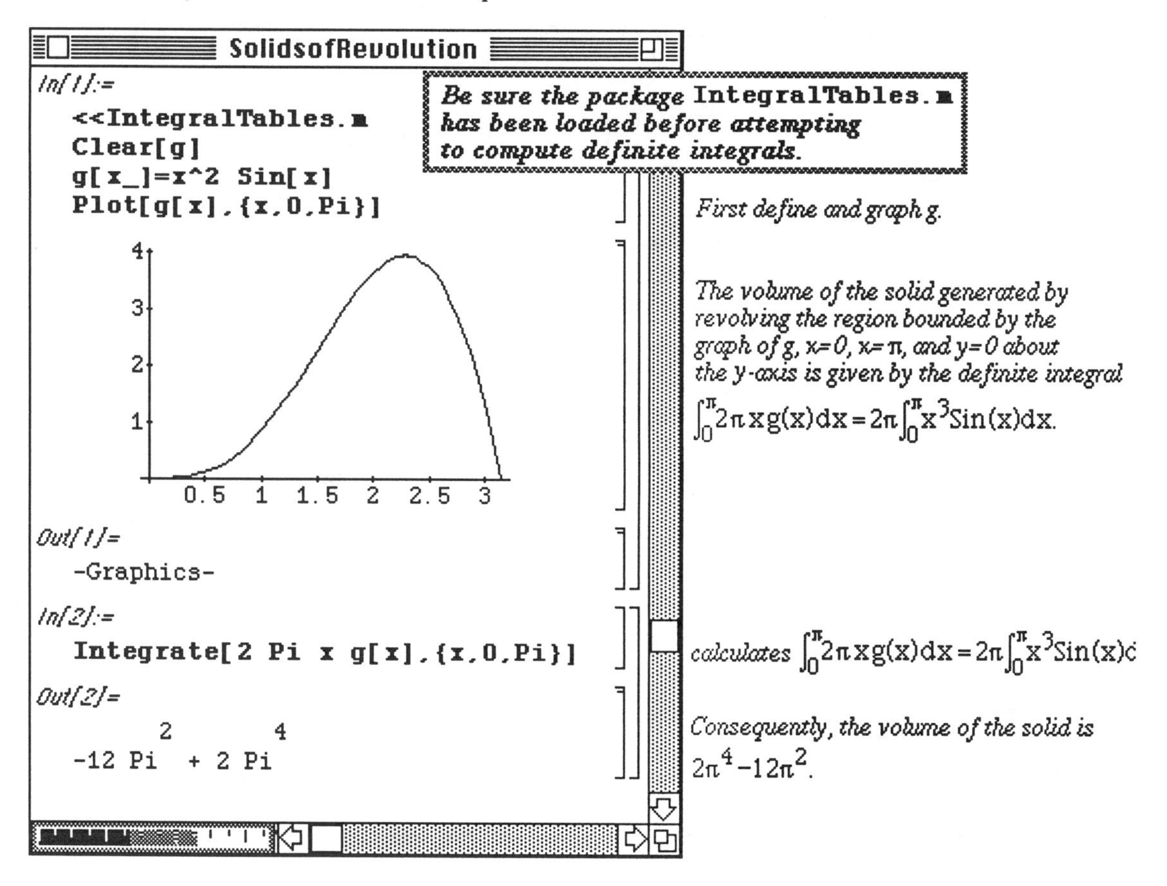

□ Example:

Use *Mathematica* to approximate the volume of the solid generated by revolving the region bounded by the graphs of

$m(x) = e^{-(x-3)^2 \text{Cos}[4(x-3)]}$, $x = 1$, $x = 5$, and $y = 0$ about the x-axis.

The solid generated by revolving the region bounded by the graphs of

$m(x) = e^{-(x-3)^2 \text{Cos}[4(x-3)]}$, $x = 1$, $x = 5$, and $y = 0$ about the x-axis can be visualized using *Mathematica*.

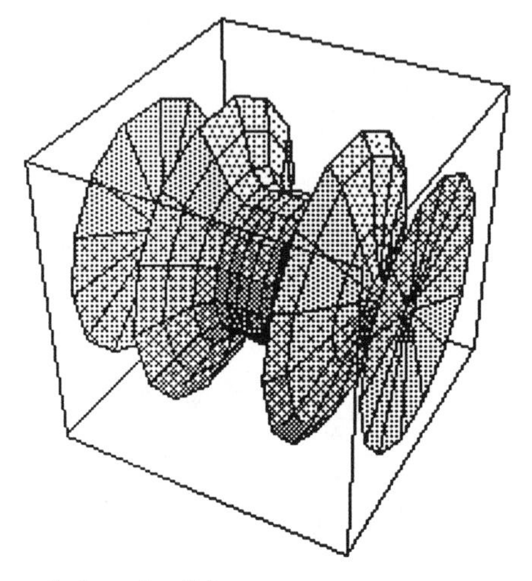

The disk method is used to determine the volume of this spool-shaped solid.

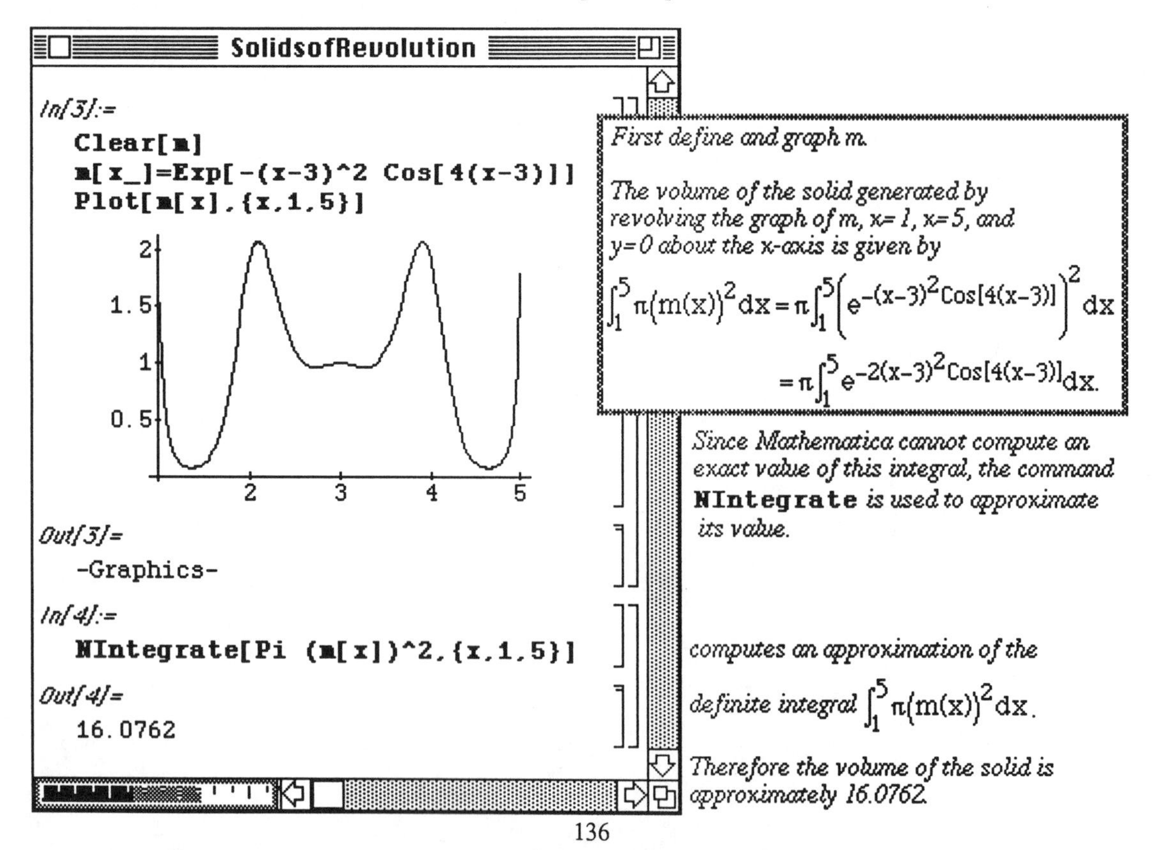

3.5 Series

Computing Power Series

Recall that a power series expansion of a function f(x) about the point x=a is given by the following expression:

$f(x)=\sum_{k=0}^{\infty}\frac{f^{(k)}(a)}{k!}(x-a)^k$. *Mathematica* computes the power series expansion of a function f(x) about the point x = **a** up to order **n** with the command: `Series[f(x),{x,a,n}]`.

Several familiar power series are computed below using this command.

Example:

Compute the first few terms of the power series expansion of f(x) about the point x=a for

(A) $f(x)=e^x$, $a=0$; (B) $f(x)=\mathrm{Sin}\,(x)$, $a=\pi$; (C) $f(x)=\mathrm{Cos}(x)$, $a=0$; and
(D) $f(x)=\mathrm{Log}(x)$, $a=1$.

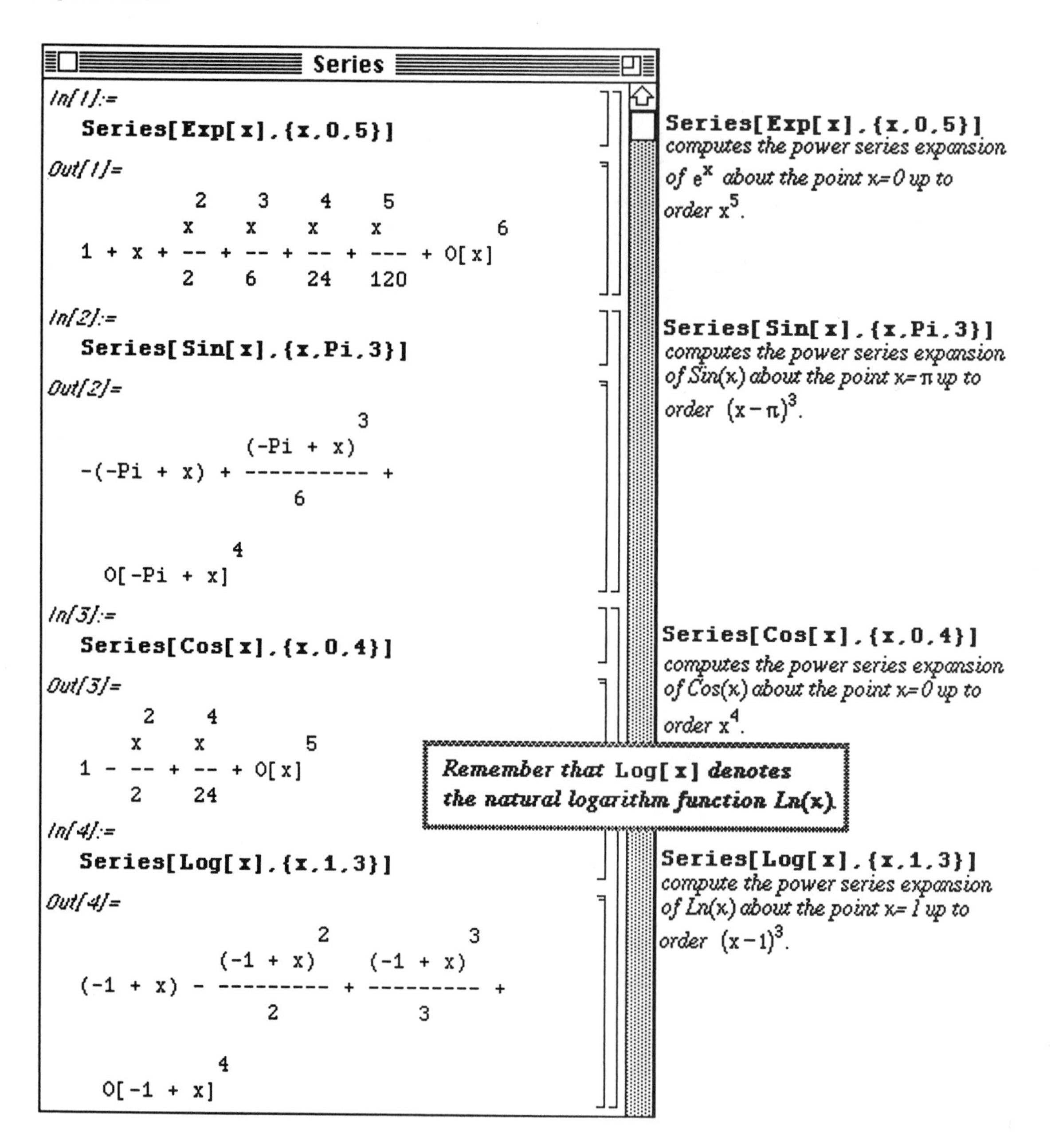
Series
In[1]:=
Series[Exp[x],{x,0,5}]
Out[1]=
1 + x + x^2/2 + x^3/6 + x^4/24 + x^5/120 + O[x]^6
In[2]:=
Series[Sin[x],{x,Pi,3}]
Out[2]=
-(-Pi + x) + (-Pi + x)^3/6 + O[-Pi + x]^4
In[3]:=
Series[Cos[x],{x,0,4}]
Out[3]=
1 - x^2/2 + x^4/24 + O[x]^5
In[4]:=
Series[Log[x],{x,1,3}]
Out[4]=
(-1 + x) - (-1 + x)^2/2 + (-1 + x)^3/3 + O[-1 + x]^4
Series[Exp[x],{x,0,5}] computes the power series expansion of e^x about the point x=0 up to order x^5.
Series[Sin[x],{x,Pi,3}] computes the power series expansion of Sin(x) about the point x=π up to order (x−π)^3.
Series[Cos[x],{x,0,4}] computes the power series expansion of Cos(x) about the point x=0 up to order x^4.
Remember that Log[x] denotes the natural logarithm function Ln(x).
Series[Log[x],{x,1,3}] compute the power series expansion of Ln(x) about the point x=1 up to order (x−1)^3.

Mathematica can also compute the general formula for the power series expansion of a function y(x) :

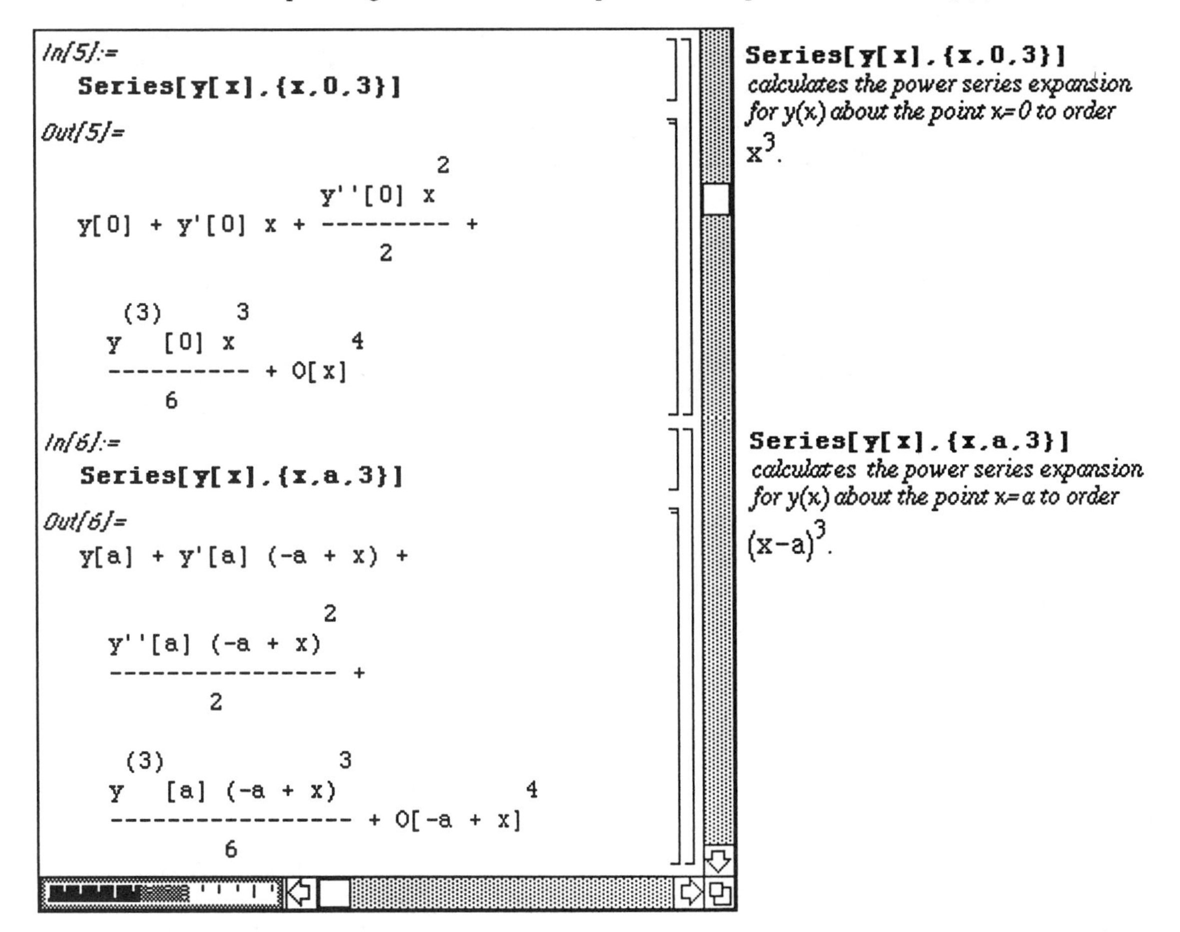

Mathematica can truncate (remove the remainder term) of the power series **`Series[f[x],{x,a,n}]`** with the command **`Normal[Series[f[x],{x,a,n}]]`**. Hence, with the **`Normal`** command, a polynomial is obtained. This polynomial serves as an approximation to the function f(x). These ideas are illustrated below :

□ Example:

Let f(x)=Sin(x) Cos(x). Compute the first 8 terms of the power series for f(x) about x=0. Use the command **Normal** to remove the remainder term from the series. Call the resulting polynomial function g(x). Compare the graphs of f(x) and g(x) on the interval [-π/2,π/2]. Graph the function |f(x)-g(x)| on the interval [-π/2,π/2].

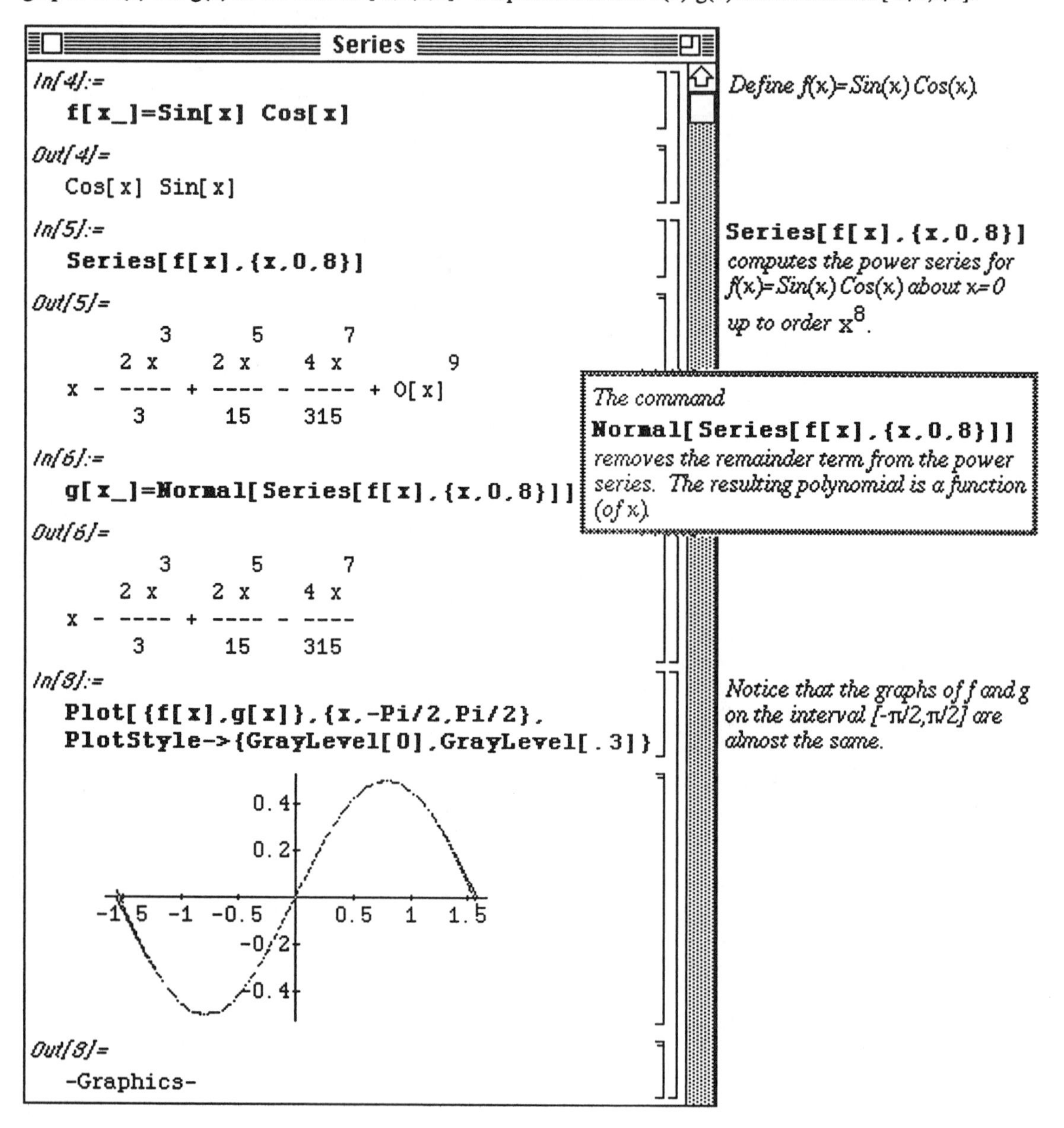

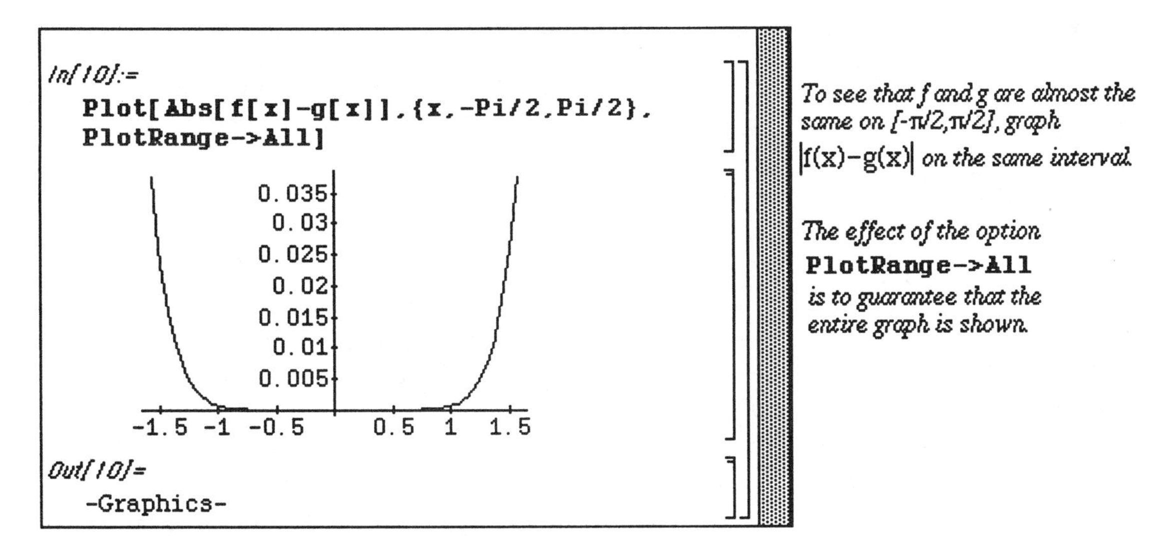

■ **Application: Approximating the Remainder**

Let f have (at least) n+1 derivatives in an interval containing a. If x is any number in the interval, then

$$f(x) = \underbrace{f(a) + \frac{f^{(1)}(a)}{1!}(x-a) + \frac{f^{(2)}(a)}{1!}(x-a)^2 + \ . \ . \ . \ + \frac{f^{(n-1)}(a)}{(n-1)!}(x-a)^{n-1} + \frac{f^{(n)}(a)}{n!}(x-a)^n}_{\text{Taylor Polynomial of degree n for f(x) at a}} +$$

$$\underbrace{\frac{f^{(n+1)}(z)}{(n+1)!}(x-a)^{n+1}}_{\text{Taylor Remainder of Taylor Polynomial of degree n for f(x) at a}}$$

where z is between a and x.

□ **Example:**

As in the above, let f(x)=Sin(x) Cos(x). Compute the Taylor Remainder of the Taylor Polynomial of degree 8, 9 and 10 for x=0.

We proceed by defining a function that symbolically computes the Taylor remainder of degree **n** for **f[x]** at 0:

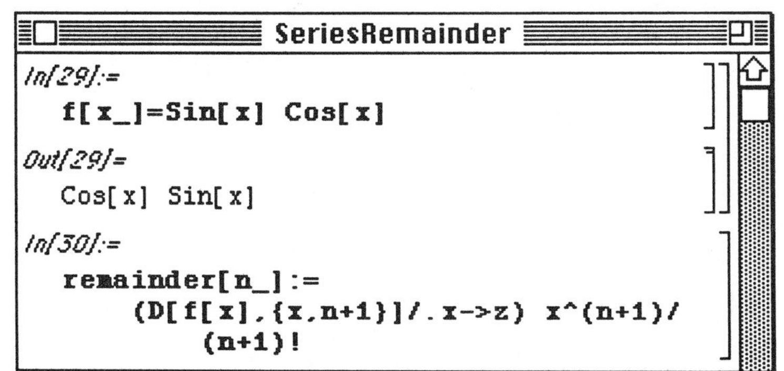

Then, **remainder[8]**, **remainder[9]**, and **remainder[10]** are the desired remainders:

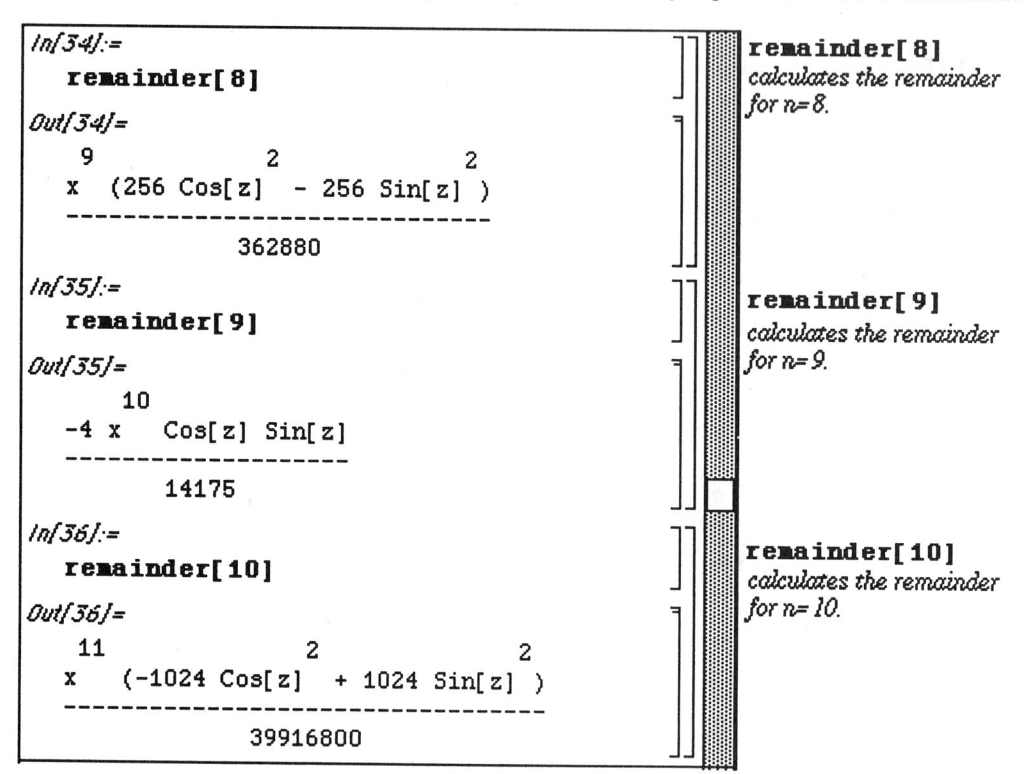

Compute the Taylor Remainder of the Taylor Polynomial of degree 11 for x=0. What is an upper bound for the remainder on the interval $[-\pi/2,\pi/2]$?

First compute the Taylor remainder of the Taylor polynomial of degree 11 for x=0 using the function **remainder** defined above:

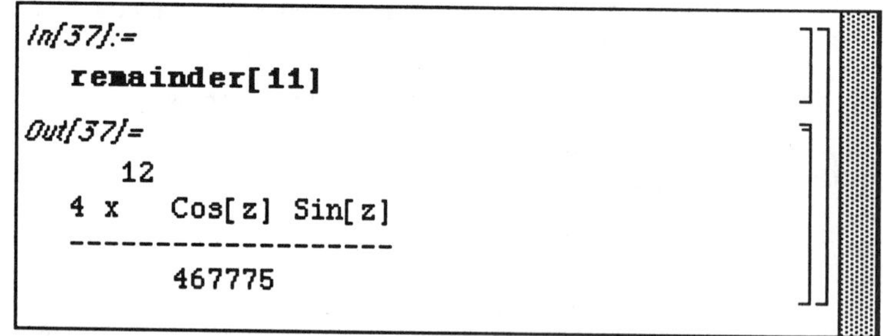

To obtain an upper bound of the remainder on the interval [-π/2,π/2], notice that for

$x \in [-\pi/2, \pi/2]$, $|x| \le \frac{\pi}{2}$ and for all values of z, $|\mathrm{Cos}(z)\mathrm{Sin}(z)| \le 1$. Thus, an upper bound on **remainder[11]** is obtained by replacing **x** by π/2 and Cos[z]Sin[z] by 1:

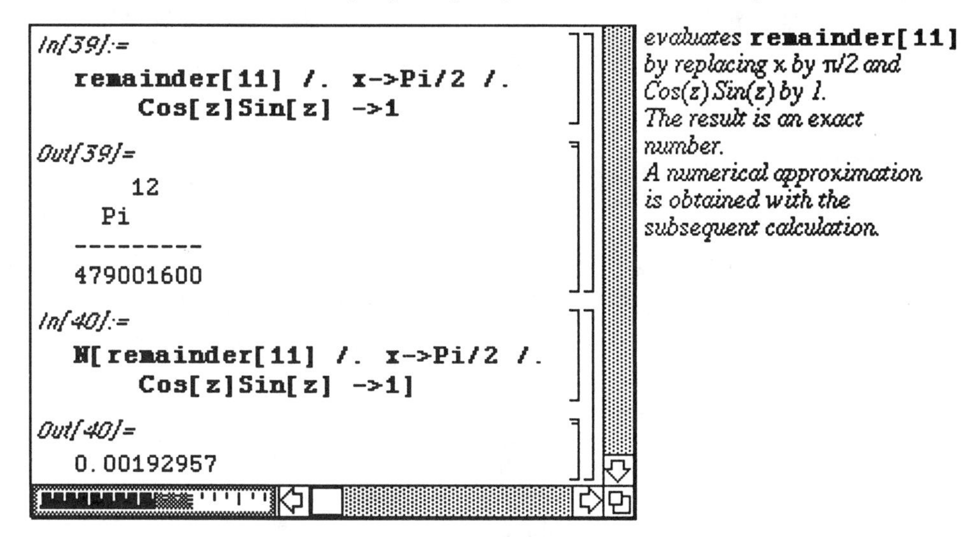

```
In[39]:=
  remainder[11] /. x->Pi/2 /.
       Cos[z]Sin[z] ->1
Out[39]=
       12
     Pi
   ---------
   479001600
In[40]:=
  N[remainder[11] /. x->Pi/2 /.
       Cos[z]Sin[z] ->1]
Out[40]=
  0.00192957
```

evaluates **remainder[11]** *by replacing* x *by* π/2 *and* Cos(z) Sin(z) *by 1. The result is an exact number. A numerical approximation is obtained with the subsequent calculation.*

■ Application: Series Solutions to Differential Equations

□ Example:

Find a function y(x) that satisfies the ordinary differential equation y"-4y'-5y=0 and the initial conditions y(0)=1 and y'(0)=5.

Remark: In **Chapter 5**, the command **DSolve** will be used to solve the differential equation y"-4y'-5y=0 subject to the initial conditions y(0)=1 and y'(0)=5.

Since the point x = 0 is an ordinary point of the differential equation, the solution y(x) is assumed to be the power

series $y(x) = \sum_{k=0}^{\infty} \frac{y^{(k)}(0)}{k!} x^k$

The power series is then substituted into the differential equation in order to determine the coefficients.

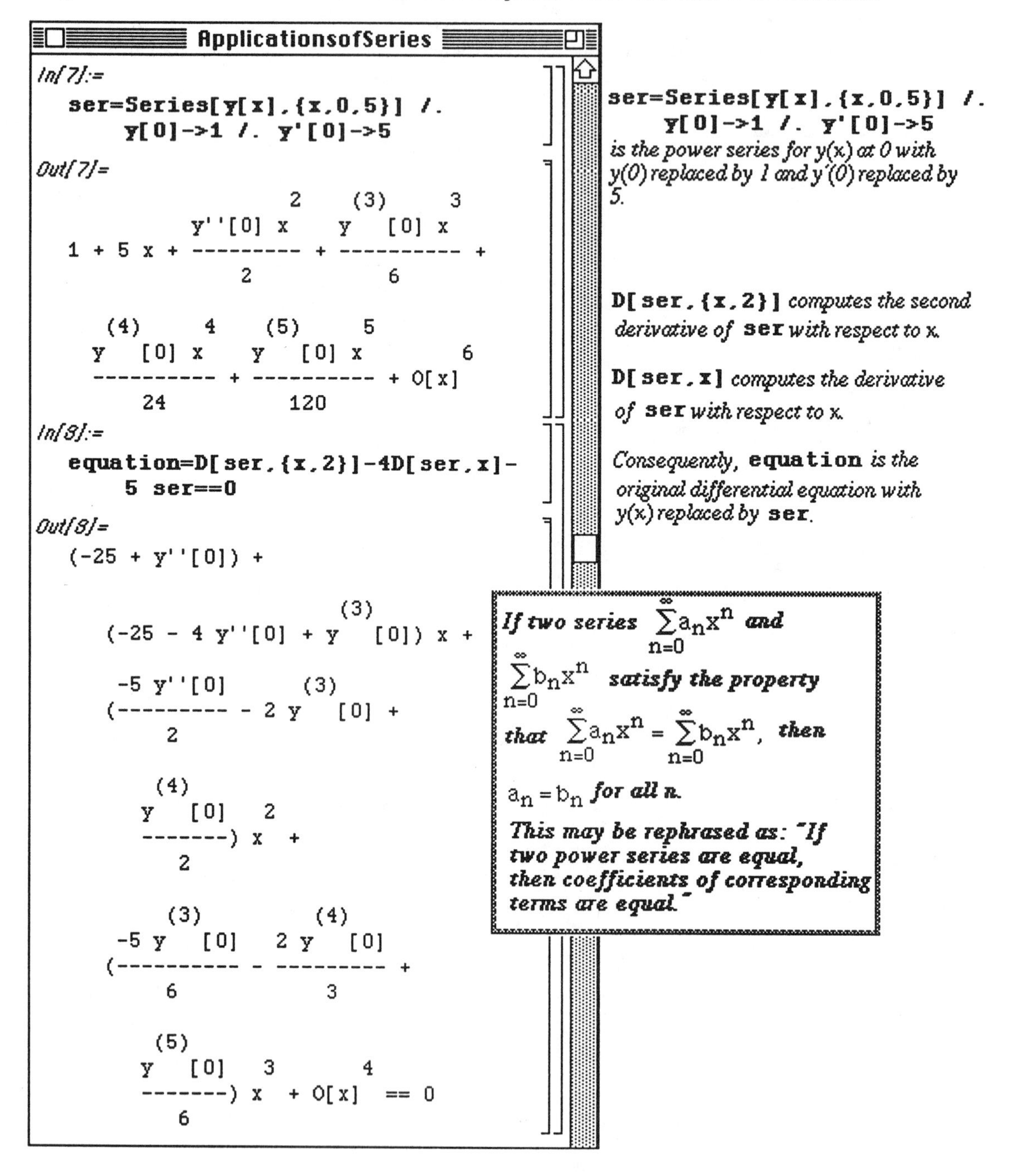

`ser=Series[y[x],{x,0,5}] /. y[0]->1 /. y'[0]->5` *is the power series for y(x) at 0 with y(0) replaced by 1 and y'(0) replaced by 5.*

`D[ser,{x,2}]` *computes the second derivative of* `ser` *with respect to x.*

`D[ser,x]` *computes the derivative of* `ser` *with respect to x.*

Consequently, `equation` *is the original differential equation with y(x) replaced by* `ser`.

If two series $\sum_{n=0}^{\infty} a_n x^n$ *and* $\sum_{n=0}^{\infty} b_n x^n$ *satisfy the property that* $\sum_{n=0}^{\infty} a_n x^n = \sum_{n=0}^{\infty} b_n x^n$, *then* $a_n = b_n$ *for all n.*

This may be rephrased as: "If two power series are equal, then coefficients of corresponding terms are equal."

Mathematica equates the coefficients of like powers of x on each side of the equation with the command **LogicalExpand[equation]**.

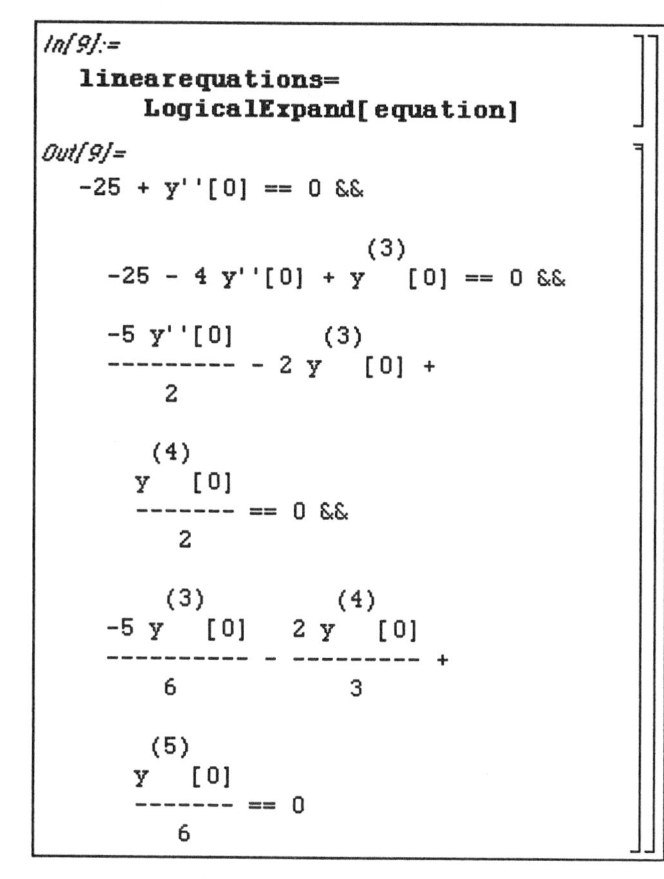

LogicalExpand[equation] *equates the corresponding coefficients in* **equation** *; the resulting system of equations is named* **linearequations**.

The above system of equations is then solved using **Solve** in order to determine the values of the higher order derivatives of y evaluated at x = 0. Once these values are determined, they are substituted back into the power series to obtain an approximate solution to the differential equation.

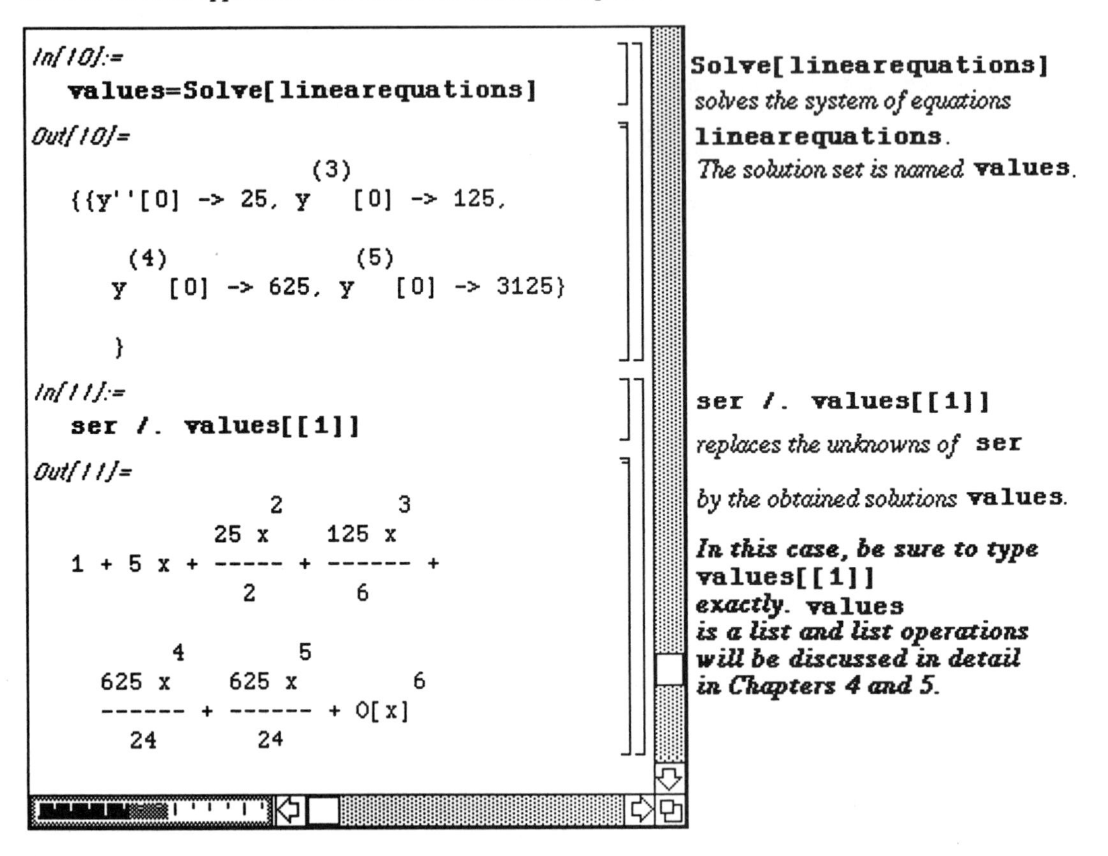

3.6 Multi-Variable Calculus

Elementary Three-Dimensional Graphics

As was mentioned in **Chapter 1**, functions of more than one variable can be defined with *Mathematica*. Of particular interest are functions of two variables. The command which plots the graph of the function f(x,y)

over the rectangular domain R where R: $xmin \le x \le xmax,\ ymin \le y \le ymax$

is: **Plot3D[f[x,y],{x,xmin,xmax},{y,ymin,ymax}]**. Since the graphs of functions of two variables are surfaces in three dimensions, the *Mathematica* command, **Plot3D**, must be introduced.

Mathematica can also plot the level curves of the function f(x,y). (Recall : **Level curves** are curves in the xy-plane which satisfy the equation f(x,y)= constant.) Level curves are plotted with the command **ContourPlot[f[x,y],{x,xmin,xmax},{y,ymin,ymax}].**

These commands are demonstrated below using Version 1.2:

- Many options are available with the Version 2.0 command **ContourPlot** that are not available in earlier editions of *Mathematica* and are discussed last.

Example:

Let $h(x,y) = \left(x^2 + y^2\right)^{1/3}$. Graph h on the rectangle $[-1,1] \times [-2,2]$.

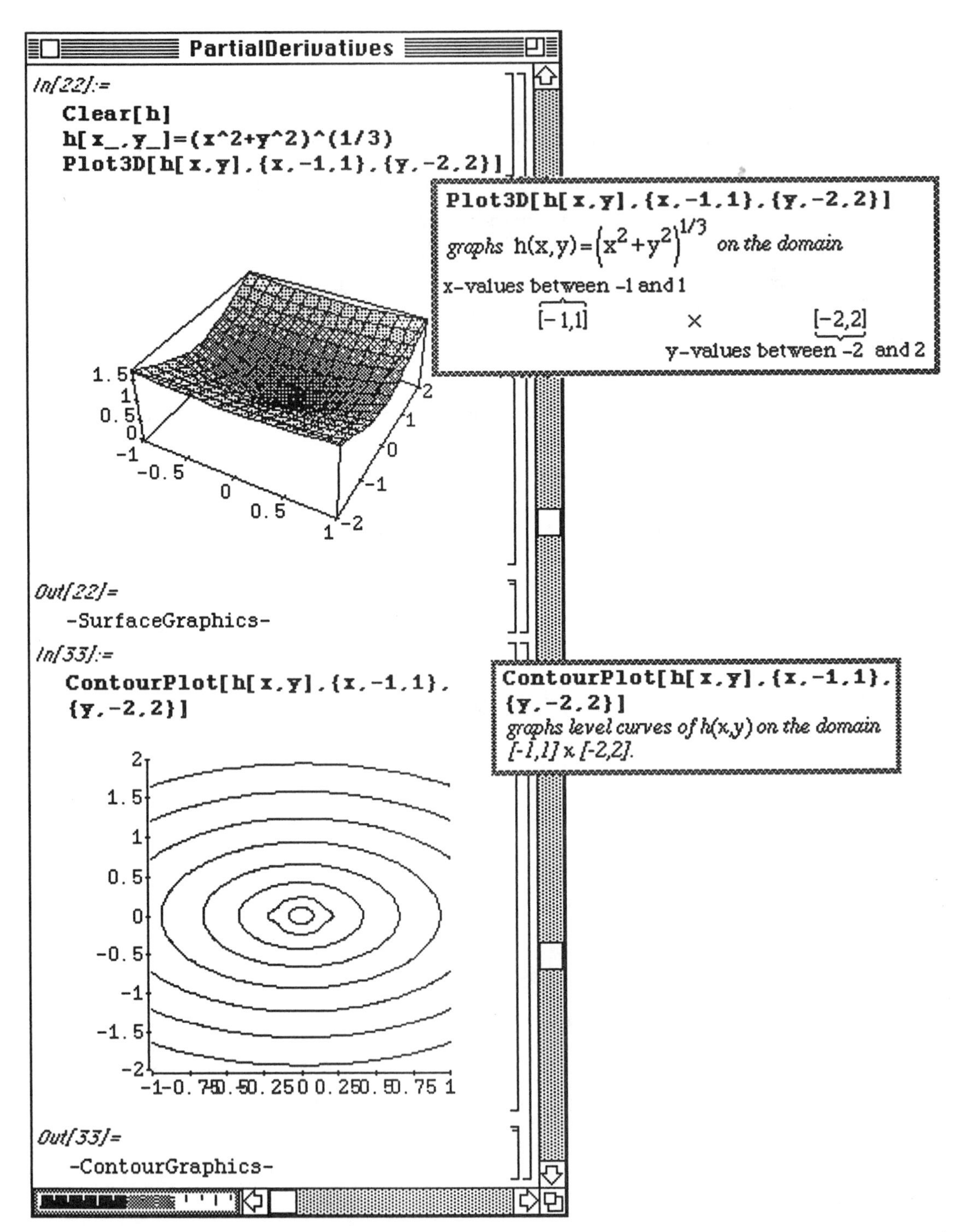

There are many options which can be included in the **Plot3D** command. The following examples illustrate the **Shading** and **PlotPoints** options.

Shading->False causes *Mathematica* to NOT shade squares in the graph.

PlotPoints->n causes *Mathematica* to evaluate the function at **n^2** points when plotting the graph. These **n^2** points are called sample points. In the command, **Plot3D[f[x,y],{x,xmin,xmax},{y,ymin,ymax}, PlotPoints->n],** the sample points are obtained by dividing each interval **[xmin,xmax]** and **[ymin,ymax]** into **n** subintervals. Hence, a larger value of **n** yields a smoother graph.

Notice the difference in the shading in the following graph of g(x,y) as opposed to the graph of h(x,y) above. The graph of g(x,y) is smoother than that of h(x,y) since the default (the value assumed when not otherwise stated) on **PlotPoints** is 15. Thus, the **Plot3D** command which plotted f(x,y) earlier automatically used 225 sample points.

□ **Example:**

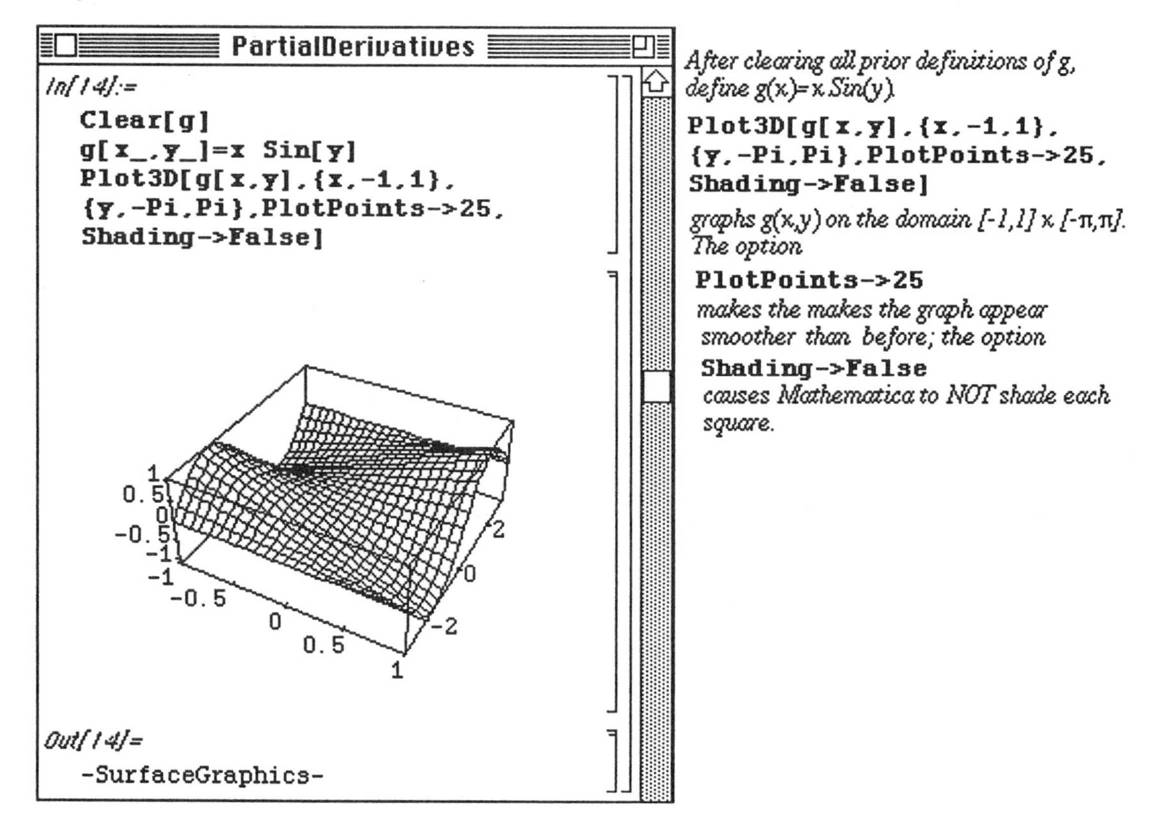

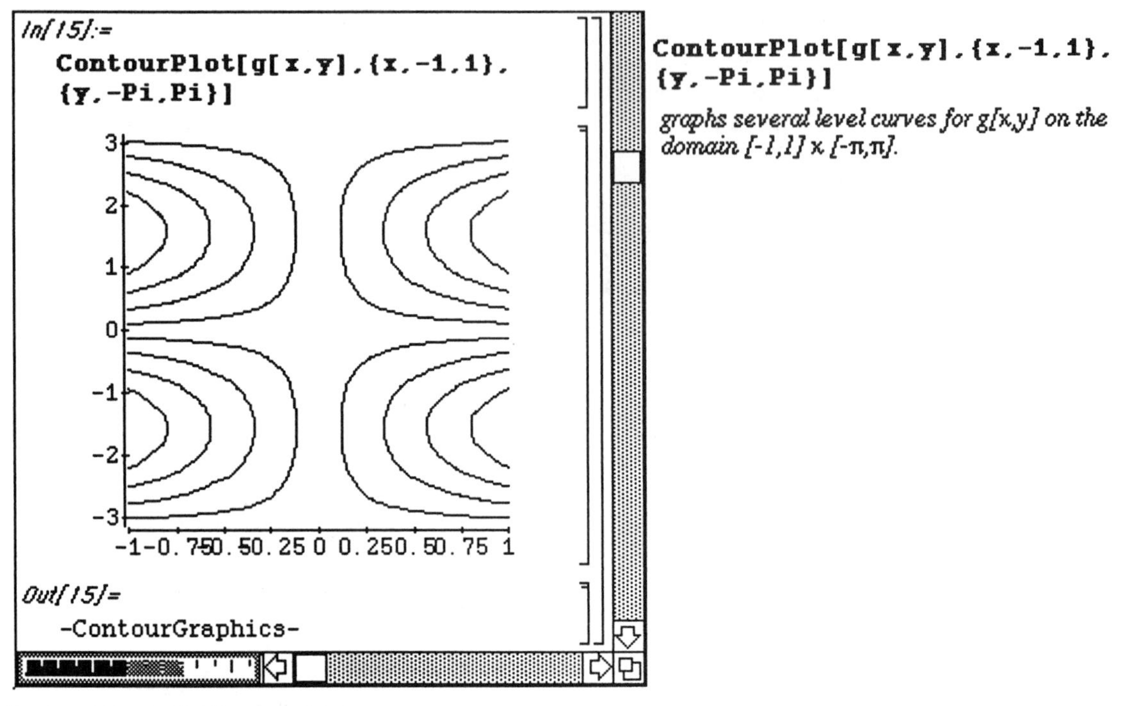

ContourPlot[g[x,y],{x,-1,1}, {y,-Pi,Pi}]

graphs several level curves for g[x,y] on the domain [-1,1] x [-π,π].

- **ContourPlot and Version 2.0**

The **ContourPlot** command is much improved in Version 2.0. Evidence of this is observed in the contour plot of the function **g[x,y]** defined below. The three-dimensional plot of this function is given in **plot3d**. Recall that contour levels represent intersections of planes of the form **g[x,y]** = constant with the surface shown in **plot3d**. **cp1** contains the contour plot of **g** which is obtained without any of the **ContourPlot** options. This plot differs greatly from that obtained in Version 1.2 in that shading is included in all contour plots unless otherwise specified.

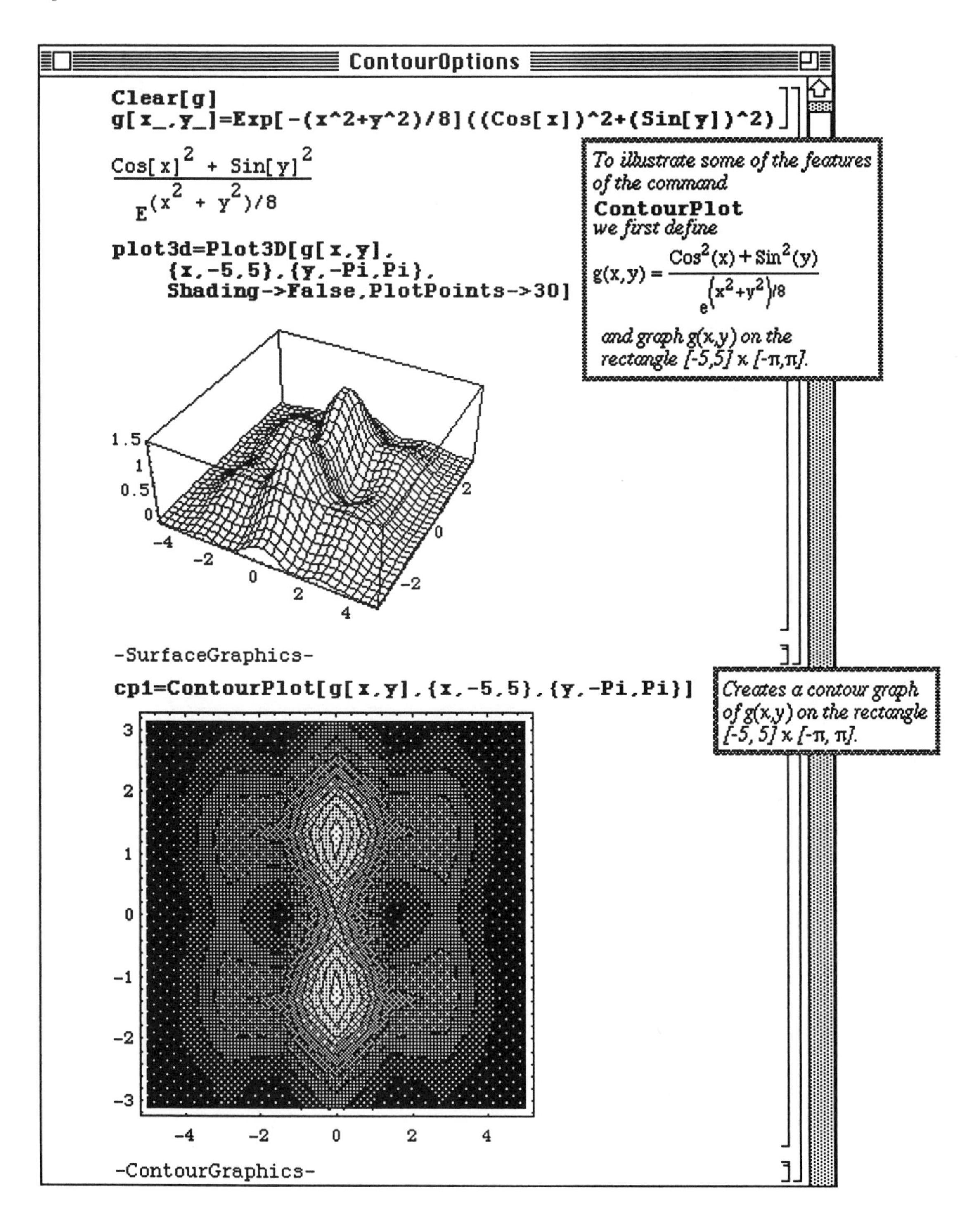
ContourOptions
Clear[g]
g[x_,y_]=Exp[-(x^2+y^2)/8]((Cos[x])^2+(Sin[y])^2)
Cos[x]^2 + Sin[y]^2 / E^((x^2 + y^2)/8)
plot3d=Plot3D[g[x,y],
{x,-5,5},{y,-Pi,Pi},
Shading->False,PlotPoints->30]
-SurfaceGraphics-
cp1=ContourPlot[g[x,y],{x,-5,5},{y,-Pi,Pi}]
-ContourGraphics-
To illustrate some of the features of the command
ContourPlot
we first define
g(x,y) = (Cos²(x) + Sin²(y)) / e^((x²+y²)/8)
and graph g(x,y) on the rectangle [-5,5] x [-π,π].
Creates a contour graph of g(x,y) on the rectangle [-5, 5] x [-π, π].

As mentioned above, all contour plots are shaded unless the **ContourShading->False** option is employed. This option is illustrated below along with **Contours->k** which instructs *Mathematica* to use **k** contour levels. (This replaces the **ContourLevels** option in Version 1.2.) In **cp2**, 25 contour levels are used and shading is suppressed. In the previous contour plot, the default value of 10 contour levels was used.

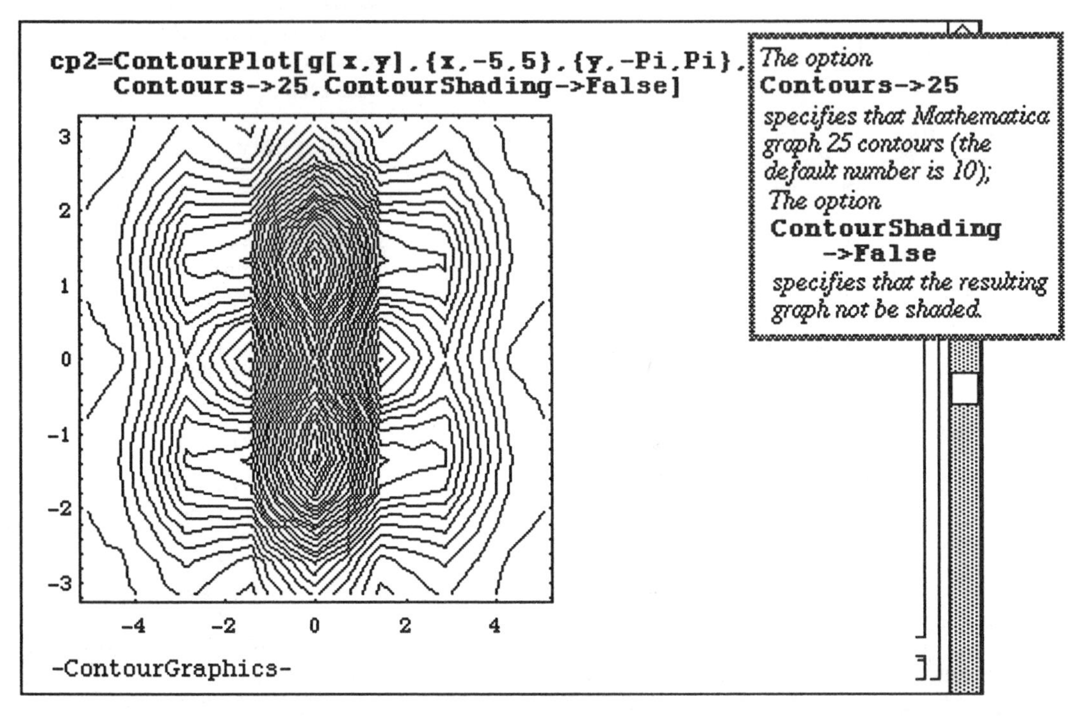

In the previous examples, *Mathematica* has selected the contour levels. However, these values can be chosen by the user with the **Contour->valuelist** option. This is demonstrated below with the table **contvals**. After this table is created, **Contours->contvals** forces *Mathematica* to use the contour levels given in **contvals**. Another option is shown below as well. The contour plot in **cp3** is redone in **cp4** with the addition of the **ContourSmoothing->Automatic** option. This causes the contour levels in **cp3** to appear smoother in **cp4**.

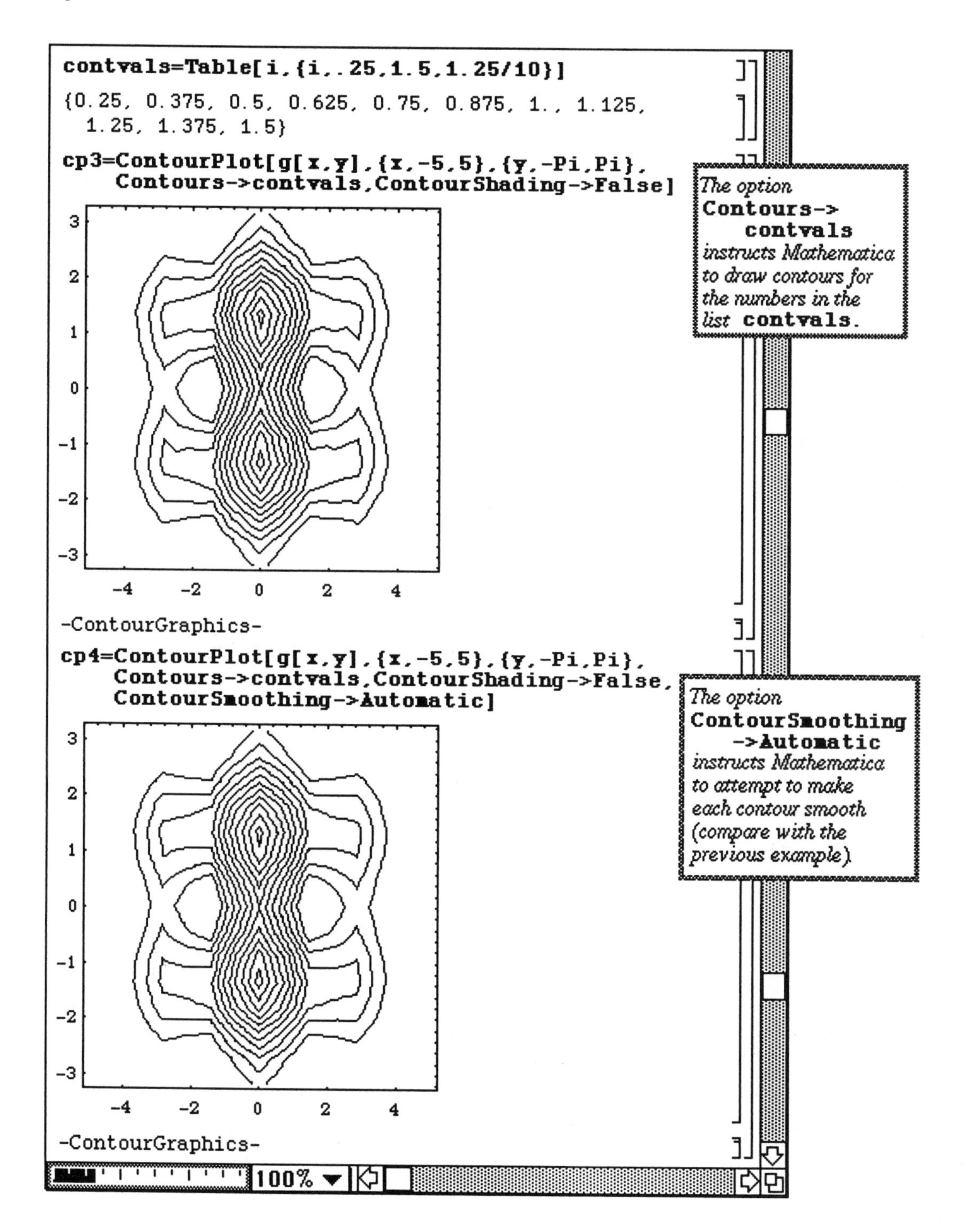

contvals=Table[i,{i,.25,1.5,1.25/10}]
{0.25, 0.375, 0.5, 0.625, 0.75, 0.875, 1., 1.125, 1.25, 1.375, 1.5}
cp3=ContourPlot[g[x,y],{x,-5,5},{y,-Pi,Pi}, Contours->contvals,ContourShading->False]
The option Contours-> contvals instructs Mathematica to draw contours for the numbers in the list contvals.
-ContourGraphics-
cp4=ContourPlot[g[x,y],{x,-5,5},{y,-Pi,Pi}, Contours->contvals,ContourShading->False, ContourSmoothing->Automatic]
The option ContourSmoothing ->Automatic instructs Mathematica to attempt to make each contour smooth (compare with the previous example).
-ContourGraphics-
100%

■ Partial Differentiation

Partial derivatives can be calculated with *Mathematica* using the command

D[f[x,y],variable]
where **f[x,y]** is differentiated with respect to **variable**.

Second order derivatives can be found using **D[f[x,y],variable1,variable2]** where **f[x,y]** is differentiated first with respect to **variable2** and then with respect to **variable1**.

□ Example:

Let $h(x,y) = \left(x^2 + y^2\right)^{1/3}$ (as above). Calculate $\frac{\partial h}{\partial x}$, $\frac{\partial h}{\partial y}$, and $\frac{\partial^2 h}{\partial y \partial x}$.

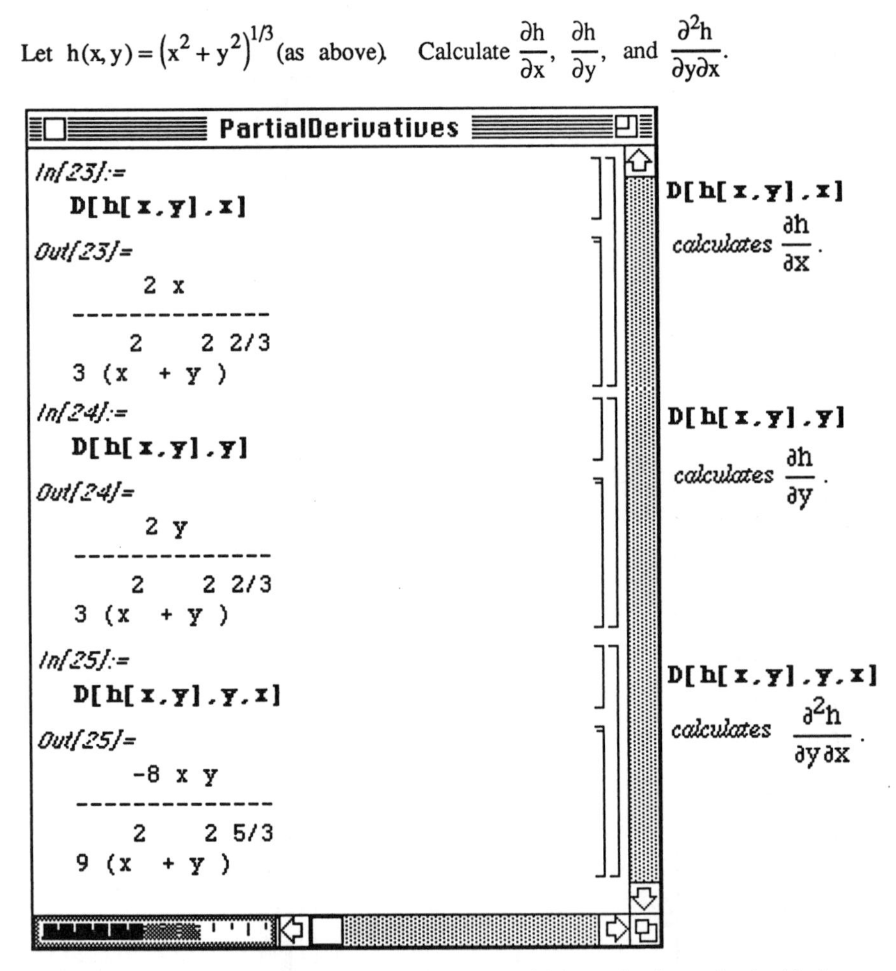

Higher order derivatives with respect to the same variable can be determined with the command

D[f[x,y],{variable,n}]. This command computes the **n**th partial derivative of **f** with respect to **variable**.

For example, **D[f[x,y],{x,n}]** computes $\frac{\partial^n f}{\partial x^n}$.

□ **Example:**

Let $h(x,y) = \left(x^2 + y^2\right)^{1/3}$ (as above). Calculate $\frac{\partial^2 h}{\partial x^2}$ and $\frac{\partial^2 h}{\partial y^2}$.

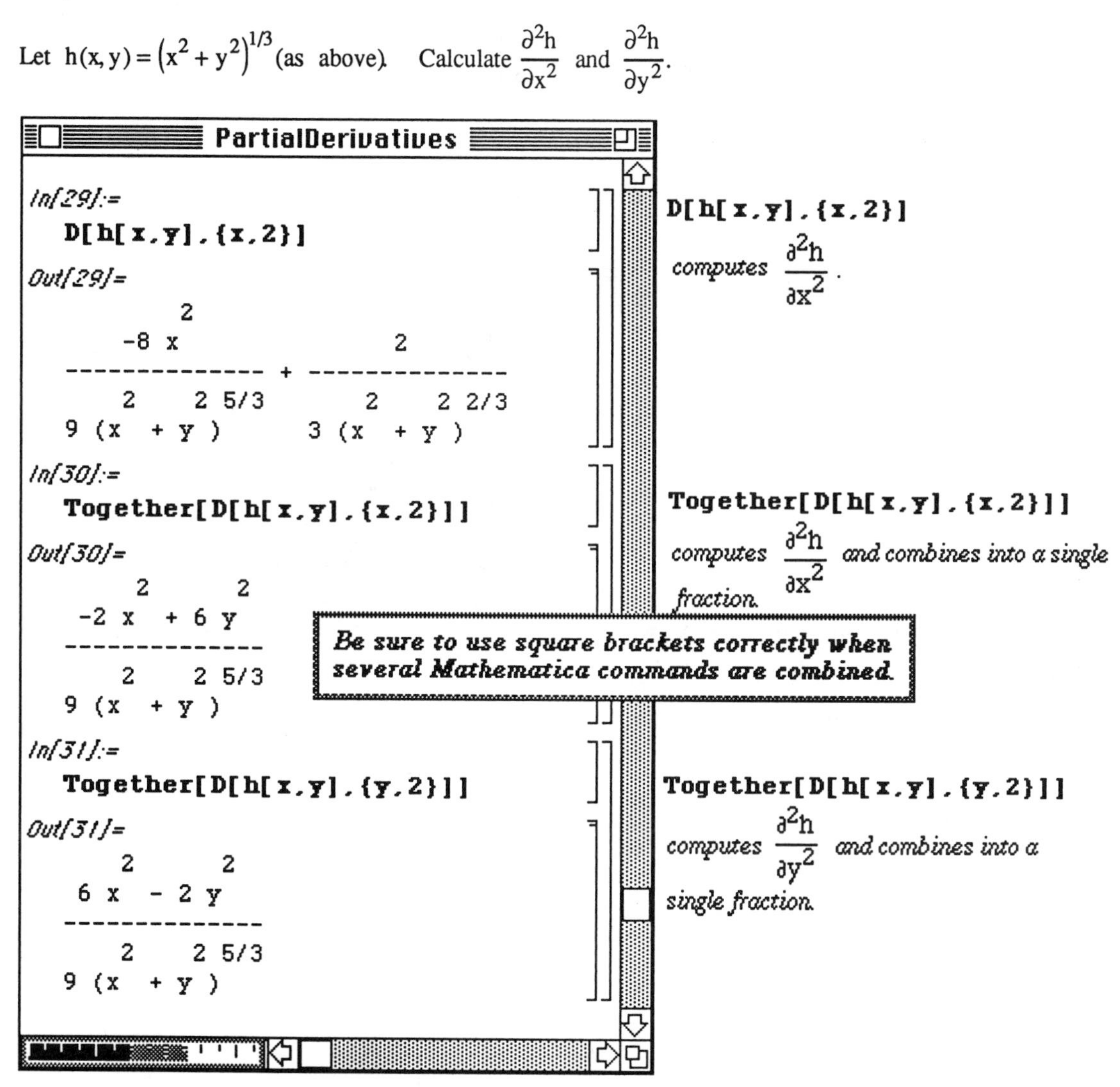

■ Other Methods of Computing Derivatives

The command **Derivative** can also be used to compute derivatives of functions. For example, if **f[x]** is a function of a single variable, the command **Derivative[1][f][a]** computes the derivative of **f** with respect to x and evaluates the result by replacing **x** by **a**; the command **Derivative[n][f][a]** computes the **n**th derivative of **f** with respect to **x** and evaluates the result by replacing **x** by **a**.

Similarly, if **f[x,y]** is a function of two variables, the command **Derivative[1,0][f][a,b]** computes the partial derivative of **f** with respect to **x** and evaluates the result by replacing **x** by **a** and **y** by **b**; the command **Derivative[0,1][f][a,b]** computes the partial derivative of **f** with respect to **y** and evaluates the result by replacing **x** by **a** and **y** by **b**; and the command **Derivative[n,m][f][a,b]** computes the **n**th partial derivative of **f** with respect to **x** and then the **m**th partial derivative of **f** with respect to **y** and evaluates the result by replacing **x** by **a** and **y** by **b**.

□ **Example:**

Use the command **Derivative** to compute $\frac{\partial g}{\partial x}, \frac{\partial g}{\partial y}, \frac{\partial^2 g}{\partial x \partial y}, \frac{\partial^2 g}{\partial x^2}, \frac{\partial^2 g}{\partial y^2}$, and $\frac{\partial^3 g}{\partial x \partial y^2}\left(\frac{\pi}{3}, \frac{\pi}{6}\right)$ if $g(x,y) = e^{-\left(x^2+y^2\right)/8}\left(\mathrm{Cos}^2(x) + \mathrm{Sin}^2(y)\right)$.

After defining **g**, we illustrate that **Derivative[1,0][g][x,y]** and **D[g[x,y],x]** both produce the same result.

PartialDerivatives

```
In[59]:=
    Clear[g]
    g[x_,y_]=Exp[-(x^2+y^2)/8]*
                (Cos[x]^2+Sin[y]^2)
Out[59]=
          2         2
    Cos[x]  + Sin[y]
    ----------------
       2    2
     (x  + y )/8
    E
In[60]:=
    gx=Derivative[1,0][g][x,y]
Out[60]=
    -2 Cos[x] Sin[x]
    ---------------- -
        2    2
      (x  + y )/8
     E
                2         2
      x (Cos[x]  + Sin[y] )
      ---------------------
              2    2
            (x  + y )/8
         4 E
In[61]:=
    D[g[x,y],x]
Out[61]=
    -2 Cos[x] Sin[x]
    ---------------- -
        2    2
      (x  + y )/8
     E
                2         2
      x (Cos[x]  + Sin[y] )
      ---------------------
              2    2
            (x  + y )/8
         4 E
```

After clearing all prior definitions of g, define

$g(x,y) = e^{-\left(x^2+y^2\right)/8}\left(\mathrm{Cos}^2(x) + \mathrm{Sin}^2(y)\right)$.

Since g is defined on two lines, be sure to include the ***** *to denote multiplication.*

Both **Derivative[1,0][g][x,y]** *and* **D[g[x,y],x]** *produce* $\frac{\partial g}{\partial x}$.

Similarly **Derivative[1,1][g][x,y]** and **Derivative[g,x,y]** produce the same result:

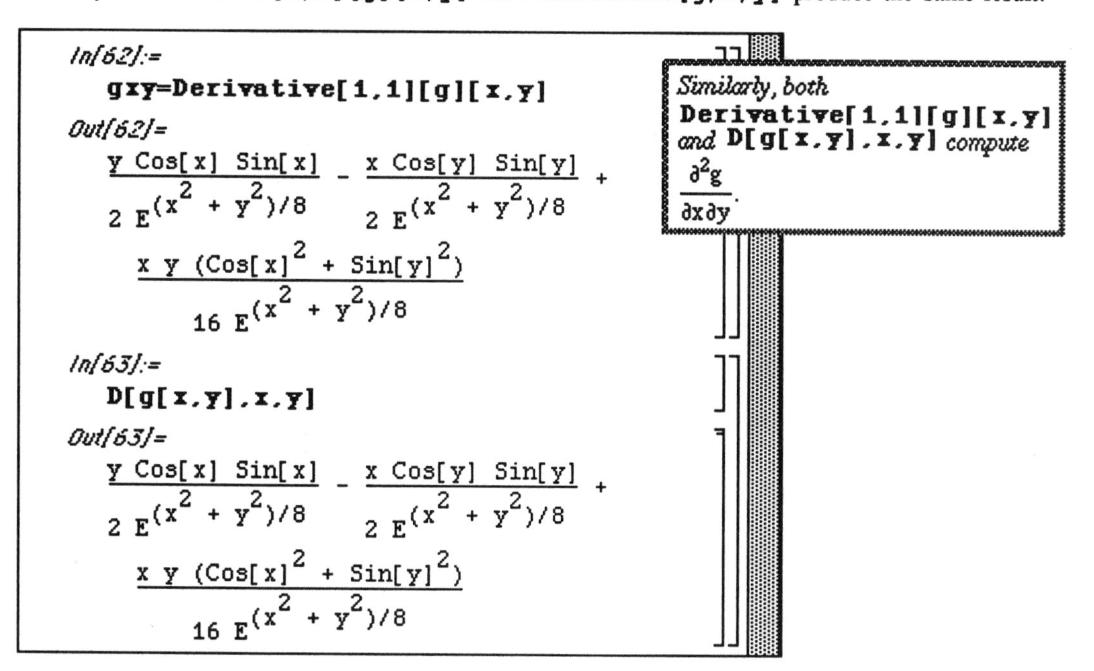

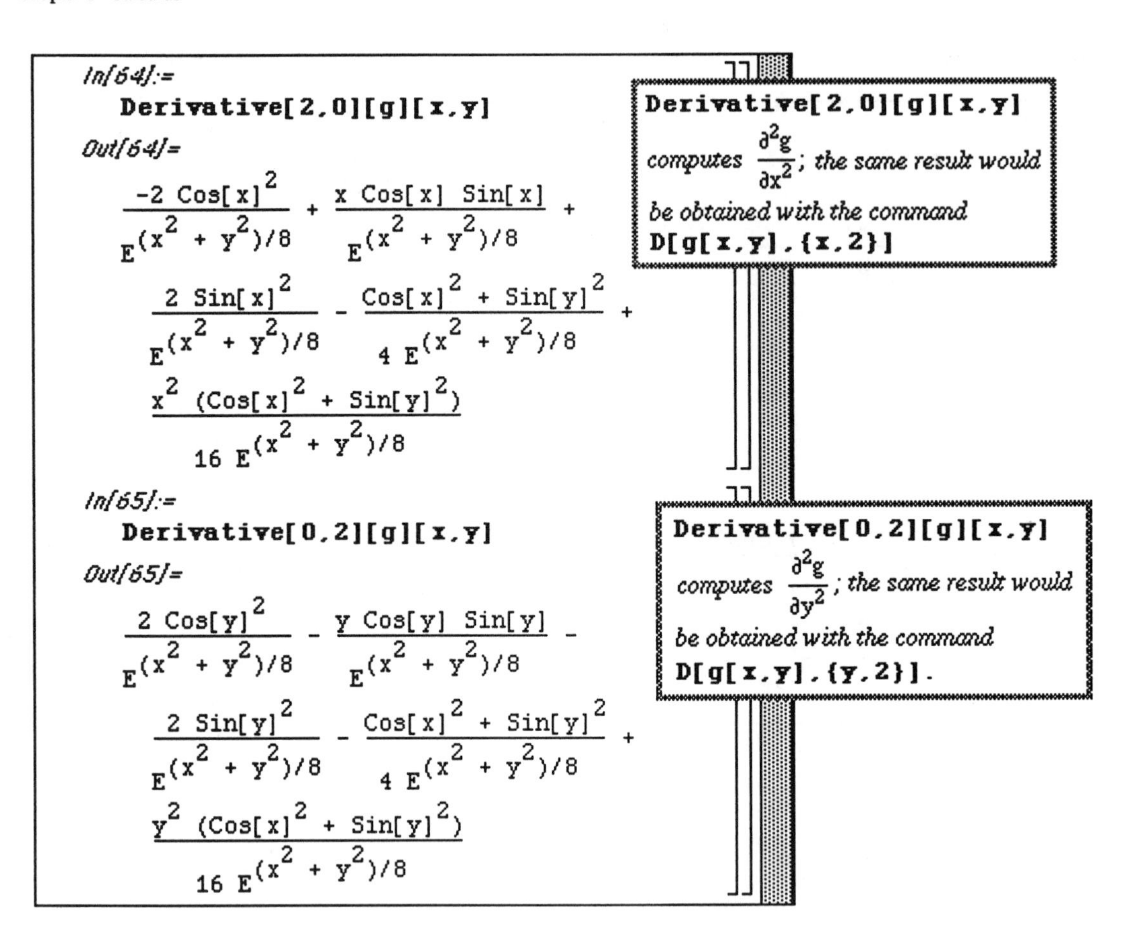

In[64]:=
Derivative[2,0][g][x,y]
Out[64]=
-2 Cos[x]^2/E^((x^2 + y^2)/8) + x Cos[x] Sin[x]/E^((x^2 + y^2)/8) +
2 Sin[x]^2/E^((x^2 + y^2)/8) - (Cos[x]^2 + Sin[y]^2)/(4 E^((x^2 + y^2)/8)) +
x^2 (Cos[x]^2 + Sin[y]^2)/(16 E^((x^2 + y^2)/8))
Derivative[2,0][g][x,y]
computes ∂²g/∂x²; the same result would
be obtained with the command
D[g[x,y],{x,2}]
In[65]:=
Derivative[0,2][g][x,y]
Out[65]=
2 Cos[y]^2/E^((x^2 + y^2)/8) - y Cos[y] Sin[y]/E^((x^2 + y^2)/8) -
2 Sin[y]^2/E^((x^2 + y^2)/8) - (Cos[x]^2 + Sin[y]^2)/(4 E^((x^2 + y^2)/8)) +
y^2 (Cos[x]^2 + Sin[y]^2)/(16 E^((x^2 + y^2)/8))
Derivative[0,2][g][x,y]
computes ∂²g/∂y²; the same result would
be obtained with the command
D[g[x,y],{y,2}].

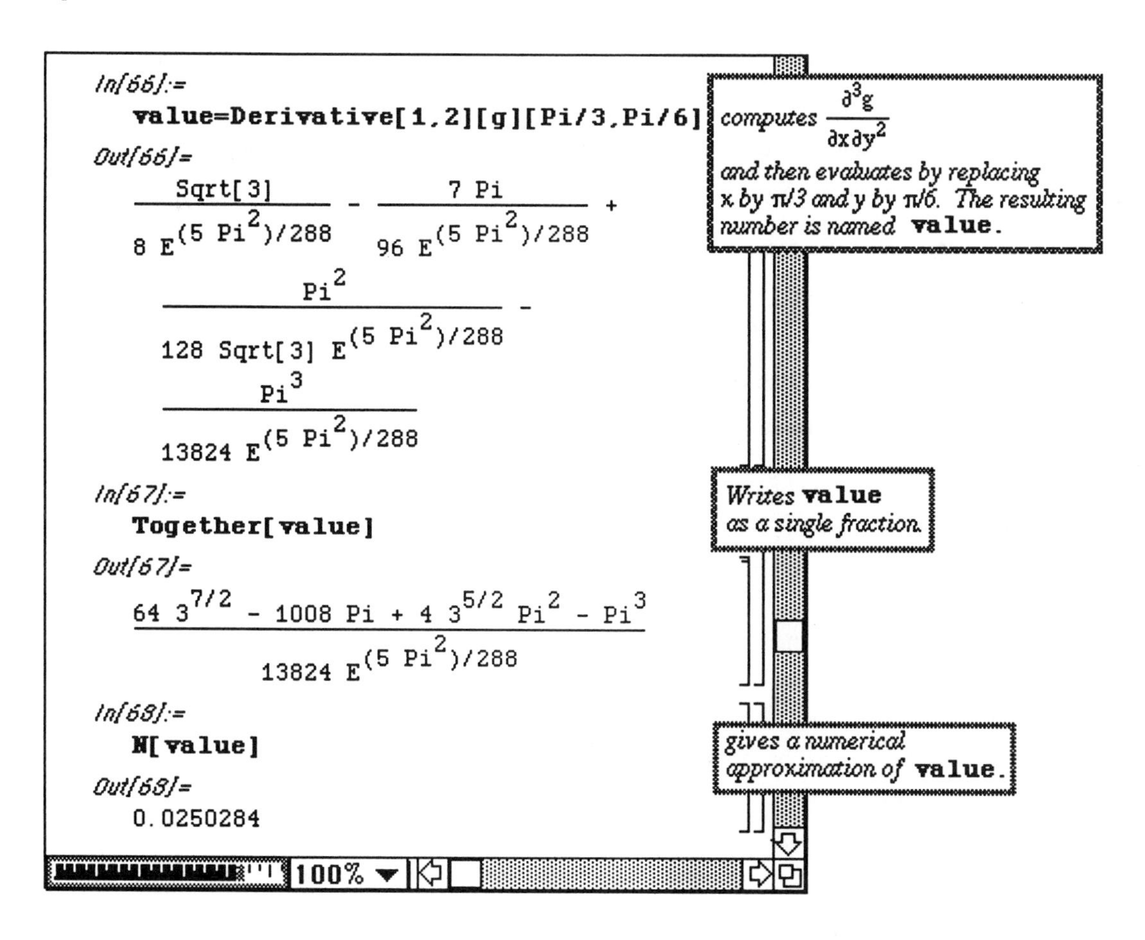
In[66]:=
value=Derivative[1,2][g][Pi/3,Pi/6]
Out[66]=
Sqrt[3] / 8 E^((5 Pi^2)/288) - 7 Pi / 96 E^((5 Pi^2)/288) +
Pi^2 / 128 Sqrt[3] E^((5 Pi^2)/288) -
Pi^3 / 13824 E^((5 Pi^2)/288)
In[67]:=
Together[value]
Out[67]=
64 3^(7/2) - 1008 Pi + 4 3^(5/2) Pi^2 - Pi^3 / 13824 E^((5 Pi^2)/288)
In[68]:=
N[value]
Out[68]=
0.0250284
100%
computes ∂³g/∂x∂y²
and then evaluates by replacing x by π/3 and y by π/6. The resulting number is named value.
Writes value as a single fraction.
gives a numerical approximation of value.

■ **Application: Classifying Critical Points**

Recall the definition of a critical point :

■ Let f(x,y) be a real-valued function with continuous second-order partial derivatives. A **critical point** of f is a point (x_0, y_0) in the interior of the domain of f for which $f_x(x_0, y_0) = 0$ and $f_y(x_0, y_0) = 0$.

□ **Remark: The following notation is used:**

$$f_x(x,y) = \frac{\partial f}{\partial x}(x,y),\ f_y(x,y) = \frac{\partial f}{\partial y}(x,y),\ f_{yx}(x,y) = \frac{\partial f}{\partial x \partial y}(x,y),\ f_{xx}(x,y) = \frac{\partial^2 f}{\partial x^2}(x,y),\ \text{and}\ f_{yy}(x,y) = \frac{\partial^2 f}{\partial y^2}(x,y)$$

Critical points are classified using the following test:

(Second-Derivative Test for Extrema) Let

$$D(f;(x_0,y_0)) = \begin{vmatrix} f_{xx}(x_0,y_0) & f_{xy}(x_0,y_0) \\ f_{xy}(x_0,y_0) & f_{yy}(x_0,y_0) \end{vmatrix} = \left(f_{xx}(x_0,y_0)\right)\left(f_{yy}(x_0,y_0)\right) - \left(f_{xy}(x_0,y_0)\right)^2.$$

(a) If $D > 0$ and $f_{xx}(x_0,y_0) > 0$, then f has a relative minimum at (x_0,y_0);
(b) If $D > 0$ and $f_{xx}(x_0,y_0) < 0$, then f has a relative maximum at (x_0,y_0);
(c) If $D < 0$, then f has a saddle point at (x_0,y_0); and
(d) If $D = 0$, no conclusion can be drawn and (x_0,y_0) is called a **degenerate critical point.**

□ Example:

Locate and classify all critical points of the function $f(x,y)=-120x^3-30x^4+18x^5+5x^6+30xy^2$.

In order to find the critical points of f(x,y), the derivatives $f_x(x,y)$ and $f_y(x,y)$

are calculated and set equal to zero. These steps are shown below. Notice how the derivatives are given names to make using them later in the problem easier.

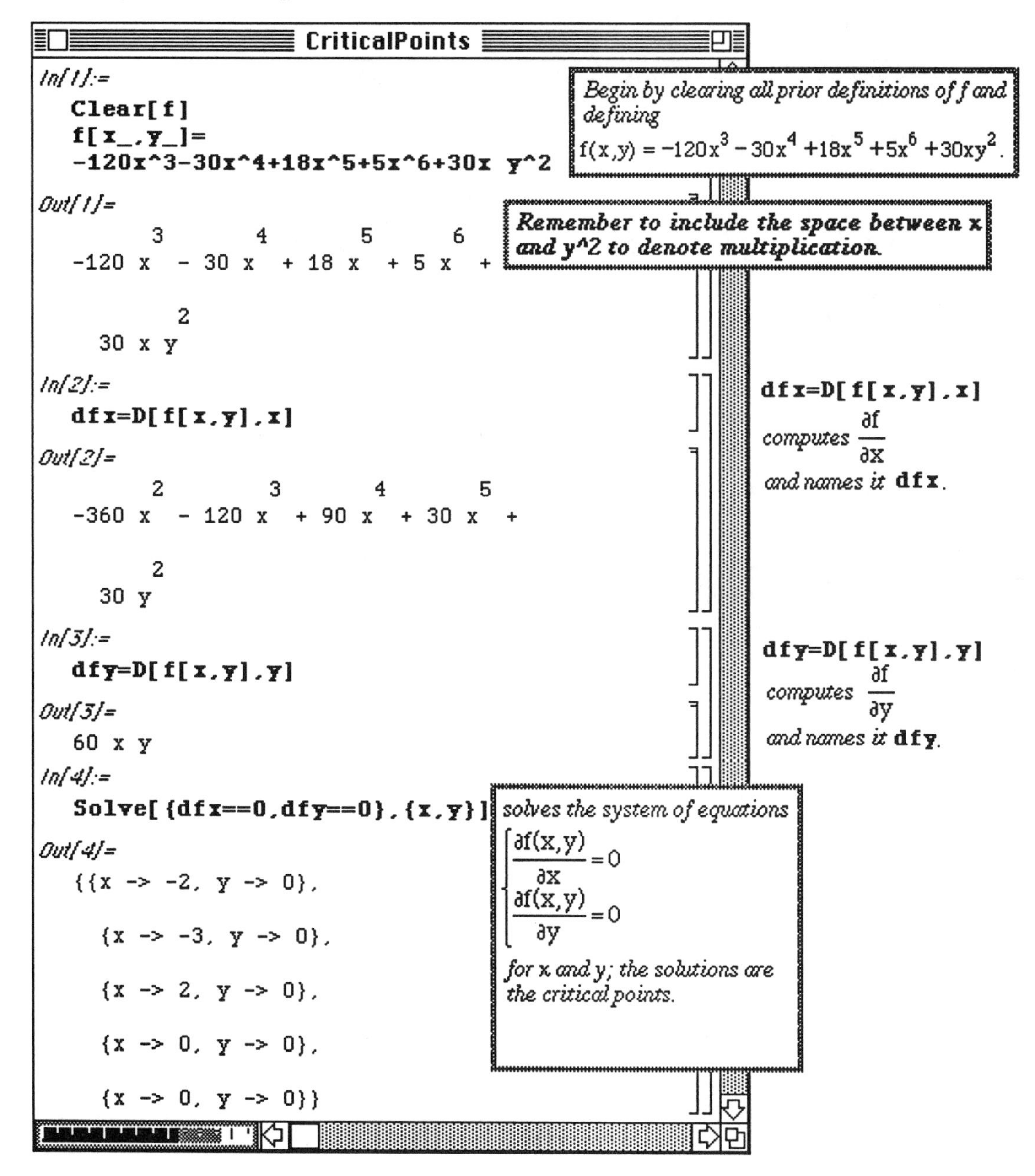

The second derivatives $\frac{\partial^2 f}{\partial x^2}$, $\frac{\partial^2 f}{\partial y^2}$, and $\frac{\partial^2 f}{\partial x \partial y}$ are computed below. Recall that if

$\frac{\partial f}{\partial x}$, $\frac{\partial f}{\partial y}$, $\frac{\partial^2 f}{\partial x \partial y}$, and $\frac{\partial^2 f}{\partial y \partial x}$ are continuous on an open set, then $\frac{\partial^2 f}{\partial x \partial y} = \frac{\partial^2 f}{\partial y \partial x}$

for each point in the set. Therefore, since f(x,y) is a polynomial and, thus, continuous for all values of x and y, only three second order derivatives need to be calculated.

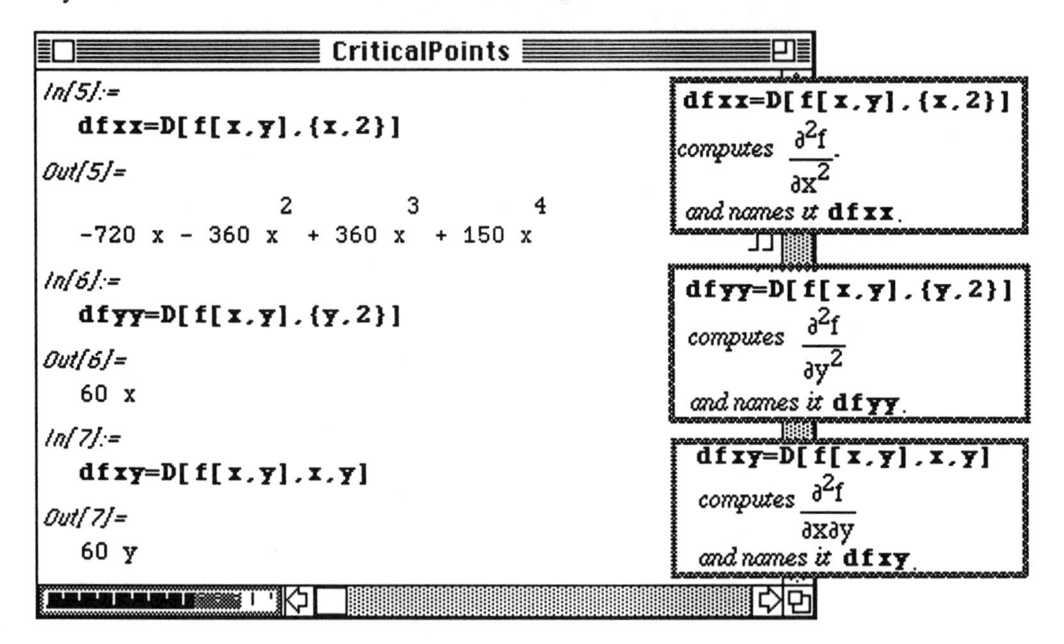

The second-order derivative $\frac{\partial^2 f}{\partial^2 x}$ is evaluated at the critical points for later use in the second derivative test:

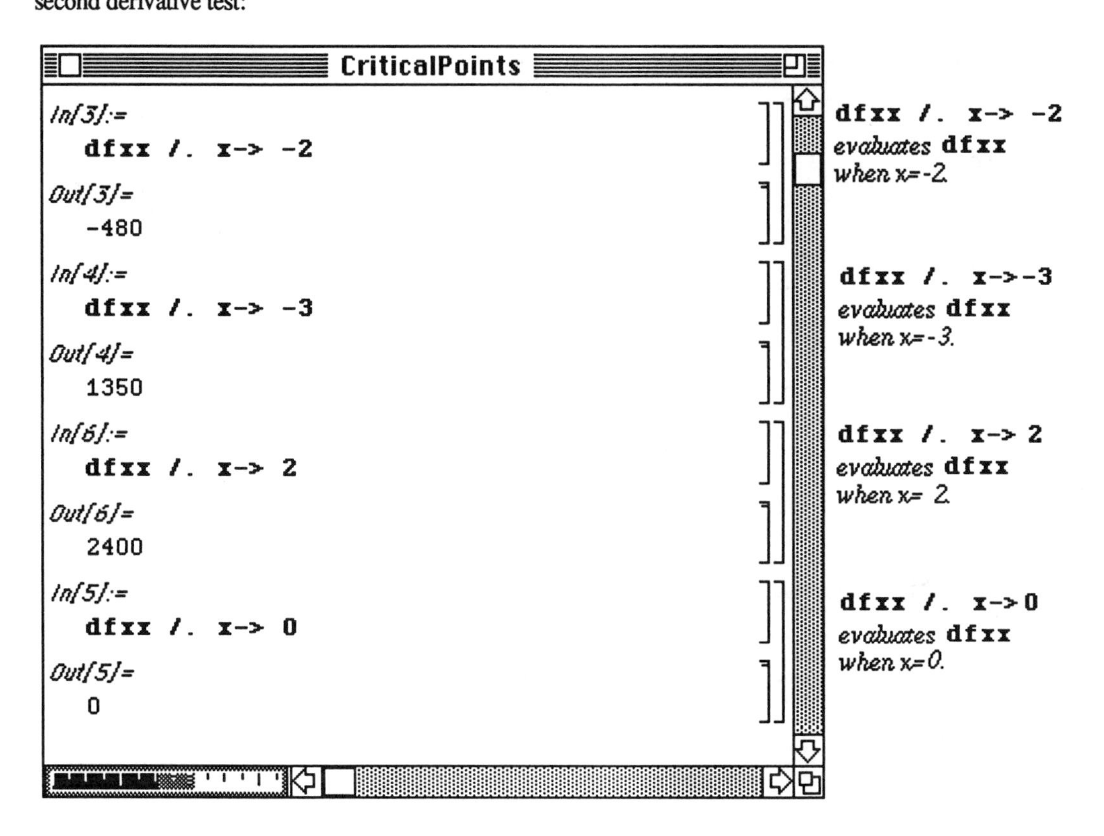

dfxx /. x-> -2 *evaluates* **dfxx** *when x=-2.*

dfxx /. x->-3 *evaluates* **dfxx** *when x=-3.*

dfxx /. x-> 2 *evaluates* **dfxx** *when x= 2.*

dfxx /. x->0 *evaluates* **dfxx** *when x=0.*

The discriminant $\frac{\partial^2 f}{\partial^2 x}(x_0,y_0)\frac{\partial^2 f}{\partial^2 y}(x_0,y_0)-\left(\frac{\partial^2 f}{\partial x \partial y}(x_0,y_0)\right)^2$.

is defined as a function so that its value at the critical points can be computed easily.

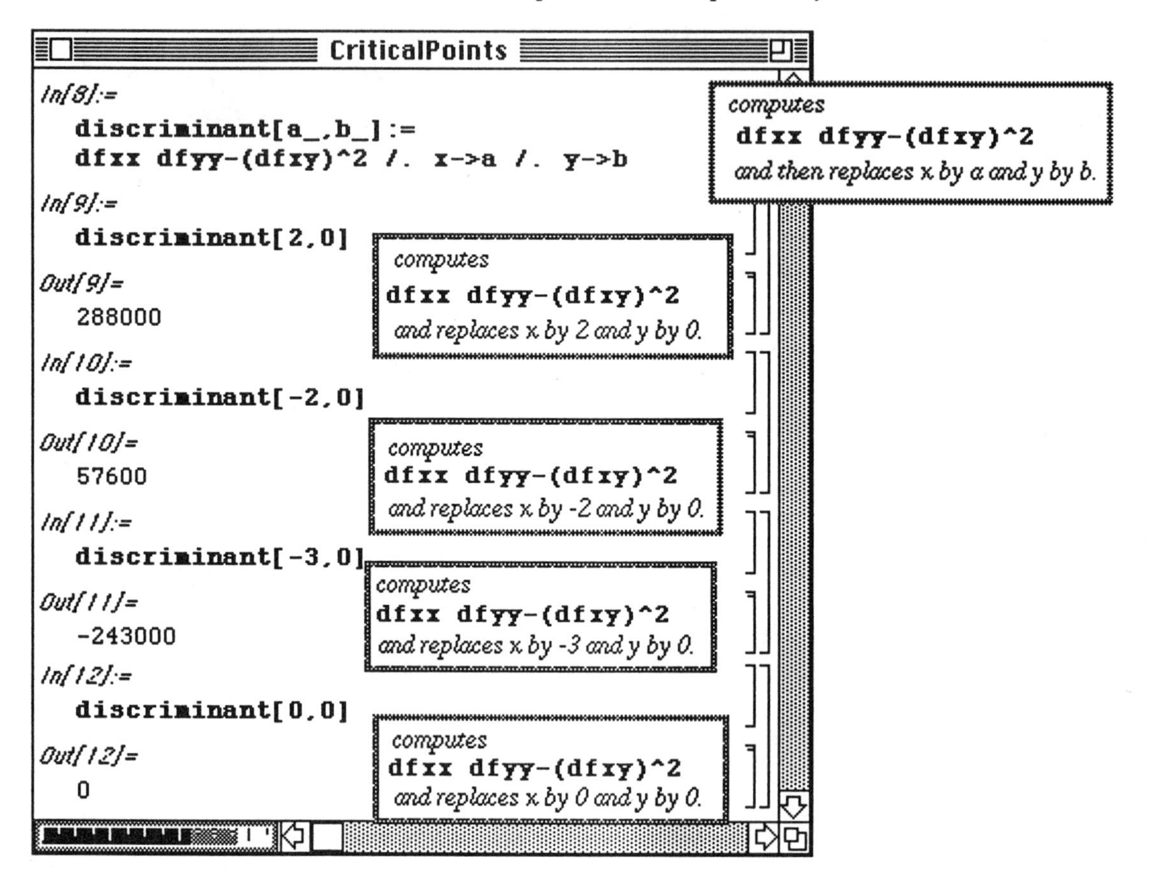

The **Plot3D** command is used to verify that the critical points have been classified correctly by plotting f(x,y) around each critical point.

These graphs also illustrate some of the **Plot3D** options.

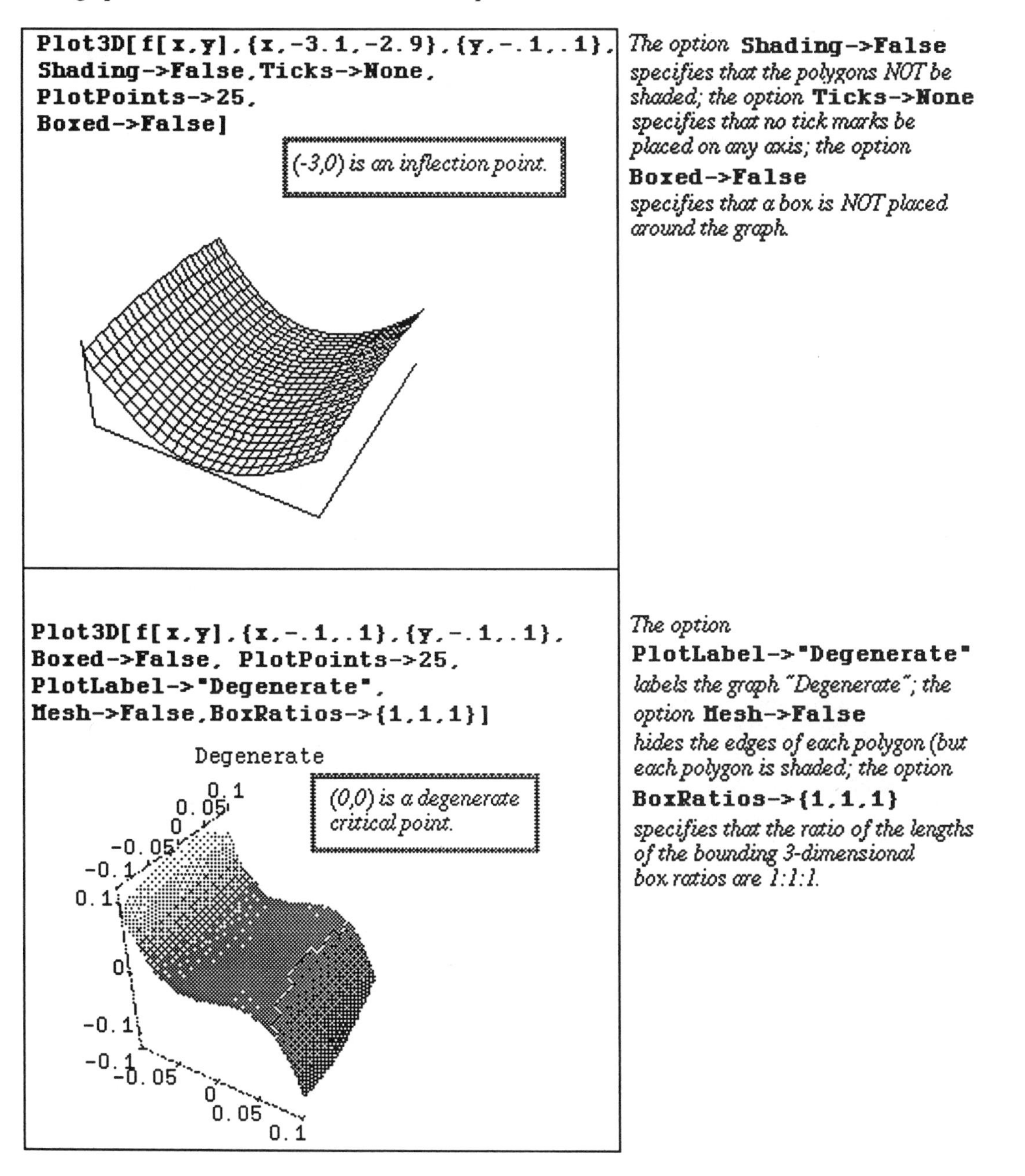


```
Plot3D[f[x,y],{x,-3.1,-2.9},{y,-.1,.1},
Shading->False,Ticks->None,
PlotPoints->25,
Boxed->False]
```

The option **Shading->False** *specifies that the polygons NOT be shaded; the option* **Ticks->None** *specifies that no tick marks be placed on any axis; the option* **Boxed->False** *specifies that a box is NOT placed around the graph.*

```
Plot3D[f[x,y],{x,-.1,.1},{y,-.1,.1},
Boxed->False, PlotPoints->25,
PlotLabel->"Degenerate",
Mesh->False,BoxRatios->{1,1,1}]
```

The option **PlotLabel->"Degenerate"** *labels the graph "Degenerate"; the option* **Mesh->False** *hides the edges of each polygon (but each polygon is shaded; the option* **BoxRatios->{1,1,1}** *specifies that the ratio of the lengths of the bounding 3-dimensional box ratios are 1:1:1.*

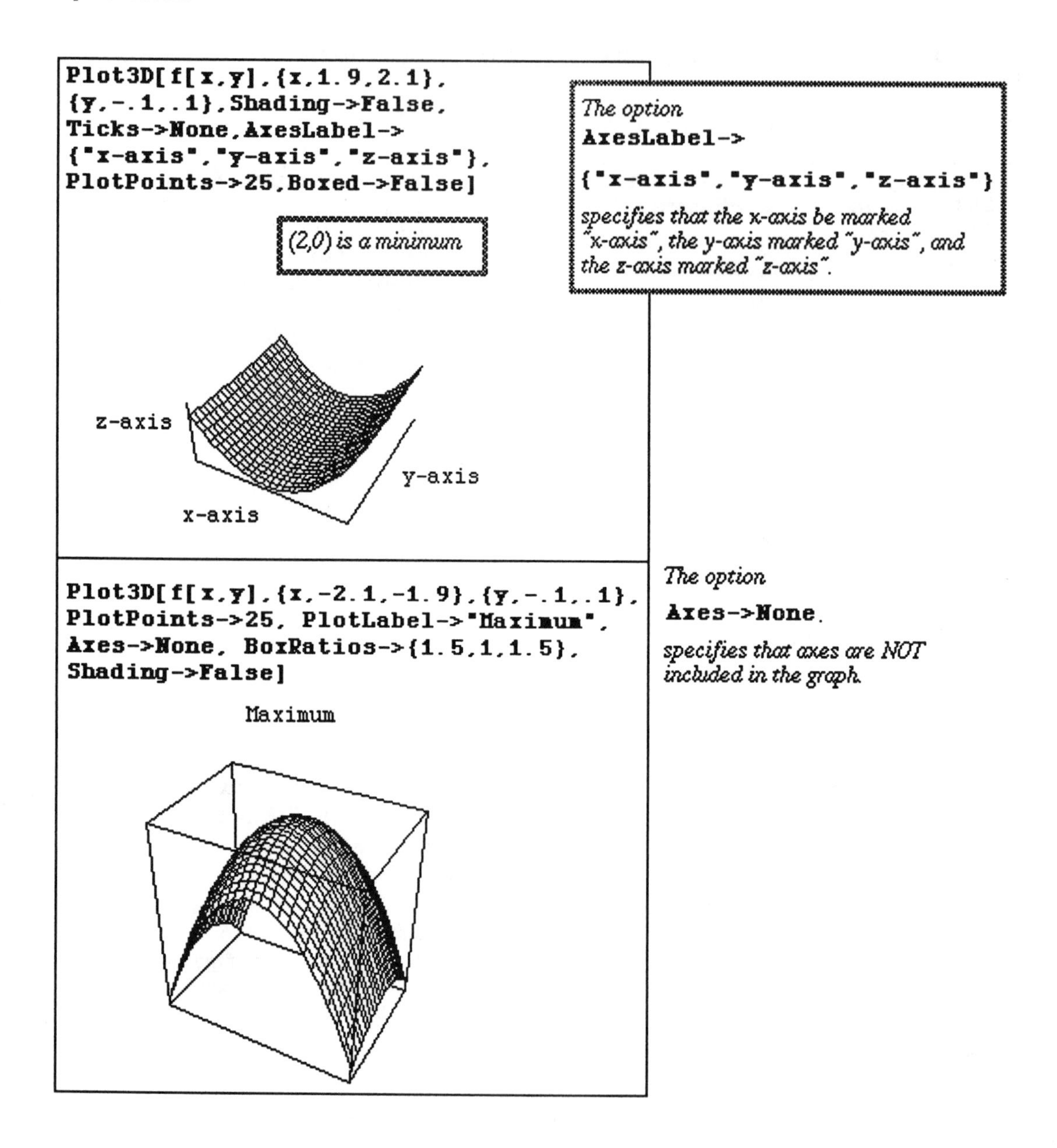
Plot3D[f[x,y],{x,1.9,2.1},
{y,-.1,.1},Shading->False,
Ticks->None,AxesLabel->
{"x-axis","y-axis","z-axis"},
PlotPoints->25,Boxed->False]
The option
AxesLabel->
{"x-axis","y-axis","z-axis"}
specifies that the x-axis be marked "x-axis", the y-axis marked "y-axis", and the z-axis marked "z-axis".
(2,0) is a minimum
z-axis
y-axis
x-axis
Plot3D[f[x,y],{x,-2.1,-1.9},{y,-.1,.1},
PlotPoints->25, PlotLabel->"Maximum",
Axes->None, BoxRatios->{1.5,1,1.5},
Shading->False]
The option
Axes->None.
specifies that axes are NOT included in the graph.
Maximum

■ Application: Tangent Planes

Let z=f(x,y). Then a function that describes the plane tangent to the graph of z=f(x,y)

at the point $(x_0, y_0, z_0 = f(x_0, y_0))$ is given by

$\text{tanplane}_{(f,x_0,y_0)}(x, y) = f_x(x_0, y_0)(x - x_0) + f_y(x_0, y_0)(y - y_0) + z_0$.

□ Example:

Find a function that describes the plane that is tangent to the graph of $f(x, y) = -6xy\,e^{-x^2-y^2}$ at the point (.5, 0).

Mathematica can be used to plot the graph of f(x,y). Using some of the options available to **Plot3D**, this graph can be plotted in a manner which will aide in visualizing the tangent plane to the graph at (.5,0).

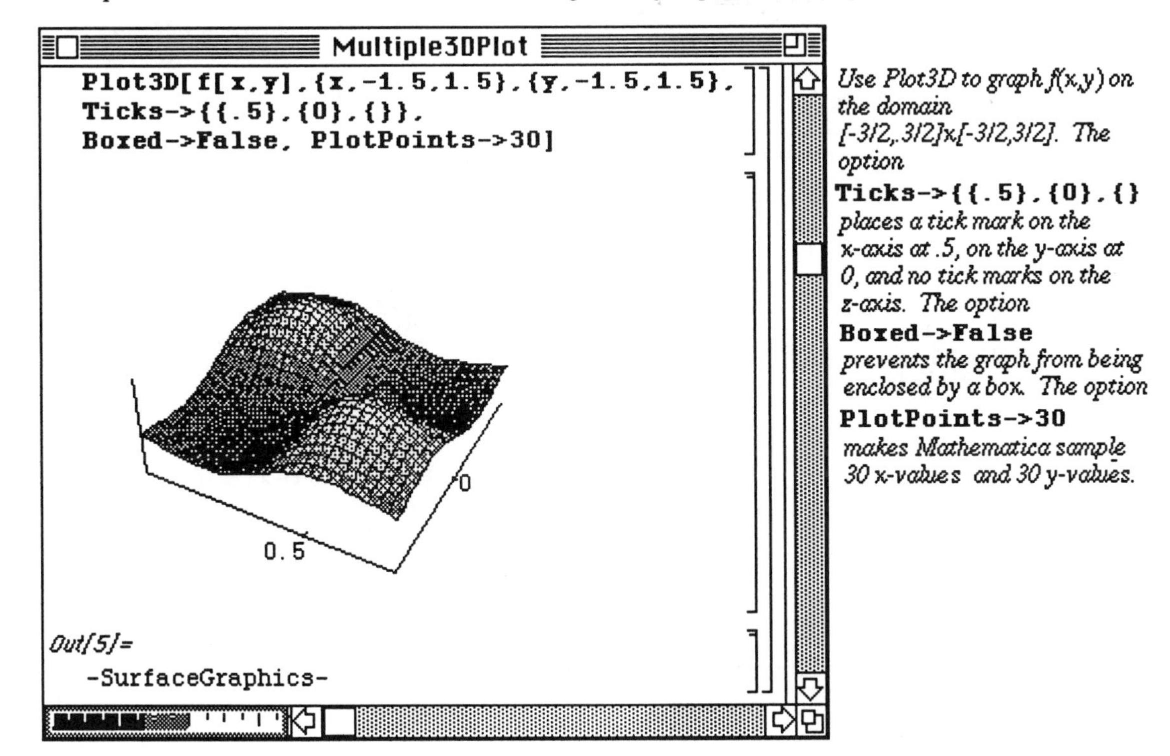

Use Plot3D to graph f(x,y) on the domain [-3/2,.3/2]x[-3/2,3/2]. The option **Ticks->{{.5},{0},{}** *places a tick mark on the x-axis at .5, on the y-axis at 0, and no tick marks on the z-axis. The option* **Boxed->False** *prevents the graph from being enclosed by a box. The option* **PlotPoints->30** *makes Mathematica sample 30 x-values and 30 y-values.*

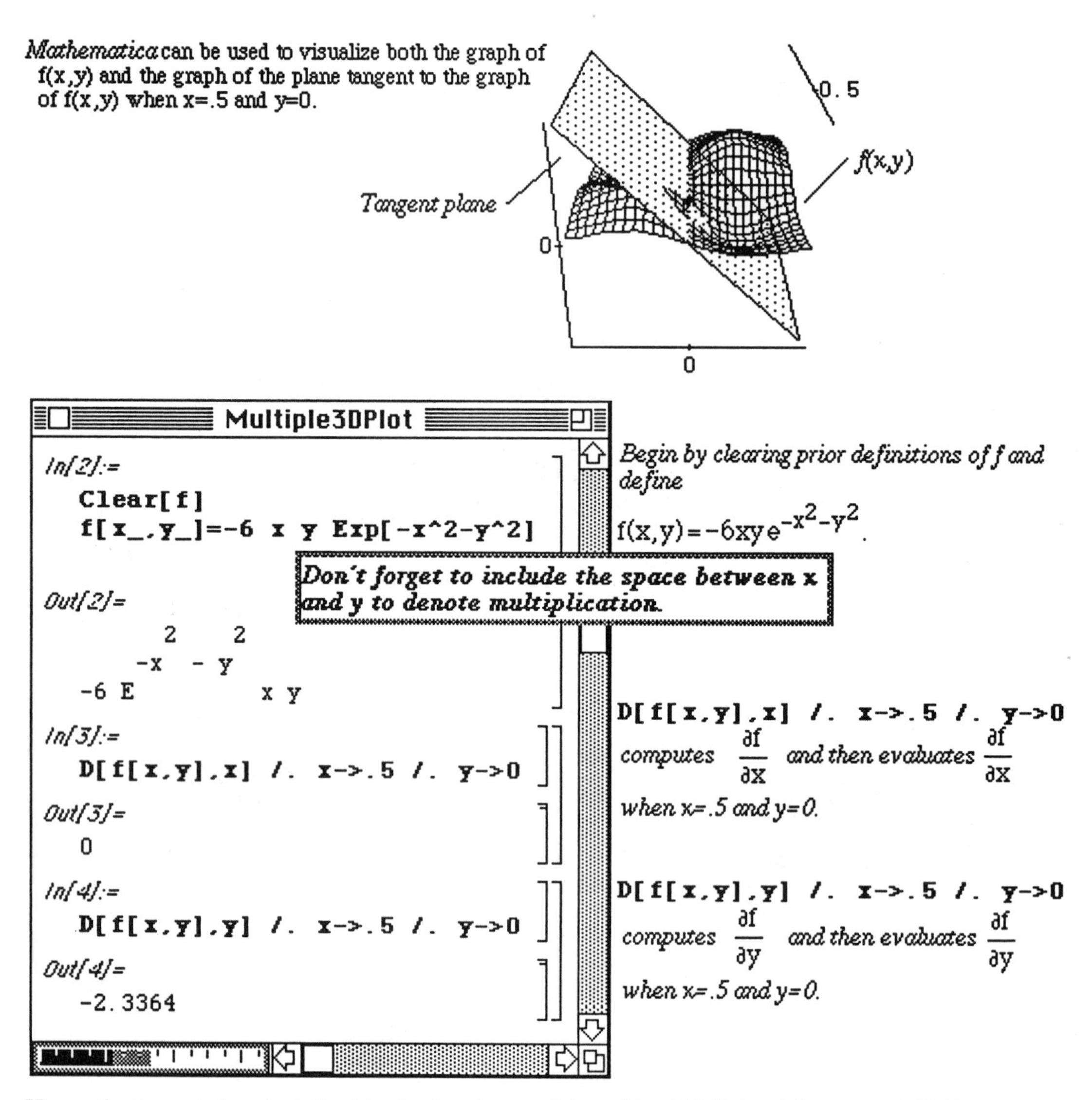

Hence, the tangent plane is defined by the function z = 0 (x - .5) - 2.3364 (y - 0) or, z = - 2.3364 y

o f and z can each be graphed with the command **Plot3D**. However, prior to the release of Version 2.0, two 3-dimension graphics objects created with **Plot3D** could not be shown simultaneously. However, Version 2.0 permits two (or more) objects created with the command **Plot3D** to be shown simultaneously with the command **Show**. In the following example, **f** is graphed on the rectangle [-3,3] x [-2,2], the resulting graph is not displayed and named **plotf**. Similarly, **z** is graphed on the rectangle [-2,2] x [-1,1], the resulting graph is not displayed and named **plotz**. The graphs are shown together (but not actually displayed) with the command **both=Show[plotf,plotz,BoxRatios->{1,1,1}]**.

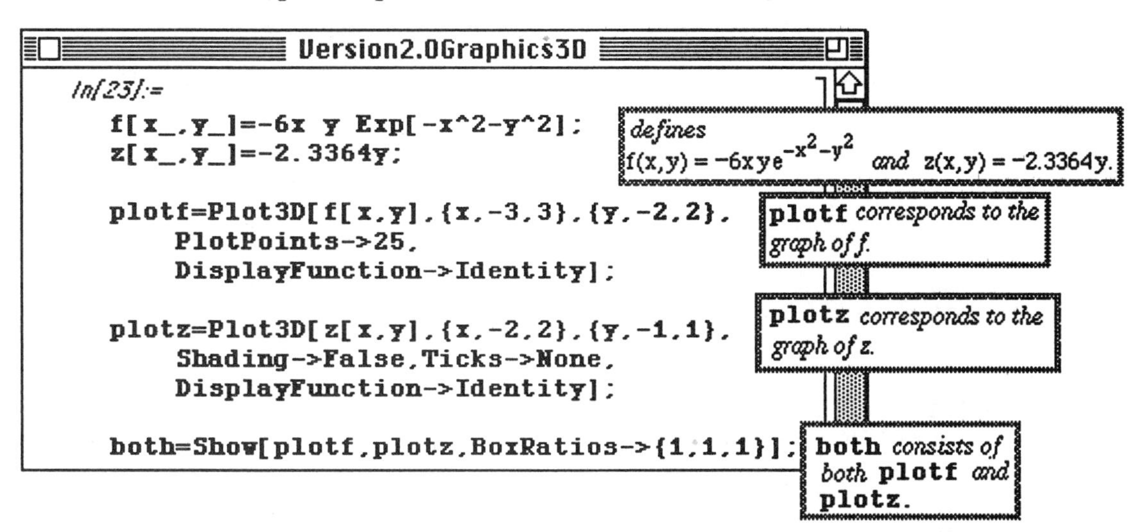

Since **plotf**, **plotz**, and **both** are graphics objects, they may be shown in a single graphics cell with the command **GraphicsArray**:

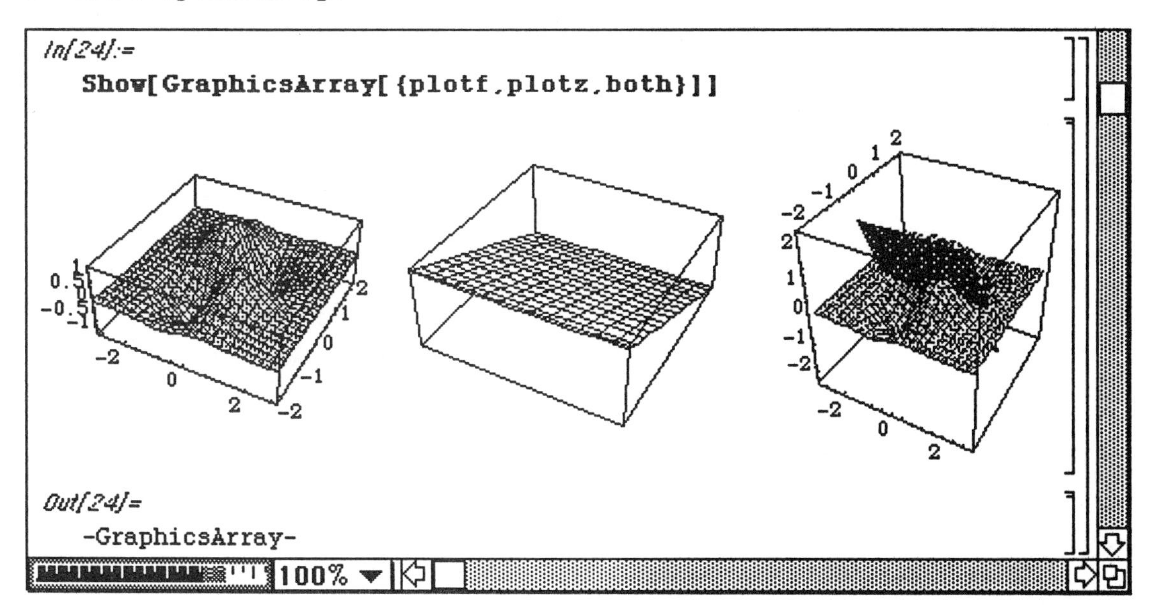

■ Application: Lagrange Multipliers

Certain types of optimization problems can be solved using the method of Lagrange multipliers. This method is based on the following theorem :

□ Lagrange's Theorem:

Let f and g have continuous partial derivatives and f have an extremum at a point

(x_0, y_0) on the smooth constraint curve $g(x,y) = c$. If $g_x(x_0, y_0) \neq 0$ and $g_y(x_0, y_0) \neq 0$, then there is a real number λ such that $f_x(x_0, y_0) = \lambda g_x(x_0, y_0)$, $f_y(x_0, y_0) = \lambda g_y(x_0, y_0)$; and $g(x_0, y_0) = 0$.

□ Example:

Find the maximum and minimum values of $f(x,y) = 4y^2 - 4xy + x^2$ subject to the constraint $x^2 + y^2 = 1$.

The commands used to create the following graphics are discussed in the **Appendix**.

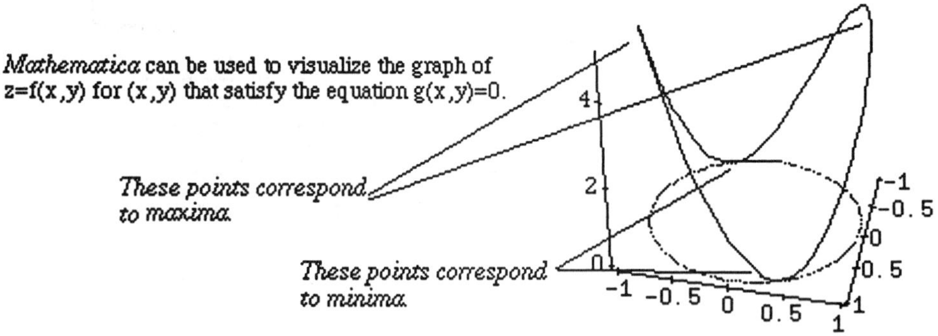

Mathematica can be used to visualize the graph of z=f(x,y) for (x,y) that satisfy the equation g(x,y)=0.

These points correspond to maxima.

These points correspond to minima.

The first order derivatives (with respect to x and y) of f and g are computed in order that Lagrange's Theorem can be applied. (The lambda in Lagrange's Theorem is represented in the calculations below as **11** .)

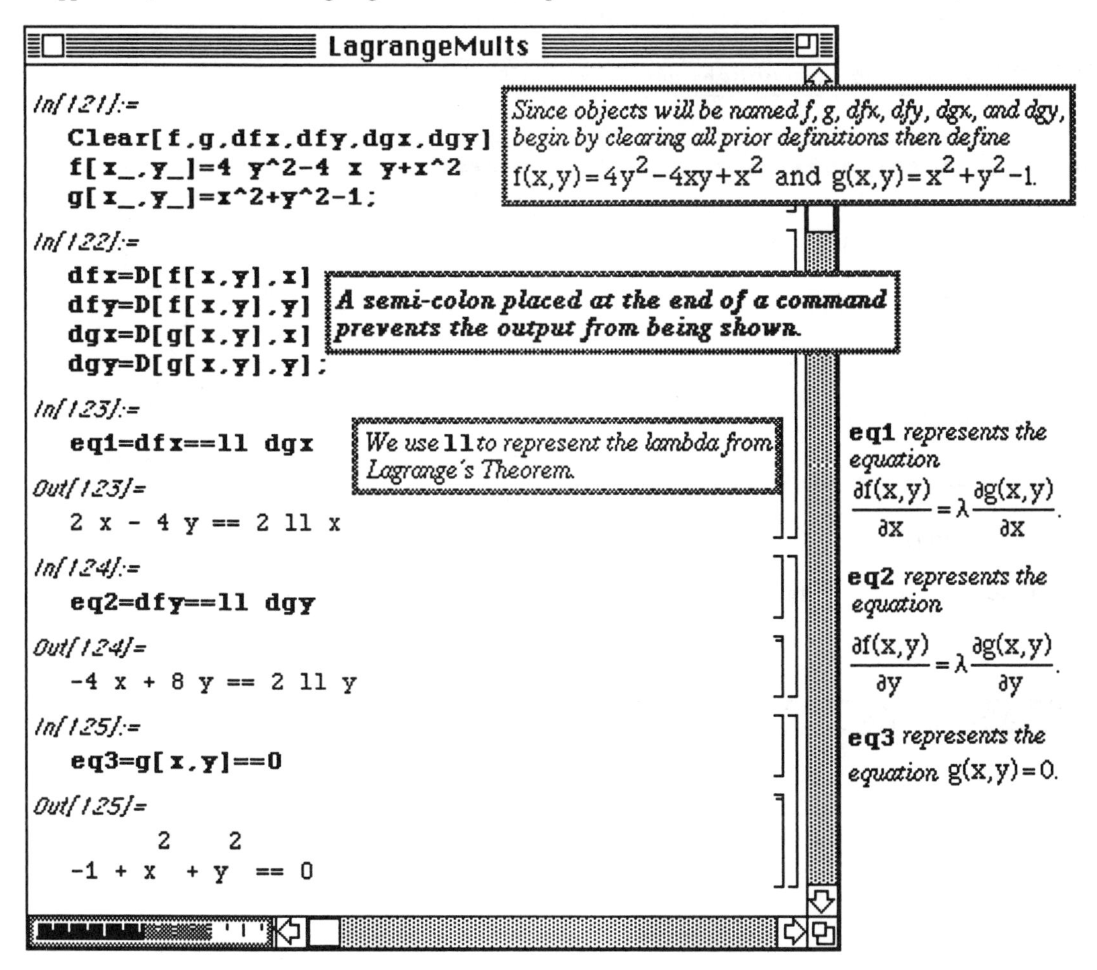

The values of x, y, and lambda which satisfy the system of three equations in Lagrange's Theorem are determined using **Solve**. The solutions of this system are ordered triples (x, y, lambda). The values of x and y in each ordered triple represent the point at which f may have a maximum or minimum value.

LagrangeMults

```
In[126]:=
  Solve[{eq1,eq2,eq3},{x,y,ll}] // N
Out[126]=
  {{ll -> 0., y -> 0.447214,
     x -> 0.894427},
    {ll -> 0., y -> -0.447214,
     x -> -0.894427},
    {ll -> 5., y -> -0.894427,
     x -> 0.447214},
    {ll -> 5., y -> 0.894427,
     x -> -0.447214}}
```

solves the system of equations for x, y and ll. The values of x and y that result in the maximum and minimum values of f subject to the constraint will occur at these points.

Note: In this case, Mathematica produces exact values for x, y, and ll. However, since they are cumbersome to manipulate, we work with numerical approximations instead.

Thus, the maximum and minimum values of f are found by substituting these points back into the function f(x,y) and comparing the resulting values of f.

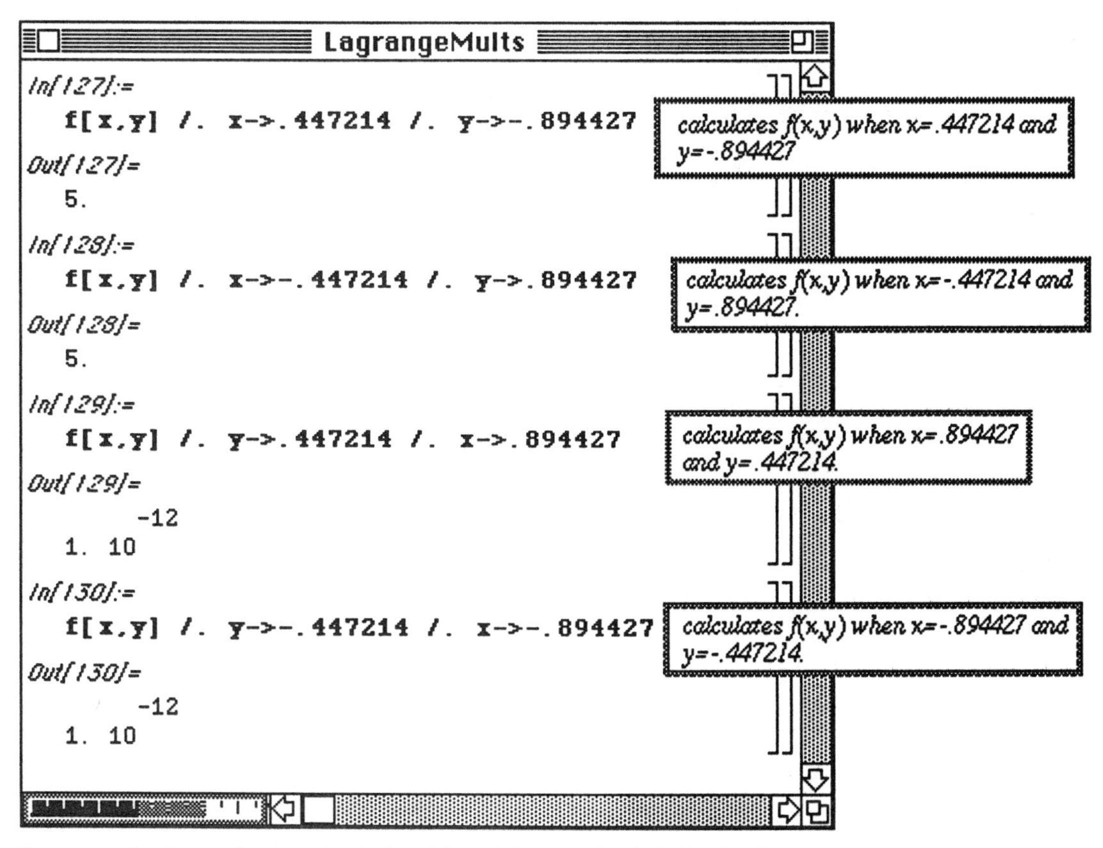

Consequently, the maximum value is 5 and the minimum value is 0. Notice that

$1.\ 10^{-12}$ is assumed to be zero.

In fact, the command `Chop[f[x,y] /. y->-.447214 /. x->-.894427]` yields zero.

◻ Example:

Find the maximum and minimum values of $k(x, y) = x^2 + 4y^3$ subject to the constraint $x^2 + 2y^2 = 1$.

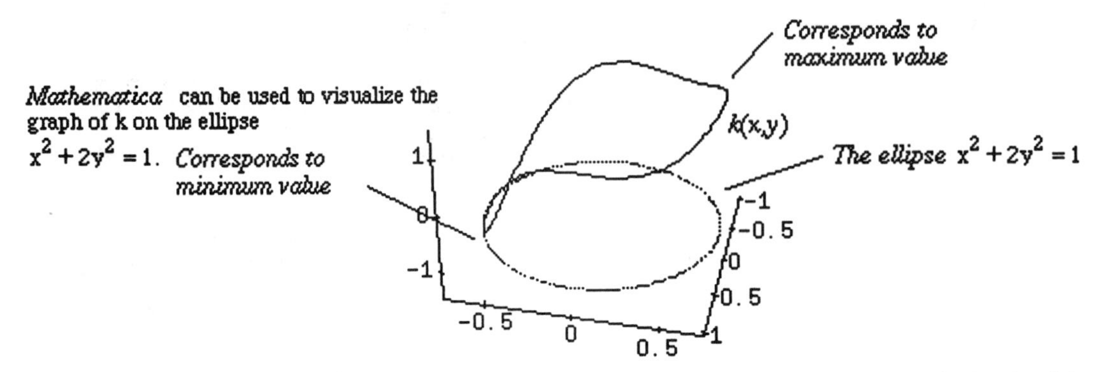

This problem is solved much like the previous example. Notice how the first derivatives are calculated and the system of equations stated in Lagrange's Theorem are established in a single command. The numerical approximation of the solutions of this system are then found.

LagrangeMults

```
In[2]:=
  Clear[k,h]
  k[x_,y_]=x^2+4y^3
  h[x_,y_]=x^2+2y^2-1;
In[3]:=
  dkx=D[k[x,y],x]
  dky=D[k[x,y],y]
  dhx=D[h[x,y],x]
  dhy=D[h[x,y],y]
  eq1=dkx==ll dhx
  eq2=dky==ll dhy
  eq3=h[x,y]==0;
In[4]:=
  Solve[{eq1,eq2,eq3},{x,y,ll}] // N
Out[4]=
  {{ll -> 1., y -> 0.333333,
    x -> 0.881917},
   {ll -> 1., y -> 0.333333,
    x -> -0.881917},
   {ll -> 1., y -> 0., x -> 1.},
   {ll -> 1., y -> 0., x -> -1.},
   {ll -> 2.12132, y -> 0.707107,
    x -> 0.}, {ll -> -2.12132,
    y -> -0.707107, x -> 0.}}
```

Notice that several commands can be evaluated if they are combined into a single input cell. The semi-colon prevents the output from being shown.

Since the goal is to find the maximum and minimum values of k(x,y) subject to the constraint $x^2+2y^2=1$ *begin by defining* $k(x,y)=x^2+4y^3$ *and* $h(x,y)=x^2+2y^2-1$.

eq1 *represents* $\frac{\partial k(x,y)}{\partial x}=\lambda\frac{\partial h(x,y)}{\partial x}$;

eq2 *represents* $\frac{\partial k(x,y)}{\partial y}=\lambda\frac{\partial h(x,y)}{\partial y}$;

and **eq3** *represents* $h(x,y)=0$.

Solve *solves eq1, eq2, and eq3 for x, y, and ll. Even though Mathematica finds the exact roots, we use numerical approximations for convenience.*

The values of x and y that maximize and minimize k must occur at these points. To determine which values maximize k and which values minimize k, evaluate k(x,y) for each set of values.

The points which satisfy the system are substituted into k(x,y) to obtain the maximum and minimum values of the function by comparing the values obtained.

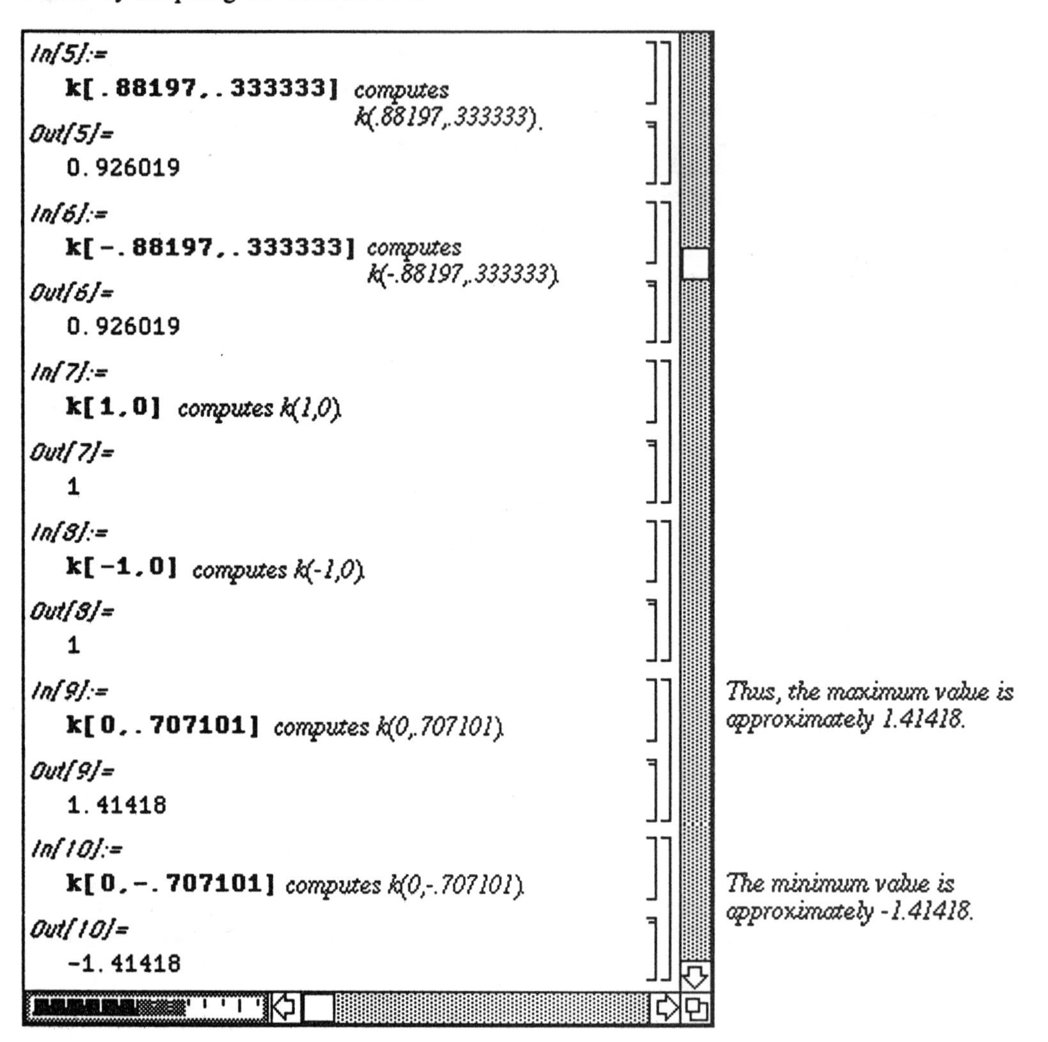

Thus, the maximum value is approximately 1.41418.

The minimum value is approximately -1.41418.

■ Multiple Integrals

Mathematica can compute multiple integrals. The command which computes the double integral

$$\int_{xmin}^{xmax}\int_{ymin}^{ymax} f(x,y)\,dy\,dx$$ is:

```
Integrate[f[x,y],{x,xmin,xmax},{y,ymin,ymax}]
```

Multiple integrals are numerically computed with the command
`NIntegrate[f[x,y],{x,xmin,xmax},{y,ymin,ymax}]`.

The first variable given (in this case, x), corresponds to the outermost integral and integration with respect to this variable is done last. Also, the inner limits of integration (in this case, **ymin** and **ymax**) can be functions of the outermost variable. Limits on the outermost integral must <u>always</u> be constants.

o When using Version 1.2, be sure to load the package **IntegralTables.m**; if using Version 2.0, loading **IntegralTables.m** is not necessary since *Mathematica* automatically loads the package.

Several examples are shown below :

□ **Example:**

MultipleIntegrals

```
In[2]:=
  <<IntegralTables.m
Out[2]=
  Integrator`
In[15]:=
  Integrate[x y^2,{y,1,2},
  {x,1-y,Sqrt[y]}]
Out[15]=
  163
  ---
  120
In[16]:=
  Integrate[y Sin[x]-x Sin[y],
  {x,0,Pi/6},{y,0,Pi/2}]
Out[16]=
    2              2
  Pi     Sqrt[3] Pi
  --- - -----------
   9         16
In[18]:=
  Integrate[Exp[x] Sin[y],
  {y,Pi/6,Pi/4},{x,0,Cos[y]}]
Out[18]=
  Sqrt[2]   Sqrt[3]      Sqrt[2]/2
  ------- - ------- - E            +
     2         2

     Sqrt[3]/2
    E
```

Be sure the package **IntegralTables.m** *has been loaded before attempting to compute definite integrals.*

Integrate[x y^2,{y,1,2},{x,1-y,Sqrt[y]}] *calculates* $\int_1^2 \int_{1-y}^{\sqrt{y}} xy^2\,dx\,dy$.

Integrate[y Sin[x]-x Sin[y],{x,0,Pi/6},{y,0,Pi/2}] *calculates* $\int_0^{\pi/6} \int_0^{\pi/2} (y\mathrm{Sin}(x)-x\mathrm{Sin}(y))\,dy\,dx$.

Integrate[Exp[x] Sin[y],{y,Pi/6,Pi/4},{x,0,Cos[y]}] *calculates* $\int_{\pi/6}^{\pi/4} \int_0^{\mathrm{Cos}(y)} e^x \mathrm{Sin}(y)\,dx\,dy$.

In cases in which the double integral cannot be computed exactly, the command

NIntegrate[f[x,y],{x,xmin,xmax},{y,ymin,ymax}]

can be used to calculate a numerical approximation of the integral

$\int_{xmin}^{xmax}\int_{ymin}^{ymax} f[x,y]\,dy\,dx.$

□ Example:

Approximate the value of the double integral $\int_0^1\int_0^1 \text{Sin}\left(e^{xy}\right)dy\,dx$.

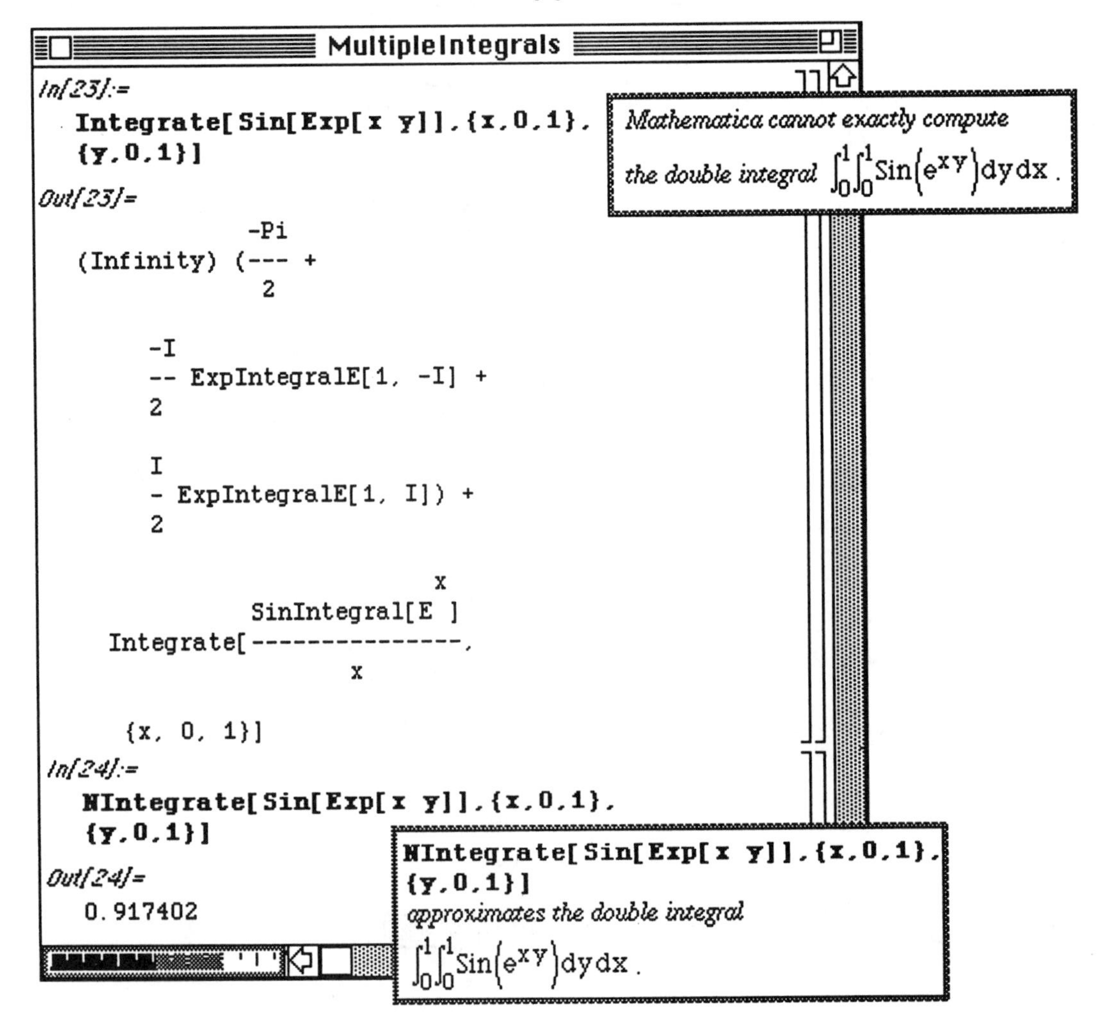

However, **NIntegrate** does not always produce the quickest result. The following example illustrates what happens when an integral which can be computed exactly is attempted using **NIntegrate**.

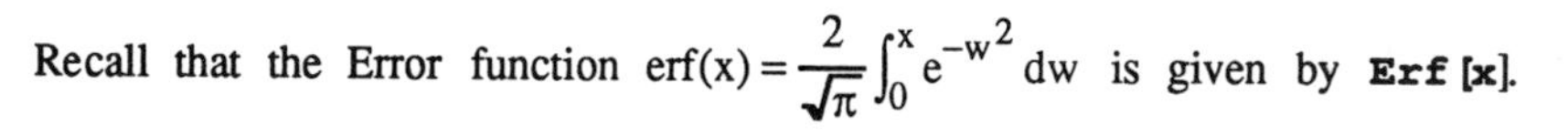

Recall that the Error function $\mathrm{erf}(x) = \frac{2}{\sqrt{\pi}} \int_0^x e^{-w^2}\, dw$ is given by **Erf[x]**.

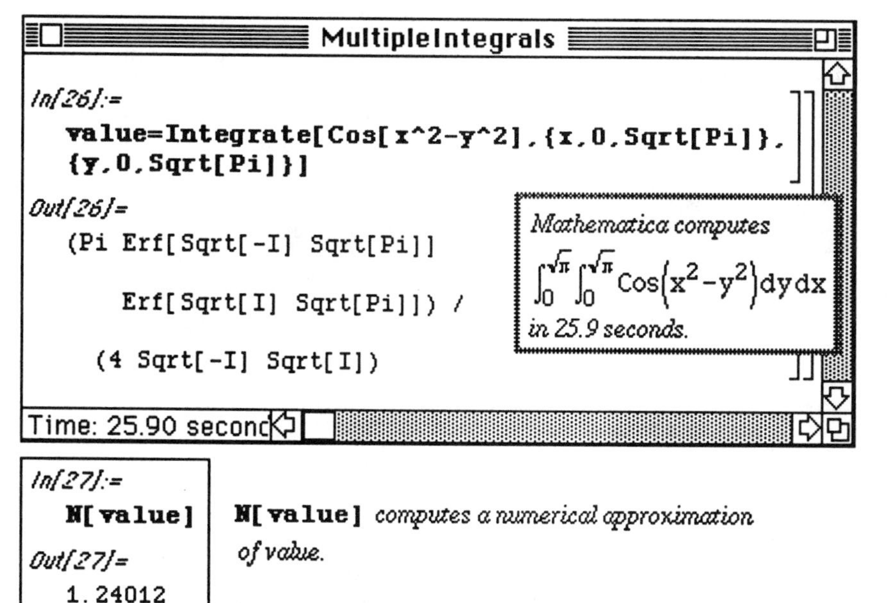

The command **NIntegrate[Cos[x^2-y^2],{x,0,Sqrt[Pi]},{y,0,Sqrt[Pi]}]** does not produce an output for an approximation of the integral after ten minutes of computing on a Macintosh IIcx.

■ Application: Volume

□ Example:

Find the volume of the region between the graphs of

$q(x, y) = \text{Cos}\left(x^2 + y^2\right) e^{-x^2}$ and $w(x, y) = 3 - x^2 - y^2$ on the domain $[-1,1] \times [-1,1]$.

The region can be viewed using *Mathematica*'s `Plot3D` command.

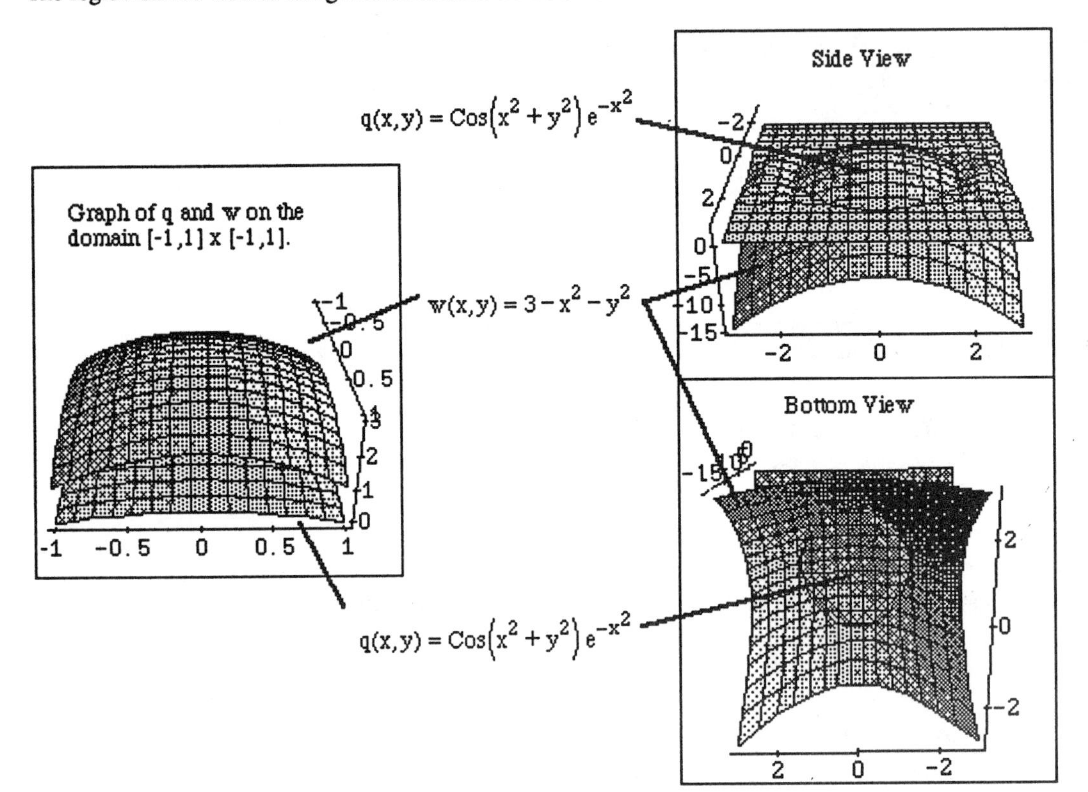

The above graphs show that the region is bounded above by w(x,y) and below by q(x,y). Hence, the volume is determined as follows :

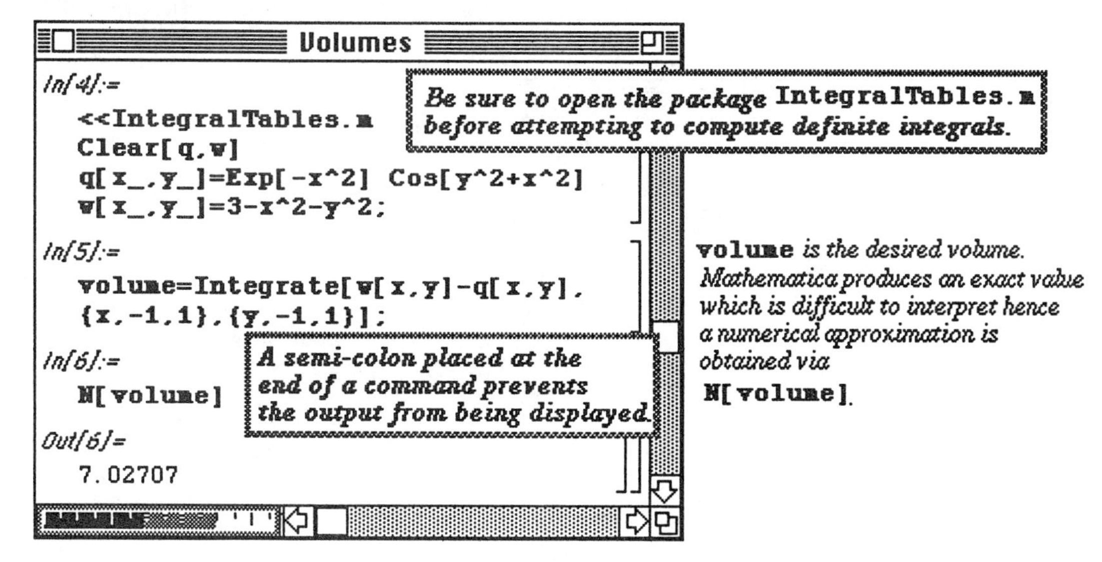

volume is the desired volume. *Mathematica* produces an exact value which is difficult to interpret hence a numerical approximation is obtained via N[volume].

□ **Example:**

Find the volume of the solid bounded by the graphs of $f(x,y)=1-x-y$ and $g(x,y)=2-x^2-y^2$. The graph is used to determine that the region is bounded above by the paraboloid and below by the plane.

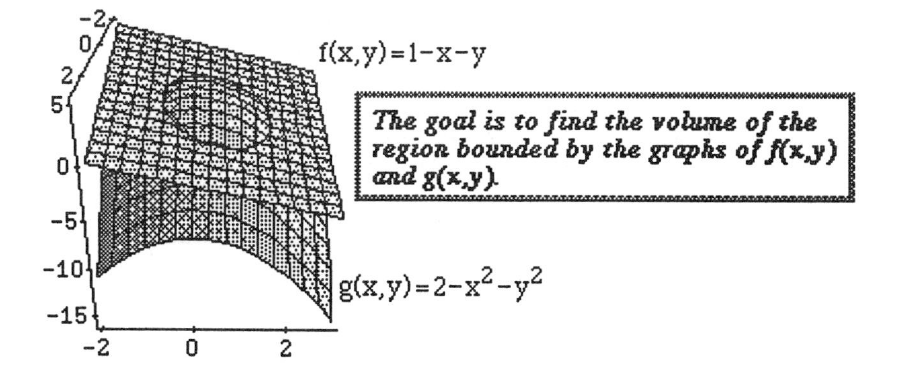

This problem is more difficult than the first example since the limits of integration must be determined. However, using **Solve** to find the values of x and y such that f(x,y) = g(x,y), these limits are found easily. First, the equation f(x,y) = g(x,y) is solved for y (in terms of x). To facilitate the use of these y-values, they are given the names **y1** and **y2**.

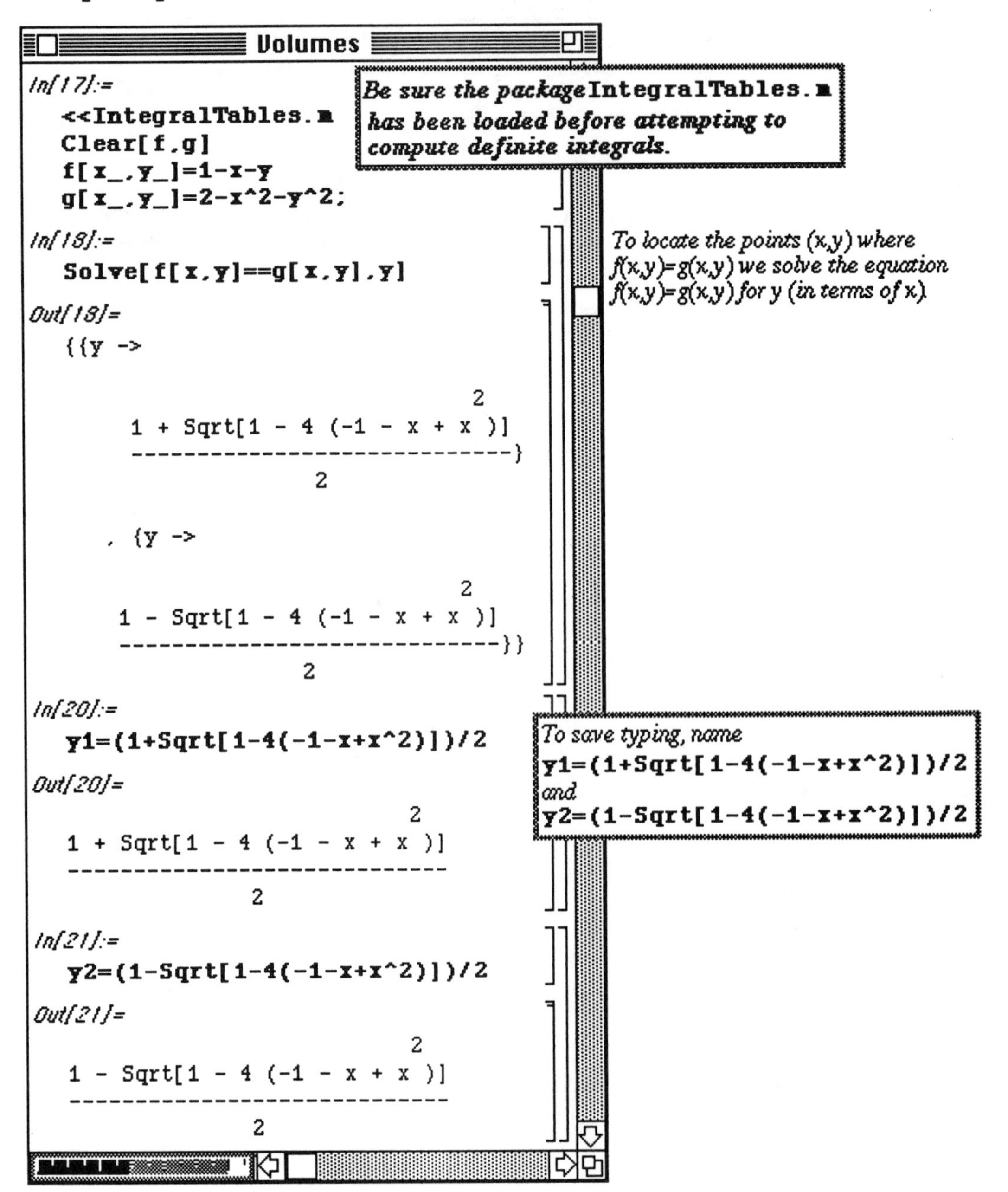

Next, the limits on the x-coordinate are determined :

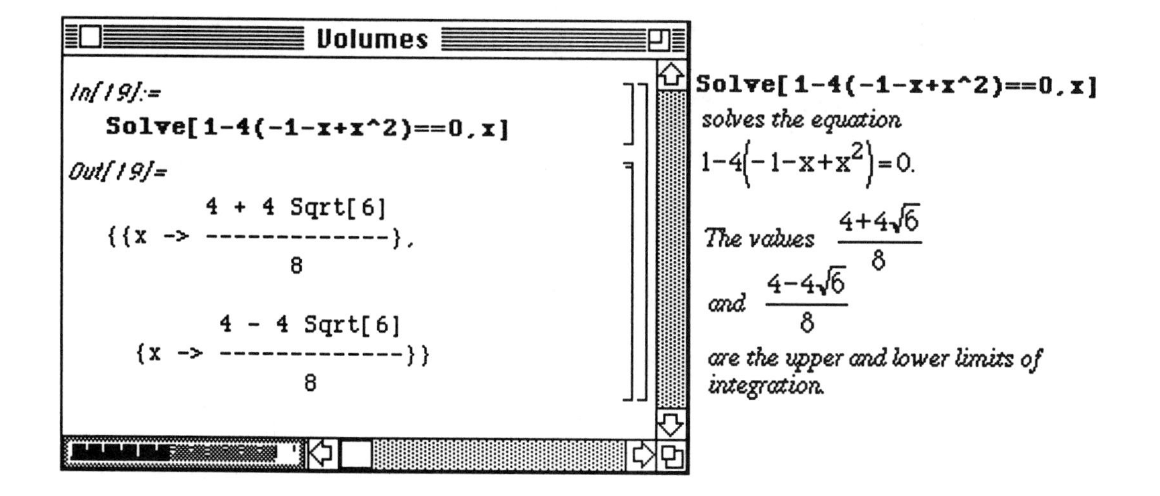

Finally after the limits of integration have been determined, the volume is found :

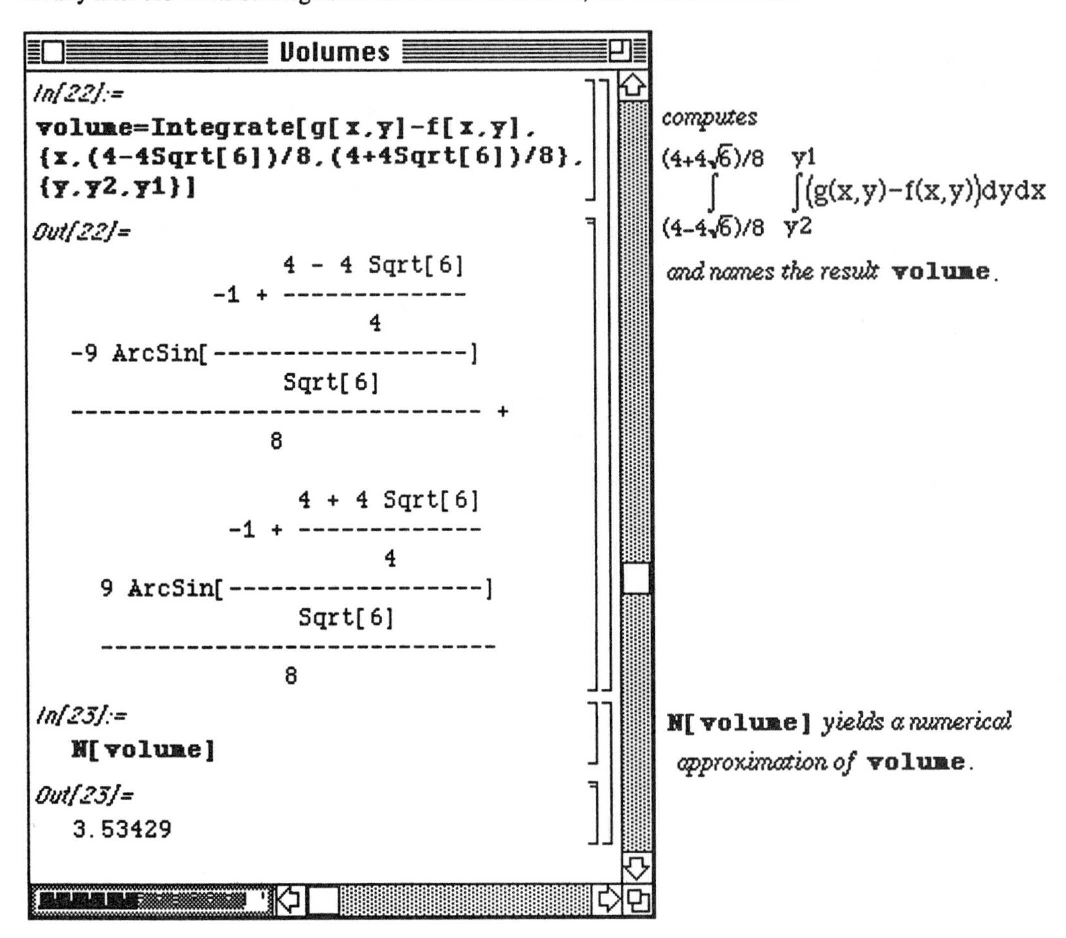

Volumes

```
In[22]:=
volume=Integrate[g[x,y]-f[x,y],
 {x,(4-4Sqrt[6])/8,(4+4Sqrt[6])/8},
 {y,y2,y1}]
Out[22]=
                  4 - 4 Sqrt[6]
            -1 + -------------
                        4
  -9 ArcSin[------------------]
                  Sqrt[6]
  ----------------------------- +
                 8

                   4 + 4 Sqrt[6]
             -1 + -------------
                         4
    9 ArcSin[------------------]
                   Sqrt[6]
    -----------------------------
                  8
In[23]:=
  N[volume]
Out[23]=
  3.53429
```

computes

$$\int_{(4-4\sqrt{6})/8}^{(4+4\sqrt{6})/8} \int_{y2}^{y1} (g(x,y)-f(x,y))\,dy\,dx$$

and names the result **volume**.

N[volume] *yields a numerical approximation of* **volume**.

■ Series in More than One Variable

In the same manner as **Integrate**, **NIntegrate**, and **D** generalize to functions of more than one variable, the command **Series** also computes power series of funtions of more than one variable.

The command **Series[f[x,y],{x,x0,n},{y,y0,m}]** computes a power series expansion of **f[x,y]** about **y0** up to order **m** and then computes a power series expansion about **x0** up to order **n**.

The result of using the **Series** command always includes a remainder term and hence cannot be treated as a function. In order to remove the remainder term, use the command **Normal**.

□ Example:

Define $f(x,y) = \text{Cos}(x - \pi \text{Sin}(y))$. (i) Generate the first two terms of the powers series for f about $y = 1$; (ii) generate the first two terms of the power series of f about $y = 1$ and then generate the first three terms of the power series about $x = 0$; and (iii) generate the first three terms of the power series of f about $x = 0$ and then generate the first two terms of the power series about $y = 1$.
After removing the remainder term, graph and compare the results.

We begin by defining f and then computing the first two terms of the power series for f about y=1:

```
SeriesinTwoVariables

In[3]:=
  f[x_,y_]=Cos[x-Pi Sin[y]]
Out[3]=
  Cos[x - Pi Sin[y]]
In[7]:=
  sery=Series[f[x,y],{y,1,2}]
Out[7]=
  Cos[x - Pi Sin[1]] + Pi Cos[1] Sin[x - Pi Sin[1]] (-1 + y) +

              2       2
      -(Pi  Cos[1]  Cos[x - Pi Sin[1]])
     (--------------------------------- -
                         2

        Pi Sin[1] Sin[x - Pi Sin[1]]              2           3
        ----------------------------) (-1 + y)  + O[-1 + y]
                       2
```

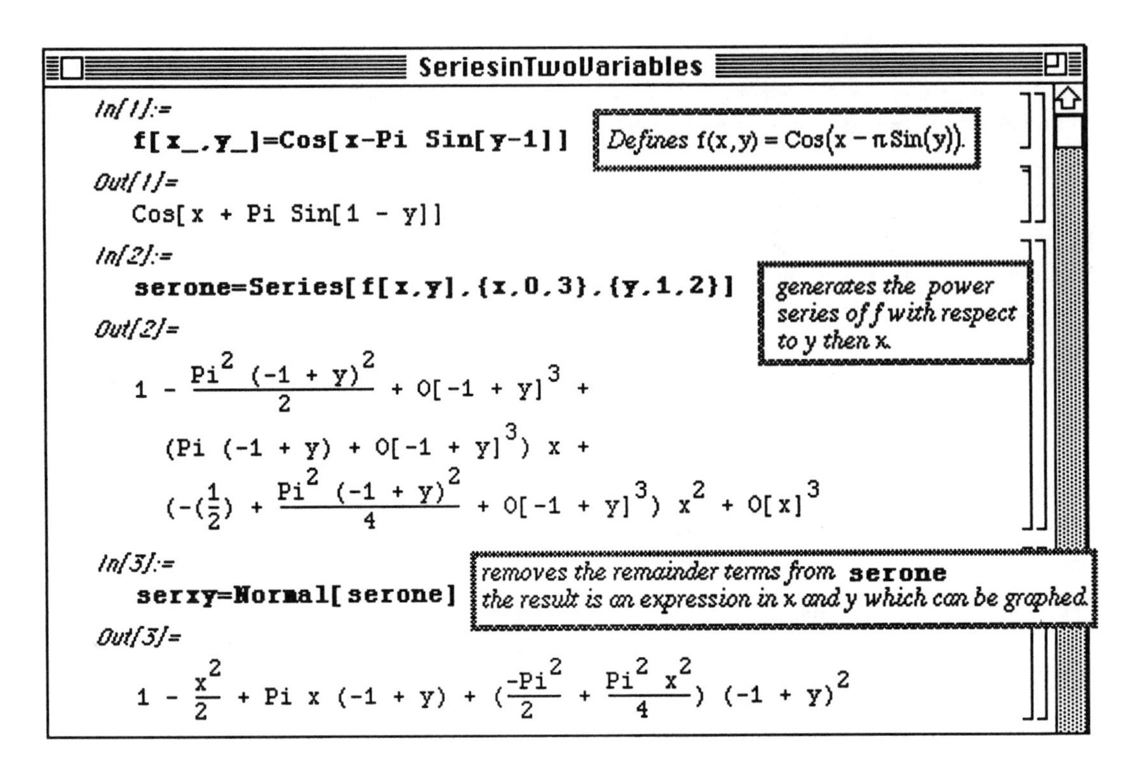
SeriesinTwoVariables
In[1]:=
f[x_,y_]=Cos[x-Pi Sin[y-1]]
Defines f(x,y) = Cos(x − π Sin(y)).
Out[1]=
Cos[x + Pi Sin[1 - y]]
In[2]:=
serone=Series[f[x,y],{x,0,3},{y,1,2}]
generates the power series of f with respect to y then x.
Out[2]=
In[3]:=
serxy=Normal[serone]
removes the remainder terms from serone
the result is an expression in x and y which can be graphed.
Out[3]=

The command **seryx=Series[f[x,y],{y,1,2},{x,0,3}]//Normal** computes the series with respect to x then y and removes the remainder terms.

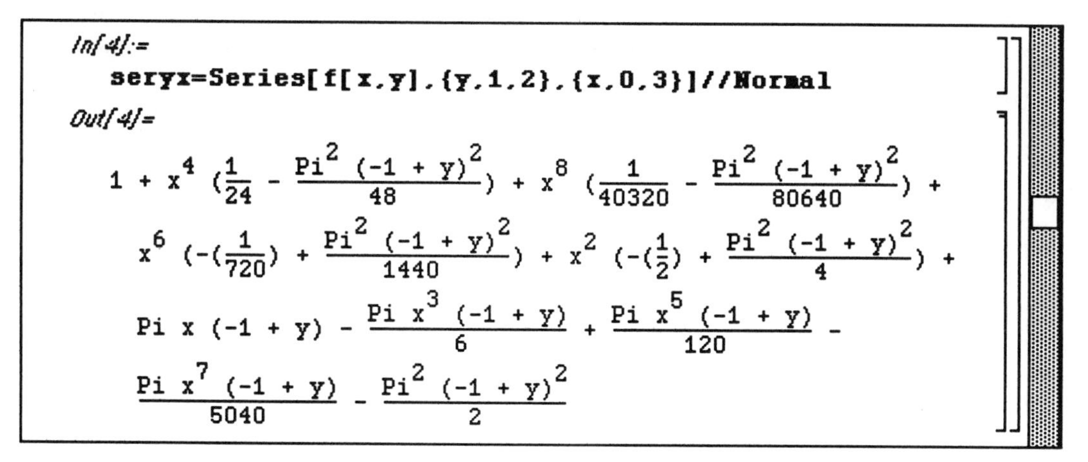

```
In[4]:=
seryx=Series[f[x,y],{y,1,2},{x,0,3}]//Normal
Out[4]=
```

$$1 + x^4 \left(\frac{1}{24} - \frac{Pi^2 (-1 + y)^2}{48}\right) + x^8 \left(\frac{1}{40320} - \frac{Pi^2 (-1 + y)^2}{80640}\right) + x^6 \left(-\left(\frac{1}{720}\right) + \frac{Pi^2 (-1 + y)^2}{1440}\right) + x^2 \left(-\left(\frac{1}{2}\right) + \frac{Pi^2 (-1 + y)^2}{4}\right) + Pi\ x (-1 + y) - \frac{Pi\ x^3 (-1 + y)}{6} + \frac{Pi\ x^5 (-1 + y)}{120} - \frac{Pi\ x^7 (-1 + y)}{5040} - \frac{Pi^2 (-1 + y)^2}{2}$$

We create the graphs with Version 2.0 to take advantage of Version 2.0's improved graphics features. In particular, we graph **f[x,y]**, **serxy**, and **seryx** with **Plot3D** and then show all three simultaneously with the **Show** command. Finally, all four graphics objects are displayed in a single cell with the command **GraphicsArray**.

```
In[9]:=
plotf=Plot3D[f[x,y],{x,0,2Pi},{y,0,2Pi},
        PlotPoints->25,Boxed->False,
        Ticks->{{0,Pi,2Pi},None,{-1,0,1}},
        DisplayFunction->Identity,
        Shading->False];

plotxy=Plot3D[serxy,{x,0,2Pi},{y,0,2Pi},
        PlotPoints->25,Boxed->False,
        DisplayFunction->Identity,
        Ticks->{None,{0,Pi,2Pi},{-1,0,1}},
        Shading->False];

plotyx=Plot3D[seryx,{x,0,2Pi},{y,0,2Pi},
        PlotPoints->25,Boxed->False,
        Ticks->{{0,Pi,2Pi},{0,Pi,2Pi},None},
        DisplayFunction->Identity,
        Shading->False];

allthree=Show[plotxy,plotyx,plotf];

Show[GraphicsArray[{{plotf,plotxy},
    {plotyx,allthree}}]]
```

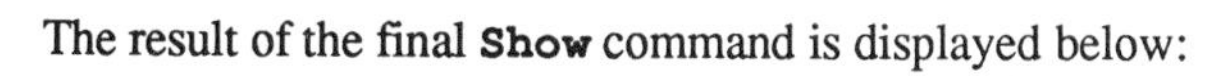

The result of the final **Show** command is displayed below:

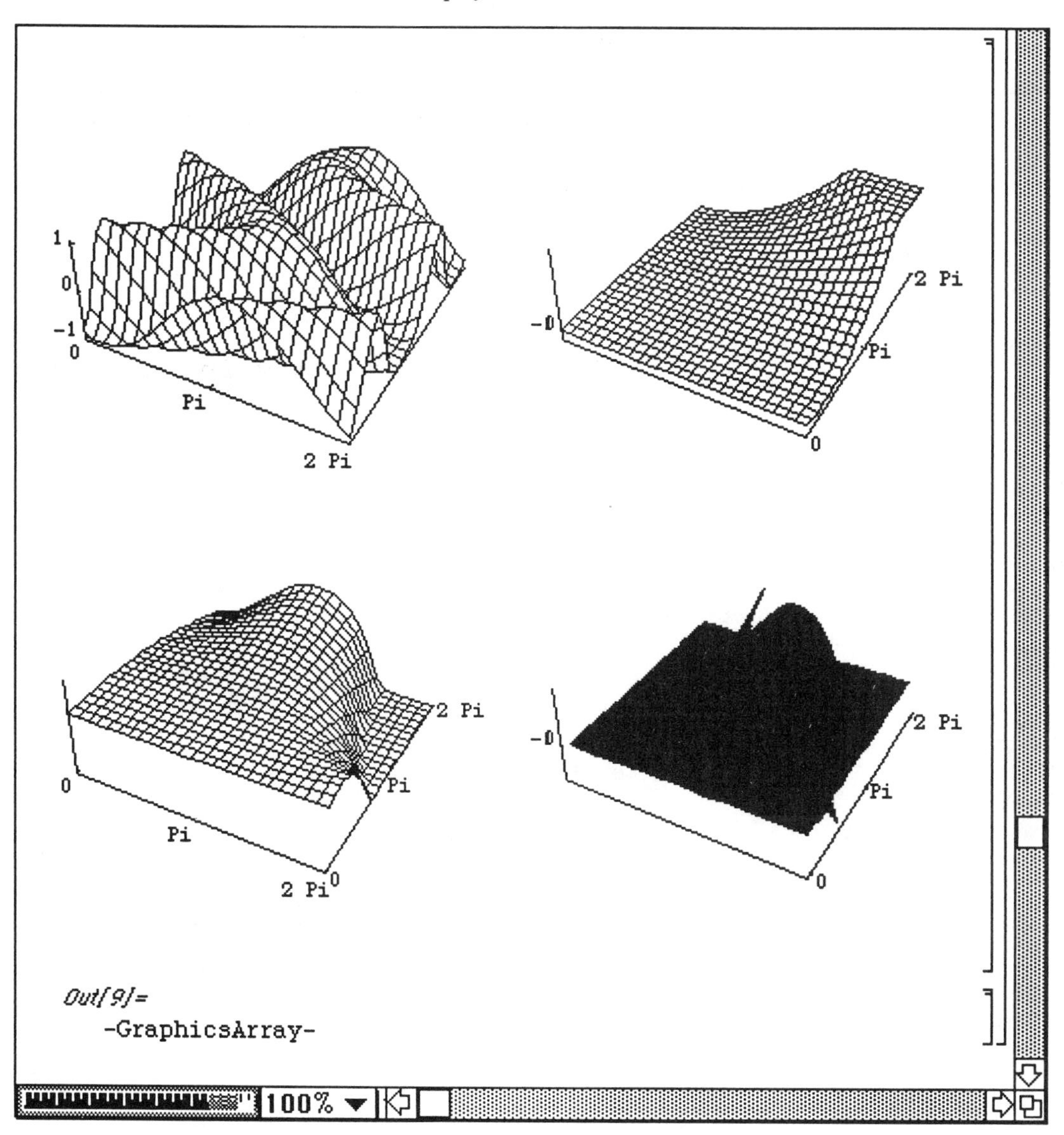

Chapter 4
Introduction to Lists and Tables

- **Chapter 4** introduces elementary operations on lists and tables. **Chapter 4** is a prerequisite for **Chapter 5** which discusses nested lists and tables in detail. The examples used to illustrate the various commands in this chapter are taken from calculus, business, and engineering applications.

- Commands introduced and discussed in this chapter from **Version 1.2** are:

Operations on Tables and Lists:

```
Table[function[index],{index,start,finish,step]
TableForm[table]
table // TableForm
MatrixForm[table]
table // MatrixForm
Prime[positiveinteger]
table[[positiveinteger]]
Short[expression]
Length[list]
Last[list]
First[list]
Apply[function,list]
Map[function,list]
Release[table]
Fit[data,{linearlyindependentfunctionsofvariable},variable]
Sum[f[j],{j,jstart,jstop,jstep}]
//
/@
```

Graphics Operations on Lists:

```
ListPlot[list,options]
ParametricPlot[{x[t],y[t]},{t,tmin,tmax},options]
```

Options:

```
PlotJoined->True
Ticks->None
AspectRatio->positivenumber
PlotRange->All
```

Other Operations:

```
Do[statement[i],{i,istart,istop,istep}]
Print[InputForm[expression]]
HermiteH[n,x]
LaguerreL[n,x]
Plus
Times
Random[type,{min,max}]
```

- Commands introduced and discussed in this chapter from **Version 2.0** include:

```
SymbolicSum[rationalfunction, {k, kmin, kmax}]
AccountingForm[number]
```

■ Applications in this chapter include:

□ Business:
Creating Business Tables:
Interest, Annuities, and Amortization

□ Calculus:
Calculating and Graphing Lists of Functions
Graphing Equations
Tangent Lines and Animations

□ Engineering:
Curve Fitting
Introduction to Fourier Series:
Calculating Fourier Series
The Heat Equation

4.1 Defining Lists

A list is a *Mathematica* object of the form
{element[1], element[2], . . . ,element[n-1], element[n]} where **element[i]** is called the ith element of the list. Elements of a list are separated by commas. Notice that lists are always enclosed in "curly" brackets **{}** and each element of a list may be (almost any) *Mathematica* object; even other lists. Since lists are *Mathematica* objects, they can be named. For easy reference, we will usually name lists.

Lists may be defined in a variety of ways. Lists may be completely typed in or they may be created by the **Table** command. For a function **f** with domain non-negative integers and a positive integer **n**, the command **Table[f[i],{i,0,n}]** creates the list
{f[0], f[1], . . . ,f[n-1], f[n]}.

Example:

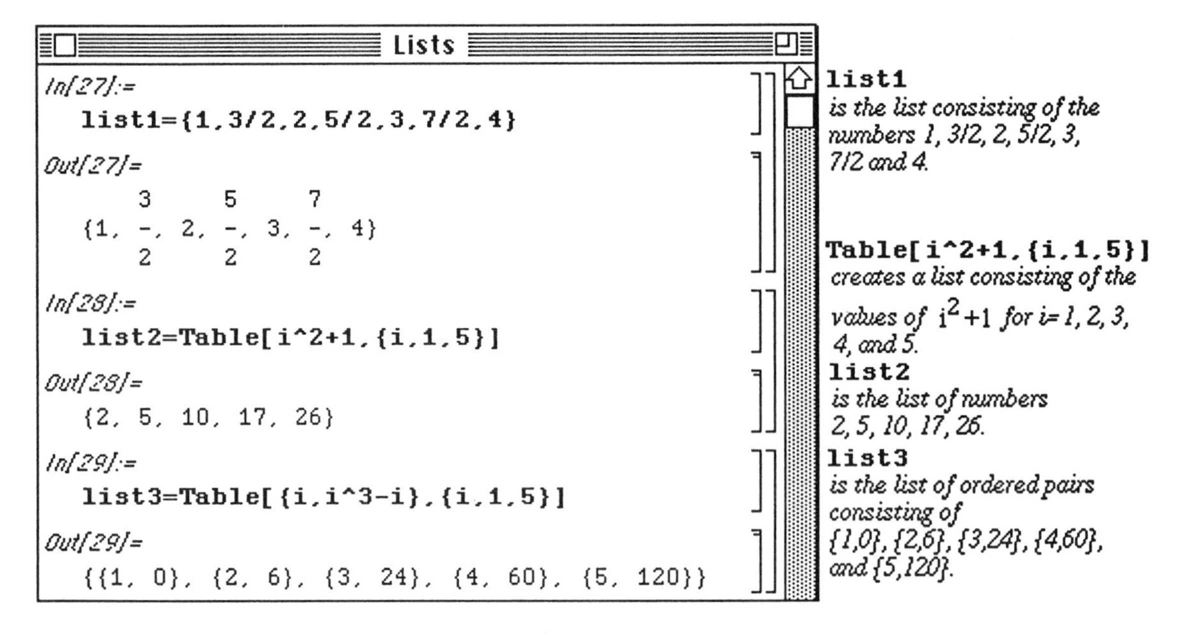

list1 *is the list consisting of the numbers 1, 3/2, 2, 5/2, 3, 7/2 and 4.*

Table[i^2+1,{i,1,5}] *creates a list consisting of the values of* i^2+1 *for i= 1, 2, 3, 4, and 5.*

list2 *is the list of numbers 2, 5, 10, 17, 26.*

list3 *is the list of ordered pairs consisting of {1,0}, {2,6}, {3,24}, {4,60}, and {5,120}.*

Mathematica will display a list, like other output, on successive lines which may sometimes be difficult to read or interpret. The commands **`TableForm`** and **`MatrixForm`** are used to display lists in traditional row/column form. In the following example, **`list4`** is the list of ordered triples **`{i,Sqrt[i] // N,Sin[i]}`** for **`i=1, 2, 3, 4, 5`**. Notice that each ordered triple is a list. This will be discussed in more detail in **Chapter 5**.

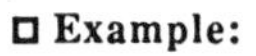

Example:

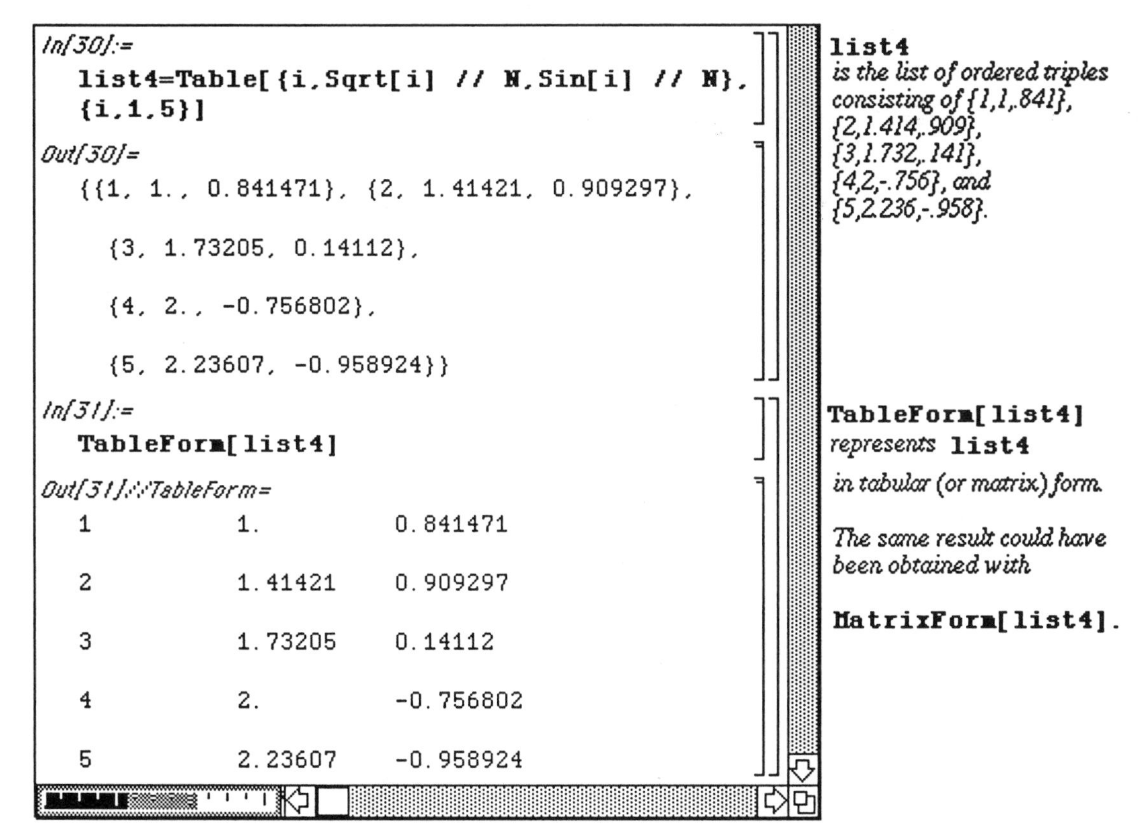

`list4` *is the list of ordered triples consisting of {1,1,.841}, {2,1.414,.909}, {3,1.732,.141}, {4,2,-.756}, and {5,2.236,-.958}.*

`TableForm[list4]` *represents* **`list4`** *in tabular (or matrix) form.*

The same result could have been obtained with

`MatrixForm[list4]`.

As indicated above, elements of lists can be numbers, ordered pairs, functions, and even other lists. For example, *Mathematica* has built-in definitions of many commonly used special functions. Consequently, lists of special functions can be quickly created.

□ **Example:**

The **Hermite polynomials**, $H_n(x)$, satisfy the differential equation $y''-2xy'+2ny=0$.

The built-in command **HermiteH[n,x]** yields the Hermite polynomial $H_n(x)$.

Create a table of the first five Hermite polynomials and name the resulting list **hermitetable**.

ListsofFunction

```
In[6]:=
  hermitetable=Table[HermiteH[n,x],{n,1,5}]
Out[6]=
                      2                3
  {2 x, -2 + 4 x , -12 x + 8 x ,
 
               2         4               3        5
    12 - 48 x  + 16 x , 120 x - 160 x  + 32 x }
```

creates a table of polynomials **HermiteH[1,x],...., HermiteH[5,x].** *The resulting list is named* **hermitetable.**

```
In[7]:=
  TableForm[hermitetable]
Out[7]//TableForm=

  2 x

            2
  -2 + 4 x

               3
  -12 x + 8 x

            2         4
  12 - 48 x  + 16 x

                  3        5
  120 x - 160 x  + 32 x
```

displays **hermitetable** *as a column.*

In fact, lists can be evaluated at certain numbers so that lists of numbers are created:

```
In[18]:=
  hermitetable /. x->1
Out[18]=
  {2, 2, -4, -20, -8}
```

calculates the value of by replacing x by 1.

```
In[19]:=
  values=Table[
      N[hermitetable /.x->j],
      {j,0,2,2/10}];
  TableForm[values]
Out[19]//TableForm=
```

computes a table of approximations of the value of **hermitetable** *by replacing x by j for j=0, 2/10, 4/10, ... , 18/10,2. The resulting list of numbers is named*

The semi-colon at the end of the command suppresses the resulting output. Instead **values** *is displayed in a tabular form.*

0.	-2.	0.	12.	0.
0.4	-1.84	-2.336	10.1056	22.7302
0.8	-1.36	-4.288	4.7296	38.0877
1.2	-0.56	-5.472	-3.2064	39.9283
1.6	0.56	-5.504	-12.1664	24.5658
2.	2.	-4.	-20.	-8.
2.4	3.76	-0.576	-23.9424	-52.8538
2.8	5.84	5.152	-20.6144	-98.9363
3.2	8.24	13.568	-6.0224	-127.816
3.6	10.96	25.056	24.4416	-112.458
4.	14.	40.	76.	-16.

Moreover, operations, such as differentiation, can be performed lists:

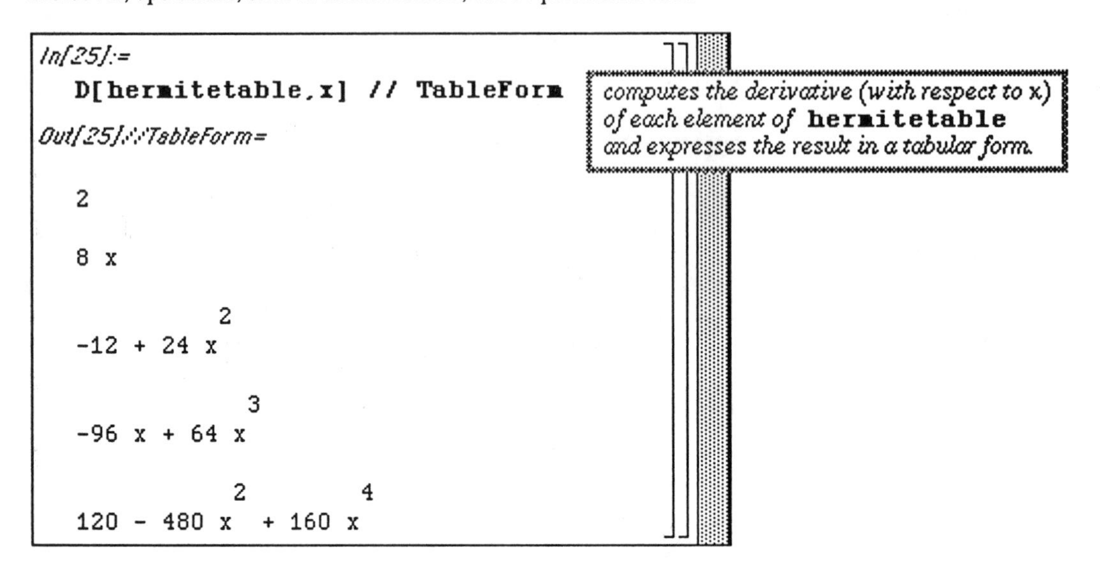

4.2 Operations on Lists

Extracting Elements of Lists

Individual elements of lists are obtained using double-square brackets. For example if **table** is a list, then **table[[2]]** is the second element of the list **table**. The **j**th element of **table** is **table[[j]]**.

Example:

The *Mathematica* function **Prime** can be used to calculate prime numbers. **Prime[1]** yields 2; **Prime[2]** yields 3; and, in general, **Prime[k]** yields the **k**th prime number.

Make a table of the first fifteen prime numbers. What is the third prime number? the thirteenth prime number?

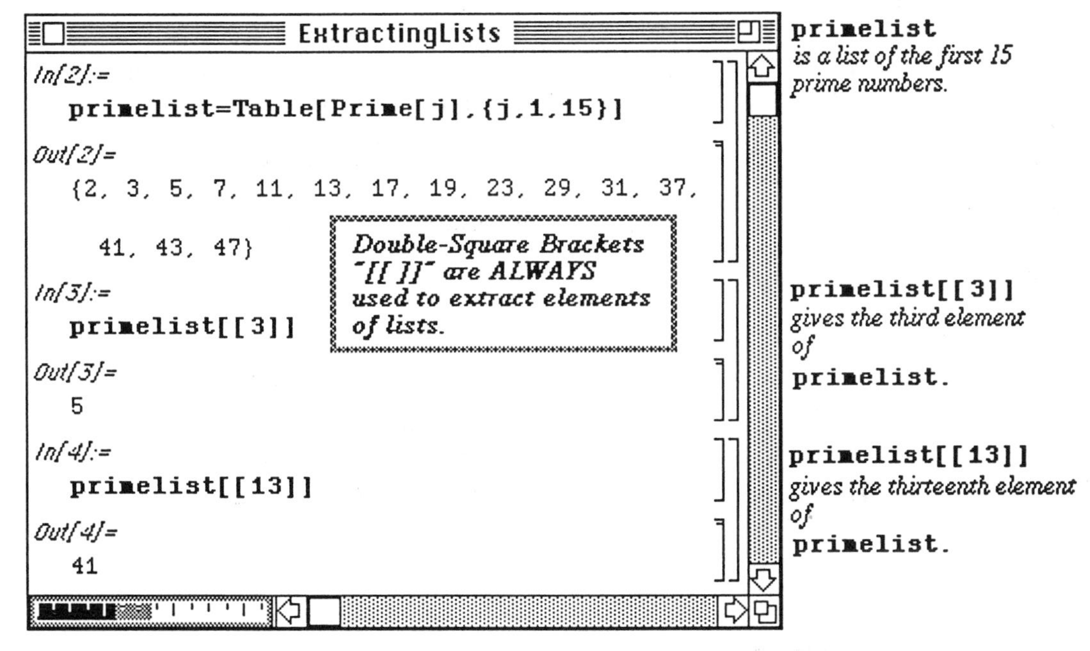

primelist *is a list of the first 15 prime numbers.*

primelist[[3]] *gives the third element of* **primelist.**

primelist[[13]] *gives the thirteenth element of* **primelist.**

Graphing Lists and Lists of Functions

If **list={a[1], a[2], ... , a[n]}** is a list of numbers, **ListPlot[list]** plots the points {1,a[1]}, {2,a[2]}, {3,a[3]}, ... , {n,a[n]}. In general, the command **ListPlot** has the same options as the command **Plot**.

Sometimes it is desirable to suppress the output of lists; particularly when long lists are used. In general, a semi-colon "**;**" placed at the end of a command suppresses the resulting output.

□ **Example:**

The following example demonstrates how **ListPlot[list]** is used to plot the points **{1,f[1]}, {2,f[2]},{3,f[3]},...,{1000,f[1000]}** where **f[x] = Sin[x]//N**.

Let f(x)=Sin (x). First, make a table of the values Sin (1), Sin (2), ... , Sin (10); then make a table of the values Sin(i) for i=1, 2, ... , 1000. Graph the points (i,Sin(i)) for i=1, ... , 1000.

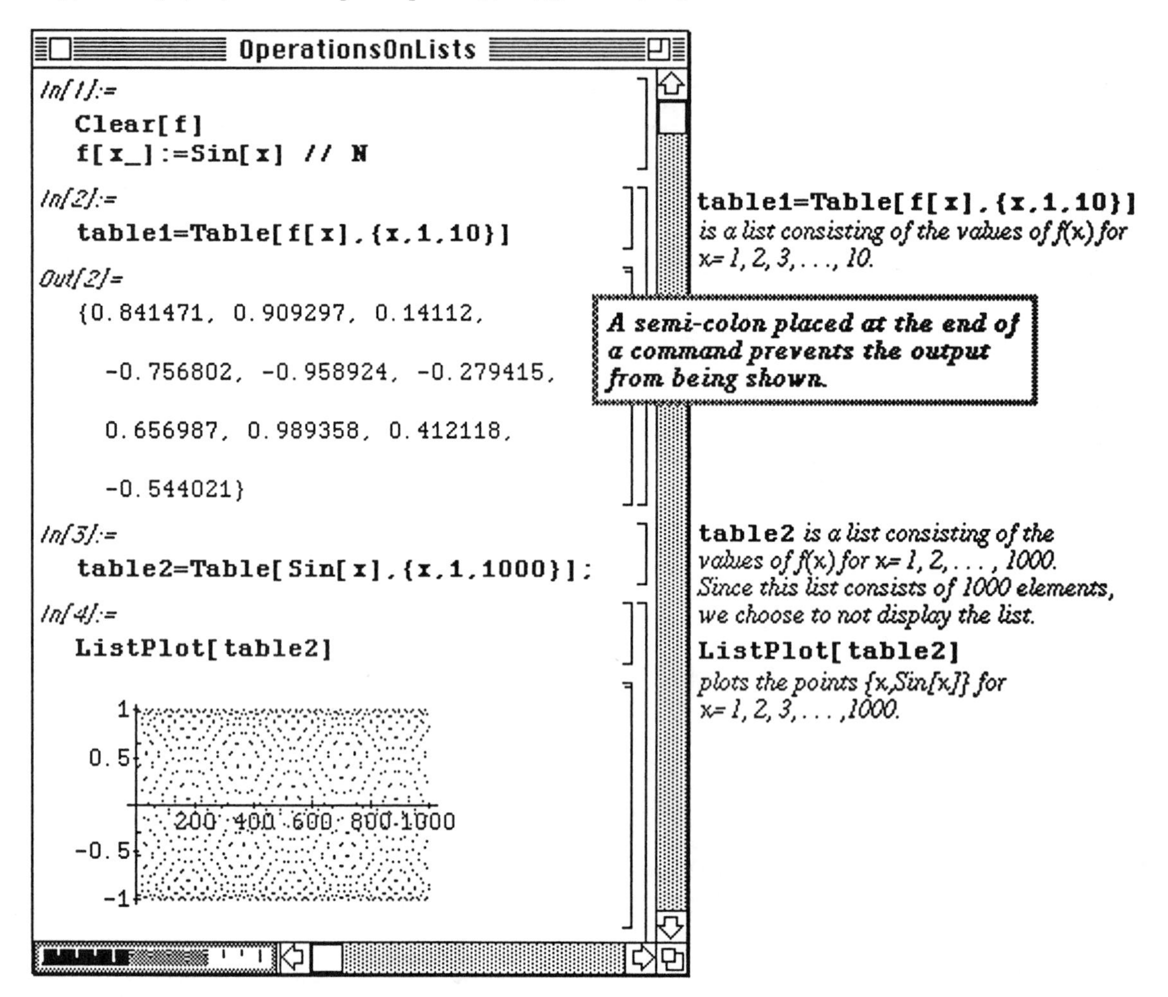

Both tables of numbers and tables of functions can be graphed. In the following example, we graph the elements of **hermitetable** (created above) on the interval [-3,3].

□ **Example:**

Use *Mathematica* to graph the first five Hermite polynomials on the interval [–3,3].

Notice that within the **Plot** command, **hermitetable** is enclosed by the command **Release**.

Release[hermitetable] allows the elements of **hermitetable** to be evaluated for the values of x on [-3,3] instead of recreating the table for each value of x.

o Version 1.2 (or earlier) users must always enclose the table to be graphed by **Release**. In Version 2.0, the command **Release** has been replaced by the command **Evaluate**. Hence, when graphing tables of functions with Version 2.0, be sure to enclose the table by **Evaluate**.

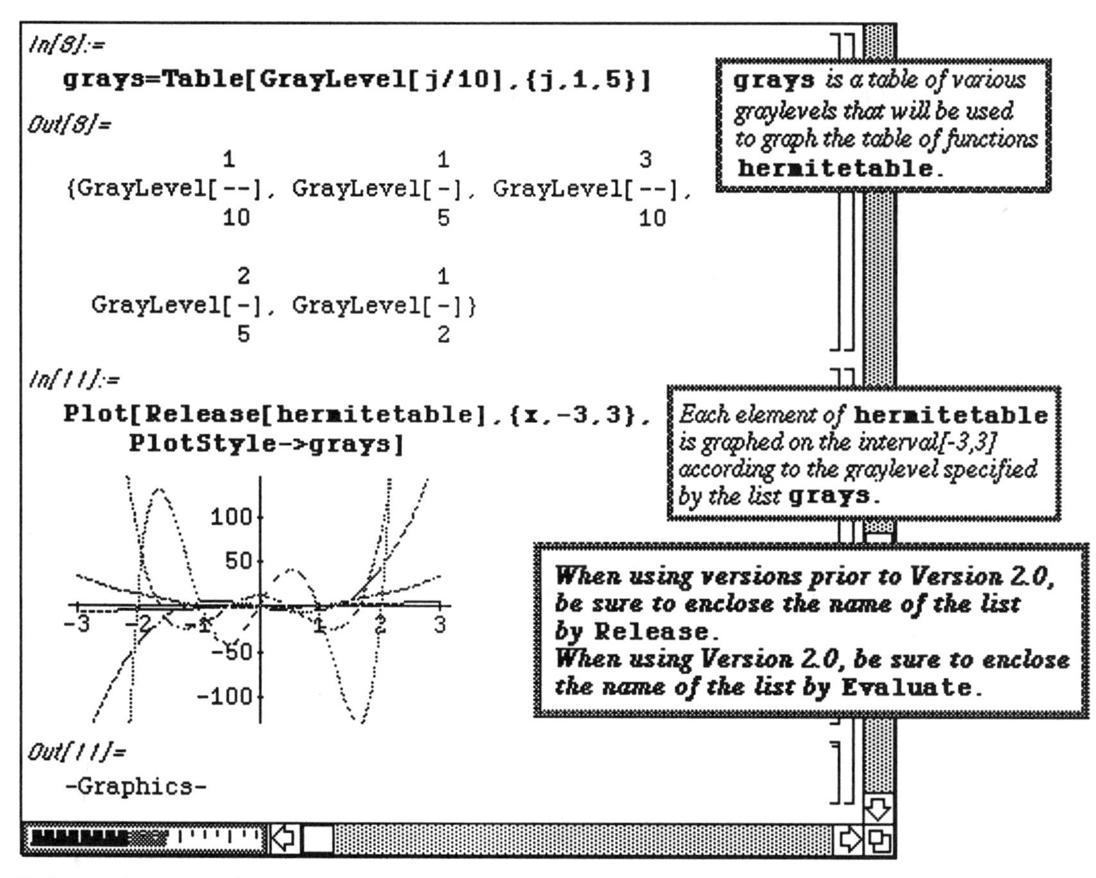

In the previous examples, the domain of the functions has been the set of natural numbers. This does not have to be the case, however. The command

Table[f[x] , {x,xmin, xmax, xstep}]

creates a list by evaluating **f** at values of **x** from **x** = **xmin** to **x** = **xmax** using a stepsize of **xstep**. If **n** is a positive integer, the command **Table[f[x],{x,xmin,xmax,(xmax-xmin)/n}]** creates the list, containing the n+1 elements:

$$\left\{\texttt{f[xmin]},\ \texttt{f}\left[\texttt{xmin}+\frac{\texttt{xmax}-\texttt{xmin}}{\texttt{n}}\right],\ \texttt{f}\left[\texttt{xmin}+2\frac{\texttt{xmax}-\texttt{xmin}}{\texttt{n}}\right],\ \ldots\ ,\ \texttt{f[xmax]}\right\}.$$

When dealing with a long *Mathematica* object expression, another useful *Mathematica* command is **Short[expression]**. This command produces an abbreviated, one-line output of **expression.** If **list** is a table, the command **Short[list]** produces a one-line output of **list**. If **n** is a positive integer greater than one, **Short[list,n]** produces an abbreviated **n**-line output of **list**. This abbreviated list includes an element of the form << n >> which indicates the number of elements of **list** that are omitted in the abbreviated output.

■ Evaluation of Lists by Functions

Another helpful command is **Map[f,list]**.
This command creates a list consisting of elements obtained by evaluating **f** for each element of **list**, provided that each member of **list** is an element of the domain of **f**.

■ *Notice:*

To avoid errors, be sure to check that each element of **list** is in the domain of **f** prior to executing the command **Map[f,list]**.

If **list** is a table of n numbers and each element of **list** is in the domain of **f**, recall that **list[[i]]** denotes the ith element of **list**. The command **Map[f,list]** produces the same list as the command **Table[f[list[[i]]],{i,1,n}]**. Using **Map** in conjunction with **ListPlot** yields an alternative approach to graphing functions.

◻ Example:

Let $g(x) = \text{Cos}(x) - 2\text{Sin}(x)$. First create a table of the values from π to 5π in steps of $\frac{4\pi}{40}$. Name the list **table 3**. This is the same as creating the table of values

$$\left\{\pi, \frac{44\pi}{40}, \ldots, \pi + n\frac{4\pi}{40}, \ldots, 5\pi\right\} = \left\{\pi + j\frac{4\pi}{40} : j = 0, 1, 2, \ldots, 40\right\}.$$

Evaluate g for each element in `table3`.

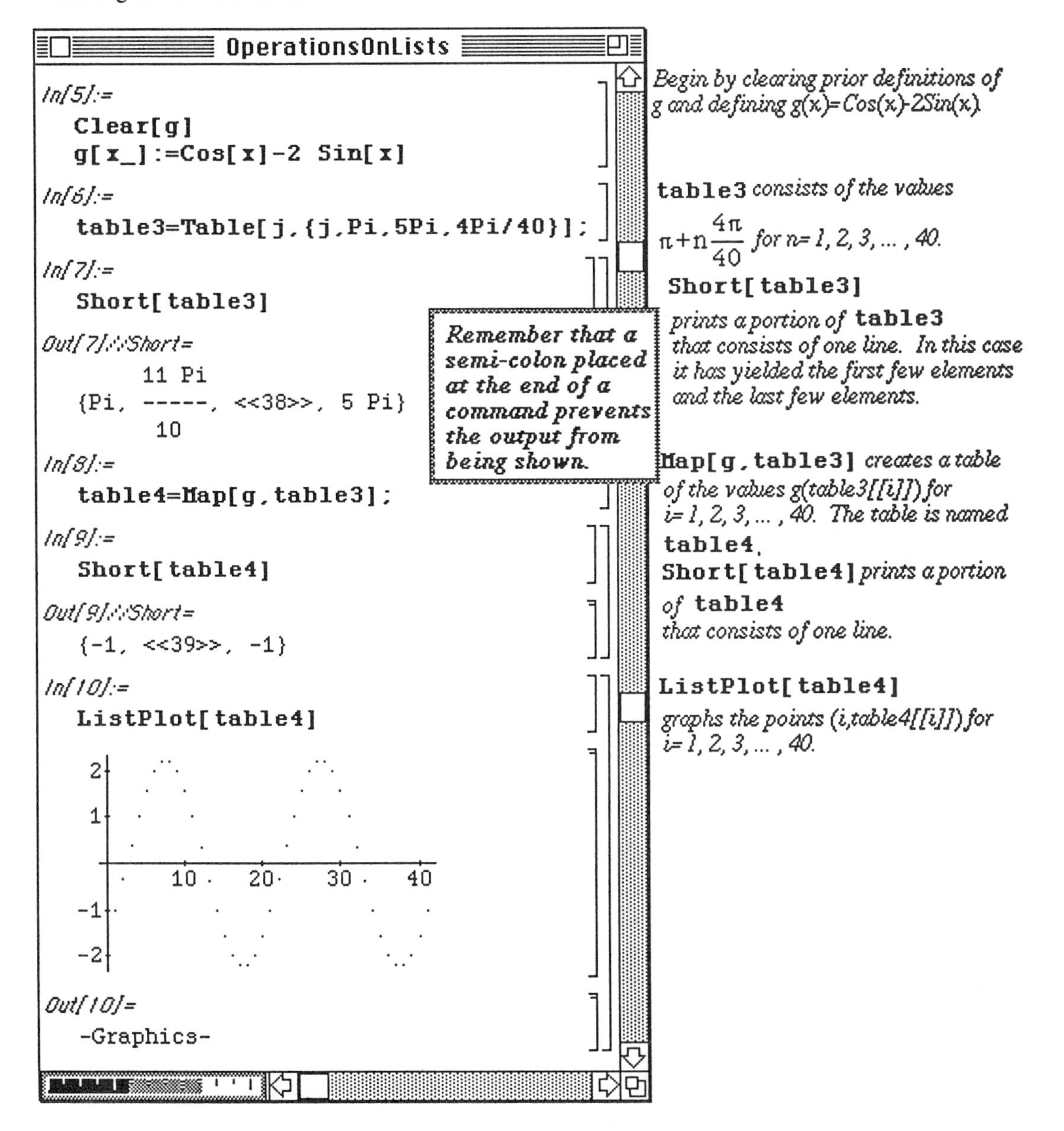

In general, **ListPlot** and **Plot** share many of the same options. However, since the **ListPlot** command graphs a set of points, and is NOT connected, a connected graph is obtained by using the **PlotJoined** option:

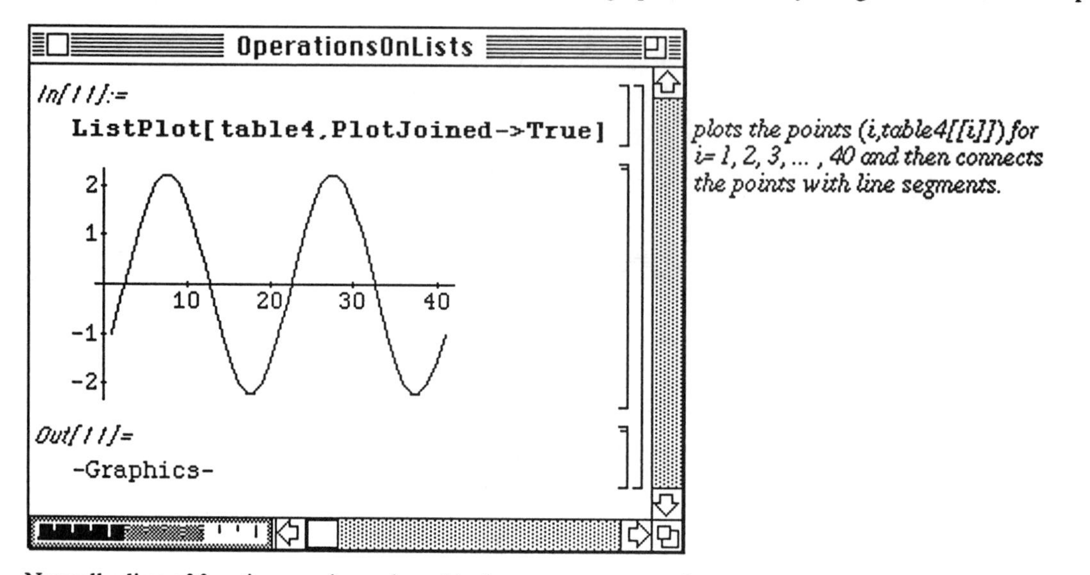

plots the points (i,table4[[i]]) for i= 1, 2, 3, ... , 40 and then connects the points with line segments.

Naturally, lists of functions can be evaluated in the same manner as above:

□ **Example:**

The **Laguerre polynomials** are defined recursively by the relationship $L_0(x) = 1$ and

$L_n(x) = \frac{e^x}{n!}\frac{d^n\left(x^n e^{-x}\right)}{dx^n}$. For each n, $L_n(x)$ satisfies the differential equation $x y'' + (1-x)y' + n y = 0$.

$L_n(x)$ is computed with the built-in function **LaguerreL [n,x]**.

For $n = 1, 2, 3, 4$, use *Mathematica* to verify that $L_n(x)$ satisfies $x y'' + (1-x)y' + n y = 0$.

First a table of the first four Laguerre polynomials is created and then a function **f** is defined as follows:

For a given ordered pair **{n,poly}**, **f[{n,poly}]** returns an ordered quadruple given by

```
{xD[poly,{x,2}],
 (1-x)D[poly,x],
  n*poly,D[poly,
  {x,2}]+(1-x)D[poly,x]n  poly
}.
```

Remember that **D[poly,{x,n}]** computes the **n**th derivative of **poly** with respect to **x**.

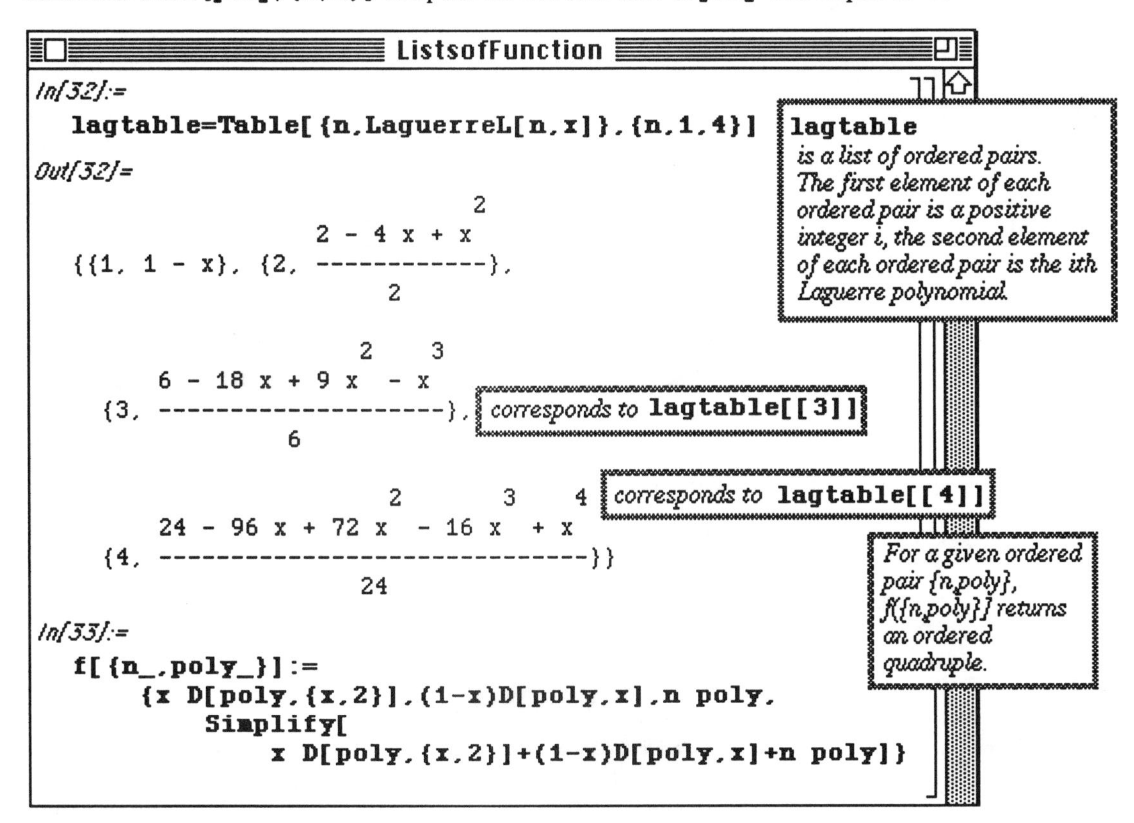

Map[f,lagtable] evaluates **f** for each element of **lagtable**. The same result would have been obtained with the command **Table[f[lagtable[[i]],{i,1,4}]**.

Note that the fourth component in each ordered quadruple below is zero. Hence, each member of **lagtable** is a solution.

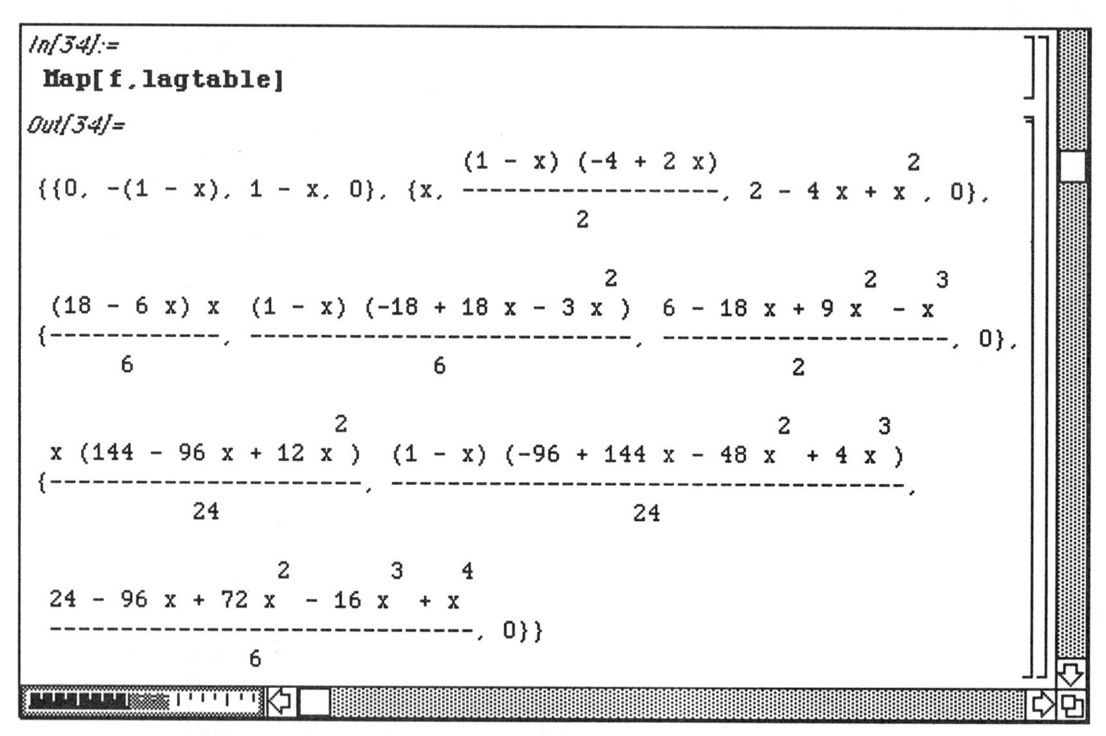

```
In[34]:=
Map[f,lagtable]
Out[34]=
{{0, -(1 - x), 1 - x, 0}, {x, ((1 - x) (-4 + 2 x))/2, 2 - 4 x + x^2, 0},
{((18 - 6 x) x)/6, ((1 - x) (-18 + 18 x - 3 x^2))/6, (6 - 18 x + 9 x^2 - x^3)/2, 0},
{(x (144 - 96 x + 12 x^2))/24, ((1 - x) (-96 + 144 x - 48 x^2 + 4 x^3))/24,
(24 - 96 x + 72 x^2 - 16 x^3 + x^4)/6, 0}}
```

■ Other List Operations

A specific operation can be applied to the elements of a list through the command **Apply[operation, list]**. Of course, in order to use this command, the given **operation** must be defined for the elements of **list**.

For example, if **numbers** is a list of real numbers, then the command **Apply[Plus,numbers]** adds together all the elements of **numbers**.

Some other *Mathematica* commands used with lists are:
Length[list], which gives the number of elements in **list**;
First[list], which gives the first element of **list**; and
Last[list], which gives the last member of **list**.

Several examples of these commands are shown below.

Also notice that the definition of a vector-valued function $f:\Re \to \Re^n$

which maps the real numbers to n-space can be made using a list. This is done below in the following manner:

f[x_]$:=\{f_1(x), f_2(x), f_3(x), \cdots, f_n(x)\}$ where $f_k(x):\Re \to \Re$ for $1 \le k \le n$.

□ Example:

Define $f(x)=\left\{\left(\frac{x}{100}\right)^2, 36-\left(\frac{x}{100}\right)^2\right\}$. Create a table of the numbers 10^2, 20^2, ..., 50^2, and 60^2. Evaluate f for each number in the table Name the resulting table **list 2**.

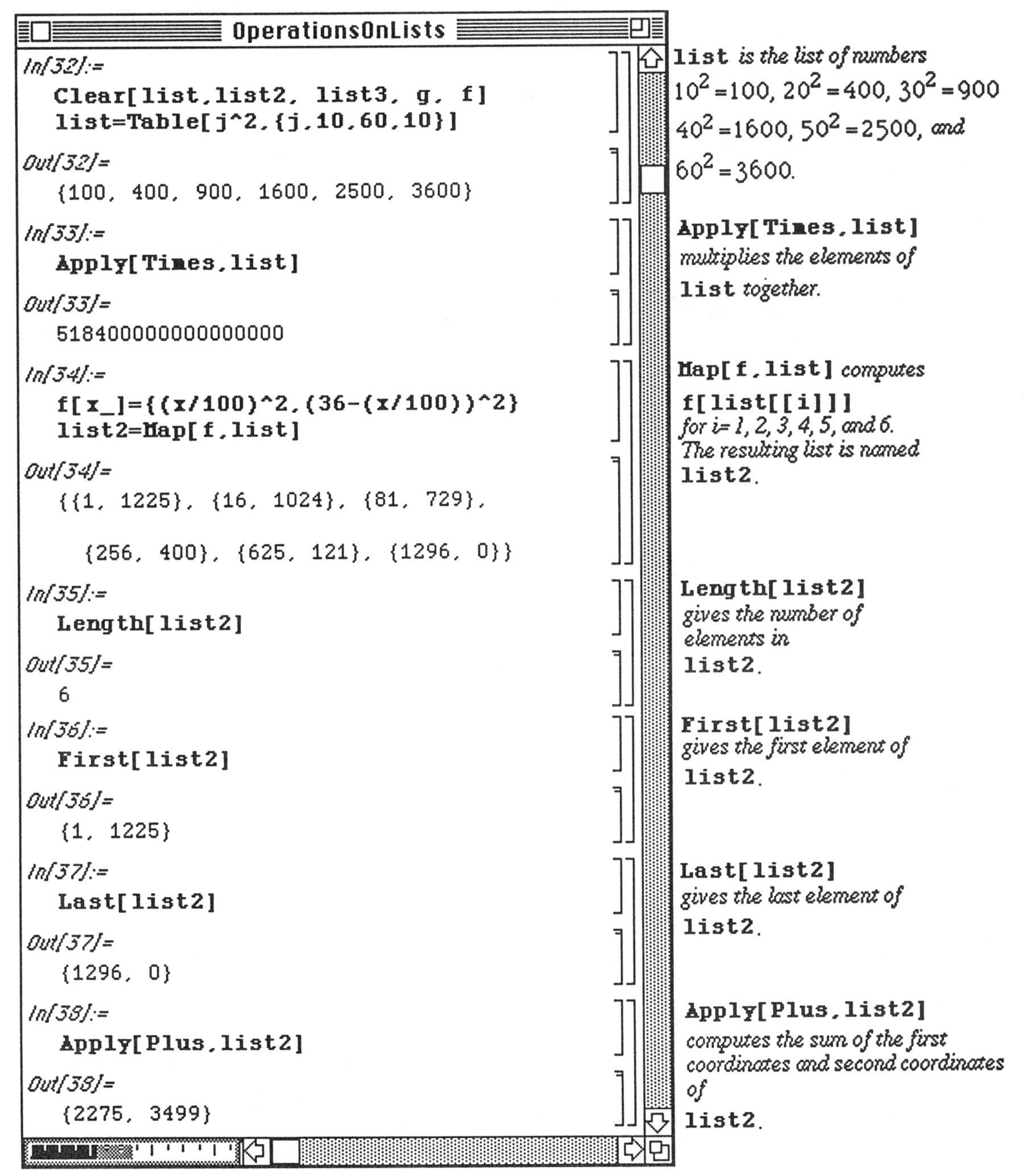

list *is the list of numbers* $10^2=100$, $20^2=400$, $30^2=900$ $40^2=1600$, $50^2=2500$, *and* $60^2=3600$.

Apply[Times,list] *multiplies the elements of* **list** *together.*

Map[f,list] *computes* **f[list[[i]]]** *for i= 1, 2, 3, 4, 5, and 6. The resulting list is named* **list2**.

Length[list2] *gives the number of elements in* **list2**.

First[list2] *gives the first element of* **list2**.

Last[list2] *gives the last element of* **list2**.

Apply[Plus,list2] *computes the sum of the first coordinates and second coordinates of* **list2**.

In the following example, **Map** is used with the function $g:\Re^2 \to \Re$ defined via $g(x,y)=\sqrt{x^2+y^2}$.

□ Example:

Evaluate g for each element of **list2**; call the resulting table **list3**. Add up the elements of **list3**; multiply together the elements of **list3**.

As has been the case with other examples, a numerical approximation of each member of the list is obtained using **//N**. Otherwise, exact values are given.

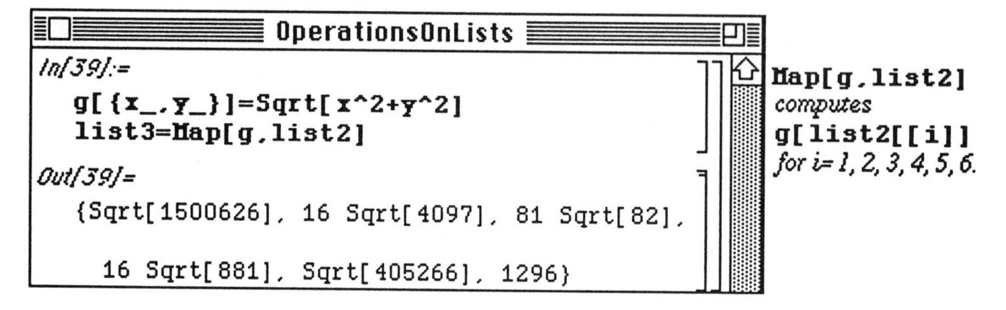

We use the **Apply** command to compute the desired sum and product. *Mathematica* gives exact results unless otherwise requested:

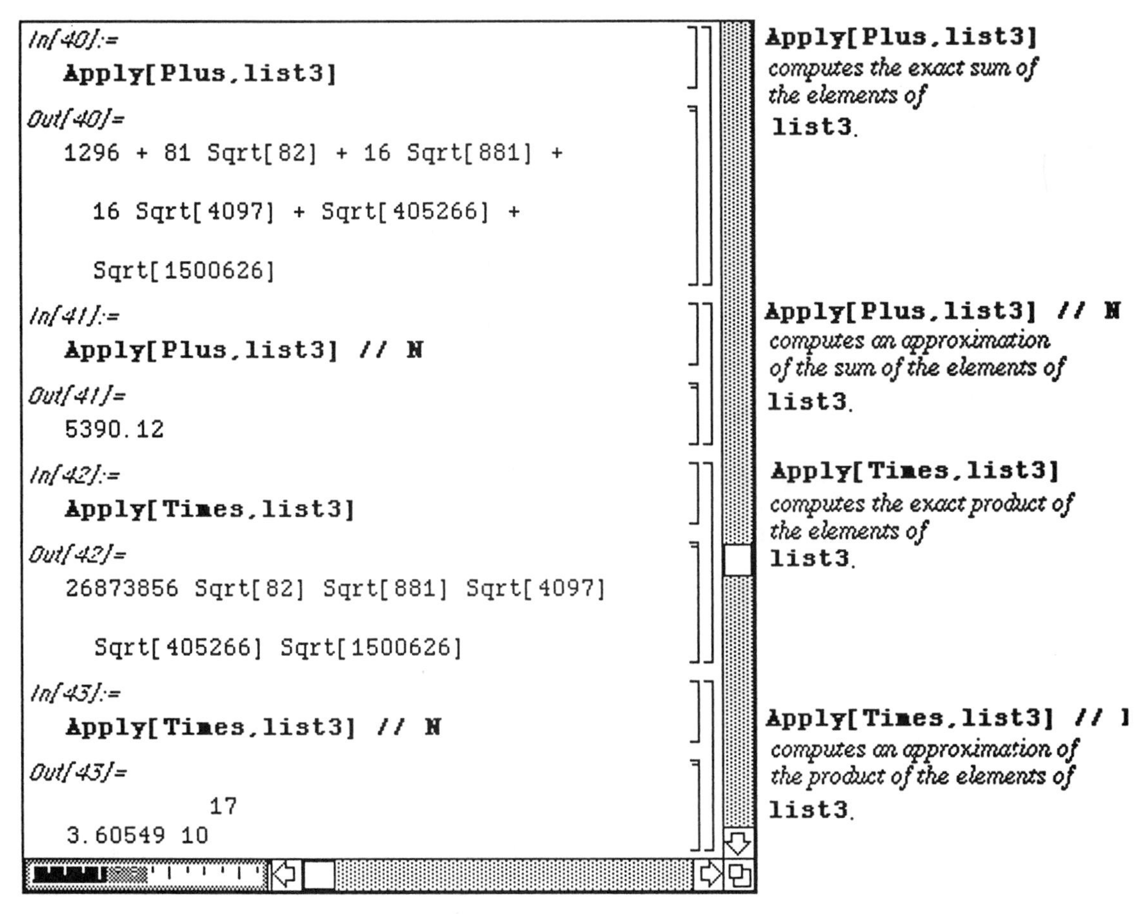

■ Alternative Way to Evaluate Lists by Functions

□ Example:

A table consisting of ten random real numbers on the interval **{0,5}** is found with **Table[Random[Real,{0,5}],{10}]** and is called **t1**. A function **g[x]=Mod[x,1]** is then defined; hence, **g** is merely x Modulo 1. In the same manner as above, the command **Map[g,t1]** evaluates **g** at each element in **t1**. However, the same result is obtained with **t1//g** and **g/@t1** as illustrated below:

■ Note that since the command **Random** is used, if you enter the following sequence of calculations, **t1** will differ each time.

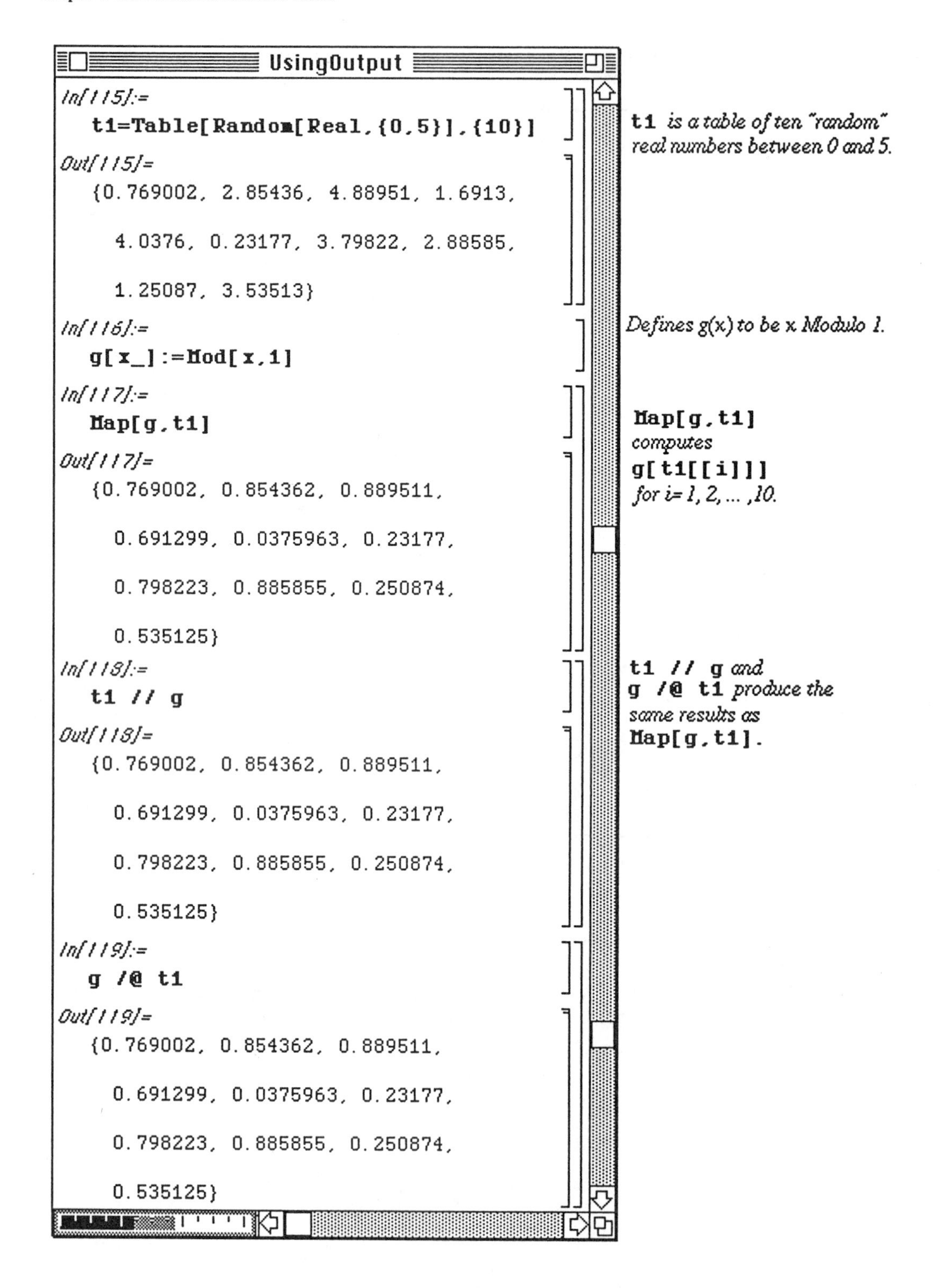

UsingOutput

```
In[115]:=
  t1=Table[Random[Real,{0,5}],{10}]
Out[115]=
  {0.769002, 2.85436, 4.88951, 1.6913,
    4.0376, 0.23177, 3.79822, 2.88585,
    1.25087, 3.53513}
In[116]:=
  g[x_]:=Mod[x,1]
In[117]:=
  Map[g,t1]
Out[117]=
  {0.769002, 0.854362, 0.889511,
    0.691299, 0.0375963, 0.23177,
    0.798223, 0.885855, 0.250874,
    0.535125}
In[118]:=
  t1 // g
Out[118]=
  {0.769002, 0.854362, 0.889511,
    0.691299, 0.0375963, 0.23177,
    0.798223, 0.885855, 0.250874,
    0.535125}
In[119]:=
  g /@ t1
Out[119]=
  {0.769002, 0.854362, 0.889511,
    0.691299, 0.0375963, 0.23177,
    0.798223, 0.885855, 0.250874,
    0.535125}
```

t1 *is a table of ten "random" real numbers between 0 and 5.*

Defines g(x) to be x Modulo 1.

Map[g,t1] *computes* **g[t1[[i]]]** *for i= 1, 2, ... ,10.*

t1 // g *and* **g /@ t1** *produce the same results as* **Map[g,t1]**.

◻ Example:

The sum of the squares of the first 100 positive integers is computed by several different methods below. **First,** table `t2` of the squares of the first 100 positive integers is created. The commands `Sum[i^2,{i,1,100}]`, `Apply[Plus,t2]`, and `Plus@@t2` all achieve the correct sum of 338350.

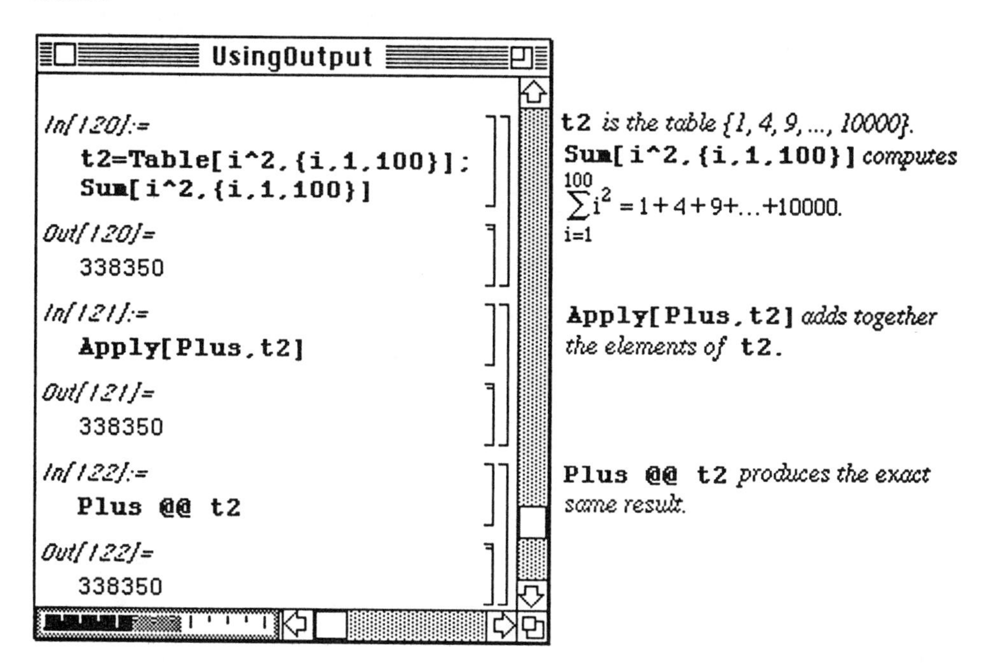

4.3 Applications

■ Application: Interest, Annuities, and Amortization

The use of lists and tables are quite useful in economic applications which deal with interest rates, annuities, and amortization. *Mathematica* is, therefore, of great use in these types of problems through its ability to show the results of problems in tabular form. Also, if a change is made in the problem, *Mathematica* can easily recompute the results.

A common problem in economics is the determination of the amount of interest earned from an investment. Consider the following: If P dollars are invested for t years at an annual interest rate of r% compounded m times per year, the compound amount A(t) at time t is given by: $A(t) = P\left(1+\frac{r}{m}\right)^{mt}$.

A specific example is shown below where the amount of money accrued at time t represents the sum of the original investment and the amount of interest earned on that investment at time t.

□ **Example:**

Suppose $12,500 is invested at an annual rate of 7% compounded daily. How much money has accumulated at the end of each five year period for t = 5, 10, 15, 20, 25, 30?

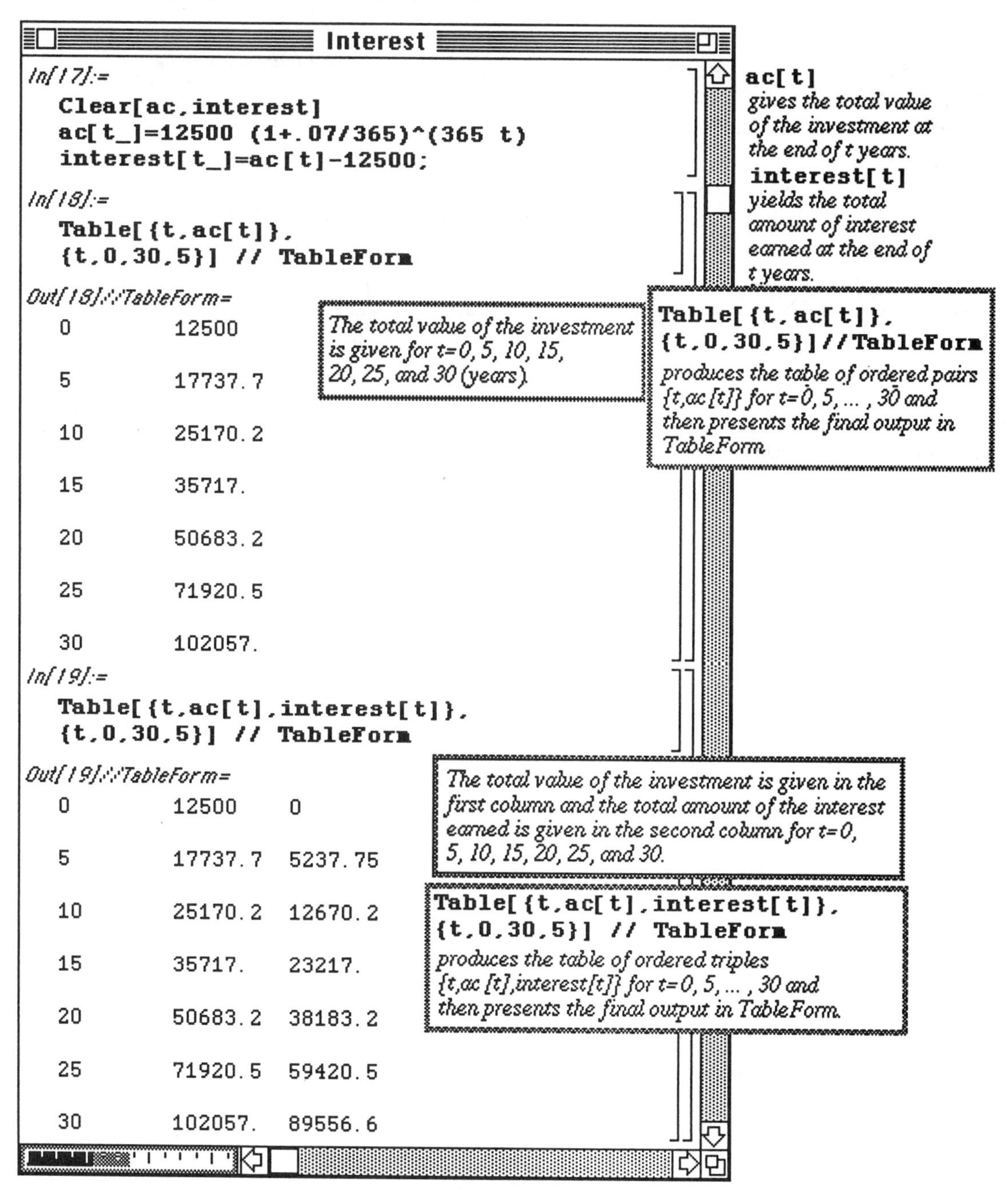

Interest

In[17]:=

```
Clear[ac,interest]
ac[t_]=12500 (1+.07/365)^(365 t)
interest[t_]=ac[t]-12500;
```

In[18]:=

```
Table[{t,ac[t]},
{t,0,30,5}] // TableForm
```

Out[18]//TableForm=

```
0     12500
5     17737.7
10    25170.2
15    35717.
20    50683.2
25    71920.5
30    102057.
```

In[19]:=

```
Table[{t,ac[t],interest[t]},
{t,0,30,5}] // TableForm
```

Out[19]//TableForm=

```
0     12500     0
5     17737.7   5237.75
10    25170.2   12670.2
15    35717.    23217.
20    50683.2   38183.2
25    71920.5   59420.5
30    102057.   89556.6
```

ac[t] *gives the total value of the investment at the end of t years.*

interest[t] *yields the total amount of interest earned at the end of t years.*

The total value of the investment is given for t=0, 5, 10, 15, 20, 25, and 30 (years).

Table[{t,ac[t]},{t,0,30,5}]//TableForm *produces the table of ordered pairs {t,ac[t]} for t=0, 5, ... , 30 and then presents the final output in TableForm*

The total value of the investment is given in the first column and the total amount of the interest earned is given in the second column for t=0, 5, 10, 15, 20, 25, and 30.

Table[{t,ac[t],interest[t]},{t,0,30,5}] // TableForm *produces the table of ordered triples {t,ac [t],interest[t]} for t=0, 5, ... , 30 and then presents the final output in TableForm.*

The problem can be redefined for arbitrary values of **t**, **P**, **r**, and **n** as follows :

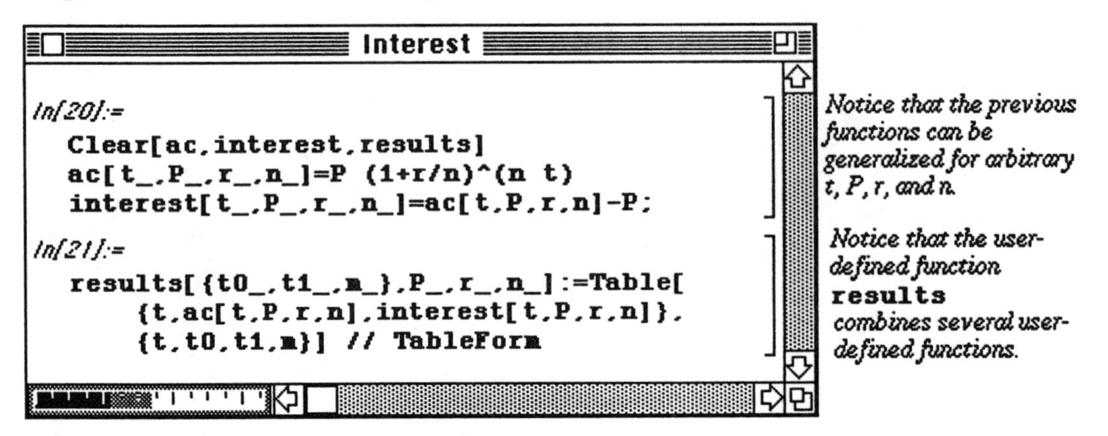

Notice that the previous functions can be generalized for arbitrary t, P, r, and n.

Notice that the user-defined function **results** *combines several user-defined functions.*

Hence, any problem of this type can be worked using the functions defined above.

□ **Example:**

For example, suppose \$10,000 is invested at an interest rate of 12% compounded daily. Create a table consisting of the total value of the investment and the interest earned at the end of 0, 5, 10, 15, 20, and 25 years.

In this case, we use the function **results** defined above. Here, **t0=0**, **t1=25**, **m=5**, **P=10000**, **r=.12**, and **n=365**:

Notice that if the conditions are changed to **t0=0, t1=30, m=10, P=15000, r=.15**, and **n=365**, the desired table can be quickly calculated:

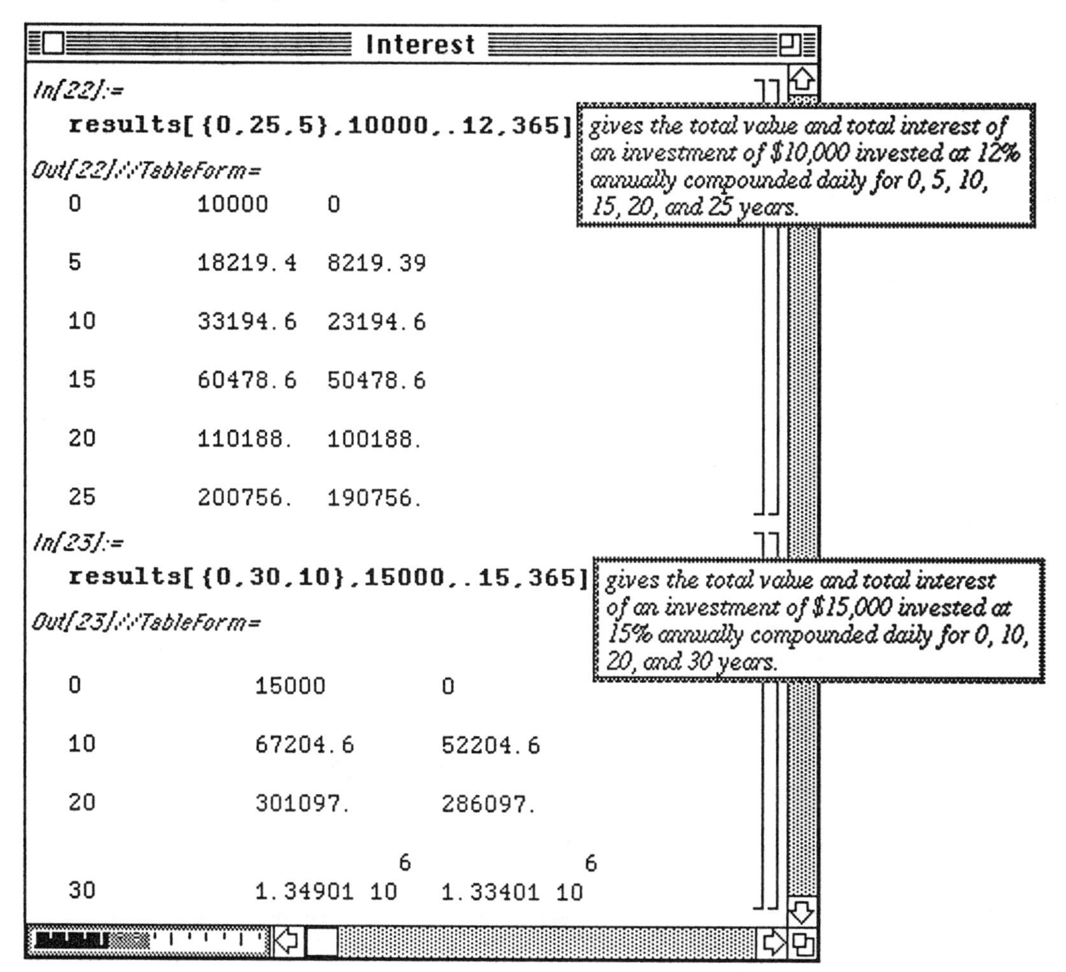

Interest

In[22]:=

```
results[{0,25,5},10000,.12,365]
```

gives the total value and total interest of an investment of $10,000 invested at 12% annually compounded daily for 0, 5, 10, 15, 20, and 25 years.

Out[22]//TableForm=

0	10000	0
5	18219.4	8219.39
10	33194.6	23194.6
15	60478.6	50478.6
20	110188.	100188.
25	200756.	190756.

In[23]:=

```
results[{0,30,10},15000,.15,365]
```

gives the total value and total interest of an investment of $15,000 invested at 15% annually compounded daily for 0, 10, 20, and 30 years.

Out[23]//TableForm=

0	15000	0
10	67204.6	52204.6
20	301097.	286097.
30	$1.34901\ 10^{6}$	$1.33401\ 10^{6}$

The problem of calculating the interest earned on an investment is altered if the interest is <u>compounded continuously</u>. The formula used in this case is as follows :

If P dollars are invested for t years at an annual interest rate of r% compounded continuously, the **compound amount** A(t) at time t is given by: $A(t) = Pe^{rt}$.

□ **Example: (Future Value)**

Consider the following :

If R dollars are deposited at the **end** of each period for n periods in an annuity hat earns interest at a rate of j per period, the **future value** of the annuity is given by: $S_{future} = R\left[\frac{(1+j)^n - 1}{j}\right]$.

A function which calculates the future value of the annuity and several examples using this function are given below :

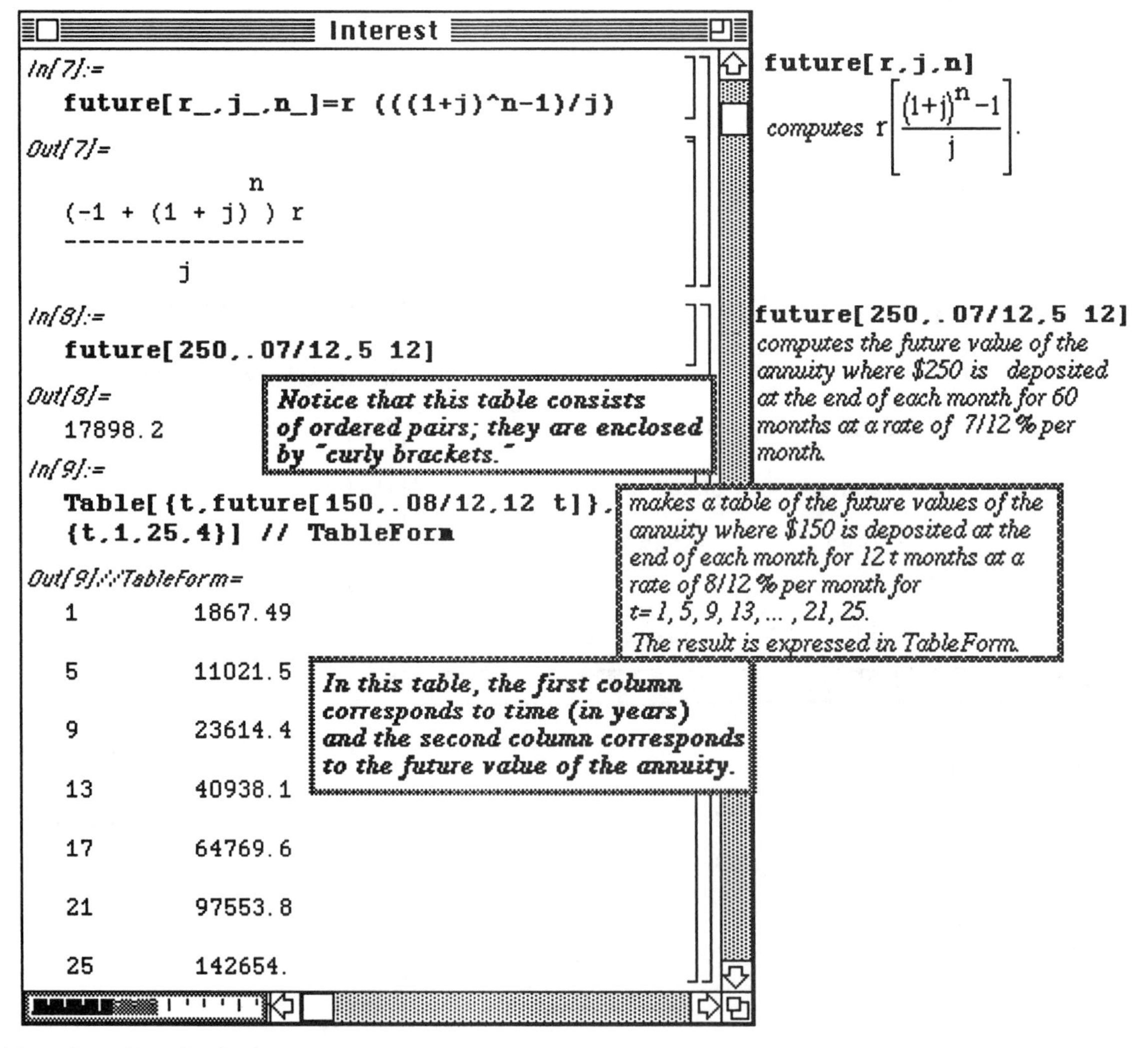

□ **Example: (Annuity Due)**

Another type of annuity is as follows :

If R dollars are deposited at the **beginning** of each period for n periods in an annuity with interest that earns interest at a rate of j per period, the **annuity due** is given by: $S_{due} = R\left[\frac{(1+j)^{n+1}-1}{j}\right] - R$.

Again, the function to determine the amount due is defined below with accompanying examples.

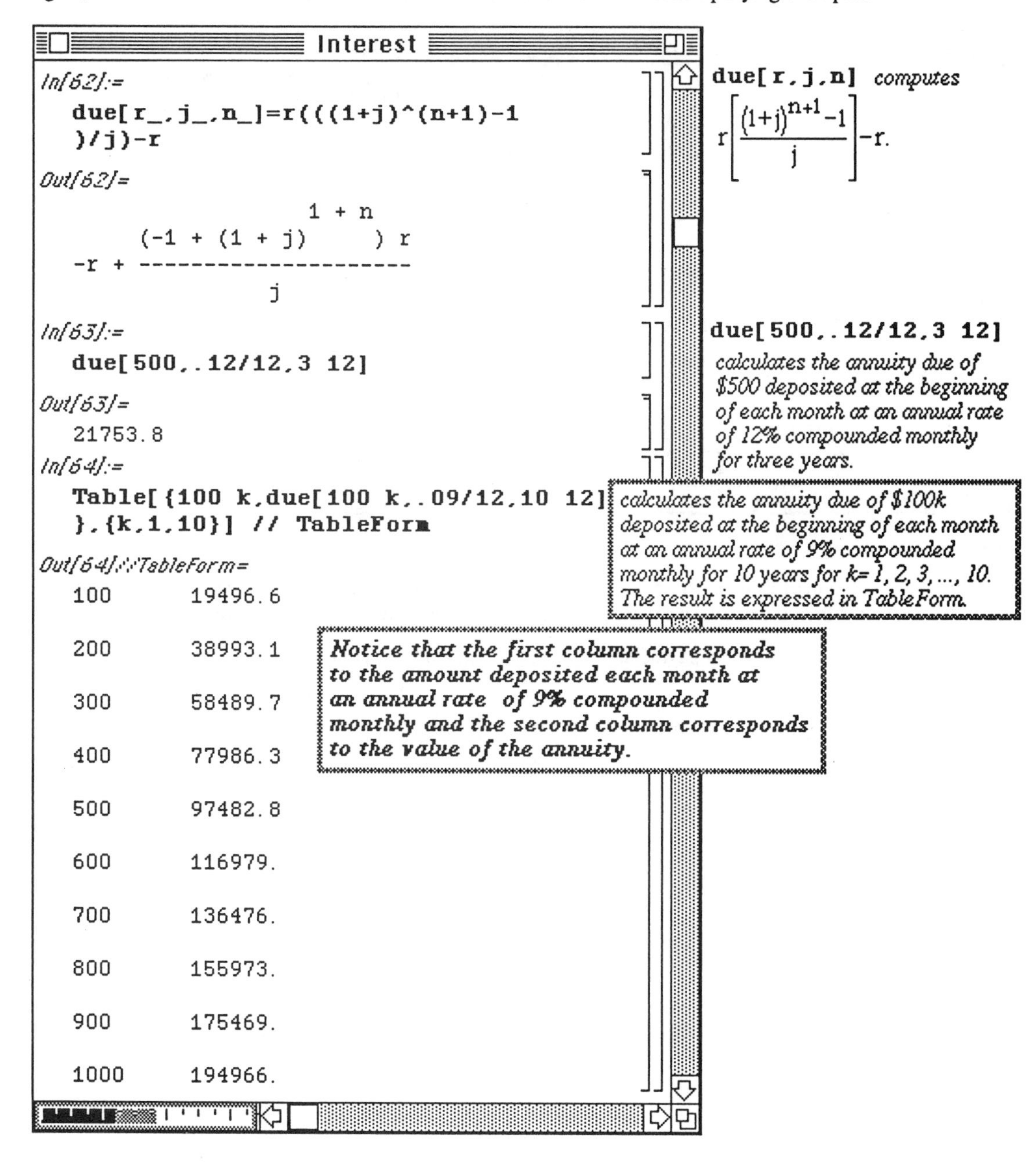

Interest

```
In[62]:=
  due[r_,j_,n_]=r(((1+j)^(n+1)-1
  )/j)-r
Out[62]=
                        1 + n
        (-1 + (1 + j)      ) r
  -r + ---------------------
                  j
In[63]:=
  due[500,.12/12,3 12]
Out[63]=
  21753.8
In[64]:=
  Table[{100 k,due[100 k,.09/12,10 12]
  },{k,1,10}] // TableForm
Out[64]//TableForm=
```

100	19496.6
200	38993.1
300	58489.7
400	77986.3
500	97482.8
600	116979.
700	136476.
800	155973.
900	175469.
1000	194966.

due[r,j,n] *computes* $r\left[\frac{(1+j)^{n+1}-1}{j}\right]-r.$

due[500,.12/12,3 12] *calculates the annuity due of $500 deposited at the beginning of each month at an annual rate of 12% compounded monthly for three years.*

calculates the annuity due of $100k deposited at the beginning of each month at an annual rate of 9% compounded monthly for 10 years for k= 1, 2, 3, ..., 10. The result is expressed in TableForm.

Notice that the first column corresponds to the amount deposited each month at an annual rate of 9% compounded monthly and the second column corresponds to the value of the annuity.

□ Example:

The following table compares the annuity due on a $100 k monthly investment at an annual rate of 8% compounded monthly for t= 5, 10,15, 20; and k = 1, 2, 3, 4, 5. This type of table can prove to quite useful in the analysis of investments. (Note that the values of k and t were later added to the table.)

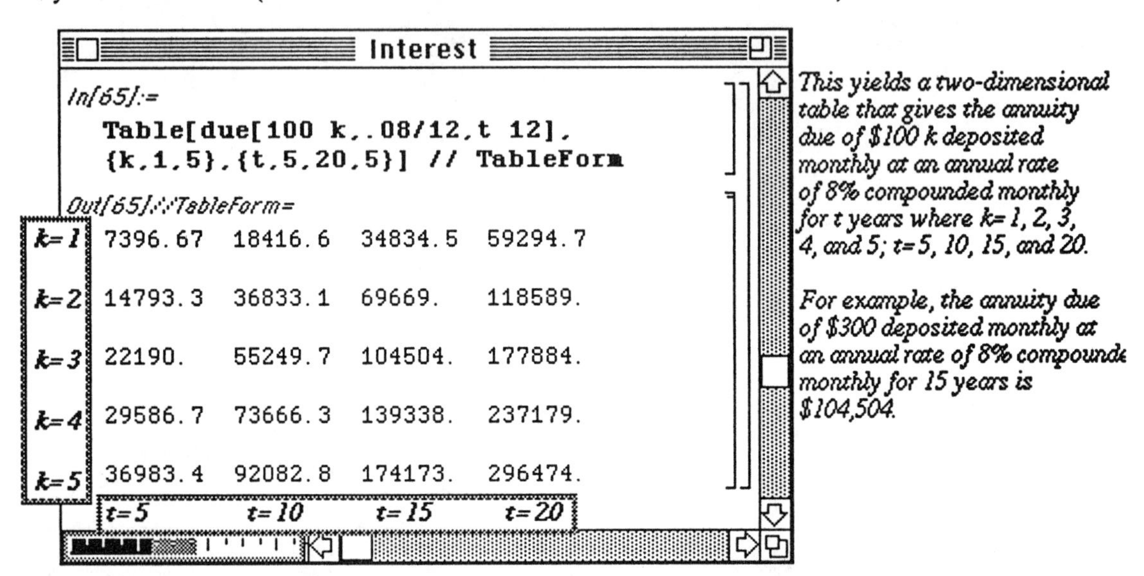

Interest

In[65]:=

```
Table[due[100 k,.08/12,t 12],
{k,1,5},{t,5,20,5}] // TableForm
```

Out[65]//TableForm=

	t=5	t=10	t=15	t=20
k=1	7396.67	18416.6	34834.5	59294.7
k=2	14793.3	36833.1	69669.	118589.
k=3	22190.	55249.7	104504.	177884.
k=4	29586.7	73666.3	139338.	237179.
k=5	36983.4	92082.8	174173.	296474.

This yields a two-dimensional table that gives the annuity due of $100 k deposited monthly at an annual rate of 8% compounded monthly for t years where k= 1, 2, 3, 4, and 5; t=5, 10, 15, and 20.

For example, the annuity due of $300 deposited monthly at an annual rate of 8% compounde monthly for 15 years is $104,504.

□ Example: (Present Value)

Yet another type of problem deals with determining the amount of money which must be invested in order to insure a particular return on the investment over a certain period of time. This is given with the following :

The **present value P** of an annuity of n payments of R dollars each at the end of consecutive interest periods with interest compounded at a rate of interest j per period is given by: $P = R\left[\frac{1-(1+j)^{-n}}{j}\right]$.

This problem is illustrated below :

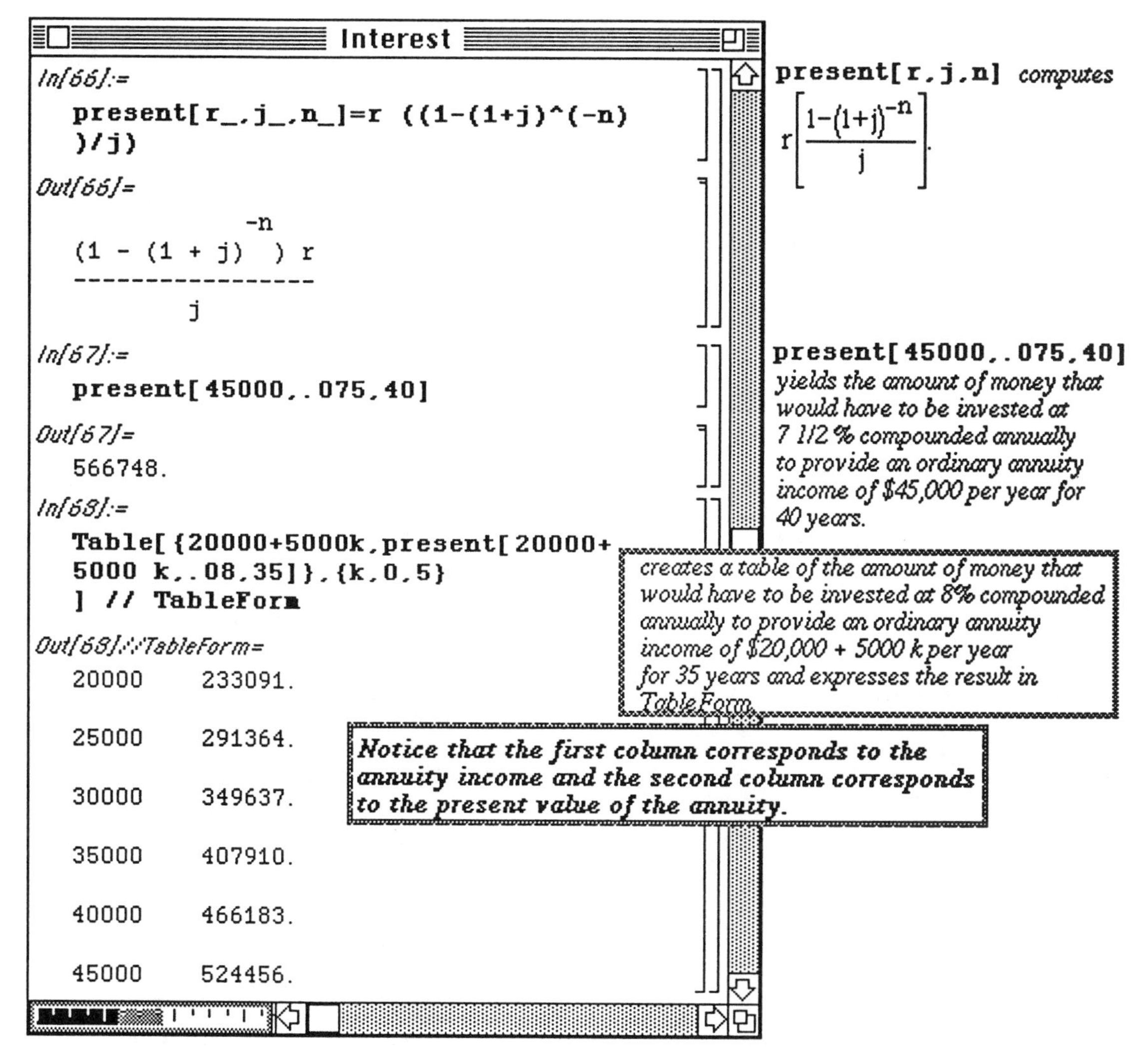

□ Example: (Deferred Annuities)

Deferred annuities can also be considered :

The <u>present value</u> of a deferred annuity of R dollars per period for n periods deferred for k periods with interest rate j per period is given by: $P_{def} = R\left[\frac{1-(1+j)^{-(n+k)}}{j} - \frac{1-(1+j)^{-k}}{j}\right]$.

The function which computes the present value of a deferred annuity is given below where
r = the amount of the deferred annuity,
n= the number of years in which in annuity is received,
k = the number of years in which the lump sum investment is made, and
j = the interest rate.

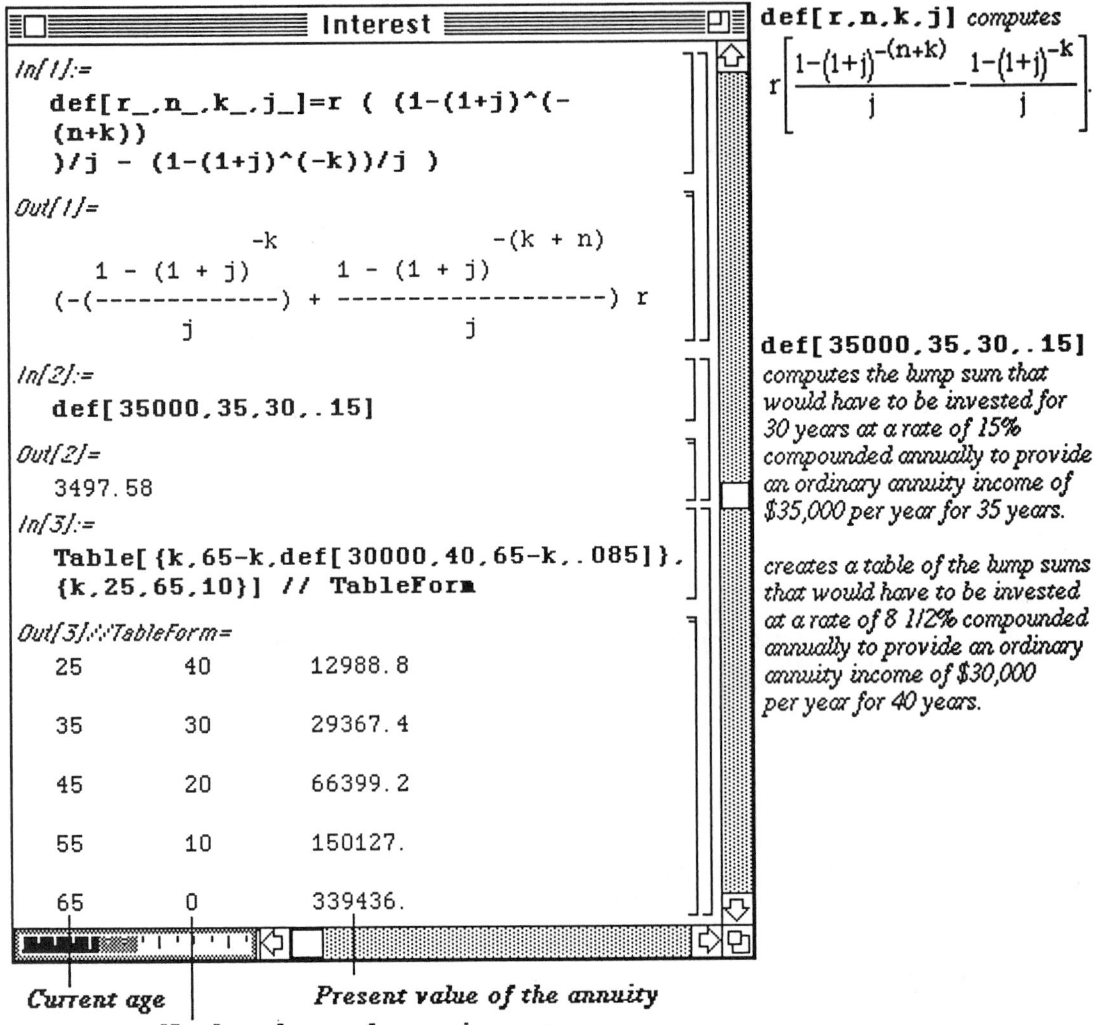

$$r\left[\frac{1-(1+j)^{-(n+k)}}{j}-\frac{1-(1+j)^{-k}}{j}\right].$$

□ **Example:** **(Amortization)**

A loan is **amortized** if both the principal and interest are paid by a sequence of equal periodic payments. A loan of P dollars at interest rate j per period may be amortized in n equal periodic payments of R dollars made at the end of each period, where $R=\dfrac{Pj}{1-(1+j)^{-n}}$.

The function, **amort[p,j,n]**, defined below determines the monthly payment needed to amortize a loan of **p** dollars with an interest rate of **j** % compounded monthly over **n** months. A second function, **totintpaid[p,j,n]**, calculates the total amount of interest paid to amortize a loan of **p** dollars with an interest rate of **j** % compounded monthly over **n** months.

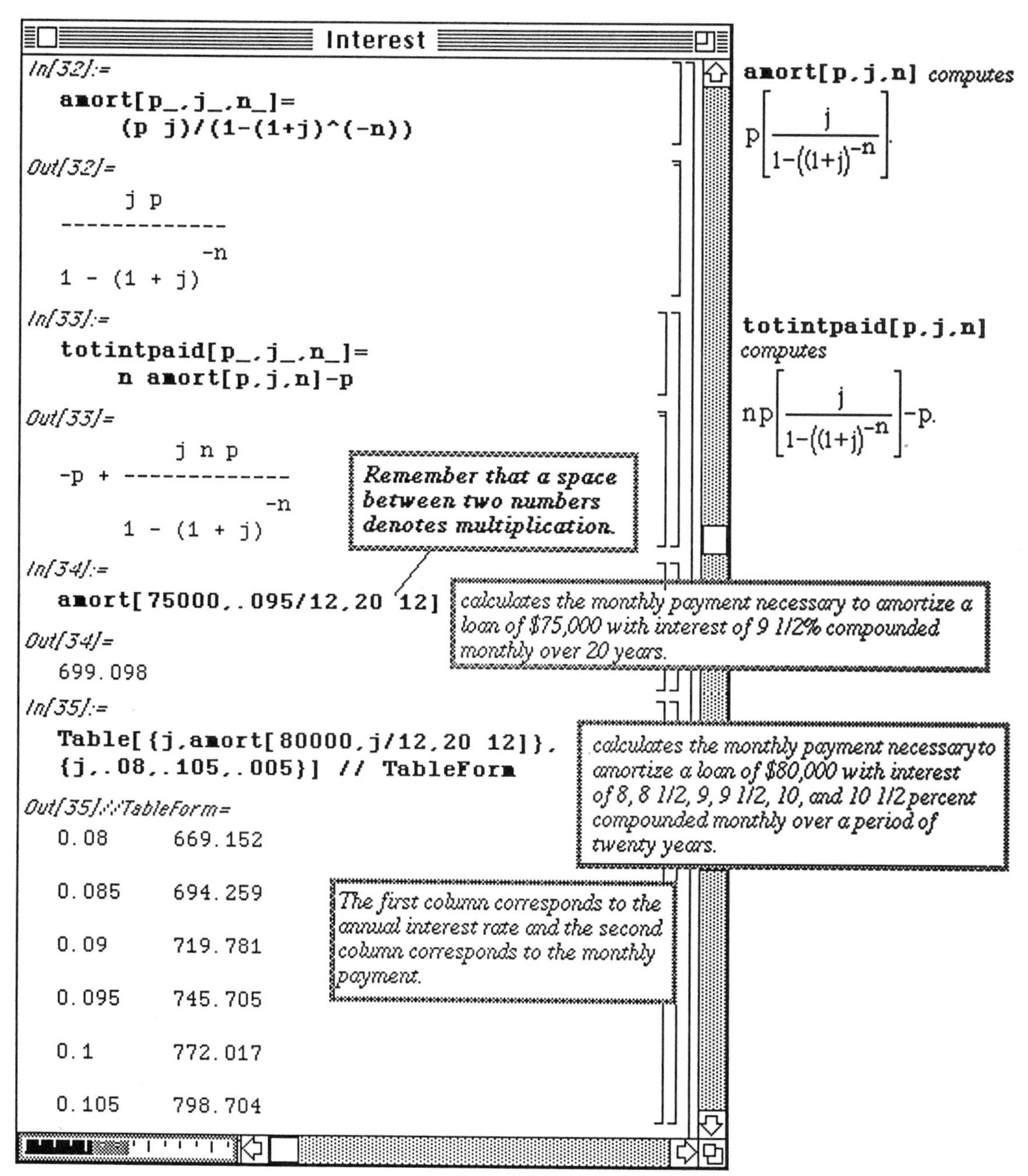

☐ Example:

The first calculation below determines the total amount paid on a loan of $75,000 at a rate of 9.5% compounded monthly over twenty years while the second shows how much of this amount was paid towards the interest.

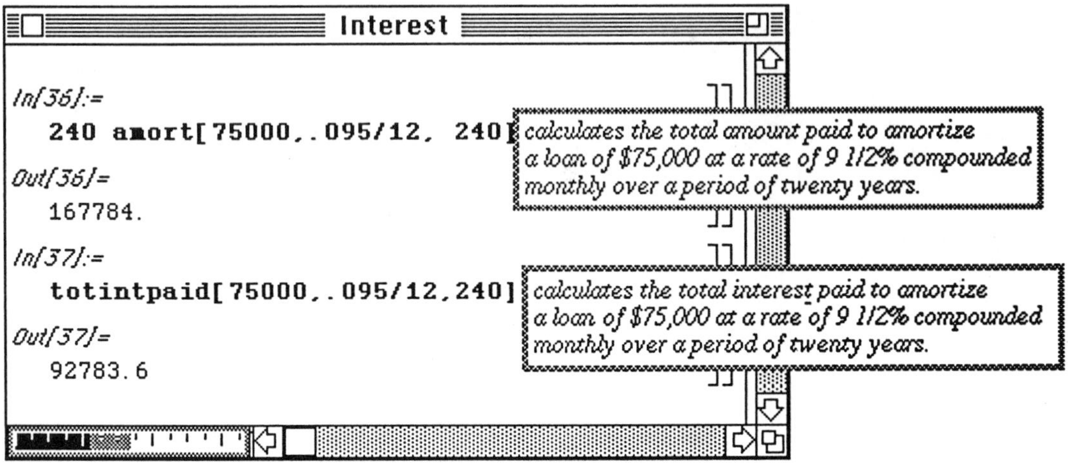

```
Interest

In[36]:=
  240 amort[75000,.095/12, 240]
Out[36]=
  167784.
In[37]:=
  totintpaid[75000,.095/12,240]
Out[37]=
  92783.6
```

calculates the total amount paid to amortize a loan of $75,000 at a rate of 9 1/2% compounded monthly over a period of twenty years.

calculates the total interest paid to amortize a loan of $75,000 at a rate of 9 1/2% compounded monthly over a period of twenty years.

In many cases, the amount paid towards the principle of the loan and the total amount which remains to be paid after a certain payment need to be computed. This is easily accomplished with the functions **unpaidbalance** and **curprinpaid** defined below using the function **amort[p,j,n]** that was previously defined:

Remark: *Mathematica* does not retain definitions of functions from previous *Mathematica* sessions. This means that in order to use a function definition from a previous *Mathematica* session, the definition must be re-entered.

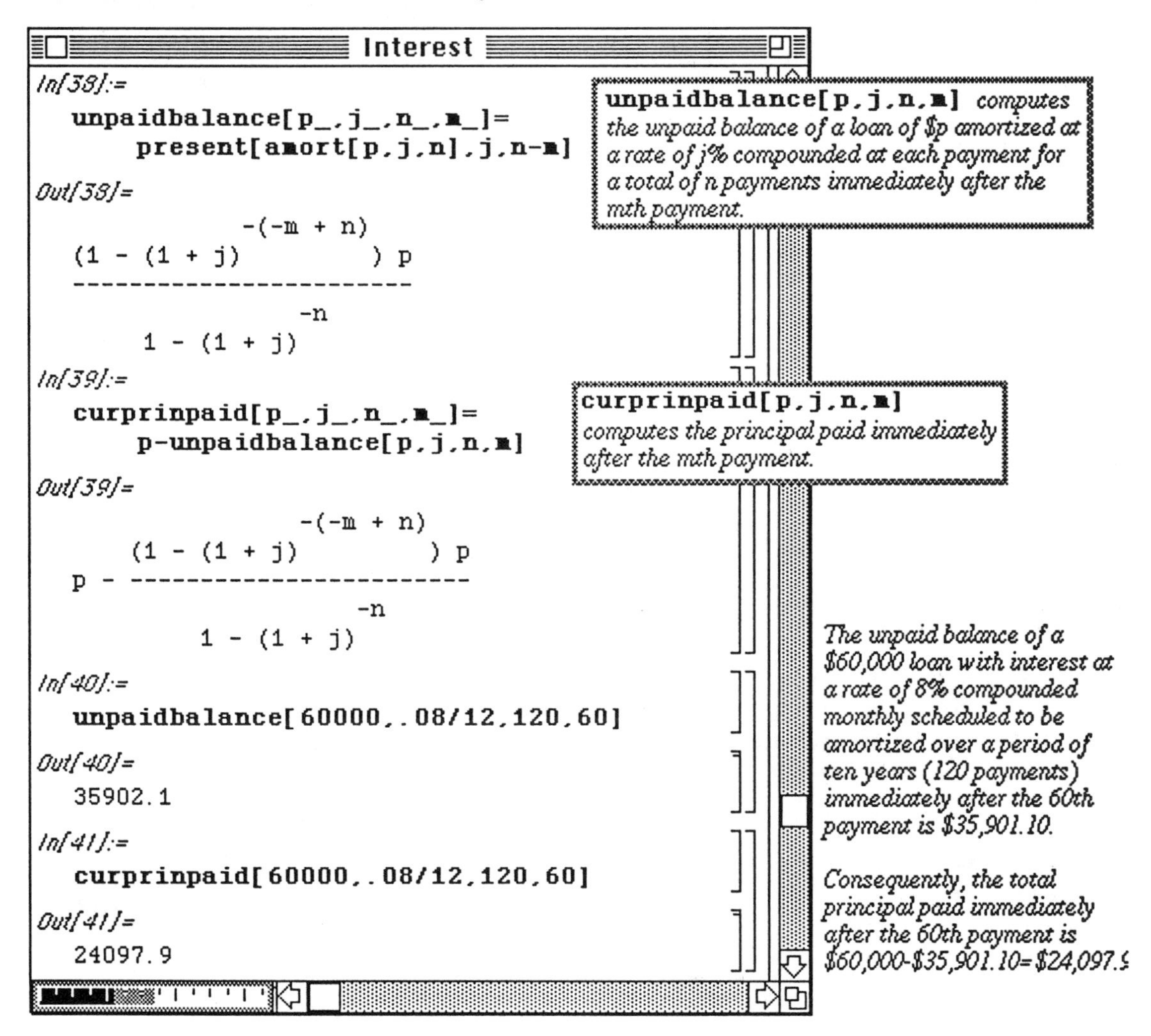

Mathematica can also be used to determine the total amount of interest paid on a loan using the following function :

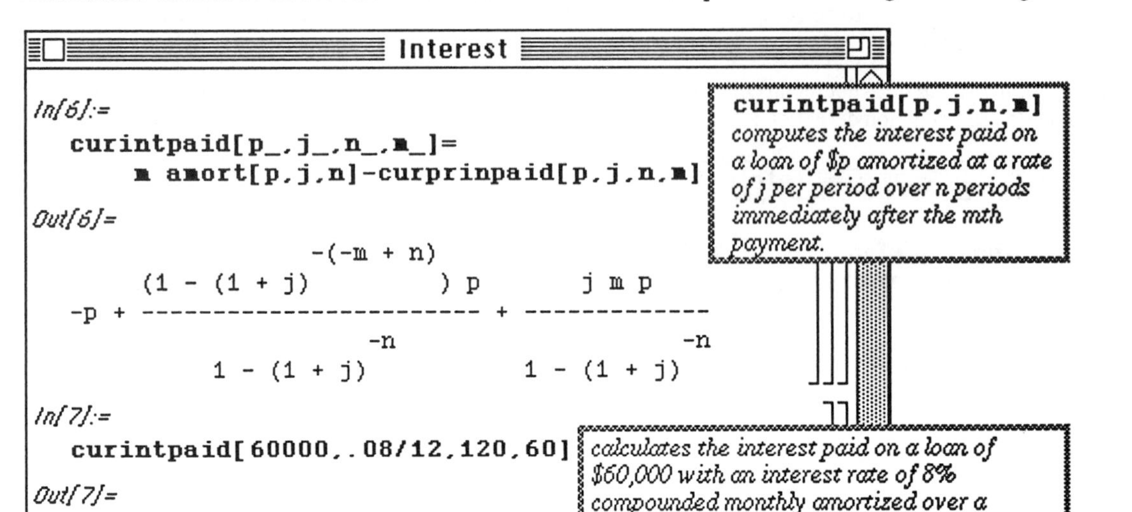

Tables can be created which show a breakdown of the payments made on a loan (i.e., how much of the total amount paid is allotted to the principle and how much to the interest.) An example is given below :

Interest

In[8]:=

```
amort[45000,.07/12,15 12]
```

calculates the monthly payment necessary to amortize a loan of $45,000 with interest rate 7% compounded monthly over a period of 15 years (15 12= 180 months).

Out[8]=

404.473

Don't forget that a space between two numbers denotes multiplication.

In[9]:=

```
Table[{t,curprinpaid[45000,.07/12,15 12,12 t],
curintpaid[45000,.07/12,15 12,12 t]},
{t,0,15,3}] // TableForm
```

This table shows the interest paid and principle paid at the end of 0, 3, 6, 9, 12 and 15 years.

Out[9]//TableForm=

0	0.	0.
3	5668.99	8892.03
6	12658.4	16463.6
9	21275.9	22407.2
12	31900.6	26343.5
15	45000	27805.1

Column 1 represents number of years, column 2 represents principle paid, and column 3 represents interest paid.

Thus, at the end of twelve years, $31,900.60 of the principle has been paid; $26,343.50 in interest has been paid.

Since **curintpaid[p,j,n,y]** computes the interest paid on a loan of $**p** amortized at a rate of **j** per period over **n** periods immediately after the **y**th payment, and **curintpaid[p,j,n,y-12]** computes the interest paid on a loan of $**p** amortized at a rate of **j** per period over **n** periods immediately after the (**y**-12)th payment,
curintpaid[p,j,n,y]-curintpaid[p,j,n,y-12] yields the amount of interest paid on a loan of $**p** amortized at a rate of **j** per period over **n** periods between the (**y**-12)th and **y**th payment.

Consequently, the interest paid and the amount of principle paid over a year can also be computed :

```
Interest

In[10]:=
  annualintpaid[p_,j_,n_,y_]:=
      curintpaid[p,j,n,y]-curintpaid[p,j,n,y-12];
  annualprinpaid[p_,j_,n_,y_]:=
      curprinpaid[p,j,n,y]-curprinpaid[p,j,n,y-12];

In[11]:=
  Table[{t,annualintpaid[45000,.07/12,15 12, 12 t],
  annualprinpaid[45000,.07/12,15 12, 12t]},{t,1,5,1}]
                          // TableForm
Out[11]//TableForm=
   1          3094.26   1759.41

   2          2967.08   1886.6

   3          2830.69   2022.98

   4          2684.45   2169.22

   5          2527.64   2326.03
```

Column 1 represents the number of years the loan has been held; column 2 represents the interest paid on the loan during the year; and column 3 represents the amount of the principle that has been paid.

• Additional Output Features of Version 2.0

□ Example:

Suppose an investor begins investing at a rate of d dollars per year at an annual rate of j%. Each year the investor increases the amount invested by i%. How much has the investor accumulated after m years?

The following table illustrates the amount invested each year and the value of the annual investment after m years:

Year	Rate of Increase	Annual Interest	Annual Investment	Value after m Years
0		j%	d	$(1+j\%)^m d$
1	i%	j%	$(1+i\%)d$	$(1+i\%)(1+j\%)^{m-1} d$
2	i%	j%	$(1+i\%)^2 d$	$(1+i\%)^2(1+j\%)^{m-2} d$
3	i%	j%	$(1+i\%)^3 d$	$(1+i\%)^3(1+j\%)^{m-3} d$
4	i%	j%	$(1+i\%)^4 d$	$(1+i\%)^4(1+j\%)^{m-4} d$
5	i%	j%	$(1+i\%)^5 d$	$(1+i\%)^5(1+j\%)^{m-5} d$
k	i%	j%	$(1+i\%)^k d$	$(1+i\%)^k(1+j\%)^{m-k} d$
m	i%	j%	$(1+i\%)^m d$	$(1+i\%)^m d$

It follows that the total value of the amount invested for the first k years after m years is given by:

Year	Total Investment
0	$(1+j\%)^m d$
1	$(1+j\%)^m d+(1+i\%)(1+j\%)^{m-1} d$
2	$(1+j\%)^m d+(1+i\%)(1+j\%)^{m-1} d+(1+i\%)^2(1+j\%)^{m-2} d$
3	$\sum_{n=0}^{3}(1+i\%)^n(1+j\%)^{m-n} d$
4	$\sum_{n=0}^{4}(1+i\%)^n(1+j\%)^{m-n} d$
5	$\sum_{n=0}^{5}(1+i\%)^n(1+j\%)^{m-n} d$
k	$\sum_{n=0}^{k}(1+i\%)^n(1+j\%)^{m-n} d$
m	$\sum_{n=0}^{m}(1+i\%)^n(1+j\%)^{m-n} d$

o The package **SymbolicSum.m**, contained in the folder **Algebra**, can be used to find a closed form of the sums $\sum_{n=0}^{k}(1+i)^n(1+j)^{m-n} d$ and $\sum_{n=0}^{m}(1+i)^n(1+j)^{m-n} d$.

`SymbolicSum.m` is included with Version 2.0 but not prior versions of *Mathematica*:

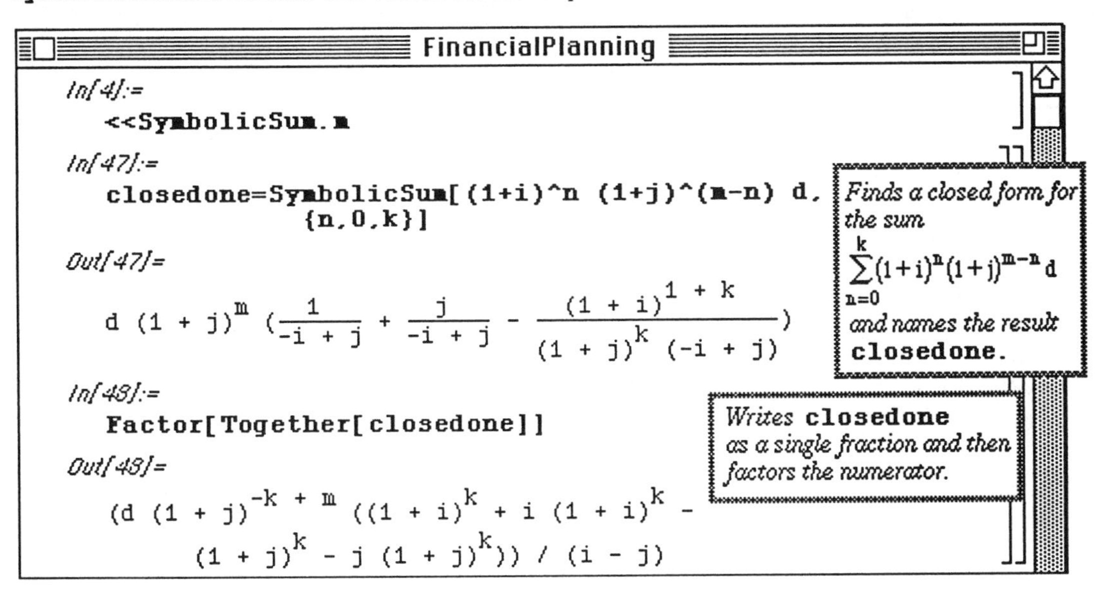

In the exact same manner a closed form was found and simplified for $\sum_{n=0}^{k}(1+i)^n(1+j)^{m-n}\,d$,

`SymbolicSum` is used to find a closed form for $\sum_{n=0}^{m}(1+i)^n(1+j)^{m-n}\,d$.

In this case, however, the final result is displayed in a print cell in input form with the command `Print[InputForm[%]]`.

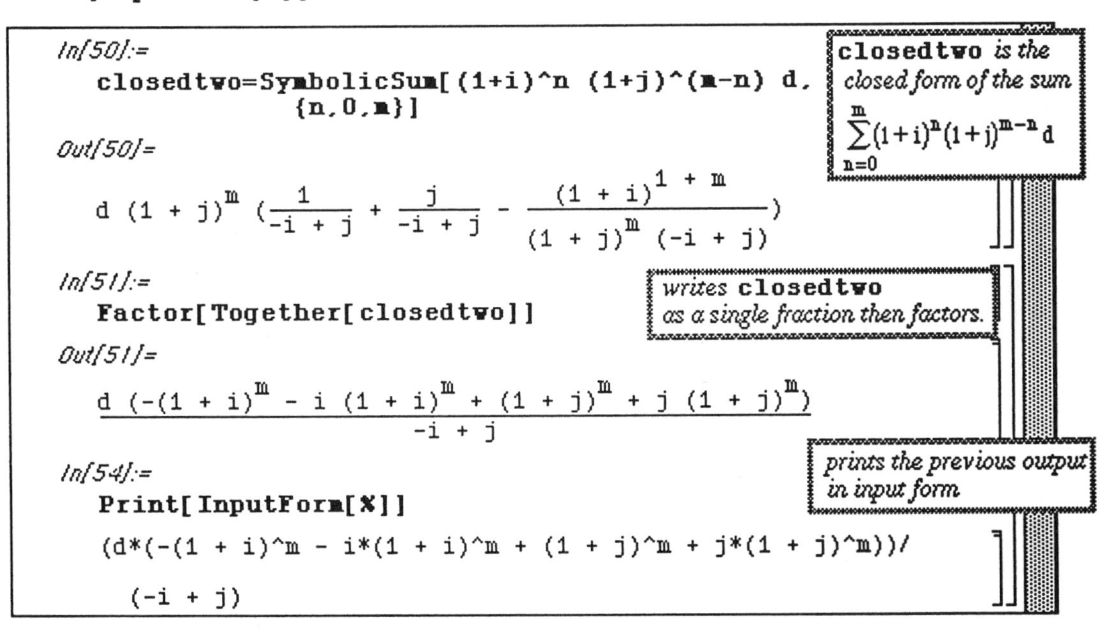

The above results are used to define the functions **investment[{d,i,j},{k,m}]** and **investmenttot[{d,i,j},m]**. In the second case, notice that print cells can be edited like any other input or text cell. Consequently, we use Macintosh editing features to copy and paste the above result when we define the the function **investmenttot**.

```
investment[{d_,i_,j_},{k_,m_}]=
    (d*(1 + j)^(-k + m)*((1 + i)^k + i*(1 + i)^k -
    (1 + j)^k - j*(1 + j)^k))/(i - j)
Out[41]=
```

$$(d\ (1 + j)^{-k + m}\ ((1 + i)^k + i\ (1 + i)^k - (1 + j)^k - j\ (1 + j)^k))\ /\ (i - j)$$

```
In[72]:=
investmenttot[{d_,i_,j_},m_]=
    (d*(-(1 + i)^m - i*(1 + i)^m +
        (1 + j)^m + j*(1 + j)^m))/(-i + j)
Out[72]=
```

$$\frac{d\ (-(1 + i)^m - i\ (1 + i)^m + (1 + j)^m + j\ (1 + j)^m)}{-i + j}$$

Finally, **investment** and **investmenttot** are used to illustrate various financial scenarios. In the first example, **investment** is used to compute the value after twenty-five years of investing $6500 the first year and then increasing the amount invested 5% per year is shown for 5, 10, 15, 20, and 25 years assuming a 15% rate of interest on the amount invested. Version 2.0 contains the built-in function **AccountingForm** which can be used to convert numbers expressed in exponential notation to ordinary notation. In the second example, **investmenttot** is used to compute the value after twenty-five years of investing $6500 the first year and then increasing the amount invested 5% per year is shown assuming various rates of interest. The results are displayed in **AccountingForm**:

```
In[64]:=
results=Table[
     {t,investment[{6500,.05,.15},{t,25}]},
     {t,5,25,5}] // TableForm
Out[64]//TableForm=
5      1.03506 10^6
10     1.55608 10^6
15     1.88668 10^6
20     2.09646 10^6
25     2.22957 10^6
TableForm[AccountingForm[results]]
Out[65]//TableForm=
5      1035065.
10     1556078.
15     1886680.
20     2096460.
25     2229573.
In[82]:=
scenes=Table[{i,investmenttot[{6500,.05,i},25]},
     {i,.08,.20,.02}];
AccountingForm[TableForm[scenes]]
Out[82]//AccountingForm=
0.08     832147.
0.1      1087126.
0.12     1437837.
0.14     1921899.
0.16     2591636.
0.18     3519665.
0.2      4806524.
```

The command **Accounting Form** *can be used to convert numbers expressed in exponential notation to ordinary digit form.*

This table illustrates the total value of investing $6500 the first year and then increasing the amount invested by 5% per year for 25 years for various rates of interest.

100%

■ Application: Graphing Parametric Equations with `ListPlot` and `ParametricPlot`

If `list={{x[1],y[1]},{x[2],y[2]}, ... , {x[n],y[n]}}` is a list of ordered pairs, `ListPlot[list]` graphs the set of ordered pairs in `list`. The commands `ListPlot` and `Plot` share the same options.

The following example demonstrates how `ListPlot` is used to create a parametric plot. In this case, both coordinates depend on the variable t. A list is produced by evaluating the function `f` at values of t running from

$t=0$ to $t=3\pi$ using increments of $\frac{3\pi}{150}$.

The ordered pairs obtained are then plotted using `ListPlot`.

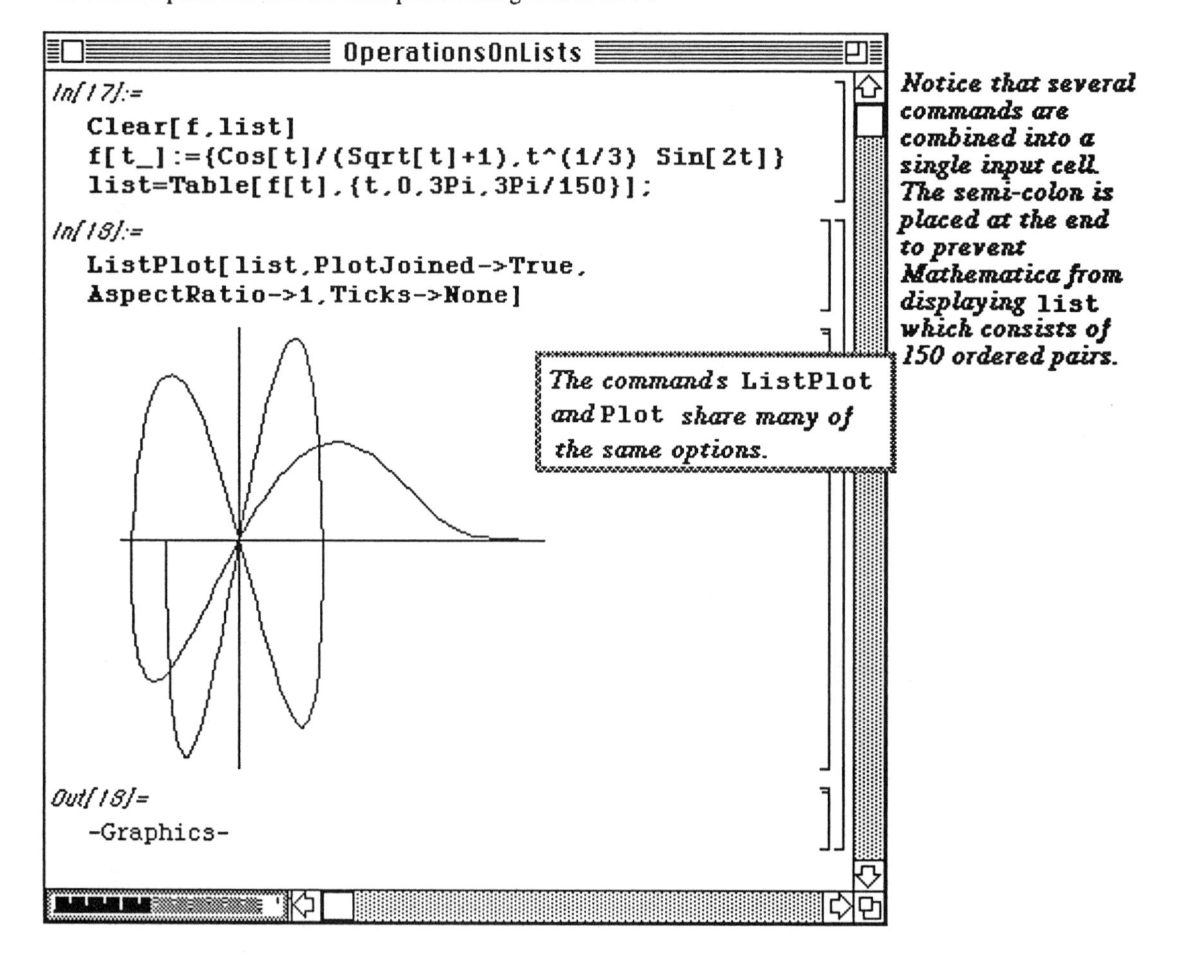

Two-dimensional parametric functions can also be graphed with the built-in function **ParametricPlot**.

Example:

The unit circle is given by the parametric equation $\{\mathrm{Cos}(t), \mathrm{Sin}(t)\}$, $0 \le t \le 2\pi$. To graph the unit circle, proceed as follows:

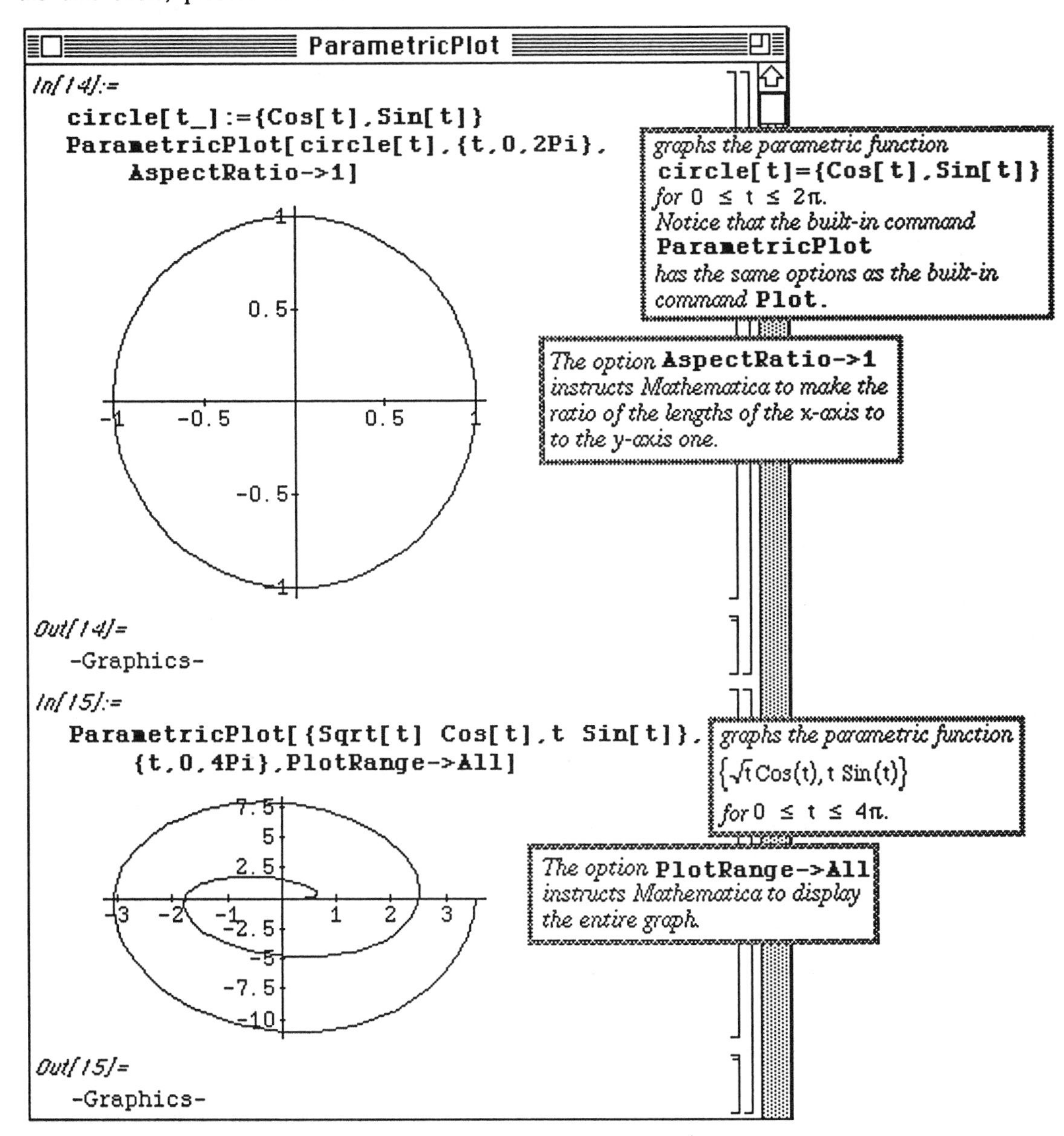

We can obtain essentially the same result with **ParametricPlot** as we obtained above with **ListPlot**:

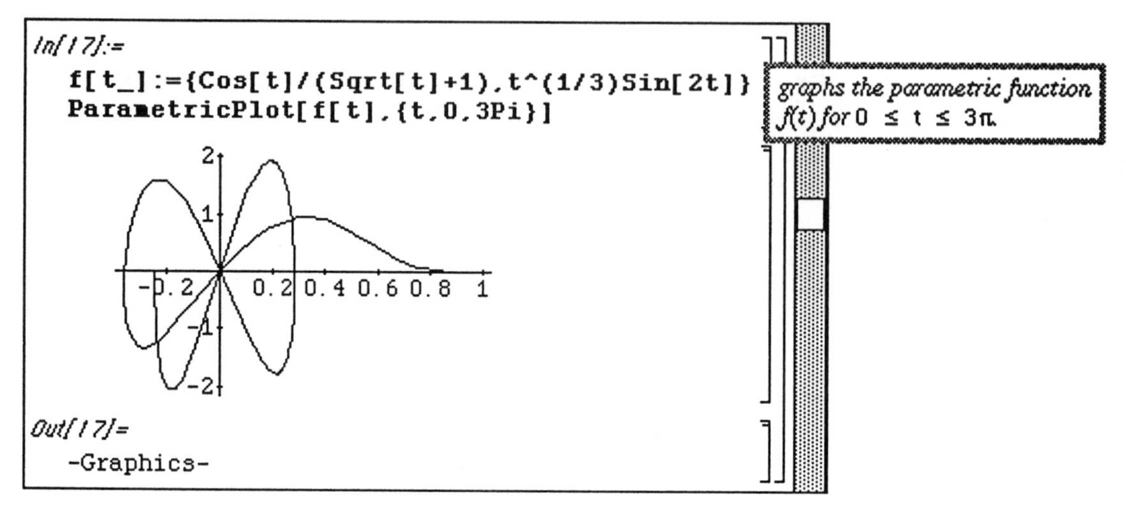

■ Application:

Given a function f, the following example illustrates how to create a table of the first, second, third, ... , and nth derivatives (provided they all exist) of f and then graph the resulting table.

□ Example:

Compute and graph the first three derivatives of $f(x) = xe^x$.

Mathematica can produce a table of these derivatives rather easily. This is accomplished through several commands. After defining f, a list of f and its first three derivatives in simplified form is obtained with the command

Table[Simplify[D[f[x],{x,n}]],{n,0,3}].

This list is then placed in the form of a table using the command

TableForm[list].

Lists can be useful in plotting the graphs of functions. Instead of entering the **GrayLevel** for each function in a multiple plot, these **GrayLevel** assignments can be made with a list. This approach to plotting several functions is shown below :

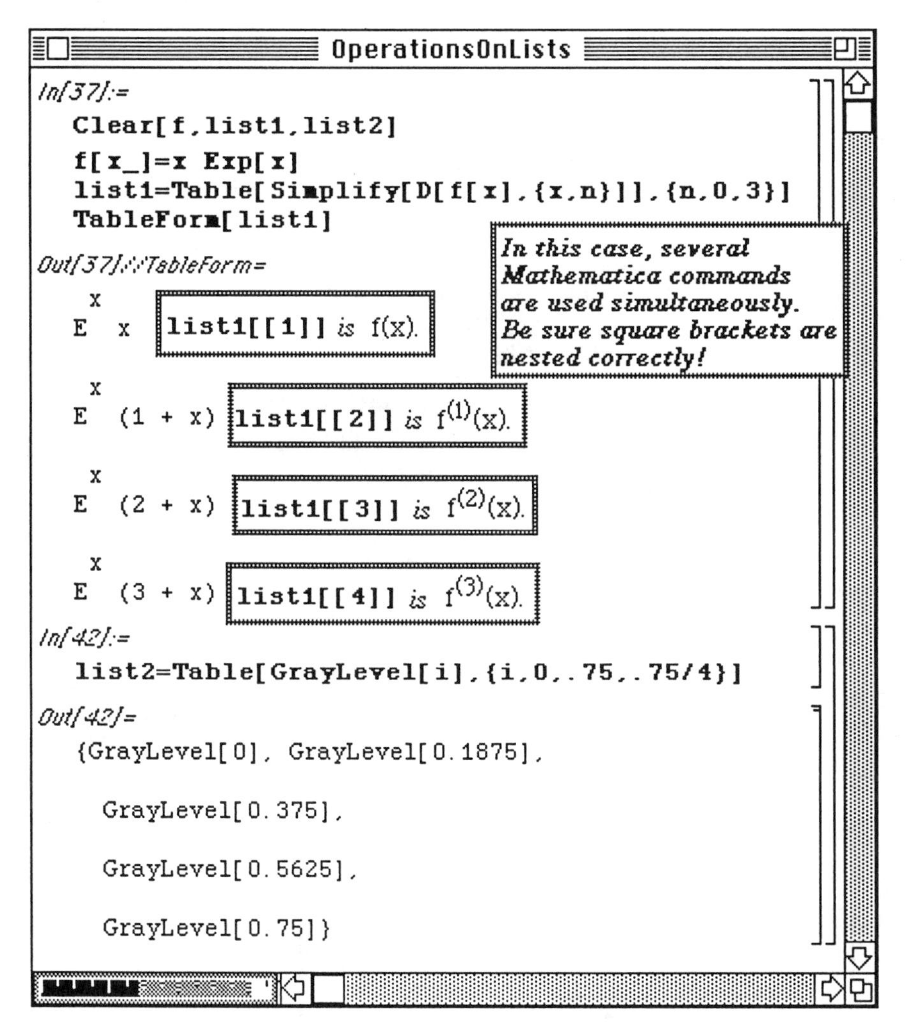

Notice that the following result could have been accomplished with the command
Plot[{f[x], f'[x],f''[x],f'''[x]},{x,-1,1},PlotStyle->{GrayLevel[0], GrayLevel[.1875],GrayLevel[.375],GrayLevel[.5625}].

However, since the use of lists simplifies the commands needed, this alternate approach is used. In order to make use of **list1** and **list2** given above, the **Release** command must be used. The command **Release[argument]** causes **argument** to be evaluated immediately.

Hence, the command
Plot[Release[list1],{x,-1,1}]
given below causes *Mathematica* to first produce the list of functions in **list1** and then evaluate the functions in the list at the values of x between **-1** and **1** in order to plot the functions. Otherwise, a new list would be created for each value of x. **Release** is used similarly with the list of **GrayLevel** values.

o In Version 2.0, **Release** has been replaced by the command **Evaluate**. Consequently, when using Version 2.0, be sure to use **Evaluate** instead of **Release**.

A list of functions can be created and plotted in a single command. The second example below shows how the lines
y=mx (where **m** varies from **m** = -4 to **m** = 2 in increments of 6/5) are plotted using the same **GrayLevel** list used in the first example.

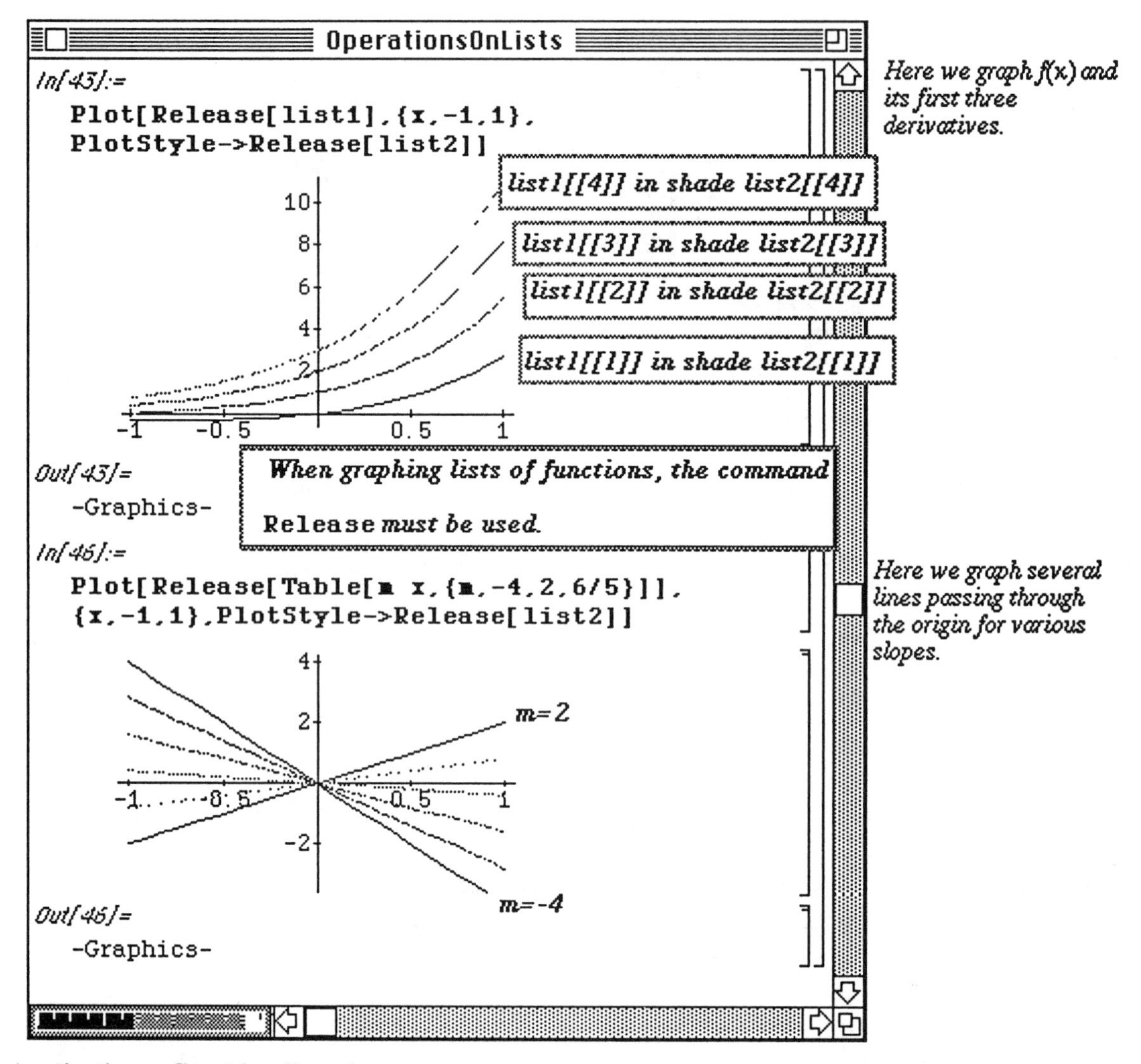

■ **Application: Graphing Equations**

Often when working problems, the ability to extract a particular element from a list is quite useful. The following example considers the equation $4x^2 + 9y^2 = 81$.

Solving this equation for y yields the two solutions $y = \pm\frac{\sqrt{81-4x^2}}{3}$.

Using *Mathematica* , these y-values appear in the form of a list. Notice below how each element of a list can be extracted for later use. Since many results are rather complicated, this technique can save a great deal of time used on typing and limit careless mistakes.

□ Example:

Graph the equation $4x^2 + 9y^2 = 81$.

Begin by defining **equation** to be the equation **4x^2+9y^2==81**. ***Be sure*** to include the double equals sign so that *Mathematica* interprets **equation** as a mathematical equation.

Notice that the elements of the list **ycoords** are lists. Notice that if **table={list1,list2,...,listn}**, where **list1**, ... , **listn** are lists, **table[[1]]** yields the first element of **table** which is **list1**; **table[[3,2]]** yields the second element of the third element of **table**. In general, **table[[i,j]]** yields the jth element of the ith element of **table**. Lists of lists, or equivalently, nested lists will be discussed in further detail in **Chapter 4**.

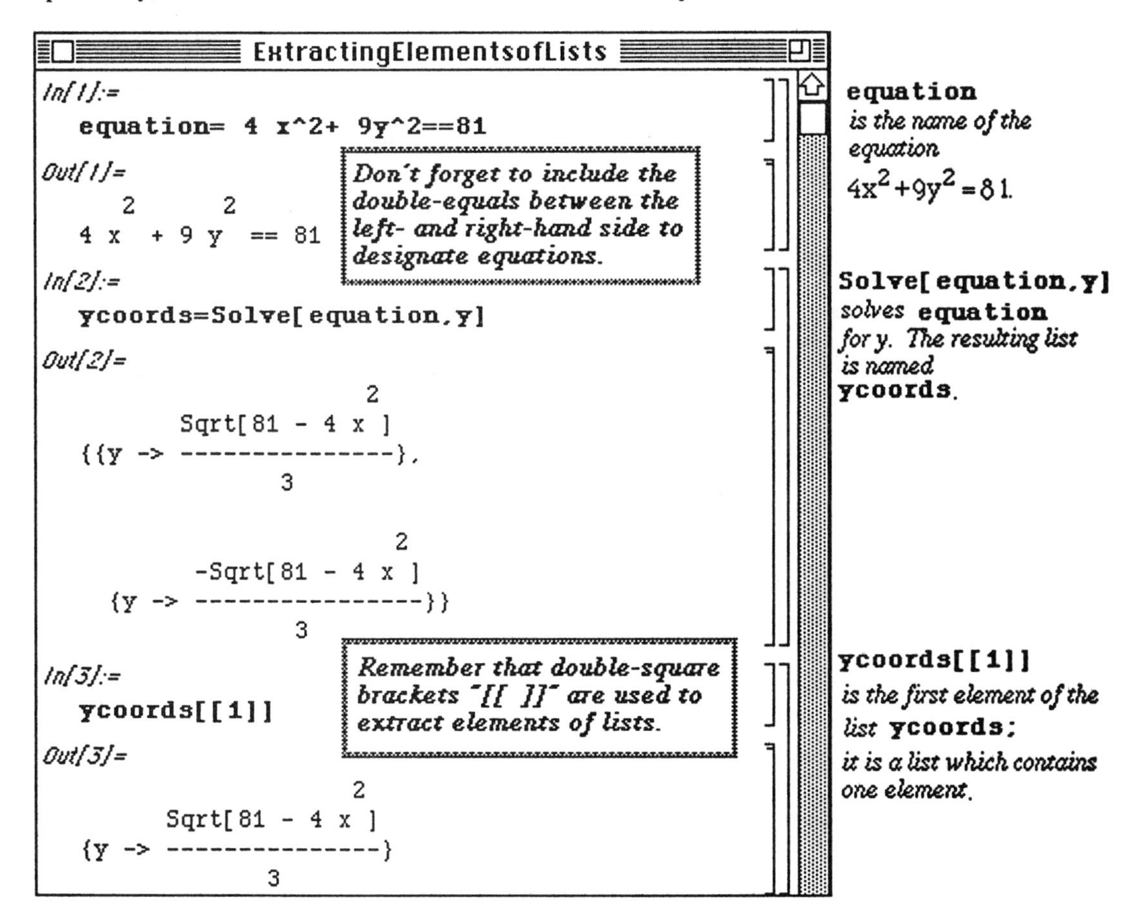

The first (and only) element of **ycoords[[1]]** is found below with **ycoords[[1,1]]**. This expression is made up of two parts: the part in front of the arrow and the part following the arrow. Therefore, to obtain the desired formula (the second part), **ycoords[[1,1,2]]** is used:

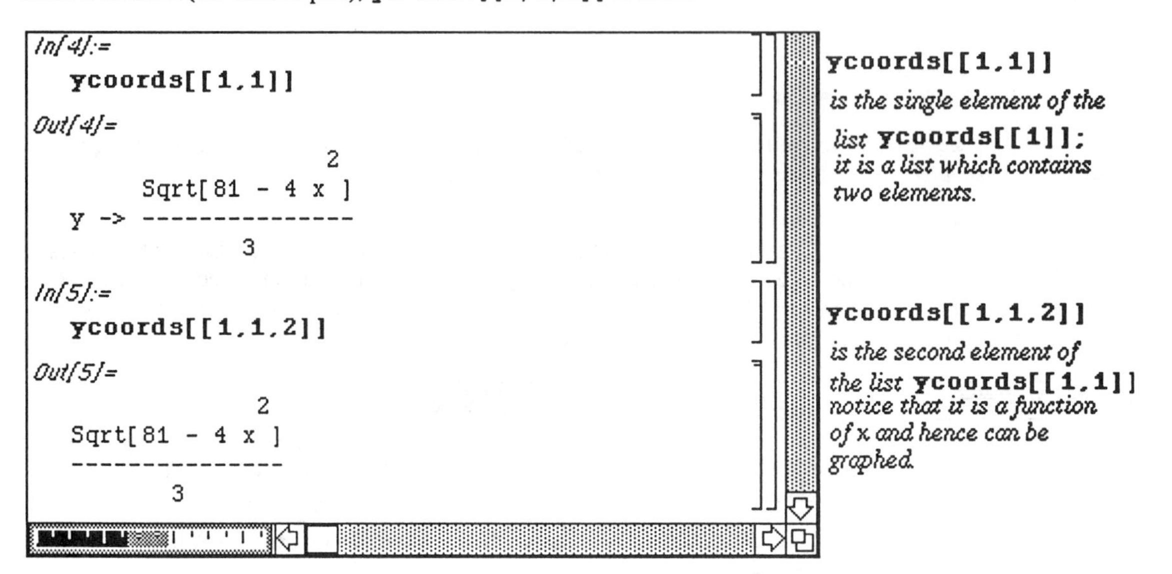

After extracting the appropriate elements from the list, tehy can be used below in other commands to determine where the curve intersects the x-axis and then plot the curve.

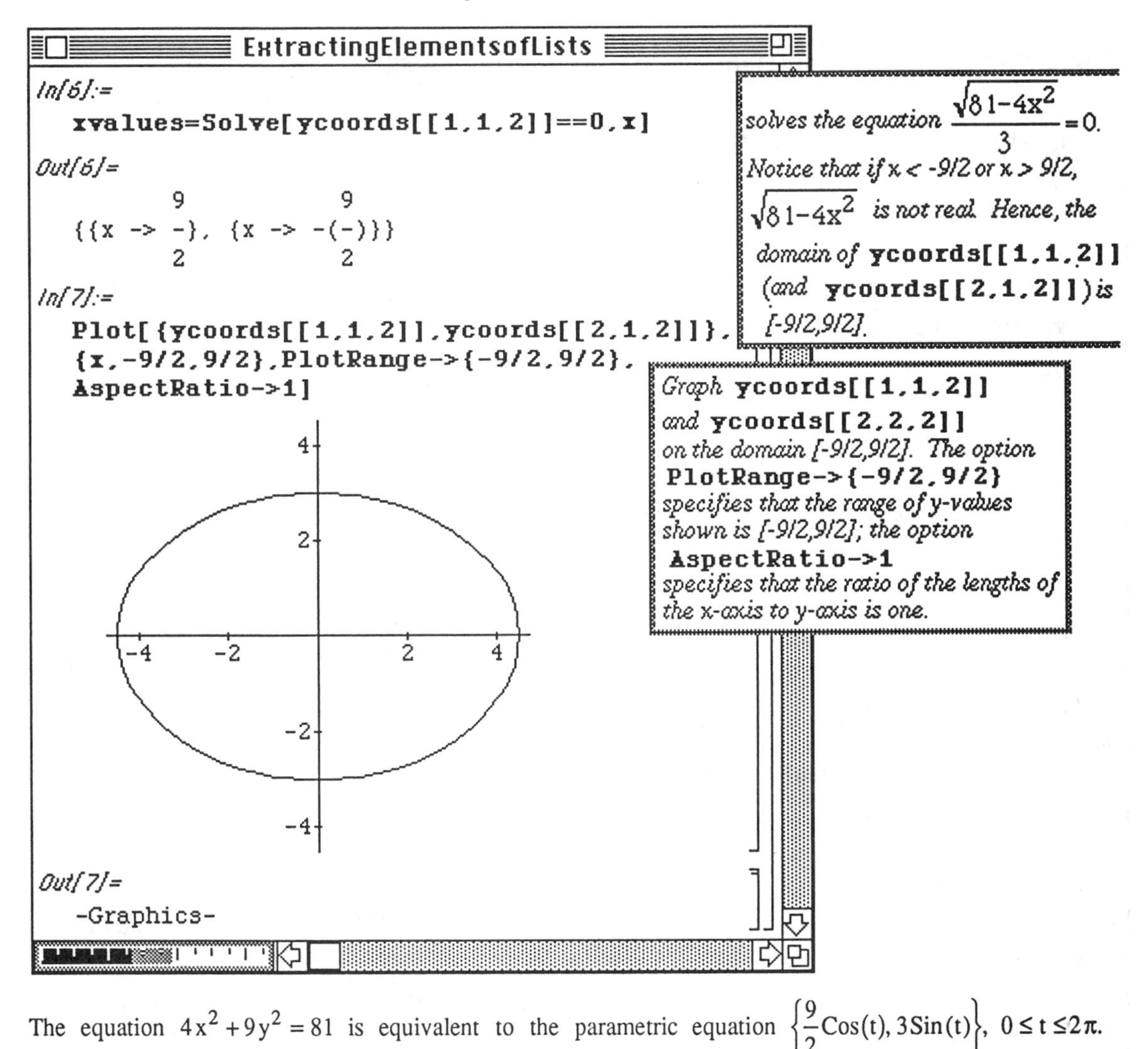

The equation $4x^2+9y^2=81$ is equivalent to the parametric equation $\left\{\frac{9}{2}\text{Cos}(t), 3\text{Sin}(t)\right\}$, $0 \le t \le 2\pi$.

Consequently, the same result could have been obtained with the command
`ParametricPlot[{9/2  Cos[t],3Sin[t]},{t,0,2Pi}]`.

o Moreover, Version 2.0 includes the package **`ImplicitPlot.m`** which contains the command **`ImplicitPlot`**. The command **`ImplicitPlot`** can be used to graph the previous example as well as the next example. For additional information on **`ImplicitPlot`**, see **Chapter 8**. Unfortunately, **`ImplicitPlot`** is not available in versions of *Mathematica* released prior to Version 2.0.

The following example deals with a slightly more complicated curve :

□ Example:

Graph the curve $y^2 - x^4 + 2x^6 - x^8 = 0$.

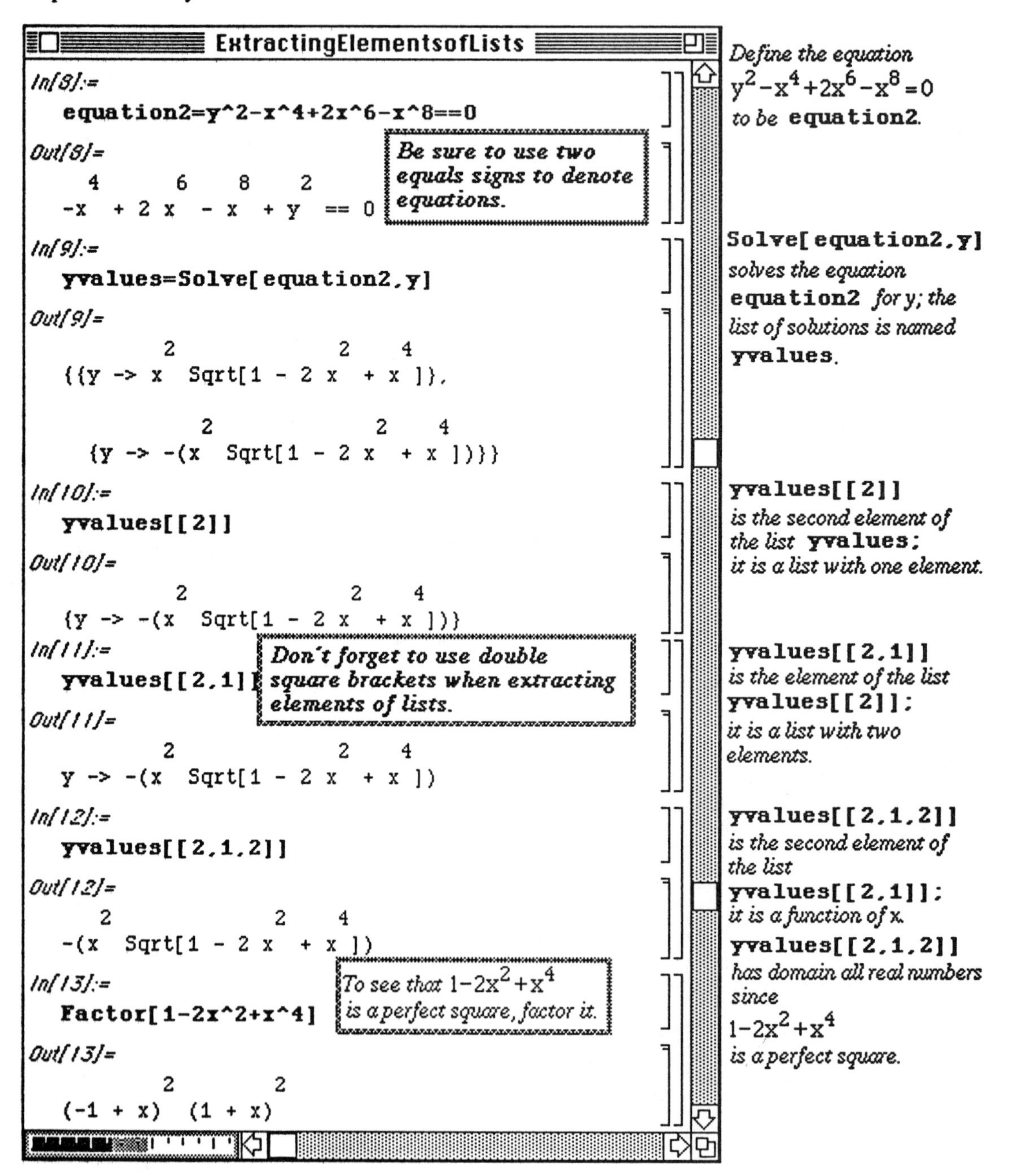

Define the equation $y^2-x^4+2x^6-x^8=0$ *to be* **equation2**.

Solve[equation2,y] *solves the equation* **equation2** *for y; the list of solutions is named* **yvalues**.

yvalues[[2]] *is the second element of the list* **yvalues**; *it is a list with one element.*

yvalues[[2,1]] *is the element of the list* **yvalues[[2]]**; *it is a list with two elements.*

yvalues[[2,1,2]] *is the second element of the list* **yvalues[[2,1]]**; *it is a function of x.* **yvalues[[2,1,2]]** *has domain all real numbers since* $1-2x^2+x^4$ *is a perfect square.*

The curve is then easily plotted using the elements of the **yvalues** list obtained above. Notice that this curve passes through the x-axis at x = -1 and x = 1 as expected from the results of the previous command: **Factor[1-2x^2+x^4]**.

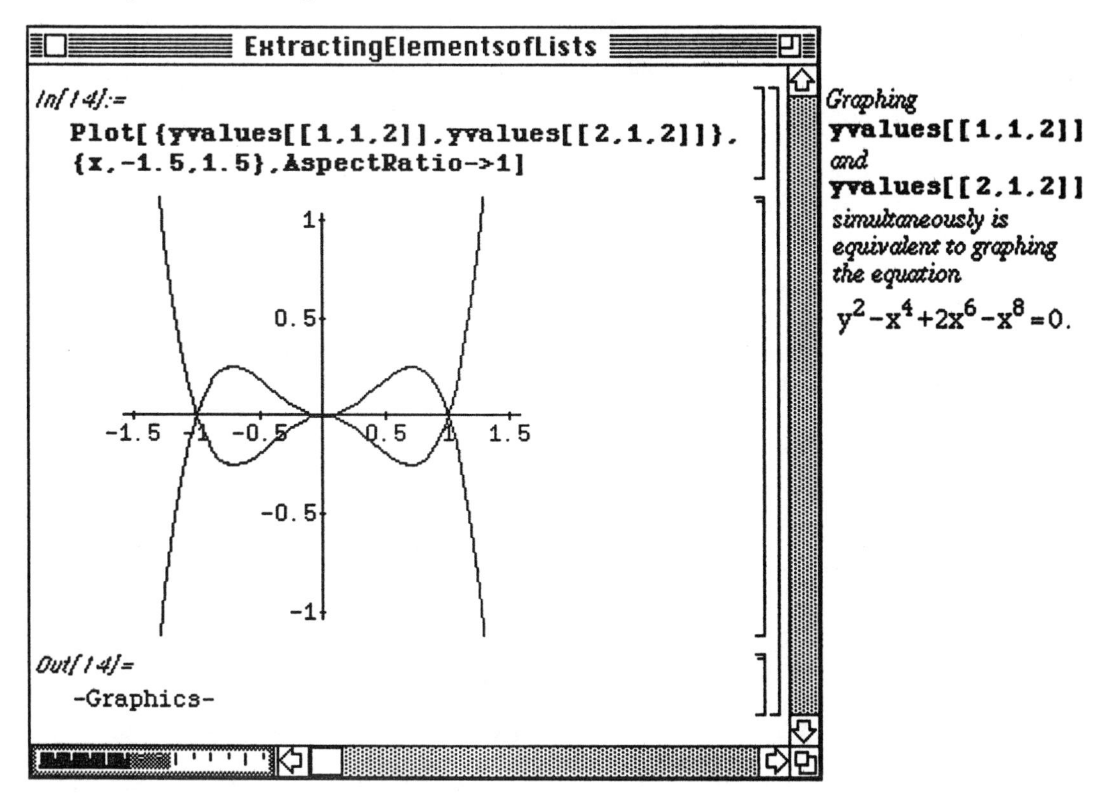

■ **Application: Tangent Lines and Animations**

The following example illustrates how the tangent line to the graph of $f(x)=x^3-\frac{9}{2}x^2+\frac{23}{4}x-\frac{15}{8}$ at many values of x on the interval [0,3] can be determined and plotted through the use of lists. Provided f is differentiable when x=a, the line tangent to the graph of f at x=a is given by y-f(x)=f'(x)(x-a).

Hence, this line line can be defined as a function of x and a with

```
tanline [x_,a_]:= f'[a](x-a)+f[a]
```

The function **tangraph[a]** defined below plots the tangent line to f at x = **a** for values of x between 0 and 3. In plotting these lines, a list of values of **a** is needed. This list is created in **table** below :

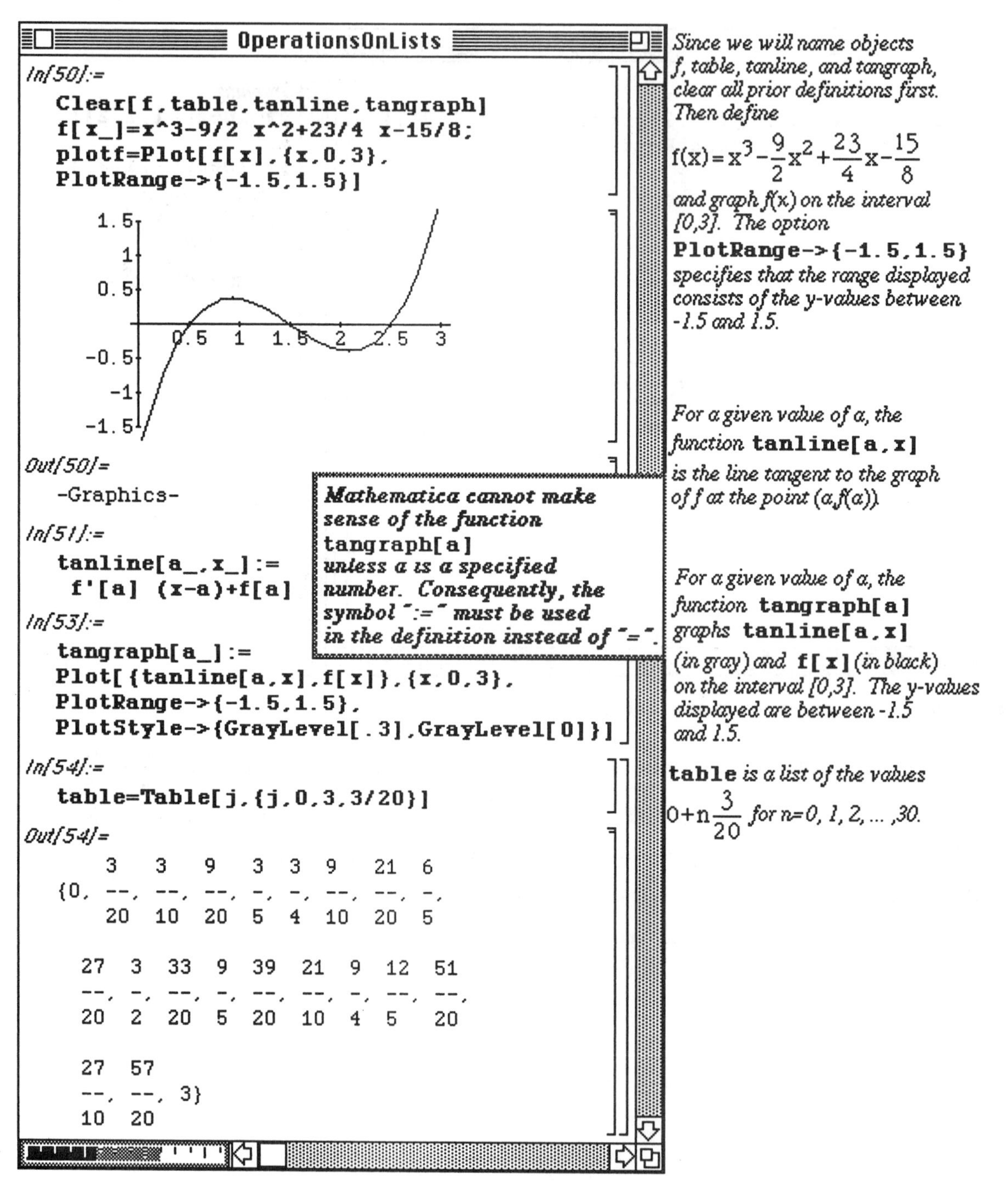

OperationsOnLists

```
In[50]:=
  Clear[f,table,tanline,tangraph]
  f[x_]=x^3-9/2 x^2+23/4 x-15/8;
  plotf=Plot[f[x],{x,0,3},
  PlotRange->{-1.5,1.5}]
```



```
Out[50]=
  -Graphics-

In[51]:=
  tanline[a_,x_]:=
   f'[a] (x-a)+f[a]

In[53]:=
  tangraph[a_]:=
  Plot[{tanline[a,x],f[x]},{x,0,3},
  PlotRange->{-1.5,1.5},
  PlotStyle->{GrayLevel[.3],GrayLevel[0]}]

In[54]:=
  table=Table[j,{j,0,3,3/20}]

Out[54]=
      3   3   9   3  3  9   21  6
  {0, --, --, --, -, -, --, --, -,
      20  10  20  5  4  10  20  5

   27  3  33  9  39  21  9  12  51
   --, -, --, -, --, --, -, --, --,
   20  2  20  5  20  10  4  5   20

   27  57
   --, --, 3}
   10  20
```

***Mathematica* cannot make sense of the function tangraph[a] unless a is a specified number. Consequently, the symbol ":=" must be used in the definition instead of "=".**

Since we will name objects f, table, tanline, and tangraph, clear all prior definitions first. Then define

$$f(x)=x^3-\frac{9}{2}x^2+\frac{23}{4}x-\frac{15}{8}$$

and graph f(x) on the interval [0,3]. The option **PlotRange->{-1.5,1.5}** *specifies that the range displayed consists of the y-values between -1.5 and 1.5.*

For a given value of a, the function **tanline[a,x]** *is the line tangent to the graph of f at the point (a,f(a)).*

For a given value of a, the function **tangraph[a]** *graphs* **tanline[a,x]** *(in gray) and* **f[x]** *(in black) on the interval [0,3]. The y-values displayed are between -1.5 and 1.5.*

table *is a list of the values* $0+n\frac{3}{20}$ *for n=0, 1, 2, ... ,30.*

Once the list of **a** values is established in **table**, **Map[tangraph, table]** evaluates **tangraph** at each value of **a** in **table** . Hence, a list of the graphs of the lines tangent to the graph of f when x = **a** is produced. Only the first graph in this list is shown below. The others are hidden but can be seen by double-clicking on the outer ("half-arrow") cell containing the first graph. These graphs can also be viewed using *Mathematica*'s animation capabilities.

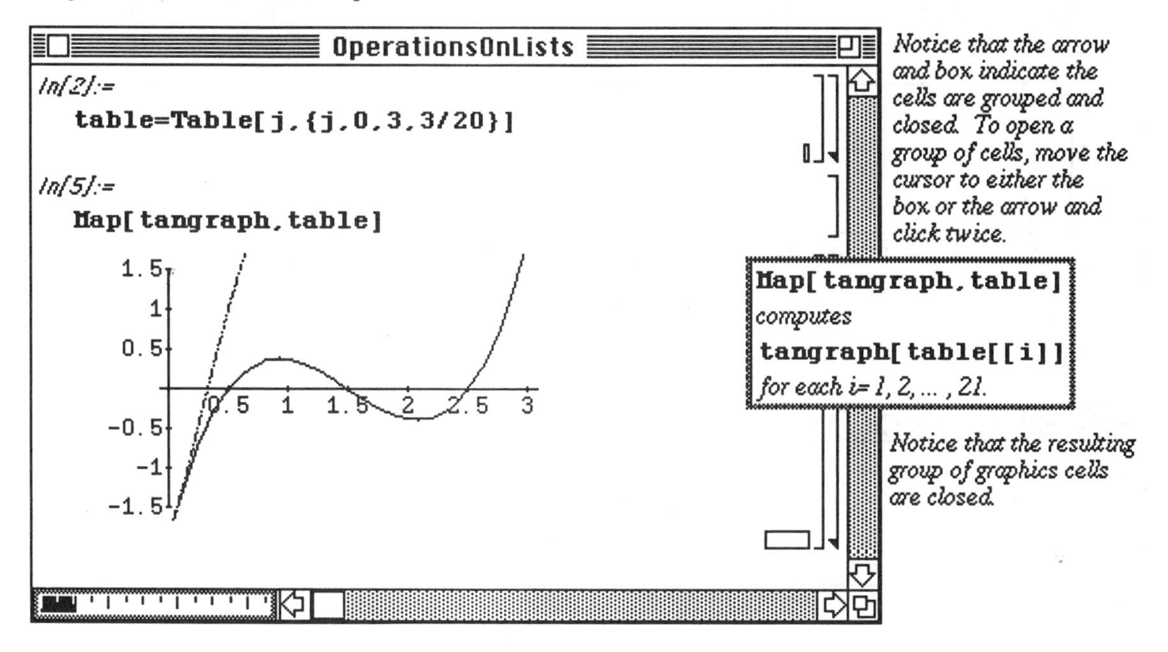

◻ Animation :

Through using animation, many graphs can be seen in succession. Hence, the graphs can be compared quite easily. The first step towards animation is to click once on the outer cell which encloses all of the graphs. This selects all of the graphs contained in the list. Next, select **Animate Selected Graphics** found under the **Graph** heading in the Menu at the top of the screen. Animation can be halted by clicking once anywhere on the screen. Some of the animation options are demonstrated below :

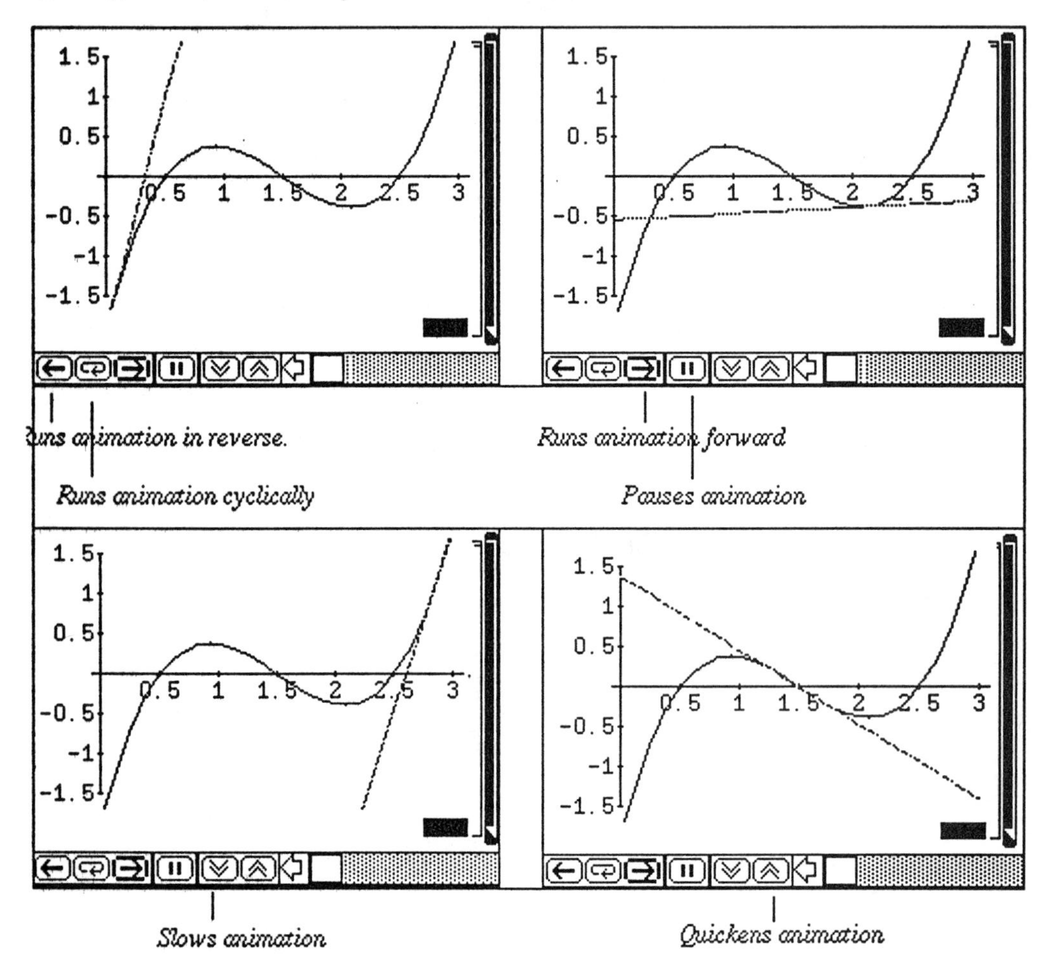

■ Application: Approximating Lists with Functions

Another interesting application of lists is that of curve-fitting. The command **Fit[data,functionset, variables]** fits a list of data points found in **data** using the functions in **functionset** by the method of least-squares. The functions in **functionset** are functions of the variables listed in **variables**.

An example is shown below which gives a quadratic fit to the data points in **datalist**.

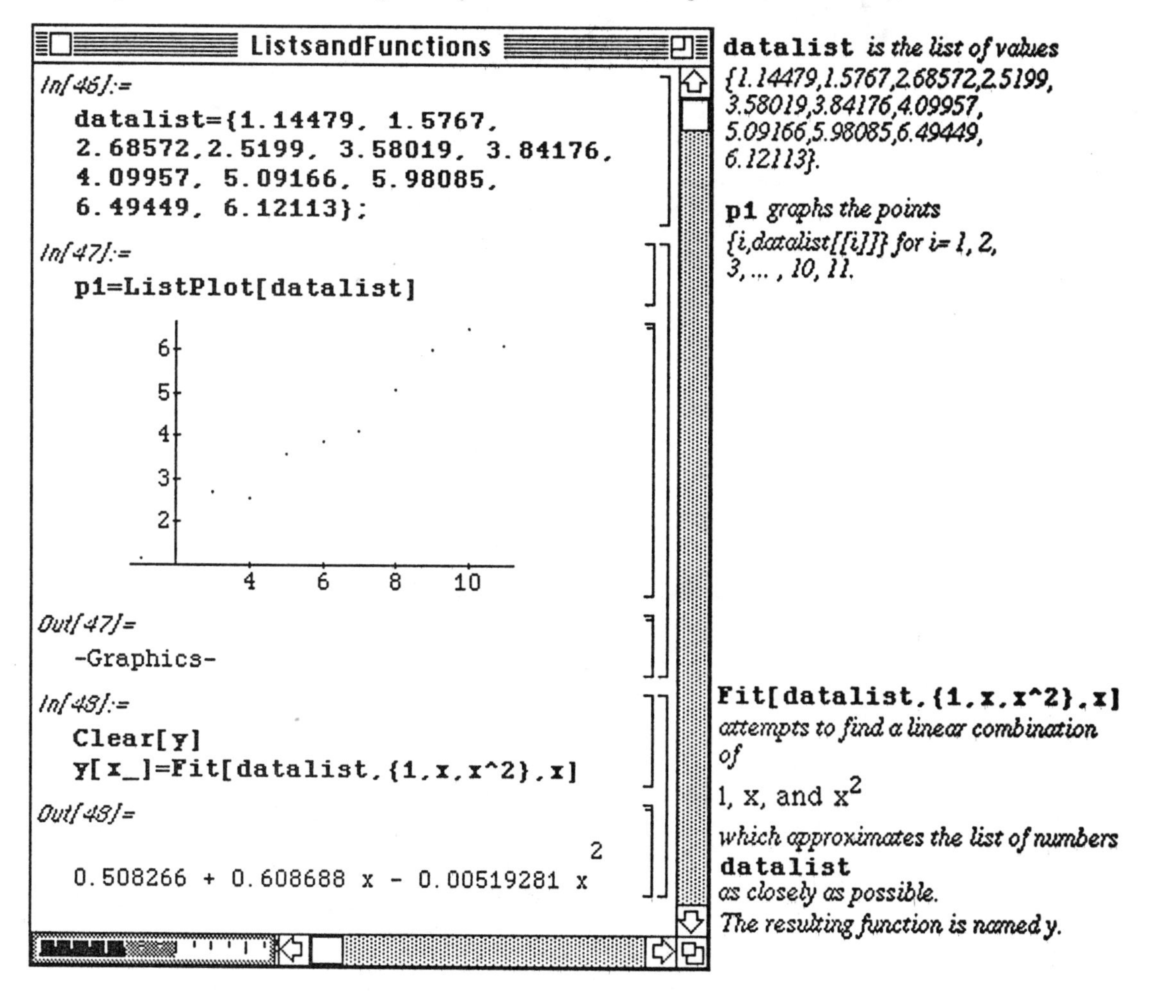

The approximating function obtained above via the least-squares method can be plotted along with the data points. This is demonstrated below. Notice that many of the data points are not very close to the approximating function. Hence, a better approximation is obtained below using a polynomial of higher degree (4).

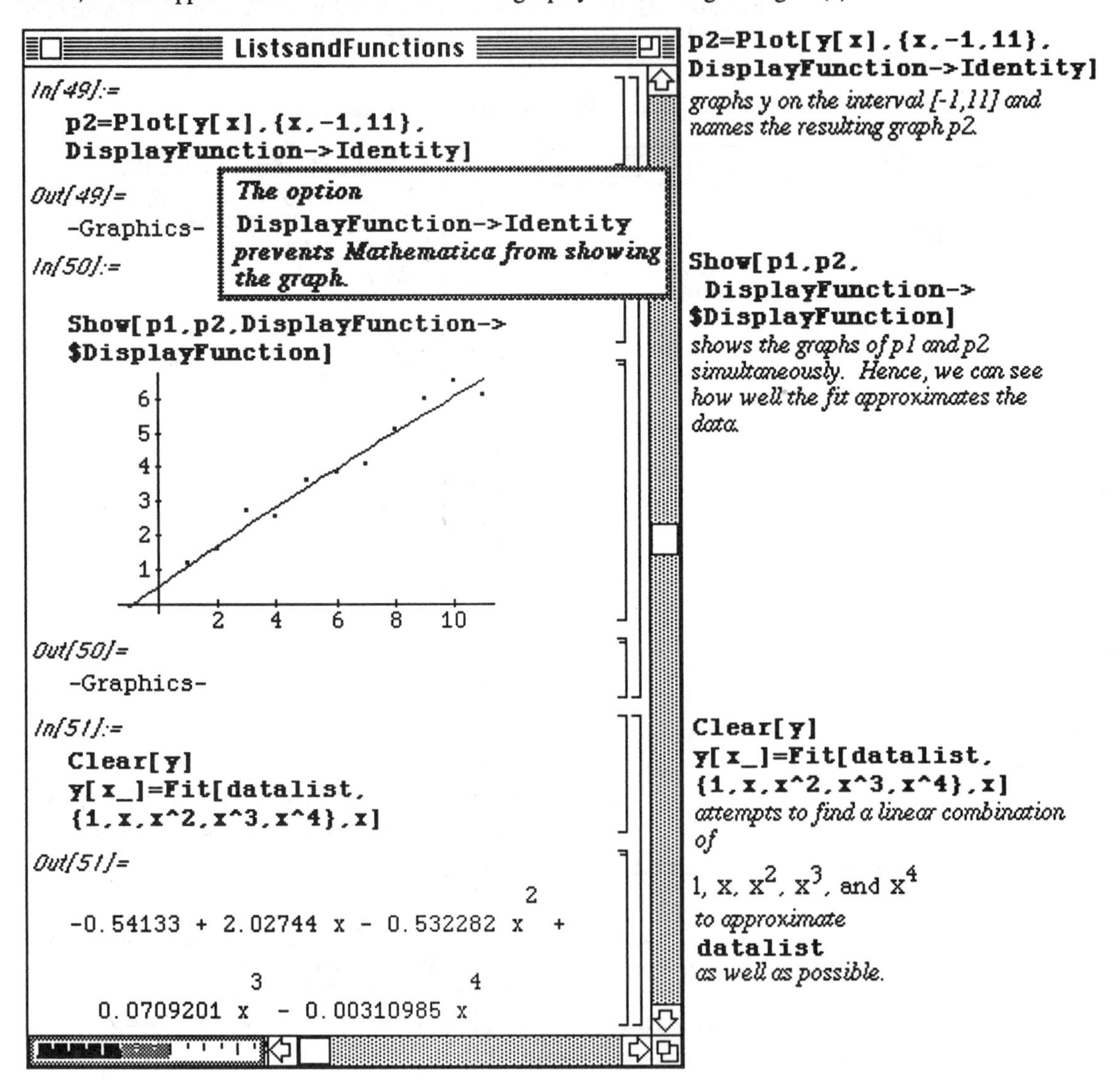

p2=Plot[y[x],{x,-1,11}, DisplayFunction->Identity] *graphs y on the interval [-1,11] and names the resulting graph p2.*

Show[p1,p2, DisplayFunction-> $DisplayFunction] *shows the graphs of p1 and p2 simultaneously. Hence, we can see how well the fit approximates the data.*

Clear[y] y[x_]=Fit[datalist, {1,x,x^2,x^3,x^4},x] *attempts to find a linear combination of*

1, x, x^2, x^3, and x^4

to approximate **datalist** *as well as possible.*

To check its accuracy, this second approximation is simultaneously with the data points.

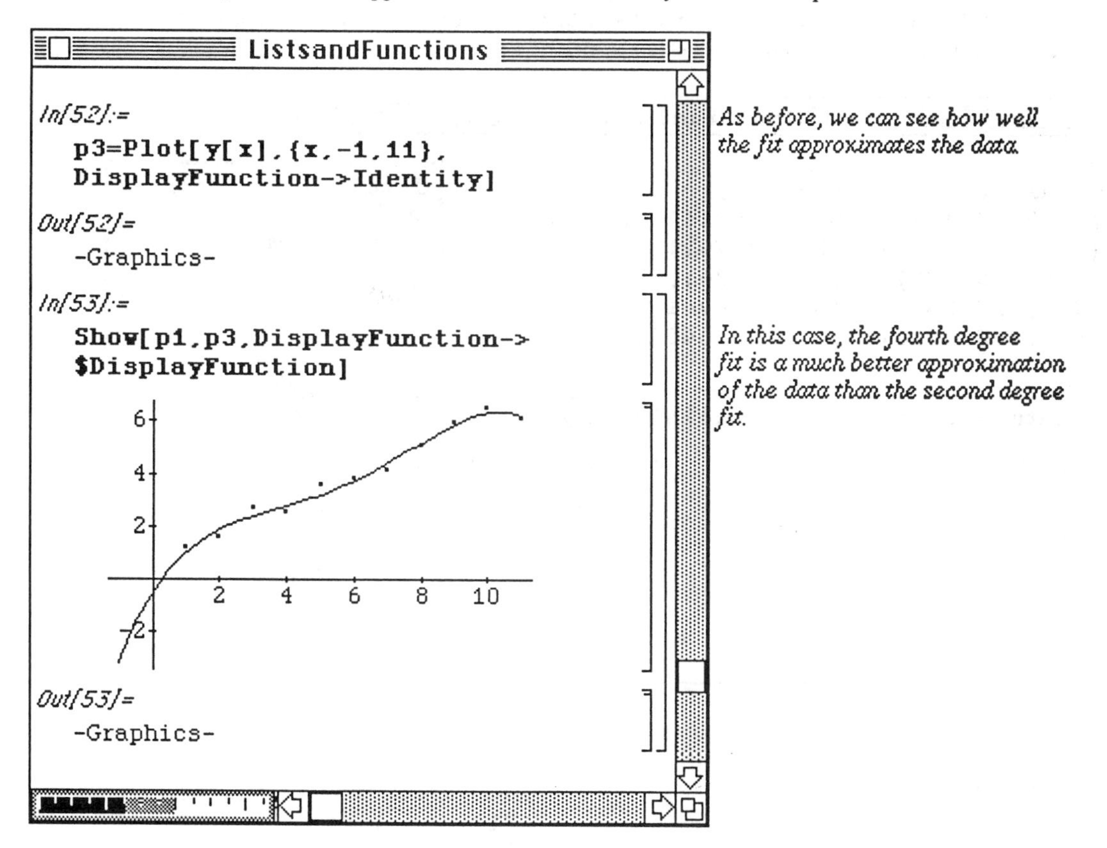

As before, we can see how well the fit approximates the data.

In this case, the fourth degree fit is a much better approximation of the data than the second degree fit.

Next, consider a list of data points made up of ordered pairs. These points are plotted below with **ListPlot**, and fitted with a polynomial of degree 3 using **Fit**. (Note that, in this case, since the data is given as ordered pairs, **ListPlot** plots the points as they are given in **datalist**.)

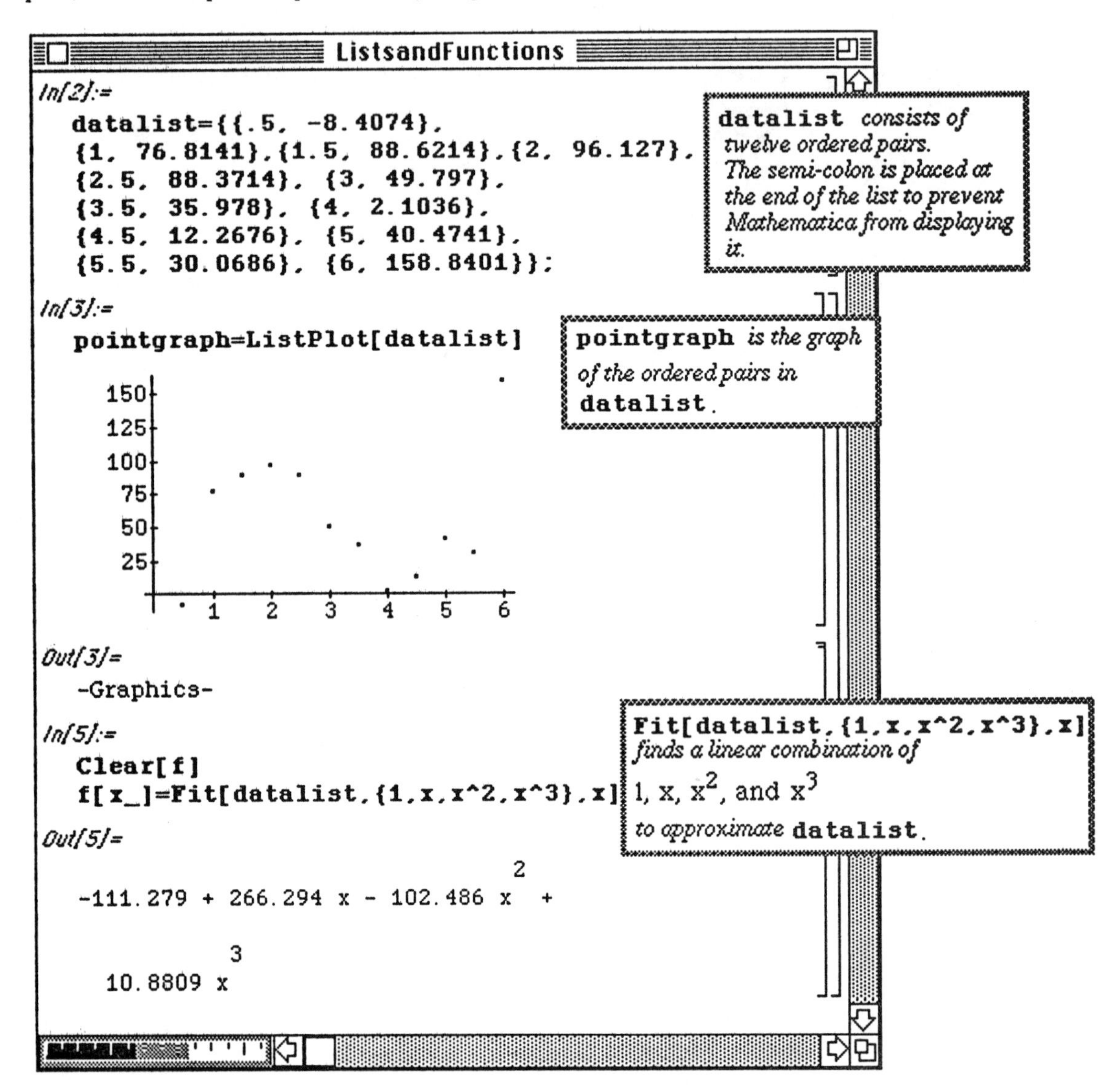

The list of data and the approximating curve f(x) are plotted together to check the accuracy. **Then, a polynomial of** degree 4 is used to fit the points.

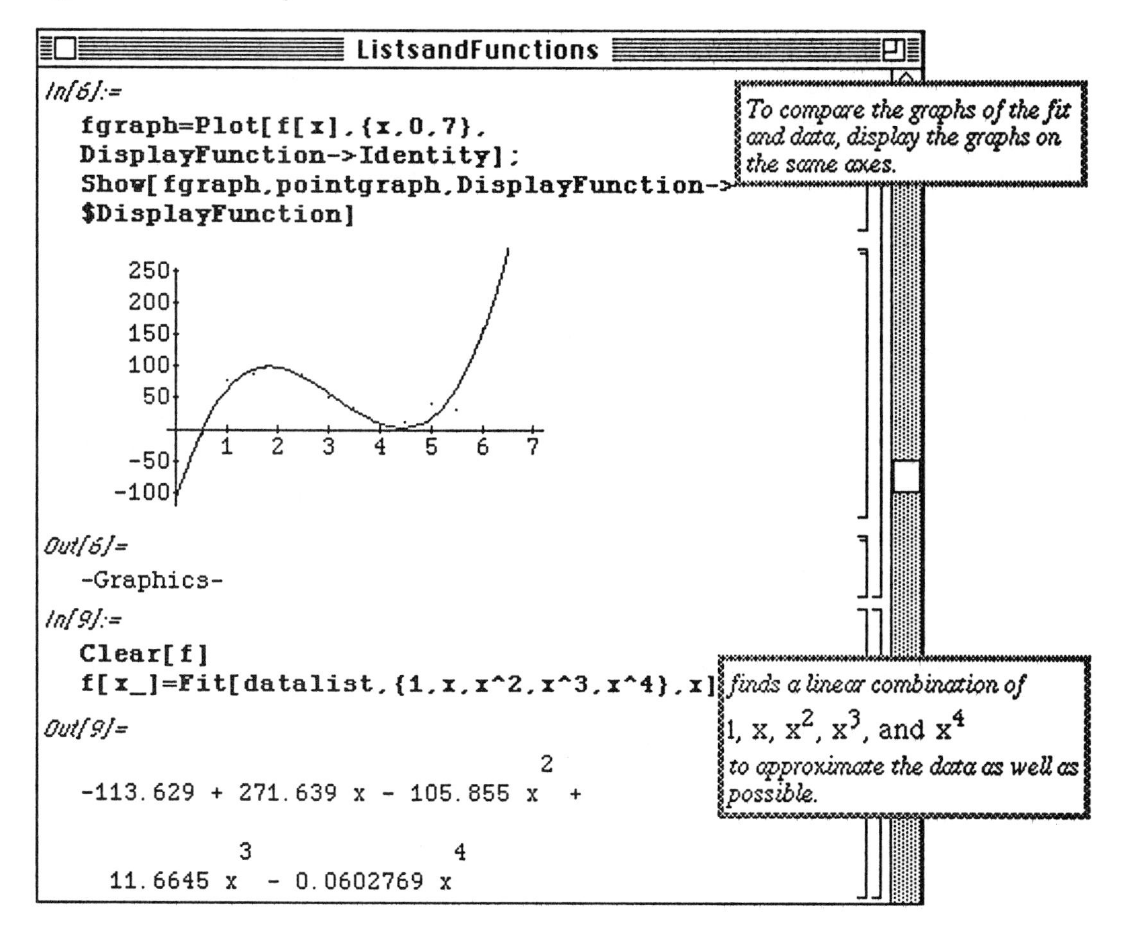

In this case, the fit resulting from `Fit[datalist,{1,x,x^2,x^3,x^4},x]` does not appear to be much more accurate than the fit resulting from `Fit[datalist,{1,x,x^2,x^3},x]`:

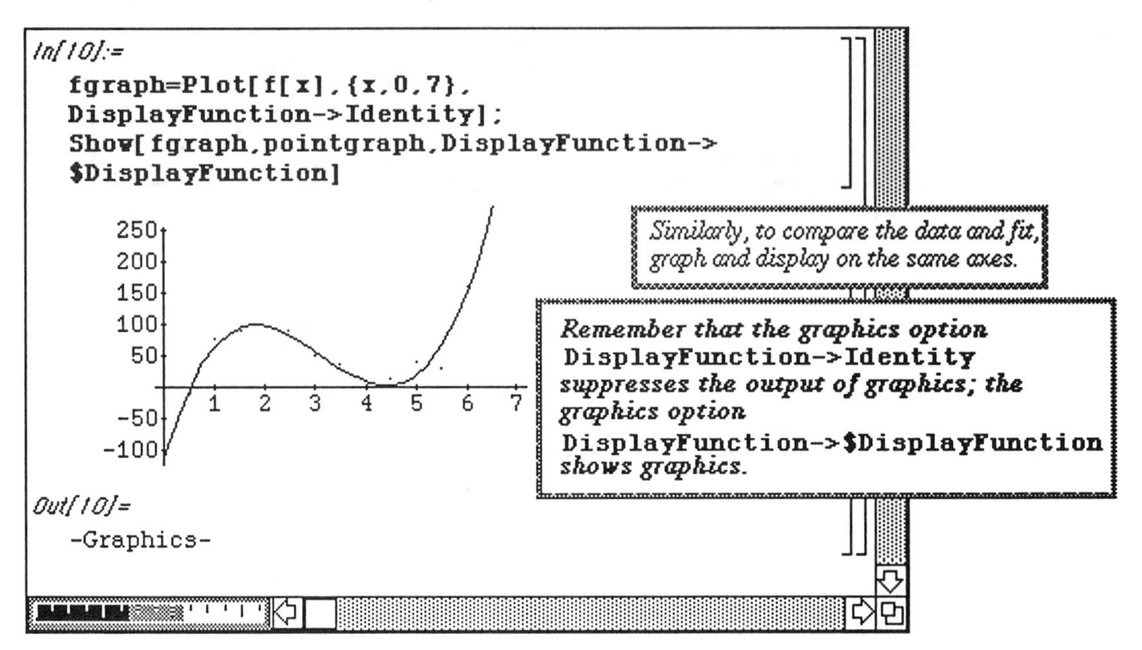

Again, consider a set of data points composed of ordered pairs. These points are listed and plotted below using `ListPlot`. In addition to curve-fitting with polynomials, *Mathematica* can also fit the data with trigonometric functions of the form $C_1 + C_2 \cos x + C_3 \sin x + C_4 \cos 2x + C_5 \sin 2x + \cdots$

The approximating function, called g, using the first three terms of the above expression is determined below:

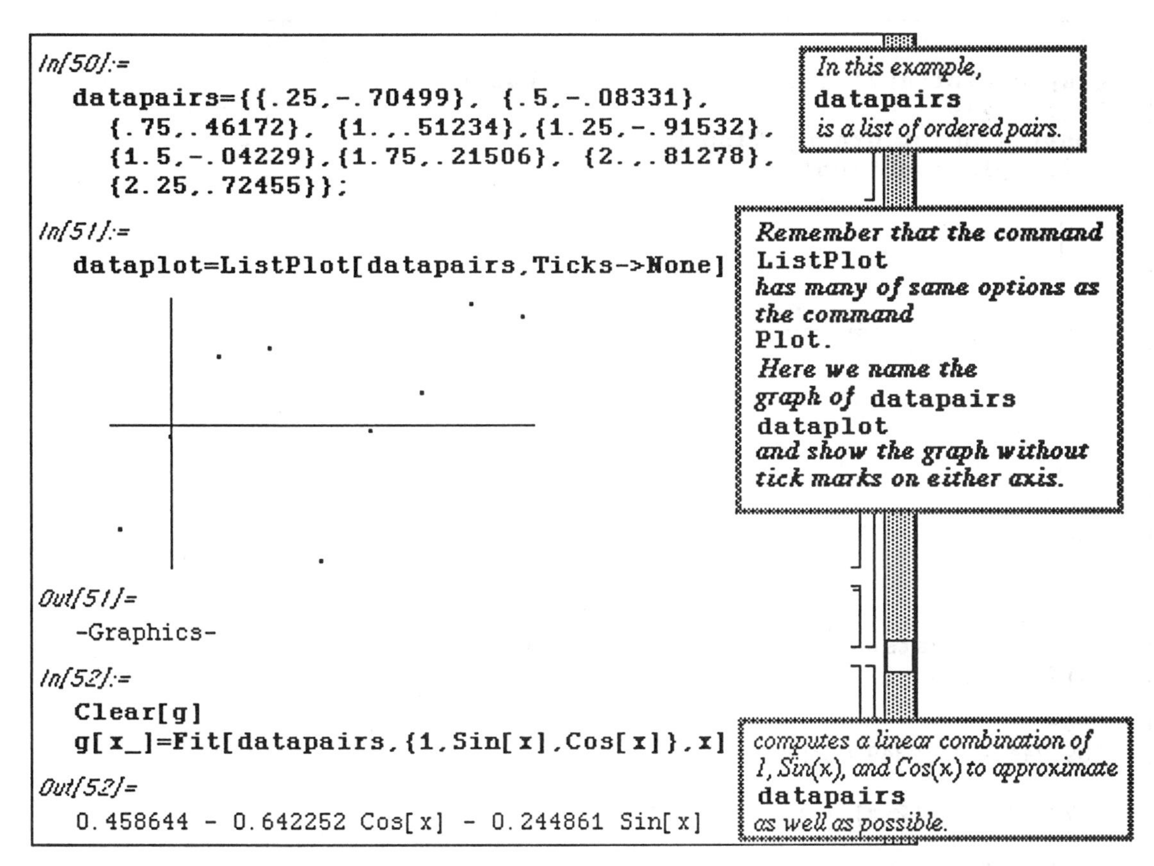

The trigonometric function g and the data are then plotted on the same graph. Afterwards, another approximation is calculated using a function of the form

$C_1 + C_2\text{Cos}x + C_3\text{Sin}x + C_4\text{Cos}2x + C_5\text{Sin}2x + C_6\text{Cos}3x + C_7\text{Sin}3x.$

This function is later plotted together with the set of data points and proves to be a much better fit.

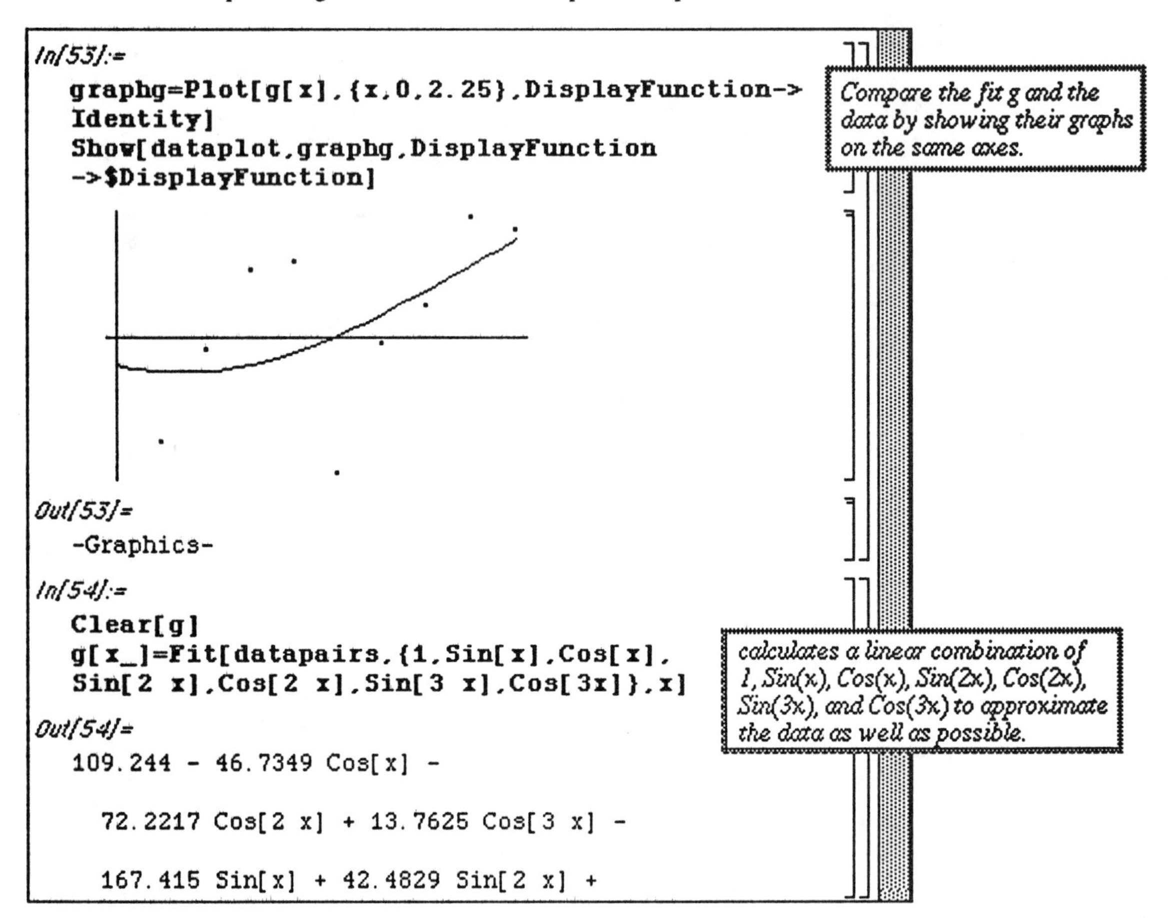

In this case, the fit, g, resulting from
`Fit[datapairs,{1,Sin[x],Cos[x],Sin[2x],Cos[2x],Sin[3x],Cos[3x]},x]` is much better than the fit resulting from `Fit[datapairs,{1,Sin[x],Cos[x]},x]`:

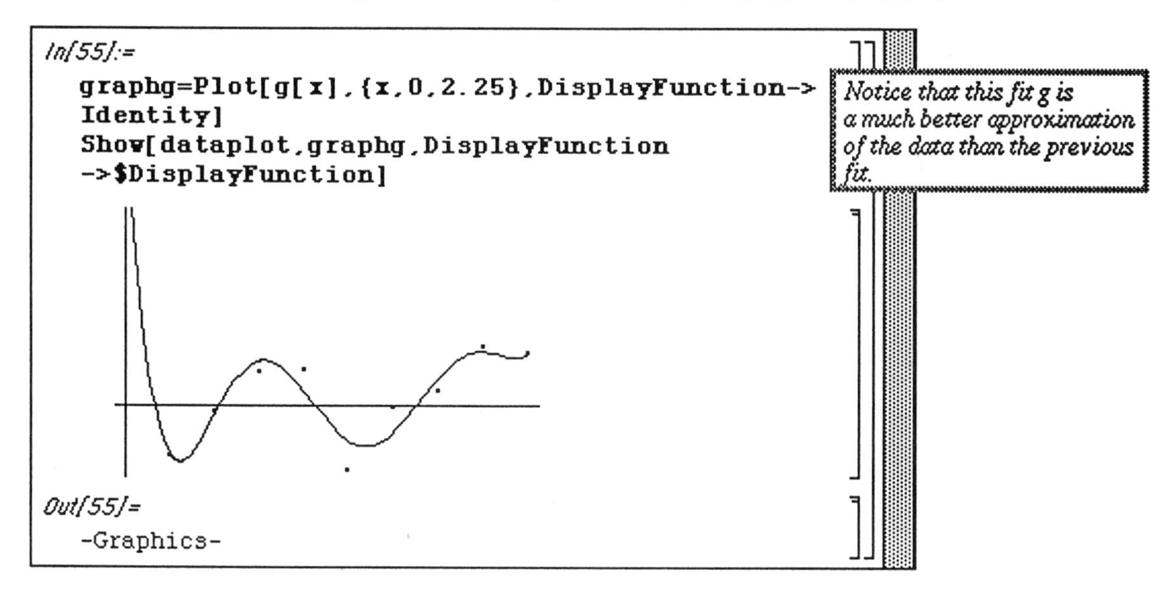

Mathematica supplies several packages which can be used to fit data using different techniques. For additional information regarding the different packages, see **Chapter 7.**

■ Application: Introduction to Fourier Series

Many problems in applied mathematics are solved through the use of Fourier series. *Mathematica* assists in the computation of these series in several ways. First, recall the definition :

The Fourier series of a periodic function f(x) with period 2L is the trigonometric series

$$a_0+\sum_{n=1}^{\infty}\left[a_n \operatorname{Cos}\left(\frac{n\pi x}{L}\right)+b_n \operatorname{Sin}\left(\frac{n\pi x}{L}\right)\right],\quad \text{where } a_0=\frac{1}{2L}\int_{-L}^{L} f(x)dx,\quad a_n=\frac{1}{L}\int_{-L}^{L} f(x)\operatorname{Cos}\left(\frac{n\pi x}{L}\right)dx,$$

and $b_n=\frac{1}{L}\int_{-L}^{L} f(x) \operatorname{Sin}\left(\frac{n\pi x}{L}\right)dx.$

The kth term of the Fourier series

$$a_0+\sum_{n=1}^{\infty}\left[a_n \operatorname{Cos}\left(\frac{n\pi x}{L}\right)+b_n \operatorname{Sin}\left(\frac{n\pi x}{L}\right)\right] \text{ is } a_k \operatorname{Cos}\left(\frac{k\pi x}{L}\right)+b_k \operatorname{Sin}\left(\frac{k\pi x}{L}\right)$$

The kth partial sum of the Fourier series

$$a_0+\sum_{n=1}^{\infty}\left[a_n \operatorname{Cos}\left(\frac{n\pi x}{L}\right)+b_n \operatorname{Sin}\left(\frac{n\pi x}{L}\right)\right] \text{ is } a_0+\sum_{n=1}^{k}\left[a_n \operatorname{Cos}\left(\frac{n\pi x}{L}\right)+b_n \operatorname{Sin}\left(\frac{n\pi x}{L}\right)\right].$$

It is a well-known theorem that if f(x) is a periodic function with period 2L and f'(x) is continuous on [-L,L] except at most finitely many points, then at each point x the Fourier series corresponding to f converges and

$$a_0 + \sum_{n=1}^{\infty}\left[a_n \operatorname{Cos}\left(\frac{n\pi x}{L}\right) + b_n \operatorname{Sin}\left(\frac{n\pi x}{L}\right)\right] = \frac{\lim_{y\to x+} f(y) + \lim_{y\to x-} f(y)}{2}.$$

In fact, if the series $\sum_{n=1}^{\infty}\left(|a_n| + |b_n|\right)$ converges, then the Fourier series $a_0 + \sum_{n=1}^{\infty}\left[a_n \operatorname{Cos}\left(\frac{n\pi x}{L}\right) + b_n \operatorname{Sin}\left(\frac{n\pi x}{L}\right)\right]$ converges uniformly on $\Re$.

Mathematica simplifies the process of determining the coefficients as well as assists in verifying the convergence of the series. Additional applications discussing Fourier Series will be discussed in **Chapter 5**. Consider the following example.

o Version 2.0 includes the package **`FourierTransform.m`** in the **`Calculus`** folder. **`FourierTransform.m`** contains several commands which can be used to compute exact or approximate Fourier series of some functions. The package **`FourierTransform.m`** is discussed in more detail in **Chapter 7**.

Example:

Let $f(x)=\begin{cases} 1 \text{ if } 0 \le x < 1 \\ -x \text{ if } -1 \le x < 0 \end{cases}$ and let g be the periodic extension of f of period 2.

Notice how the piecewise-defined function f is entered. The functions f and g are then plotted. The graph of g is named **graphg** for later use.

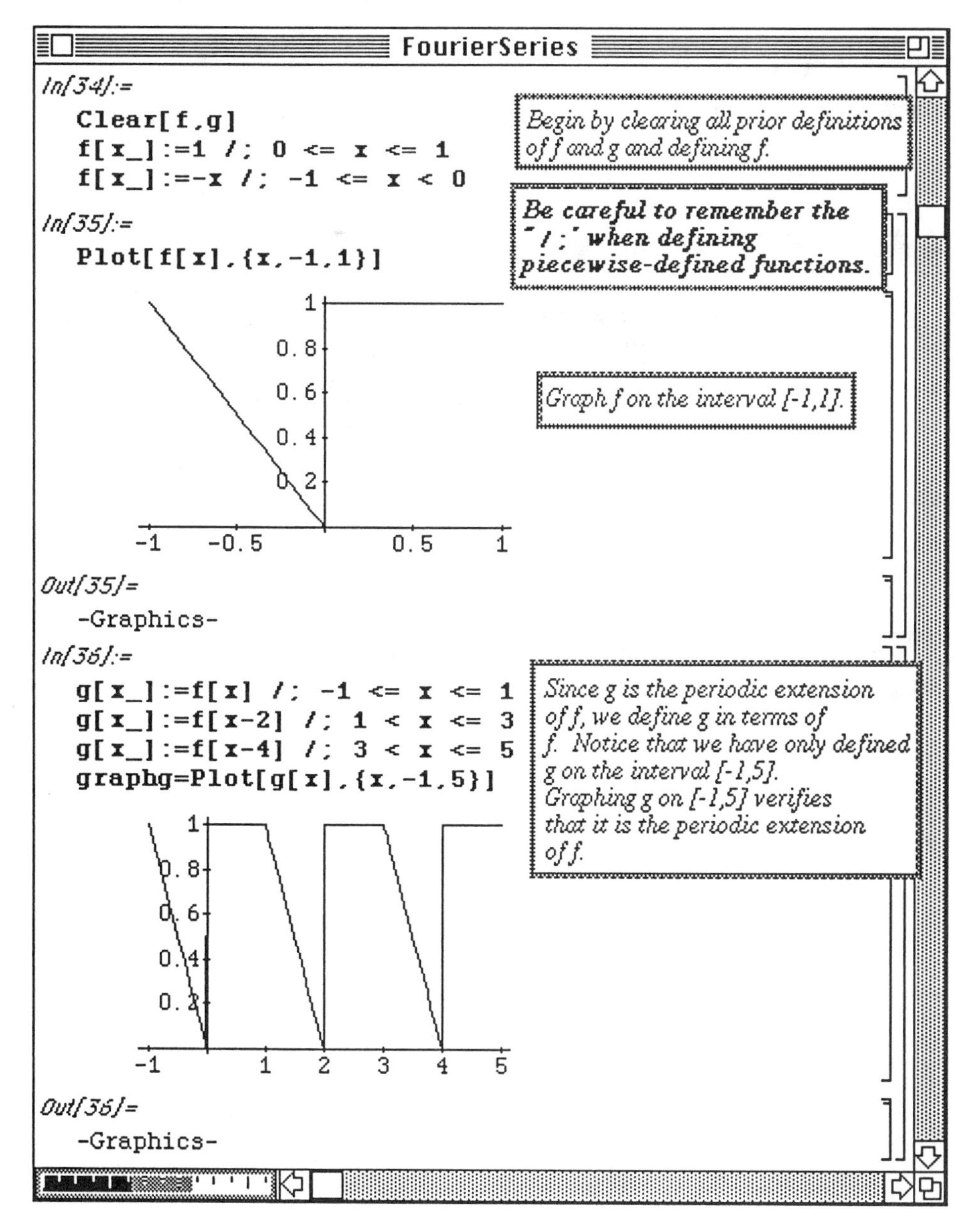

The Fourier series coefficients are computed with the integral formulae given earlier. Executing the command
L=1
a[0]=1/(2L) NIntegrate[f[x],{x,-L,L}] defines **L** to be 1 and **a[0]** to be an approximation of

the integral $\frac{1}{2L}\int_{-L}^{L} f(x)dx$. Executing the command

```
a[n_]:=1/L  NIntegrate[f[x]  Cos[n  Pi  x/L],{x,-L,L}]
b[n_]:=1/L  NIntegrate[f[x]  Sin[n  Pi  x/L],{x,-L,L}]
```

defines **a[n]** to be an approximation of the integral

$\frac{1}{L}\int_{-L}^{L} f(x)\mathrm{Cos}\left(\frac{n\pi x}{L}\right)dx$ and **b[n]** to be an approximation of the integral $\frac{1}{L}\int_{-L}^{L} f(x)\mathrm{Sin}\left(\frac{n\pi x}{L}\right)dx$.

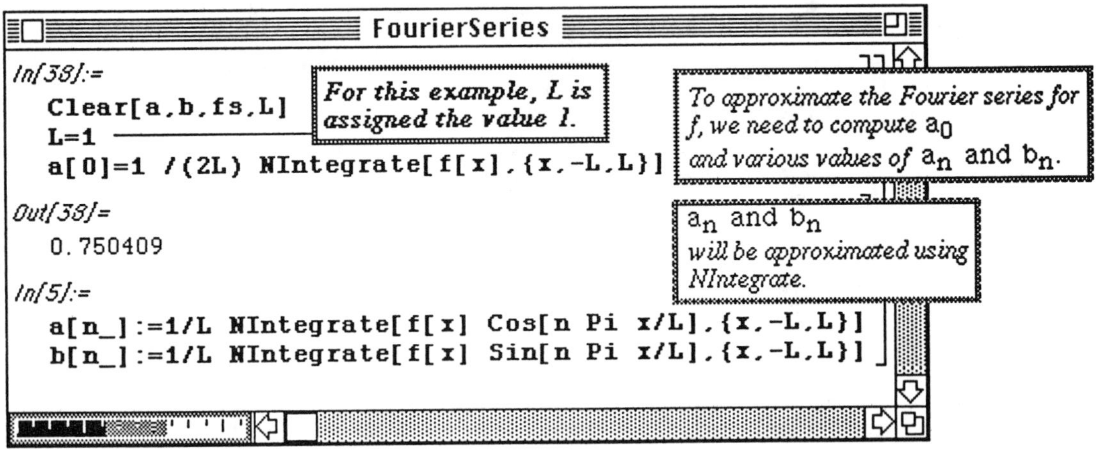

FourierSeries

In[38]:=

```
Clear[a,b,fs,L]
L=1
a[0]=1 /(2L) NIntegrate[f[x],{x,-L,L}]
```

For this example, L is assigned the value 1.

To approximate the Fourier series for f, we need to compute a_0 *and various values of* a_n and b_n.

Out[38]=

```
0.750409
```

a_n and b_n *will be approximated using NIntegrate.*

In[5]:=

```
a[n_]:=1/L NIntegrate[f[x] Cos[n Pi x/L],{x,-L,L}]
b[n_]:=1/L NIntegrate[f[x] Sin[n Pi x/L],{x,-L,L}]
```

A table containing the coefficients **a[i]** and **b[i]** for **i = 1, 2, 3,..., 10** is created. Notice how the elements of the table are extracted using double brackets with **coeffs**, the name of the table :

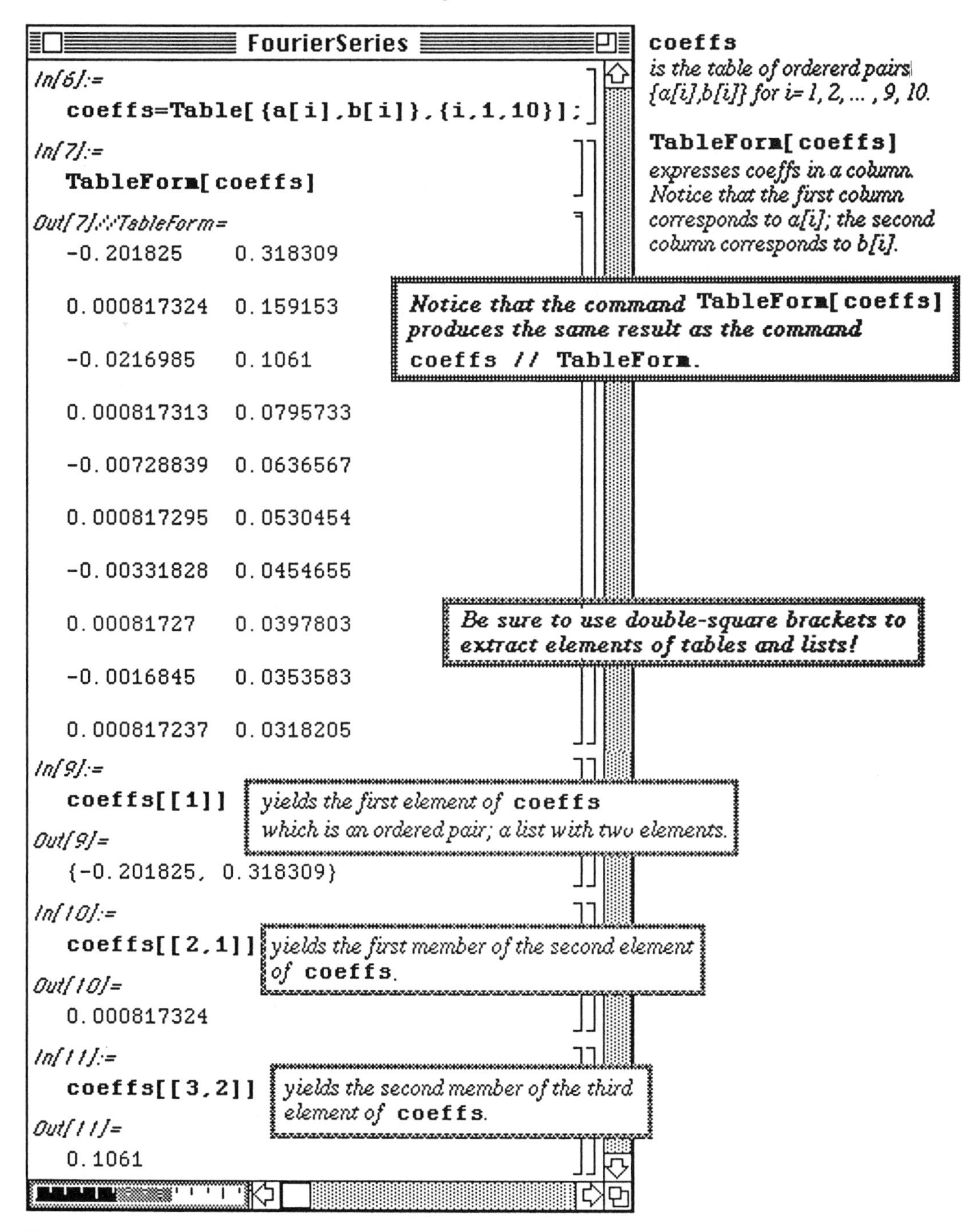

The command **Sum[f[i],{i,1,n}]** computes the sum

$$\texttt{f[1]+f[2]+...+f[n-1]+f[n]} = \sum_{i=1}^{n} \texttt{f[i]}.$$

Once the coefficients are calculated, the kth partial sum of the Fourier series is obtained with **Sum**.

The kth term of the series is:

```
fs[k_,x_] := coeffs[[k,1]] Cos[k Pi x] + coeffs[[k,2]] Sin[k Pi x]
```

where $k = 1, 2, \cdots$, $a_k =$ **coeffs [[k,1]]**, and $b_k =$ **coeffs [[k,2]]**.

Therefore, the nth partial sum (the sum of the first n terms) of the Fourier series is obtained by summing the **fs[k,x]** over k from k = 1 to k = n and adding to this summation the coefficient **a[0]**:

$$\text{nth partial sum} = \texttt{a[0]+fs[1,x]+...+fs[n-1,x]+fs[n,x]} = \texttt{a[0]} + \sum_{k=1}^{n} \texttt{fs[k,x]}.$$

This is defined below with the function **fourier[n,x]**. Several examples of this function are then given. Note that the largest value of **n** which can be used is **n** = 10 since the coefficients for the Fourier series have not been calculated for larger values of **n**.

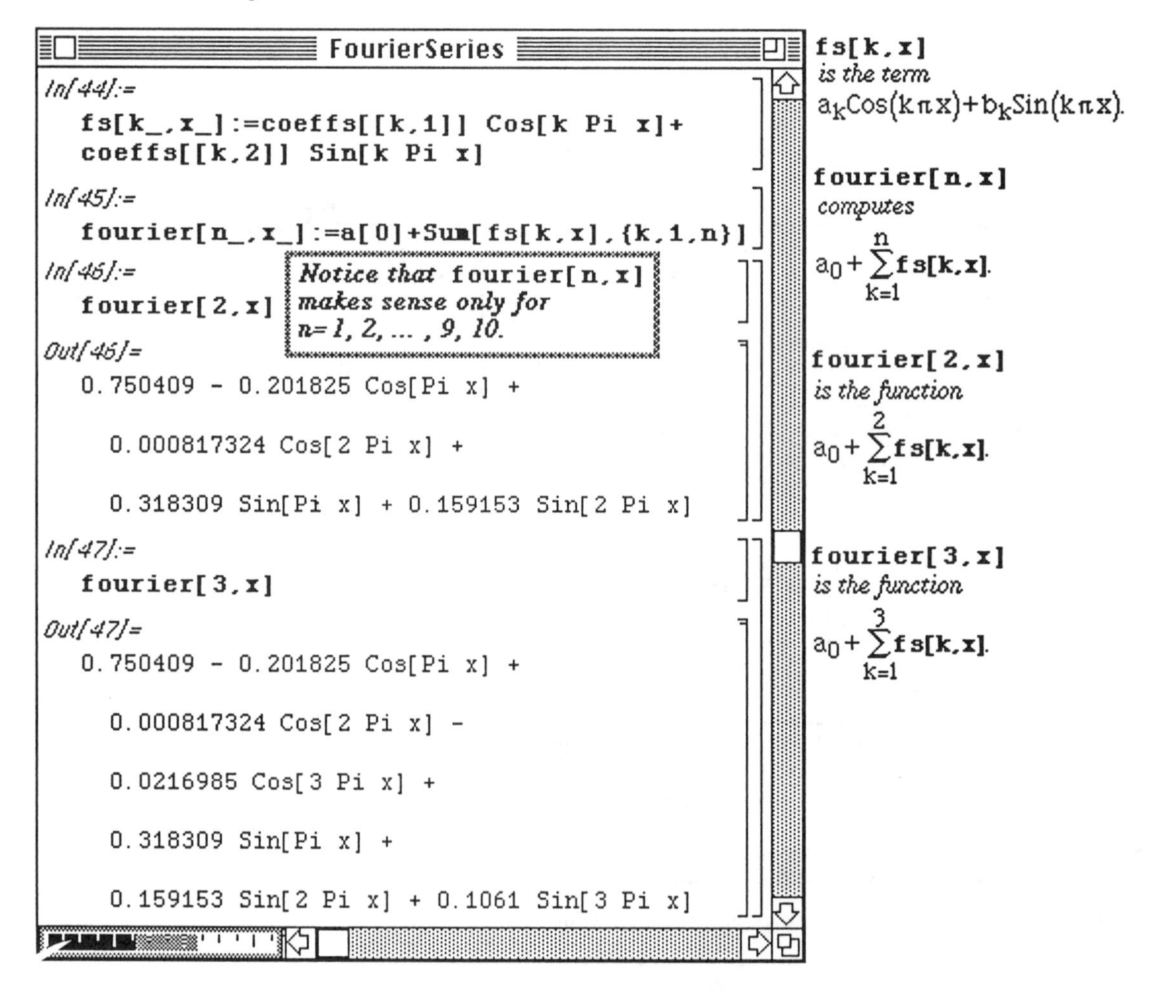

To see how the Fourier series approximates the function, the function g is plotted together with the graph of the Fourier series with n = 2. This does not appear to be a very good approximation. (The resulting graph is named **`graphone`**.)

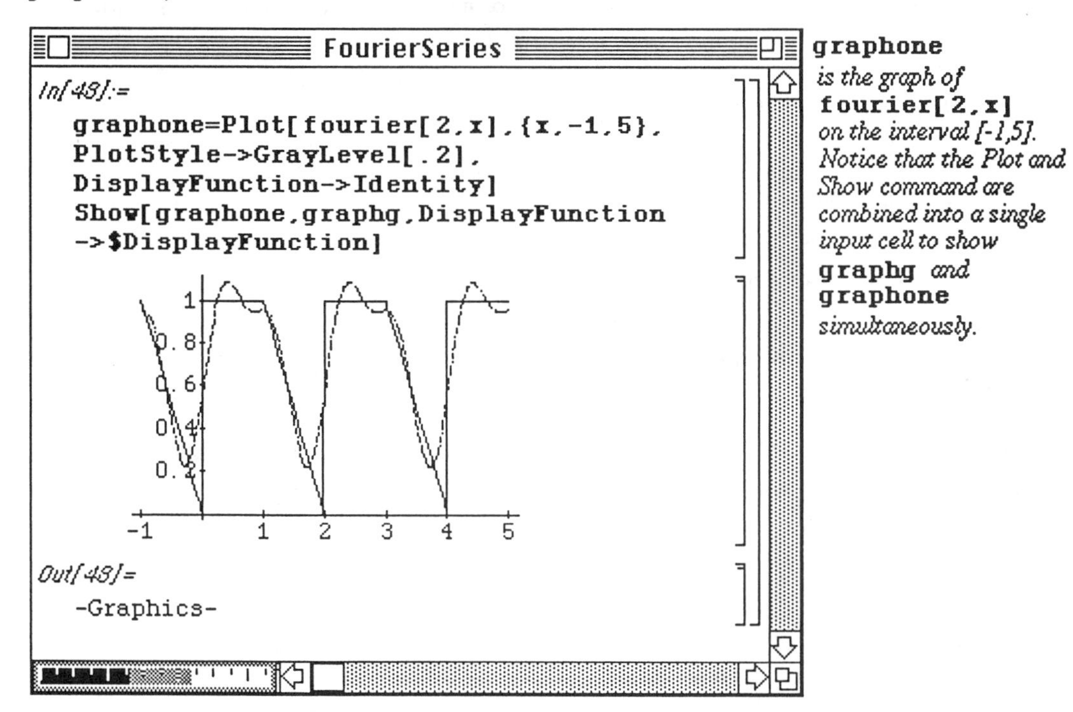

The use of more terms in the Fourier series yields a more accurate approximation of g. Graphs are shown for n = 5 and n = 9. These are named **`graphfive`** and **`graphnine`**, respectively.

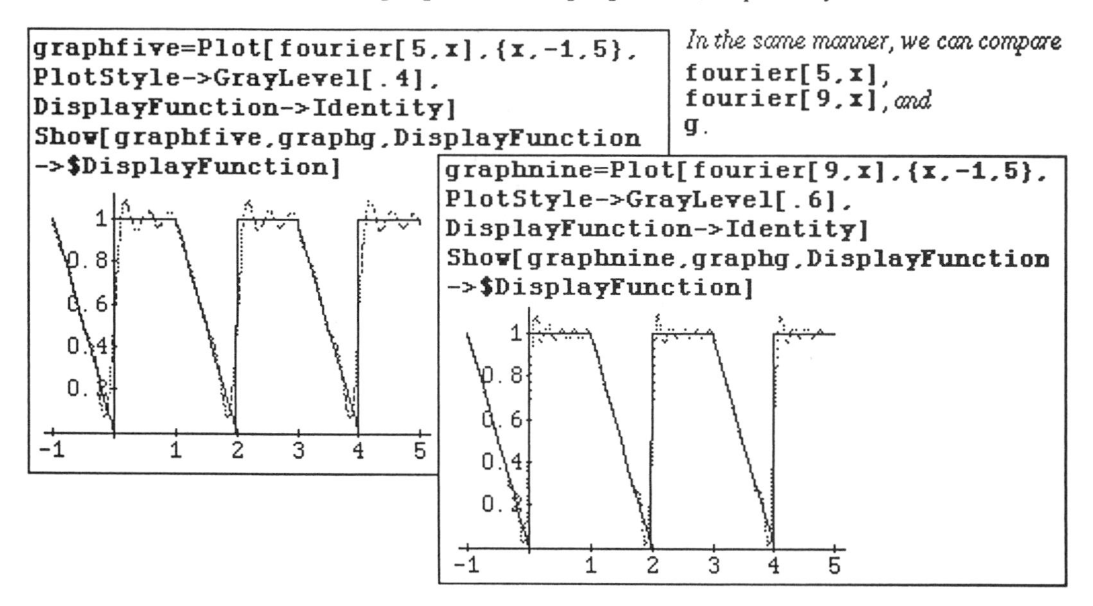

□ **Example: (One-Dimensional Heat Equation)**

A typical problem in applied mathematics which involves the use of Fourier series is that of the one-dimensional heat equation. This initial-value problem which describes the temperature in a uniform rod with insulated surface is given by:

(1) $k\frac{\partial^2 u}{\partial x^2}=\frac{\partial u}{\partial t}, \quad 0 < x < a,\ t > 0;$

(2) $u(0,t)=T_0; \quad t > 0;$

(3) $u(a,t)=T_a; \quad t > 0;$ and

(4) $u(x,0)=f(x), \quad 0 < x < a.$

The solution to the problem is well-known:

$$u(x,t)=\underbrace{T_0+\frac{x(T_a-T_0)}{a}}_{v(x)}+\sum_{n=1}^{\infty} b_n \mathrm{Sin}(\lambda_n x)e^{-\lambda_n^2 kt}, \text{ where } \lambda_n=\frac{n\pi}{a} \text{ and } b_n=\frac{2}{a}\int_0^a [f(x)-v(x)]\mathrm{Sin}\left(\frac{n\pi x}{a}\right)dx$$

and is obtained through separation of variables techniques The coefficient b_n in the solution $u(x,t)$ is the Fourier series coefficient b_n of the function $f(x)-v(x)$, where $v(x)$ is the steady-state temperature.

Consider the heat equation with initial temperature distribution $f(x) = -(x-1)\,\mathrm{Cos}(\pi x)$.

The steady−state temperature for this problem is $v(x)=1-x$, and the eigenvalue, λ_n, is given by $\frac{n\pi}{2}$.

The function f is defined and plotted below. Also, the steady-state temperature, v(x), and the eigenvalue are defined. Finally, **NIntegrate** is used to define a function which will be used to calculate the coefficients of the solution.

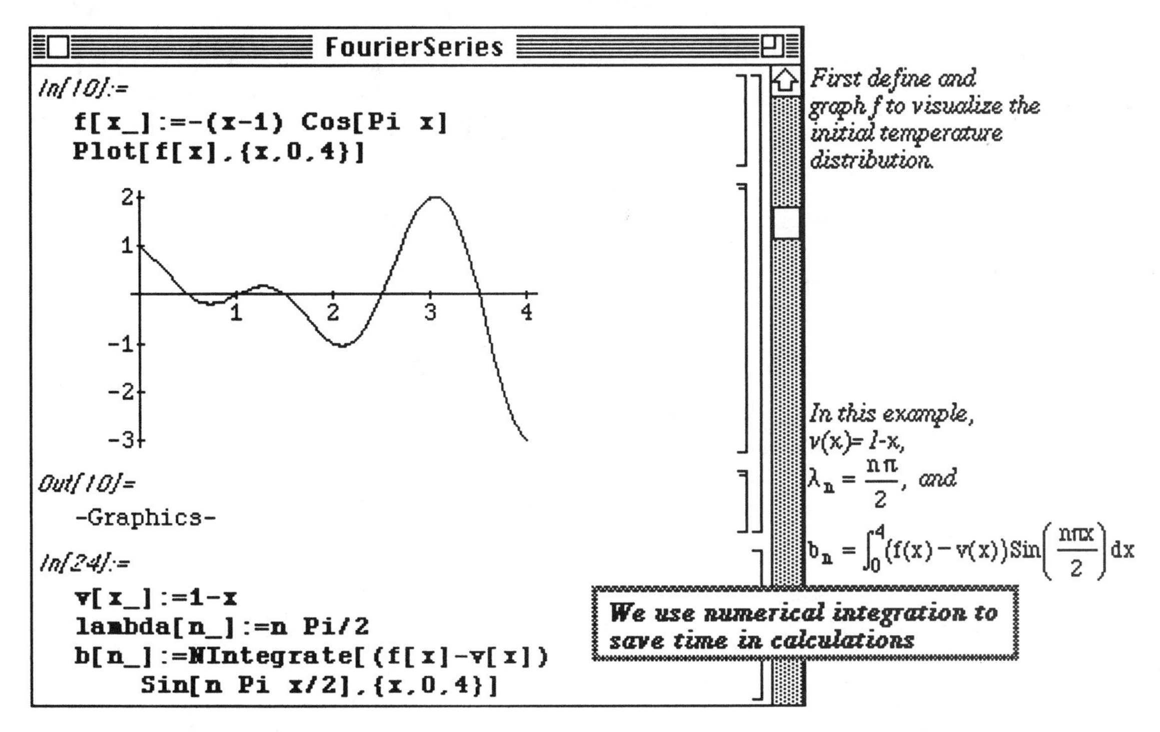

Let $S_m = b_m \mathrm{Sin}(\lambda_m x) e^{-\lambda_m{}^2 kt}$. Then the desired solution u(x,t) is given by $u(x,t) = v(x) + \sum_{m=1}^{\infty} S_m$.

Let $u(x,t,n) = v(x) + \sum_{m=1}^{n} S_m$.

Notice that $u(x,t,n) = u(x,t,n-1) + S_n$. Consequently approximations of the solution to the heat equation are obtained recursively. The solution is first defined for n = 1, **u[x,t,1]**.

Subsequent partial sums, **u[x,t,n]**, are obtained by adding the nth term of the series,

$S_n = b_n \mathrm{Sin}(\lambda_n x) e^{-\lambda_n{}^2 kt}$ to **u[x,t,n −1]**.

```
In[25]:=
u[x_,t_,1]:=v[x]+b[1] Sin[lambda[1] x
    ]Exp[-lambda[1]^2 1/4 t]
u[x_,t_,n_]:=u[x,t,n-1]+b[n] Sin[
    lambda[n] x]Exp[-lambda[n]^2 1/4 t]
```

Notice u(x,t,n) is obtained via adding the nth term to u(x,t,n-1). Hence, we take advantage of Mathematica's ability to compute recursively.

By defining the solution in this manner a table can be created which includes the partial sums of the solution :

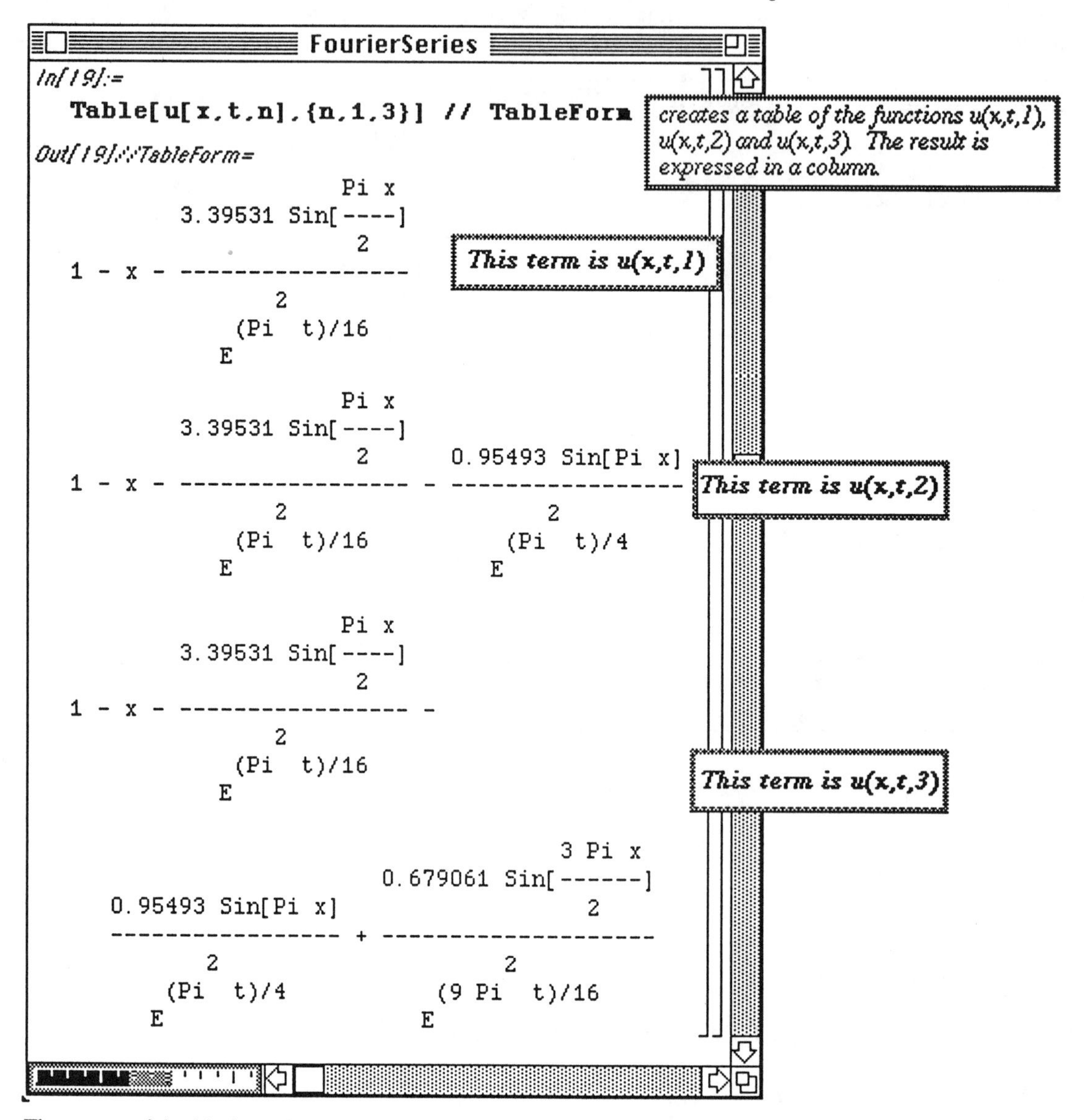

The syntax of the *Mathematica* command **`Do`** is similar to the syntax of the command **`Table`**. The command **`Do[statement[i],{i,istart,istop,istep}]`** instructs *Mathematica* to execute **`statement[i]`** for values of **`i`** beginning with **`istart`** and continuing through **`istop`** in increments of **`istept`**.

The solution with n = 8 is plotted below for t = 0 to t = 6 using a step-size in t of 6/20. Remember that **u[x,t,n]** is determined with a **Table** command. Therefore, **Release** must be used in the **Do** command below so that *Mathematica* first computes the solution u and then evaluates u at the particular values of **x**. Otherwise, u is recalculated for each value of x.

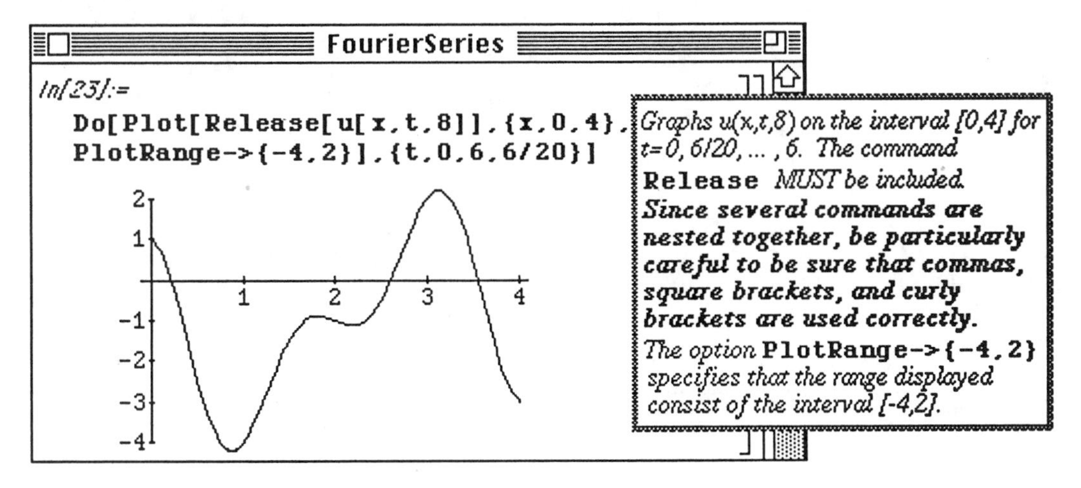

The plots of the solution obtained above can be animated :

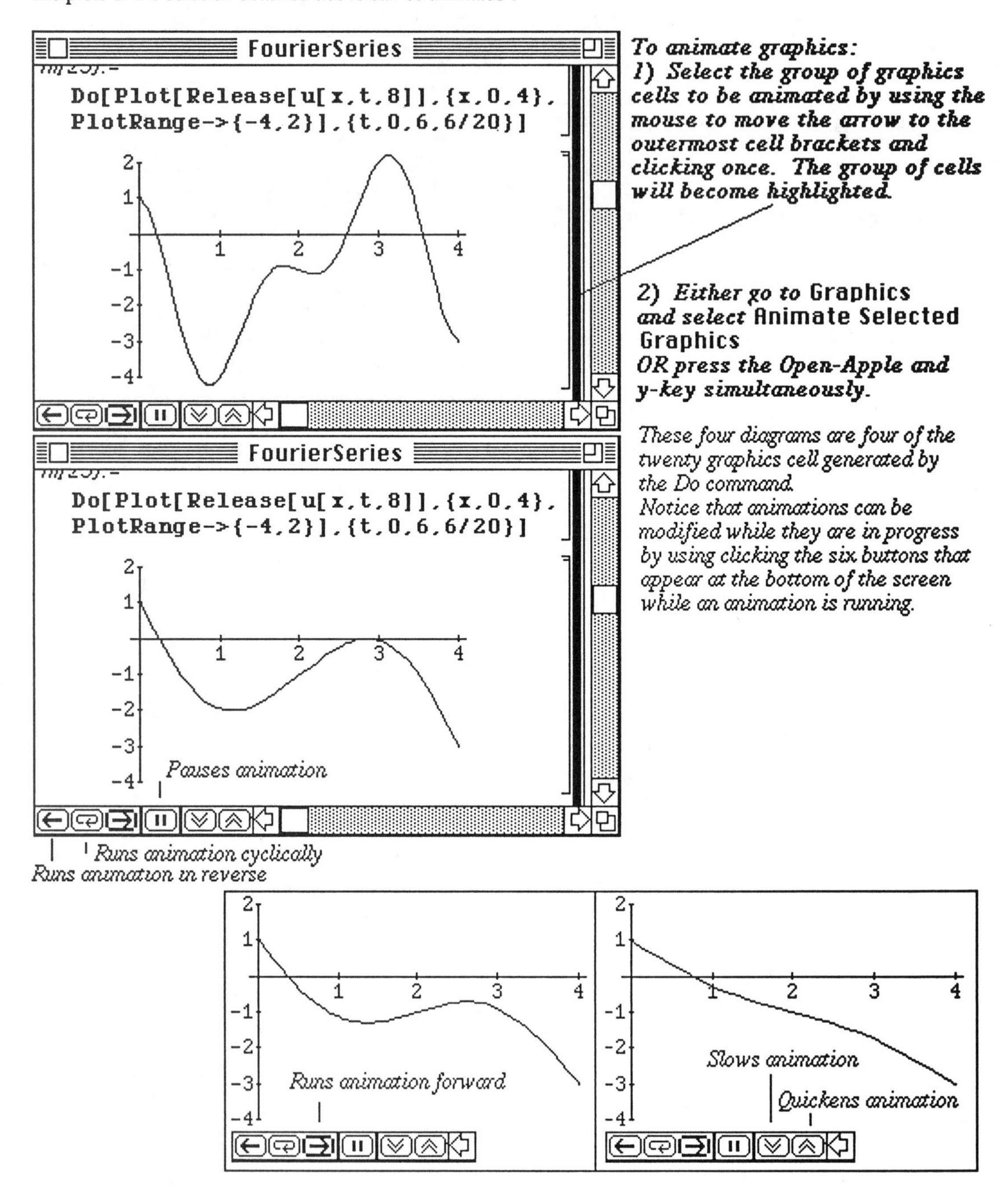

Chapter 5
Introduction to Nested Lists: Matrices and Vectors

- **Chapter 5** discusses operations on matrices and vectors, including vector calculus and systems of equations. Several linear programming examples are discussed.

- Commands introduced and discussed in this chapter from **Version 1.2** include:

Operations on Matrices

```
matrixa+matrixb
matrixa.matrixb
Det[matrix]
Eigenvalues[matrix]
Eigenvectors[matrix]
IdentityMatrix[positiveinteger]
Inverse[matrix]
Transpose[matrix]
LinearSolve[matrix,vector]
MatrixPower[matrix,positiveinteger]
MatrixForm[matrix]
```

Operations to Create Lists and Tables

```
Table[expression,{positiveinteger}]
Array[variable,positiveinteger]
```

Vector Calculus

```
<<VectorAnalysis.m
Grad[scalarfield,coordinatesystem]
Laplacian[scalarfield,coordinatesystem]
Div[vectorfield,coordinatesystem]
Curl[vectorfield,coordinatesystem]
SetCoordinates[System]
```

Linear Programming

```
ConstrainedMin[function,{inequalities},{variables}]
ConstrainedMax[function,{inequalities},{variables}]
LinearProgramming[vectorc,matrixa,vectorb]
```

Other Commands

```
TrigExpand[expression]
BesselJ[alpha,x]
Print[expression]
```

Saving and Appending Output for Future Mathematica Sessions

```
>>filename
>>>filename
```

Commands introduced and discussed in this chapter from **Version 2.0** include:

```
Expand[Trig->True]
```

- Applications in this chapter include linear programming, vector calculus, and saving results for future *Mathematica* sessions.

5.1 Nested Lists: Introduction to Matrices, Vectors and Matrix Operations

Defining Matrices and Vectors

Matrix algebra can be performed with *Mathematica*. Before introducing the operations involved in matrix algebra, the method by which a matrix is entered must first be discussed. In *Mathematica*, a matrix is simply a list of lists where each list represents a row of the matrix. Therefore, the m x n matrix

$$A = \left[a_{i,j}\right] = \begin{bmatrix} a_{1,1} & a_{1,2} & a_{1,3} & \cdots & a_{1,n} \\ a_{2,1} & a_{2,2} & a_{2,3} & \cdots & a_{2,n} \\ a_{3,1} & a_{3,2} & a_{3,3} & \cdots & a_{3,n} \\ \vdots & \vdots & \vdots & & \vdots \\ a_{m,1} & a_{m,2} & a_{m,3} & \cdots & a_{m,n} \end{bmatrix}$$ is entered in the following manner:

$A = \{\{a_{1,1}, a_{1,2}, a_{1,3}, \cdots, a_{1,n}\}, \{a_{2,1}, a_{2,2}, a_{2,3}, \cdots, a_{2,n}\}, \cdots, \{a_{m,1}, a_{m,2}, a_{m,3}, \cdots, a_{m,n}\}\}$.

For example, to use *Mathematica* to define **m** to be the matrix $\begin{bmatrix} a[1,1] & a[1,2] \\ a[2,1] & a[2,2] \end{bmatrix}$, execute the command **m = {{a [1,1], a [1,2]},{a [2,1], a [2,2]}}**.

Another way to create a matrix is to use the command **Array**. The command **m=Array[a, {2,2}]** produces the same result as above.

The following examples illustrate the definition of a 3 x 3 matrix.

□ **Example:**

`matrixa=Table[a[i,j],{i,1,3},{j,1,3}]` and `matrixaprime=Array[a,{3,3}]` produce the same result.

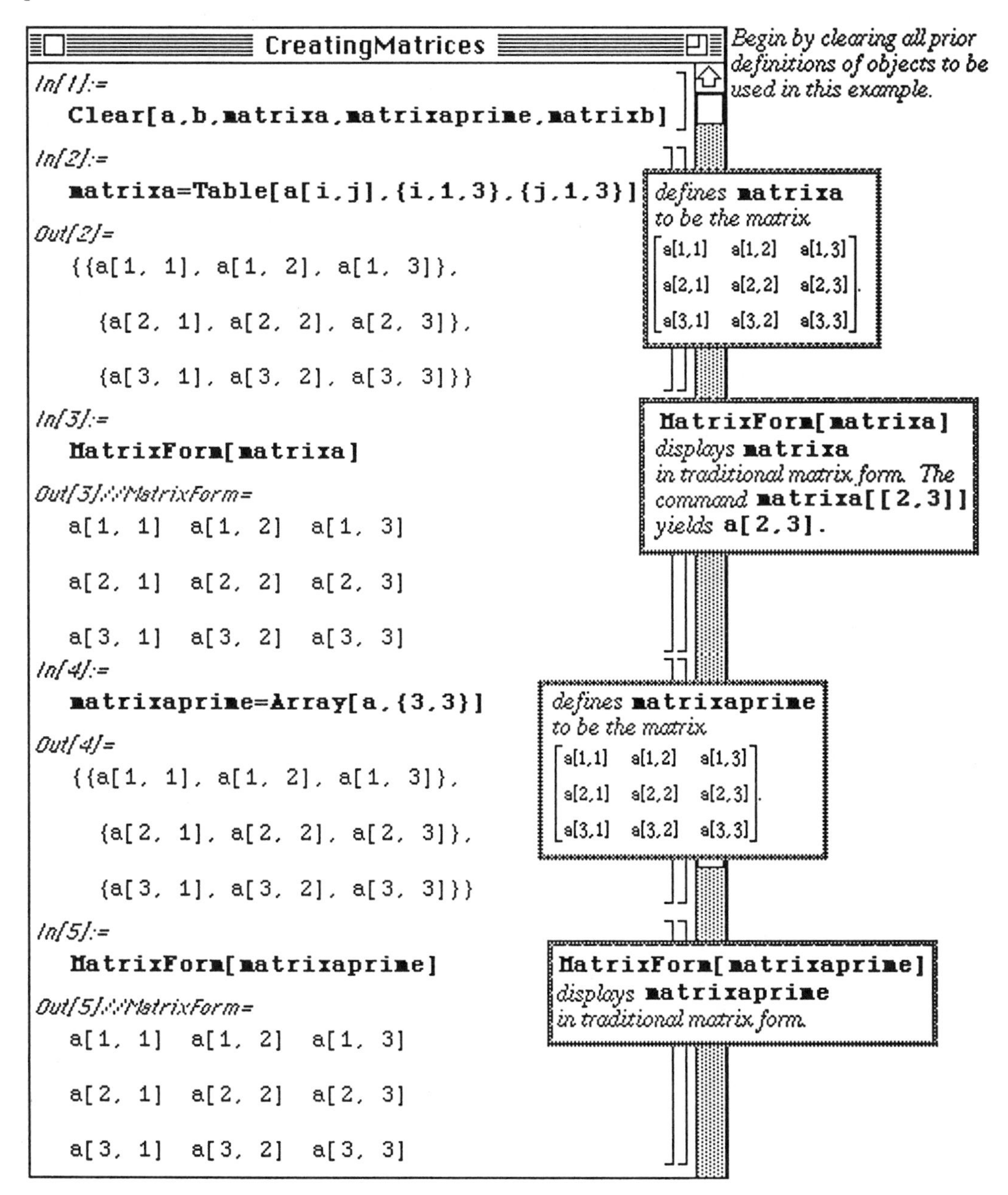

In addition, non-square matrices may also be defined using *Mathematica*. For example,

to define **matrixb** to be the matrix $\begin{bmatrix} b[1,1] & b[1,2] & b[1,3] & b[1,4] \\ b[2,1] & b[2,2] & b[2,3] & b[2,4] \end{bmatrix}$, enter either

matrixb=Table[b[i,j],{i,1,2},{j,1,4}] or **matrixb=Array[b,{2,4}]**.

```
In[6]:=
  matrixb=Array[b,{2,4}]
Out[6]=
  {{b[1, 1], b[1, 2], b[1, 3], b[1, 4]},

    {b[2, 1], b[2, 2], b[2, 3], b[2, 4]}}
In[7]:=
  MatrixForm[matrixb]
Out[7]//MatrixForm=
  b[1, 1]  b[1, 2]  b[1, 3]  b[1, 4]

  b[2, 1]  b[2, 2]  b[2, 3]  b[2, 4]
```

Array[b,{2,4}]
yields the 2 x 4 matrix
$\begin{bmatrix} b[1,1] & b[1,2] & b[1,3] & b[1,4] \\ b[2,1] & b[2,2] & b[2,3] & b[2,4] \end{bmatrix}$.
The same result would have been obtained using the command
Table[b[i,j],{i,1,2},{j,1,4}]

□ Example:

Use *Mathematica* to create the matrix **matrixc**,

$$\begin{bmatrix} c[1,1] & c[1,2] & c[1,3] & c[1,4] \\ c[2,1] & c[2,2] & c[2,3] & c[2,4] \\ c[3,1] & c[3,2] & c[3,3] & c[3,4] \end{bmatrix},$$ where $c(i,j)$ is the numerical value of $\operatorname{Cos}\left(j^2-i^2\right)\operatorname{Sin}\left(i^2-j^2\right)$.

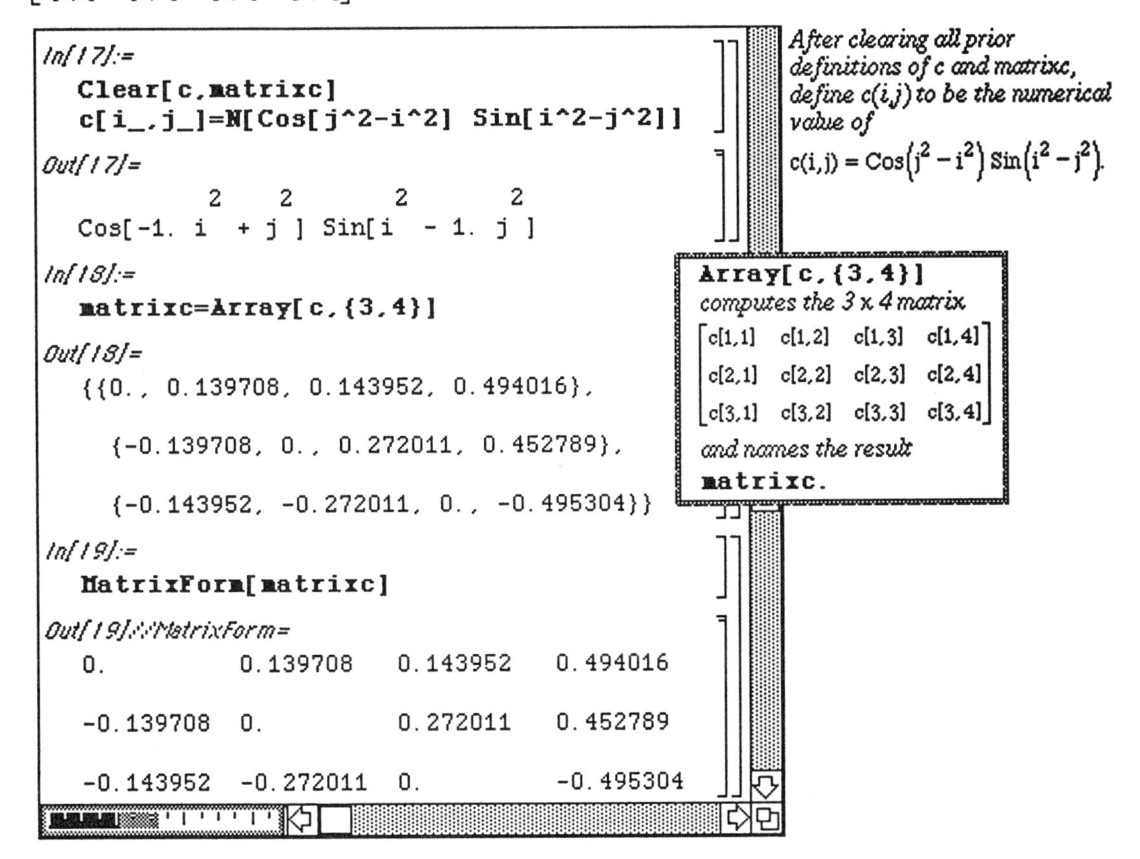

A matrix is a nested list. For the 2 x 2 matrix **m={{a[1,1],a[1,2]},{a[2,1],a[2,2]}}** defined earlier, **m[[1]]** yields the first element of matrix **m** which is the list **{a[1,1],a[1,2]}; m[[2,1]}** yields the first element of the second element of matrix **m** which is **a[2,1]**. In general, if **matrixm** is an m x n matrix, **matrixm[[i,j]]** yields the unique element in the ith row and jth column.

Once a matrix has been entered, it can be placed in the usual form (with rows and columns) using the command **MatrixForm[A]**.

In *Mathematica*, a <u>vector</u> is a list of numbers. For example, to use *Mathematica*

to define the row vector **vectorv** to be $\left[v[1], \ v[2], \ v[3]\right]$,

enter **vectorv={v[1],v[2],v[3]}**.

Similarly, to define the column vector **vectorv** to be $\begin{bmatrix} v[1] \\ v[2] \\ v[3] \end{bmatrix}$,

enter **vectorv={v[1],v[2],v[3]}**. Thus, *Mathematica* does not distinguish between row and column vectors. Nevertheless, *Mathematica* performs computations with vectors and matrices correctly.

■ Extracting Elements of Matrices:

Once **matrixa** has been defined via either

```
matrixa=
{{a[1,1],...,a[1,n]},{a[2,1],...,a[2,n]},...,{a[m,1],...a[m,n]}}
```

or **matrixa=Array[a,{m,n}]**, the unique element of **matrixa** in the **i**th row and **j**th column is obtained with **matrixa[[i,j]]**.

□ Example:

In the previous examples, **mb** was defined to be the matrix $\begin{bmatrix} 10 & -6 & -9 \\ 6 & -5 & -7 \\ -10 & 9 & 12 \end{bmatrix}$.

mb[[i,j]] yields the (unique) number in the **i**th row and **j**th column of **mb**. The determinant of **mb** can be calculated with **Det[mb]**. Observe how various components of **mb** (rows,elements) can be extracted and how **mb** is placed in **MatrixForm**.

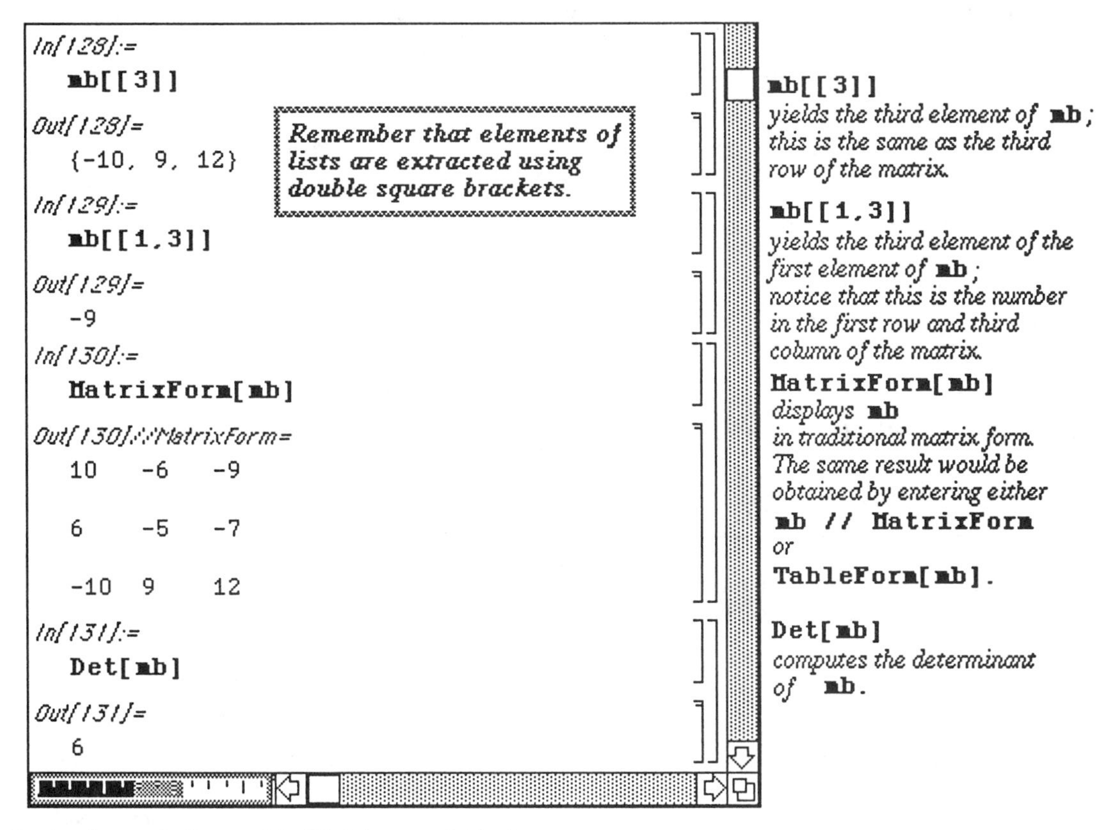

■ Basic Computations with Matrices and Vectors:

Mathematica performs all of the usual operations on matrices. Matrix addition (**A+B**), scalar multiplication (**kA**), matrix multiplication (**A.B**), and combinations of these operations are all possible. In addition, the transpose of a matrix **A** is found with the built-in command **Transpose[A]**.

If A and B are n x n matrices satisfying $AB = BA = I$. Then <u>B is the inverse of A</u> and B is denoted by

A^{-1}. Then $AA^{-1} = A^{-1}A = I$.

The inverse of a matrix **A**, provided it exists, is found with the built-in command **Inverse[A]**.

Recall that if $A = [a_{i,j}]$ then the **transpose** of A is denoted A^t where $A^t = [a_{j,i}]$.

Equivalently, the transpose of A is obtained by interchanging the rows and columns of A.

□ **Example:**

Use *Mathematica* to define matrix **ma** to be $\begin{bmatrix} 3 & -4 & 5 \\ 8 & 0 & -3 \\ 5 & 2 & 1 \end{bmatrix}$ and matrix **mb** to be $\begin{bmatrix} 10 & -6 & -9 \\ 6 & -5 & -7 \\ -10 & 9 & 12 \end{bmatrix}$.

Compute (i) **ma+mb**; (ii) **mb-4ma**; (iii) the inverse of **ma.mb**; and (iv) the transpose of **(ma-2mb).mb**.

As described above, we enter **ma** and **mb** as nested lists where each element corresponds to a row of the matrix:

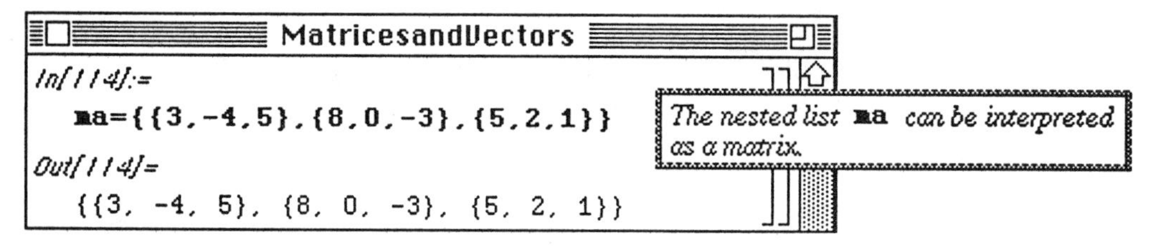

mb is defined similarly:

```
In[127]:=
  Clear[mb]
  mb={{10,-6,-9},{6,-5,-7},{-10,9,12}}
Out[127]=
  {{10, -6, -9}, {6, -5, -7}, {-10, 9, 12}}
```

Define **mb** *to be the nested list {{10,-6,-9},{6,-5,-7},{-10,9,12}}. This is interpreted as the matrix*

$$\begin{bmatrix} 10 & -6 & -9 \\ 6 & -5 & -7 \\ -10 & 9 & 12 \end{bmatrix}.$$

Matrices can be expressed in traditional matrix form using either the command **MatrixForm** or **TableForm**. The operations are performed and the resulting matrix is expressed in traditional matrix form:

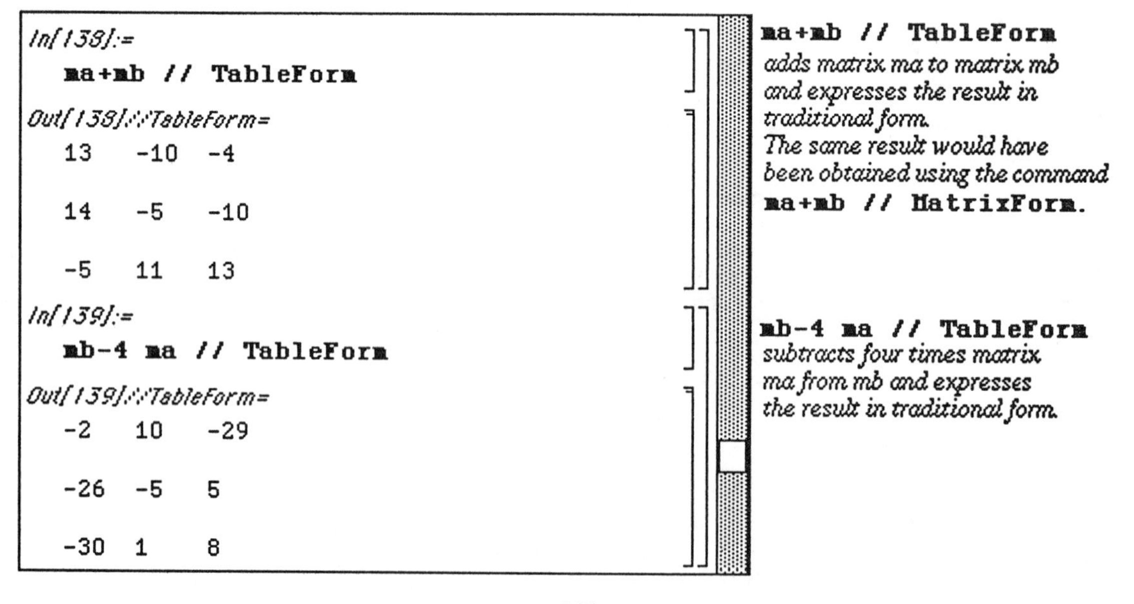

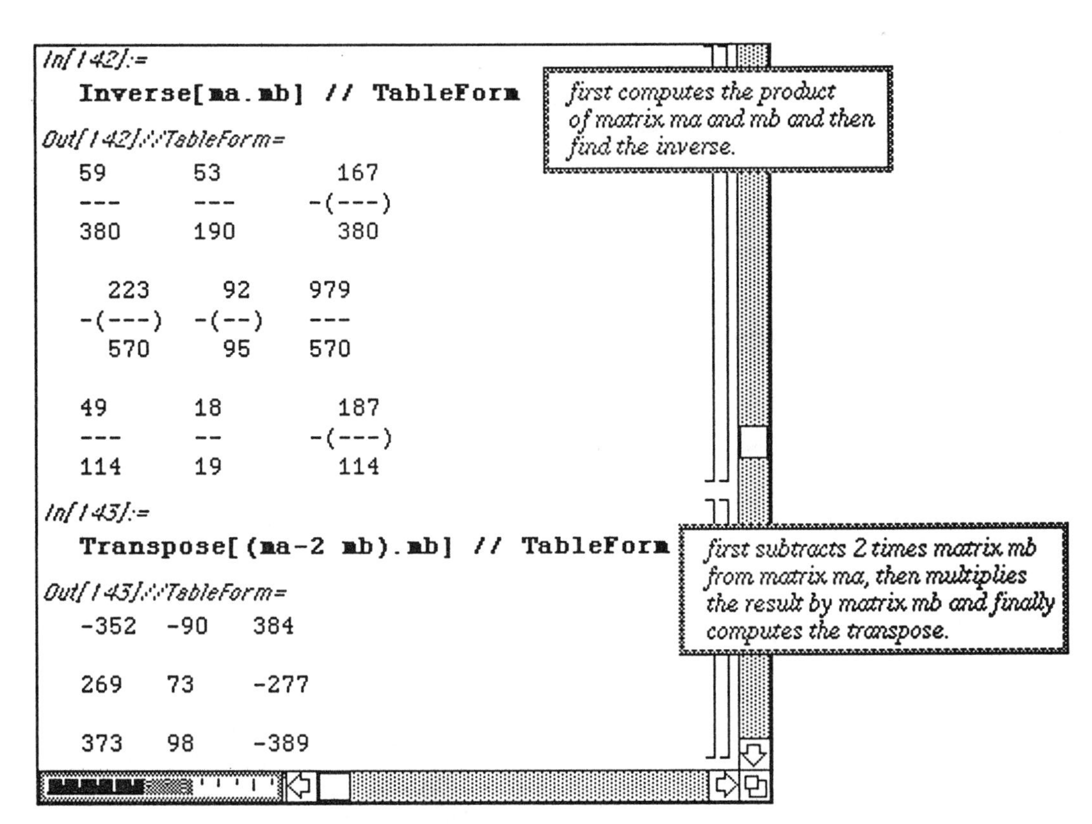

All of the basic operations on matrices can be performed with *Mathematica* . These include the determinant, **`Det[A]`**, as shown earlier, as well as the computation of the eigenvalues and corresponding eigenvectors of the matrix.

Recall that a nonzero vector x is an <u>eigenvector</u> for the square matrix A means there exists

a scalar λ such that $Ax = \lambda x$. λ is called an **eigenvalue** (or **characteristic value**) for A.

The command **`Eigenvalues[m]`** gives a list of the eigenvalues of the square matrix **`m`**.

The command **`Eigenvectors[m]`** gives a list of the eigenvectors of the square matrix **`m`**.
Several examples are shown below. (Notice that by naming the matrix **`ma`** , the matrix operations involving **`ma`** are easier to perform.)

□ **Example:**

Use *Mathematica* to compute (i) the determinant of $\begin{bmatrix} 3 & -4 & 5 \\ 8 & 0 & -3 \\ 5 & 2 & 1 \end{bmatrix}$;

(ii) numerical approximations of the eigenvalues; and

(iii) numerical approximations of the eigenvectors

In the previous example, *Mathematica* was used to define **ma** to be $\begin{bmatrix} 3 & -4 & 5 \\ 8 & 0 & -3 \\ 5 & 2 & 1 \end{bmatrix}$.

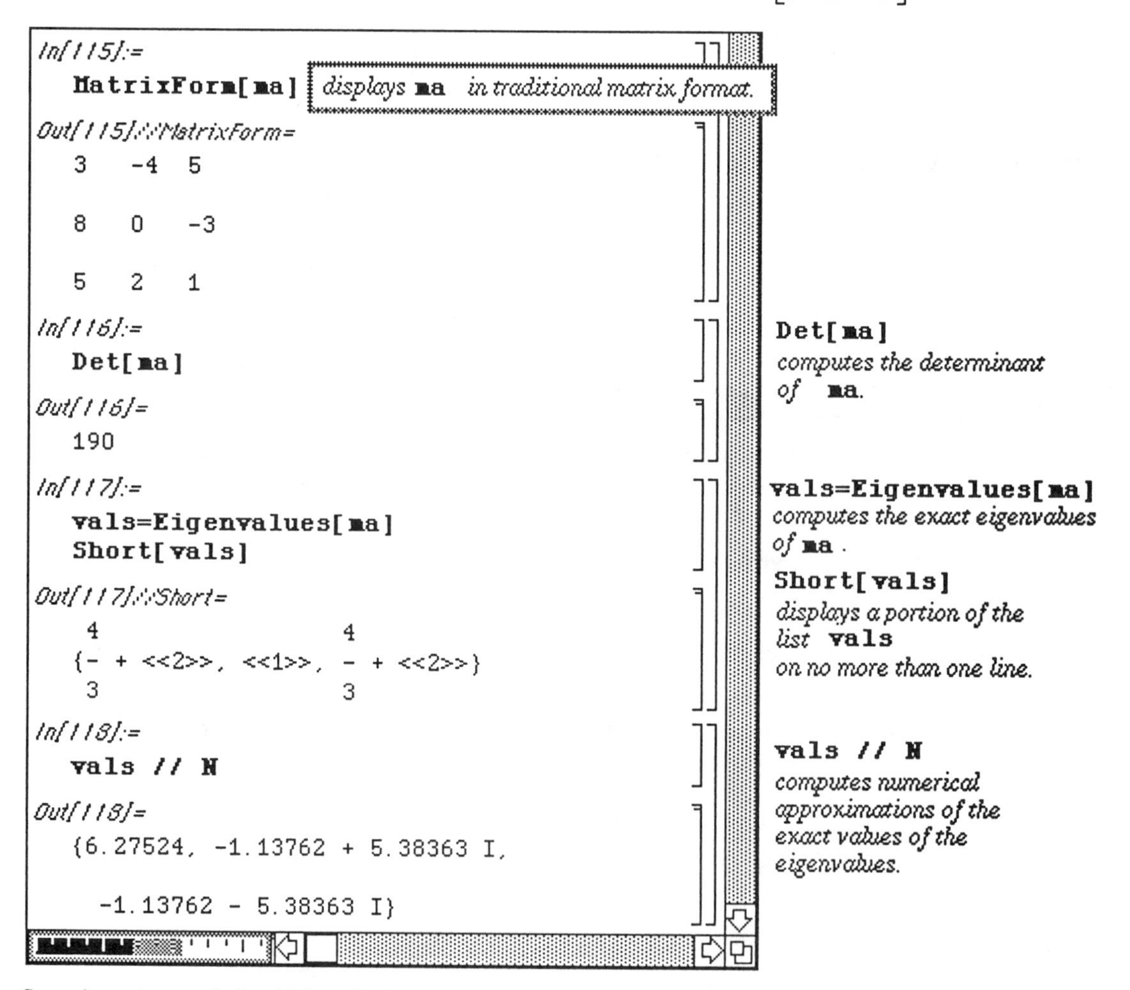

Sometimes the matrix in which each element is numerically approximated is more useful than the matrix in its original form. This is obtained below for the matrix **ma** with **N[ma]**. *Mathematica* can also be used to compute the eigenvectors of the matrix with **Eigenvectors[ma]**.

Notice that this command results in a list of eigenvectors as was the case with eigenvalues. Hence, individual eigenvectors can be extracted from the list of eigenvectors, **vector** , using **vector[[i]]** which gives the **i**th member of the list of eigenvectors. A similar command yields individual eigenvalues. (This method of extracting elements of lists was discussed earlier in **Chapter 4**.) Once obtained from the list of eigenvectors, certain operations can be performed with the extracted eigenvector. For example, the **i**th eigenvector can be multiplied by the matrix, **ma** . This is accomplished with the command
ma.vector[[i]].

The above command can be used with **N[vals[[j]]] vector[[i]]**
to verify that the numerical approximation of the eigenvalue **vals[[j]]** corresponds to the eigenvector **vector[[i]].** If this pair corresponds, then
ma.vector[[i]]= N[vals[[j]]] vector[[i]] according to the definition given above (i.e., $Ax=\lambda x$).

Notice that the lists of eigenvalues and eigenvalues are not given in corresponding order. That is to say, the **i**th eigenvalue in the list of eigenvalues does not necessarily correspond with the **i**th eigenvector in the list of eigenvectors.
Examples of these operations are demonstrated below :

□ Example:

Compute numerical approximations of the eigenvectors of **ma** $=\begin{bmatrix}3 & -4 & 5\\ 8 & 0 & -3\\ 5 & 2 & 1\end{bmatrix}$.

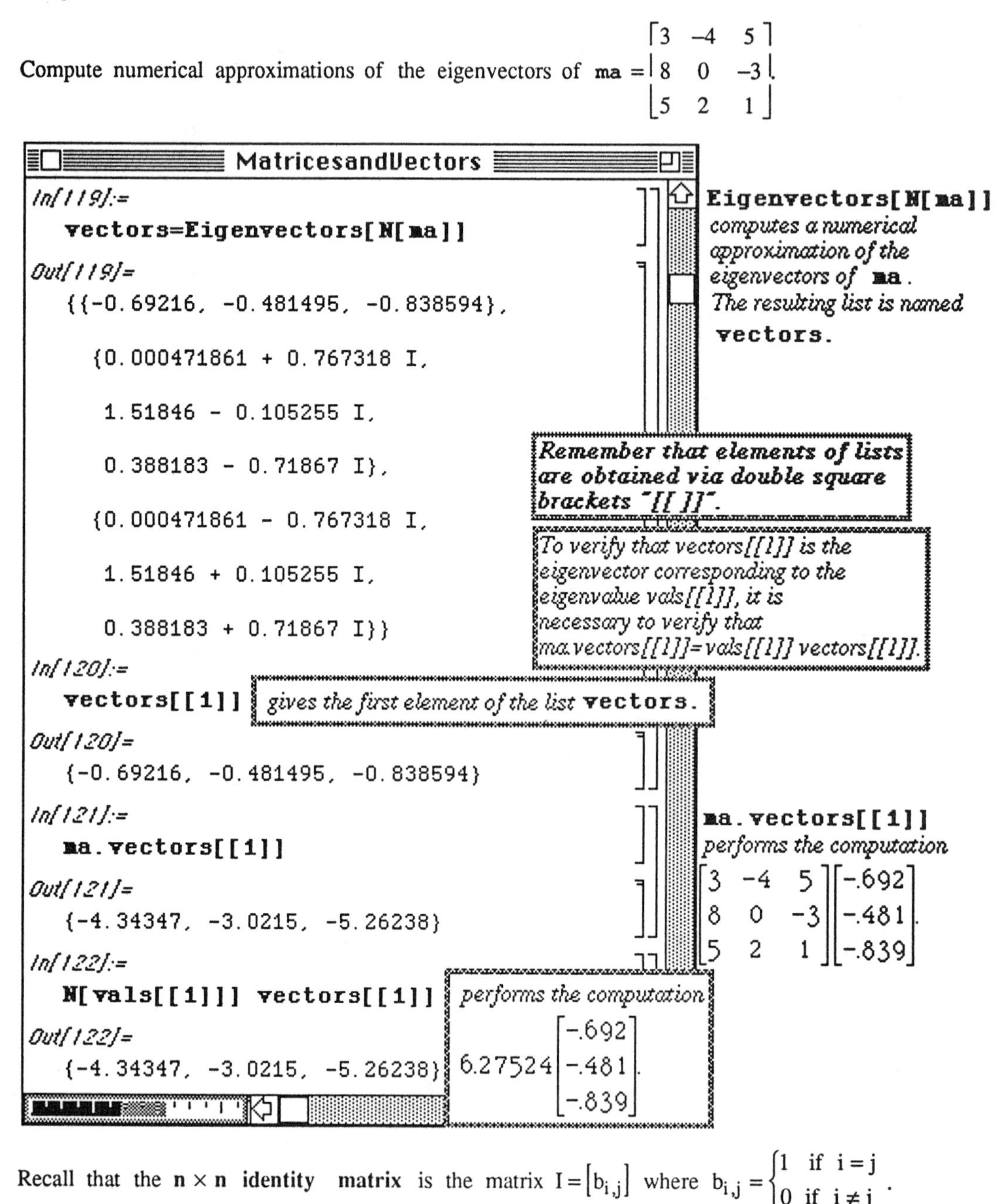

Recall that the **n × n identity matrix** is the matrix $I=[b_{i,j}]$ where $b_{i,j}=\begin{cases}1 & \text{if } i=j\\ 0 & \text{if } i\neq j\end{cases}$.

The command `IdentityMatrix[n]` yields the n x n identity matrix .

□ **Example:**

Compute the inverse, $\mathbf{ma}^{-1}$, of the matrix $\mathbf{ma} = \begin{bmatrix} 3 & -4 & 5 \\ 8 & 0 & -3 \\ 5 & 2 & 1 \end{bmatrix}$ and verify that $\mathbf{ma}^{-1}\mathbf{ma} = \begin{bmatrix} 1 & 0 & 0 \\ 0 & 1 & 0 \\ 0 & 0 & 1 \end{bmatrix}$.

The inverse of the matrix **ma** is found with the command **Inverse[ma]** and is named **mai** for easier use. The matrix **ma** is then multiplied by its inverse **mai** and placed in **Tableform** to verify that the identity matrix is obtained.

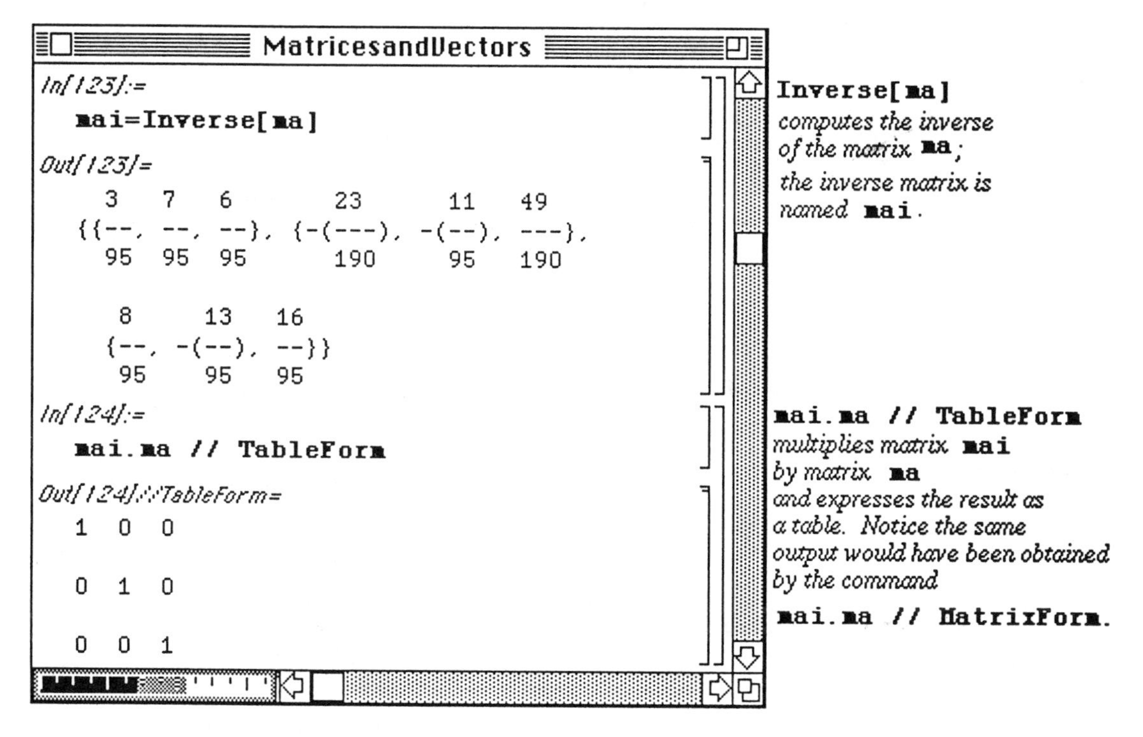

▫ Example:

Define **matrixb** to be the matrix $\begin{bmatrix} -2 & 3 & 4 & 0 \\ -2 & 0 & 1 & 3 \\ -1 & 4 & -6 & 5 \\ 4 & 8 & 11 & -4 \end{bmatrix}$. Compute (i) $\text{Det}\begin{bmatrix} -2 & 3 & 4 & 0 \\ -2 & 0 & 1 & 3 \\ -1 & 4 & -6 & 5 \\ 4 & 8 & 11 & -4 \end{bmatrix}$;

(ii) $\begin{bmatrix} -2 & 3 & 4 & 0 \\ -2 & 0 & 1 & 3 \\ -1 & 4 & -6 & 5 \\ 4 & 8 & 11 & -4 \end{bmatrix}^2$; (iii) $\begin{bmatrix} -2 & 3 & 4 & 0 \\ -2 & 0 & 1 & 3 \\ -1 & 4 & -6 & 5 \\ 4 & 8 & 11 & -4 \end{bmatrix}^3$; and (iv) $\begin{bmatrix} -2 & 3 & 4 & 0 \\ -2 & 0 & 1 & 3 \\ -1 & 4 & -6 & 5 \\ 4 & 8 & 11 & -4 \end{bmatrix}^4$.

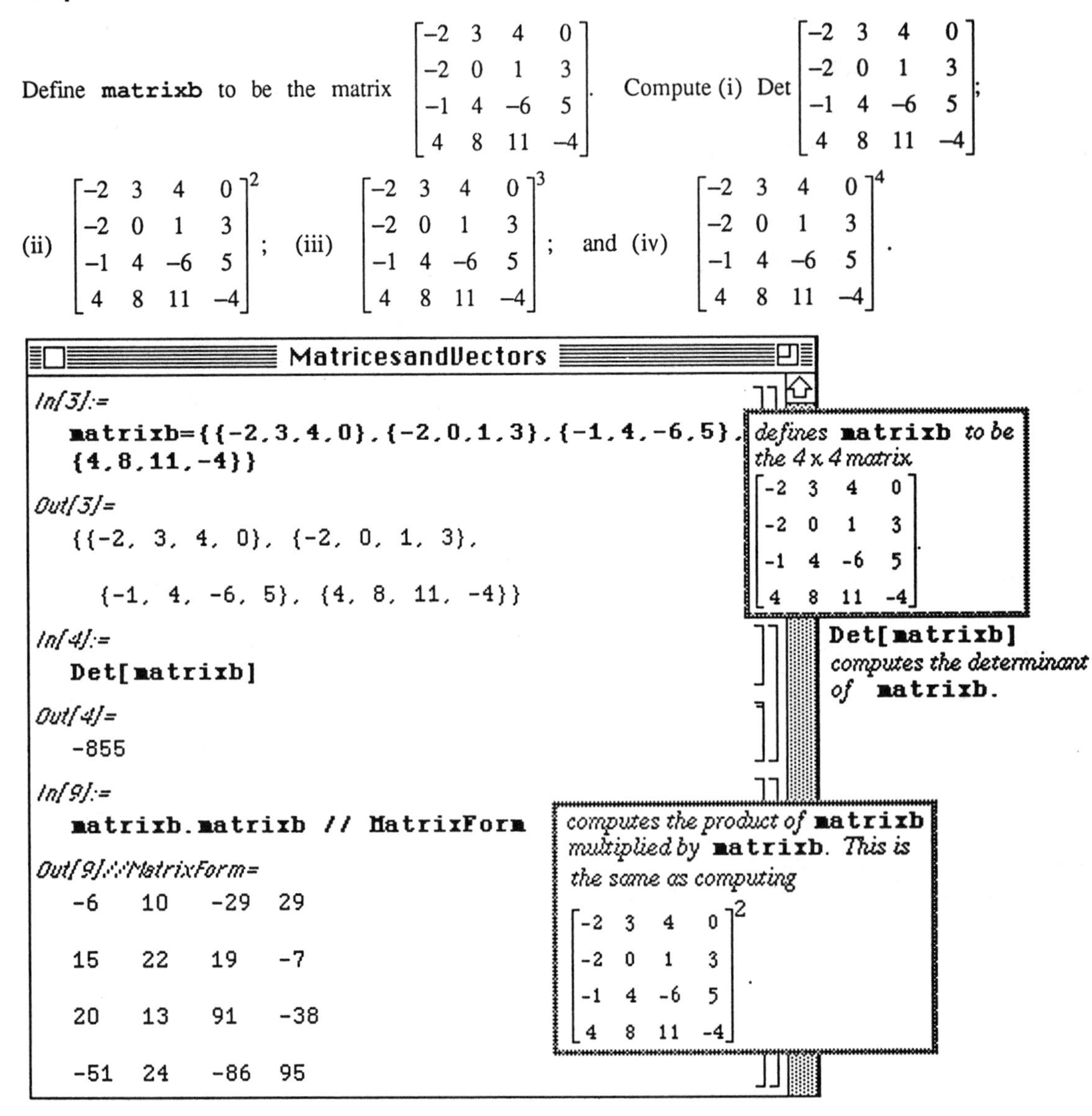

Special attention must be given to the notation which must be used in taking the product of a matrix with itself. The following example illustrates how *Mathematica* interprets the expression **(matrixb)^3**. Usually, the matrix product
matrixb matrixb matrixb is represented as **(matrixb)^3** .
However, the command **(matrixb)^3** cubes each element of the matrix **matrixb**.

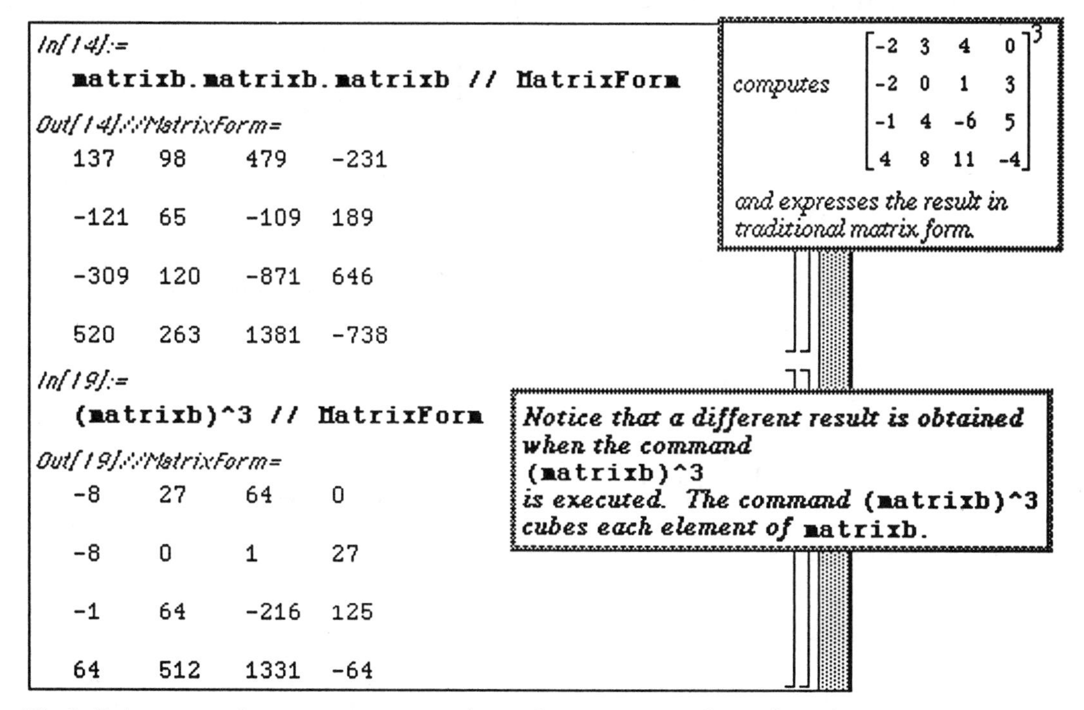

The built-in command **MatrixPower** may be used to compute products of matrices:

MatrixPower[matrix,n] computes $\underbrace{\texttt{matrix} \bullet \texttt{matrix} \bullet \ldots \bullet \texttt{matrix}}_{n-\text{times}}$.

However, to illustrate *Mathematica*'s recursive abilities, we define a function **matrixpower** that performs the same calculations as the built-in function **MatrixPower**.

We define the function **matrixpower** by first defining the matrix with **matrixpower[a_,1]** and then the matrix product **matrixpower[a_,n_]:= a.matrixpower[a,n-1]**.

Hence, *Mathematica* computes the desired power of the matrix.

MatricesandVectors

```
In[26]:=
  matrixpower[a_,1]:=a
  matrixpower[a_,n_]:=a.matrixpower[a,n-1]

In[28]:=
  matrixpower[matrixb,3]

Out[28]=
  {{137, 98, 479, -231},

    {-121, 65, -109, 189},

    {-309, 120, -871, 646},

    {520, 263, 1381, -738}}

In[29]:=
  matrixpower[matrixb,4] // MatrixForm

Out[29]//MatrixForm=
  -1873   479    -4769    3613

  977     713    2314     -1106

  3833    757    11216    -6579

  -5899   1180   -14061   10646
```

Using Mathematica's recursion ability, for a given matrix a, define **matrixpower[a,1]=a** *and define, for n a positive integer,* **matrixpower[a,n]=a.matrixpower[a,n-1]** *Then,* **matrixpower[a,n]** *computes* $\underbrace{a.a.....a}_{n \text{ times}}$.

matrixpower[matrixb,3] *computes*

$$\begin{bmatrix} -2 & 3 & 4 & 0 \\ -2 & 0 & 1 & 3 \\ -1 & 4 & -6 & 5 \\ 4 & 8 & 11 & -4 \end{bmatrix}^3.$$

computes

$$\begin{bmatrix} -2 & 3 & 4 & 0 \\ -2 & 0 & 1 & 3 \\ -1 & 4 & -6 & 5 \\ 4 & 8 & 11 & -4 \end{bmatrix}^4$$

and expresses the result in matrix form.

5.2 Linear Systems of Equations

Calculating Solutions of Linear Systems of Equations

To solve the system of linear equations Ax=b, where A is the coefficient matrix, b is a known vector and x is the unknown vector, we proceed in the usual manner: If A^{-1} exists, then $A^{-1}Ax = A^{-1}b$ so $x = A^{-1}b$.

Mathematica offers several commands for solving systems of linear equations, however, which do not depend on the computation of the inverse of A. These commands are discussed in the following examples.

Example:

Solve the system of three equations $\begin{cases} x-2y+z=-4 \\ 3x+2y-z=8 \\ -x+3y+5z=0 \end{cases}$ for x, y, and z.

In order to solve an n x n system of equations (n equations and n unknown variables), the command
`Solve[{eqn1,eqn2,...,eqnn},{var1,var2,...,varn}]`
is used. In other words, the equations as well as the variables are entered as lists. If one wishes to solve for all variables that appear in a system, the command **`Solve[{eqn1,eqn2,...eqnn}]`** attempts to solve **`eqn1`**, **`eqn2`**, ..., **`eqnn`** for all variables that appear in them. The system given above with 3 equations and 3 unknowns is solved below. (**Remember** that a double equals sign must be used in each equation.) The time required to perform the calculation is also displayed. The steps necessary to have the time displayed are given later in this section.

In this case, entering either **`Solve[{x-2y+z==-4,3x+2y==8,-x+3y+5z==0}]`** or **`Solve[{x-2y+z,3x+2y,-x+3y+5z}=={-4,8,0}]`** yield the same result.

Remark: Be sure to include the double equals signs between the left- and right-hand sides of each equation.

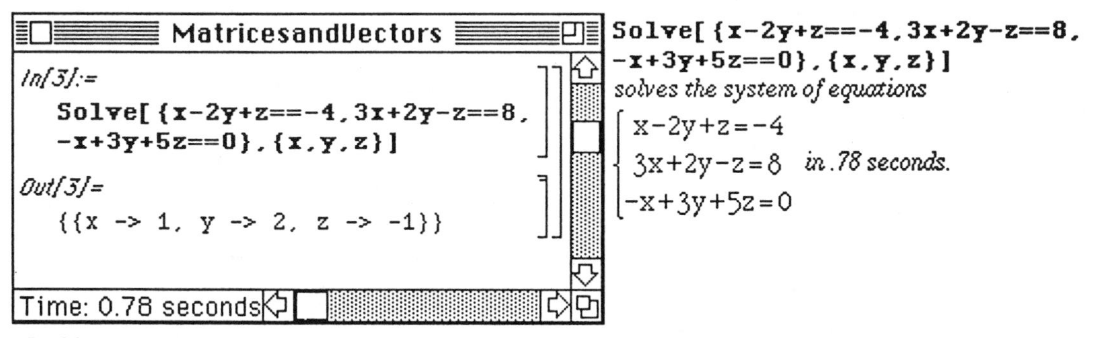

In this case, Mathematica displays the time it takes to perform the calculation.

Another way to solve systems of equations is based on the matrix form of the system of equations, **Ax=b**. The matrix of coefficients in the previous example is entered as **matrixa** along with the vector of right-hand side values **vectorb** . After defining the vector of variables, **vectorx**, the system **Ax=b** is solved explicitly with the command **Solve[matrixa.vectorb==vectorb,vectorx]** .
Compare the computation times for each calculation.

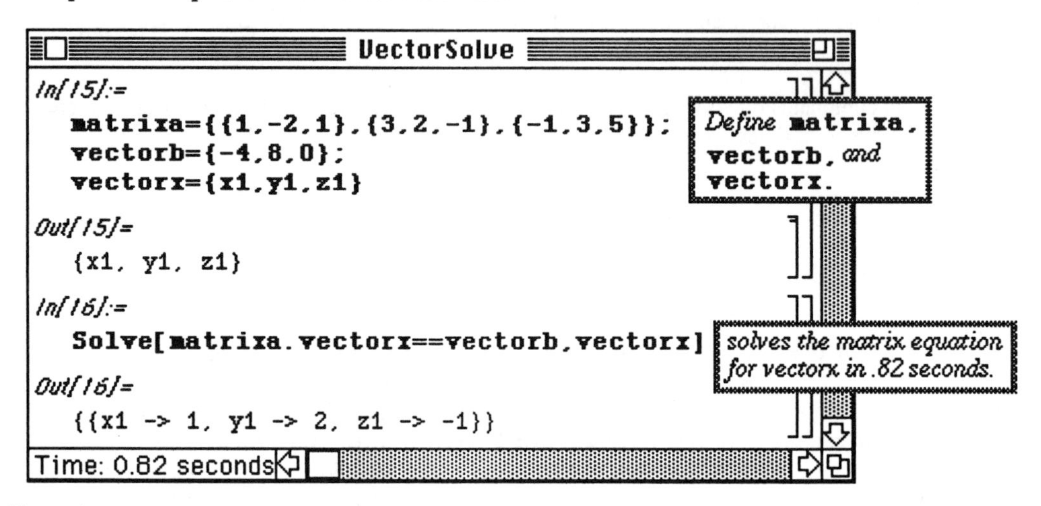

□ **Example:**

Next, the system $\begin{cases} 2x - 4y + z = -1 \\ 3x + y - 2z = 3 \\ -5x + y - 2z = 4 \end{cases}$ is solved in a similar manner. Notice that exact values are given with **Solve**. This system takes longer to solve than the first example.

MatricesandVectors

```
In[19]:=
  Solve[{2x-4y+z==-1,3x+y-2z==3,
  -5x+y-2z==4},{x,y,z}]
Out[19]=
          1            15            51
  {{x -> -(-), y -> -(--), z -> -(--)}}
          8            56            28
Time: 1.10 seconds
```

Solve[{2x-4y+z==-1,3x+y-2z==3, -5x+y-2z==4},{x,y,z}] *solves the system of equations* $\begin{cases} 2x-4y+z=-1 \\ 3x+y-2z=3 \\ -5x+y-2z=4 \end{cases}$ *for x, y, and z in 1.10 seconds.*

As before, consider the alternate approach to solving this system using matrices and vectors.

MatricesandVectors

In[20]:=

```
matrixa={{2,-4,1},{3,1,-2},
{-5,1,-2}};
vectorb={-1,3,4}
```

Out[20]=

```
{-1, 3, 4}
```

Be sure to nest brackets appropriately.

matrixa={{2,-4,1},{3,1,-2},{-5,1,-2}};
vectorb={-1,3,4}
defines **matrixa** *to be* $\begin{bmatrix} 2 & -4 & 1 \\ 3 & 1 & -2 \\ -5 & 1 & -2 \end{bmatrix}$ *and* **vectorb** *to be* $\begin{bmatrix} -1 \\ 3 \\ 4 \end{bmatrix}$.

Time: 0.32 seconds

Notice that executing the command takes .32 seconds.

The command
LinearSolve[A,b]
calculates the solution **x** of the system **Ax=b**.
Comparing the computation times for each, **LinearSolve** performs the task more quickly (.57 secs.) than does **Solve**. If the time needed to enter the matrix **A** and vector **b** (.32 secs.) is considered, however, the total time (.89 secs.) for this calculation was slightly larger than the original(.78 secs.).

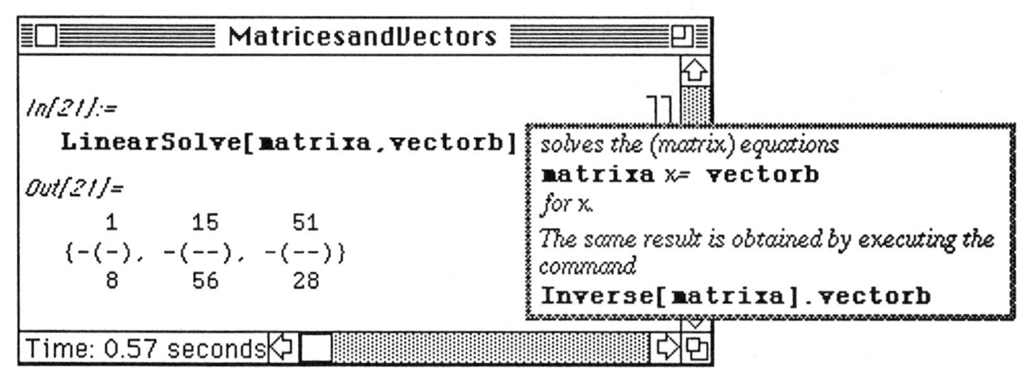

In this case, the calculation takes only .57 seconds.

□ To Display Time After Each Evaluations:

In the above examples, the time to perform each calculation is displayed instead of the usual thermometer indicating memory used. To display the time clock after each evaluation, proceed as follows:

1) Use the Mouse to move the cursor to **Edit** and select **Settings**; then select **Action**;
2) The **Action Settings** window will appear;
3) Click the box **Display clock timing after each evaluation**; and

4) Click the [OK] button.

Action Settings in Version 1.2; the Action Settings in Version 2.0 are the same.

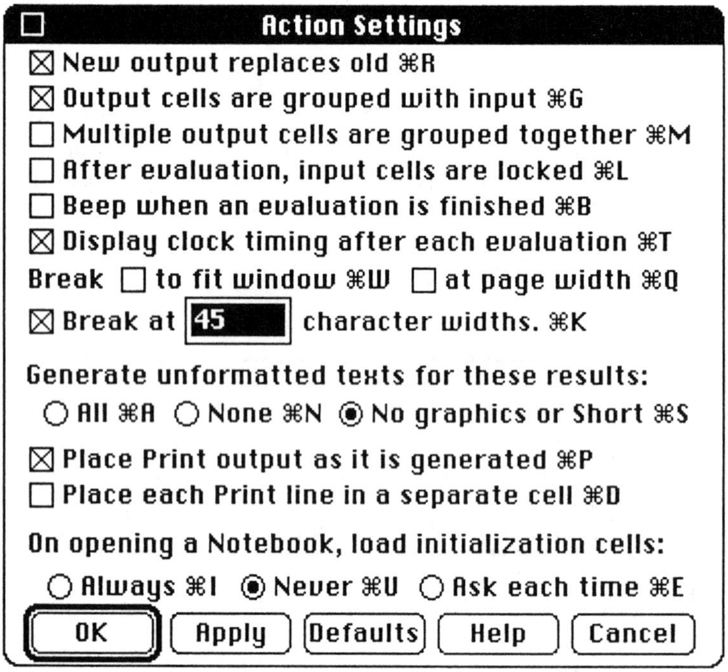

Action Settings

☒ New output replaces old ⌘R
☒ Output cells are grouped with input ⌘G
☐ Multiple output cells are grouped together ⌘M
☐ After evaluation, input cells are locked ⌘L
☐ Beep when an evaluation is finished ⌘B
☒ Display clock timing after each evaluation ⌘T
Break ☐ to fit window ⌘W ☐ at page width ⌘Q
☒ Break at [45] character widths. ⌘K

Generate unformatted texts for these results:
○ All ⌘A ○ None ⌘N ◉ No graphics or Short ⌘S

☒ Place Print output as it is generated ⌘P
☐ Place each Print line in a separate cell ⌘D

On opening a Notebook, load initialization cells:
○ Always ⌘I ◉ Never ⌘U ○ Ask each time ⌘E

[OK] [Apply] [Defaults] [Help] [Cancel]

When using the `Solve` command, the equations may be entered in several different ways as the following example shows. For example, if `equations` is a list of equations and `variables` is a list of variables, then *Mathematica* attempts to solve `equations` in terms of `variables` when the command `Solve[equations,variables]` is entered; *Mathematica* attempts to solve `equations` in terms of all variables that appear in `equations` when `Solve[equations]` is entered:

□ Example:

Solve the system of equations $\begin{cases} 4x_1+5x_2-5x_3-8x_4-2x_5=5 \\ 7x_1+2x_2-10x_3-x_4-6x_5=-4 \\ 6x_1+2x_2+10x_3-10x_4+7x_5=-7 \\ -8x_1-x_2-4x_3+3x_5=5 \\ 8x_1-7x_2-3x_3+10x_4+5x_5=7 \end{cases}$ for x_1, x_2, x_3, x_4, and x_5.

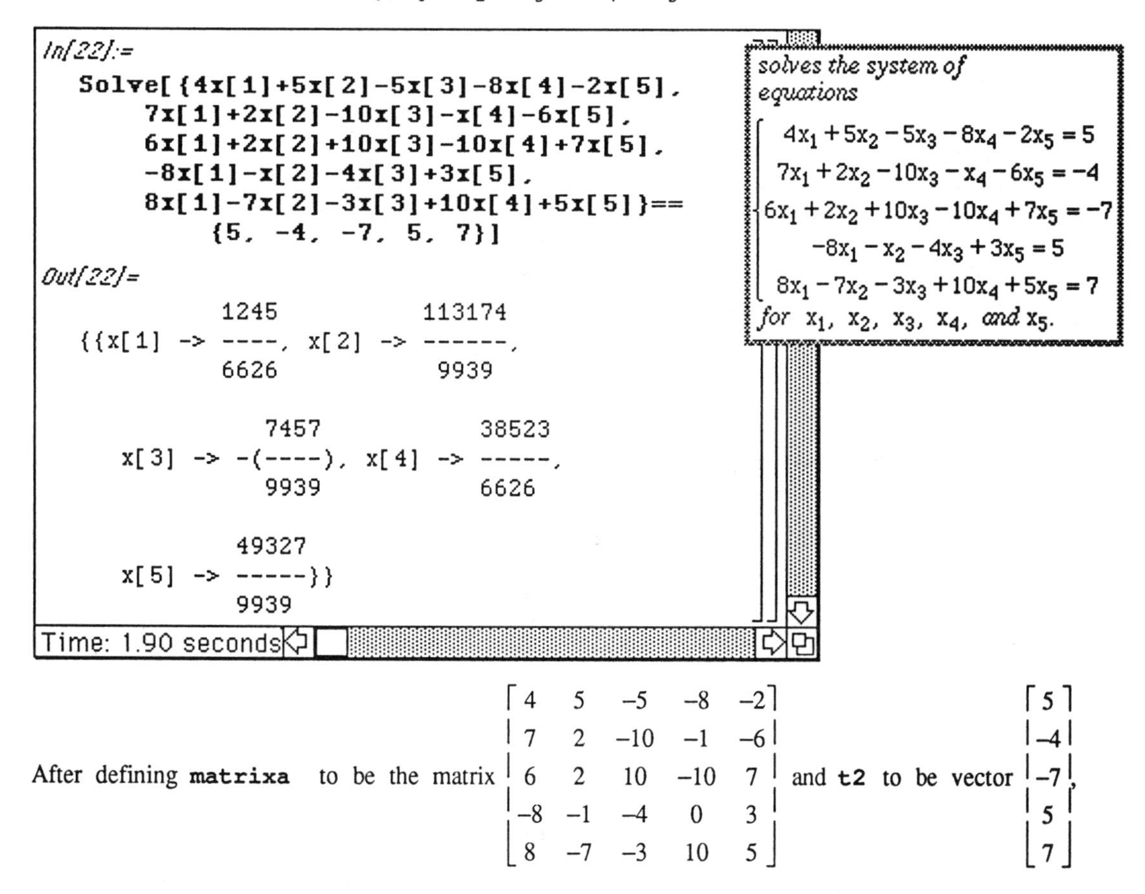

After defining **matrixa** to be the matrix $\begin{bmatrix} 4 & 5 & -5 & -8 & -2 \\ 7 & 2 & -10 & -1 & -6 \\ 6 & 2 & 10 & -10 & 7 \\ -8 & -1 & -4 & 0 & 3 \\ 8 & -7 & -3 & 10 & 5 \end{bmatrix}$ and **t2** to be vector $\begin{bmatrix} 5 \\ -4 \\ -7 \\ 5 \\ 7 \end{bmatrix}$,

LinearSolve is used to solve the same system much faster:

```
In[21]:=
  matrixa={{4,5,-5,-8,-2},{7,2,-10,-1,-6},
           {6,2,10,-10,7},{-8,-1,-4,0,3},
           {8,-7,-3,10,5}}
  t2={5,-4,-7,5,7}
  xarray=Array[x,{5}];

In[24]:=
  LinearSolve[matrixa,t2]

Out[24]=
   1245  113174     7457    38523  49327
  {----, ------, -(----), -----, -----}
   6626    9939     9939   6626    9939

Time: 0.88 seconds
```

■ Application: Characteristic and Minimal Polynomials

The characteristic polynomial of the n x n matrix A is the polynomial

$p_A(x) = \mathrm{Det}[x I_n - A]$, where I_n is the $n \times n$ identity matrix

It is well-known that the eigenvalues of A are the roots of the characteristic polynomial of A.

The **trace** of an $n \times n$ matrix $A = [a_{i,j}]$ is $a_{1,1} + a_{2,2} + \ldots + a_{n,n} = \sum_{k=1}^{n} a_{k,k}$.

If A is a matrix with non-zero determinant and characteristic polynomial

$p_A(x) = x^n + c_{n-1}x^{n-1} + c_{n-2}x^{n-2} + \ldots + c_1 x + c_0$, then $\mathrm{Det}[A] = (-1)^n c_0$

and $\mathrm{Trace}[A] = a_{1,1} + \ldots + a_{n,n} = \sum_{k=1}^{n} a_{k,k} = -c_{n-1}$.

Let $p_A(x)$ be the characteristic polynomial of A and let

$p_A(x) = (p_1(x))^{n1}(p_2(x))^{n2} \bullet \ldots \bullet (p_m(x))^{nm}$ be the factorization of $p_A(x)$.

The minimal polynomial q(x) of A is the monic polynomial of least degree satisfying q(A)=0. It is well-known

that $p_1(x)p_2(x) \bullet \ldots \bullet p_m(x)$ divides q(x); hence, $p_A(A) = 0$.

□ **Example:**

Find the trace, characteristic polynomial, and minimal polynomial of the matrix $\begin{bmatrix} 0 & 6 & 3 \\ -2 & -8 & -2 \\ 0 & 0 & -2 \end{bmatrix}$

The process begins by entering the matrix, **matrixa**, and then the associated matrix, **assoca**, where **assocaa=x IdentityMatrix[4]-matrixa.**

The characteristic polynomial is then determined using **Det[assoca].**

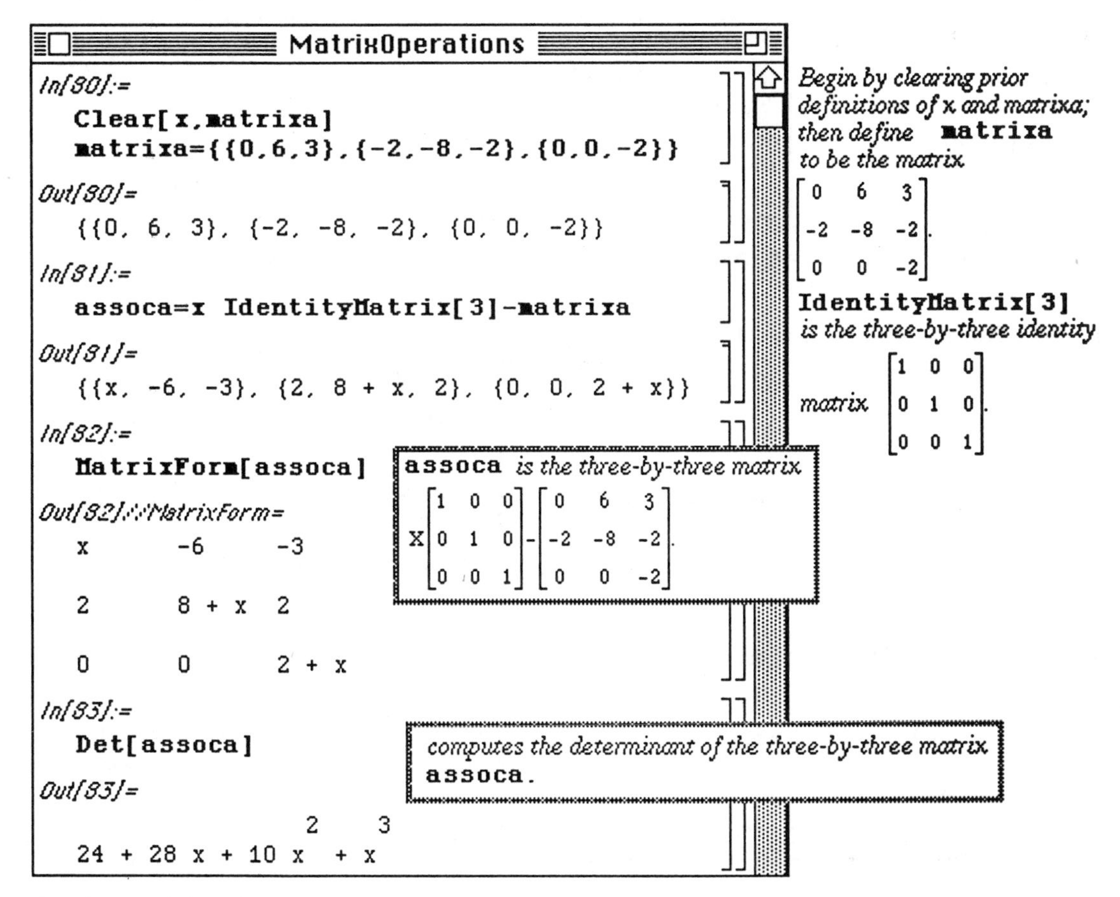

The characteristic polynomial is then easily factored using **Factor[Det[assoca]].**

This yields $(2+x)^2(6+x)$.

Since the minimal polynomial divides the characteristic polynomial, the minimal polynomial must either be

$(2+x)^2(6+x)$ or $(2+x)\ (6+x)$.

Now, in order to determine which is the minimal polynomial, the definition given earlier must be employed. Since this definition involves substituting **matrixa** into these two polynomials to see which gives the zero matrix (the matrix of all zeros), determining the powers of **matrixa** is necessary. Therefore, the function **matrixpower** discussed earlier is redefined and used to square and cube **matrixa**. For easier use, the 3x3 identity matrix is named **ident**, the square of **matrixa** is called **matrixs** , and the cube of **matrixa** is called **matrixc** .

Note that the same result can be accomplished using the built-in function **MatrixPower**.

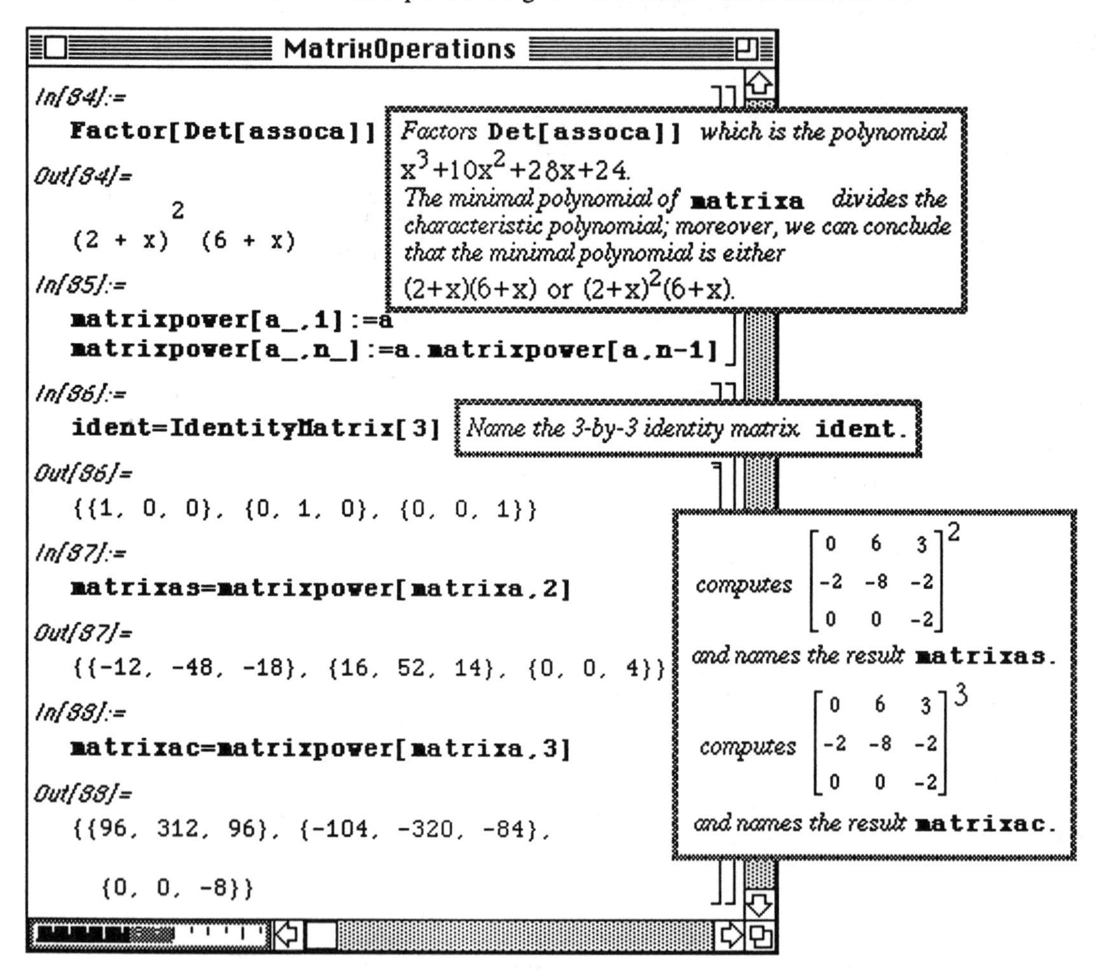

Substitution of **matrixa** into the possible minimal polynomial (2+x)(6+x) is done as follows:

```
(2 ident + matrixa).(6 ident + matrixa)
```

Notice that since **matrixa** is a matrix as opposed to a scalar, then each constant in the polynomial must be converted to a matrix by multiplying by the identity matrix. Otherwise, the command would not be defined. Substitution into the other polynomial is done in a similar manner.
Since substitution into the second polynomial yields the zero matrix, the minimal polynomial is

$(2+x)^2(6+x)$ or $x^3+10x^2+28x+24$.

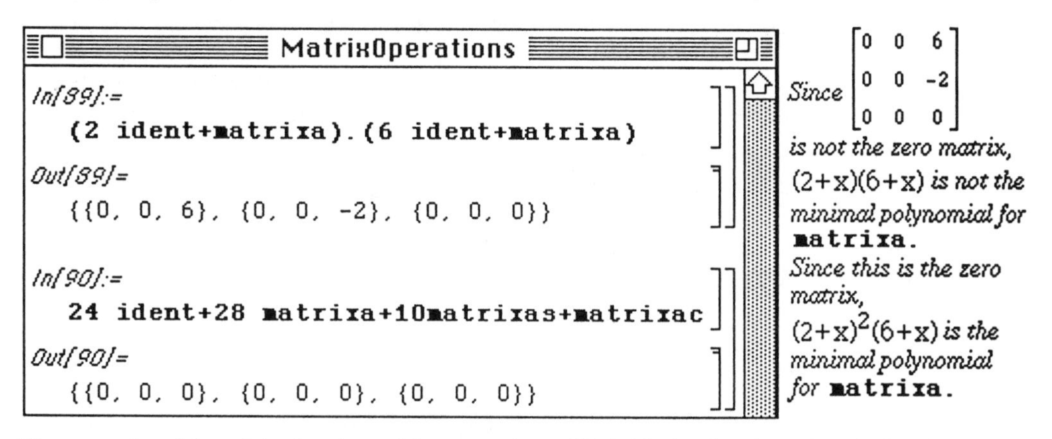

The properties of the minimal polynomial can now be verified. Notice that the order of this polynomial is three. Since the general formula for the minimal polynomial of order three is

$q(x) = x^3 + c_2x^2 + c_1x + c_0$ we have in this case, $c_0 = 24$, $c_1 = 28$, and $c_2 = 10$.

Hence, the trace of A can be computed with following formula:

$$\text{Trace}[A] = a_{1,1} + a_{2,2} + a_{3,3} = \sum_{k=1}^{3} a_{k,k} = -c_{3-1} = -c_2 = -10$$

as well as the determinant: $\text{Det}[A] = (-1)^3 c_0 = -24$.

These results agree with the calculations shown below using *Mathematica* :

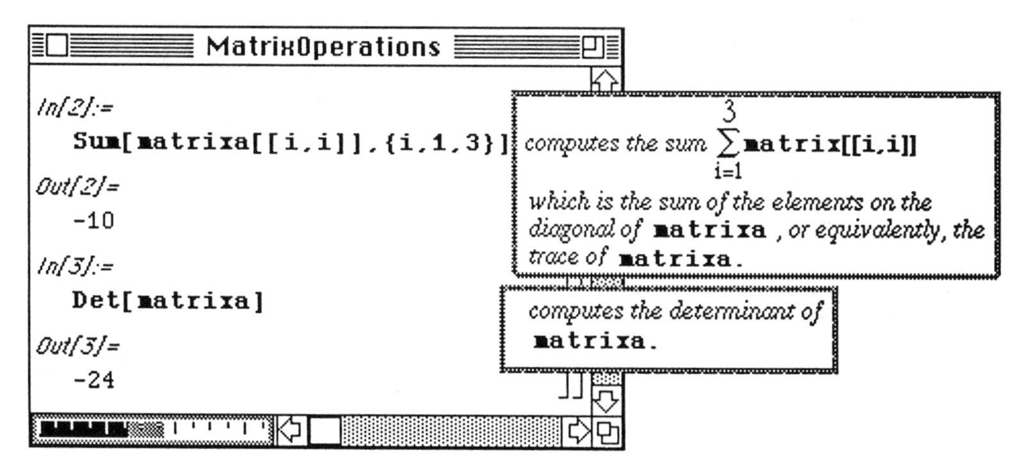

□ Example:

Find the trace, characteristic polynomial, and minimal polynomial of the matrix $\begin{bmatrix} 2 & -3 & 1 & 2 \\ 3 & 7 & -5 & 2 \\ 0 & 1 & 9 & -5 \\ -2 & 3 & 4 & 3 \end{bmatrix}$.

This problem is solved in a manner similar to that of the previous example. First, the matrix is entered and named **matrixc**. Then, the associated matrix **assocc** is defined. Note that in this case the 4x4 identity must be used. The characteristic polynomial is then found with **Det[assocc]** and called **char** for later use. From **char**, the numerical approximation of the roots of the characteristic polynomial (the eigenvalues) of **matrixc** can be determined with **NRoots[char==0,x]** . In this case, the characteristic polynomial is irreducible.

MoreMatrices

```
In[15]:=
  matrixc={{2,-3,1,2},{3,7,-5,2},
  {0,1,9,-5},{-2,3,4,3}}
Out[15]=
  {{2, -3, 1, 2}, {3, 7, -5, 2},

    {0, 1, 9, -5}, {-2, 3, 4, 3}}
In[16]:=
  assocc=x IdentityMatrix[4]-matrixc
Out[16]=
  {{-2 + x, 3, -1, -2}, {-3, -7 + x, 5, -2},

    {0, -1, -9 + x, 5}, {2, -3, -4, -3 + x}}
In[17]:=
  char=Det[assocc]
Out[17]=
                         2        3    4
  1665 - 848 x + 181 x  - 21 x  + x
In[18]:=
  NRoots[char==0,x]
Out[18]=
  x == 3.59859 - 5.14335 I ||

    x == 3.59859 + 5.14335 I ||

    x == 4.58305 || x == 9.21977
```

Define **matrixc** *to be the matrix* $\begin{bmatrix} 2 & -3 & 1 & 2 \\ 3 & 7 & -5 & 2 \\ 0 & 1 & 9 & -5 \\ -2 & 3 & 4 & 3 \end{bmatrix}$.

Then define **assocc** *to be the matrix* $x\begin{bmatrix} 1 & 0 & 0 & 0 \\ 0 & 1 & 0 & 0 \\ 0 & 0 & 1 & 0 \\ 0 & 0 & 0 & 1 \end{bmatrix} - \begin{bmatrix} 2 & -3 & 1 & 2 \\ 3 & 7 & -5 & 2 \\ 0 & 1 & 9 & -5 \\ -2 & 3 & 4 & 3 \end{bmatrix}$.

The characteristic polynomial of **matrixc** *is the determinant of* **assocc**.

Although Mathematica can compute the exact roots of the characteristic polynomial, it is much faster to obtain numerical approximations of them.

A numerical approximation of the eigenvalues can also be found with **`Eigenvalues[N[matrixc]]`**. This is demonstrated below.
The method by which **`assocc`** is raised to a power is slightly different in this example. Instead of making use of the user-defined function **`matrixpower`** seen in the example above, we choose to take advantage of the built-in *Mathematica* function **`MatrixPower[matrix,n]`**. (Note the capital letters.) This function determines the matrix obtained when **`matrix`** is raised to the power **`n`** as did **`matrixpower`**. The matrices obtained by using **`MatrixPower`** to raise **`matrixc`** to the powers 2, 3, and 4 are necessary in determining the minimal polynomial. These are calculated and named **`matrix2`**, **`matrix3`**, and **`matrix4`**, respectively. (Note that only the output of the last command in the second input cell is displayed.) To see that the characteristic polynomial is the minimal polynomial, **`matrixc`** is substituted into the characteristic polynomial, **`char`**, to yield the zero matrix.

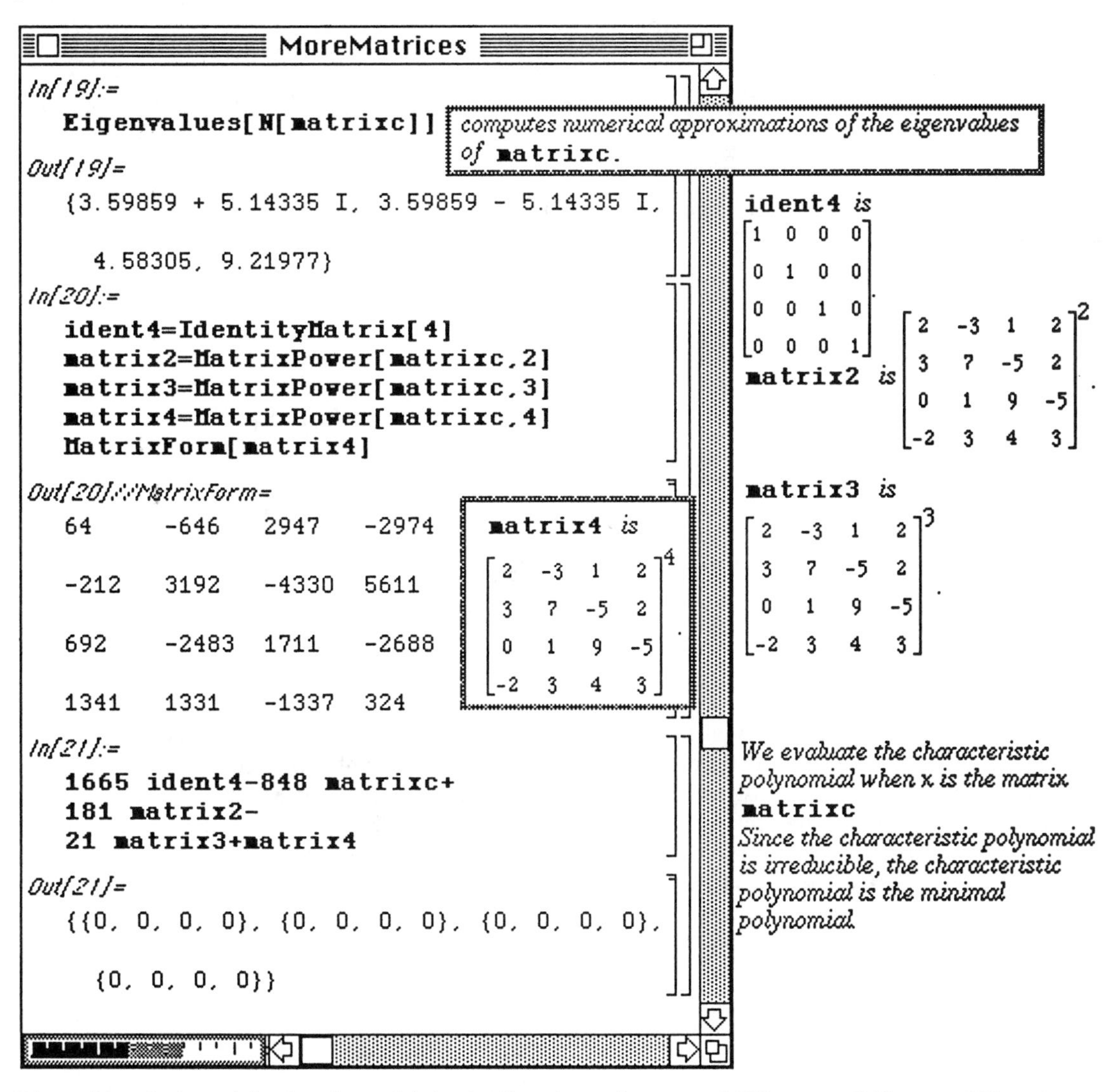

MoreMatrices

```
In[19]:=
  Eigenvalues[N[matrixc]]
Out[19]=
  {3.59859 + 5.14335 I, 3.59859 - 5.14335 I,
    4.58305, 9.21977}
In[20]:=
  ident4=IdentityMatrix[4]
  matrix2=MatrixPower[matrixc,2]
  matrix3=MatrixPower[matrixc,3]
  matrix4=MatrixPower[matrixc,4]
  MatrixForm[matrix4]
Out[20]//MatrixForm=
  64      -646    2947    -2974
  -212    3192    -4330   5611
  692     -2483   1711    -2688
  1341    1331    -1337   324
In[21]:=
  1665 ident4-848 matrixc+
  181 matrix2-
  21 matrix3+matrix4
Out[21]=
  {{0, 0, 0, 0}, {0, 0, 0, 0}, {0, 0, 0, 0},
    {0, 0, 0, 0}}
```

computes numerical approximations of the eigenvalues of **`matrixc`**.

`ident4` *is*
$$\begin{pmatrix}1&0&0&0\\0&1&0&0\\0&0&1&0\\0&0&0&1\end{pmatrix}.$$

`matrix2` *is*
$$\begin{pmatrix}2&-3&1&2\\3&7&-5&2\\0&1&9&-5\\-2&3&4&3\end{pmatrix}^2.$$

`matrix3` *is*
$$\begin{pmatrix}2&-3&1&2\\3&7&-5&2\\0&1&9&-5\\-2&3&4&3\end{pmatrix}^3.$$

`matrix4` *is*
$$\begin{pmatrix}2&-3&1&2\\3&7&-5&2\\0&1&9&-5\\-2&3&4&3\end{pmatrix}^4.$$

We evaluate the characteristic polynomial when x is the matrix **`matrixc`**
Since the characteristic polynomial is irreducible, the characteristic polynomial is the minimal polynomial.

The order of the minimal polynomial is 4. Therefore, since $c_0 = 1665$, $c_1 = -848$, $c_2 = 181$, and $c_3 = 21$, we have $\text{Trace}[A] = -c_{4-1} = -c_3 = -21$.

■ Application: Maxima and Minima Using Linear Programming

We call the linear programming problem of the form:

Minimize $\underbrace{Z = c_1 x_1 + c_2 x_2 + \ldots + c_n x_n}_{\text{function}}$, subject to the restrictions

$$\text{inequalities}\begin{cases} a_{1,1} x_1 + a_{1,2} x_2 + \ldots + a_{1,n} x_n \geq b_1 \\ a_{2,1} x_1 + a_{2,2} x_2 + \ldots + a_{2,n} x_n \geq b_2 \\ \vdots \\ a_{m,1} x_1 + a_{m,2} x_2 + \ldots + a_{m,n} x_n \geq b_m \end{cases}, \text{ and } x_1 \geq 0,\ x_2 \geq 0, \ldots, x_n \geq 0$$

the **standard form** of the linear programming problem.

The *Mathematica* command
ConstrainedMin[function,{inequalities},{variables}] solves the standard form of the linear programming problem.

Similarly, the *Mathematica* command
ConstrainedMax[function,{inequalities},{variables}] solves the linear programming problem

Maximize $\underbrace{Z = c_1 x_1 + c_2 x_2 + \ldots + c_n x_n}_{\text{function}}$, subject to the restrictions

$$\text{inequalities}\begin{cases} a_{1,1} x_1 + a_{1,2} x_2 + \ldots + a_{1,n} x_n \geq b_1 \\ a_{2,1} x_1 + a_{2,2} x_2 + \ldots + a_{2,n} x_n \geq b_2 \\ \vdots \\ a_{m,1} x_1 + a_{m,2} x_2 + \ldots + a_{m,n} x_n \geq b_m \end{cases}, \text{ and } x_1 \geq 0,\ x_2 \geq 0, \ldots, x_n \geq 0.$$

□ Example:

Maximize $z(x_1, x_2, x_3) = 4x_1 - 3x_2 + 2x_3$ subject to the constraints

$3x_1 - 5x_2 + 2x_3 \leq 60,$

$x_1 - x_2 + 2x_3 \leq 10,$

$x_1 + x_2 - x_3 \leq 20,$ and x_1, x_2, x_3 all non – negative

In order to solve a linear programming problem with *Mathematica* , the variables **{x1,x2,x3}** and objective function **z[x1,x2,x3]** must first be defined. In an effort to limit the amount of typing required to complete the problem, the set of inequalities is assigned the name **ineqs** while the set of variables is called **vars** . Notice that the symbol "<=", obtained by typing the "<" key and then the "=" key, represents "less than or equal to" and is used in **ineqs** . Hence, the maximization problem is solved with the command **ConstrainedMax[z[x1,x2,x3],ineqs,vars]**.

The solution gives the maximum value of **z** subject the given constraints as well as the values of **x1**, **x2**, and,**x3** which maximize **z** . These steps are shown below :

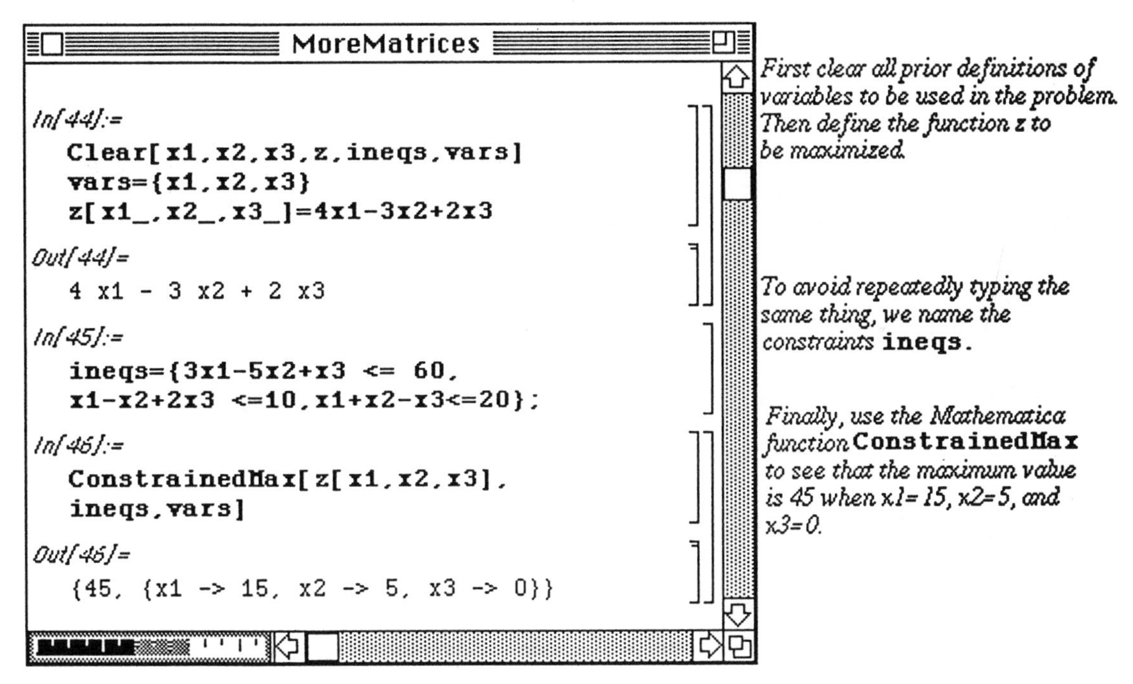

Minimization problems are solved in a similar manner. Consider the following :

□ **Example:**

Minimize $z(x_1, x_2, x_3) = 4x_1 - 3x_2 + 2x_3$ subject to the constraints

$3x_1 - 5x_2 + 2x_3 \leq 60,$

$x_1 - x_2 + 2x_3 \leq 10,$

$x_1 + x_2 - x_3 \leq 20,$ and x_1, x_2, x_3 all non-negative

After clearing all previously used names of functions and variable values, the variables, objective function, and set of constraints for this problem are defined and entered as they were in the first example. By using
`ConstrainedMin[z[x1,x2,x3], ineqs,vars]`
the minimum value of the objective function is obtained as well as the variable values which give this minimum.

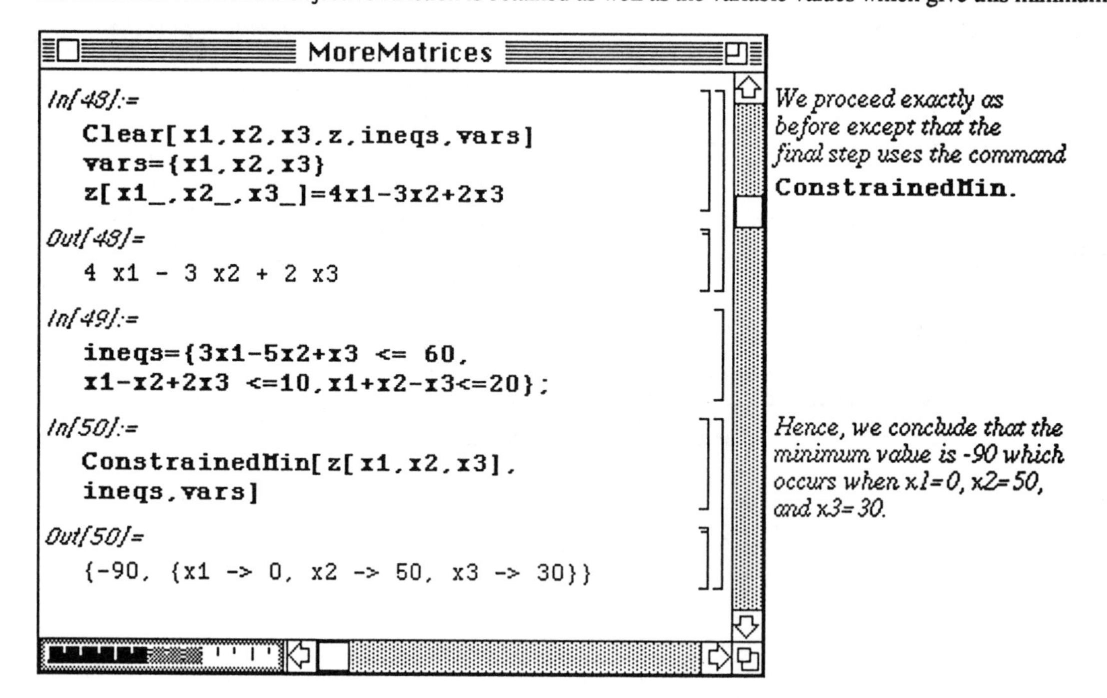

■ **The Dual Problem**

Given the standard form linear programming problem

Minimize $Z = \sum_{j=1}^{n} c_j\, x_j$ subject to the constraints $\sum_{j=1}^{n} a_{i,j}\, x_j \geq b_i$ for $i = 1, 2, \ldots, m$

and $x_j \geq 0$ for $j = 1, 2, \ldots, n$, the **dual problem** is

Maximize $Y = \sum_{i=1}^{m} b_i\, y_i$ subject to the constraints $\sum_{i=1}^{m} a_{i,j}\, y_i \leq c_j$ for $j = 1, 2, \ldots, n$

and $y_i \geq 0$ for $i = 1, 2, \ldots, m$.

Similarly, for the problem

Maximize $Z = \sum_{j=1}^{n} c_j x_j$ subject to the constraints $\sum_{j=1}^{n} a_{i,j} x_j \leq b_i$ for $i = 1, 2, \ldots, m$

and $x_j \geq 0$ for $j = 1, 2, \ldots, n$, the **dual problem** is

Minimize $Y = \sum_{i=1}^{m} b_i y_i$ subject to the constraints $\sum_{i=1}^{m} a_{i,j} y_i \geq c_j$ for $j = 1, 2, \ldots, n$

and $y_i \geq 0$ for $i = 1, 2, \ldots, m$.

□ **Example:**

Maximize $Z = 6x_1 + 8x_2$ subject to the constraints $5x_1 + 2x_2 \leq 20$, $x_1 + 2x_2 \leq 10$, $x_1 \geq 0$, and $x_2 \geq 0$.

State the dual problem and find its solution.
First, the original (primal) problem is solved. The objective function for this problem is represented by **`zx`** while that of the dual is given by **`zy`**. The set of variables **`{x[1],x[2]}`** of the primal are called **`valsx`**. Similarly, those of the dual **`{y[1],y[2]}`** are assigned the name **`valsy`**. Finally, the set a of inequalities for the primal and dual are **`ineqsx`** and **`ineqsy`**, respectively. Using the command **`ConstrainedMax[zx,ineqsx,{x[1],x[2]}]`**, the maximum value of**`zx`** is found to be 45.

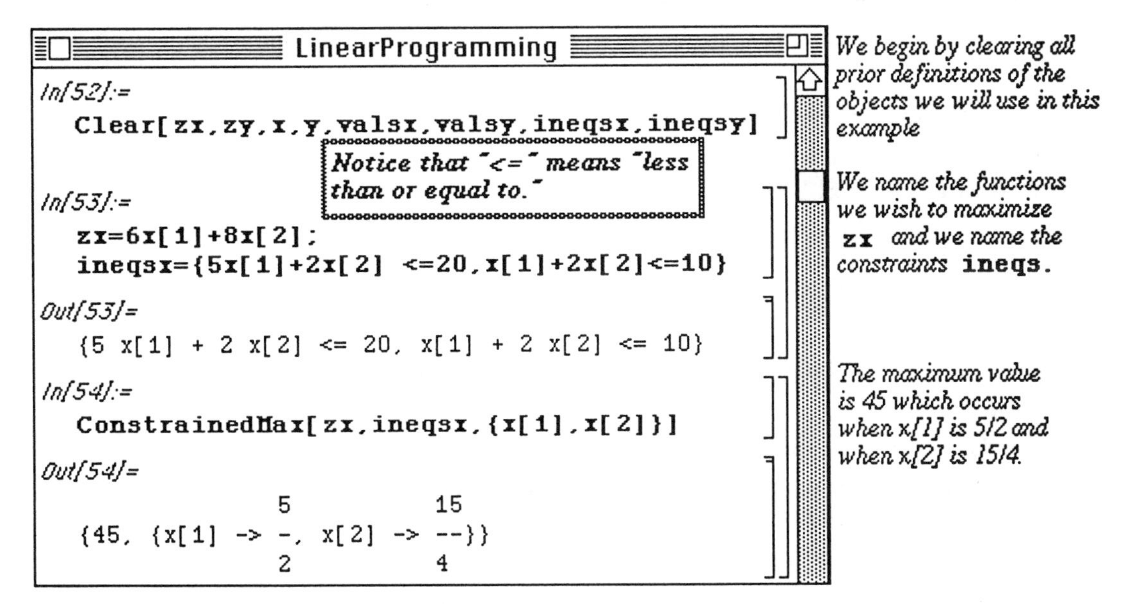

Since in this problem, we have $c_1 = 6$, $c_2 = 8$, $b_1 = 20$, and $b_2 = 10$, the dual problem is

Minimize $Z = 20y_1 + 10y_2$ subject to the constraints $5y_1 + y_2 \geq 6$, $2y_1 + 2y_2 \geq 8$, $y_1 \geq 0$, and $y_2 \geq 0$.

The dual is solved in a similar fashion by defining the objective function **zy** and the collection of inequalities **ineqsy**. The minimum value obtained by **zy** subject to the constraints **ineqsy** is 45 which agrees with the result of the primal and is found with
ConstrainedMin[zy,ineqsy,{y[1] , y[2]}] .

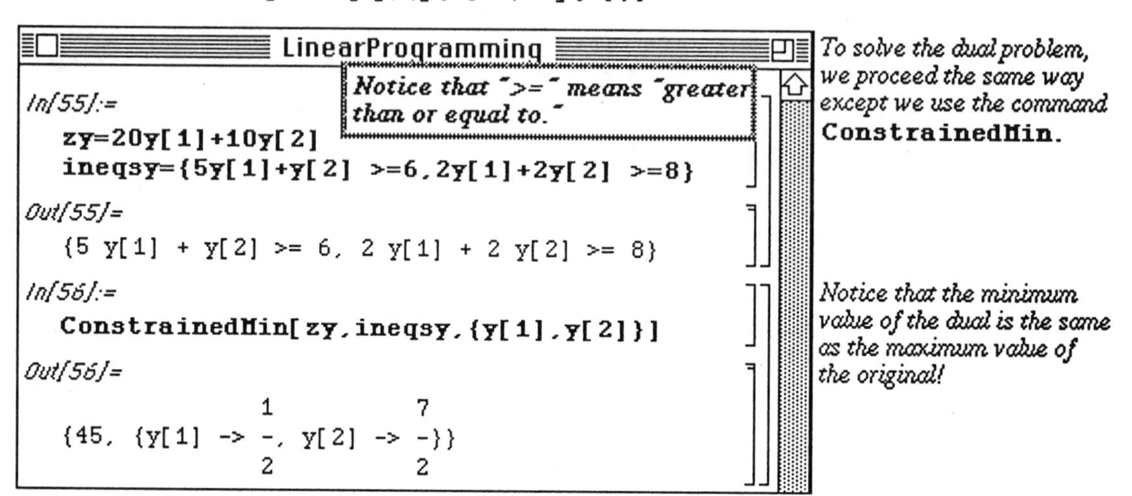

Of course, linear programming models can involve numerous variables. Consider the following :
Given the standard form linear programming problem

Minimize $Z = c_1 x_1 + c_2 x_2 + \ldots + c_n x_n$, subject to the restrictions

$$a_{1,1} x_1 + a_{1,2} x_2 + \ldots + a_{1,n} x_n \geq b_1$$
$$a_{2,1} x_1 + a_{2,2} x_2 + \ldots + a_{2,n} x_n \geq b_2$$
$$\vdots$$
$$a_{m,1} x_1 + a_{m,2} x_2 + \ldots + a_{m,n} x_n \geq b_m, \text{ and } x_1 \geq 0,\ x_2 \geq 0, \ldots, x_n \geq 0$$

let $x = \begin{bmatrix} x_1 \\ \vdots \\ x_n \end{bmatrix}$, $b = \begin{bmatrix} b_1 \\ \vdots \\ b_m \end{bmatrix}$, $c = [c_1, \ldots, c_n]$, and A denote the $m \times n$ matrix $A = [a_{i,j}]$.

Then the standard form of the linear programming problem is equivalent to find the vector x that maximizes the number Z= c.x subject to the restrictions A.x $\geq$ b and x $\geq$ 0. The **dual problem** of ***Maximize the number Z=c.x subject to the restrictions A.x ≥ b and x ≥ 0*** is ***Minimize the number Y=y.b subject to the restrictions y.A ≤ c and y ≥ 0.***

The *Mathematica* command **LinearProgramming[c,A,b]** finds the vector x which minimizes the quantity Z=c.x subject to the restrictions A.x $\geq$ b and x $\geq$ 0. This command does not yield the minimum value of Z as did **ConstrainedMin** and **ConstrainedMax**. This value must be determined from the resulting vector.

□ **Example:**

Maximize $Z = 5x_1 - 7x_2 + 7x_3 + 5x_4 + 6x_5$ subject to the constraints

$2x_1 + 3x_2 + 3x_3 + 2x_4 + 2x_5 \leq 10,$

$6x_1 + 5x_2 + 4x_3 + x_4 + 4x_5 \leq 30,$

$-3x_1 - 2x_2 - 3x_3 - x_4 \leq -5,$

$-x_1 - x_2 - x_4 \leq -10,$ and $x_i \geq 0$ for $i = 1, 2, 3, 4, 5.$

For this problem, $x = \begin{bmatrix} x_1 \\ x_2 \\ x_3 \\ x_4 \\ x_5 \end{bmatrix}$, $b = \begin{bmatrix} 10 \\ 30 \\ -5 \\ -10 \\ 0 \end{bmatrix}$, $c = [5, \ -7, \ 7, \ 5, \ 6]$, and $A = \begin{bmatrix} 2 & 3 & 3 & 2 & 2 \\ 6 & 5 & 4 & 1 & 4 \\ -3 & -2 & -3 & -4 & 0 \\ -1 & -1 & 0 & -1 & 0 \\ 0 & 0 & 0 & 0 & 0 \end{bmatrix}$.

Clearly, *Mathematica*'s ability to perform matrix algebra will be advantageous in the completion of this type of problem. First, the vectors **c** and **b** are entered.

Remark: Notice that *Mathematica* does NOT make a distinction between row and column vectors; it "interprets" the vector correctly and consequently performs the calculation correctly

The matrix A is entered and named **matrixa**.
A helpful function which can be used in this problem is **zerovec = Table[0,{5}]**
which creates a list of five zeros. This can be used instead of typing a vector made up of 5 zeros and is used in defining **matrixa** below. In general, the command **Table[expression,{n}]** produces a list of **n** copies of **expression**.

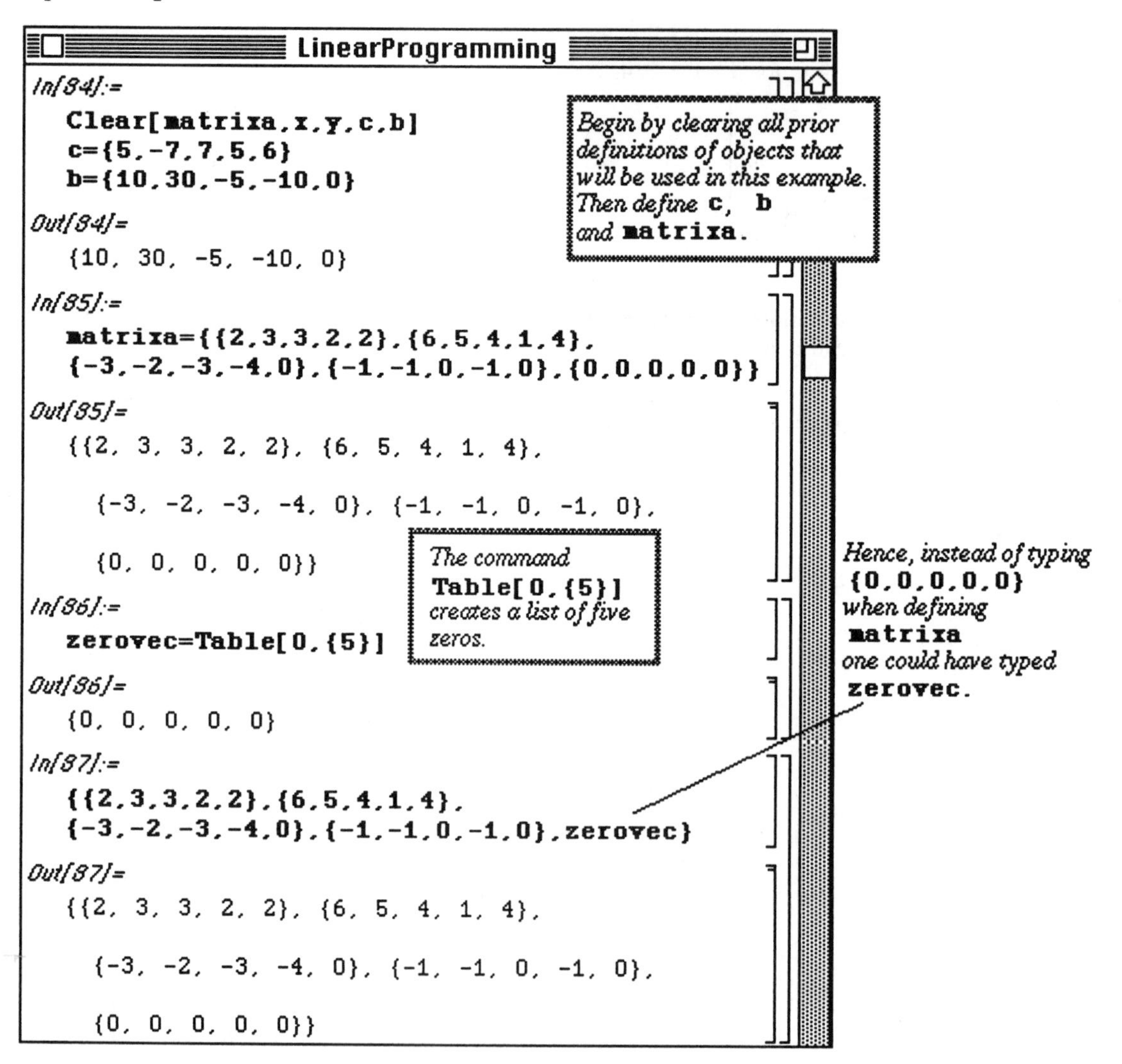

Another useful *Mathematica* command is **Array[x, n]** which creates the list of **n** elements **{x[1],x[2],... ,x[n]}**. This command is used below to define the list of variables **xvec**. Similarly, the command **Table[x[i],{i,1,n}]** yields the same list. These variables must be defined before attempting to solve this linear programming problem.

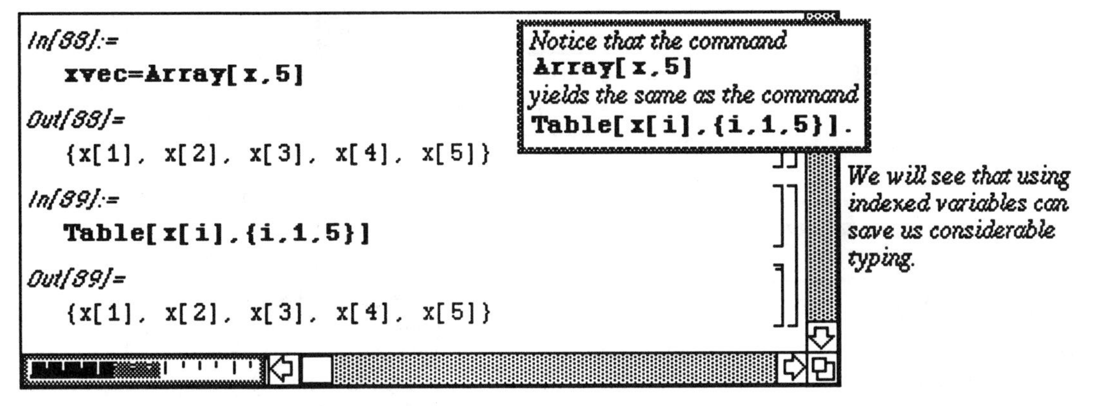

After entering the objective function coefficients with the vector **c**, the matrix of coefficients from the inequalities with **matrixa**, and the right-hand side values found in **b**; the problem is solved with **LinearProgramming[c,matrixa,b]**.
The solution is called **xvec**. Hence, the maximum value of the objective function is obtained by evaluating the objective function at the variable values which yields a maximum. Since these values are found in **xvec**, the maximum is determined with the product of the vector **c** and the vector **xvec**. (Recall that this product is entered as **c.xvec**.) This value is found to be 35/4.

LinearProgramming

```
In[90]:=
  xvec=LinearProgramming[c,matrixa,b]
Out[90]=
       5          35
  {0, -, 0, 0, --}
       2          8
In[91]:=
  c.xvec
Out[91]=
  35
  --
  4
```

solves the linear programming problem.

Thus, the maximum value of z subject to the given constraints occurs when x[1]=0, x[2]=5/2, x[3]=0, x[4]=0, and x[5]=35/8.

c.xvec *computes the maximum value which is 35/4.*

State the dual problem. What is its solution?

Since the dual of the problem is Minimize the number Y=y.b subject to the restrictions y.A $\leq$ c and y $\geq$ 0, we use *Mathematica* to calculate y.b and y.A:

Remark: Notice that *Mathematica* does NOT make a distinction between row and column vectors; it interprets the vector correctly and consequently performs the calculation properly.

A list of the dual variables **{y[1],y[2],y[3],y[4],y[5]}** is created with **Array[y,5]** . This list includes 5 elements since there are five constraints in the original problem. The objective function of the dual problem is, therefore, found with **yvec.b**, and the left-hand sides of the set of inequalities are given with **yvec.matrixa** .

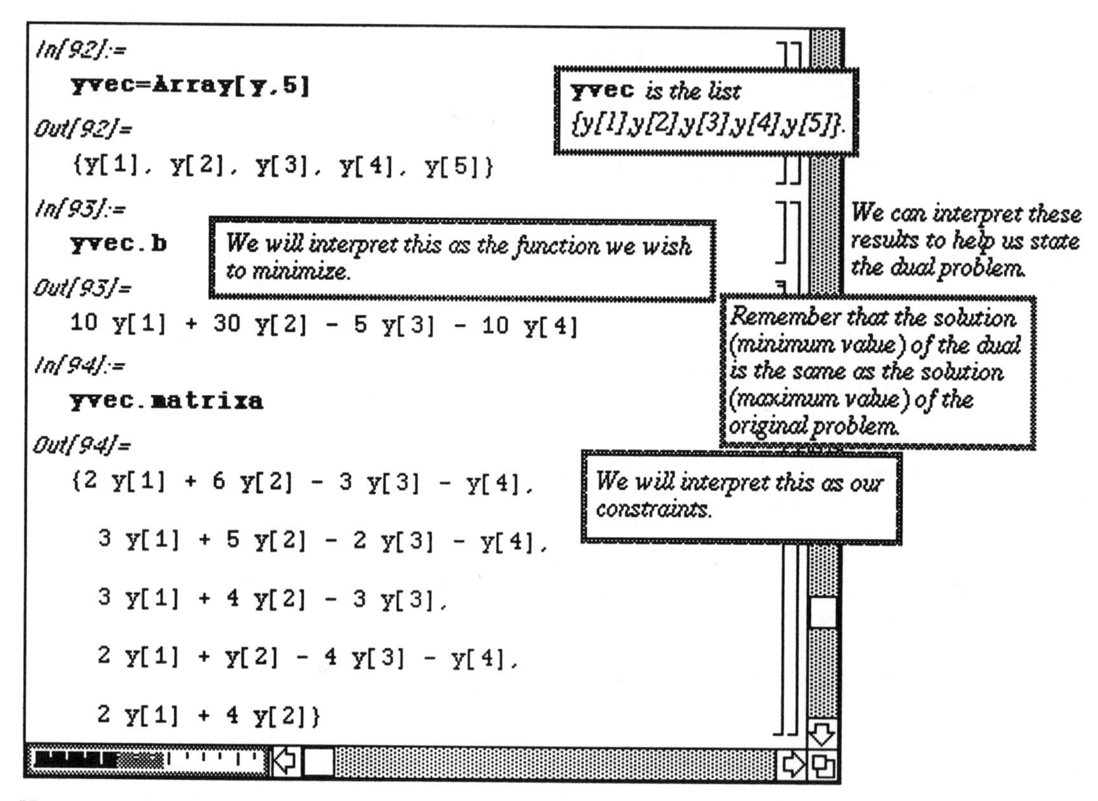

Hence, we may state the dual problem as follows:

Minimize $Y = 10y_1 + 30y_2 - 5y_3 - 10y_4$ subject to the constraints

$2y_1 + 6y_2 - 3y_3 - y_4 \leq 5$,

$3y_1 + 5y_2 - 2y_3 - y_4 \leq -7$,

$3y_1 + 4y_2 - 3y_3 \leq 7$,

$2y_1 + y_2 - 4y_3 - y_4 \leq 5$,

$2y_1 + 4y_2 \leq 6$, and $y_i \geq 0$ for $i = 1, 2, 3,$ and 4.

■ Application: A Transportation Problem

A certain company has two factories, F 1 and F 2, each producing two products, Product 1 and Product 2, that are to be shipped to three distribution centers, Dist 2, Dist 2, and Dist 3.

The following table illustrates the cost associated with shipping each product from the factory to the distribution center, the minimum number of each product each distribution center needs, and the maximum output of each factory.

	F1 /P 1	F 1/P 2	F 2/P 1	F 2/P 2		Minimum
Dist 1/Product 1	$0.75		$0.80			500
Dist 1/Product 2		$0.50		$0.40		400
Dist 2/Product 1	$1.00		$0.90			300
Dist 2/Product 2		$0.75		$1.20		500
Dist 3/Product 1	$0.90		$0.85			700
Dist 3/Product 2		$0.80		$0.95		300
Maximum Output	1000	400	800	900		

How much of each product should be shipped from each plant to each distribution center to minimize the total shipping costs?

Let x_1 denote the number of units of Product 1 shipped from F 1 to Dist 1;
x_2 denote the number of units of Product 2 shipped from F 1 to Dist 1;
x_3 denote the number of units of Product 1 shipped from F 1 to Dist 2;
x_4 denote the number of units of Product 2 shipped from F 1 to Dist 2;
x_5 denote the number of units of Product 1 shipped from F 1 to Dist 3; and
x_6 denote the number of units of Product 2 shipped from F 1 to Dist 3.

Let x_7 denote the number of units of Product 1 shipped from F 2 to Dist 1;
x_8 denote the number of units of Product 2 shipped from F 2 to Dist 1;
x_9 denote the number of units of Product 1 shipped from F 2 to Dist 2;
x_{10} denote the number of units of Product 2 shipped from F 2 to Dist 2;
x_{11} denote the number of units of Product 1 shipped from F 2 to Dist 3; and
x_{12} denote the number of units of Product 2 shipped from F 2 to Dist 3.

Then it is necessary to **minimize** the number:

$$Z = .75x_1 + .5x_2 + x_3 + .75x_4 + .9x_5 + .8x_6 + .8x_7 + .4x_8 + .9x_9 + 1.2x_{10} + .85x_{11} + .95x_{12}$$

subject to the constraints

$x_1 + x_3 + x_5 \leq 1000$; $x_2 + x_4 + x_6 \leq 400$; $x_7 + x_9 + x_{11} \leq 800$;
$x_8 + x_{10} + x_{12} \leq 900$; $x_1 + x_7 \geq 500$; $x_3 + x_9 \geq 300$; $x_5 + x_{11} \geq 700$;
$x_2 + x_8 \geq 400$; $x_4 + x_{10} \geq 500$; $x_6 + x_{12} \geq 300$; and $x_i \geq 0$ for $i = 1, \ldots, 12$.

In order to solve this linear programming problem, the objective function which computes the total cost, the 12 variables, and set of inequalities must be entered. The coefficients of the objective function are given in the vector **c**. Using the command **Array[x, 12]** illustrated in the previous example to define the list of 12 variables **{x[1], x[2], . . . , x[12]}**, the the objective function is given by the product **z=xvec.c** where **xvec** is the name assigned to the list of variables.

LinearProgramming

```
In[26]:=
   Clear[xvac,z,constraints,vars,c]
   c={.75,.5,1,.75,.9,.8,.8,.4,.9,1.2,.85,.95}
Out[26]=
   {0.75, 0.5, 1, 0.75, 0.9, 0.8, 0.8, 0.4,

     0.9, 1.2, 0.85, 0.95}
```

In order to define z, first define the table c so that the ith element of c is the coefficient of x_i.

```
In[27]:=
   xvec=Array[x,12]
Out[27]=
   {x[1], x[2], x[3], x[4], x[5], x[6], x[7],

     x[8], x[9], x[10], x[11], x[12]}
```

Array[x,12] *creates the list of twelve elements x[1], ... , x[12]; we will interpret x[i] as* x_i.

```
In[28]:=
   z=xvec.c
Out[28]=
   0.75 x[1] + 0.5 x[2] + x[3] + 0.75 x[4] +

     0.9 x[5] + 0.8 x[6] + 0.8 x[7] +

     0.4 x[8] + 0.9 x[9] + 1.2 x[10] +

     0.85 x[11] + 0.95 x[12]
```

Notice that **xvec.c** *produces the desired quantity, z, we wish to minimize. Hence, in defining z we have avoided considerable typing by using the command* **Array**.

The set of constraints are then entered and named **constraints** for easier use. Therefore, the minimum cost and the value of each variable which yields this minimum cost are found with the command **ConstrainedMin[z,constraints,xvec]**.

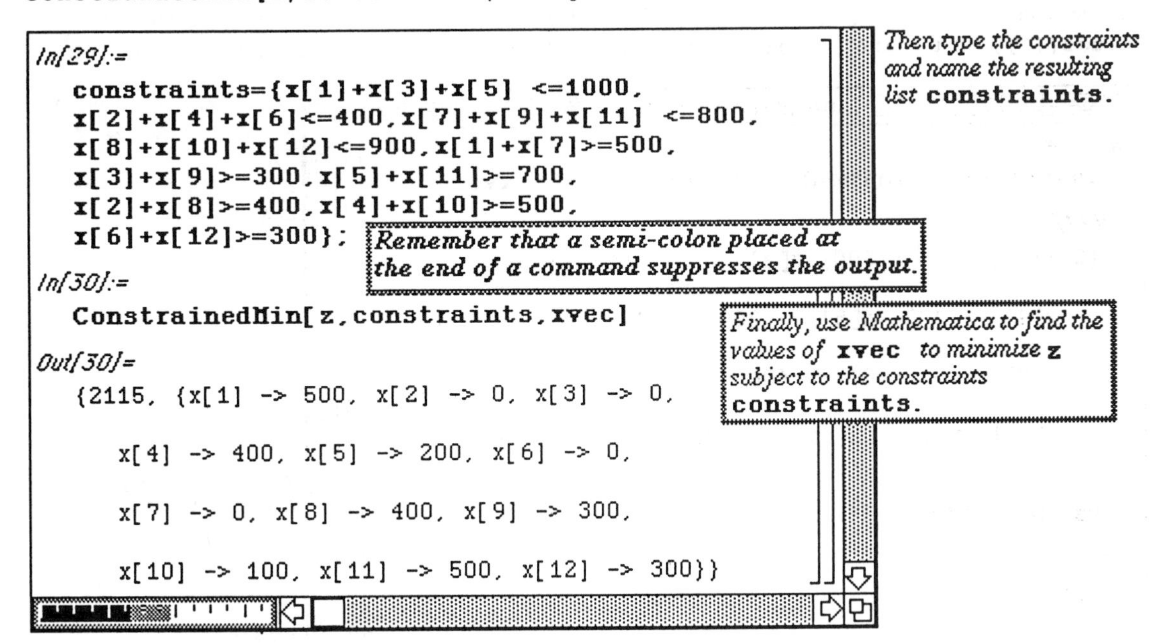

The elements of the list which results from the command `ConstrainedMin[z,constraints,xvec]` can be extracted if this list is assigned a name. Therefore, the name **values** is given to this list. Notice that **values** is a list made up of two elements, the minimum value of the cost function, 2115, and the list of the variable values `{x[1]->500,x[2]->0, ...}` Hence, the minimum cost is obtained with the command `values[[1]]` and the list of variable values which yield the minimum cost is extracted with `values[[2]]`.

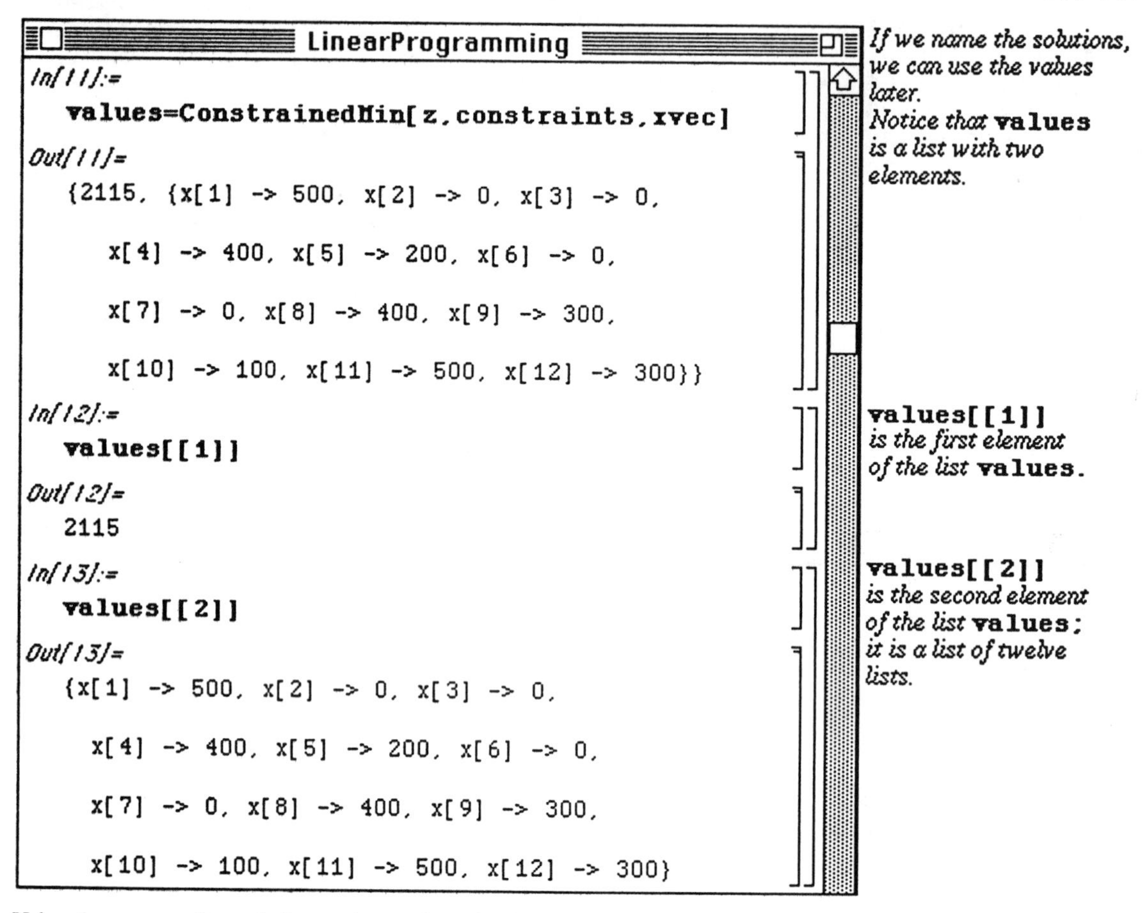

Using these extraction techniques, the number of units produced by each factory can be computed. Since

x_1 denotes the number of units of Product 1 shipped from F1 to Dist 1;;
x_3 denotes the number of units of Product 1 shipped from F1 to Dist 2;
and x_5 denotes the number of units of Product 1 shipped from F1 to Dist 3,

then the total number of units of Product 1 produced by Factory 1 is given by **x[1]+x[3]+x[5]**. The command

x[1]+x[3]+x[5] /. values[[2]]

evaluates this sum at the values of x[1], x[3], and x[5] given in the list **values[[2]]**. Similarly, the number of units of each product that each factory produces can be calculated. These results are shown below :

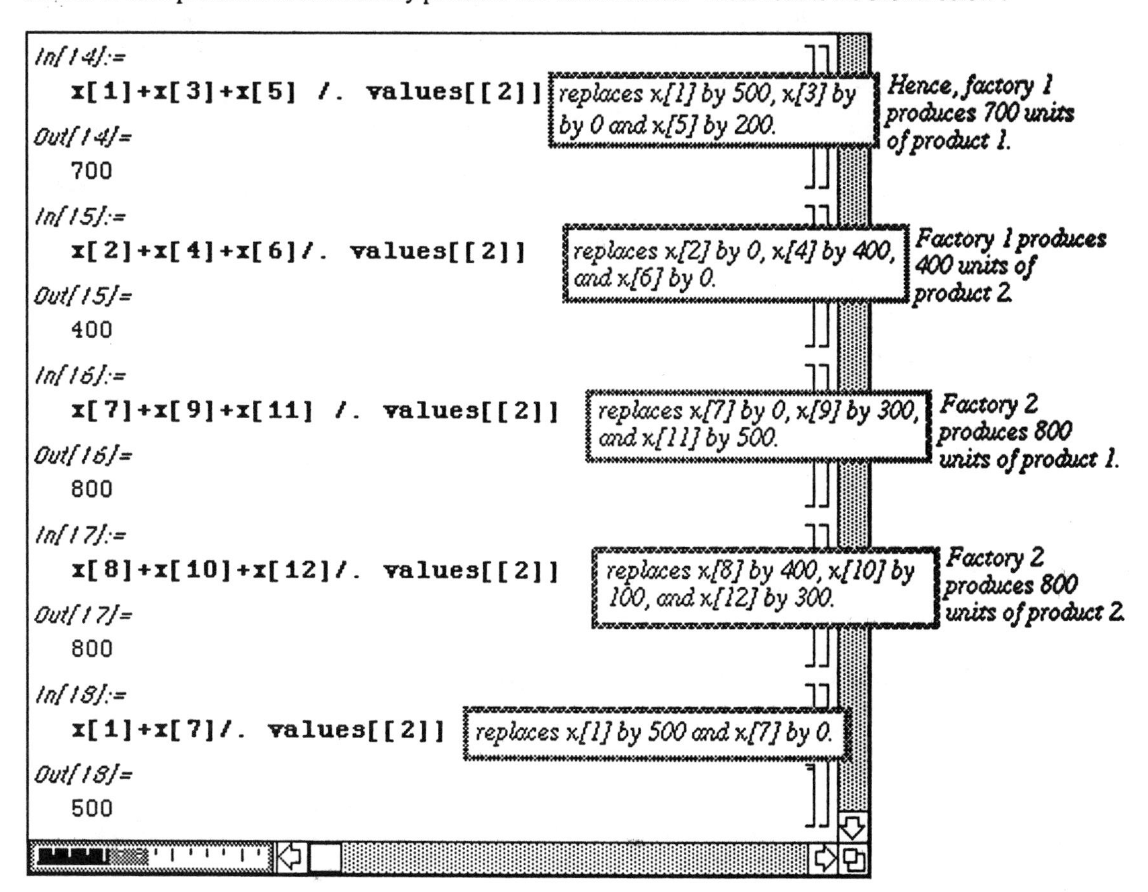

Also, the number of units of Products 1 and 2 received by each distribution center can be computed. The command
```
x[3]+x[9]/.values[[2]]
```
gives the total amount of Product 1 received at Dist 1 since
x[3]= amount of Product 1 received by Dist 2 from F1 and
x[9]= amount of Product 1 received by Dist 2 from F2 .
Notice that this amount is the minimum number of units (300) of Product 1 requested by Dist 1. The amount of Products 1 and 2 received at each distribution center are calculated in a similar manner and illustrated below :

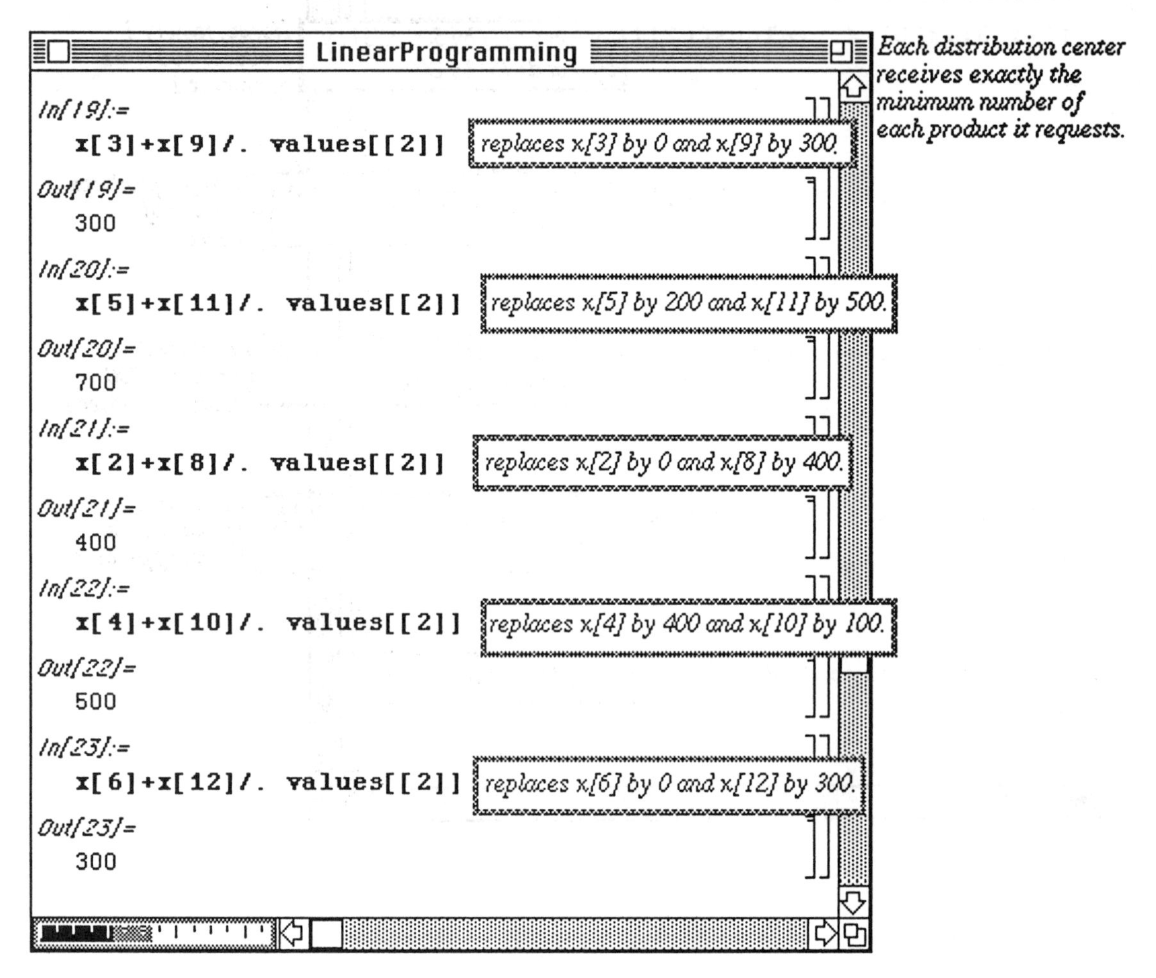

5.3 Vector Calculus

Review of Definitions and Notation

The terminology and notation used in Mathematica by Example is standard. Nevertheless, we review basic definitions briefly.

A scalar field is a function with domain a set of ordered-triples and range a subset of the real numbers:

$f:U \to V$ is a scalar field means $U \subseteq \Re^3$ and $V \subseteq \Re$.

The gradient of the scalar field f is defined to be the vector

$$\text{grad } f = \frac{\partial f}{\partial x} i + \frac{\partial f}{\partial y} j + \frac{\partial f}{\partial z} k = \left\langle \frac{\partial f}{\partial x}, \frac{\partial f}{\partial y}, \frac{\partial f}{\partial z} \right\rangle = \langle f_x, f_y, f_z \rangle, \text{ where } i = \langle 1,0,0 \rangle,\ j = \langle 0,1,0 \rangle, \text{ and } k = \langle 0,0,1 \rangle.$$

A vector field f is a vector-valued function:

$f:V \to U$, $U \subseteq \Re^3$ and $V \subseteq \Re$, is a vector field means that f can be written in the form $f(x,y,z) = f_1(x,y,z)i + f_2(x,y,z)j + f_3(x,y,z)k = \langle f_1(x,y,z), f_2(x,y,z), f_3(x,y,z) \rangle$ for each (x,y,z) in the domain of f.

A conservative vector field f is a vector field that is the gradient of a scalar field:

f is a conservative vector field means there is a scalar field g satisfying

$f = \nabla^2 g$. In this case, g is usually called a **potential function** for f.

The divergence of the vector field f is defined to be the scalar

$$\begin{aligned} \text{div } f = \text{div } f(x,y,z) &= \text{div}\langle f_1(x,y,z), f_2(x,y,z), f_3(x,y,z) \rangle \\ &= \frac{\partial f_1(x,y,z)}{\partial x} + \frac{\partial f_2(x,y,z)}{\partial y} + \frac{\partial f_3(x,y,z)}{\partial z} \\ &= \nabla \bullet f. \end{aligned}$$

The laplacian of the scalar field f is defined to be div(grad f)

$$\text{laplacian }(f) = \nabla^2 f = \Delta f = \frac{\partial^2 f}{\partial x^2} + \frac{\partial^2 f}{\partial y^2} + \frac{\partial^2 f}{\partial z^2} = f_{xx} + f_{yy} + f_{zz}.$$

For three-dimensional vector analysis, the package **`VectorAnalysis.m`** contains the commands **`Grad`**, **`Div`**, **`Curl`**, and **`Laplacian`**.

Be sure to load the package **`VectorAnalysis.m`** prior to using these functions. **`VectorAnalysis.m`** is contained in the folder **`Calculus`**.

Because *Mathematica* recognizes Cartesian (x,y,z), Cylindrical (r,phi,z), and Spherical (r,theta,phi) coordinates, and because the operations discussed in this section differ in the various coordinate systems, the desired coordinate system must be indicated. This is accomplished with
`SetCoordinates[System]` where
`System` is usually one of **`Cartesian`** , **`Cylindrical`**, or **`Spherical`** .

However, the available coordinate systems are:

Cartesian, Cylindrical, Spherical, Parabolic, ParabolicCylinder, ProlateEllipsoidal, EllipticCylinder, OblateEllipsoidal, Toroidal, Elliptic, and **Bipolar.**

The examples illustrated below are done in cartesian coordinates.
After the function **f[x,y,z]** has been defined, the gradient is found with :
Grad[f[x,y,z]].
Since this is a function of (x,y,z), it is denoted **gradientf[x,y,z]** for later use.
The laplacian of f is determined using
Laplacian[f[x,y,z]],
and the divergence is found with
Div[gradient[f[x,y,z]]].

□ Example:

Let f(x,y,z) = Cos(x y z). Compute ∇f, ∇^2f, and Div(∇f).

Note: When defining f, be sure to include the space between the variables.

Notice tha. `Laplacian[f[x,y,z]]` yields the same result as `Div[gradient[x,y,z]]`.

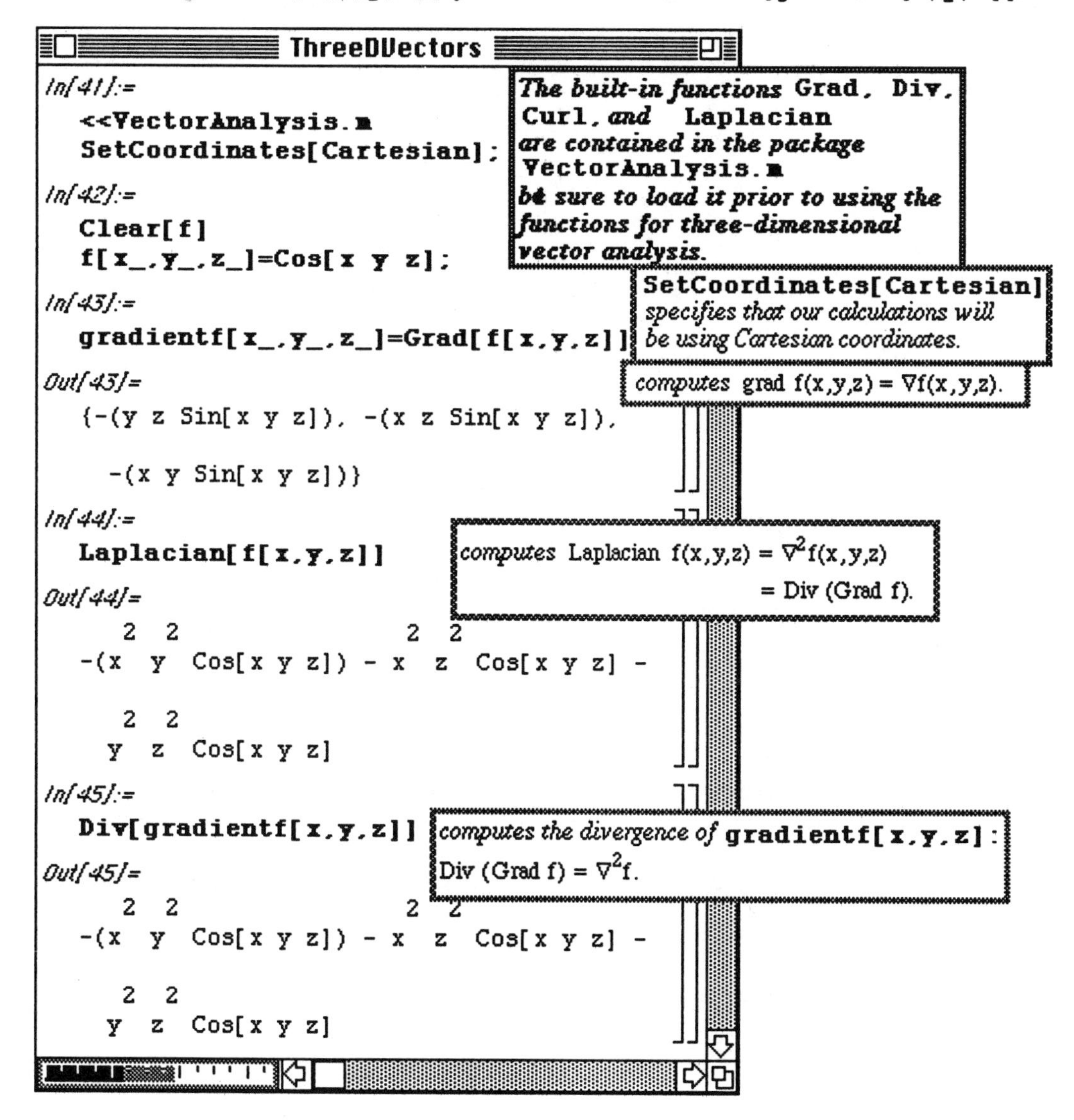

If S is the graph of f(x,y) and g(x,y,x)=z - f(x,y), then the gradient $\nabla g(x,y,z)$ is a normal vector to the graph of $g(x,y,z)=0$.
At the point (x,y,z) a **unit normal vector** n can be obtained via:

$$n=\frac{\nabla g(x,y,z)}{\|\nabla g(x,y,z)\|}=\frac{-f_x(x,y)i-f_y(x,y)j+k}{\sqrt{[f_x(x,y)]^2+[f_y(x,y)]^2+1}}=\frac{1}{\sqrt{[f_x(x,y)]^2+[f_y(x,y)]^2+1}}\langle -f_x(x,y),-f_y(x,y),1\rangle.$$

The **curl of the vector field f** is defined to be the vector field

$$\begin{aligned}\text{curl } f = \text{curl } f(x,y,z) &= \text{curl}\langle f_1(x,y,z),f_2(x,y,z),f_3(x,y,z)\rangle \\ &= \left(\frac{\partial f_3}{\partial y}-\frac{\partial f_2}{\partial z}\right)i+\left(\frac{\partial f_1}{\partial z}-\frac{\partial f_3}{\partial x}\right)j+\left(\frac{\partial f_2}{\partial x}-\frac{\partial f_1}{\partial y}\right)k \\ &= \begin{vmatrix} i & j & k \\ \frac{\partial}{\partial x} & \frac{\partial}{\partial y} & \frac{\partial}{\partial z} \\ f_1 & f_2 & f_3 \end{vmatrix}.\end{aligned}$$

□ **Example:**

Let $f(x,y,z)=xy\,i+xz^2y\,j-e^{2z}k=\{xy,xz^2y,-e^{2z}\}$. Compute

$$\text{Curl} f=\left\{\left(\frac{\partial}{\partial y}\left(-e^{2z}\right)-\frac{\partial}{\partial z}\left(xz^2y\right)\right),\left(\frac{\partial}{\partial z}(xy)-\frac{\partial}{\partial x}\left(-e^{2z}\right)\right),\left(\frac{\partial}{\partial x}\left(xz^2y\right)-\frac{\partial}{\partial y}(xy)\right)\right\};$$

$\text{Div} f=\frac{\partial}{\partial x}(xy)+\frac{\partial}{\partial y}\left(xz^2y\right)+\frac{\partial}{\partial z}\left(-e^{2z}\right)$; $\text{Laplacian}(\text{Div} f)=\nabla^2(\text{Div} f)$;

$\text{Grad}(\text{Laplacian}(\text{Div} f))=\text{Grad}\left(\nabla^2(\text{Div} f)\right)$; and

$\text{Laplacian}(\text{Grad}(\text{Laplacian}(\text{Div} f)))=\nabla^2(\text{Grad}(\text{Laplacian}(\text{Div} f)))$.

The first step towards solving this problem is to enter the unit vectors in Cartesian coordinates `i={1,0,0}`, `j={0,1,0}`, and `k={0,0,1}`. The vector-valued function `f[x,y,z]` can then be defined using these three unit vectors as follows:
`f[x_,y_,z_]=x y i+x z^2 y j-Exp[2z] k` (remembering to place appropriate spaces between variables for multiplication).

Notice that the coordinate system has not been set in this problem. However, the correct system can be indicated in each command. For example, the curl of f in Cartesian coordinates is determined with **`Curl[f[x,y,z],Cartesian]`**. The curl could similarly be obtained in the other systems by replacing **`Cartesian`** with **`Cylindrical`**, **`Spherical`**, or one of the other available coordinate systems, in the command above.

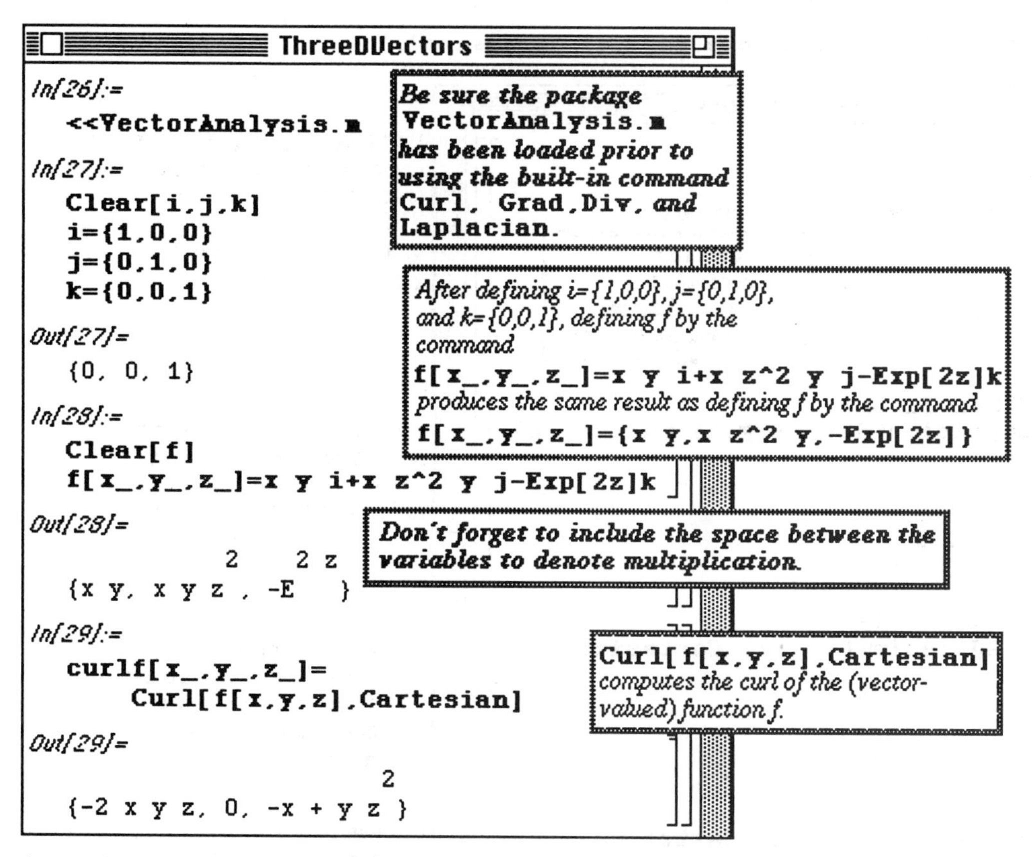

As was the case with computing the curl of f, the divergence of f can be calculated in Cartesian coordinates with **`Div[f[x,y,z],Cartesian]]`**. Again, since the divergence is a function of (x,y,z), it is named **`divf[x,y,z]`** for later use. Hence, the Laplacian of the divergence of f is computed with **`Laplacian[divf[x,y,z],Cartesian]]`**. This function is called **`ladiv[x,y,z]`** so that

$\text{Grad}(\text{Laplacian}(\text{Div}\, f)) = \text{Grad}(\nabla^2(\text{Div}\, f))$

can be found with **Grad[ladivf[x,y,z],Cartesian]]**. The resulting function is then named **grad[x,y,z]** so that $\text{Laplacian}(\text{Grad}(\text{Laplacian}(\text{Div}\, f))) = \nabla^2(\text{Grad}(\text{Laplacian}(\text{Div}\, f)))$

can be computed with **Laplacian[grad[x,y,z],Cartesian]**.

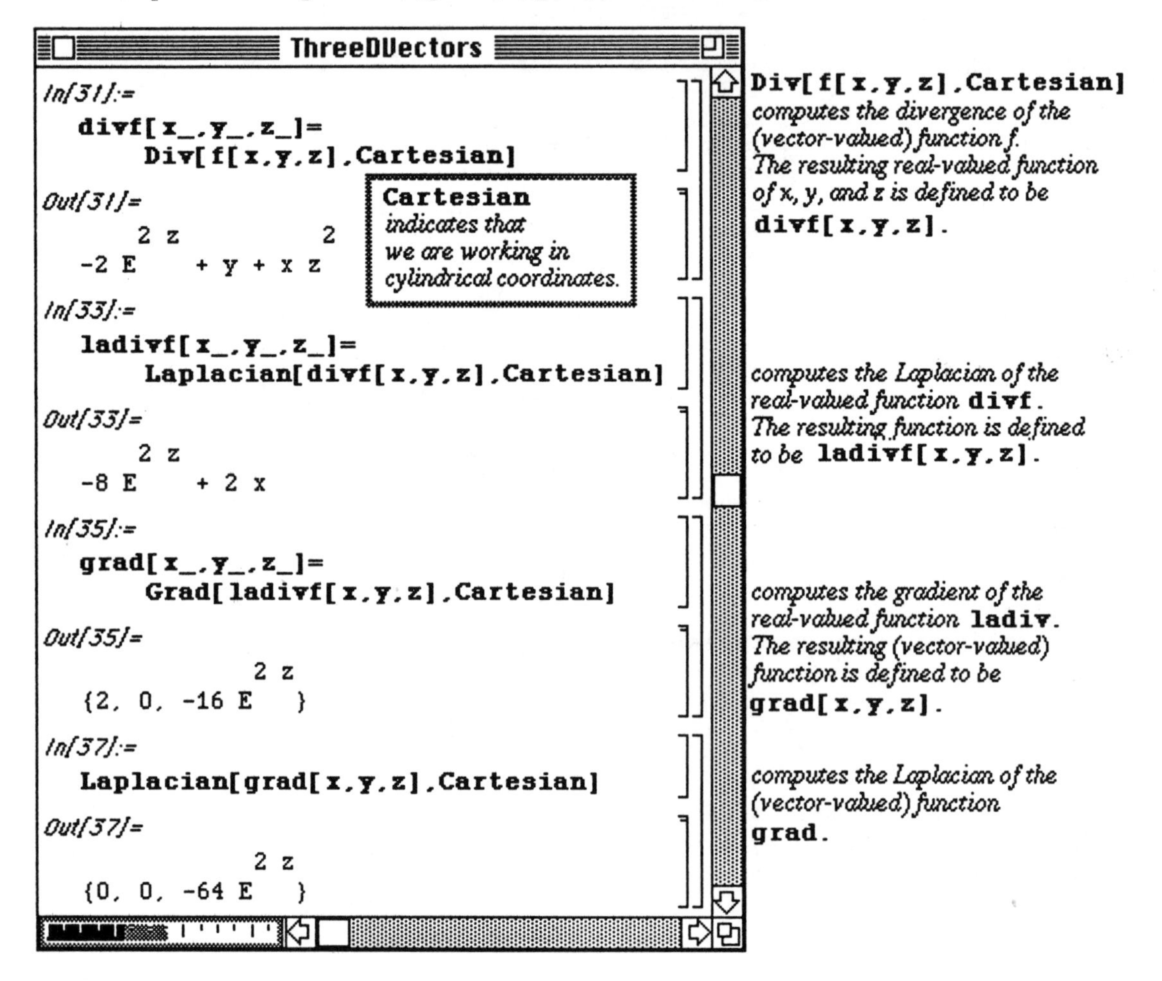

□ **Example:**

Let $w(x,y) = \mathrm{Cos}\left(4x^2 + 9y^2\right)$. Let $n_{(x,y)}$ denote a unit vector normal to the graph of w at the point (x,y,w(x,y)). Find a formula for n.

In order to visualize the unit normal vector at points (x,y,w(x,y)) to the surface w(x,y), this function is plotted using several of the options available with `Plot3D`. These options are discussed below :

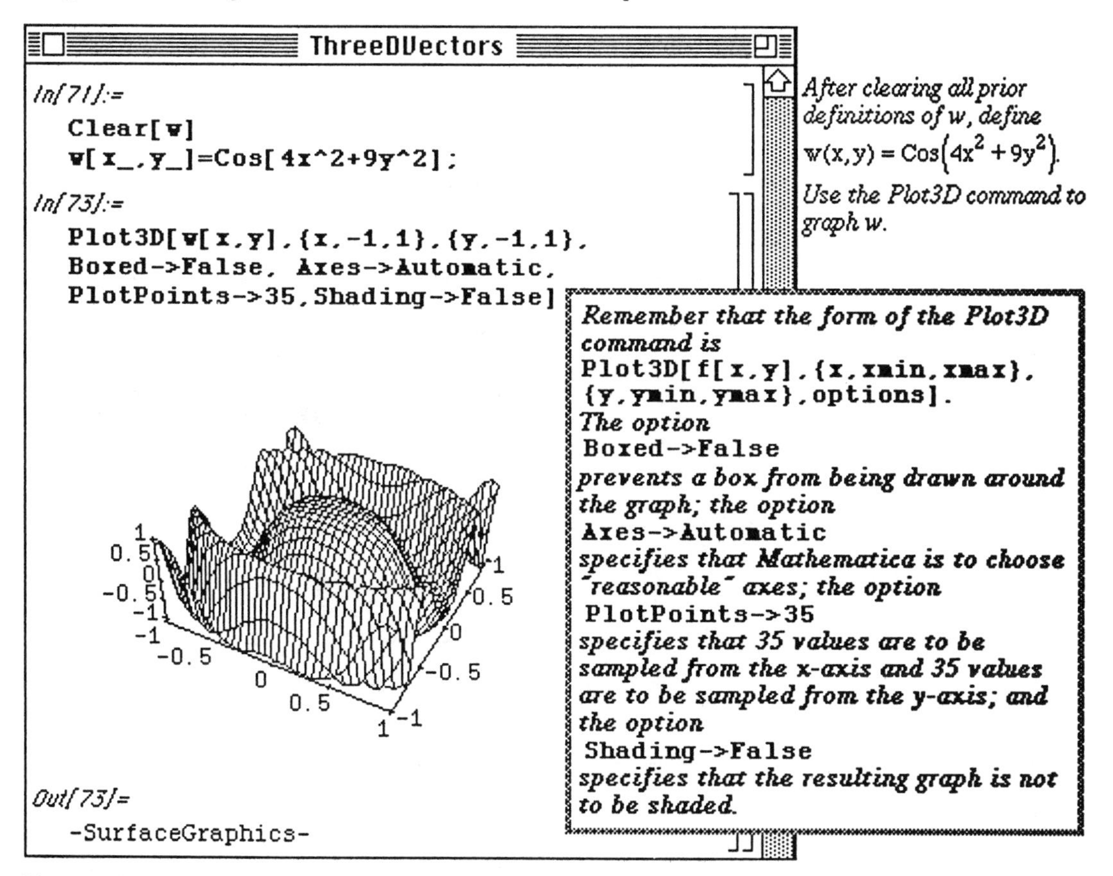

The equation z = w(x,y) is written as z - w(x,y) = 0. The left-hand side of this equation is a function of x, y, and z and is defined as `wz[x_,y_,z_]=z-w[x,y]`.

Since the partial derivative of **wz** with respect to **z** is -1, the gradient of **wz** is a function of x and y only. Hence, the gradient of **wz** is named **gw[x_,y_]** and is computed with **Grad[wz[x,y,z],Cartesian]**. The length of the gradient of **wz** which is necessary in determining the unit normal vector is the square root of the dot product of the gradient of **wz** with itself. This product is computed with **gw[x,y].gw[x,y]**.

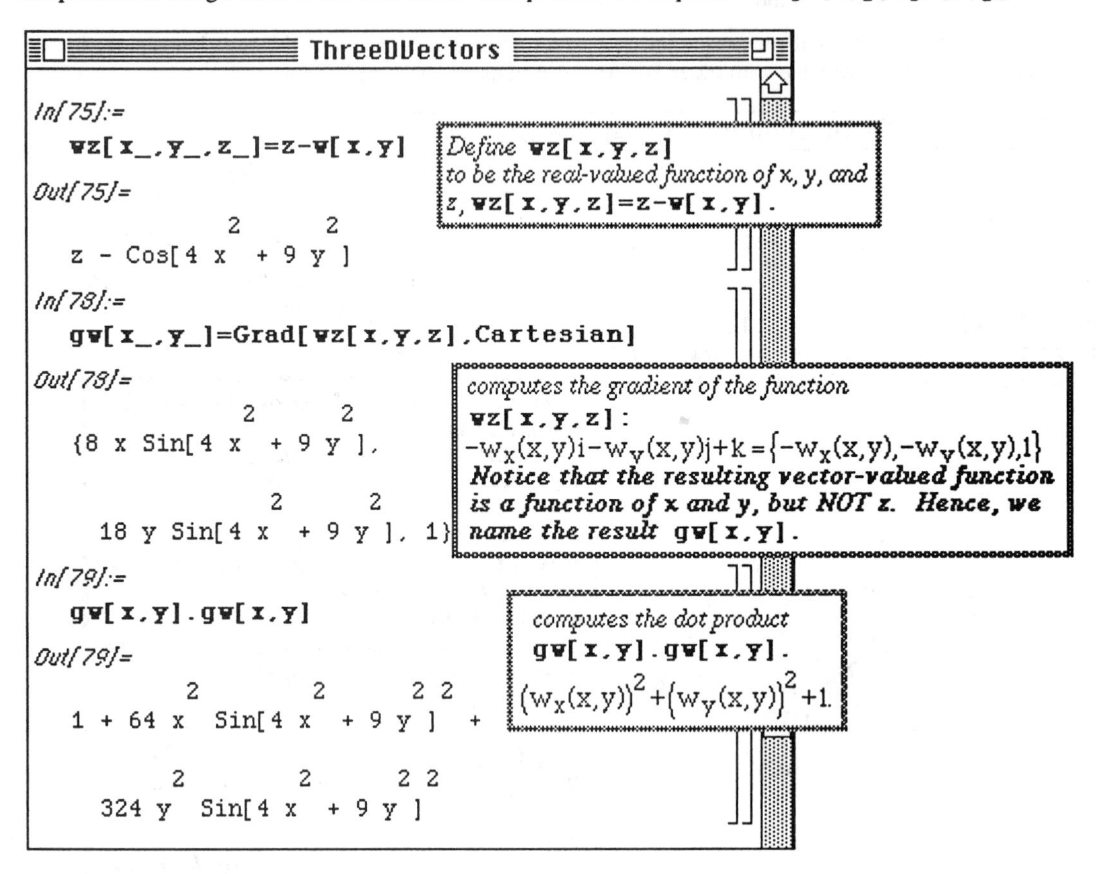

Therefore, the unit normal vector is the gradient of g, **gw[x,y]**, divided by the the square root of **gw[x,y].gw[x,y]** as shown below. This is also a function of the variables x and y since the unit normal vector differs from point to point on the surface. Hence, this vector is assigned the name **normalw[x_,y_]** so that the unit vector at any point (x,y,w(x,y)) can be easily determined by evaluating **normalw[x_,y_]** at any point (x,y).

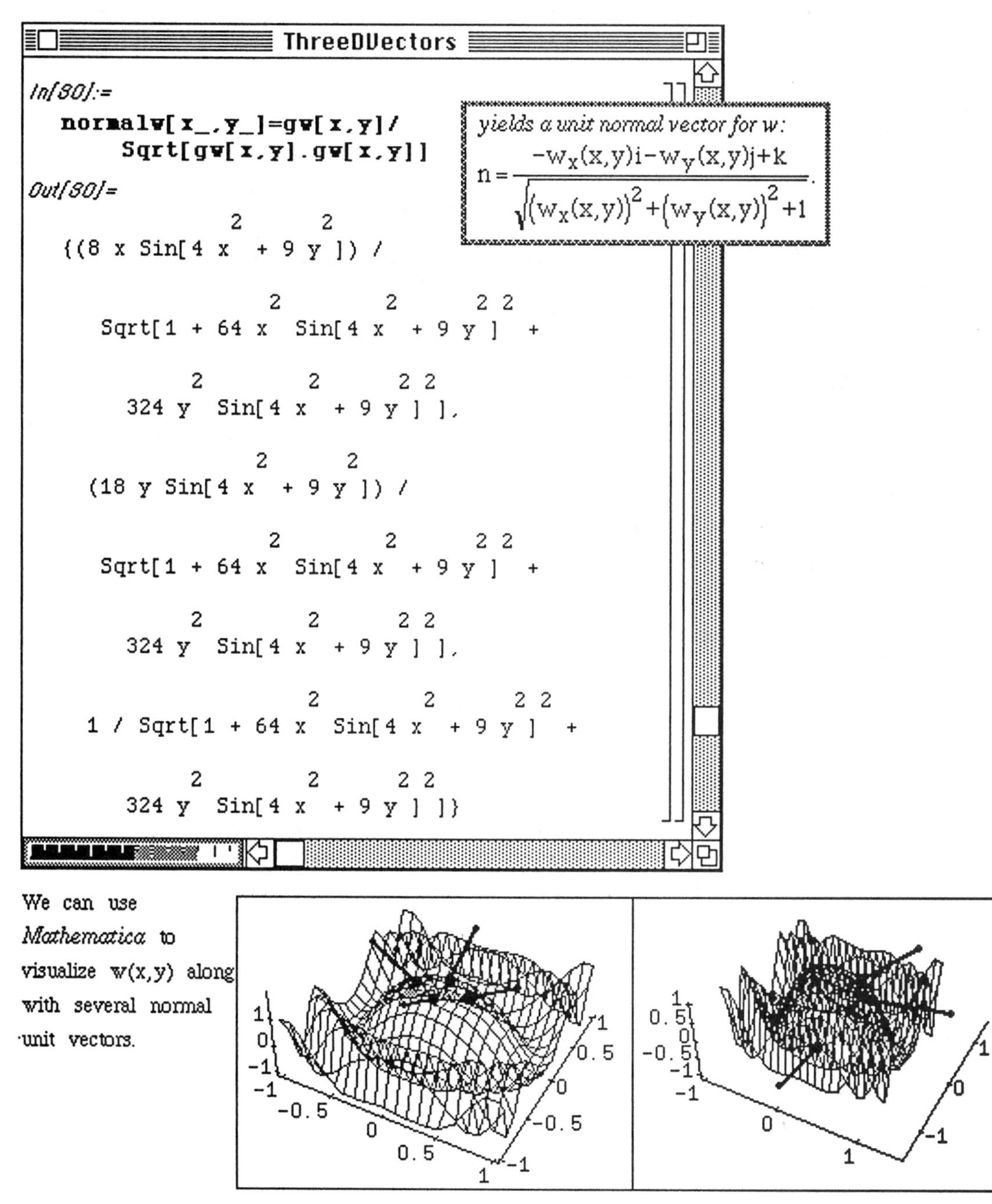

ThreeDVectors

```
In[80]:=
   normalw[x_,y_]=gw[x,y]/
        Sqrt[gw[x,y].gw[x,y]]
Out[80]=
                     2       2
   {(8 x Sin[4 x  + 9 y ]) /

                  2          2        2 2
       Sqrt[1 + 64 x  Sin[4 x  + 9 y ]  +

                2          2       2 2
         324 y  Sin[4 x  + 9 y ] ],

                      2       2
      (18 y Sin[4 x  + 9 y ]) /

                  2          2        2 2
       Sqrt[1 + 64 x  Sin[4 x  + 9 y ]  +

                2          2       2 2
         324 y  Sin[4 x  + 9 y ] ],

                        2          2       2 2
      1 / Sqrt[1 + 64 x  Sin[4 x  + 9 y ]  +

                2          2       2 2
         324 y  Sin[4 x  + 9 y ] ]}
```

yields a unit normal vector for w:

$$n = \frac{-w_x(x,y)i - w_y(x,y)j + k}{\sqrt{(w_x(x,y))^2 + (w_y(x,y))^2 + 1}}.$$

We can use *Mathematica* to visualize w(x,y) along with several normal unit vectors.

■ Application: Green's Theorem

Green's Theorem: Let C be a piecewise smooth simple closed curve and let R be the region consisting of C and its interior. If f and g are functions that are continuous and have continuous first partial derivatives throughout an open region D containing R, then $\oint_C (m(x,y)dx + n(x,y)dy) = \iint\limits_R \left(\frac{\partial n}{\partial x} - \frac{\partial m}{\partial y}\right) dA.$

□ Example:

Use Green's Theorem to evaluate $\int\limits_C \left(x + e^{\sqrt{y}}\right)dx + (2y + \mathrm{Cos}(x))dy$ where

C is the boundary of the region enclosed by the parabolas $y = x^2$ and $x = y^2$.

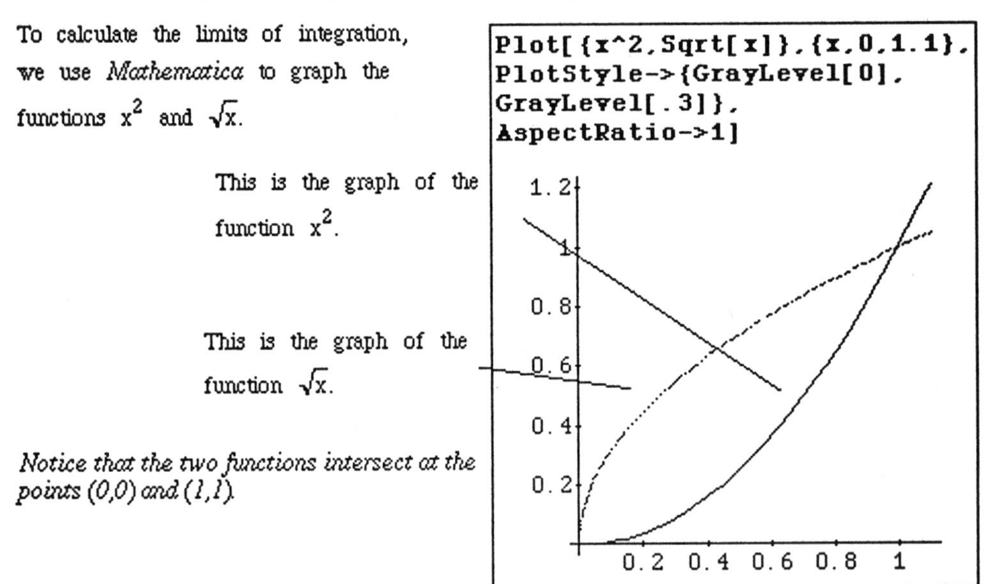

In this example,

$m(x,y) = x + e^{\sqrt{y}}$ and $n(x,y) = 2y + \mathrm{Cos}(x)$. Therefore, applying Green's Theorem,

$$\int\limits_C \left(x + e^{\sqrt{y}}\right)dx + (2y + \mathrm{Cos}(x))dy = \int\limits_C m(x,y)dx + n(x,y)dy$$

$$= \iint\limits_R \left(\frac{\partial n}{\partial x} - \frac{\partial m}{\partial y}\right) dA = \int_0^1 \int_{x^2}^{\sqrt{x}} \left(\frac{\partial n}{\partial x} - \frac{\partial m}{\partial y}\right) dy dx.$$

Therefore we will use *Mathematica* to define m(x,y), n(x,y), and to compute $\frac{\partial n}{\partial x}$, $\frac{\partial m}{\partial y}$, and $\int_0^1 \int_{x^2}^{\sqrt{x}} \left(\frac{\partial n}{\partial x} - \frac{\partial m}{\partial y}\right) dy\, dx.$

First, the functions m(x,y) and n(x,y) are defined. Recall that in computing partial derivatives, the variable of differentiation must be given. Therefore, the partial of **n[x, y]** with respect to **x** is given by **D[n[x, y], x]**. This derivative is then named **nx** for later use. Similarly, the partial derivative of **m[x, y]** with respect to **y** is found with **D[m[x, y], y]** and named **my**.

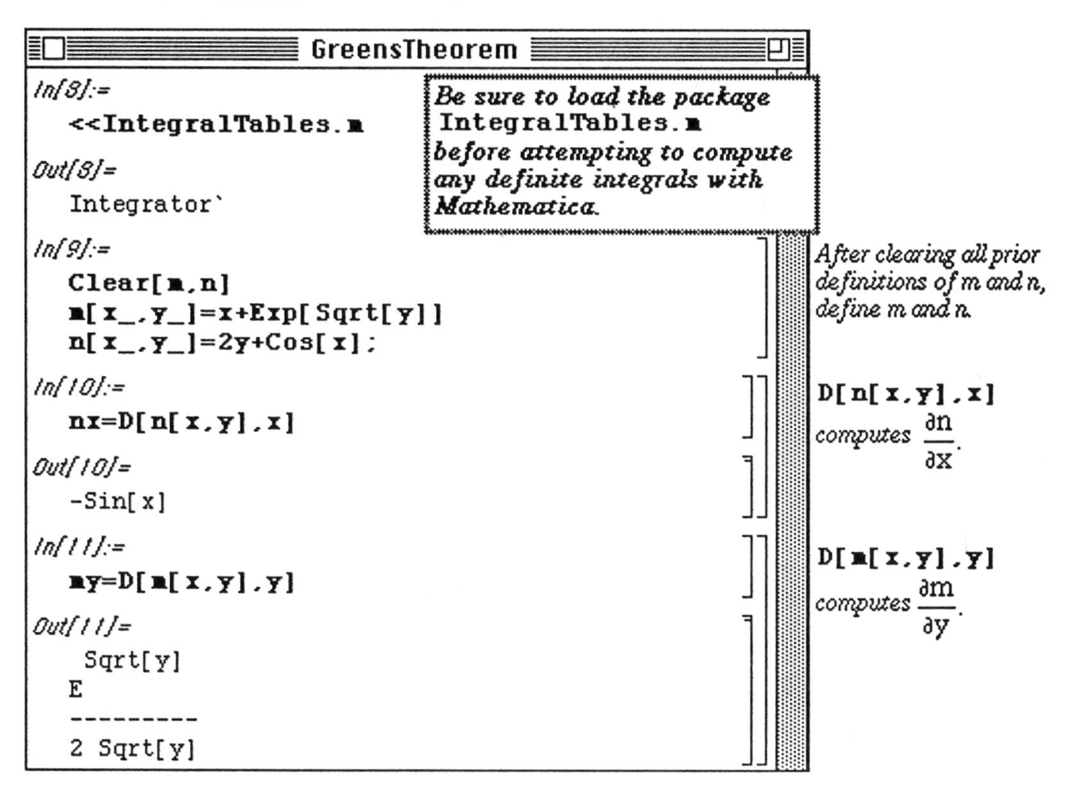

Mathematica computes the exact value of the double integral. To obtain a more meaningful value, we approximate the value of the double integral using the **N** command. In general, the command **N[%]** produces a numerical approximation of the previous output.

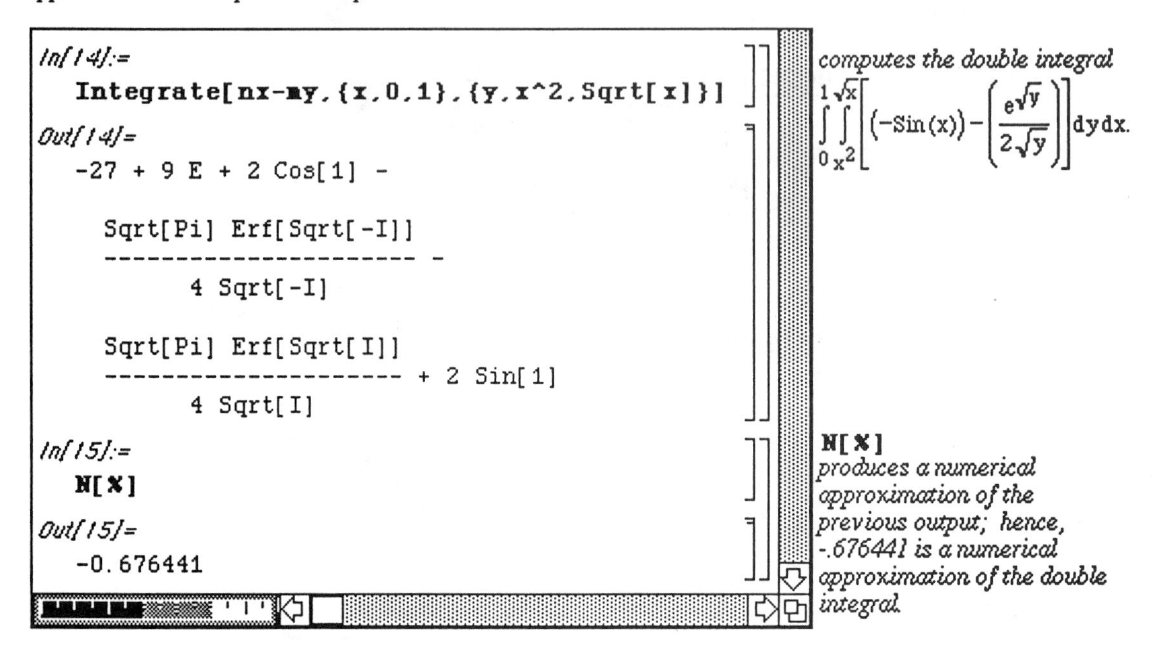

■ **Application:** **The Divergence Theorem**

The Divergence Theorem: Let Q be any domain with the property that each straight line through any interior point of the domain cuts the boundary in exactly two points, and such that the boundary S is a piecewise-smooth, closed, oriented surface with unit outer normal n. If f is a vector field that has continuous partial derivatives on Q, then

$$\iint_S \mathbf{f} \bullet \mathbf{n}\, dS = \iiint_Q \operatorname{div} \mathbf{f}\, dV = \iiint_Q \nabla \bullet \mathbf{f}\, dV.$$

$\iint_S \mathbf{f} \bullet \mathbf{n}\, dS$ is called the **outward flux** of the vector field f across the surface S.

If S is a portion of the level curve g(x,y,z)=c for some g, then a unit normal vector n may be taken to be either

$$n = \frac{\nabla g}{\|\nabla g\|} \quad \text{or} \quad n = -\frac{\nabla g}{\|\nabla g\|}.$$

■ Recall the following formulas for the evaluation of surface integrals:

Let S be the graph of z=f(x,y) (y=h(x,z) or x=k(y,z)) and let

R_{xy} (R_{xz} or R_{yz}) be the projection of S on the xy (xz or yz)−plane. Then,

$$\iint_S g(x,y,z)\,dS = \begin{cases} \displaystyle\iint_{R_{xy}} g(x,y,f(x,y))\sqrt{(f_x(x,y))^2+(f_y(x,y))^2+1}\,dA \\ \displaystyle\iint_{R_{xz}} g(x,h(x,z),z)\sqrt{(h_x(x,z))^2+(h_z(x,z))^2+1}\,dA\,. \\ \displaystyle\iint_{R_{yz}} g(k(y,z),y,z)\sqrt{(k_y(y,z))^2+(k_z(y,z))^2+1}\,dA \end{cases}$$

□ **Example:**

Use the Divergence Theorem to compute the outward flux of the field

$$vf(x,y,z)=\{xy+x^2yz,\,yz+xy^2z,\,xz+xyz^2\}=(xy+x^2yz)i+(yz+xy^2z)j+(xz+xyz^2)k$$

through the surface of the cube cut from the first octant by the planes x=2, y=2, and z=2.

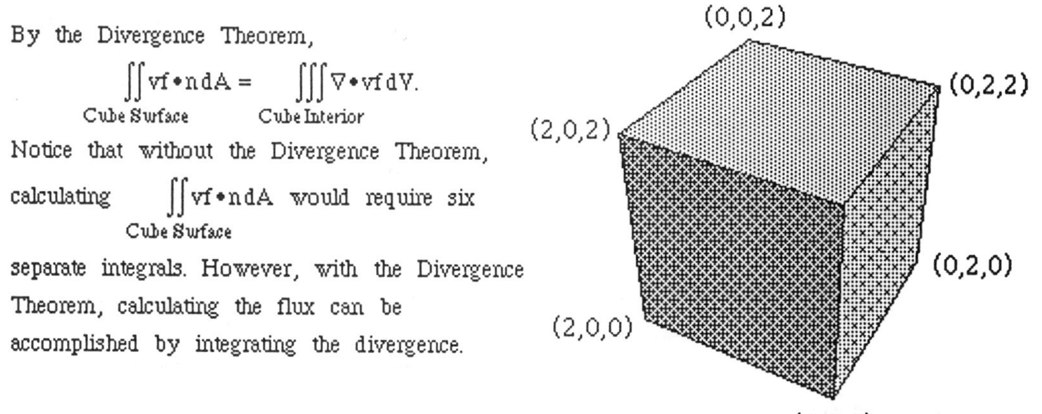

After loading the **IntegralTables.m** and **VectorAnalysis.m** packages, the vector field **vf** is defined as a list of three elements, the x, y, and z components. Since the volume integral is that of a cube, the logical choice for a coordinate system is cartesian coordinates. Hence the divergence of **vf** is calculated with **Div[vf[x,y,z],Cartesian]** . Since the divergence is a function of (x,y,z), it is defined as the function **divvf** for later use in the volume integral. Therefore, the outward flux of the field **vf** through the surface of the cube is found to be 72 with **Integrate[divvf[x,y,z],{x,0,2},{y,0,2},{z,0,2}]**.

DivergenceTheorem

```
In[1]:=
<<IntegralTables.m
<<VectorAnalysis.m
```

Be sure to load the packages **IntegralTables.m** *and* **VectorAnalysis.m** *before attempting to compute definite integrals of divergence of three-dimensional vector fields.*

```
In[2]:=
vf[x_,y_,z_]={x y +x^2 z y,y z+ x y^2 z,x z+x y z^2 }
Out[2]=
          2              2              2
{x y + x  y z, y z + x y  z, x z + x y z }
```

First define the function **vf** *to be the three-dimensional vector field.*

```
In[3]:=
divvf[x_,y_,z_]=Div[vf[x,y,z],Cartesian]
Out[3]=
x + y + z + 6 x y z
```

Use the command to compute the divergence of **vf** *; name the resulting function (of x, y, and z)* **divvf**.

```
In[4]:=
Integrate[divvf[x,y,z],{x,0,2},{y,0,2},{z,0,2}]
Out[4]=
72
```

computes the triple integral $\int_0^2\int_0^2\int_0^2$ **divvf[x,y,z]**dzdydx.

■ Application: Stoke's Theorem

Stokes's Theorem: Let S be an oriented surface with finite surface area, unit normal n, and boundary C. Let F be a continuous vector-field defined on S such that the component functions of F have continuous partial derivatives at each non – boundary points of S. Then, $\int_C \mathrm{F} \bullet d\mathrm{r} = \iint_S (\mathrm{Curl\ F}) \bullet \mathrm{n}\ d\mathrm{S}$.

In other words, the surface integral of the normal component of the curl of F taken over S equals the line integral of the tangential component of the field taken over C: $\oint_C \mathrm{F} \bullet \mathrm{T}\, ds = \iint_S \mathrm{curl\,F} \bullet \mathrm{n}\, d\mathrm{S}$.

In particular, if $\mathrm{F} = M\mathrm{i} + N\mathrm{j} + P\mathrm{k} = \{M, N, P\}$, then

$$\int_C (M(x,y,z)dx + N(x,y,z)dy + P(x,y,z)dz) = \iint_S (\mathrm{Curl\ F}) \bullet \mathrm{n}\ d\mathrm{S}.$$

□ Example:

Verify Stoke's Theorem for the vector field

$vf(x,y,z) = (y^2 - z)i + (z^2 + x)j + (x^2 - y)k = \{y^2 - z, z^2 + x, x^2 - y\}$ and S the paraboloid

$z = f(x,y) = 4 - (x^2 + y^2),\ z \geq 0.$

Since we must show

$\int_C vf \bullet dr = \iint_S (\text{Curl } vf) \bullet n\, dS$, we must compute Curl vf, n, $\iint_S (\text{Curl } vf) \bullet n\, dS$, r, dr, and $\int_C vf \bullet dr$.

First define the vector-field and the surface as **vf** and **f**, respectively. Compute the curl of **vf** and name it **curlvf[x,y,z]** then compute a normal vector and name it **normal[x,y,z]**:

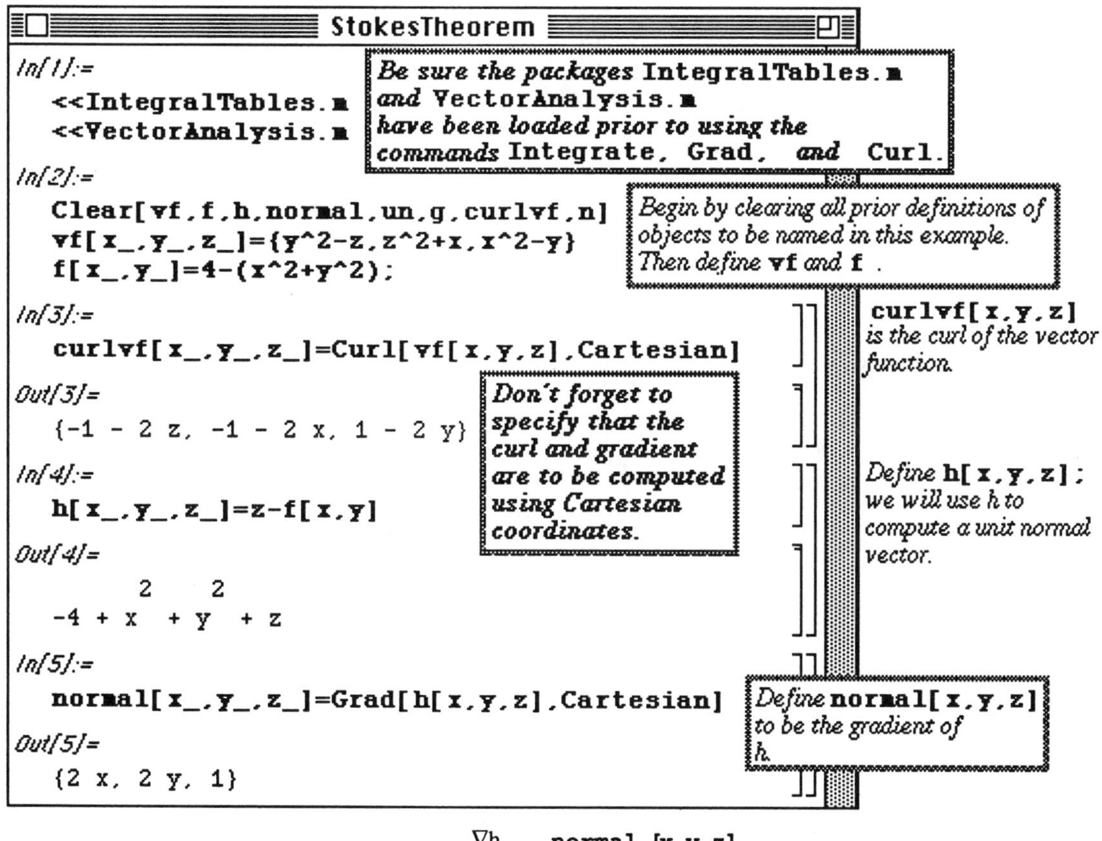

A unit normal vector n is given by $n = \frac{\nabla h}{\|\nabla h\|} = \frac{\texttt{normal [x,y,z]}}{\|\texttt{normal [x,y,z]}\|}$.

Since **normal[x,y,z]** is a list (of three elements), **normal[x,y,z][[i]]** yields the ith element of the list **normal[x,y,z]**. Therefore,
||**normal[x,y,z]**|| is given by the command
Sqrt[Sum[(normal[x,y,z][[i]])^2,{i,1,3}]]. An alternative approach is to recall that

for a vector v, $\|v\| = \sqrt{v \bullet v}$. Consequently, the command **Sqrt[normal[x,y,z].normal[x,y,z]]** yields the same result.

In order to easily use the surface integral evaluation formula, define **g[x,y,z]** to be the dot product of **curlvf[x,y,z]** and **un[x,y,z]**.

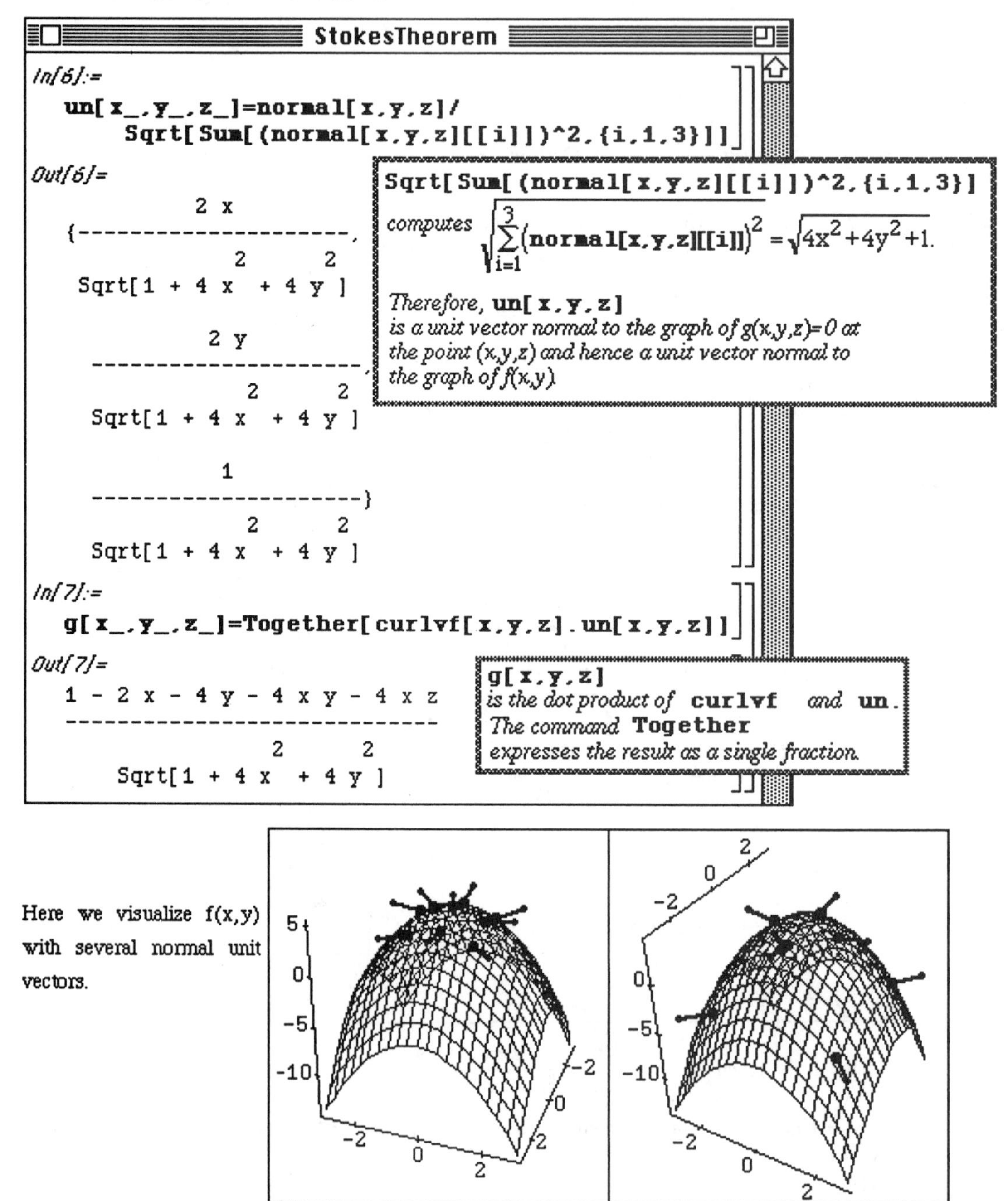

By the surface integral evaluation formula,

$$\iint_S \overbrace{(\text{Curl vf})}^{\textbf{curlvf}} \bullet \underbrace{\text{n}}_{\textbf{un}}\ dS = \iint_S g(x,y,z)dS = \iint_R g(x,y,f(x,y))\sqrt{(f_x(x,y))^2 + (f_y(x,y))^2 + 1}\ dA, \text{ where}$$

R is the projection of f(x, y) on the xy – plane. Hence, in this example, R is the region bounded by the graph of the circle $x^2 + y^2 = 4$. Thus,

$$\iint_R g(x,y,f(x,y))\sqrt{(f_x(x,y))^2 + (f_y(x,y))^2 + 1}\ dA = \int_{-2}^{2}\int_{-\sqrt{4-x^2}}^{\sqrt{4-x^2}} g(x,y,f(x,y))\sqrt{(f_x(x,y))^2 + (f_y(x,y))^2 + 1}\,dy\,dx.$$

Notice that the integral $\int_{-2}^{2}\int_{-\sqrt{4-x^2}}^{\sqrt{4-x^2}} g(x,y,f(x,y))\sqrt{(f_x(x,y))^2 + (f_y(x,y))^2 + 1}\,dy\,dx$ can be easily evaluated using polar coordinates. To do so, in $g(x,y,f(x,y))\sqrt{(f_x(x,y))^2 + (f_y(x,y))^2 + 1}$, replace each x by r Cos(t) and each y by r Sin (t).

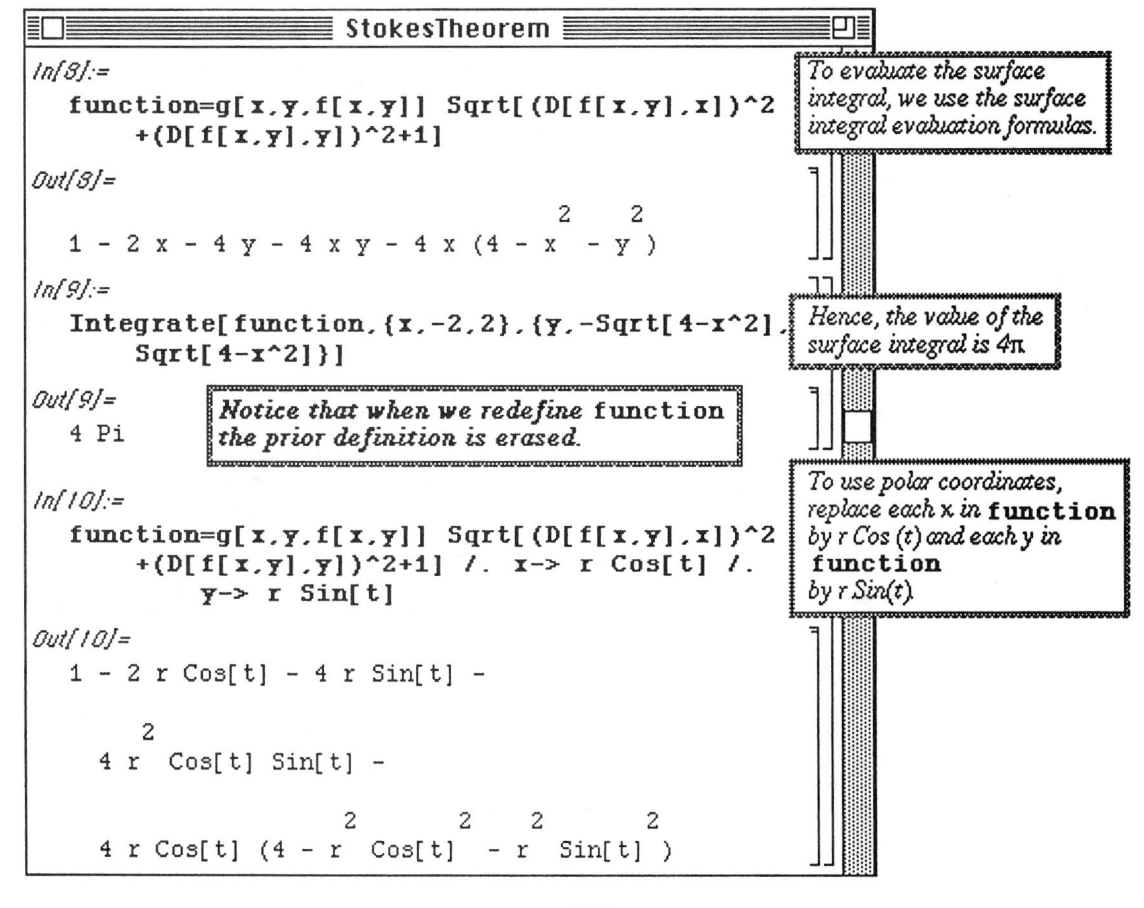

The command **TrigExpand[expression]** applies basic trigonometric identities to attempt to simplify **expression**. Finally, to evaluate the integral in polar coordinates, the limits of integration must be changed and dx dy must be replaced by r dt dr. Hence, the same value is obtained by the integral:

$$\int_0^2\int_0^{2\pi}\underbrace{\left(1-18r\,\text{Cos}(t)+4r^3\text{Cos}(t)-4r\text{Sin}(t)-2r^2\,\text{Sin}(2t)\right)}_{\text{simplified version of function}}r\,dt\,dr.$$

o Version 2.0 does not include the command **TrigExpand**. The same results as **TrigExpand[expression]** are obtained with **Expand[expression,Trig->True]**.

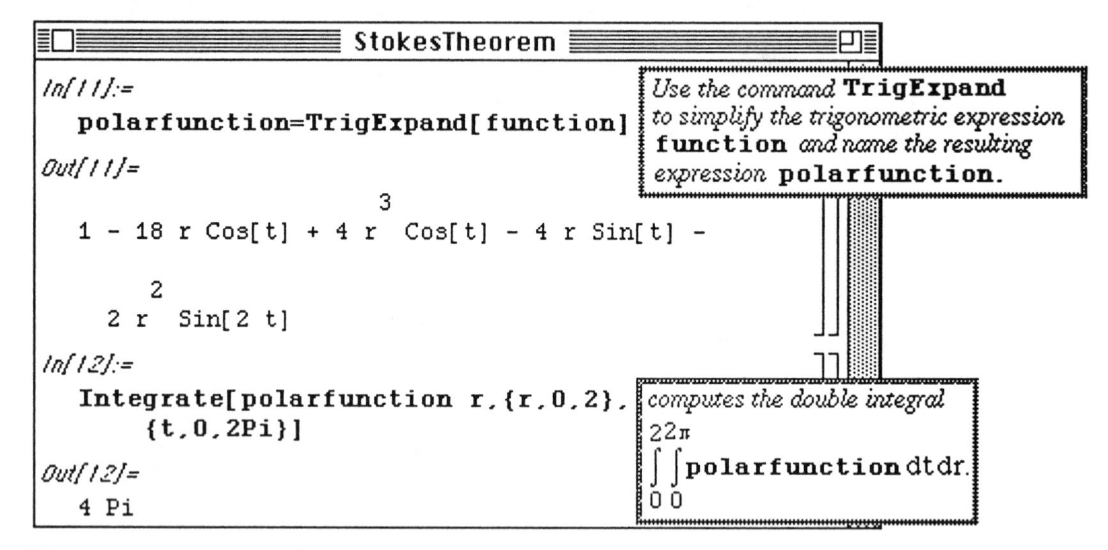
StokesTheorem

```
In[11]:=
  polarfunction=TrigExpand[function]
Out[11]=
                              3
  1 - 18 r Cos[t] + 4 r  Cos[t] - 4 r Sin[t] -

      2
   2 r   Sin[2 t]
In[12]:=
  Integrate[polarfunction r,{r,0,2},
       {t,0,2Pi}]
Out[12]=
  4 Pi
```

Use the command **TrigExpand** *to simplify the trigonometric expression* **function** *and name the resulting expression* **polarfunction.**

computes the double integral $\int_0^2\int_0^{2\pi}$ **polarfunction** dtdr.

We are able to evaluate the line integral directly by noticing that the boundary of $z=f(x,y)=4-\left(x^2+y^2\right)$, $z \geq 0$ is the circle $x^2+y^2=4$ which has parametrization $x=2\text{Cos}(s)$, $y=2\text{Sin}(s)$, and $z=0$, for $0 \leq s \leq 2\pi$.

```
In[13]:=
  pvf=vf[x,y,z] /. x-> 2 Cos[s] /.
       y->2 Sin[s] /. z->0
Out[13]=
          2                    2
  {4 Sin[s] , 2 Cos[s], 4 Cos[s]  - 2 Sin[s]}
In[14]:=
  r[s_]={2 Cos[s],2Sin[s],0}
Out[14]=
  {2 Cos[s], 2 Sin[s], 0}
```

Then replace each x in **vf** *by 2 Cos(s), each y in* **vf** *by 2 Sin (s), and each z in* **vf** *by 0.*

To evaluate the line integral directly, notice that a parametrization of C is given by r(s)={2 Cos(s),2Sin(s),0}, where $0 \leq s \leq 2\pi$.

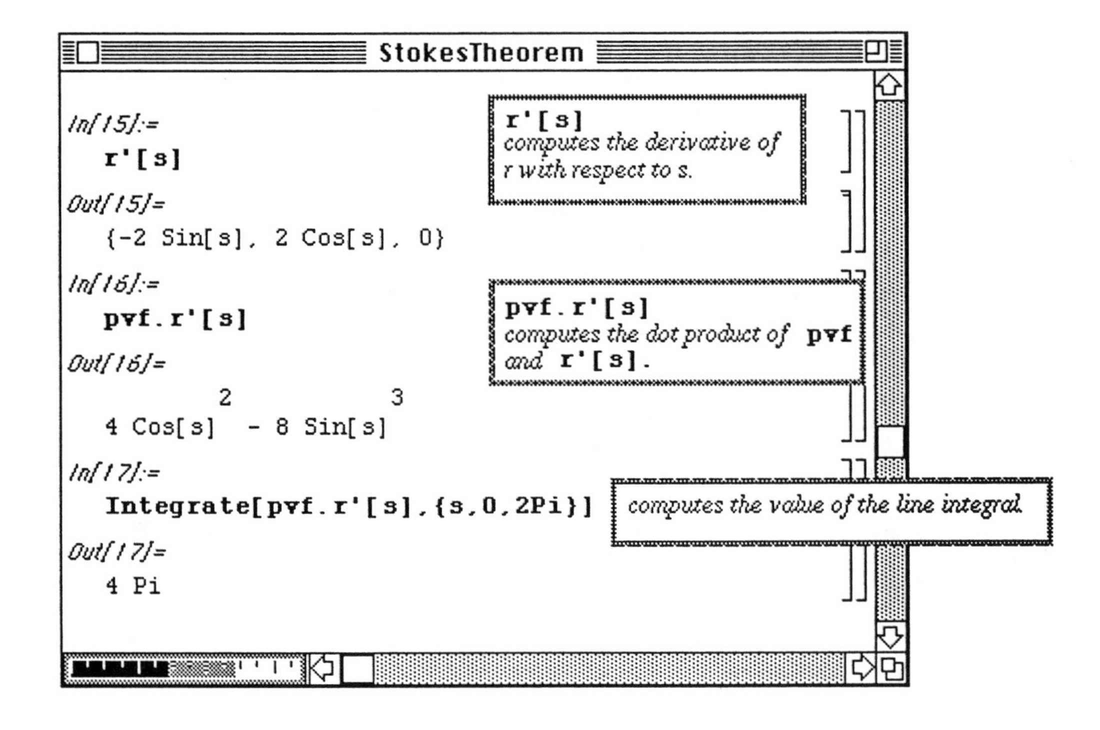
StokesTheorem
In[15]:=
r'[s]
Out[15]=
{-2 Sin[s], 2 Cos[s], 0}
In[16]:=
pvf.r'[s]
Out[16]=
4 Cos[s]^2 - 8 Sin[s]^3
In[17]:=
Integrate[pvf.r'[s],{s,0,2Pi}]
Out[17]=
4 Pi
r'[s] computes the derivative of r with respect to s.
pvf.r'[s] computes the dot product of pvf and r'[s].
computes the value of the line integral.

5.4 Saving Results for Future *Mathematica* Sessions

Beginning users of *Mathematica* quickly notice that in order to use results from a previous *Mathematica* session, they must first be re-calculated. The purpose of this example is to illustrate how results can be saved for future use.

Application: Constructing a Table of Zeros of Bessel Functions

In this example, we will create a table of the first six zeros of the Bessel functions of the first kind,

$J_0(x)$, $J_1(x)$, $J_2(x)$, $J_3(x)$, $J_4(x)$, and $J_5(x)$ then save the resulting table of numbers

in a file. Consequently, for future calculations involving zeros of Bessel functions, we need only use the table of numbers we have already created instead of re-computing the zeros. This will thus save not only substantial time but also substantial memory.

The built-in *Mathematica* command **BesselJ [alpha ,x]** represents $J_{alpha}(x)$, and, hence

will be used in the construction of this table of zeros. This command as well as Bessel functions are discussed in more detail in **Chapter 6**.

We begin by looking at a graph of **BesselJ[0,x]** on the interval [0,25] and observing the first six zeros. We create a list of six numbers corresponding to initial guesses of the first six zeros and then use **FindRoot** to approximate the first six zeros.

The list **guess[0]**, a list of initial guesses of the first six zeros of **BesselJ[0,x]** is used with **FindRoot** to approximate these zeros. Since this list is used in conjunction with **Table**, an approximation of the first zero is obtained with **guess[0][[1]]**, an approximation of the second zero is obtained with **guess[0][[2]]**, and so forth. This produces a list of approximations of the first six zeros of **BesselJ[0,x]** called **bz[0]**.

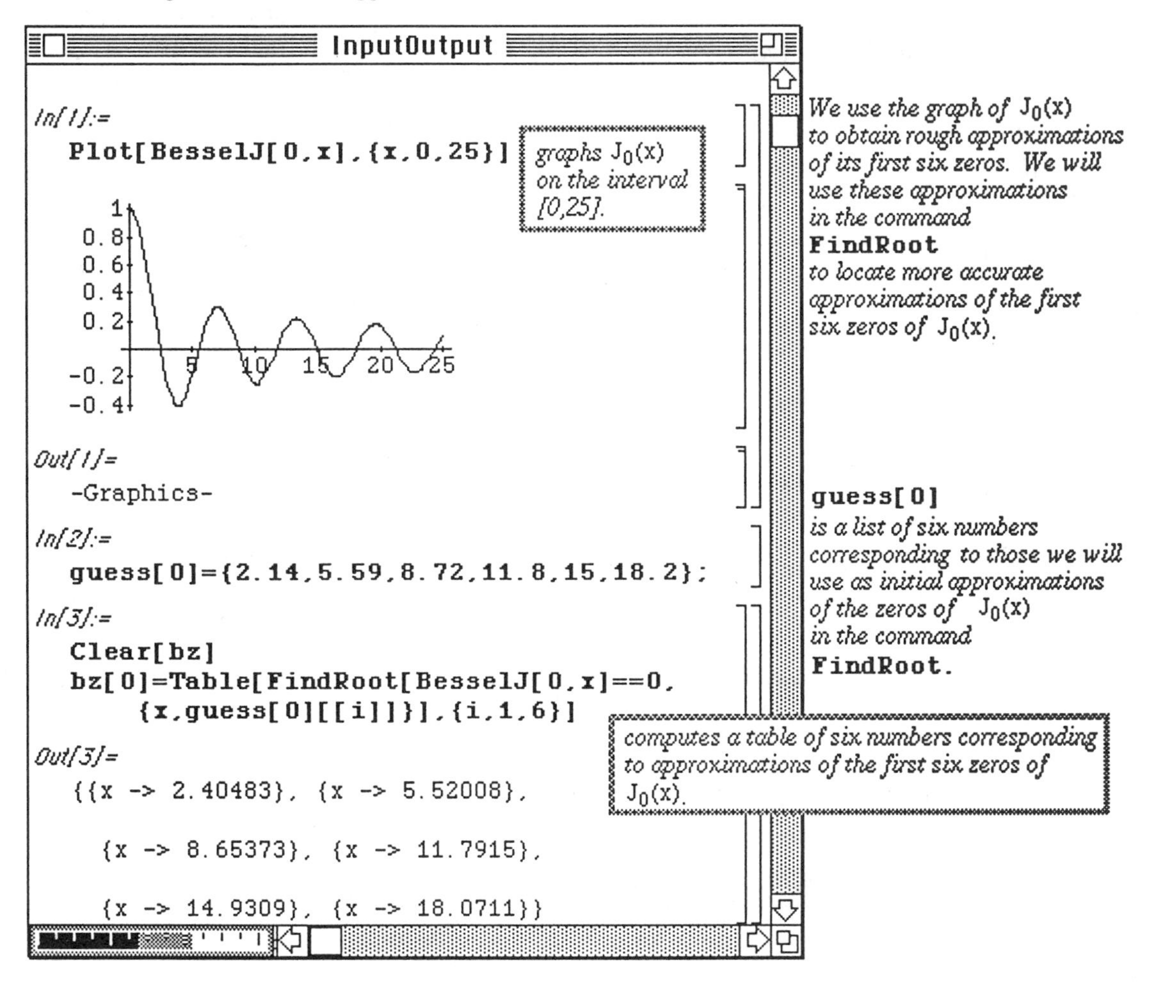

We then proceed to **BesselJ[1,x]**:

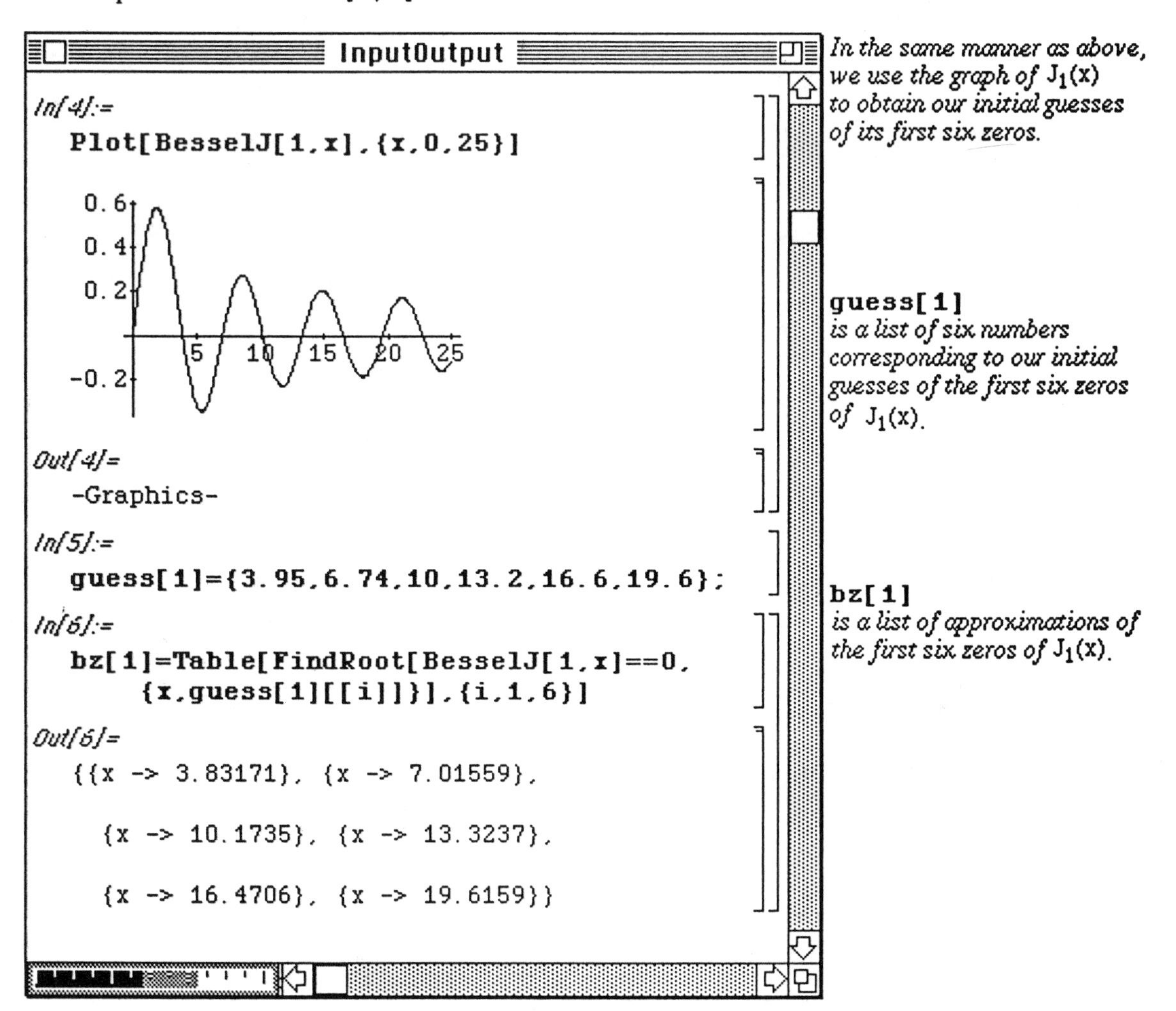

We repeat the procedure for **BesselJ[2,x]**, **BesselJ[3,x]**, **BesselJ[4,x]** and **BesselJ[5,x]**:

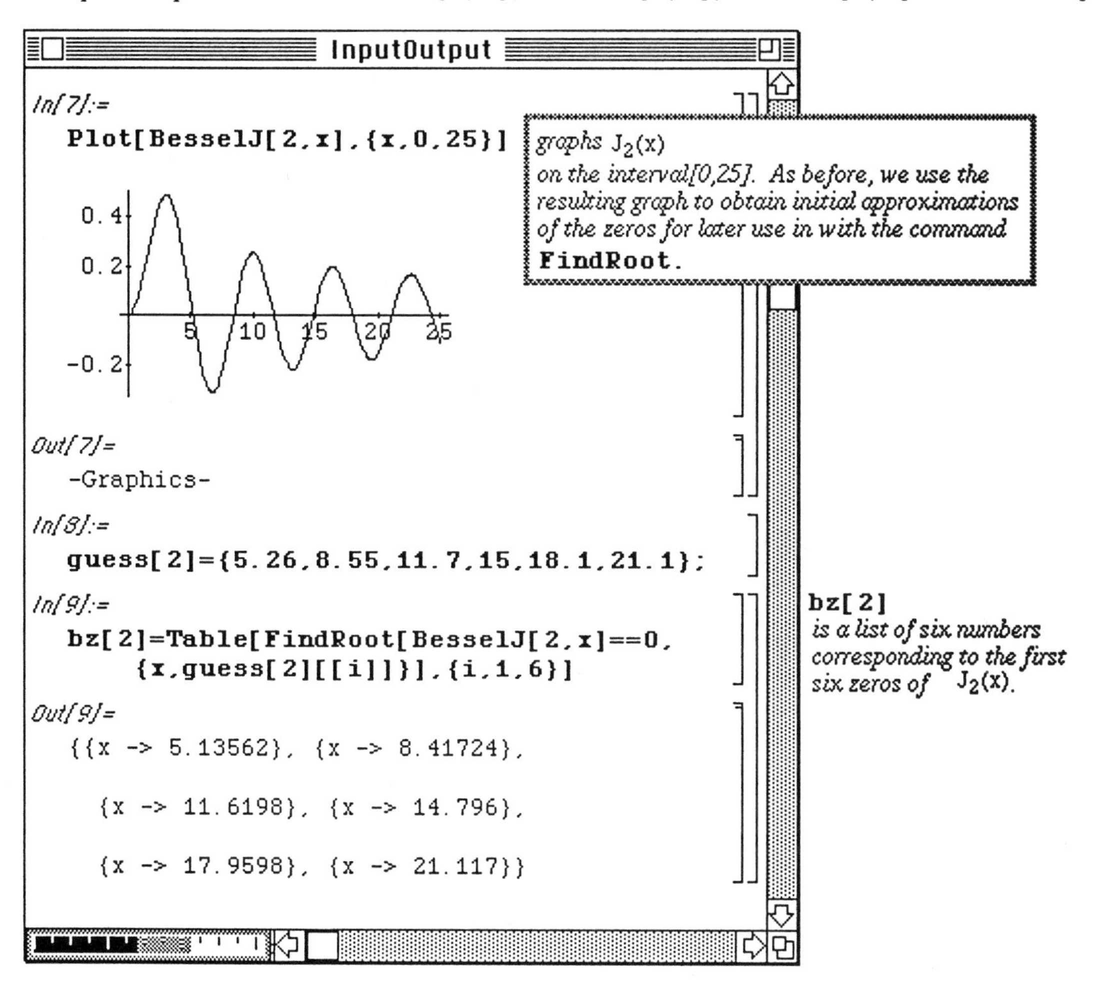

The first six zeros of **BesselJ[3,x]** are found below:

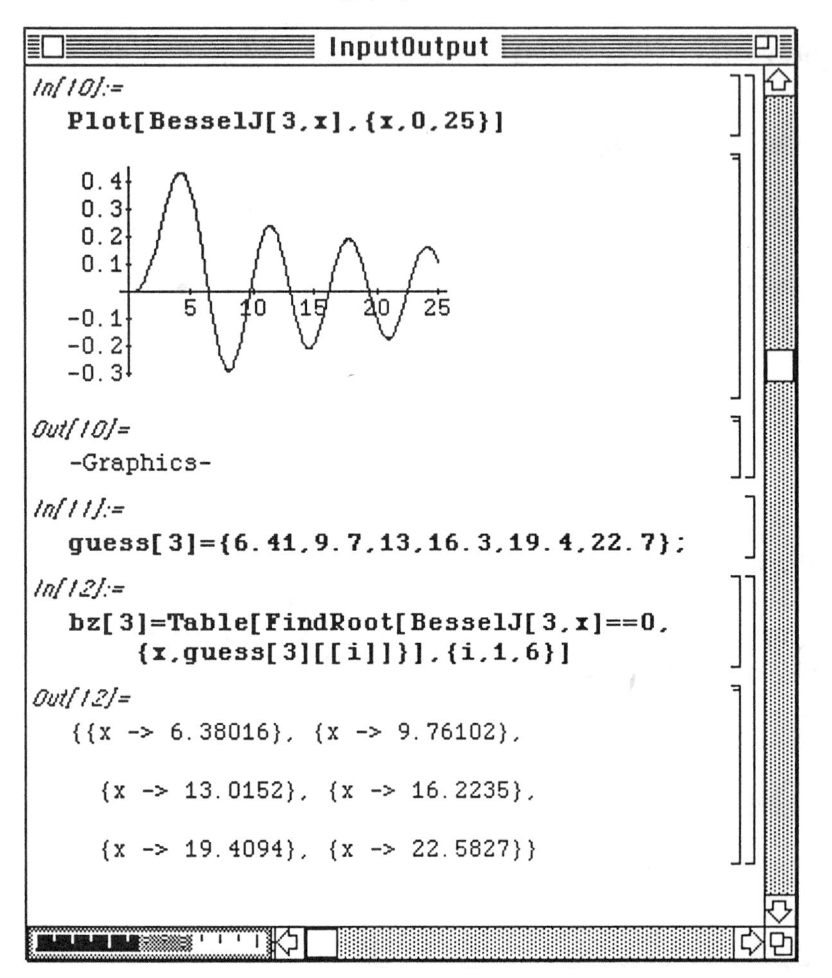

Similarly, the zeros of **BesselJ[4,x]** are found.

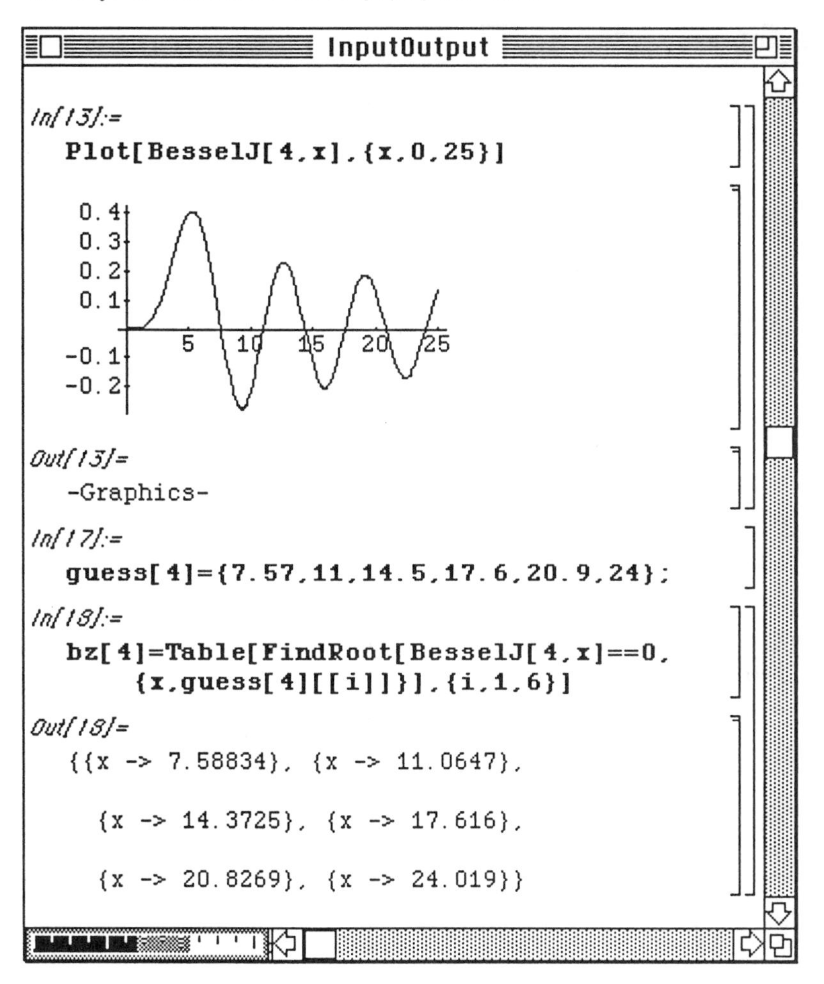

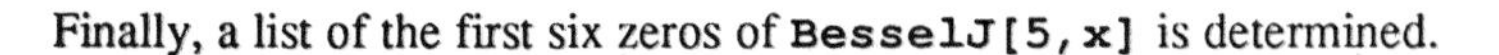

Finally, a list of the first six zeros of **BesselJ[5,x]** is determined.

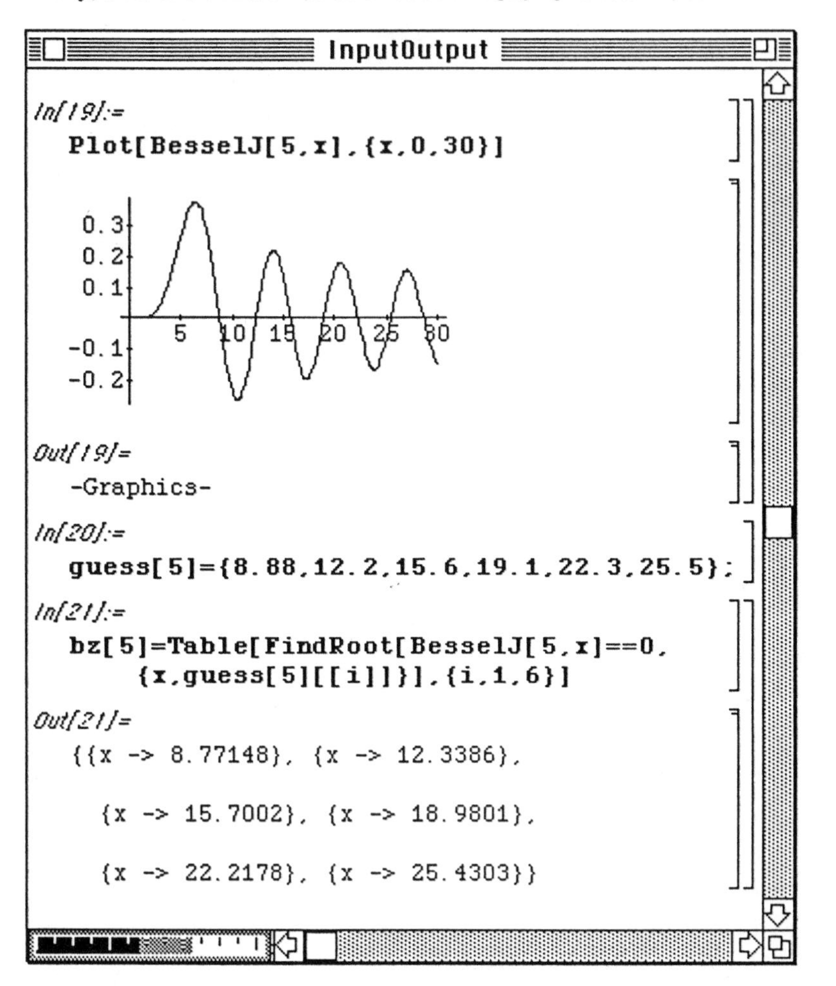

An alternative approach would have been to compile a list of initial guesses for the first six zeros by looking at the graphs of **BesselJ[0,x]**, **BesselJ[1,x]**, **BesselJ[2,x]**, **BesselJ[3,x]**, **BesselJ[4,x]** and **BesselJ[5,x]** and then to use **FindRoot**. In order to apply this alternate approach, a table of initial guesses must be compiled. This is done in **starts** below using the previously defined lists **guess[n]** for **n = 0,1,2,3,4,5**. Then, a list of the first six zeros of the Bessel functions of the first kind

$J_0(x)$, $J_1(x)$, $J_2(x)$, $J_3(x)$, $J_4(x)$, and $J_5(x)$

are computed in **besselzeros**. Notice that **besselzeros** is a list of lists, so the nth zero of the Bessel function **BesselJ[m,x]** is extracted with **besselzeros[[m+1,n,1,2]]**.

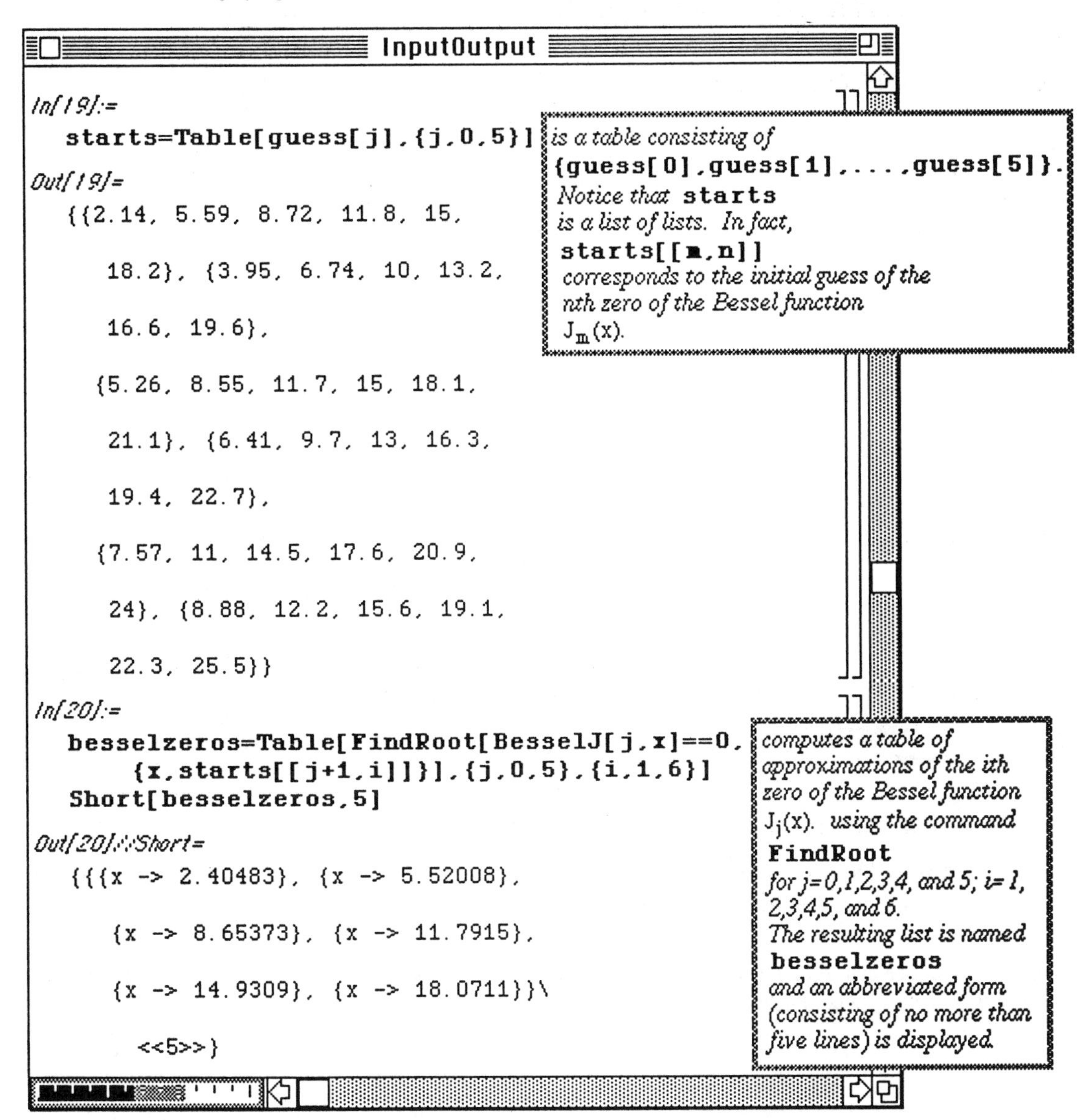

The first list in **besselzeros** is the list of zeros of **BesselJ[0,x]**. Hence, **besselzeros[[1]]** gives this list. Since the indices are shifted, **besselzeros[[j+1]]=bz[j]** where **bz[j]** is the list of the first six zeros of **BesselJ[j,x]** found earlier. We verify that the results obtained by the alternative approach are the same as those found by the previous approach.

The notation commonly used to denote the nth zero of the Bessel function $J_m(x)$ is α_{mn}.

Hence, the function **alpha[m,n]** is defined below in terms of the values in **bz[n]** for later use.

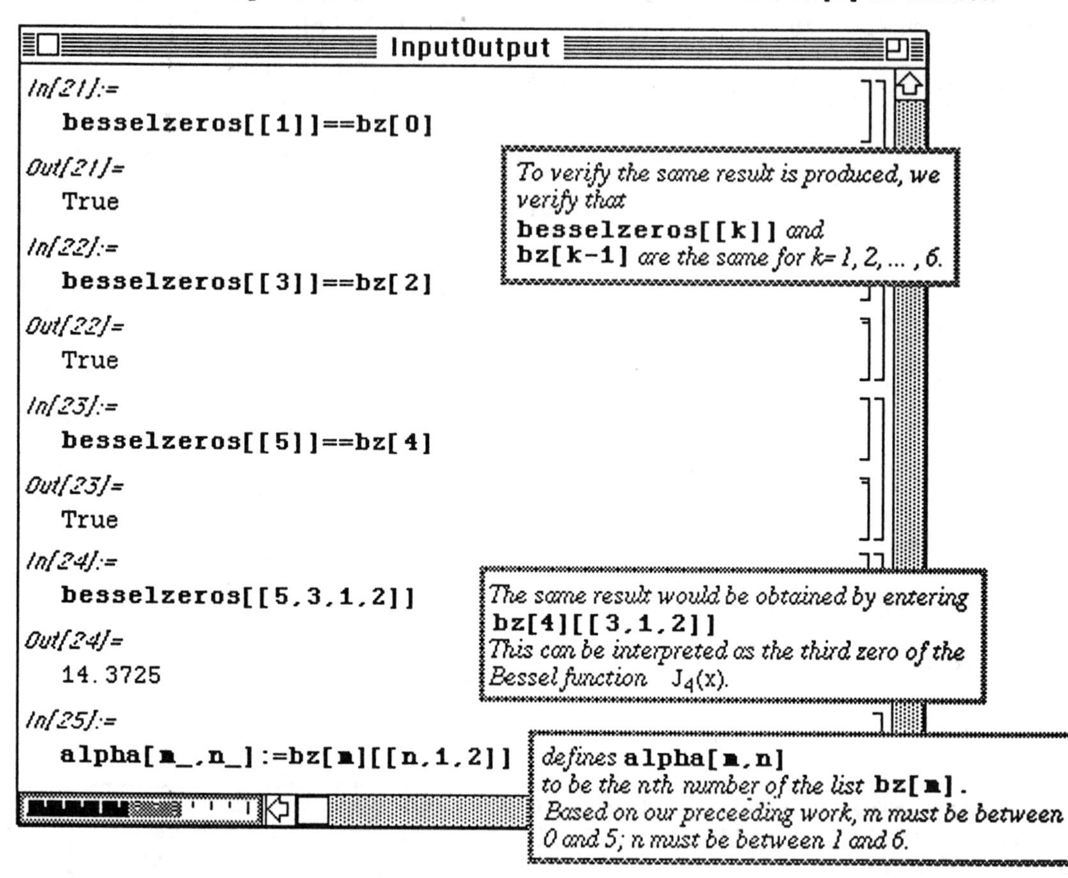

The table **zeros** is a table of the first six zeros of the first six Bessel functions.

The command **Table[alpha[m,n],{m,0,5},{n,1,6}]>>besselzeros** first computes the table of numbers **Table[alpha[m,n],{m,0,5},{n,1,6}]** and then writes the results to the file **besselzeros**.

It is important to notice that if the file **besselzeros** does not exist, it is created; if it does exist, it is written over. To **append** the results of a command to an existing file, the form of the command is **command>>>output file**.

The function **alphaalt[m,n]** can then easily find the nth zero of **BesselJ[m,x]** from the list **besselzeros**. These zeros do not have to be recomputed with this approach and are located in the table **zerosalt**. The fact that identical values are contained in **zeros** and **zerosalt** is verified.

```
InputOutput

In[26]:=
  zeros=Table[alpha[m,n],{m,0,5},{n,1,6}]
  TableForm[zeros]
```

creates a table **zeros** *and expresses the result in Table Form.*

```
Out[26]//TableForm=
  2.40483   5.52008   8.65373   11.7915

      14.9309   18.0711

  3.83171   7.01559   10.1735   13.3237

      16.4706   19.6159

  5.13562   8.41724   11.6198   14.796

      17.9598   21.117

  6.38016   9.76102   13.0152   16.2235

      19.4094   22.5827

  7.58834   11.0647   14.3725   17.616

      20.8269   24.019

  8.77148   12.3386   15.7002   18.9801

      22.2178   25.4303

In[27]:=
  Table[alpha[m,n],{m,0,5},{n,1,6}] >> besselzeros
```

creates the same table as above and writes the output to the file **besselzeros.**

```
In[28]:=
  alphaalt[m_,n_]:=besselzeros[[m+1,n,1,2]]

In[29]:=
  zerosalt=Table[alphaalt[m,n],{m,0,5},{n,1,6}];

In[30]:=
  zerosalt==zeros
```

The same results would have been obtained using the table **besselzeros.**

```
Out[30]=
  True
```

The file **besselzeros** is a nested list of numbers; it does not rely on previous calculations.

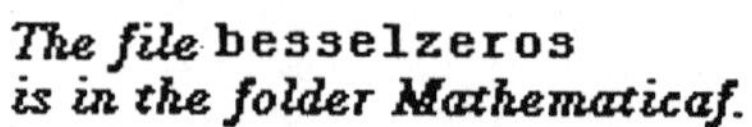


```
besselzeros
{{2.404825557624828635, 5.520078082173197342, 8
  11.79153159732223503, 14.93091770603674345,
 {3.831705251431236663, 7.01558655548211243, 10
  13.3236919362232029, 16.47063000385284817, 1
 {5.135621887451435987, 8.417243947281169003,
  14.79595114197812931, 17.95981943405801651,
 {6.380161895464166835, 9.761023129207743131,
  16.22346615704028987, 19.40941267380204939,
 {7.588342434475555476, 11.06470948793955275,
  17.61596604980382726, 20.82693295541904514,
 {8.771483762897060509, 12.33860419738229831,
  18.98013385271306593, 22.21779989413968912,
```

In this case, we modify **besselzeros** by converting the initialization cell to an ordinary input cell (by selecting **Input Cell** from **Cell Style** under the **Style** heading on the *Mathematica* **Menu**), naming the table **zbj** and defining **bzero[m,n]**. For future use, we need only open the file **besselzeros**, enter its input cells, and the function **bzero [m,n]** will give the **n**th zero of the **m**th Bessel function, $J_m(x)$.

The calculation of the zeros of the Bessel functions are important in many problems in applied mathematics, so the procedures described here can be quite useful. Notice how these values are easily obtained using **bzero[m,n]**.

In order to use the values located in **zbj**, the input cell containing **zbj** and the input cell containing the definition of **bzero** must first be entered.

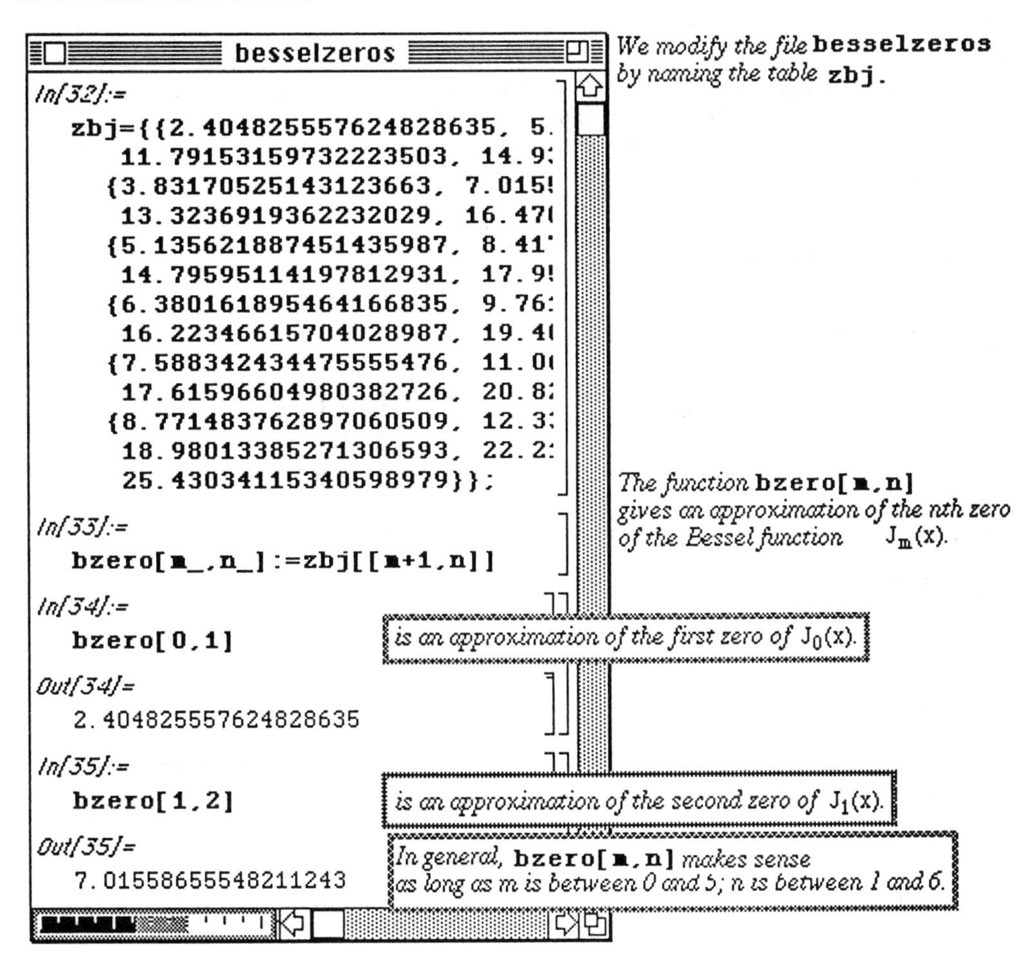

■ An Alternative Method

In the previous example, we saw how to create a table of numbers and save them in a separate file for future use. The command **Table[{Cos[j],Sin[i]},{i,1,3},{j,1,3}]>>*outputfile*** creates a table of order pairs {**Cos[j],Sin[i]**} for **i**=1,2,3 and **j**=1,2,3, creates (or erases) ***outputfile*** and places the table in ***outputfile***. Notice that in the first example below that the results are given in an output cell which cannot be accessed. Hence, if the output is to be saved for later use another approach may be more desirable. The second example illustrates such an approach. **Print** is used within the **Table** command so that the results are given in a print cell which can be accessed with the cursor. This gives the user a second method by which to save a file for future use.

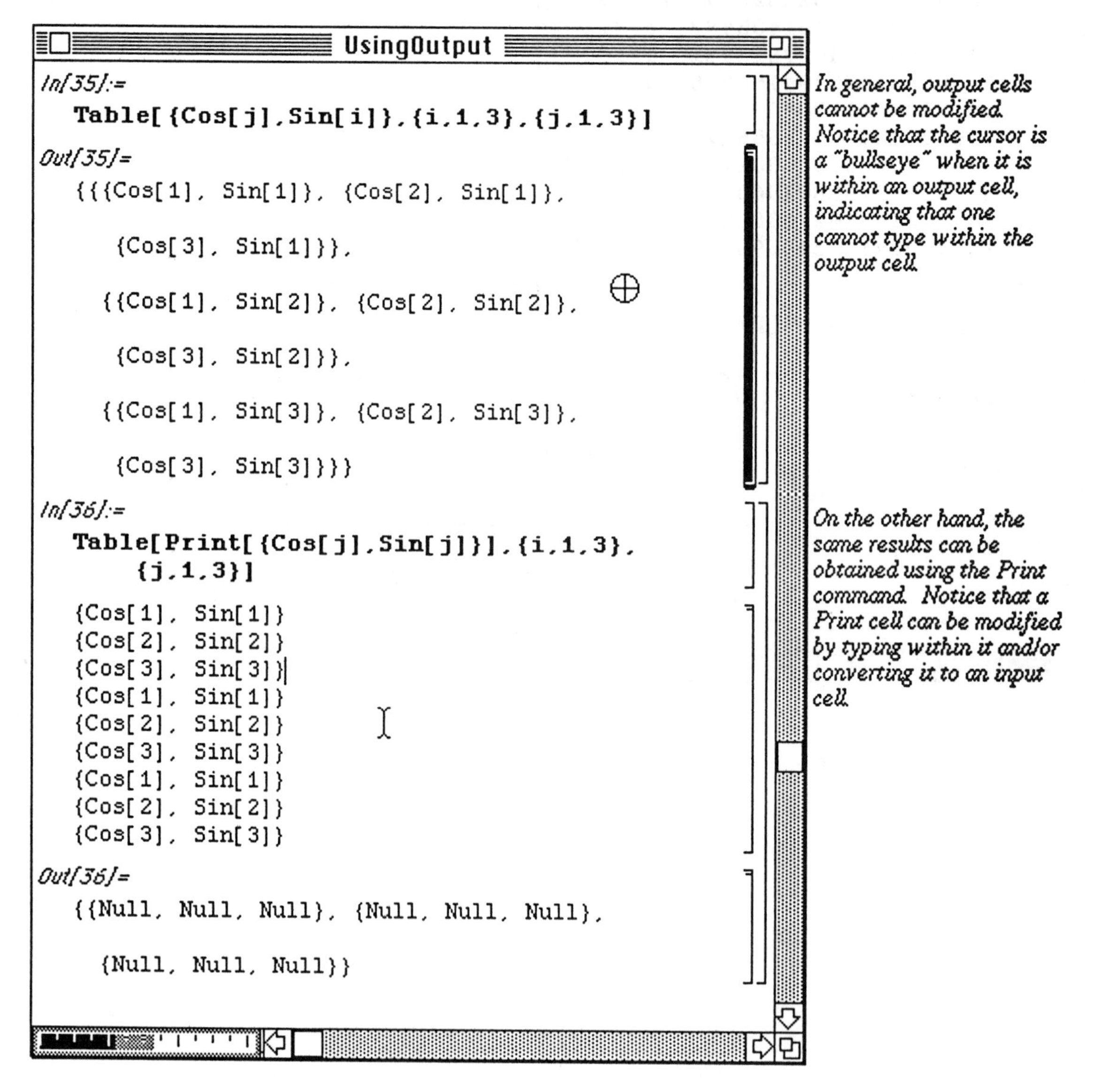

Chapter 6
Applications Related to Ordinary and Partial Differential Equations

- *Mathematica* can perform calculations necessary when computing solutions of various differential equations and, in some cases, can be used to find the exact solution of certain differential equations using the built-in command `DSolve`. In addition, Version 2.0 contains the built-in command `NDSolve` which can be used to obtain numerical solutions of other differential equations. The purpose of **Chapter 6** is to illustrate various computations *Mathematica* can perform when solving differential equations.

- Commands introduced and discussed in this chapter from **Version 1.2** include:

Differential Equations:
```
DSolve[differentialequation, function, variable]
DSolve[{differentialequations},{functions},variable]
DSolve[{de,initialcond},...]
DSolve[{des,initialconds},...]
```
Programming
```
Block[{localvariables},procedure]
```
Algebraic Operations:
```
Variables[expression]
Exponent[polynomial,variable]
Coefficient[poly,var,i]
```
Other Operations:
```
Flatten[list]
Print[expression]
Dt[function]
==
```
Special Functions:
```
BesselJ[alpha,x]
BesselY[alpha,x]
```
Trigonometric Operations:
```
TrigExpand[expression]
ComplexToTrig[expression]
```

- Commands introduced and discussed in this chapter from **Version 2.0** include:

```
NDSolve
InterpolatingFunction
Evaluate
```

Commands from previous chapters are frequently used.

- Applications discussed in this chapter include the Falling Bodies Problem, Spring Problems, Classification of Equilibrium Points, and the Wave Equation.

6.1 Linear Equations

The general solution of the linear equation $\frac{dy}{dx} + P(x)y = f(x)$, where P and f are continuous on the interval I, is $y = e^{-\int P(x)dx} \int e^{\int P(x)dx} f(x)dx + c_1 e^{-\int P(x)dx}$.

Mathematica can solve equations of this type with **DSolve**. However, since solutions of first-order linear equations obviously depend on the computation of an integral, if you are using Version 1.2, **IntegralTables.m** must be loaded before trying to solve any differential equations.

o If you are using Version 2.0, **IntegralTables.m** is automatically loaded at startup.

After this is done, the linear equation given above is solved with the command **DSolve[y'[x]+ P[x] y[x] == f[x],y[x],x]** where the functions P(x) and f(x) are usually directly entered in the **DSolve** command. Notice that the command consists of three parts : the differential equation; the dependent variable (or solution), **y[x]**; and the independent variable, **x**. Also notice that the dependent variable must be entered as **y[x]** each time it appears in the differential equation. Otherwise, **DSolve** will yield a meaningless result. Several examples are given below to illustrate the use of **DSolve**.

□ **Example:**

Solve the first order linear differential equation $(1+x^2)\frac{dy}{dx} + xy = -(x^3 + x)$.

The solution, called **sol**, is easily found below with
DSolve[(1+x^2)y'[x]+x y[x]==-(x^3+x),y[x],x].
Notice that the output for this command is a list of one element (obtained with **sol[[1]]**),

$$\{ y[x] \; > \frac{1-(1+x^2)^{3/2}}{3E^{Log[1+x^2]/2}} + \frac{C[1]}{E^{Log[1+x^2]/2}} \}.$$

This is also a list of one element (obtained with **sol[[1,1]]**),

$$y[x] -> \frac{1-(1+x^2)^{3/2}}{3E^{Log[1+x^2]/2}} + \frac{C[1]}{E^{Log[1+x^2]/2}}$$

which is composed of two parts, **y[x]** and the expression following the arrow. Of course, the second part of this element is the portion of interest since it gives the formula of the solution. Therefore, in order to extract this formula, the command **sol[[1,1,2]]** is used. (**sol[[1,1,1]]** yields y[x] .)

Note: In most instances, the ability to extract the formula for the solution will be of great importance. In order to analyze solutions to most differential equations, obtaining the formula of the solution is necessary. One alternative is to define the solution as a function once it has been found with **DSolve**. However, many solutions are rather complicated, so typing is cumbersome and mistakes are likely. Hence, the logical choice for obtaining the formula is by extracting it from the output list. At first, the technique of extracting solutions may seem difficult to understand, but it should become clearer after several examples.

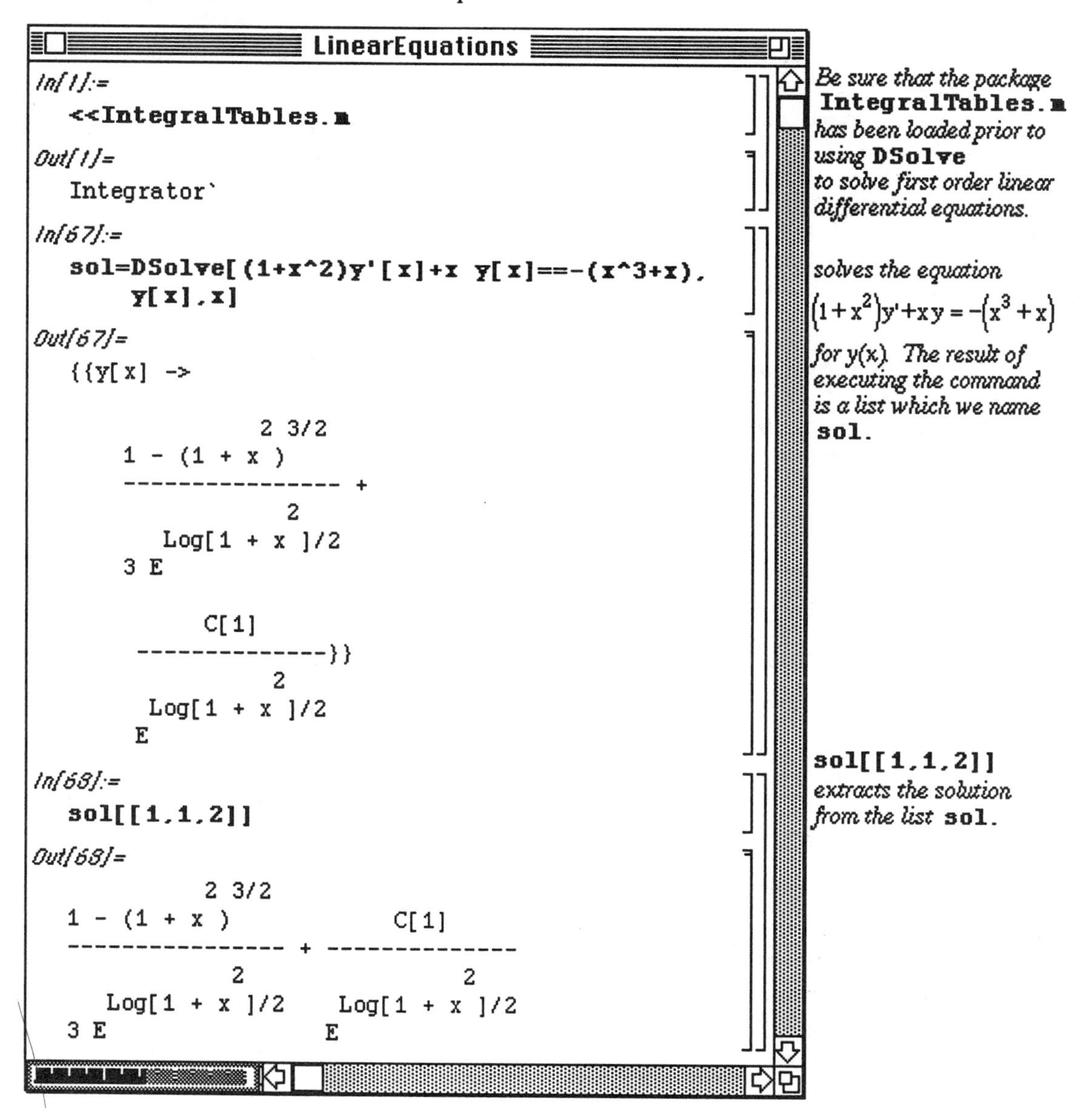

Solutions for various values of the constant `C[1]` can be graphed. First, the graphs for `C[1]` = -4, -2, 0, and 2 are requested in a single input cell using different **GrayLevel** assignments. Each plot is assigned a name so that the **Show** command can be used in the same cell to plot these four graphs simultaneously. Since all of the commands are in the same cell, only the output from the last command is shown.

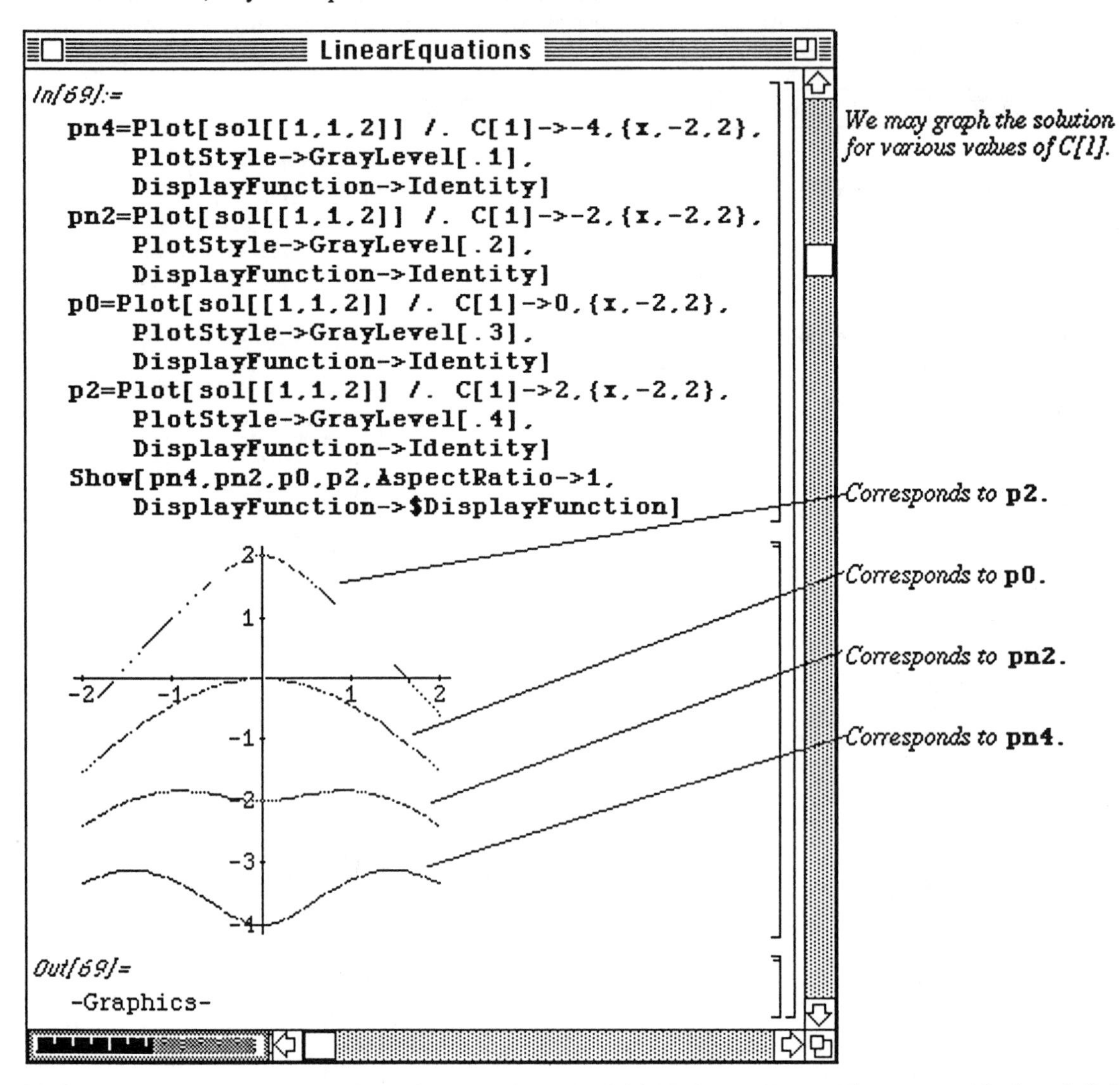

Mathematica's **DSolve** command can also be used to solve initial value problems. The command is altered slightly to include the initial condition.

□ Example:

Solve: $x\frac{dy}{dx} + (x+2)\,y = 2\,e^{-x}$, $y(1) = 0$.

The differential equation can be solved with **DSolve** as in the previous example. However, this gives the general solution to the problem which represents a family of solutions. The solution to the initial value problem the one solution which passes through the point (1,0).

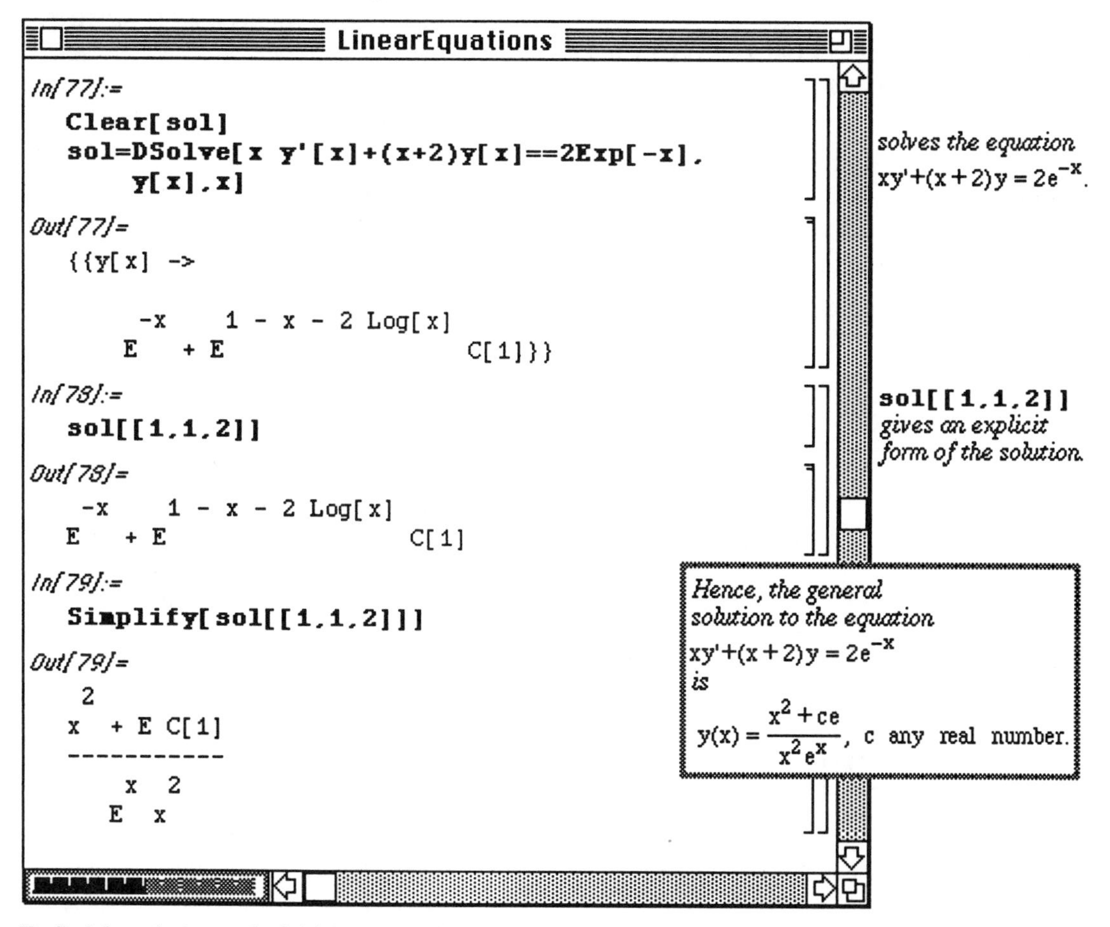

To find the solution to the initial value problem with **DSolve**, the initial condition must be entered in the **DSolve** command. This is accomplished in the following way:

```
DSolve[{x  y'[x]+(x+2)y[x]==2Exp[-x],y[1]==0},y[x],x].
```

Notice that the initial condition is placed in "curly" brackets , **{ }**, along with the differential equation. Otherwise, the command is unchanged. Also note that a double equals sign is used in the initial condition.
The solution is found below and named **sol**. The formula for y[x] is then extracted and simplified in a single command. The expression which results is called **simsol** for use in the **Plot** command which follows.

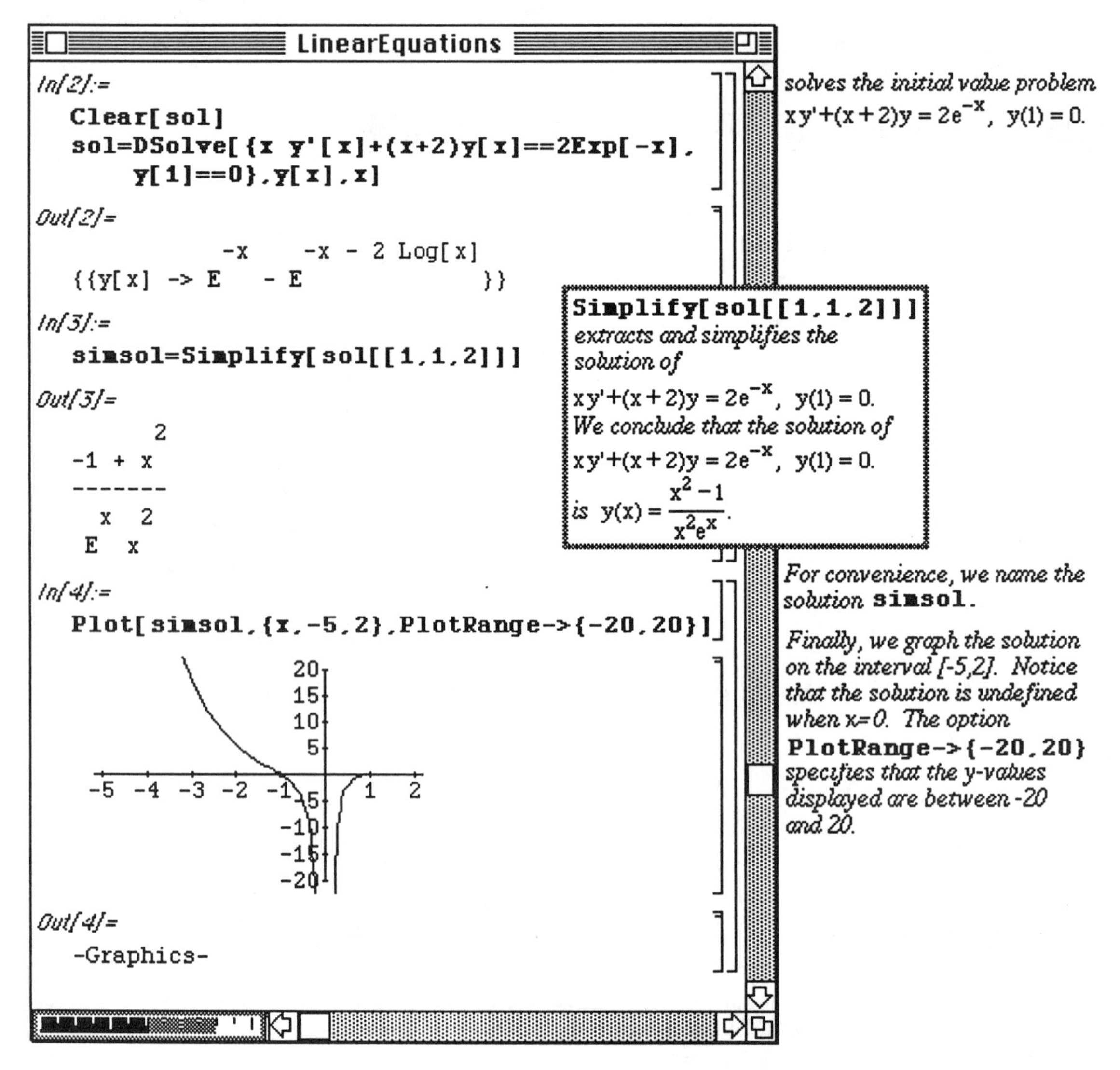

■ Application: The Falling Body Problems

A useful application of first-order differential equations is solving problems encountered in mechanics. One such problem is as follows:
A body falls through the air towards the earth. In such a circumstance, the body is subjected to a certain amount of air resistance (which in some cases is proportional to the body's velocity). The objective is to determine the velocity and the distance fallen at time t seconds.
Mathematica can be quite useful in solving problems of this type. To illustrate how these falling body problems are solved, consider the following problem.

An object weighing 32 pounds is released from rest 50 feet above the surface of a calm lake. The air resistance (in pounds) is given by 2v, where v is the velocity (in feet/sec). After the object passes beneath the surface, the water resistance (in pounds) is given by 6v. Further, the object is then buoyed up by a buoyancy force of 8 pounds. Find the velocity of the object 2 seconds after it passes beneath the surface of the lake.

This problem is made up of two parts. First, the forces acting on the object before it reaches the surface of the lake must be considered. Then, the set of forces which act upon the object beneath the lake's surface must be determined in order to solve the problem. Using Newton's second law, the initial value problem which determines the object's velocity above the surface is: $\frac{dv}{dt} = 32 - 2\,v,\ \ v(0) = 0.$

DSolve can be used to solve this initial value problem. This is done below. The velocity is then extracted from the resulting expression with **deq1[[1,1,2]]** and named **vel1** .

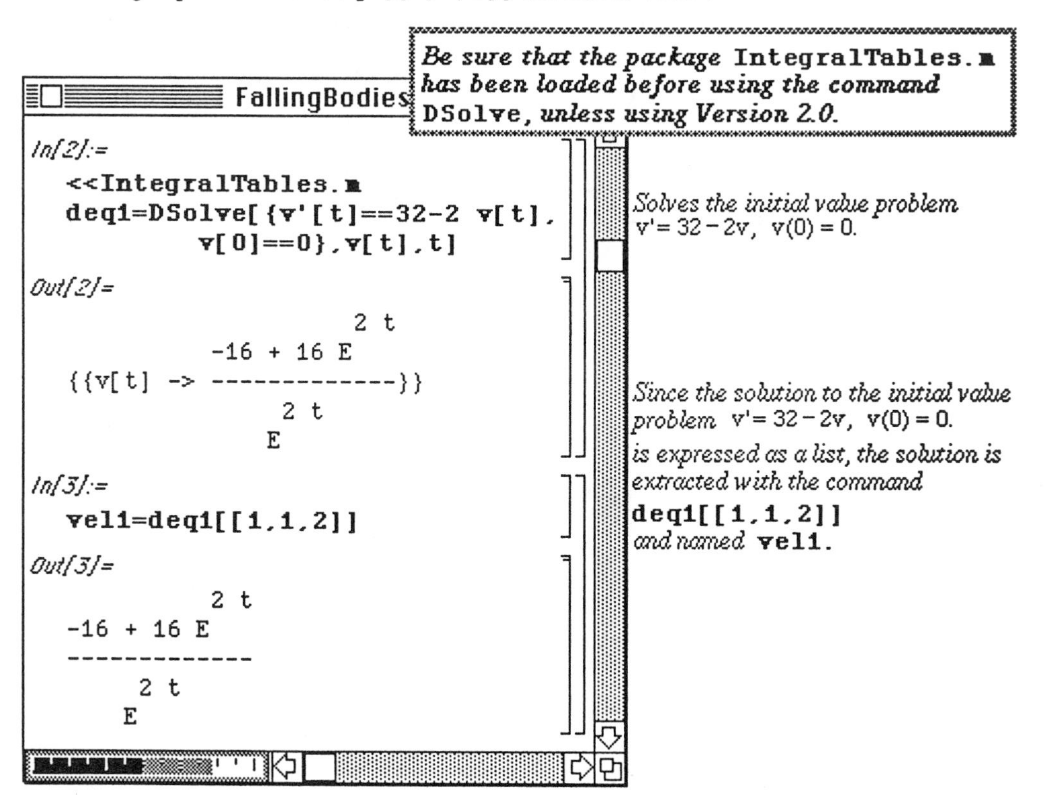

Before determining the velocity beneath the lake's surface, the object's velocity at the point of impact must be found. Therefore, the time at which the object hits the surface of the lake must be calculated by integrating the velocity v(t) to obtain the object's position function x(t) using the initial position x(0) = 0. (Recall that x'(t) = v(t)) This calculation is carried out below. Note that the position function is extracted and assigned the name **position1**.

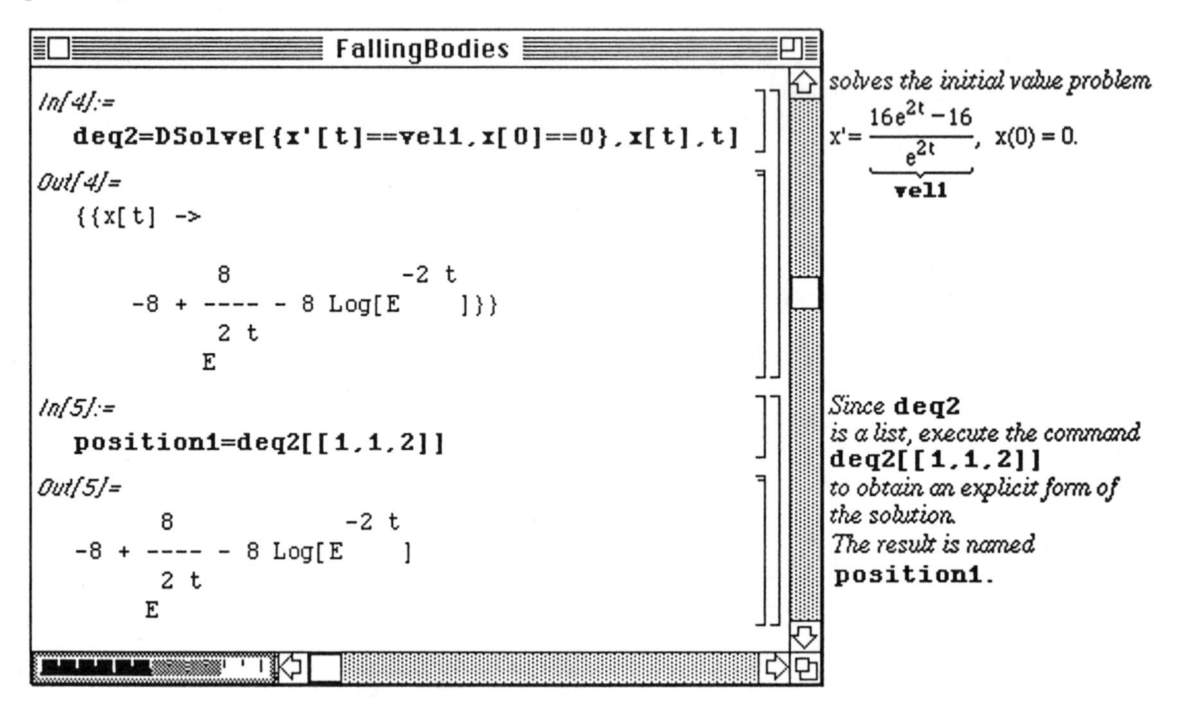

Then, the value of t when x(t) = 50 is computed (i.e., the time when the object hits the lake's surface). This is accomplished by making use of **FindRoot**. Since **FindRoot** depends on an initial guess of the solution to the equation, **position1** is graphed in order to obtain an approximate value of t when x(t) = 50. The position appears to equal 50 near t = 4. Hence, **FindRoot** is used to determine the root of the equation x(t) = 50 with the initial guess t=4. Therefore, the velocity at the point of impact is found by substituting the value obtained with **FindRoot**, t = 3.62464, into the velocity function of the object above the surface, **vel1**. This is accomplished with **vel1/.t->3.62464** , and the resulting expression is named **initvel2**.

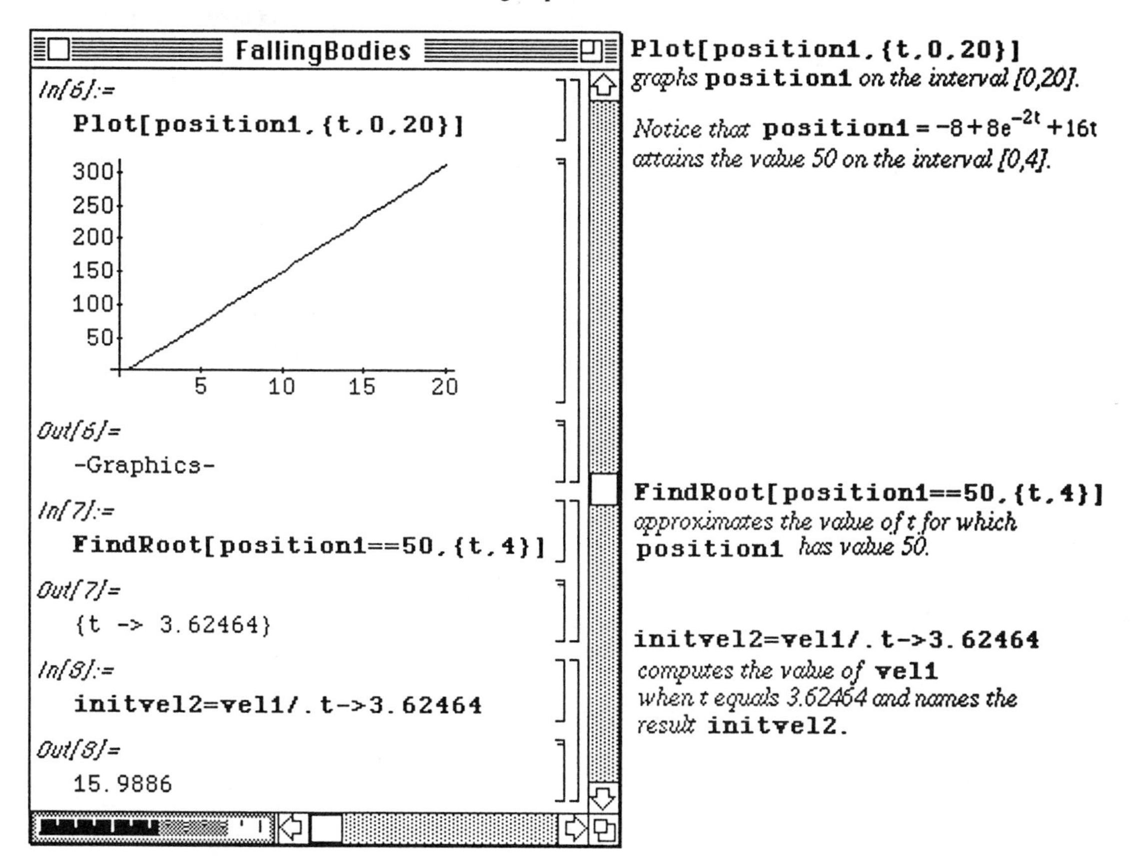

Plot[position1,{t,0,20}] *graphs* **position1** *on the interval [0,20].*

Notice that **position1** $= -8+8e^{-2t}+16t$ *attains the value 50 on the interval [0,4].*

FindRoot[position1==50,{t,4}] *approximates the value of t for which* **position1** *has value 50.*

initvel2=vel1/.t->3.62464 *computes the value of* **vel1** *when t equals 3.62464 and names the result* **initvel2**.

Now that the velocity of the object at its point of impact is known, the initial value problem to determine the velocity beneath the lake's surface is: $\frac{dv}{dt} = 24 - 6\,v,\ \ v(0) = 15.9886.$

can be solved. This problem is solved below with **DSolve**. (Note that **initvel2** is used in the initial condition instead of entering the numerical value.) Once solved, the exact value of the velocity at t=2 seconds is calculated by extracting the velocity formula from **deq3** and evaluating it at t=2 with **deq3[[1,1,2]]/.t->2**. The numerical approximation of the velocity (4.00007 seconds) is then computed.

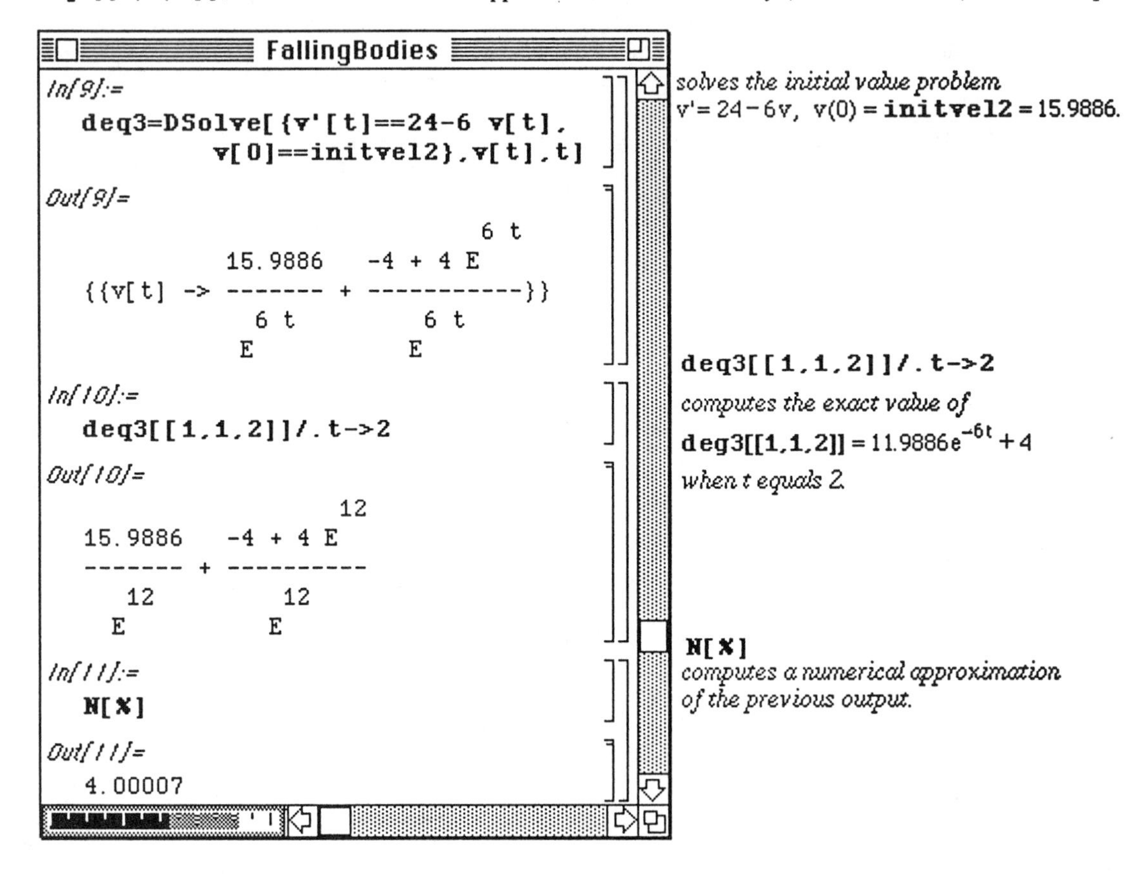

6.2 Exact Differential Equations

Certain types of nonlinear ordinary differential equations such as exact differential equations can be solved with the aid of *Mathematica*. The differential equation M(x,y) dx + N(x,y) = 0 is called an **exact differential equation** in

a domain D if there exists a function F such that $\frac{\partial F(x,y)}{\partial x} = M(x,y)$ and $\frac{\partial F(x,y)}{\partial y} = N(x,y)$

for all (x,y) in D. In order to determine if an equation is exact, the following well-known theorem can be used: Let M and N have continuous first partial derivatives at all points (x,y) in a domain D.

If $\frac{\partial M(x,y)}{\partial y} = \frac{\partial N(x,y)}{\partial x}$ for all (x,y) in D, then the differential equation $M(x,y)dx + N(x,y)dy = 0$ is exact

Hence, if the differential equation is exact the total differential of F,
dF(x,y) = M(x,y) dx + N(x,y)dy = 0.
Therefore, the solution of the exact equation is F(x,y) = Constant. The method by which F(x,y) is determined is illustrated with the following example.

□ **Example:**

Solve the initial value problem $\frac{1+8\,x\,y^{2/3}}{x^{2/3}y^{1/3}}\,dx + \frac{2x^{4/3}y^{2/3} - x^{1/3}}{y^{4/3}}\,dy = 0,\ \ y(1) = 8.$

First, we must verify that this differential equation is exact. This is done by entering the functions M(x,y) and N(x,y). (To avoid confusion with any built-in *Mathematica* function or constant, small letters are used in these definitions.) Next, the partial derivative of M(x,y) with respect to y, `D[m[x,y],y]`, must be calculated so that it can be compared with the partial derivative of N(x,y) with respect to x, `D[n[x,y],x]`. These derivatives are conveniently named **my** and **nx**.

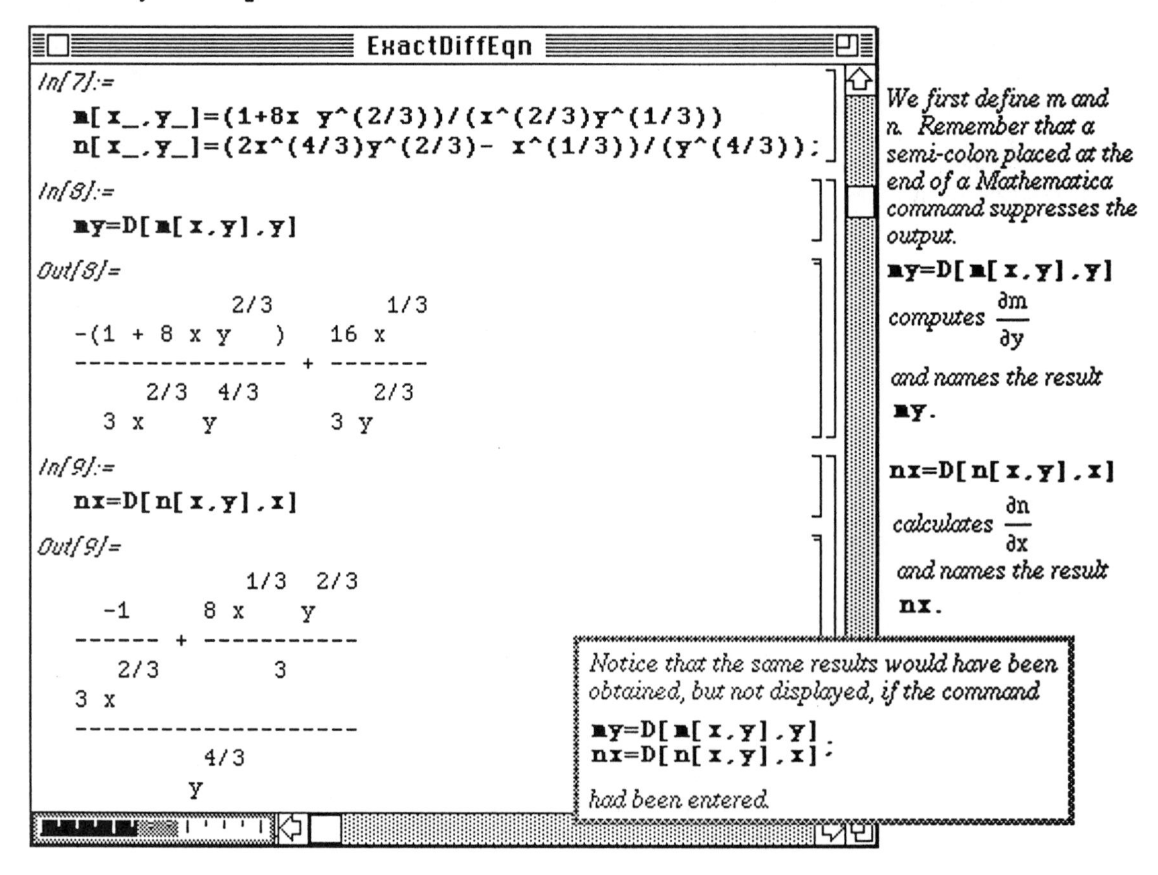

At this point **my** and **nx** do not appear to be equal. However, once simplified, equality is verified when the result of the test equality **my1 == nx1** (note the double equals sign) is `True`. (The simplified derivatives are assigned the names **my1** and **nx1**).

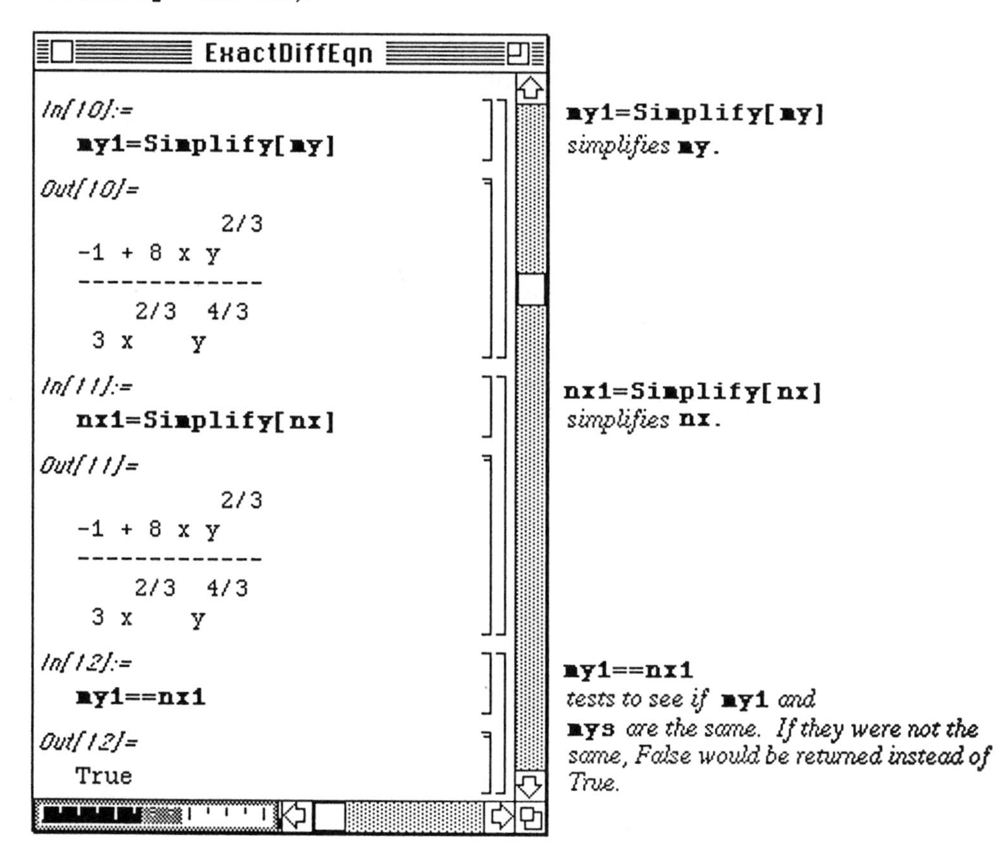

my1=Simplify[my]
simplifies **my**.

nx1=Simplify[nx]
simplifies **nx**.

my1==nx1
tests to see if **my1** *and* **mys** *are the same. If they were not the same, False would be returned instead of True.*

Now that the equation is known to be exact, the process of finding the function F(x,y) can begin. Since by definition,

$$\frac{\partial F(x,y)}{\partial x} = M(x,y), \text{ we have } F(x,y) = \int M(x,y)\,\partial x + g(y).$$

where g(y) is an arbitrary function of y. In order to determine g(y), differentiate the above equation with respect to y and, make use of the fact that $\frac{\partial F(x,y)}{\partial y} = N(x,y)$. Then, $\frac{\partial F(x,y)}{\partial y} = \frac{\partial}{\partial y}\int M(x,y)\,\partial x + g'(y) = N(x,y)$.

Therefore, $g'(y) = N(x,y) - \frac{\partial}{\partial y}\int M(x,y)\,\partial x$ so $g(y) = \int \left(N(x,y) - \frac{\partial}{\partial y} M(x,y)\,\partial x \right) dy$.

These steps are carried out with *Mathematica* in the following manner.
First, integrate M(x,y) with respect to x. The resulting expression is called **`f`**. Of course, *Mathematica* does not indicate the presence of the arbitrary function of y, g(y). However, the possibility that a nonconstant function g(y) exists must be investigated. This is accomplished by differentiating **`f`** with respect to y, naming it **`fy`**, and comparing **`fy`** to N(x,y). Since the difference of N(x,y) and **`fy`** is g'(y), their difference is computed below and named **`gprime`**. When simplified, **`gprime`** is found to be zero. Therefore, g(y) is a constant function, so the solution of this exact differential equation is **`f`** = C. The constant C is found by evaluating **`f`** at the point (1,8). This is done with the command: **`f/.x->1/.y->8`**

This shows that $C = \frac{27}{2}$. Therefore, the solution to the initial value problem is: $\frac{3x^{1/3}}{y^{1/3}} + 6\,x^{4/3}\,y^{1/3} = \frac{27}{2}$.

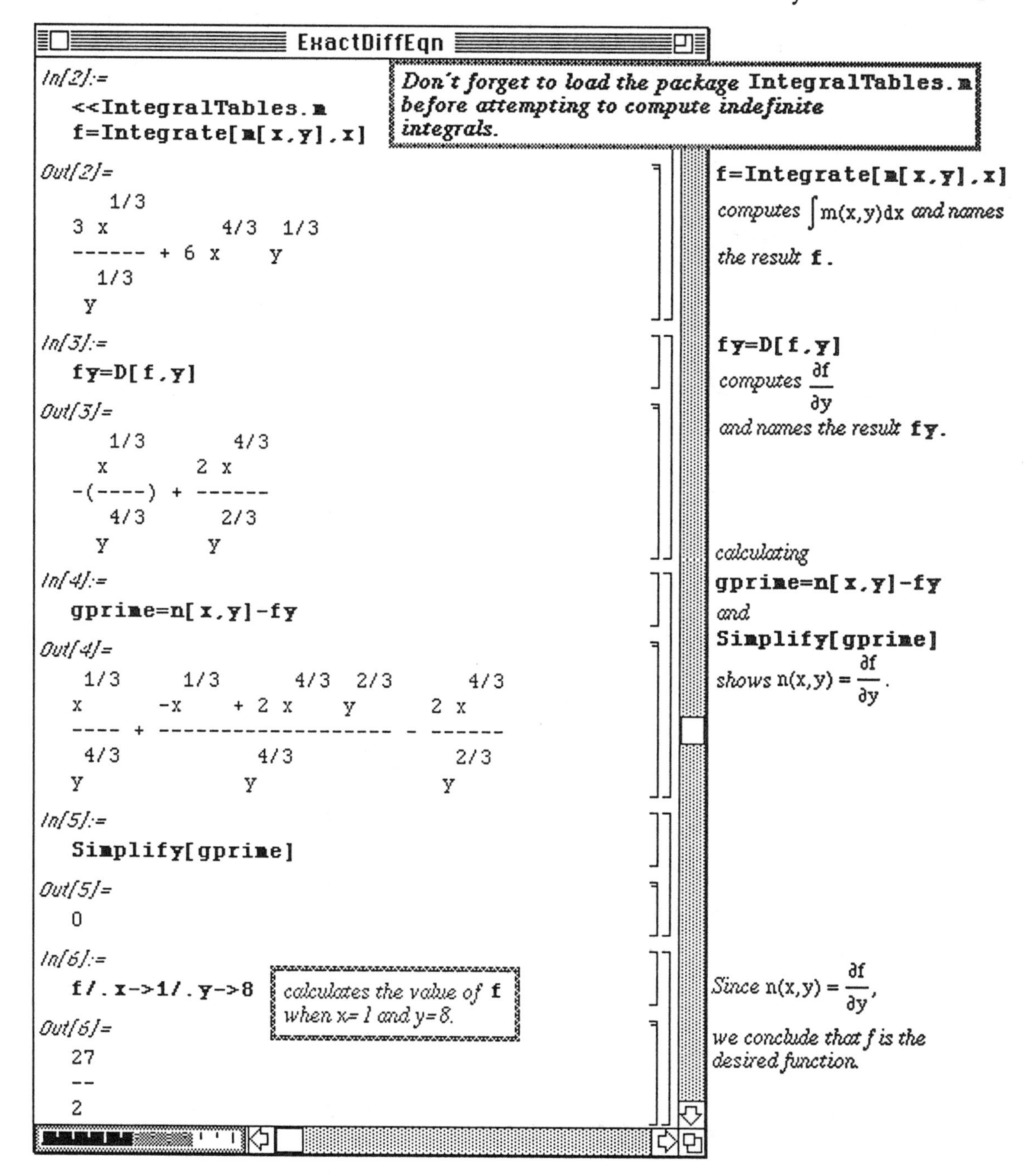

Mathematica can now be used to check that the function **f** is correct.

Dt[f] computes the total differential of **f**. This is computed below and called **totalf**. The symbols `Dt[x]` and `Dt[y]` in the output represent dx and dy, respectively. Hence, the command **Together[Coefficient[totalf,Dt[x]]]** gives the coefficient of dx and writes it as a single fraction. A similar command is used to obtain the coefficient of dy. Since the differential equation is exact, the coefficients of dx and dy should be the functions

$$M(x,y)=\frac{1+8\,x\,y^{2/3}}{x^{2/3}y^{1/3}} \quad\text{and}\quad N(x,y)=\frac{2x^{4/3}y^{2/3}-x^{1/3}}{y^{4/3}}$$

from the differential equation. The results below verify the solution.

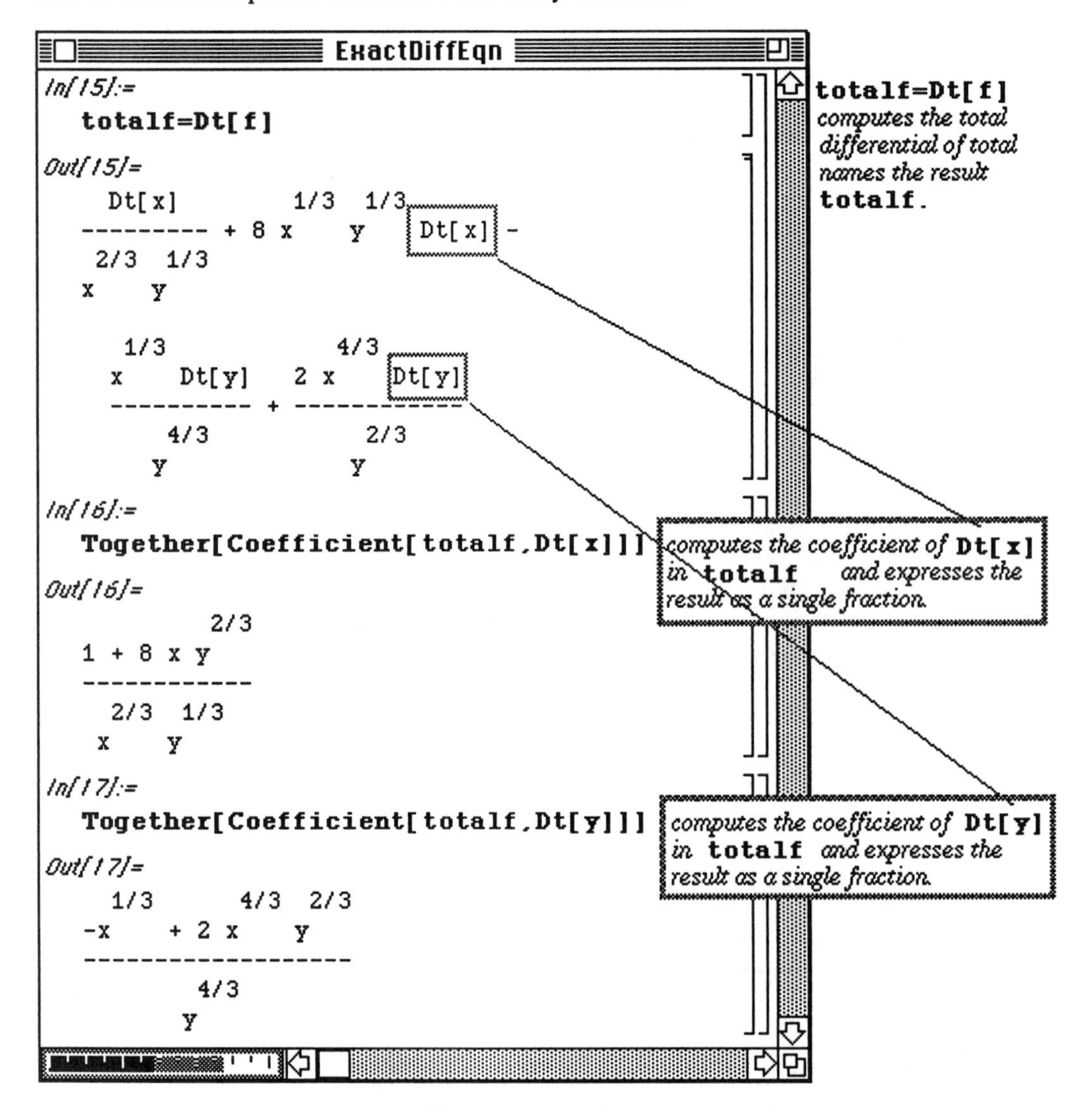

□ Example:

To illustrate that the arbitrary function g(y) may not be constant, consider the following problem:

$$\frac{3-y}{x^2}\,dx + \frac{y^2-2x}{xy^2}\,dy = 0\,,\;\; y(-1) = 2$$

Begin by clearing all previously used expression names. Then, the same steps are followed as were used in the previous example. The functions M(x,y) and N(x,y) are entered and the equation found to be exact by verifying that

$$\frac{\partial M(x,y)}{\partial y} = \frac{\partial N(x,y)}{\partial x}.$$

Note that the partial derivative of N(x,y) with respect to x must be simplified to see that it is identical to the partial of M(x,y) with respect to y.

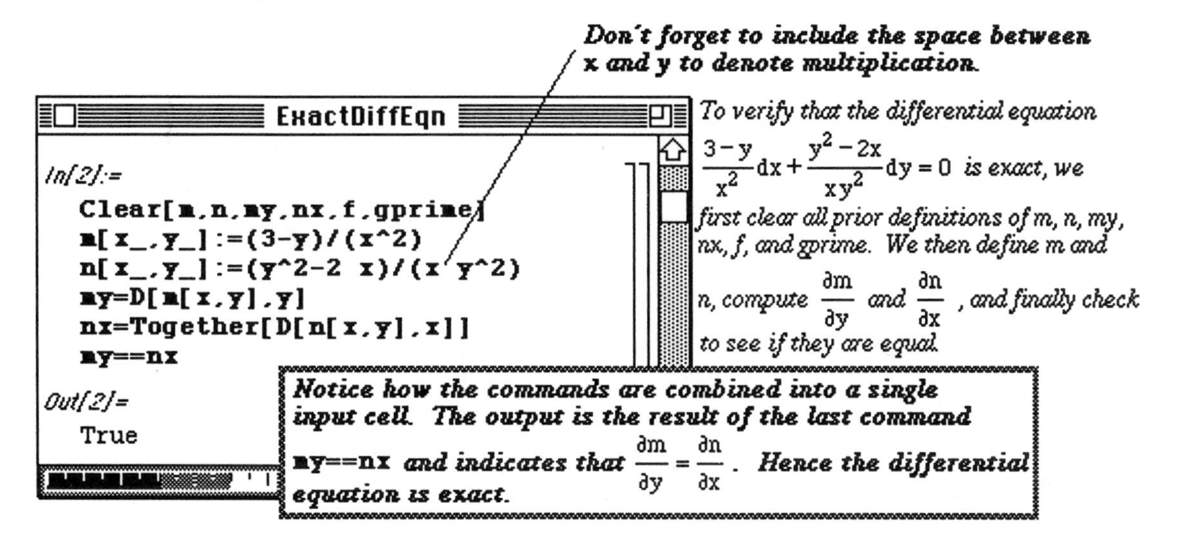

Next, the function F(x,y) is determined as in the previous example. Note that in this problem, the simplified form of **gprime** is nonconstant.

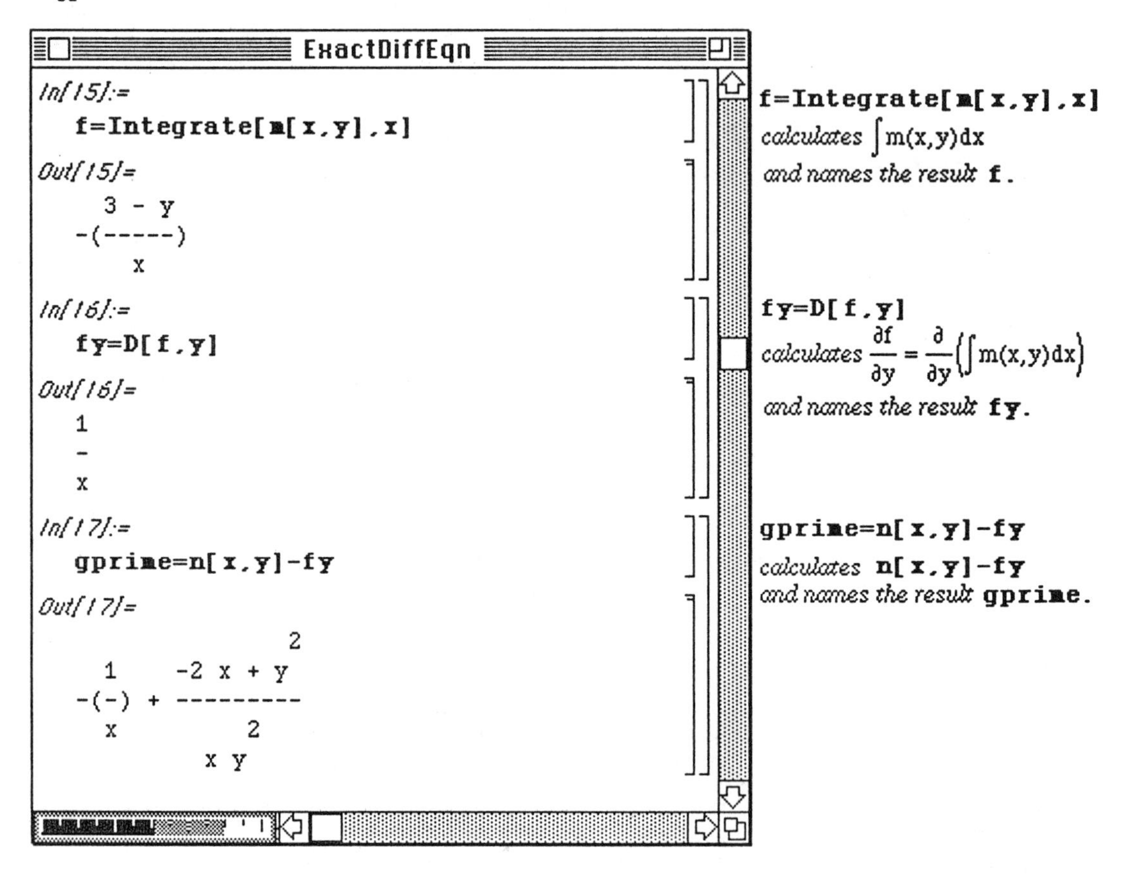

Therefore, **gprime** must be integrated with respect to y to find the formula for g(y), and the solution of the exact equation, **solution**, is the sum of the functions **f** and g(y). Finally, to solve the initial value problem, **solution** is evaluated at the point (-1,2) with the command **solution/.x->-1/.y->2** to obtain a value of 2. Hence the solution is $-\frac{3-y}{x}+\frac{2}{y}=2$

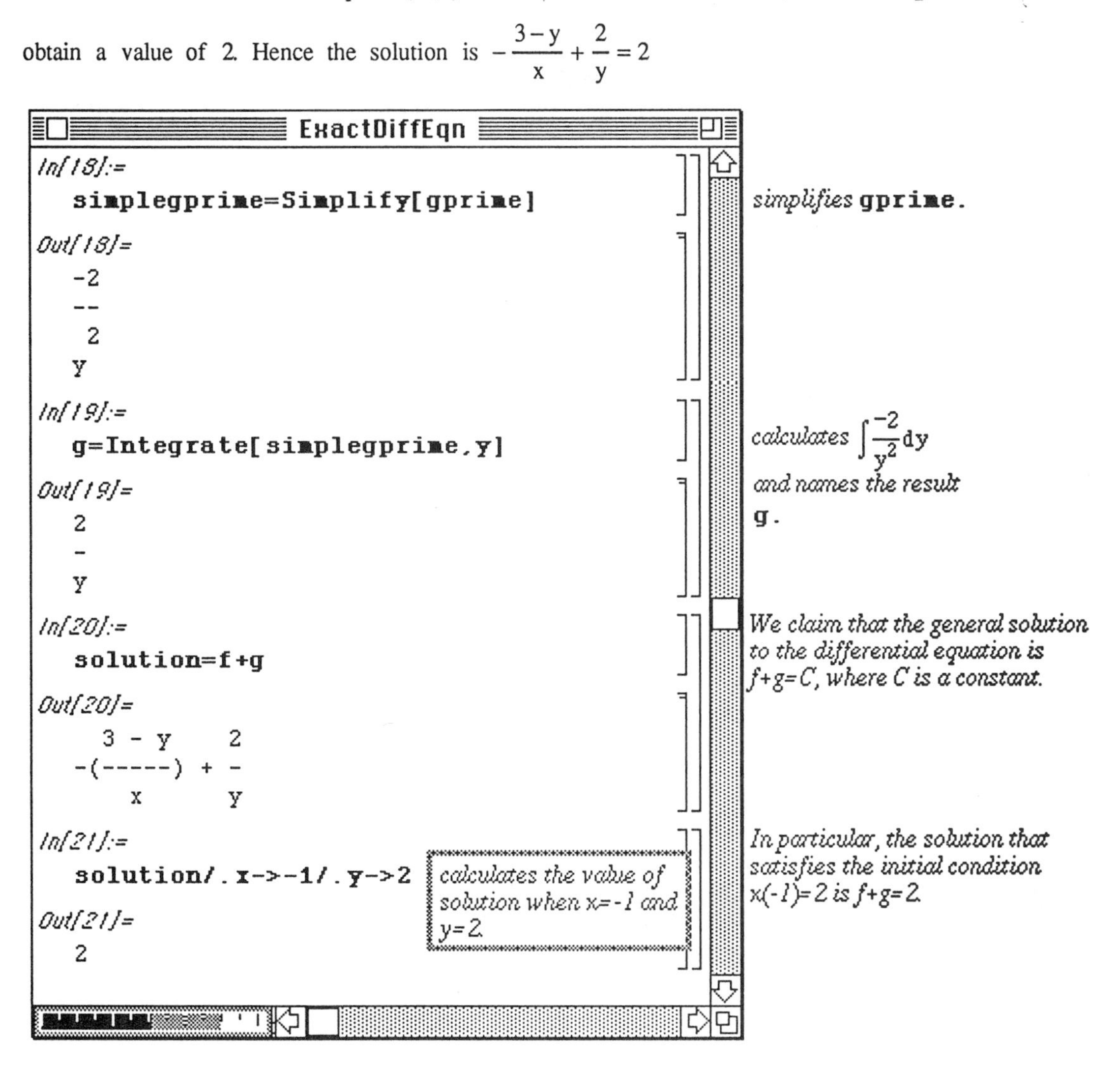

As was illustrated in the previous example, the solution can be verified by calculating the total differential, **totalsol**,of **solution** with **Dt[solution]**. Again, collecting the coefficients of **Dt[x]** and **Dt[y]** in **totalsol** and comparing these results to the functions M(x,y) and N(x,y) from the differential equation shows that the solution is correct.

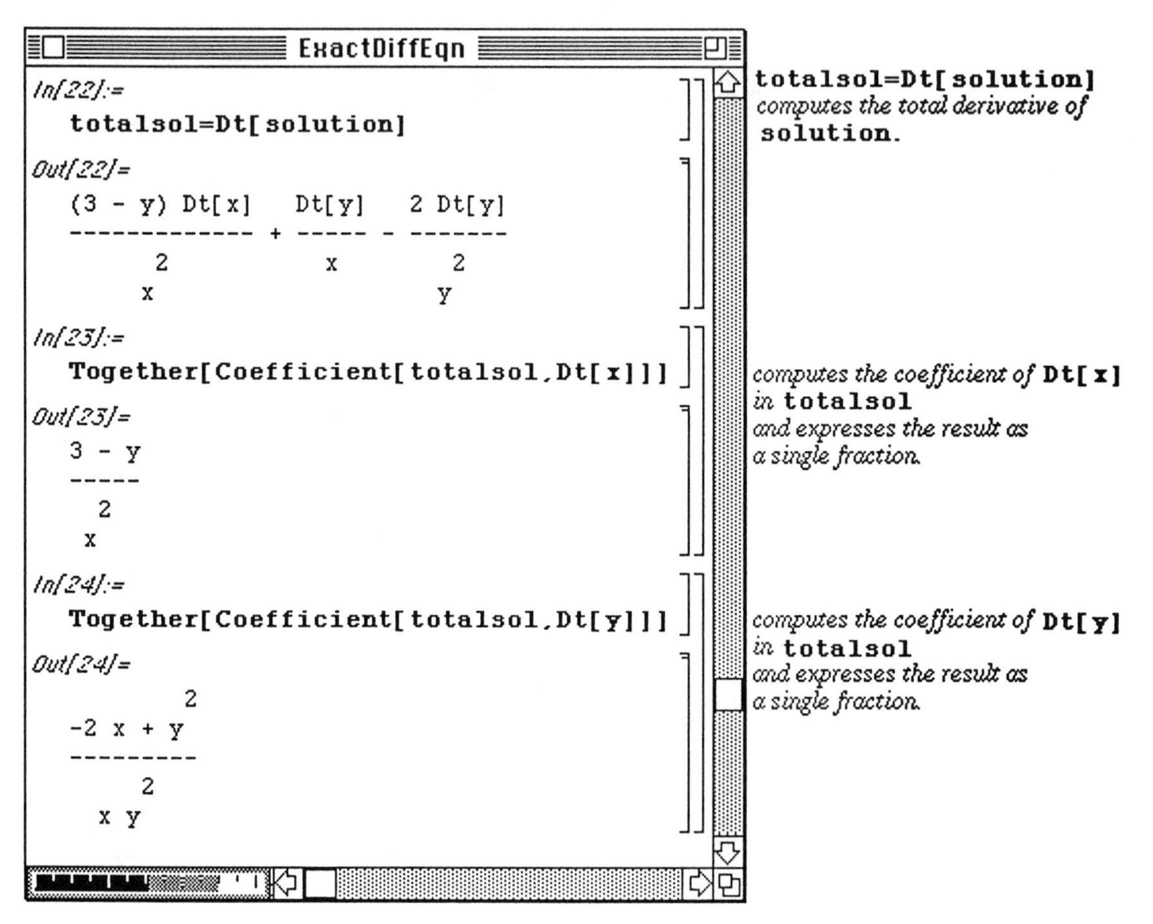

6.3 Undetermined Coefficients

Example:

Use the method of undetermined coefficients to solve the differential equation $y''-2y'+y=\text{Sin}(x)$.

Recall that the method of undetermined coefficients is used to solve nonhomogeneous linear ordinary differential equations. The general solution to the nonhomogeneous equation

$a_n y^{(n)} + a_{n-1}y^{(n-1)} + \cdots + a_1y' + a_0y = g(x)$ is $y(x) = y_h(x) + y_p(x)$, where $y_h(x)$ is the solution to the corresponding homogeneous equation $a_n y^{(n)} + a_{n-1}y^{(n-1)} + \cdots + a_1y' + a_0y = 0$, and $y_p(x)$ is a particular solution to the nonhomogeneous equation.

The following solution is constructed with Version 1.2. Notice that Version 1.2 is unable to solve the differential equation y"-2y'+y=Sin(x) with the built-in command **DSolve**.

Because **DSolve** does not solve most nonhomogeneous equations (as illustrated below) the problem must be divided into two parts. First, the homogeneous equation must be solved, and then a particular solution to the nonhomogeneous equation must be found.
Since **DSolve** can be used to solve homogeneous linear ordinary differential equations with constant coefficients of degree four or less, the homogeneous solution, $y_h(x)$, is found with **DSolve[y''[x]-2y'[x]+y[x]==0,y[x],x]** and called **sol1** for later use. Also, the nonhomogeneous function, g(x) = sinx, is assigned the name **exp**.

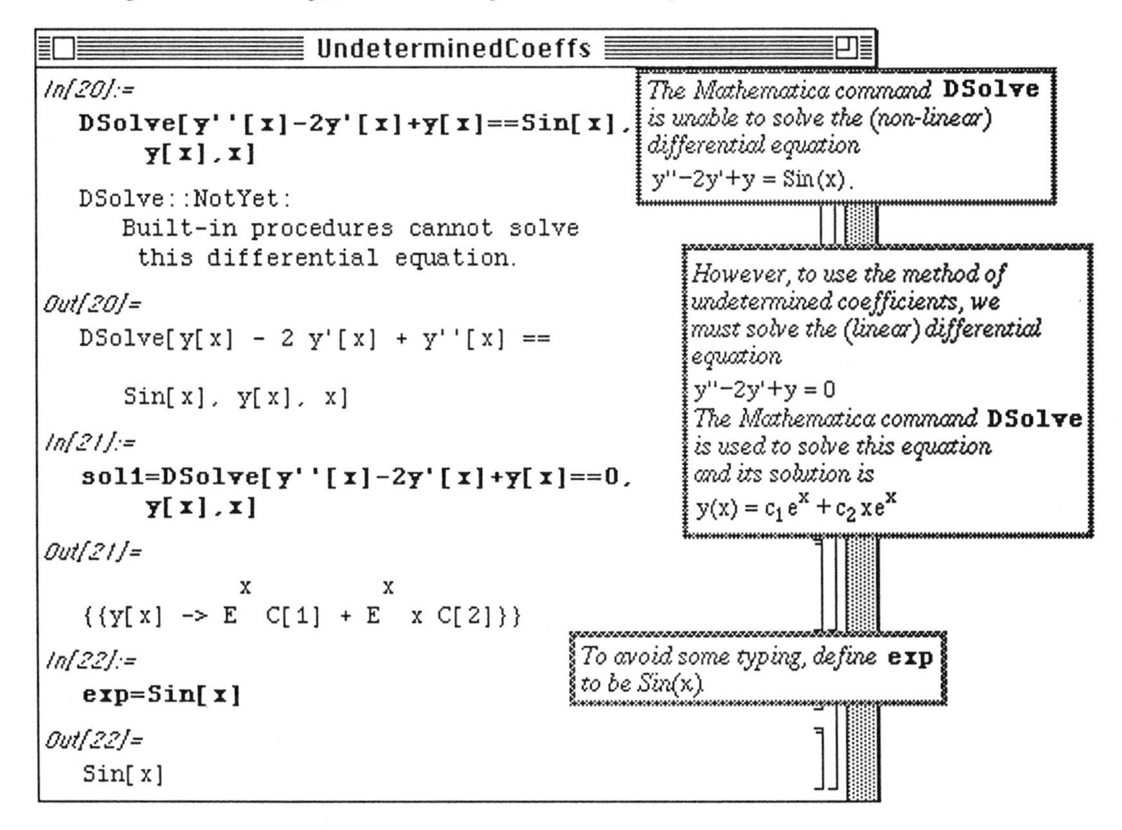

The annihilator form of the method of undetermined coefficients will be used to find a particular solution of the nonhomogeneous problem. Recall that the $n-$th order derivative of a function y is $D^n y = \dfrac{d^n y}{dx^n}$.

Thus, the linear n-th order ordinary differential equation with constant coefficients

$a_n\, y^{(n)} + a_{n-1}\, y^{(n-1)} + \cdots + a_1\, y' + a_0\, y = g(x)$ can be expressed in operator notation

$a_n D^n\, y + a_{n-1} D^{n-1}\, y + \cdots + a_1 Dy + a_0 y = g(x)$ or $(\, a_n D^n + a_{n-1} D^{n-1} + \cdots + a_1 D + a_0)\, y = g(x)$.

The expression $(\, a_n D^n + a_{n-1} D^{n-1} + \cdots + a_1 D + a_0)$ is called an **$n-$th order differential operator.**

The differential operator $(\, a_n D^n + a_{n-1} D^{n-1} + \cdots + a_1 D + a_0)$ is said to **annihilate** a function f (which possesses at least n derivatives) if $(\, a_n D^n + a_{n-1} D^{n-1} + \cdots + a_1 D + a_0)\, f(x) = 0$.

In order to solve the nonhomogeneous equation, recall the following :

(i) The differential operator D^n annihilates each of the functions $1, x, x^2, \ldots, x^{n-1}$.

(ii) The differential operator $(D-\alpha)^n$ annihilates each of the functions $e^{\alpha x}, xe^{\alpha x}, x^2 e^{\alpha x}, \ldots, x^{n-1} e^{\alpha x}$.

(iii) The differential operator $\left[D^2 - 2\alpha D + (\alpha^2 + \beta^2)\right]^n$ annihilates each of the functions

$e^{\alpha x}\cos\beta x,\; xe^{\alpha x}\cos\beta x,\; x^2 e^{\alpha x}\cos\beta x, \ldots, x^{n-1} e^{\alpha x}\cos\beta x,\; e^{\alpha x}\sin\beta x,\; xe^{\alpha x}\sin\beta x,\; x^2 e^{\alpha x}\sin\beta x, \ldots, x^{n-1} e^{\alpha x}\sin\beta x.$

Let $P(D)$ denote the differential operator $(\, a_n D^n + a_{n-1} D^{n-1} + \cdots + a_1 D + a_0)$.

Then the nonhomogeneous linear n-th order ordinary differential equation with constant coefficients equation can be expressed as $P(D)\, y = g(x)$. When $g(x)$ consists of either

(a) a constant k,
(b) a polynomial in x,
(c) an exponential function $e^{\alpha x}$,
(d) $\sin\beta x$, $\cos\beta x$

or finite sums and products of these functions, another differential operator which annihilates $g(x)$ can be determined.

Suppose that the differential operator $P_1(D)$ annihilates $g(x)$. Then applying $P_1(D)$ to the nonhomogeneous equation yields $P_1(D)\, P(D)\, y = P_1(D)\, g(x) = 0$. The form of the particular solution is found by solving the homogeneous equation $P_1(D)\, P(D)\, y = 0$.

The function **`ann[q,f]`** which verifies that the annihilating operator, **`q`** (a polynomial in **`d`**), annihilates the function **`f`** is defined below with a **`Block`**. This is accomplished by transforming the polynomial **`q`** into its equivalent differential form and applying it to the function **`f`**.

`Block[{var1,var2,...},procedure]` allows the variables to be treated locally in the procedure without affecting their value outside. Hence, each time the procedure is executed, the original value of each of the variables in the list **`{var1,var2,...}`** is saved and restored at the end of the procedure. The commands used in the definition of **`ann[q,f]`** are explained below. **`Expand[q]`** expands all products and powers in **`q`**. This is useful when the annihilator is a product of differential operators.
This expanded form of **`q`** is called **`p`** locally.

o In Version 2.0, the command `Block` has been replaced by the command `Module`.

`Variables[p]` gives a list of all of the variables in `p`. This command is used to obtain the variable, `d`, used in the annihilating operator.

`Exponent[p,var[[1]]]` yields the maximum power for which `var[[1]]` appears in `p`.
In this case, `var[[1]]` represents the variable `d`. Hence, this command determines the highest power of `d` in `p` and names it `exp`. (Since `exp` is contained in the `Block`, it should not be confused with the name of the nonhomogeneous function, g(x) = sinx, which was named `exp` outside of the `Block`.) Knowing the highest power of `d` enables the differential form of the operator to be determined by, first, finding the coefficients.

`c[i]:=Coefficient[p,var[[1]],i]` gives the coefficient of `var[[1]]^i` in `p`. In other words, it finds the coefficient of each term in `p`. (`c[1]` is the coefficient of `d`, `c[2]` is the coefficient of `d^2`, etc.) Since `c[0]` is the the constant in `p`, it can be obtained by evaluating `p` when the variable `d=0`.

The final command, `Sum[c[i] D[f,{x,i}],{i,0,exp}]`, applies the annihilator to the function by substituting `f` into the differential form of the operator.

In this example, the nonhomogeneous term in the equation is the function g(x) = sinx which falls under category (iii) given above. Therefore, the annihilator is **d^2+1** (since $\alpha = 0$, $\beta = 1$, and n = 1). This is verified with **ann[d^2+1,exp]** where g(x) = sinx was given the name **exp**.

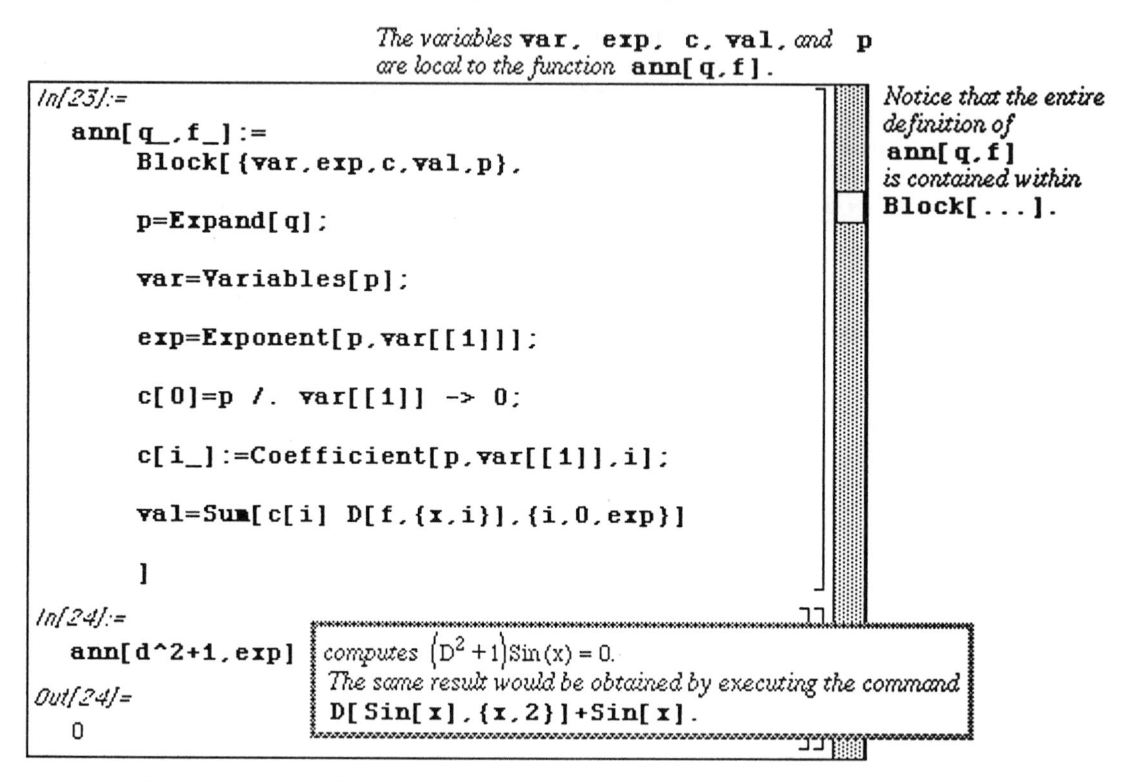

After the annihilator has been verified, the procedure to determine the particular solution to the nonhomogeneous differential equation may begin. The first step is to represent both the annihilator and the left-hand side of the differential equation in differential operator form. The annihilator is, therefore, represented as **y^2+1**, and the differential equation as **y^2-2y+y**. (The variable **y** is being used instead of **d**.) Now, applying the annihilator to the differential equation is equivalent to taking the product of these two differential operators. This is done with **Expand[(y^2+1)(y^2-2y+1)]**. Hence, a fourth-order differential operator is obtained. This operator must then be used to return to the form of a differential equation. (i.e., the fourth-order differential operator obtained must be applied to y.) The function **polytodiff** defined in the **Block** below accomplishes this task by making the

making the conversion from $(a_n D^n + a_{n-1} D^{n-1} + \cdots + a_1 D + a_0)\, y = 0$ to the differential equation $a_n\, y^{(n)} + a_{n-1}\, y^{(n-1)} + \cdots + a_1\, y' + a_0\, y = 0.$

It uses the same commands which appeared in **ann[q, f]** above to yield a fourth-order differential equation. This equation is solved with **DSolve** with the result named **sol2**.

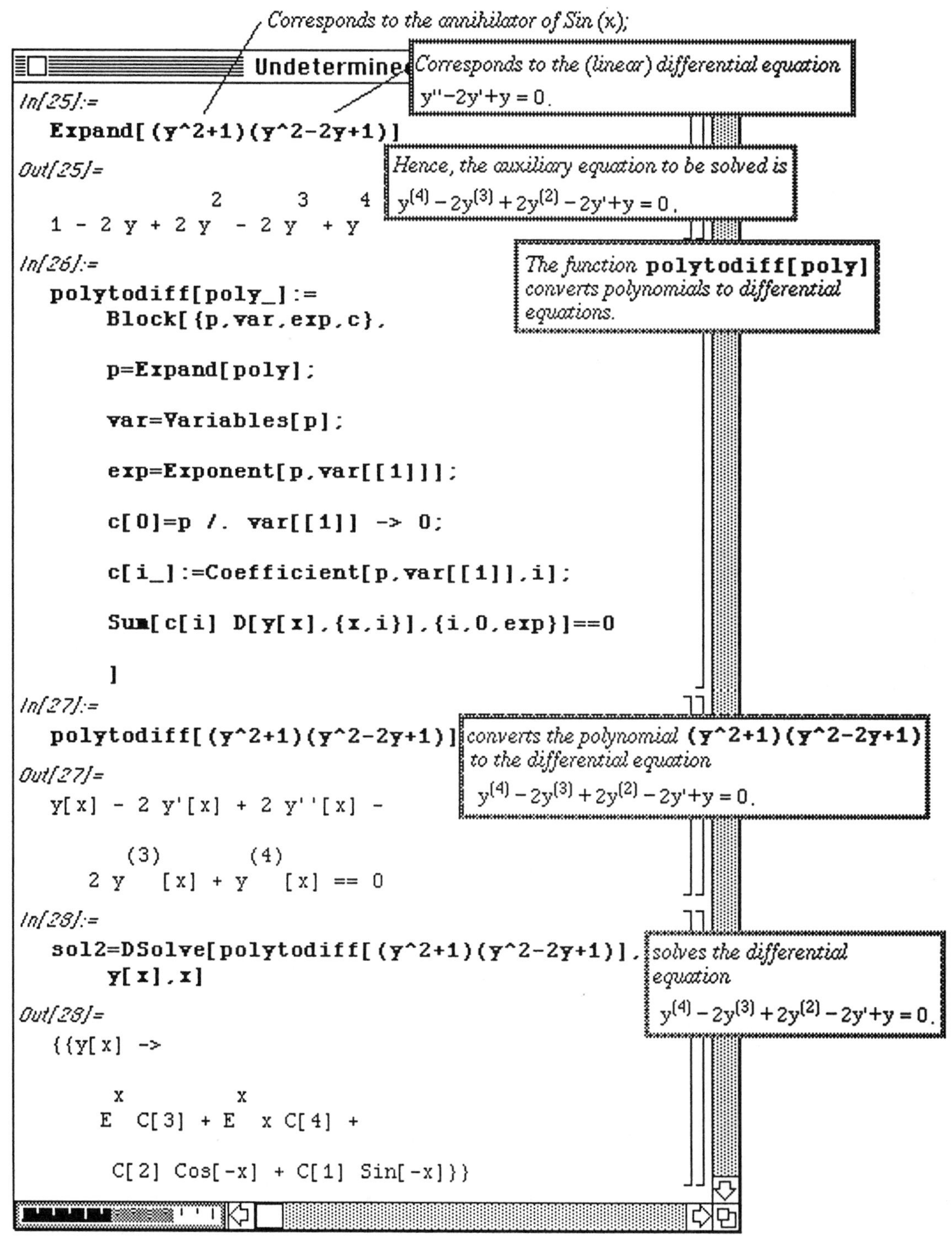

Since **sol2** is a list, the formula for y[x] must be extracted. This is done with **sol2[[1,1,2]]. Obviously,** this formula can be simplified with trigonometric identities. The command, **TrigExpand[sol2[[1,1,2]]],** does this. Recalling that the general solution to the homogeneous equation obtained earlier in the example, **sol1**, was

given by $y[x] = C[1]\,E^{x} + C[2]\,x\,E^{x}$, the form of the particular solution must be the remaining terms in the simplified version of **sol2[[1,1,2]]** given below. Therefore, the particular solution is defined as $y_p(x) = c_1 \operatorname{Cos}x + c_2 \operatorname{Sin}x$.

o In Version 2.0, **TrigExpand** is obsolete. The command **Expand[expression,Trig->True]** in Version 2.0 performs the same calculation as **TrigExpand[expression]** in Version 1.2.

Note that the constants are arbitrary, so the negative sign associated with C[1] in **sol2[[1,1,2]]** as well as the order of these constants can be ignored. The objective is to solve for the arbitrary constants found in the particular solution. Hence, the name of these constants and the signs associated with them is not a concern. In order to find the values of these two constants, the particular function is defined below as the function **yp**.

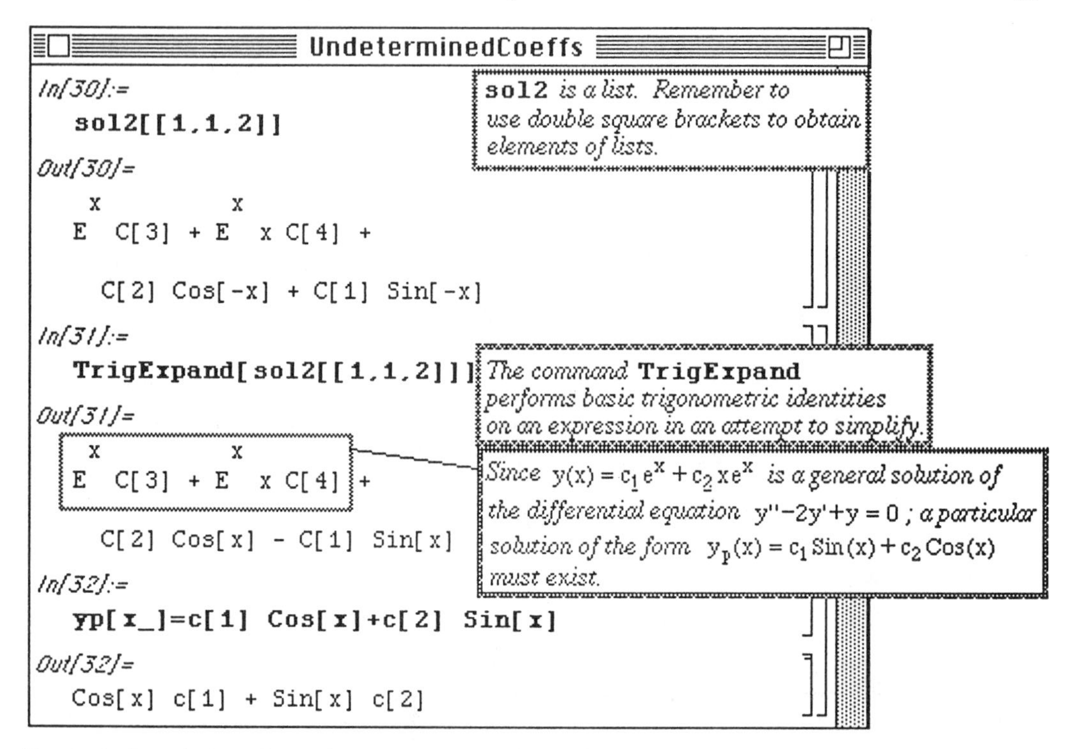

The particular solution of a nonhomogeneous equation satisfies the nonhomogeneous equation. Hence, **yp** must satisfy $y'' - 2y + y = \operatorname{Sin}(x)$. **yp** is substituted into this equation below, and the expression which results is named **eqn** so that it can be easily simplified with **Simplify[eqn]**. This simplified equation reveals that `c[1]` = 1/2 and `c[2]` = 0 by equating the coefficients of Cos[x] and Sin[x] on each side of the equation. Evaluating the function **yp** for these values of `c[1]` and `c[2]` yields the particular solution.

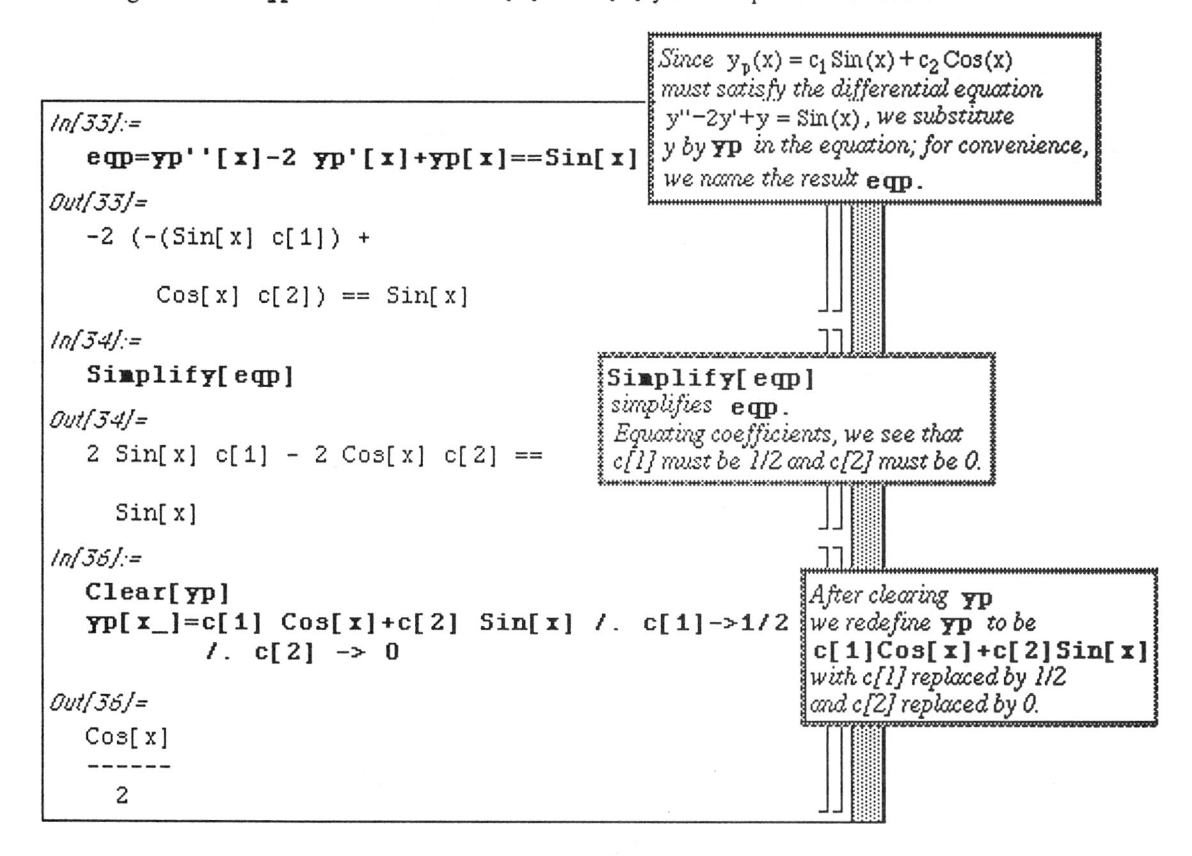

The general solution to the nonhomogeneous equation is the sum of the general solution to the homogeneous equation and the particular solution to the nonhomogeneous equation. This solution is defined below with **y**. To check that the solution is correct, it is substituted into the left-hand side of the differential equation and simplified to verify the results.

```
UndeterminedCoeffs

In[37]:=
  y[x_]=sol1[[1,1,2]]+yp[x]
Out[37]=
   x           x          Cos[x]
  E  C[1] + E  x C[2] + ------
                            2

In[40]:=
  verify=y''[x]-2y'[x]+y[x]
Out[40]=
     x          x
  2 E  C[1] + 2 E  C[2] +

      x
   2 E  x C[2] -

        x          x
   2 (E  C[1] + E  C[2] +

       x         Sin[x]
      E  x C[2] - ------)
                    2

In[41]:=
  Simplify[verify]
Out[41]=
  Sin[x]
```

Hence, every solution of $y''-2y'+y = \mathrm{Sin}(x)$ *is of the form* $y(x) = c_1 e^x + c_2 x e^x + \frac{1}{2}\mathrm{Cos}(x)$.

To verify, we compute and simplify.

o The command **DSolve** has been dramatically improved in Version 2.0. In fact, Version 2.0 computes the exact solution of y''-2y'+y=Sin (x). In general, when attempting to solve a differential equation, try **DSolve** first; if **DSolve** does not produce a solution, try other methods.

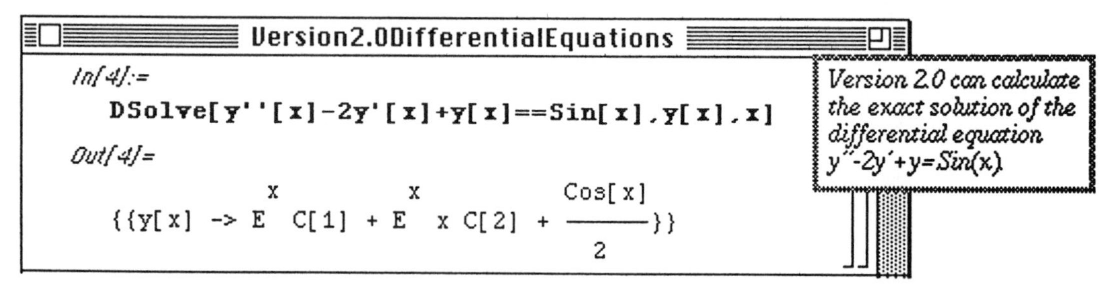

▩ 6.4 Linear n-th Order Differential Equations with Constant Coefficients

A differential equation of the form $a_n \frac{d^n y}{dx^n} + a_{n-1} \frac{d^{n-1} y}{dx^{n-1}} + \cdots + a_1 \frac{dy}{dx} + a_0 y = 0$, where each a_i is a real number, is called

a **linear n-th order homogeneous differential equation**.

The **characteristic equation** of the linear n-th order homogeneous differential equation

$a_n \frac{d^n y}{dx^n} + a_{n-1} \frac{d^{n-1} y}{dx^{n-1}} + \cdots + a_1 \frac{dy}{dx} + a_0 y = 0$ is $a_n m^n + a_{n-1} m^{n-1} + \cdots + a_1 m + a_0 = 0.$

Recall that this characteristic equation is found by assuming that the solutions of the differential equation are of the form $y(x) = e^{mx}$, and the solution is found by substituting this solution into the differential equation and solving for m. Hence, the roots of the characteristic equation are values of m which yield solutions to the differential equation. *Mathematica* is somewhat limited in its ability to solve linear n-th order homogeneous differential equations with constant coefficients directly with `DSolve` since `DSolve` only works for equations of this type of degree four or less. However, these equations can still be solved with *Mathematica* by considering the characteristic equation.

□ **Example:**

Mathematica's inability to solve the fifth-order equation

$$\frac{d^5y}{dx^5} - 3\frac{d^4y}{dx^4} - 5\frac{d^3y}{dx^3} + 15\frac{d^2y}{dx^2} + 4\frac{dy}{dx} - 12\,y = 0$$

with **DSolve** is demonstrated below by first defining the left-hand side of the equation with **lhseqn**, and then using **DSolve[lhseqn==0,y[x],x]**. Recall that the n-th derivative of **y[x]** is given with **D[y[x],{x,n}]**. Also, since y is a function of x, **y[x]** must be used in defining the left-hand side of the differential equation.
When **DSolve** fails to solve the equation, the characteristic equation is defined as **chareqn** so that an alternate method of solution can be employed.

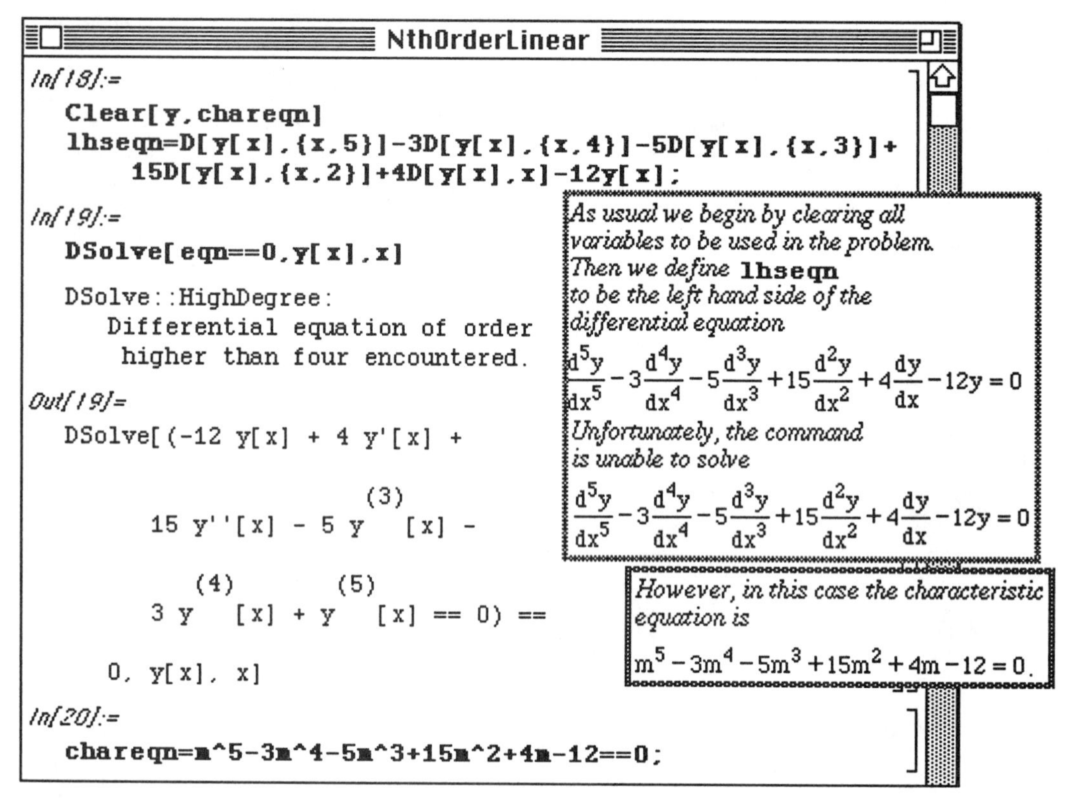

Since the characteristic equation is of degree five, the roots can be found exactly with **Solve[chareqn,m]**.

Since five distinct roots result, the functions $e^{-2x}, e^{-x}, e^{x}, e^{2x}, e^{3x}$ are five linearly independent solutions to the fifth-order differential equation. Therefore, the general solution to the differential equation is the linear combination of these five functions. The general solution is defined below as `y[x]`. It is then substituted into the left-hand side of the differential equation. Since this expression is lengthy, a shorted version is requested with `Short[lhseqn ,3]`. This expression is simplified to show that `y` satisfies the homogeneous differential equation.

```
In[21]:=
  Solve[chareqn,m]
Out[21]=
  {{m -> -1}, {m -> -2}, {m -> 3},

     {m -> 2}, {m -> 1}}
In[22]:=
  y[x_]=c[1] Exp[-2x]+c[2]Exp[-x]+c[3]Exp[x]+
        c[4]Exp[2x]+c[5]Exp[3x];
In[23]:=
  Short[lhseqn ,3]
Out[23]//Short=
  -32 c[1]
  -------- + <<8>> -
     2 x
    E

       16 c[1]
    3 (------- + <<3>> +
         2 x
        E

          3 x
      81 E    c[5])
In[24]:=
  Simplify[%]
Out[24]=
  0
```

In this case, Mathematica is able to exactly solve the characteristic equation. In any case, Mathematica is ALWAYS able to approximate the roots of the characteristic equation with the command **NRoots**.

Therefore the general solution of the differential equation is
$y(x) = c_1e^{-2x} + c_2e^{-x} + c_3e^{x} + c_4e^{2x} + c_5e^{3x}$

Remember that a semi-colon placed at the end of a command suppresses the resulting output.

After defining **y[x]** *to be the solution, we evaluate and simplify* **lhseqn** *to verify it is the general solution of the differential equation.*

If a set of initial conditions accompany the differential equation, then the constants in the general solution must be found so that the solution satisfies the differential equation and the initial conditions. Consider the following set of initial conditions:
y(0) = 0, y'(0) = 1,y''(0) = 0, y'''(0) = 3, y''''(0) = 0.
These initial conditions are entered as the list **initialconds** below. Next, a table consisting of the constants c[1],c[2],c[3],c[4],and c[5] is created with **Table[c[i],{i,1,5}]** and named **table.** Then the command **Solve[initialconds,table]** calculates the constants. The result is named **values** so that the solution of the initial value problem can be easily obtained by evaluating **y** for the values of the constants extracted from the list **values**. This is done with **y[x] /. values[[1]]**.

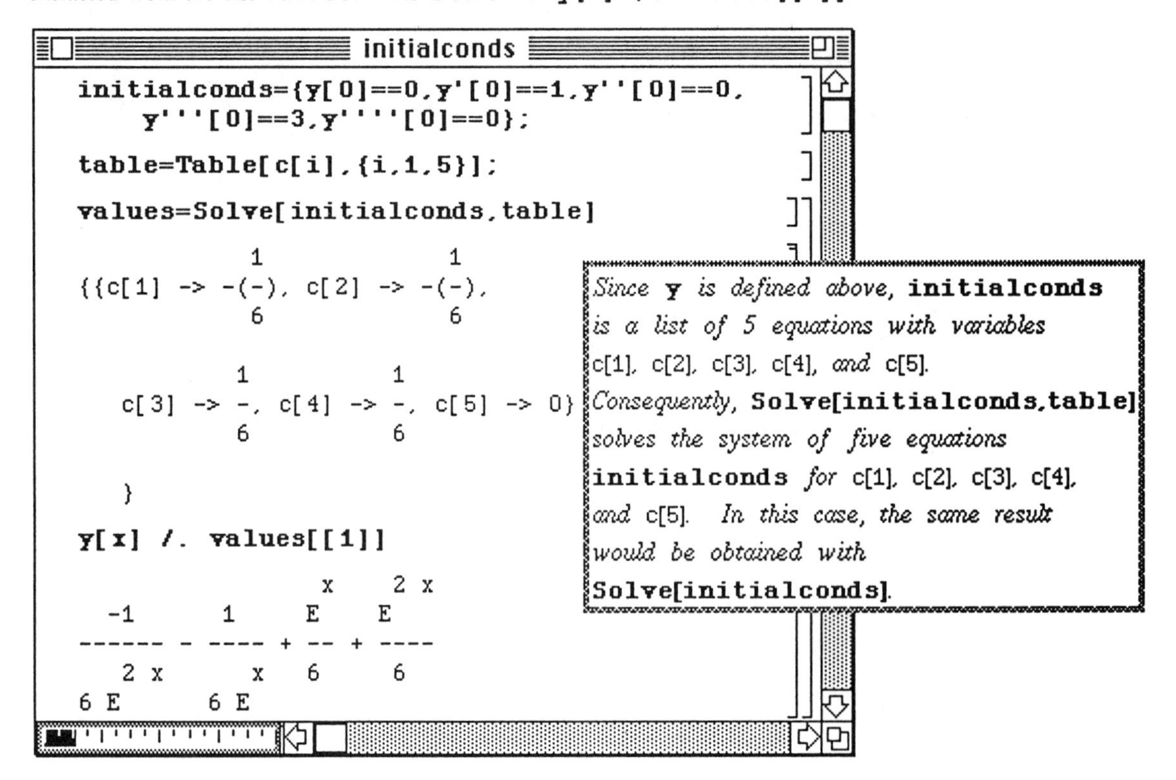

Consider another set of conditions :
$y(0) = 0,\ y'(0) = 1, y''(0) = 0,\ y'''(0) = 3,\ y''''(0) = 0.$
After clearing all previously used definitions, the problem is solved in a similar manner as the previous example. The new set of conditions are defined in a list, and the exact values of the constants are found in **values** as they were above. However, the exact values are not as meaningful as the numerical approximation to each constant. Therefore, a numerical approximation, named **nvalues**, is obtained with **N[values]**. The solution **y[x]** can then be evaluated at **nvalues[[1]]** to yield the solution to this problem.

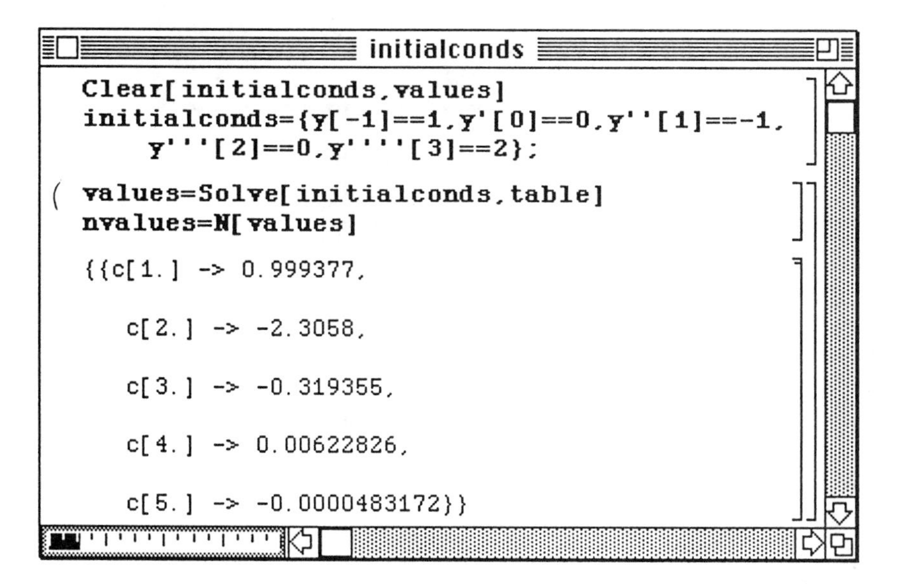

To obtain the solution, the function **y[x]** is evaluated at the elements of **nvalues** with **y[x]/.nvalues[[1]]**.

NthOrderLinear

```
values

{{c[1] ->

           6          5        6
  ((2 E   (((((-(2 E ) + 7 E ) -

              7        8         9
          2 E  - 18 E  + 26 E ) \

              10        11
          - 4 E   - 16 E   -

             12        13
          18 E   + 54 E   ) -

             14        15
          39 E   - 30 E   +

              16          17
          108 E  ) - 126 E   +

              18
          108 E  )) /

                           5
        ((((((((((-(32 E ) +
```

In this case, Mathematica calculates the exact values of c[1], c[2], c[3], c[4], and c[5]. However, numerical approximations of the exact values are probably more useful.

■ 6.5 The Cauchy-Euler Equation

A differential equation of the form $a_n x^n \frac{d^n y}{dx^n} + a_{n-1} x^{n-1} \frac{d^{n-1} y}{dx^{n-1}} + \cdots + a_1 x \frac{dy}{dx} + a_0 y = g(x)$

is called a **Cauchy-Euler equation**.

Equations of this type are solved by assuming solutions of the form $y = x^m$, substituting this solution into the differential equation, and solving for m.

□ Example:

Solve the Cauchy – Euler equation $x^2 y'' - 4xy' + 6y = 0$.

Problems of this type can be solved nicely with *Mathematica* as shown below. For convenience, the left-hand side of the equation is defined and named **exp**. After defining the solution **y[x]** as **x^m**, re-entering **exp** causes this form of the solution (**x^m**) to be substituted into the left-hand side of the equation. This results in an expression involving terms which can be simplified with **Simplify[exp]**. This simplified expression is called **exp2**.

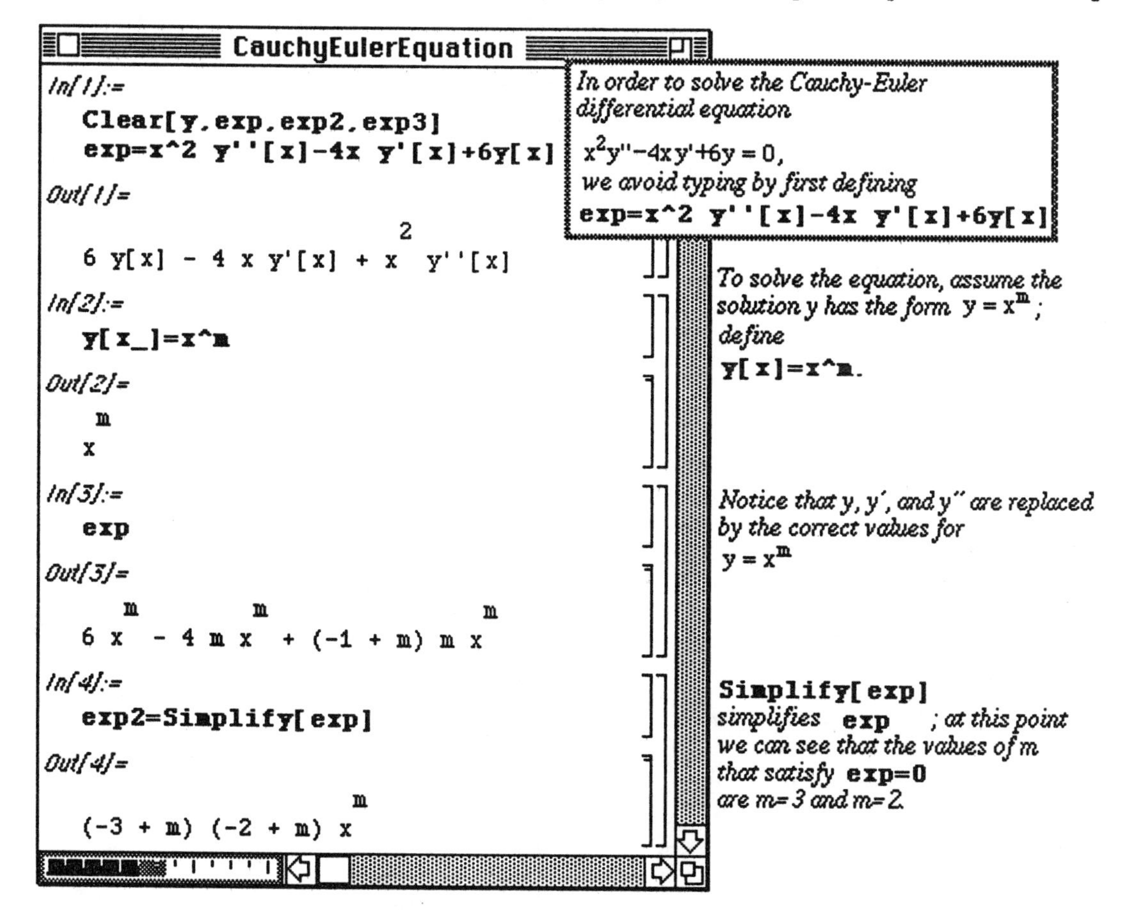

Division of **exp2** by **x^m** yields the expression **exp3** which depends only on **m**. Therefore, equating **exp3** to **0** and solving for **m**, yields the solution of the differential equation. Since these roots of are m = 3 and m = 2, the functions x^2 and x^3 are both solutions. Because these functions are linearly independent, the general solution to the differential equation is $y = c_1x^3 + c_2x^2$.

This general solution is defined as **y[x]** below. As was seen earlier, entering **exp** results in the substitution of **y[x]** into the left-hand side of the equation. Simplification reveals that **y[x]** is, in fact, the solution since a value of zero results.

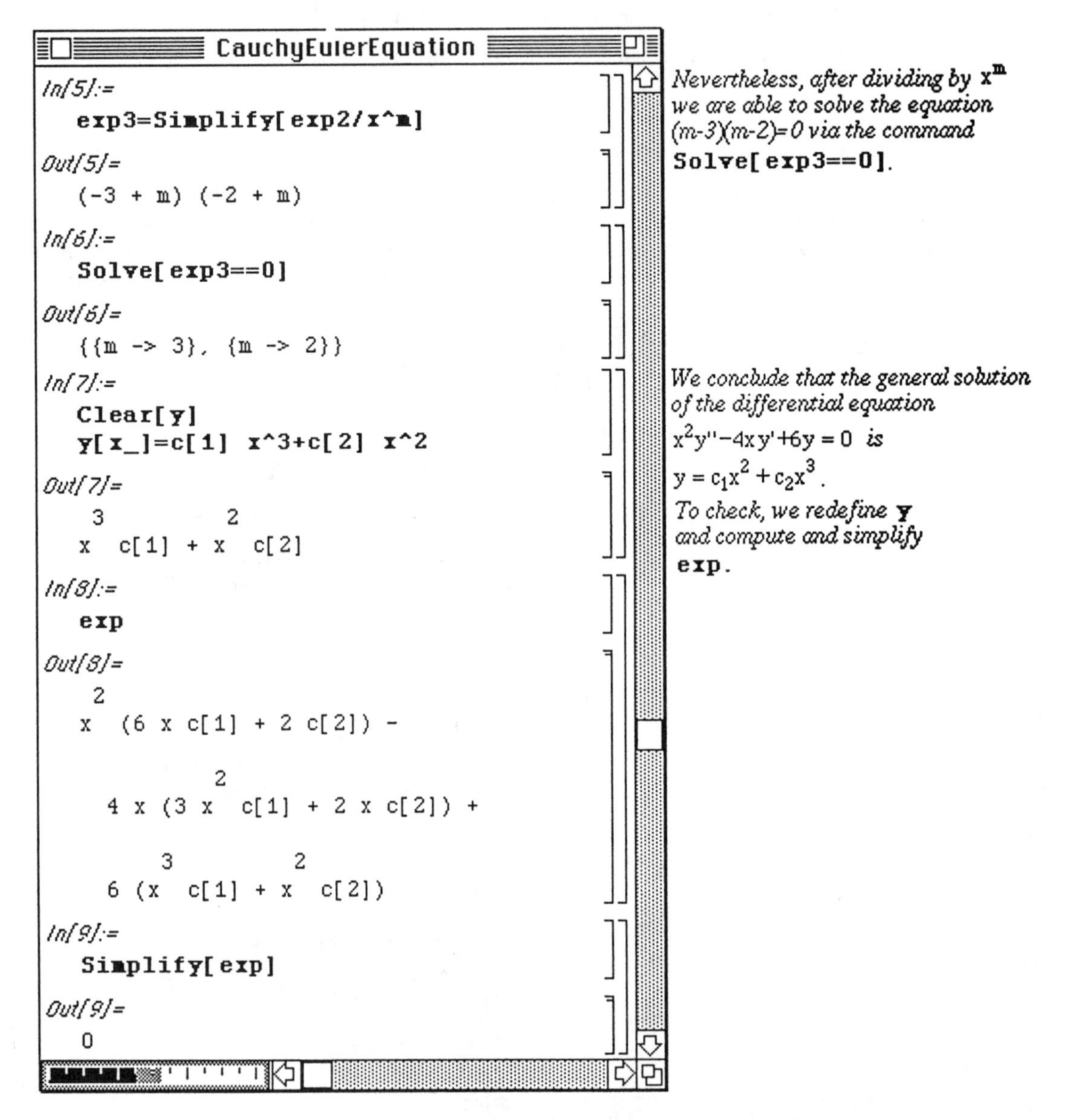

```
CauchyEulerEquation

In[5]:=
  exp3=Simplify[exp2/x^m]
Out[5]=
  (-3 + m) (-2 + m)
In[6]:=
  Solve[exp3==0]
Out[6]=
  {{m -> 3}, {m -> 2}}
In[7]:=
  Clear[y]
  y[x_]=c[1] x^3+c[2] x^2
Out[7]=
   3          2
  x  c[1] + x  c[2]
In[8]:=
  exp
Out[8]=
   2
  x  (6 x c[1] + 2 c[2]) -

               2
    4 x (3 x  c[1] + 2 x c[2]) +

          3          2
    6 (x  c[1] + x  c[2])
In[9]:=
  Simplify[exp]
Out[9]=
  0
```

Nevertheless, after dividing by x^m *we are able to solve the equation* $(m-3)(m-2)=0$ *via the command* **Solve[exp3==0]**.

We conclude that the general solution of the differential equation $x^2y''-4xy'+6y=0$ *is* $y = c_1x^2+c_2x^3$. *To check, we redefine* **y** *and compute and simplify* **exp**.

o Even though Version 1.2 is unable to solve the Cauchy-Euler equation with the command **DSolve**, Version 2.0 can solve Cauchy-Euler equations:

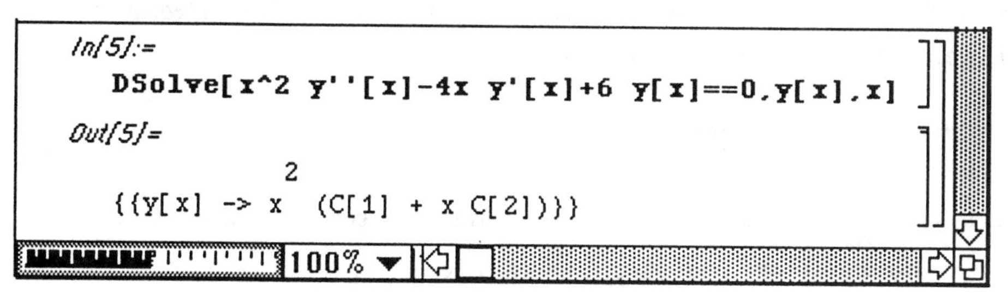

6.6 Variation of Parameters

Let y_1 and y_2 be a fundamental set of solutions on the interval I of the differential equation $y''+P(x)y'+Q(x)y=0$, where P and Q are continuous on the interval I.

The **Wronskian W** of y_1 and y_2 is $W=\det\begin{bmatrix} y_1 & y_2 \\ y_1' & y_2' \end{bmatrix}$.

Let $u_1=-\int \frac{y_2 f(x)}{W}dx$, $u_2=\int \frac{y_1 f(x)}{W}dx$, and $y_p=u_1y_1+u_2y_2$. Then the general solution of the differential equation $y''+P(x)y'+Q(x)y=f(x)$ is $y=y_c+y_p$ where $y_c=c_1y_1+c_2y_2$ is the general solution of the differential equation $y''+P(x)y'+Q(x)y=0$.

Because the method of variation of parameters depends on integration, *Mathematica* can be of great service in solving second-order linear nonhomogeneous ordinary differential equations by this method. Consider the following problem:

□ Example:

Solve $y''-\frac{4}{x}y'+\frac{6}{x^2}y=\frac{1}{x^4}$ by the method of variation of parameters

DSolve does not solve this equation, as shown below, since it is nonhomogeneous and involves variable coefficients. Once again, the problem must be solved in two parts. The homogeneous equation must be solved and a particular solution must be determined through variation of parameters.

The homogeneous equation is $y''-\frac{4}{x}y'+\frac{6}{x^2}y=0$. Multiplication by x^2 yields $x^2\,y''-4\,x\,y'+6\,y=0$.

This equation is the same as that solved the previous section on Cauchy-Euler equations. Hence, the solution to the homogeneous equation (the complimentary solution) has been found to be

$y_c(x) = c_1x^2 + c_2x^3$. This function is defined below as **yc**.

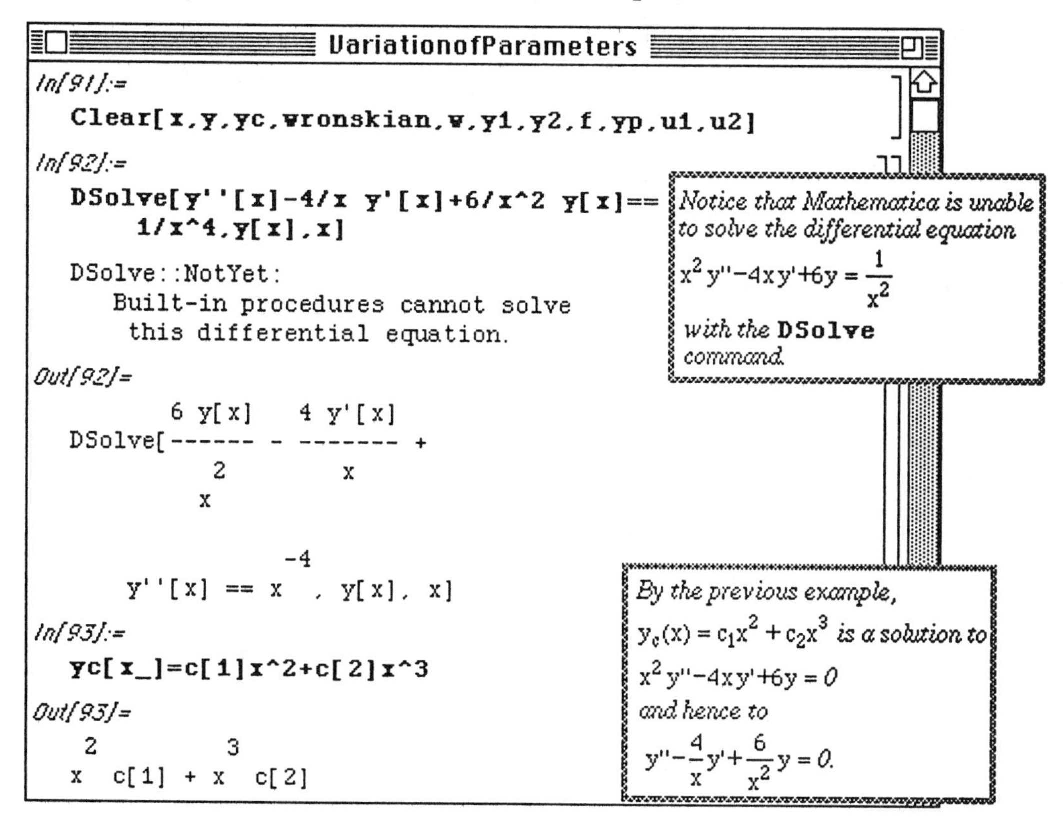

```
In[91]:=
  Clear[x,y,yc,wronskian,w,y1,y2,f,yp,u1,u2]
In[92]:=
  DSolve[y''[x]-4/x y'[x]+6/x^2 y[x]==
      1/x^4,y[x],x]

  DSolve::NotYet:
     Built-in procedures cannot solve
      this differential equation.
Out[92]=
            6 y[x]    4 y'[x]
  DSolve[------ - ------- +
              2         x
             x

                   -4
      y''[x] == x   , y[x], x]
In[93]:=
  yc[x_]=c[1]x^2+c[2]x^3
Out[93]=
   2          3
  x  c[1] + x  c[2]
```


The next step in the solution process is to find a particular solution to the nonhomogeneous problem. In order to do this by the method of variation of parameters, the Wronskian of the two solutions to the homogeneous equation must be determined. The Wronskian for two arbitrary functions **h** and **k** is defined below as a function **wronskian** of two variables, **h** and **k**. Since the solutions to the homogeneous equation are defined as **y1** and **y2**, the Wronskian is computed given by **wronskian[y1,y2]**. The Wronskian is a function of the independent variable **x** and is defined as **w[x]**.

Before attempting to evaluate the integrals involved in the formulas given earlier, Version 1.2 users must load the package **IntegralTables.m** as illustrated below.

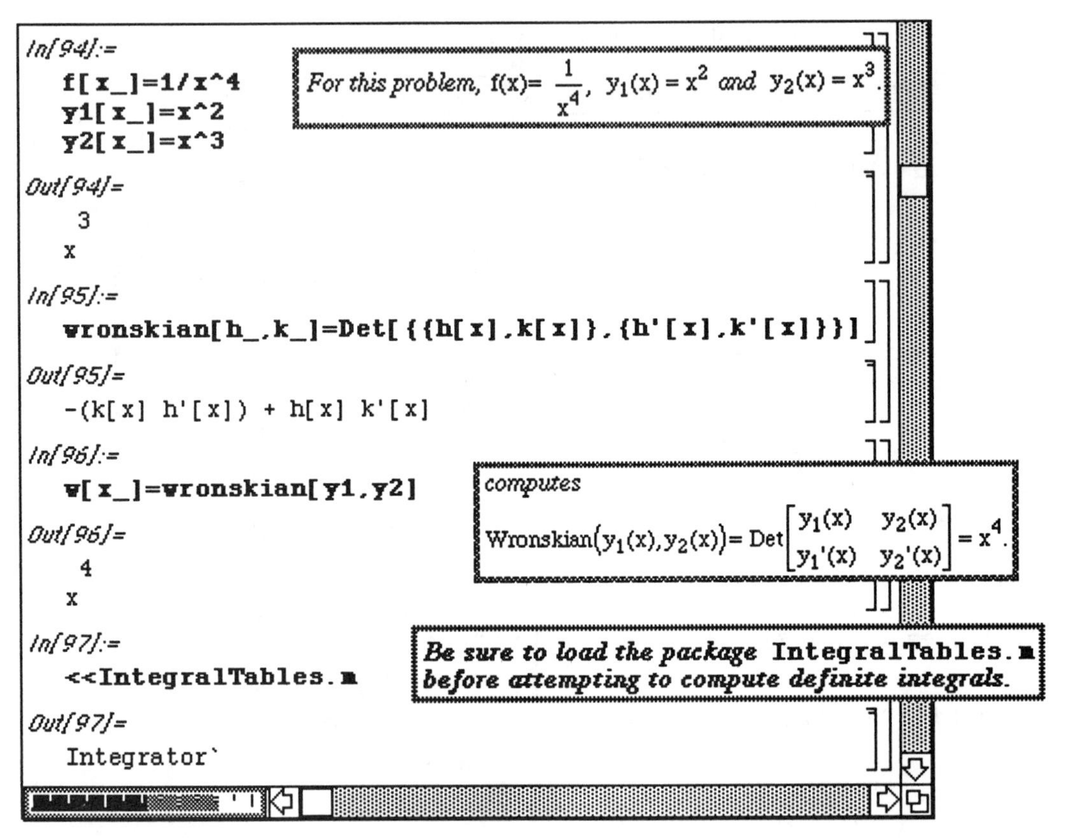

In order to find a particular solution of the form $y_p = u_1 y_1 + u_2 y_2$, the functions u_1 and u_2 must be calculated with the following integrals $u_1 = -\int \frac{y_2 f(x)}{W} dx$ and $u_2 = \int \frac{y_1 f(x)}{W} dx$.

The functions u_1 and u_2 are determined below with **Integrate**. The general solution to the nonhomogeneous problem is then defined as the sum of the complimentary solution found earlier, `yc[x]`, and the particular solution, `yp[x] = u1[x] y1[x]+ u2[x] y2[x]`. The general solution is defined as `y[x]`.

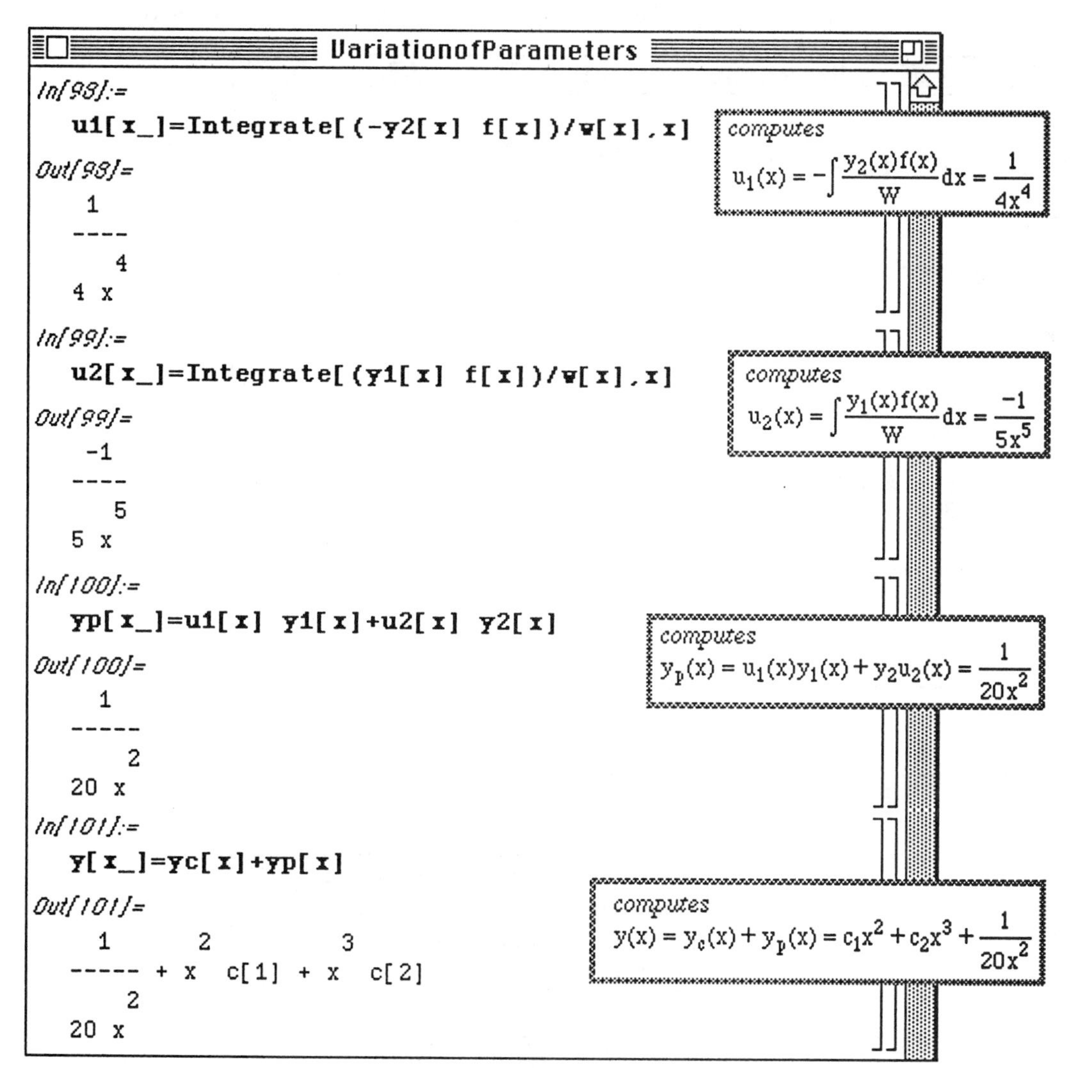

The general solution is verified by substitution into the left-hand side of the differential equation. If correct, the resulting expression is equivalent to x^{-4}. After simplification, the desired result is obtained:

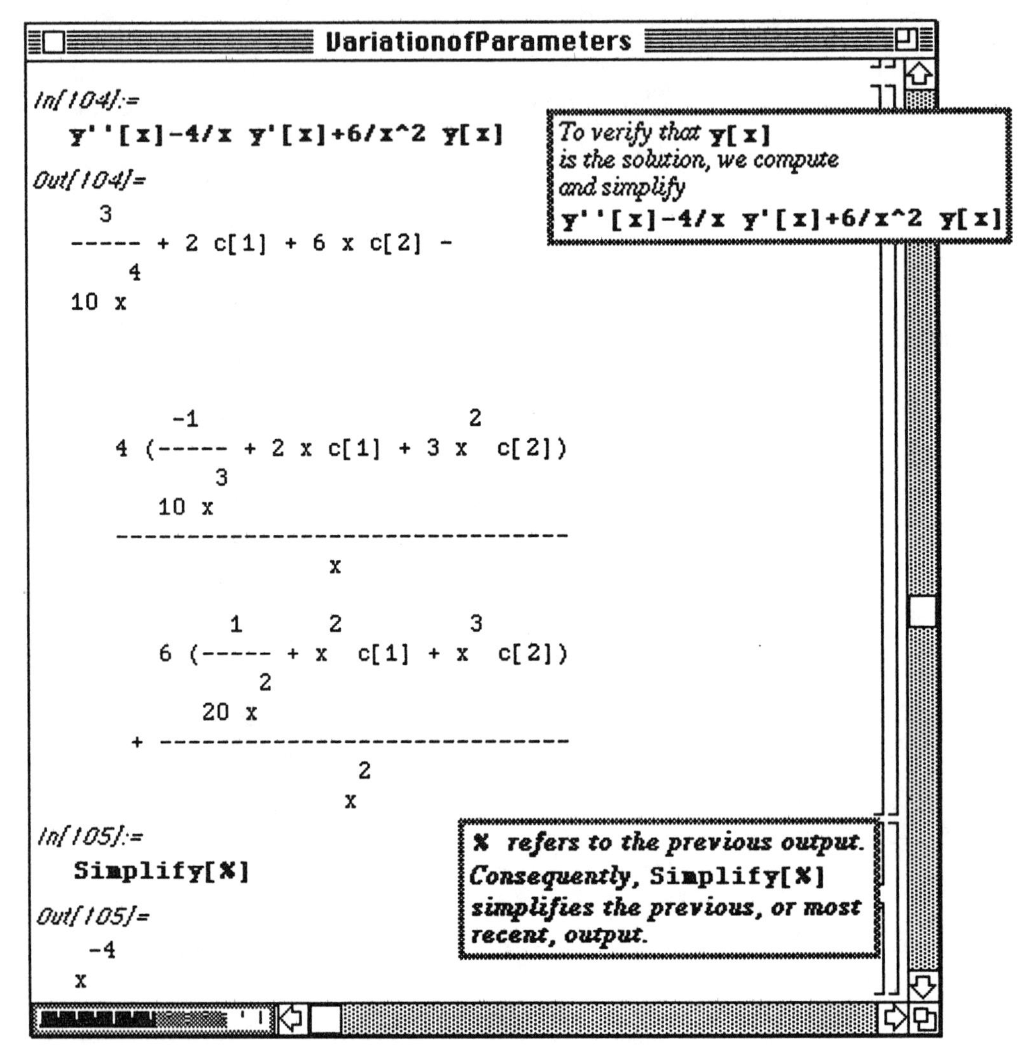

6.7 Capabilities of DSolve

Although the command **DSolve** is quite useful in Version 1.2, **DSolve** is unable to solve many standard differential equations:

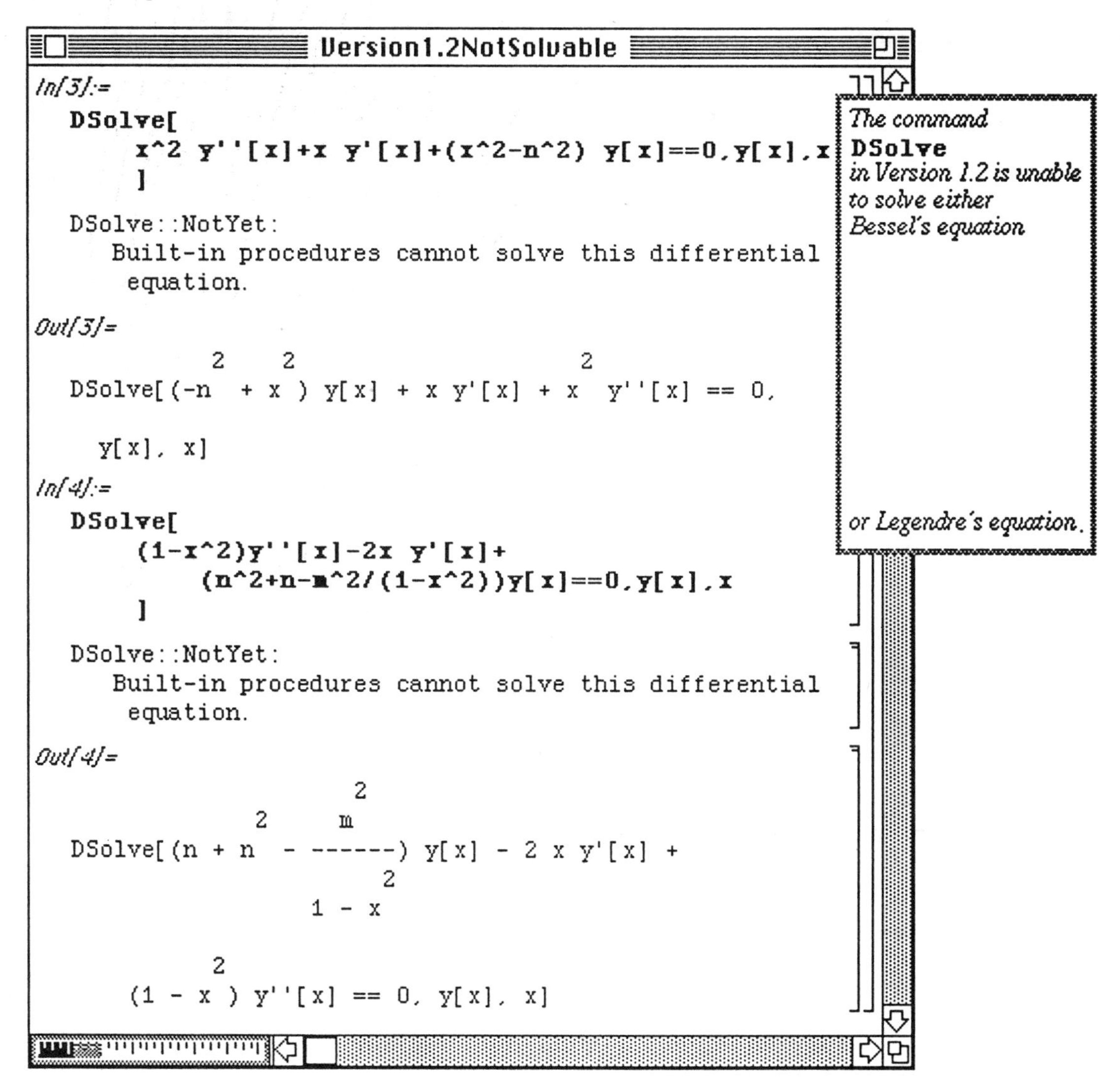

□ Example:

o In Version 2.0, the command **DSolve** has been improved. In fact, exact solutions can be computed for Bessel's equation, Legendre's equation, and Airy's equation:

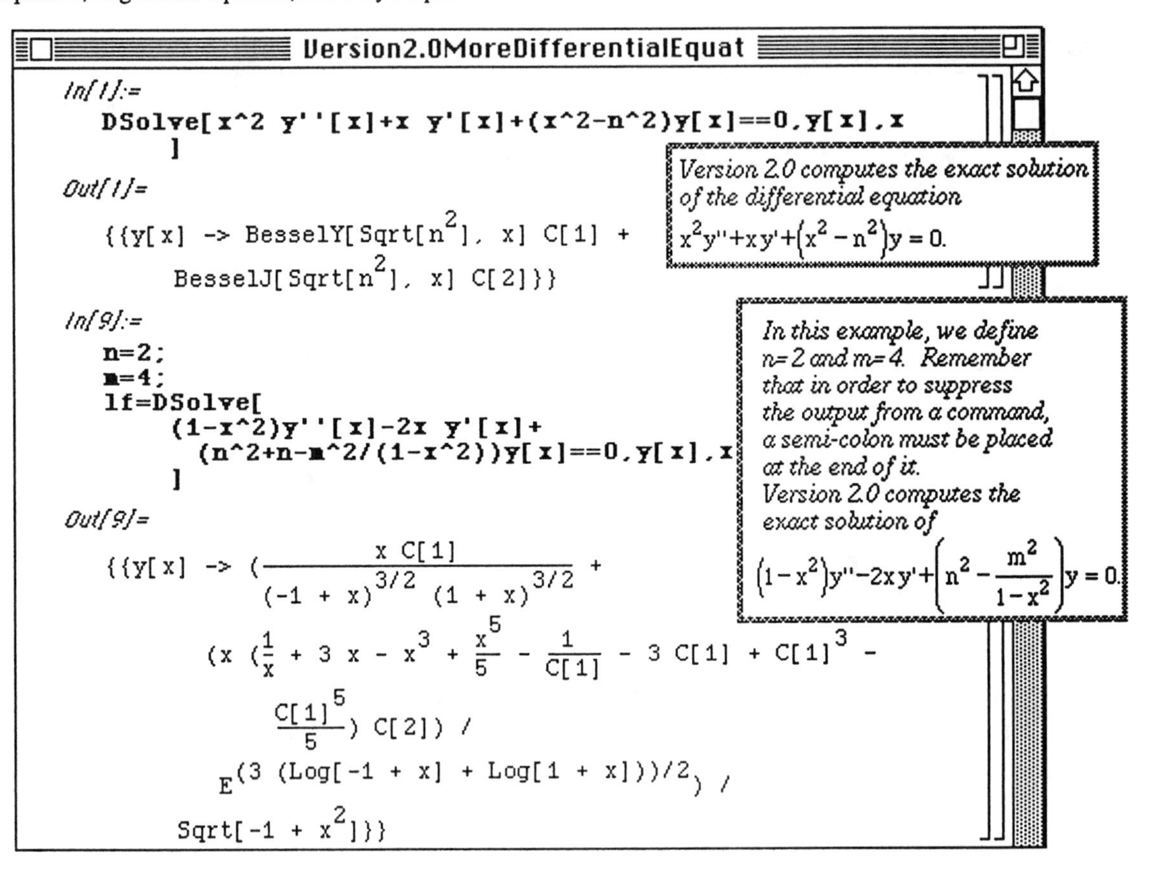

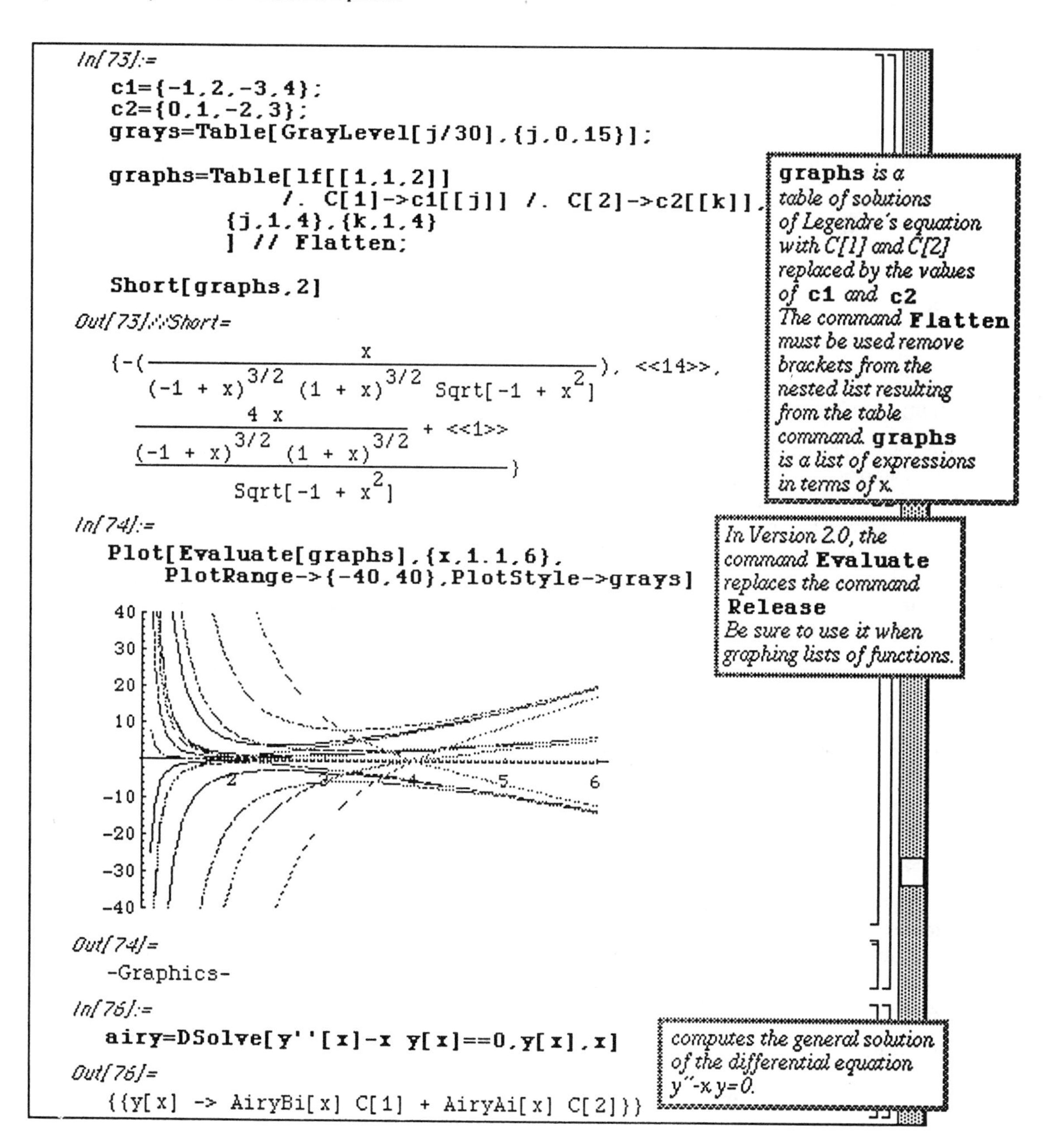

In[73]:=

```
c1={-1,2,-3,4};
c2={0,1,-2,3};
grays=Table[GrayLevel[j/30],{j,0,15}];

graphs=Table[lf[[1,1,2]]
        /. C[1]->c1[[j]] /. C[2]->c2[[k]],
    {j,1,4},{k,1,4}
    ] // Flatten;

Short[graphs,2]
```

Out[73]//Short=

```
{-(x/((-1 + x)^(3/2) (1 + x)^(3/2) Sqrt[-1 + x^2])), <<14>>,
   ((4 x)/((-1 + x)^(3/2) (1 + x)^(3/2)) + <<1>>)/Sqrt[-1 + x^2]}
```

graphs is a table of solutions of Legendre's equation with C[1] and C[2] replaced by the values of **c1** *and* **c2**. *The command* **Flatten** *must be used remove brackets from the nested list resulting from the table command.* **graphs** *is a list of expressions in terms of x.*

In[74]:=

```
Plot[Evaluate[graphs],{x,1.1,6},
    PlotRange->{-40,40},PlotStyle->grays]
```

In Version 2.0, the command **Evaluate** *replaces the command* **Release**. *Be sure to use it when graphing lists of functions.*

Out[74]=

```
-Graphics-
```

In[76]:=

```
airy=DSolve[y''[x]-x y[x]==0,y[x],x]
```

Out[76]=

```
{{y[x] -> AiryBi[x] C[1] + AiryAi[x] C[2]}}
```

computes the general solution of the differential equation y''-x y=0.

6.8 Systems of Linear Differential Equations

The general form of the first – order linear system of n dimensions is $x = A(t)\,x$ where $A(t) = [a_{i,j}(t)]$ is an $n \times n$ matrix with each $a_{i,j}$ a function of t and x(t) is a column vector of the n dependent variables The general form of the first – order linear system may be written as:

$$\begin{bmatrix} \dot{x}_1 \\ \dot{x}_2 \\ \vdots \\ \dot{x}_n \end{bmatrix} = \begin{bmatrix} a_{1,1}(t) & a_{1,2}(t) & \cdots & a_{1,n}(t) \\ a_{2,1}(t) & a_{2,2}(t) & \cdots & a_{2,n}(t) \\ \vdots & \vdots & \vdots & \vdots \\ a_{n,1}(t) & a_{n,2}(t) & \cdots & a_{n,n}(t) \end{bmatrix} \begin{bmatrix} x_1 \\ x_2 \\ \vdots \\ x_n \end{bmatrix}.$$

■ Application: Spring Problems

An application of second-order linear differential equations with constant coefficients is the differential equation of the vibrations of a mass on a spring. The problem to be solved is as follows :

A coil spring is suspended from a point on a rigid support such as a ceiling or beam. A mass is then attached to the spring and allowed to come to rest in an equilibrium position. The system is then set into motion in one of two manners: (1) the mass is pulled below (or pushed above) its equilibrium and released with a zero or nonzero initial velocity at t=0, or (2) the mass is forced out of its equilibrium position by giving it a nonzero (downward or upward) initial velocity at t=0. The problem is to determine the motion of the mass on the spring which results.

By making use of Newton's second law and Hooke's Law, and by determining the forces acting upon the mass, the differential equation for this problem is found to be $m\frac{d^2x}{dt^2} + a\frac{dx}{dt} + kx = F(t)$.

where m = the mass attached to the spring, a = the damping constant,
k = the spring constant determined with Hooke's Law, and
F(t) = the function which describes any external force acting on the spring.

□ Compare the Effects of Damping

An interesting problem to consider is that of investigating the effect that different values of the damping constant, a, have on the resulting motion of the mass on the spring. Consider the following problem :

A 3-kilogram mass is attached to a spring having spring constant, k = 12. Determine the equation of the motion which results if the motion starts at x(0) = 3 with zero initial velocity if the damping constant is (i) a = 6 and (ii) a = 12. Plot the solutions obtained.

The differential equation with initial conditions for (i) is $3x'' + 6x' + 12x = 0$, $x(0) = 3$, $x'(0) = 0$.

This initial value problem is solved below using **DSolve** . The solution is called **sol** for later use. Notice that **sol** is a list of one element which is a list made up of two parts. Therefore, **sol[[1,1,2]]** is used to extract the solution from **sol** . The name **eq1** is given to this expression for convenience when plotting.

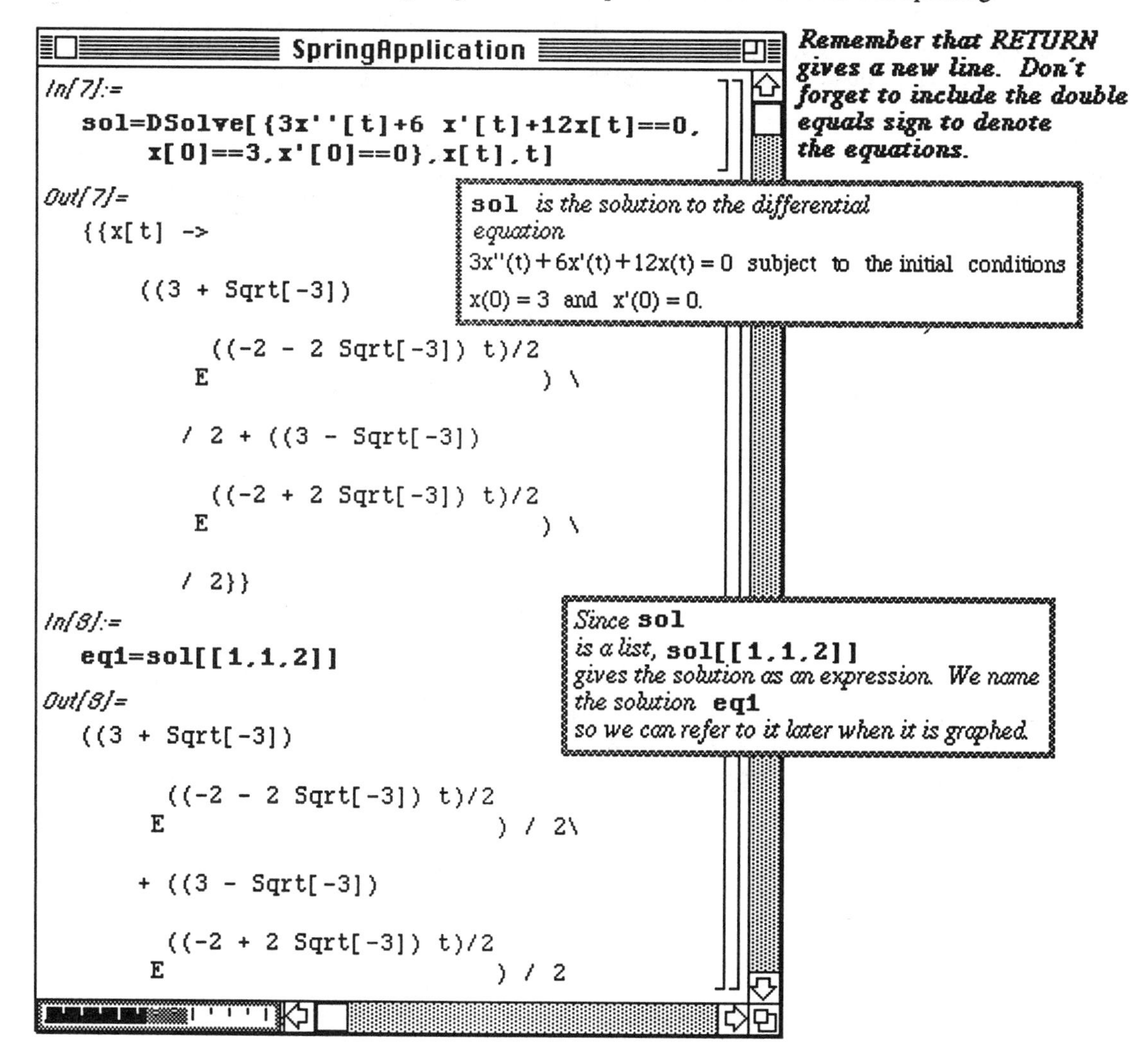

The differential equation for (ii) is similar to that of (i) with the exception that the damping constant, a = 12. Hence the coefficient of x' is 12 in this case instead of 6. The initial conditions are the same as those used in the calculations for (i). After the solution of (ii) is obtained in an identical manner as (i), the solutions to (i) and (ii) are plotted simultaneously. Note that the graph of the solution with a = 6 is the darker of the two curves since it has a **GrayLevel[0]** as compared to **GrayLevel[0.3]** for the problem with a = 12. The graph illustrates that the spring approaches its equilibrium more quickly when the damping constant is increased.

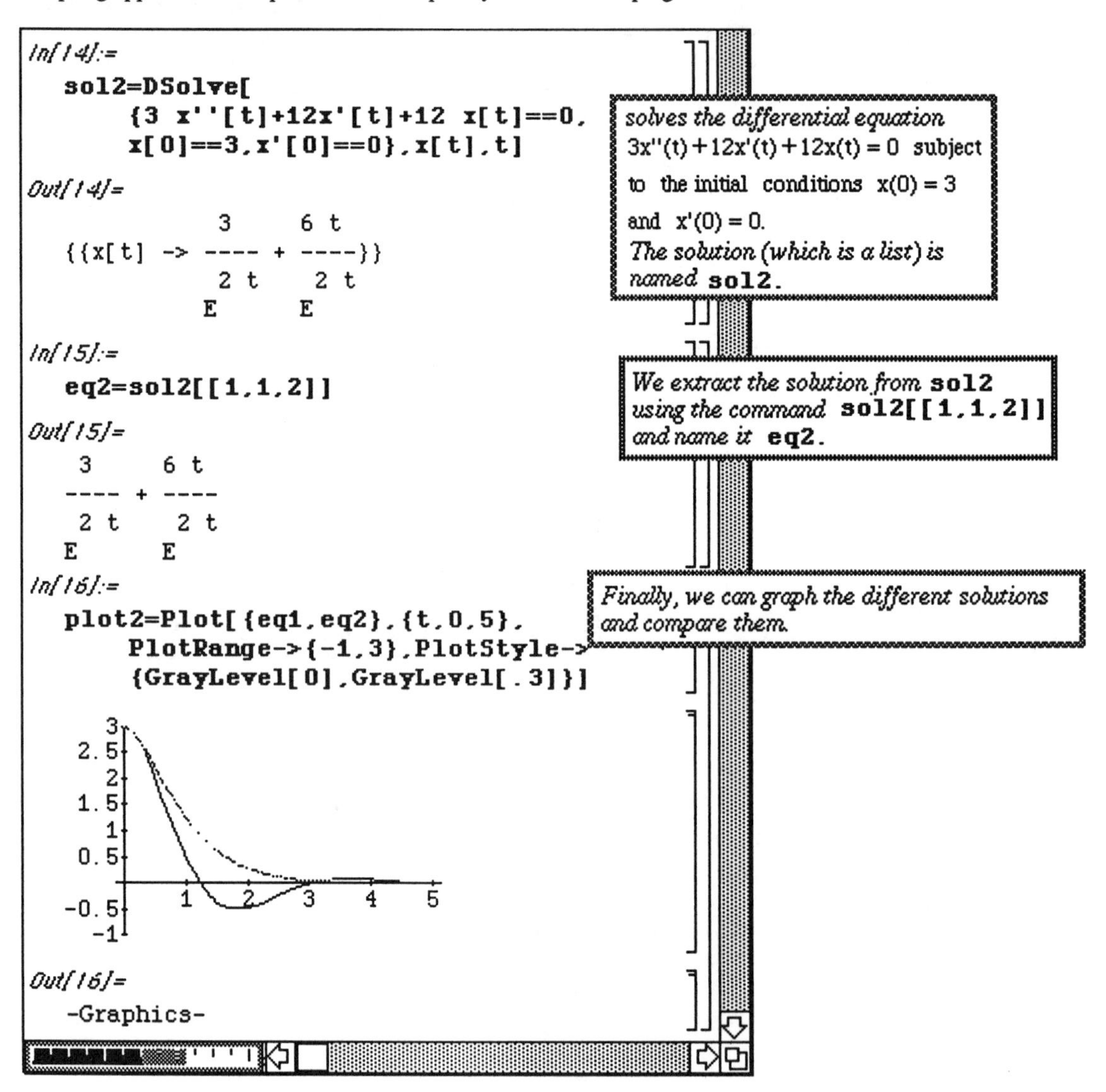

□ Compare Different Initial Velocities

The effect that a change in initial velocity has on the subsequent motion is yet another problem of interest. Again, consider the problem stated above for the damping problem with a = 6. In this case, however, consider the problem in which the damping and initial position, x(0)= 3, remain unchanged while the initial velocity is varied. First, determine the motion when the initial velocity is x'(0) = 2 and then again when x'(0) = -2. Plot the solutions to compare the effects that a change in initial velocity has on the motion of the mass on the spring.

The method of solution is similar to that of the previous problem. The initial value problem which must be solved when x'(0) = 2 is as follows :

$$3x'' + 6x' + 12x = 0, \quad x(0) = 3, \quad x'(0) = 2.$$

This problem is solved below with **DSolve** , and the resulting expression is named **soln1** for later use. As in the previous example, the equation of the solution is extracted from **soln1** with **soln1[[1,1,2]]**. This equation is assigned the name **eq1** .

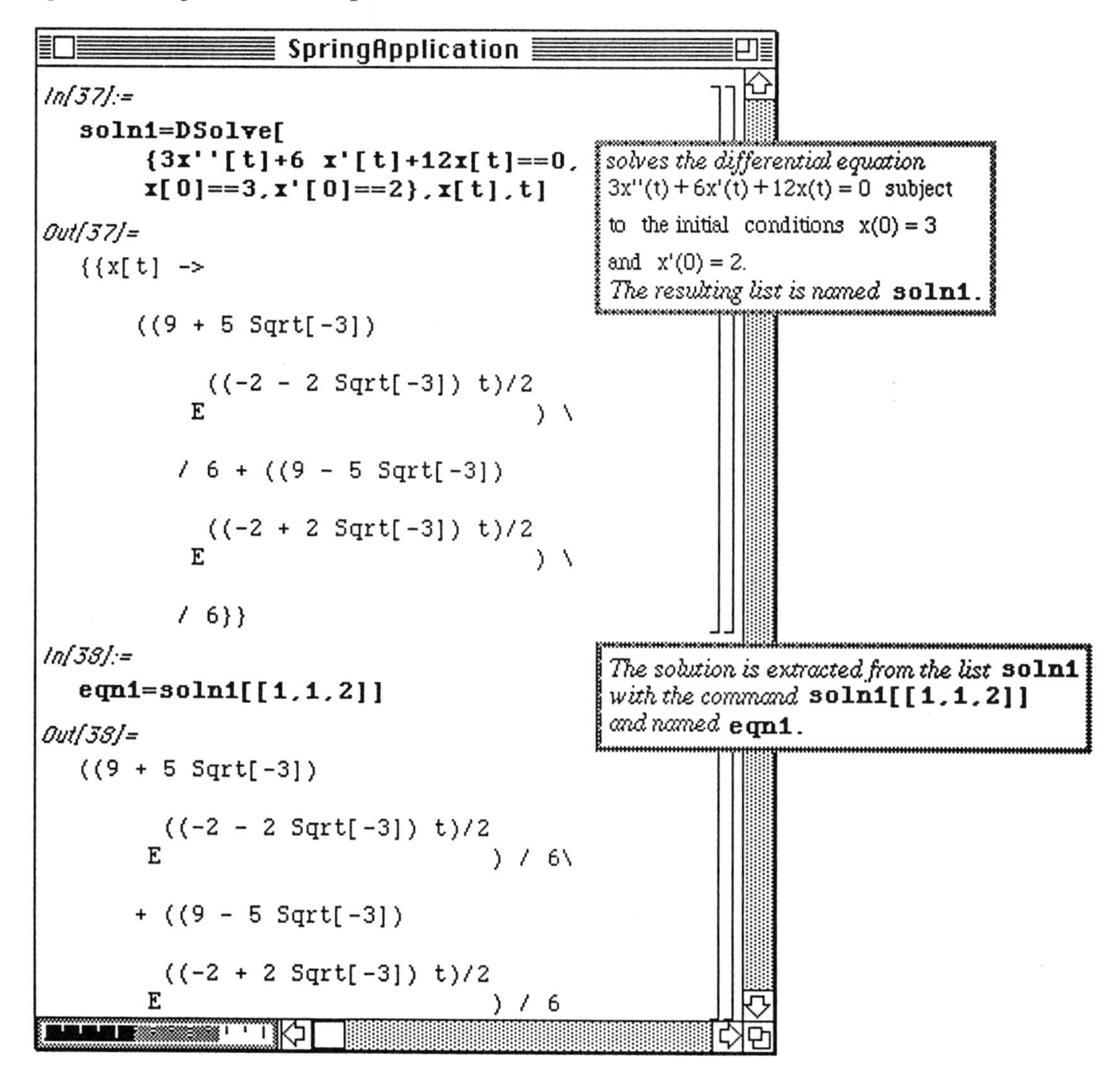

The solution to the initial value problem in which x'(0) = -2 is found in a similar manner. The value of the initial velocity is simply changed and the same commands executed.

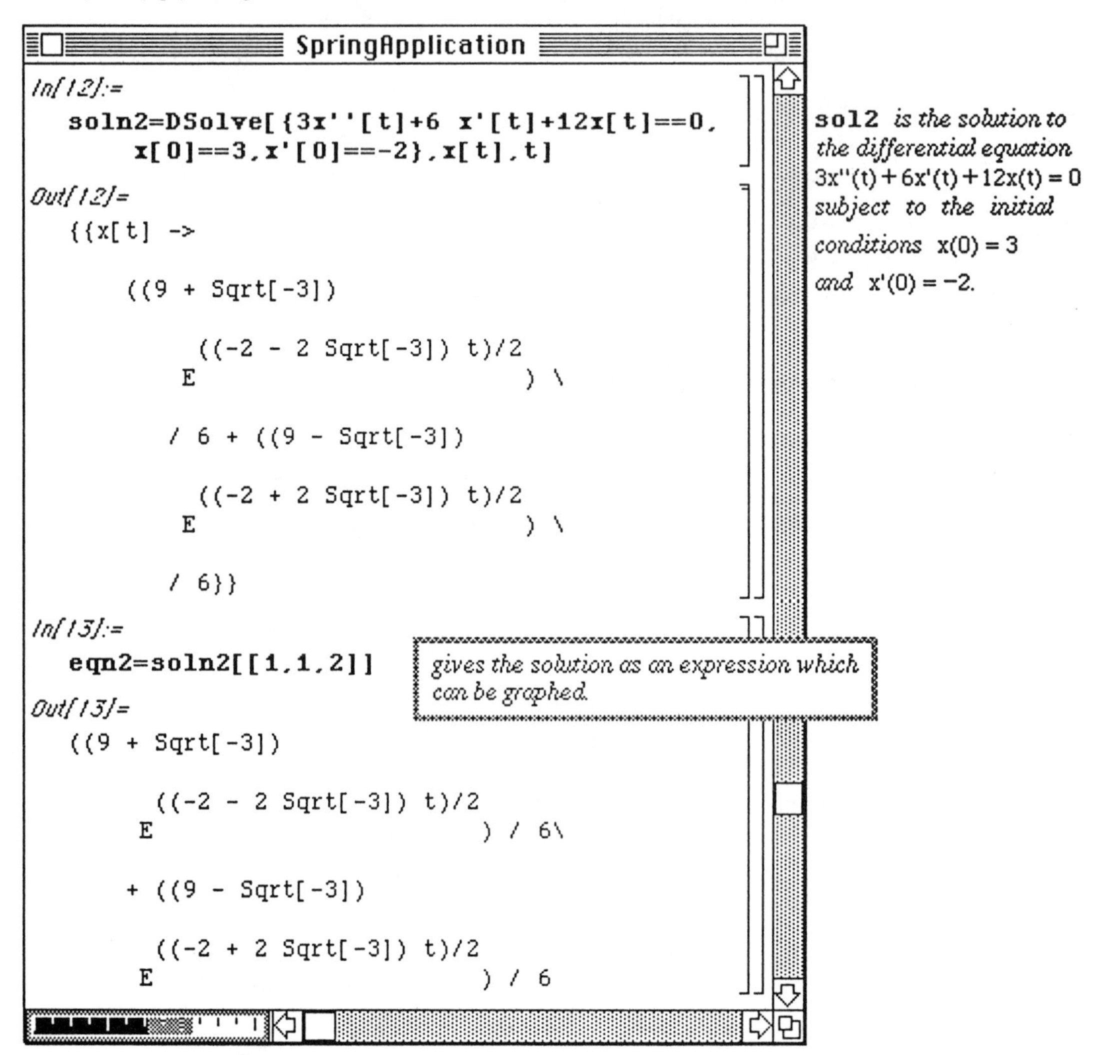

When both problems have been solved, the solutions can be compared by observing their graphs. These solutions are shown below with the solution to the problem in which x'(0) = 0 (which was solved in (i) in the previous example). Note how a positive initial velocity (x'(0) = 2) affects the motion as compared to an initial velocity in the negative direction. (Recall, **eq1** is the solution to the problem with zero initial velocity. It is the darkest of the three curves below.) Note also that the change in initial velocity eventually has little effect on the motion of the mass. (i.e., the three curves appear to overlap for larger values of t.)

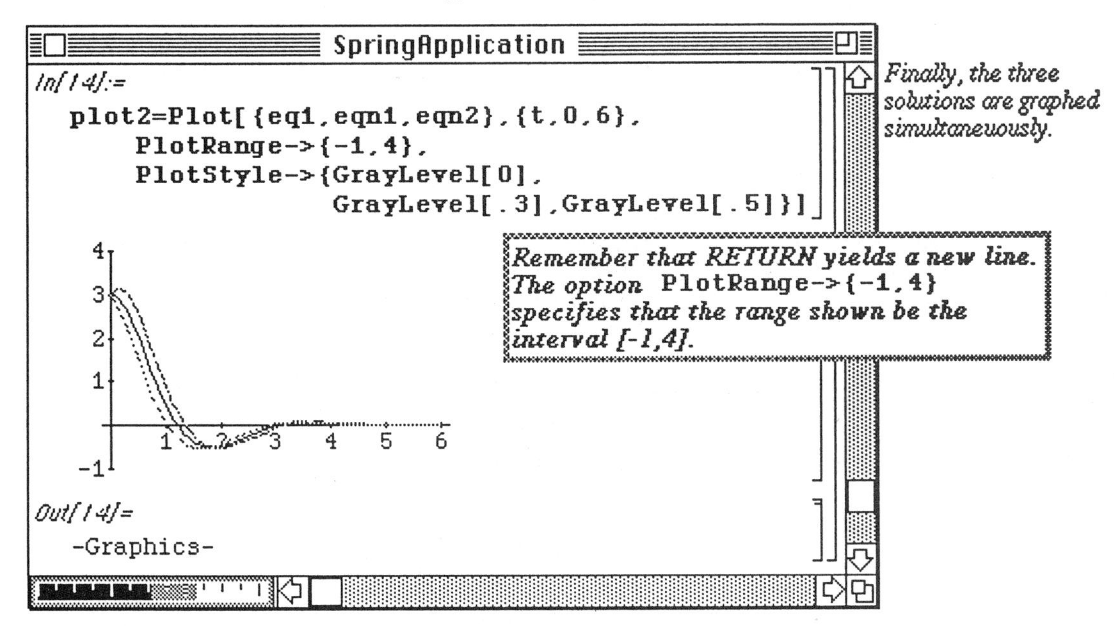

■ **Application:** Classification of Equilibrium Points

The **equilibrium points** of the general first-order system

$\dot{x} = X(x,y),\ \dot{y} = Y(x,y)$ are the points where $X(x,y) = 0$ and $Y(x,y) = 0$. If the system

$\dot{x} = X(x,y),\ \dot{y} = Y(x,y)$ has an equilibrium point (x_0, y_0) the

linear approximation to the system in the neighborhood of the equilibrium point is defined as the system

$\dot{x} = ax + by,\ \dot{y} = cx + dy$, where $a = \frac{\partial X}{\partial x}(x_0, y_0)$, $b = \frac{\partial X}{\partial y}(x_0, y_0)$, $c = \frac{\partial Y}{\partial x}(x_0, y_0)$, and $d = \frac{\partial Y}{\partial y}(x_0, y_0)$.

The eigenvalues of the system $\dot{x} = ax + by,\ \dot{y} = cx + dy$ are the

eigenvalues of the matrix $\begin{bmatrix} a & b \\ c & d \end{bmatrix}$. If the eigenvalues of the

system are complex, the phase diagram in a neighborhood of the equilibrium point is an unstable spiral. If the eigenvalues are real, distinct, and have the same sign, the equilibrium point is a node; if the eigenvalues are real, distinct, and have opposite signs, the equilibrium point is a saddle. If the eigenvalues are the same, no conclusion can be drawn.

□ **Example:**

Classify the equilibrium points of the system $\dot{x} = -x - 5y,\ \dot{y} = x + 3y$.

First, the critical points of the system must be located. This is accomplished by solving the linear system

$-x - 5y = 0$

$x + 3y = 0$

for x and y using **Solve[-x-5y==0,x+3y==0},{x,y}]**. Once the critical point (0,0) is found, it can be

classified by investigating the eigenvalues of the matrix of coefficients $\begin{bmatrix} -1 & -5 \\ 1 & 3 \end{bmatrix}$.

Recall that a matrix is represented as a list with *Mathematica* where each element of the list is a row of the matrix. Hence, the matrix of coefficients, **matrixxy**, is **{{-1,-5},{1,3}}**. The eigenvalues are easily determined with **Eigenvalues[matrixxy]**. The eigenvalues are complex conjugates with positive real part. Therefore, (0,0) is classified as an unstable spiral.

EquilibriumPoints

```
In[1]:=
    Solve[{-x-5y==0,x+3y==0},{x,y}]
Out[1]=
    {{x -> 0, y -> 0}}
In[2]:=
    matrixxy={{-1,-5},{1,3}}
Out[2]=
    {{-1, -5}, {1, 3}}
In[3]:=
    Eigenvalues[matrixxy]
Out[3]=
    {1 + I, 1 - I}
```

To locate the equilibrium points, we solve the system $\begin{cases} -x-5y=0 \\ x+3y=0 \end{cases}$.

Thus the equilibrium point is (0,0).

The eigenvalues of the system $\begin{cases} \dot{x}=-x-5y \\ \dot{y}=x+3y \end{cases}$ *are the eigenvalues of the matrix* $\begin{bmatrix} -1 & -5 \\ 1 & 3 \end{bmatrix}$.

Thus, we first define **matrixxy** *and then we use the command* **Eigenvalues[matrixxy]** *to compute the eigenvalues.*

Notice that we may solve the system using the command **DSolve**:
Each equation is entered separately. Remember that since x and y both depend on the variable t, they must be defined as functions of t in the two equations using square brackets. For convenience, the equations are assigned the names, **eq1** and **eq2**, respectively. Hence, the command
DSolve[{eq1,eq2},{x[t],y[t]},t]
solves the system for **x[t]** and **y[t]**. The resulting expression is named **complexsol** for later use. By entering the command **Short[complexsol,4]** in the same input cell, only a portion of the solution (four lines) is displayed on the screen.

In order to simplify this solution which involves complex numbers, some of the *Mathematica* commands found in the package **Trigonometry.m** must be used. This package is located in the **Algebra** folder under **Packages** and must be located by the user when loading.

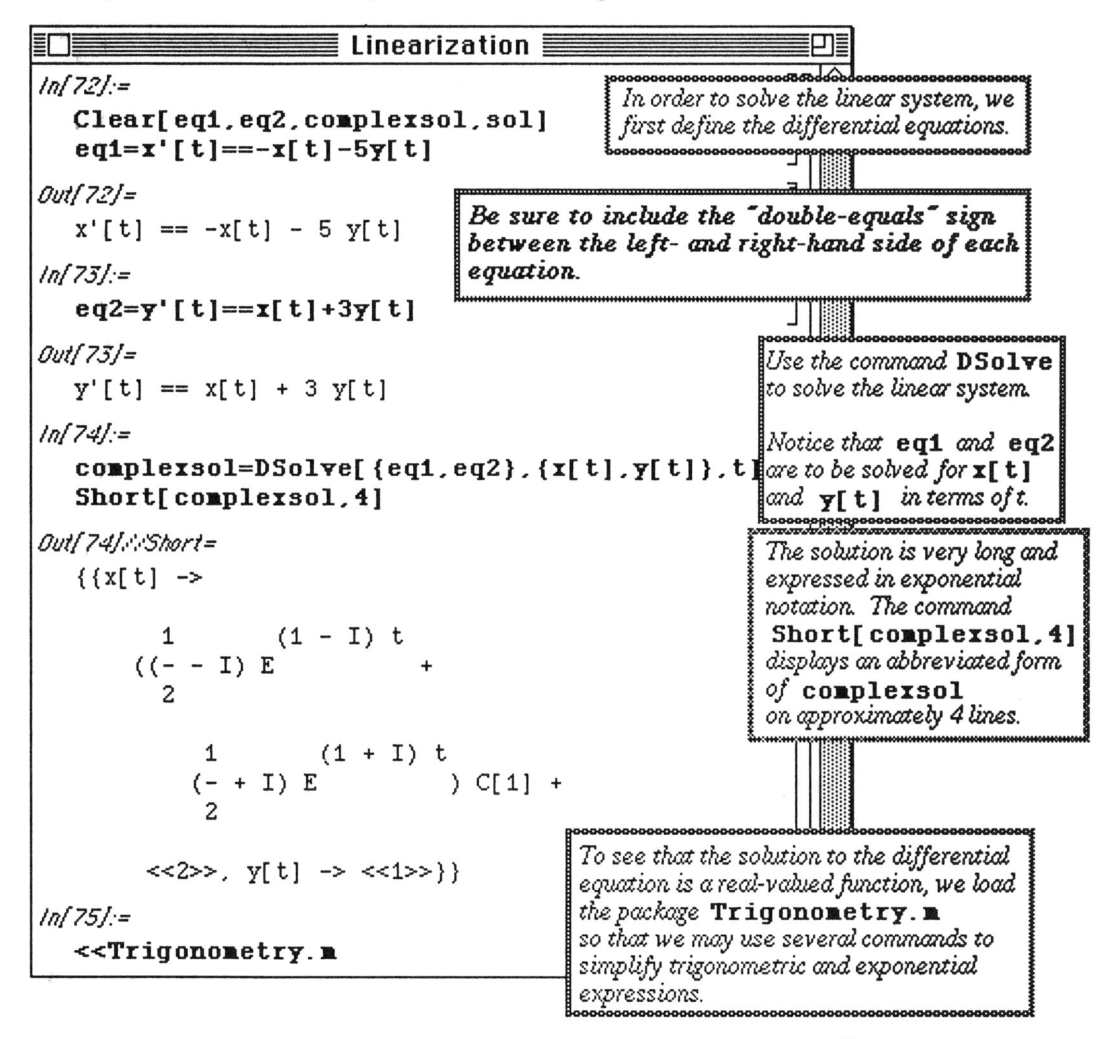

The command **ComplexToTrig [complexsol]** uses Euler's formula $e^{i\theta} = \cos\theta + i\sin\theta$ to eliminate complex exponents from the solution. Again, a shortened form of the result is requested with **Short[solution,5]**. The solution is simplified further with **Simplify[solution]**. This yields a real-valued solution. Recall that if the complex function $\begin{bmatrix} f_1(t) + i\,g_1(t) \\ f_2(t) + i\,g_2(t) \end{bmatrix}$ is a solution of a linear system of first-order differential equations,

the real part $\begin{bmatrix} f_1(t) \\ f_2(t) \end{bmatrix}$ and the imaginary part $\begin{bmatrix} g_1(t) \\ g_2(t) \end{bmatrix}$ are linearly independent solutions of the system. Therefore, the general solution is $\begin{bmatrix} x \\ y \end{bmatrix} = C[1]\begin{bmatrix} f_1(t) \\ f_2(t) \end{bmatrix} + C[2]\begin{bmatrix} g_1(t) \\ g_2(t) \end{bmatrix}$.

Thus, **Simplify[solution]** collects the real and imaginary parts of **solution** and expresses the general solution in the appropriate manner.

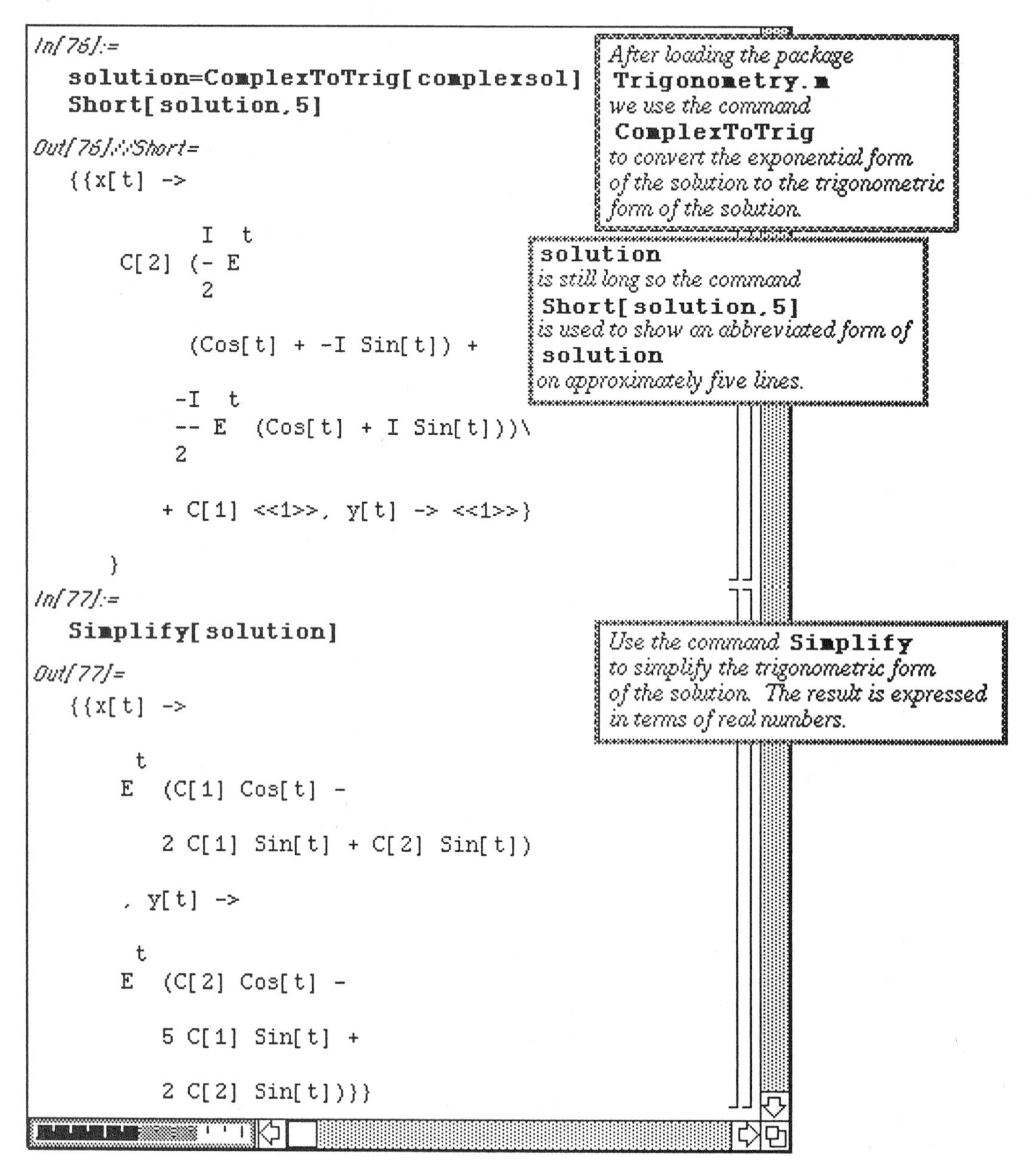

Various values of C[1] and C[2] may be substituted into the solution and then graphed to verify that the equilibrium point is an unstable spiral.

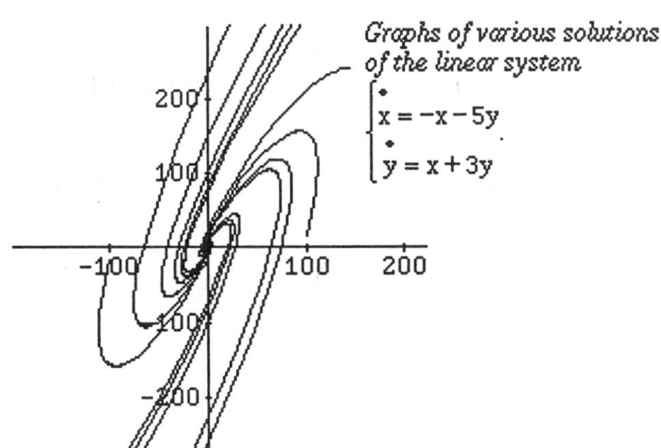

Example:

Locate and classify the equilibrium points of the (non – linear) system $\begin{cases}\dot{x} = 2x - y^3 \\ \dot{y} = 2 - 3xy\end{cases}$.

First, **Solve** is used to find the equilibrium points of the system by setting each equation equal to zero and solving for x and y. Remember, equilibrium points are real solutions to the system, so complex values are disregarded. Since the solution set of this system was named **eqpts**, the two real points (the first and third elements in **eqpts**) can be extracted from the list with **eqpts[[1]]** and **eqpts[[3]**. After locating the

two equilibrium points, the nonlinear system $\dot{x} = X(x,y)$, $\dot{y} = Y(x,y)$ must be linearized about these points using the Jacobian matrix $\begin{bmatrix}\dfrac{\partial X(x,y)}{\partial x} & \dfrac{\partial X(x,y)}{\partial y} \\ \dfrac{\partial Y(x,y)}{\partial x} & \dfrac{\partial Y(x,y)}{\partial y}\end{bmatrix}$.

The easiest way to obtain this matrix is to assign names to the functions X(x,y) and Y(x,y). In this problem, $X(x,y) = 2x - y^3$ and $Y(x,y) = 2 - 3xy$ so the assignment of

the names **xt** and **yt** to these functions, respectively, allows for the Jacobian to be determined with **{{D[xt,x],D[xt,y]},{D[yt,x],D[yt,y]}**.
This matrix is named **matrix** for later use and displayed in **MatrixForm** to verify that the desired matrix has been obtained.

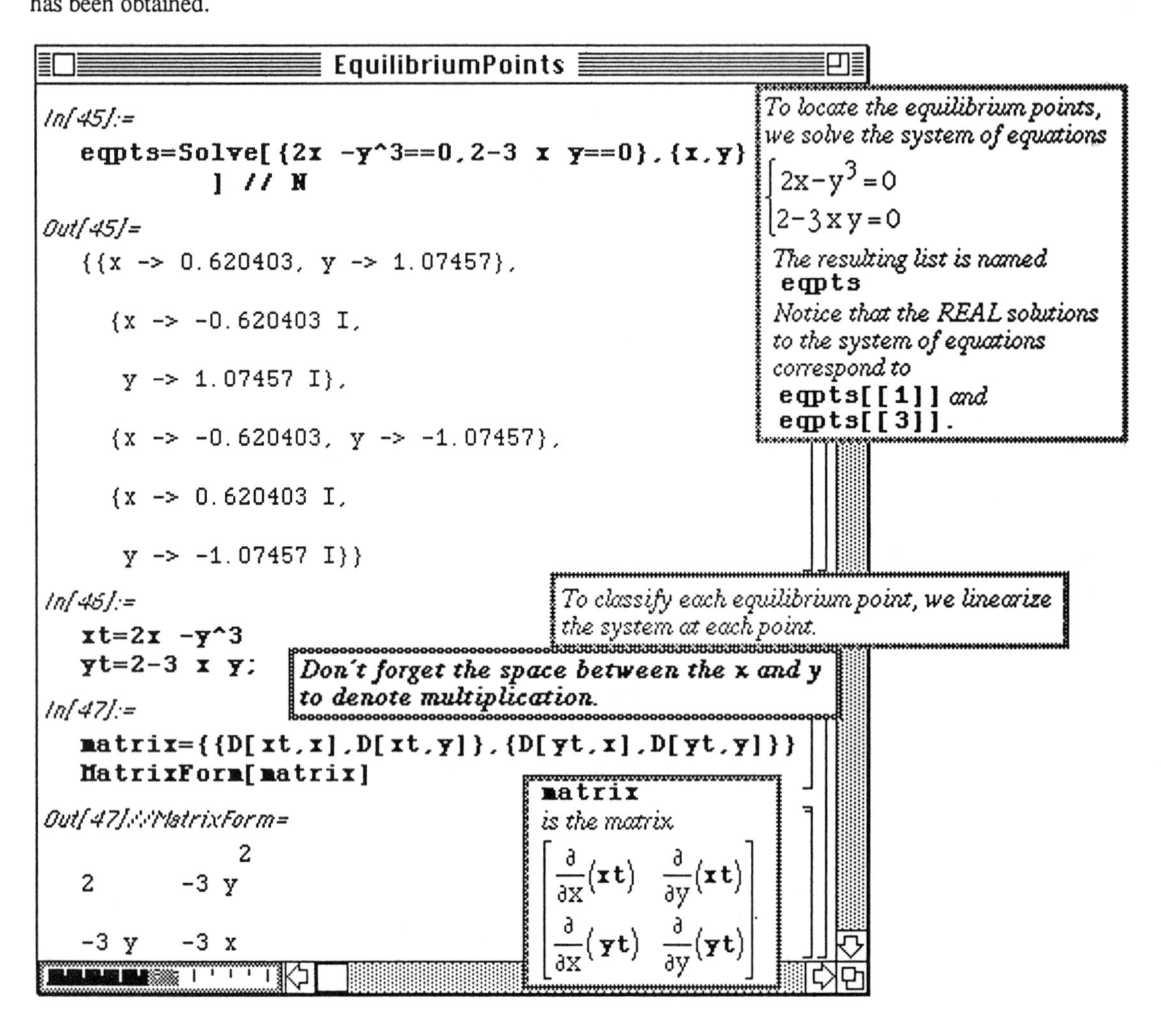

In order to linearize the nonlinear system about each equilibrium point, **matrix** must be evaluated at each point. Then, the eigenvalues for each system can be found with **Eigenvalues** to classify these equilibrium points. The matrix of coefficients for the linearized system about the first equilibrium point is easily found with **matrix/.eqpts[[1]]**.
These values are real and have opposite signs. Hence, this point is a saddle.
The linearized system about the second equilibrium point is found similarly. The eigenvalues for this system are complex with positive real part. Therefore, the second equilibrium point is classified as an unstable node.

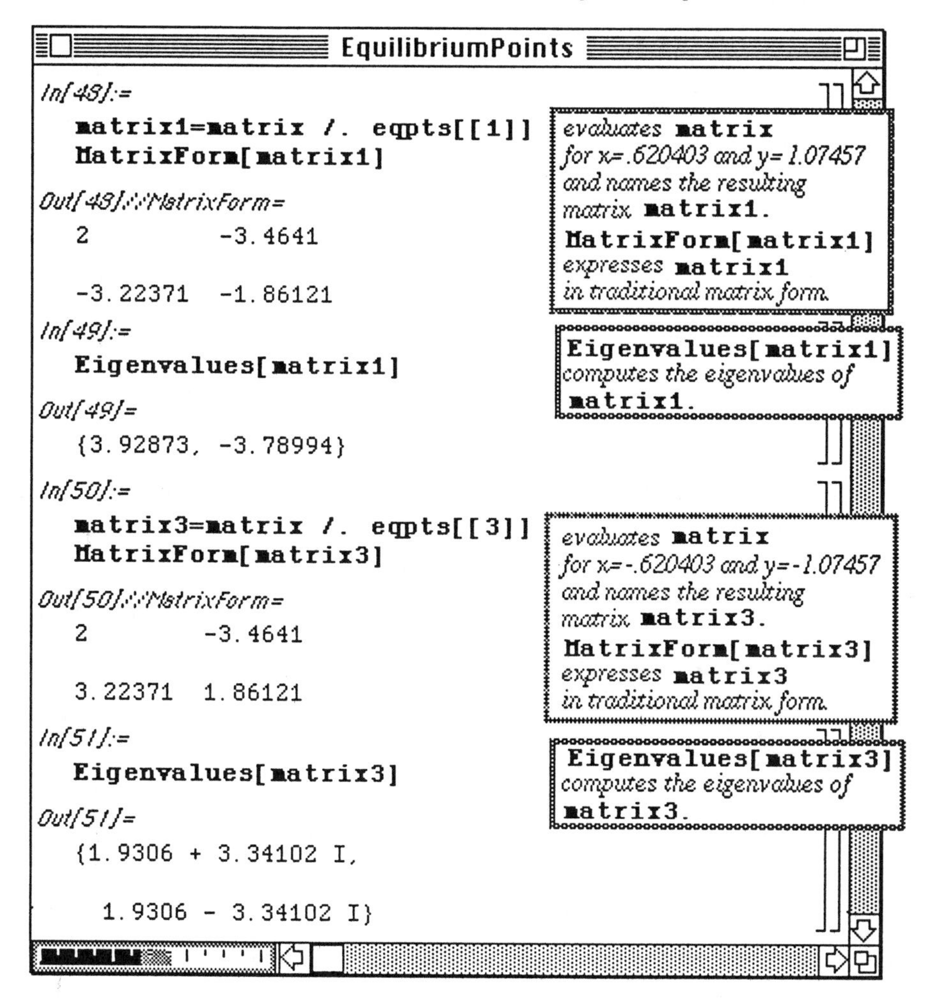

6.9 Series Solutions to Ordinary Differential Equations

As was demonstrated in **Chapter 3**, *Mathematica* is very useful in the computation of power series solutions to ordinary differential equations.

Recall that if $x = x_0$ is an ordinary point of a differential equation, then a power series solution of the form $\sum_{n=0}^{\infty} c_n (x-x_0)^n$ is assured.

Hence, the solution is obtained by finding the coefficient of each term in the series. Several examples are given below which illustrate how *Mathematica* is used to determine these values.

When using *Mathematica*'s **Series** command, the solution is assumed to have the form

$$y = \sum_{n=0}^{\infty} \frac{y^{(n)}(x_0)}{n!}(x-x_0)^n.$$

Example:

Use power series to compute an approximation of the solution to the initial value problem

$$\left(x^3+1\right)y''+3\,\mathrm{Sin}(x)y'=0, \quad y(0)=1 \text{ and } y'(0)=2.$$

Mathematica is not able to solve this differential equation with **DSolve**, so an alternate approach must be taken. The problem is solved below using power series. When solving problems of this type by hand, the first step is to compute the appropriate derivatives of the assumed power series solution and substitute the solution and its derivatives into the differential equation. Then, the coefficients are determined by collecting like powers of x. This usually involves changing the indices in one or more terms in the equation. To better understand the method of solution given here, a brief reminder of *Mathematica*'s **Series** command is given:
Series[f[x],{x,x0,n}] computes the Taylor series expansion of the function **f** about the point **x=x0** of order at most **n**. For example,
Series[y[x],{x,0,5}] yields the first 5 terms of the Taylor series expansion about $x = 0$:

$$y[0] + y'[0]\,x + \frac{y''[0]\,x^2}{2} + \frac{y^{(3)}[0]\,x^3}{6} + \frac{y^{(4)}[0]\,x^4}{24} + \frac{y^{(5)}[0]\,x^5}{120} + O[x]^6.$$

Therefore, the command
Series[(x^3+1)y''[x]+3Sin[x] y'[x],{x,0,5}]
accomplishes many steps at once. It computes the series expansions of **y[x]** and **Sin[x]**, performs the necessary multiplication, and collects like terms to compute the series expansion which results from substitution into the left-hand side of the differential equation. This series is called **serapprox**.

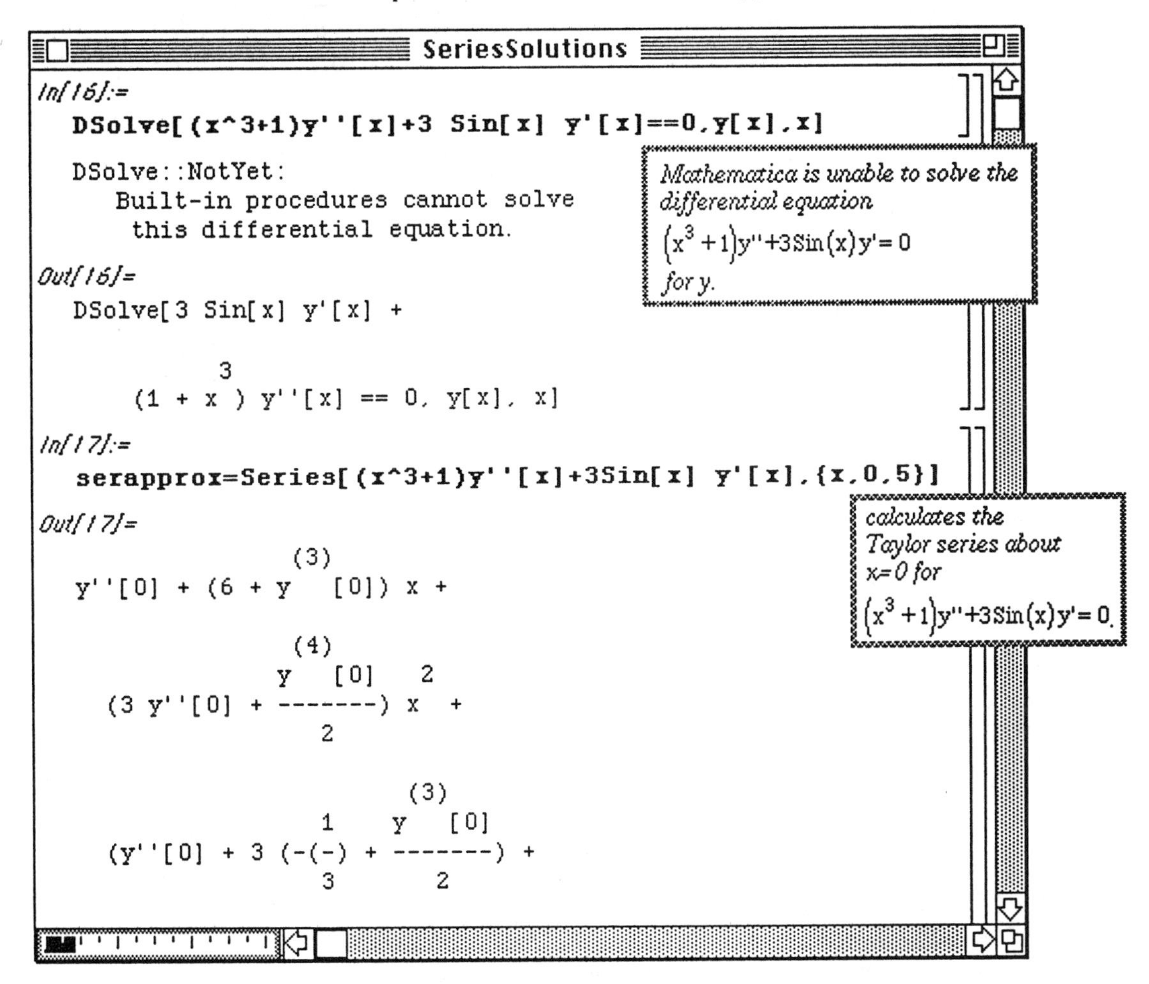

The next step in the solution process is to equate the coefficients of **serapprox** to the corresponding coefficients of the series on the right-hand side of the equation by matching like powers of x. In this case, each coefficient on the right is 0 . This is done below with **LogicalExpand**. The resulting expression is named **equations** since each term is, in fact an equation.

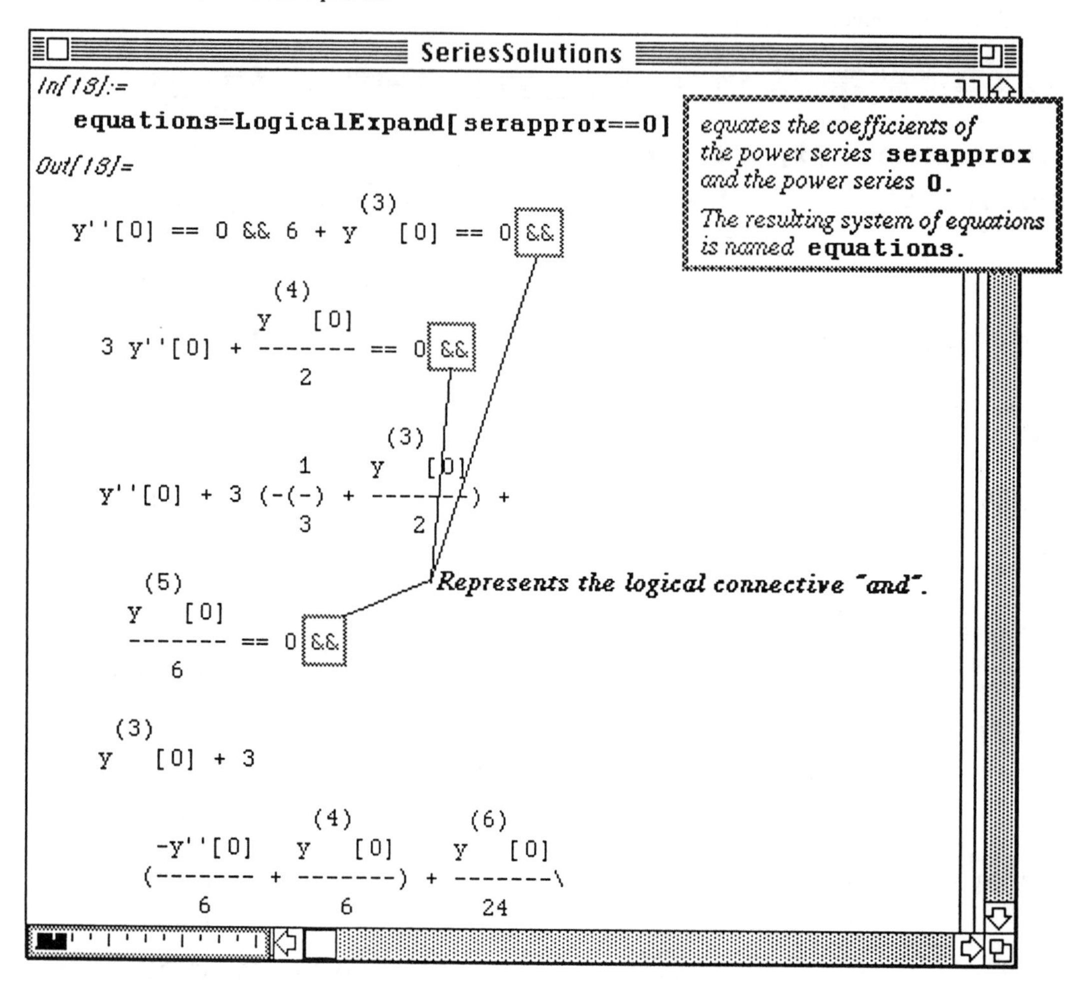

The initial conditions are entered so that **equations** can be solved for the unknown quantities. When **equations** is solved for

$$y''[0], y^{(3)}[0], y^{(4)}[0], \ldots,$$

the solution is determined by substituting these values into the assumed solution

$$y(x) = \sum_{n=0}^{\infty} \frac{y^{(n)}[0]}{n!} x^n.$$

Notice that $y^{(7)}[0]$ is the last element in the list **values** . Hence, the solution can be approximated with the series expansion of at most order 7. **values [[1]]** extracts the appropriate list from **values** . Therefore, the approximate solution, a series of order 7, is found with
Series [y[x],{x,0,7}]/.values [[1]].

The expression which results from this command cannot be considered a function since it contains a remainder term. Hence, the remainder term is eliminated with **Normal[Series[y[x],{x,0,7}/.values[[1]]**. This function is called **yapprox**.

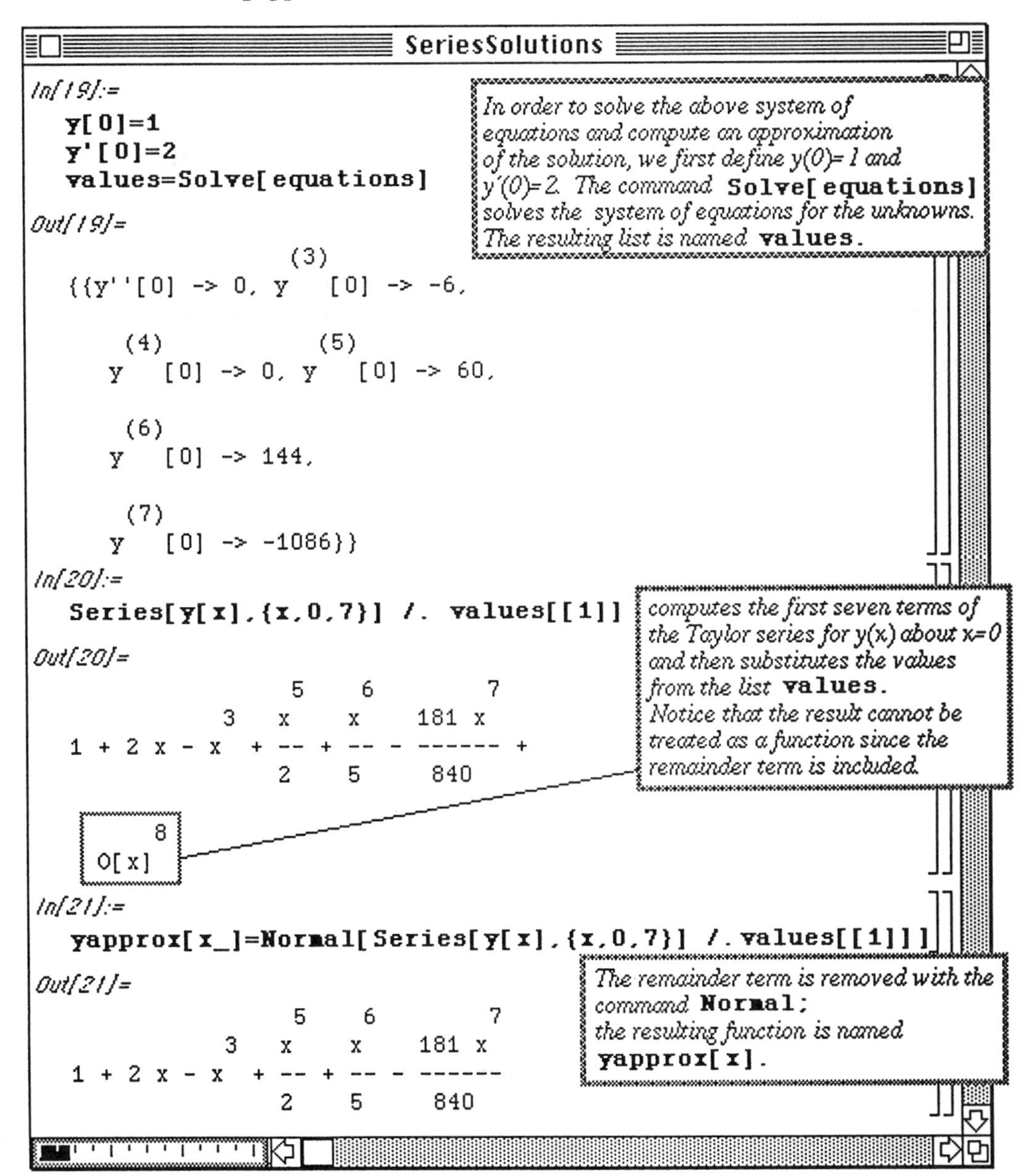

The approximate solution can then be graphed.

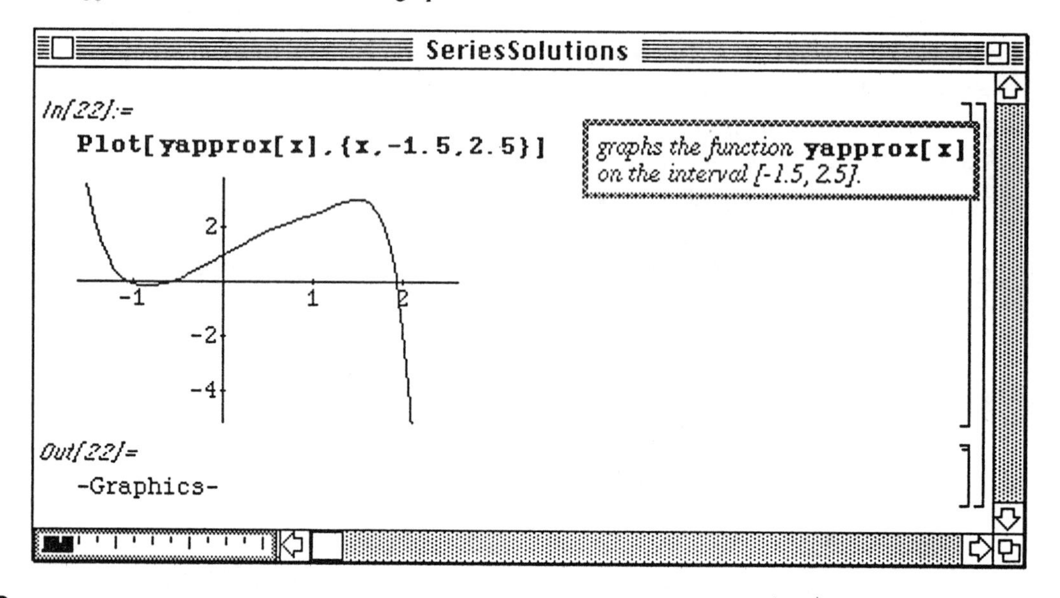

o Although **DSolve** cannot be used to find an explicit solution of $(x^3+1)y''+3\text{Sin}(x)y'=0$,

Version 2.0 users can use the built-in command DSolve to compute a numerical solution. For example, the command `sol=NDSolve[{x^3+1)y''[x]+3Sin[x]y'[x]==0,y[0]==1,y'[0]==2}, y[x],{x,-3,3}]`

computes a numerical solution of $(x^3+1)y''+3\text{Sin}(x)y'=0$ satisfying $y(0)=1$ and $y'(0)=2$ on the interval [-3,3]. The resulting **interpolating function** is named `sol` and is then graphed on the interval [-3,3] with the command `Plot[y[x] /. sol,{x,-3,3}]`.

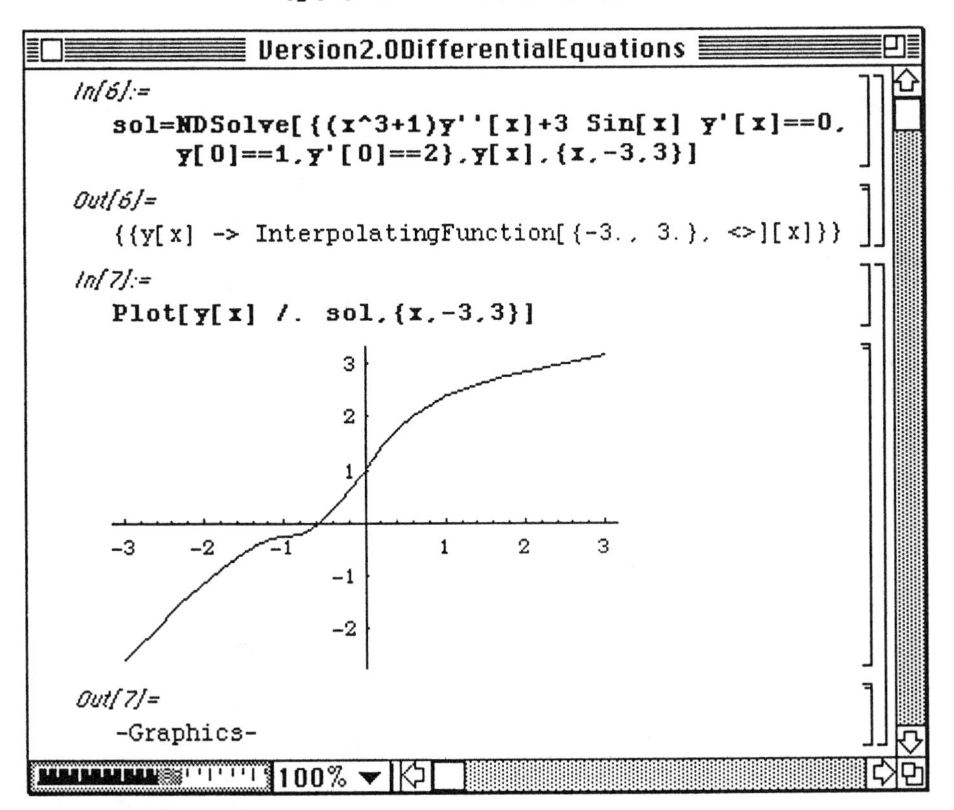

Problems which involve arbitrary initial conditions can also be considered. With the help of *Mathematica*, approximate solutions can be found in terms of these conditions and plotted for various values. The following example illustrates this idea by using the procedures found in the previous problem.

□ **Example:**

Use power series to compute approximations of the solution to the initial value problems

$y''-4x^2y'-4y = xe^x$, $y(0)=i$ and $y'(0)=j$, i and j both integers, $-2 \le i \le 2$, $-2 \le j \le 2$.

Again, this problem cannot be solved with **DSolve**. Hence, another method of solution must be used. Clearly, x = 0 is an ordinary point of this differential equation, so a power series solution can be assumed. The fifth-order power series expansions of the left and right-hand sides of the equation are computed with **Series[y''[x]-4x^2y'[x]-4y[x],{x,0,5}]** and **Series[x Exp[x], {x,0,5}]**, respectively. These expressions are named **ser1** and **ser2**, so the command **LogicalExpand[ser1==ser2]** equates the coefficients of like powers of x from **ser1** and **ser2** . This gives a sequence of equations which is called **equations**.

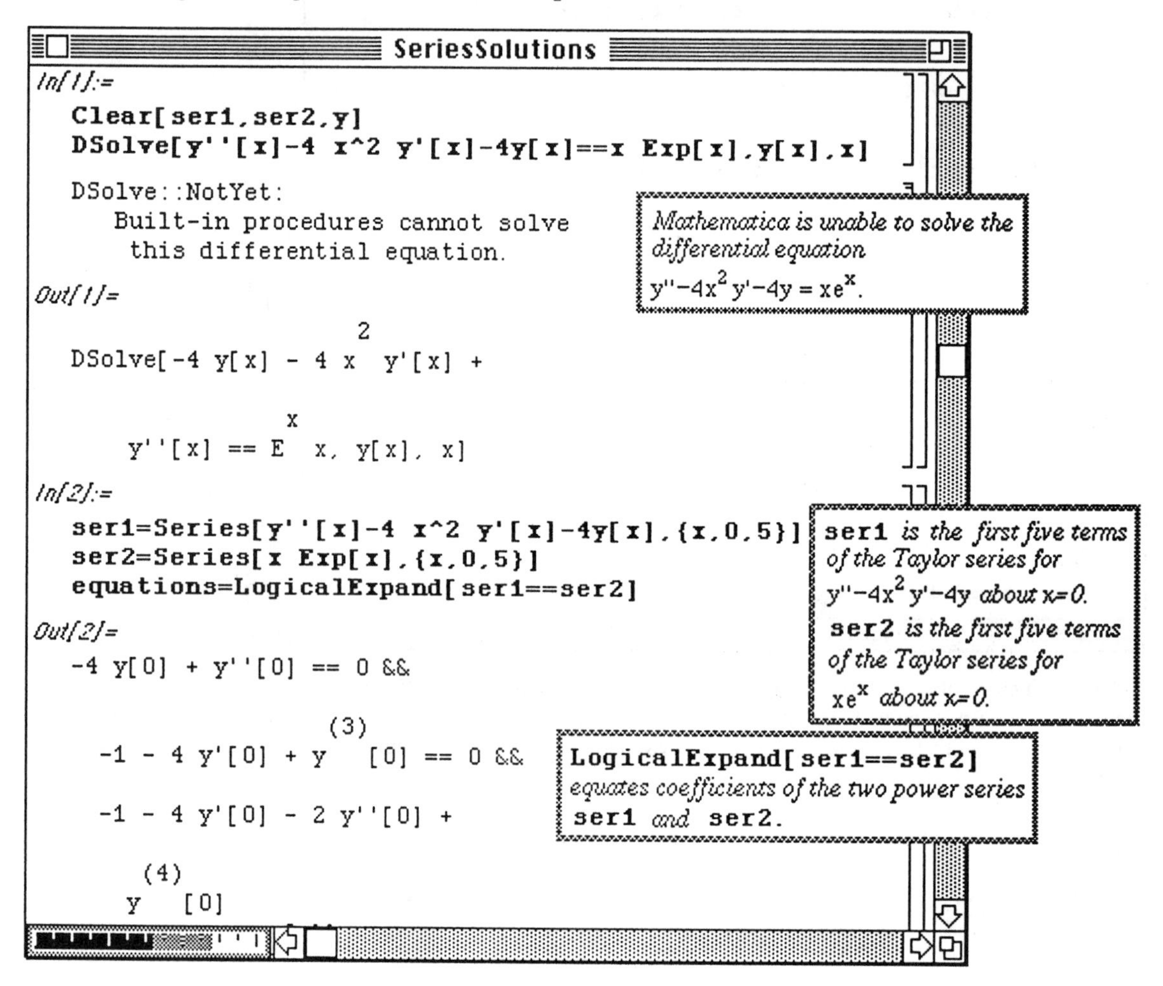

The following function, called **solutions**, is defined with a **Block** and solves **equations** to determine the coefficients of the power series solution using the initial conditions **y[0]=i** and **y'[0]=j**. It also computes the approximate solution of order five by eliminating the remainder term. A table of functions, **conditions**, is then created for several values of **i** and **j** using **solutions[i,j]**.

o In Version 2.0, the command **Block** has been replaced by the command **Module**.

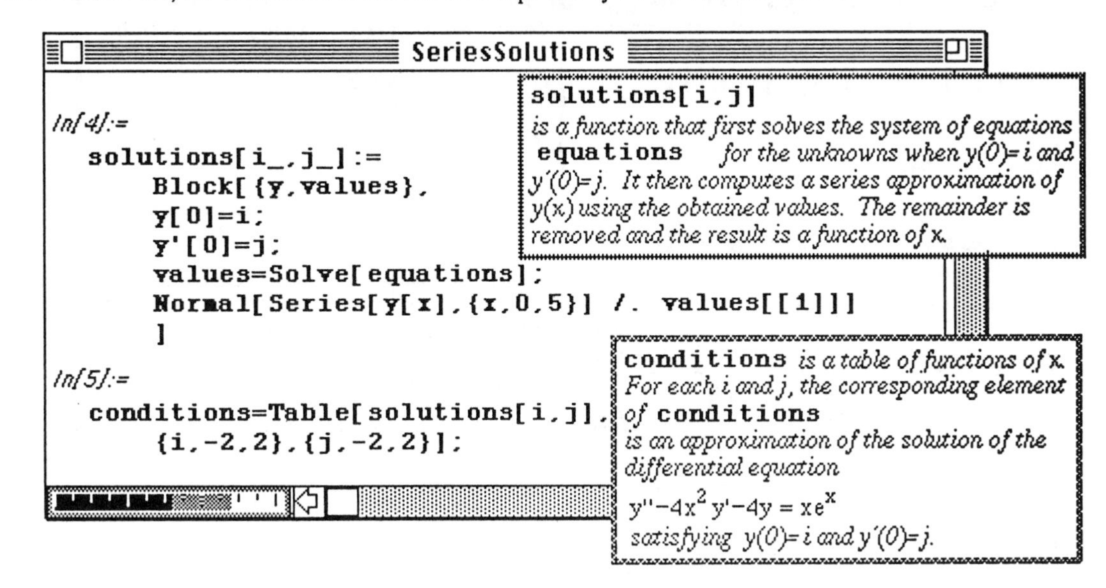

Since **conditions** is a list of lists, the *Mathematica* command **Flatten** can be used to obtain a list of functions which is needed to complete the problem. A simple example is given to illustrate this command. Hence, **Flatten[conditions]** yields a list of functions called **solutionlist**. (Only a portion of this list is shown below.)

SeriesSolutions

In[6]:=

```
Flatten[{{1,2},{3,4},{5,6}}]
```

Out[6]=

```
{1, 2, 3, 4, 5, 6}
```

In[7]:=

```
solutionlist=Flatten[conditions]
```

Out[7]=

```
                               3        4
                     2    7 x     23 x
{-2 - 2 x - 4 x  - ---- - ----- -
                           6       12

      5                        3
  217 x                 2     x
  ------, -2 - x - 4 x  - -- -
   120                        2

     4       5
  19 x    67 x
  ----- - -----,
   12      40

                  3      4       5
            2    x    5 x    37 x
  -2 - 4 x  + -- - ---- - -----,
                 6     4      24

                      3       4
                2   5 x    11 x
  -2 + x - 4 x  + ---- - ----- -
                      6      12
```

conditions *is actually a list of lists. To convert* **conditions** *from a nested list of functions to a list of functions, we use the command* **Flatten**. *Notice that* **Flatten[{{1,2},{3,4},{5,6}}]** *produces the list* **{1,2,3,4,5,6}**.

converts the list of lists of functions **conditions** *to a list of functions by removing curly brackets. The resulting list of functions is named* **solutionlist**.

In order to plot these functions, a **GrayLevel** table is useful. So that a **GrayLevel** can be assigned to each function in the list, the length of **solutionlist** must be known. Once this length is found to be 25, a table of 25 **GrayLevel** assignments is created and called **graylist**. Hence, the list of approximate solutions can be plotted and identified by referring to the **GrayLevel** of each curve.

o In Version 2.0, the command **Release** has been replaced by the command **Evaluate**.

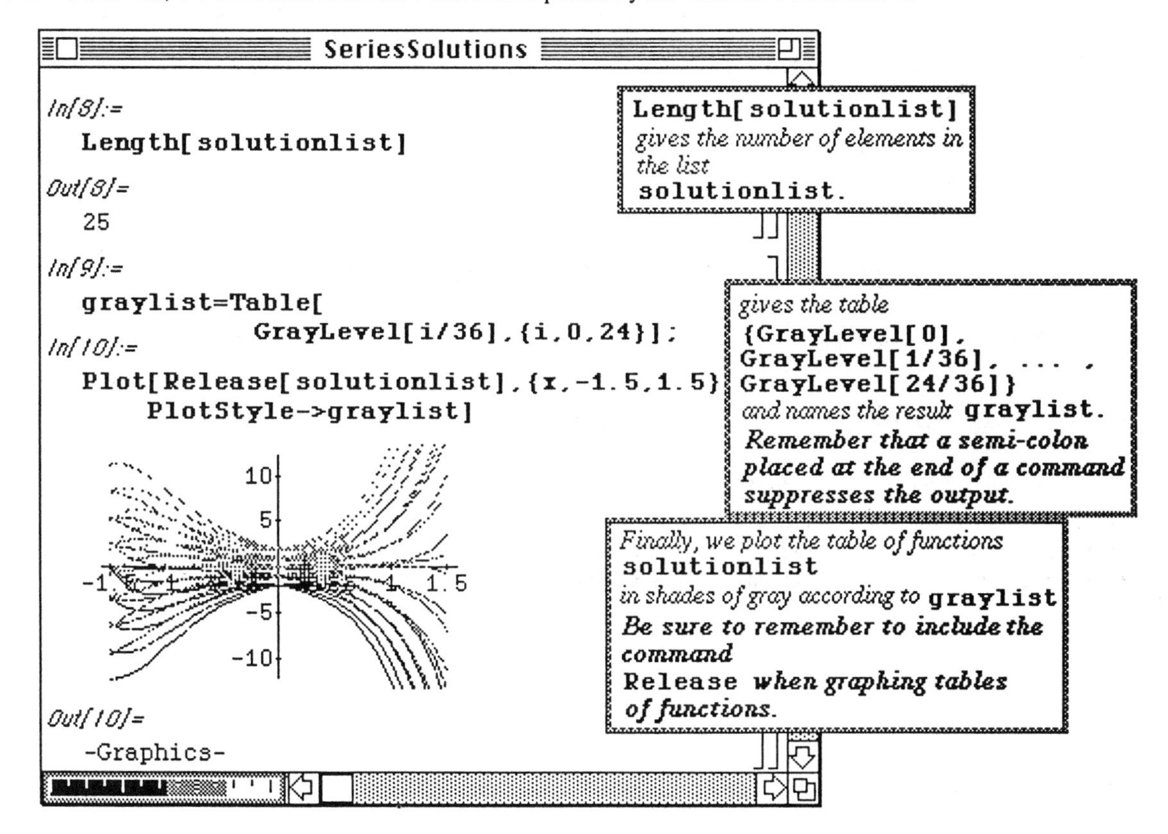

6.10 Series Solutions to Partial Differential Equations

Application: Two-Dimensional Wave Equation in a Circular Region

The **Bessel equation** is the differential equation $x^2y''+xy'+\left(x^2-\alpha^2\right)y=0$.

If α is a positive integer, the **Bessel Function** $J_\alpha(x)$ **of the first kind of order** α is defined by the series $J_\alpha(x)=\sum_{n=0}^{\infty}\frac{(-1)^n}{n!\;(n+\alpha)!}\left(\frac{x}{2}\right)^{2n+\alpha}$.

The Bessel Functions of the first kind have the following properties:

(i) $\frac{d}{dx}\left(x^{-\mu}J_\mu(x)\right)=-x^{-\mu}J_{\mu+1}(x)$; and (ii) $\frac{d}{dx}\left(x^{\mu}J_\mu(x)\right)=x^{\mu}J_{\mu-1}(x)$.

The **Bessel Function** $Y_\alpha(x)$ **of the second kind of order** α is defined by the integral $Y_\alpha(x)=J_\alpha(x)\int\frac{dx}{x(J_\alpha(x))^2}$.

It is well-known that the general solution of Bessel's equation, $x^2y''+xy'+\left(x^2-\alpha^2\right)y=0$, is given by $y_\alpha(x)=C_1J_\alpha(x)+C_2Y_\alpha(x)$.

The *Mathematica* function for the Bessel Function of the first kind of order **alpha** as a function of **x** is given by the command **BesselJ[alpha,x]**; the *Mathematica* function for the Bessel Function of the second kind of order **alpha** as a function of **x** is given by **BesselY[alpha,x]**.

□ **Example:**

Graph $J_0(x)$, $J_2(x)$, $J_4(x)$, $J_6(x)$ and $J_8(x)$ on the interval [0,15].

This is accomplished below by creating a table of the 5 Bessel functions in **tableb** as well as a table of GrayLevel assignments in **tablec**.

o If using Version 2.0, use **Evaluate** instead of **Release**.

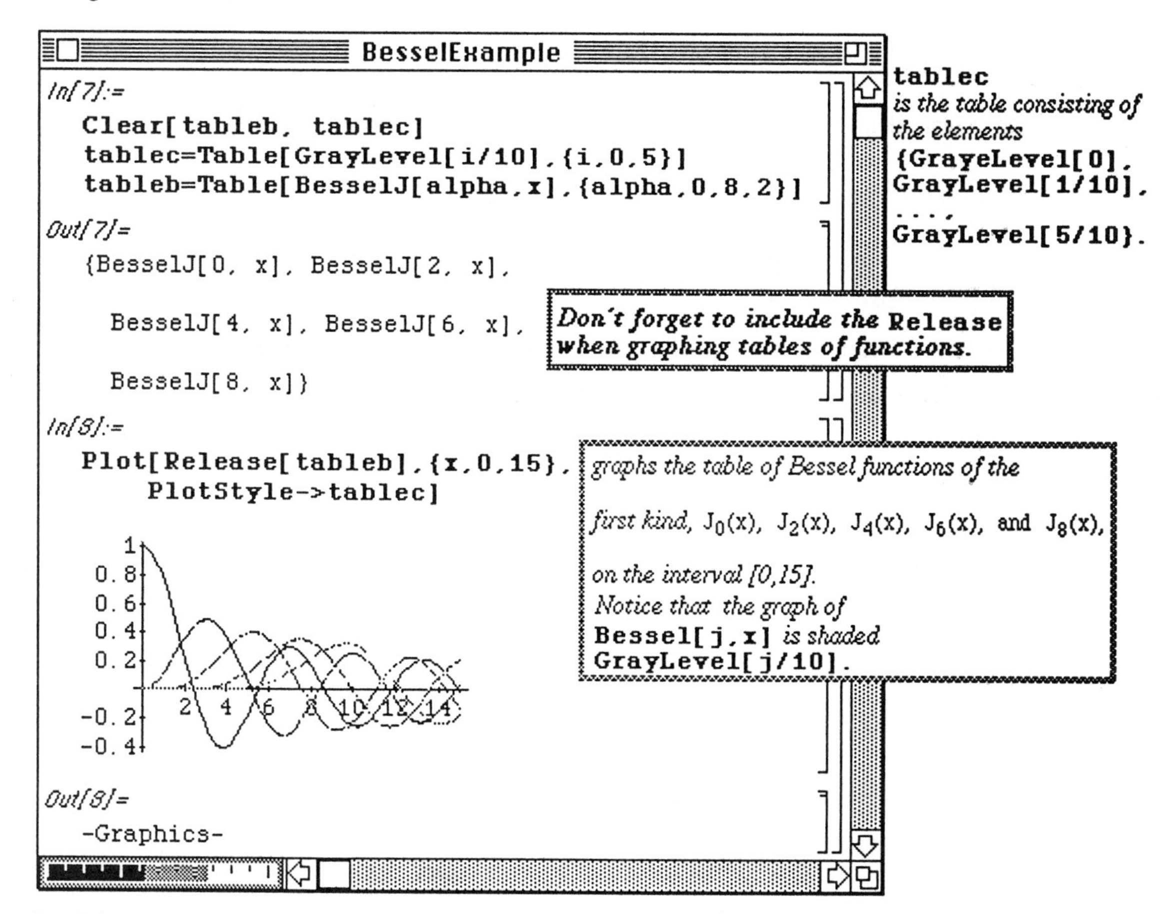

In solving numerous problems in applied mathematics, polar or cylindrical coordinate systems are often convenient to use. For example, the wave equation in a circular membrane lends itself quite naturally to the use of polar coordinates. The two-dimensional wave equation in a circular region which is radially symmetric (no dependence on θ) with boundary and initial conditions is easily expressed in polar coordinates as follows :

(i) $\dfrac{\partial^2 u}{\partial t^2} = c^2 \left(\dfrac{\partial^2 u}{\partial r^2} + \dfrac{1}{r}\dfrac{\partial u}{\partial r} \right) \quad 0 < r < R, \ 0 < t;$

(ii) $u(R, t) = 0 \quad 0 < t;$

(iii) $|u(0, t)|$ bounded, $\quad 0 < t;$

(iv) $u(r, 0) = f(r) \quad 0 < r < R;$ and

(v) $\dfrac{\partial u}{\partial t}(r, 0) = g(r) \quad 0 < r < R.$

Using the method of separation of variables with $u(r,t) = F(r)\,W(t)$ leads to the two ordinary differential equations :

$\dfrac{\partial^2 G}{\partial t^2} + \lambda^2 G = 0$ where $\lambda = ck$; and $\dfrac{\partial^2 W}{\partial r^2} + \dfrac{1}{r}\dfrac{dW}{dr} + k^2 W = 0$ where $-k^2$ is the constant of separation. $\dfrac{\partial^2 W}{\partial r^2} + \dfrac{1}{r}\dfrac{dW}{dr} + k^2 W = 0$ is Bessel's equation of order zero and, thus, has solutions of the form $W(r) = C_1 J_0(k\,r) + C_2 Y_0(k\,r)$ where $J_0(k\,r)$ and $Y_0(k\,r)$ are Bessel functions of order zero of the first and second kind, respectively:

In order to determine the constants C_1 and C_2 in W(r), the following graphs are necessary:

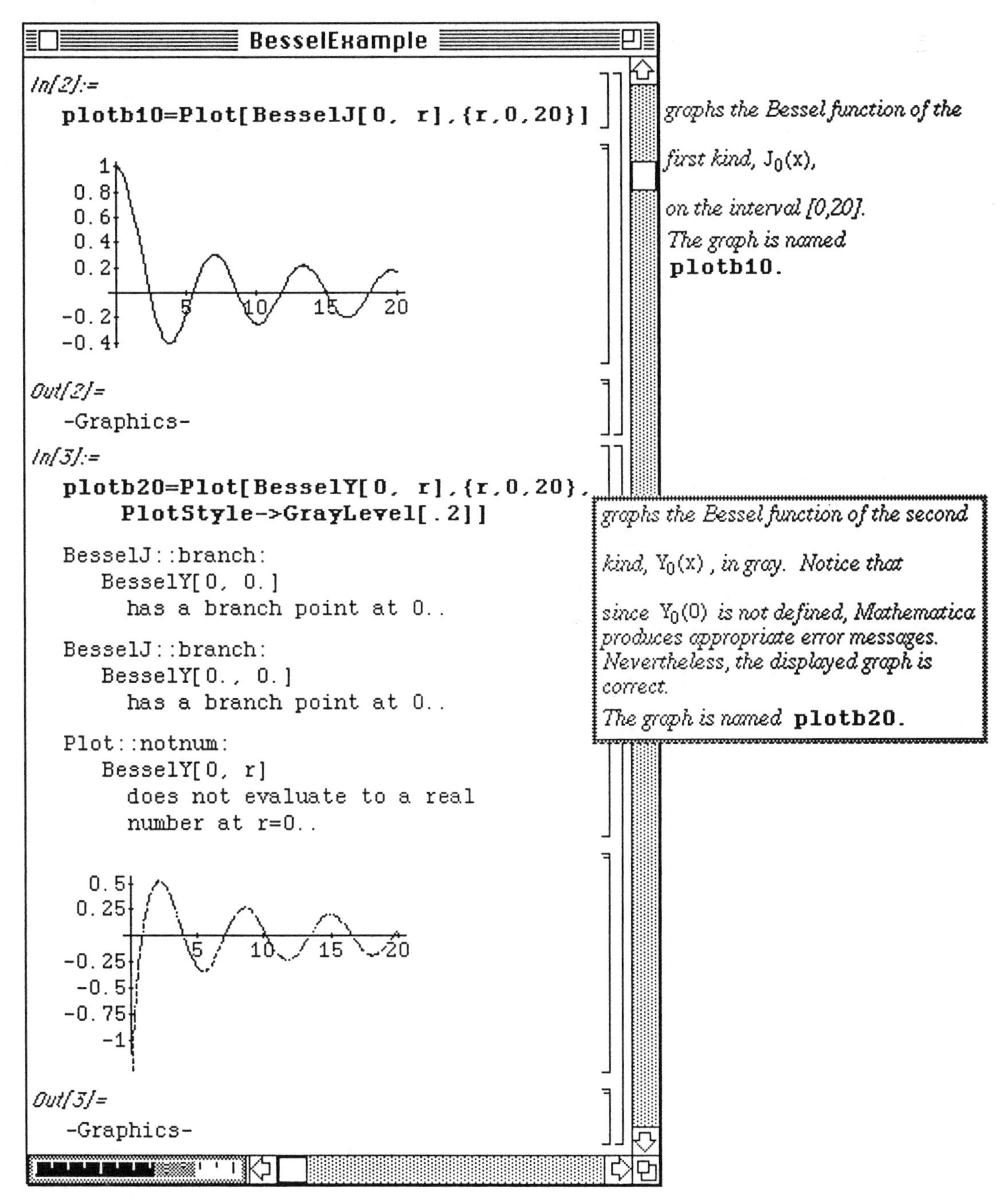

Notice the error messages which accompany the second plot. These, of course, are due to the fact that the Bessel function of the second kind is unbounded near r = 0.

These two functions can be plotted simultaneously (Note that the interval for the variable r is chosen to avoid r = 0. The darker function is the Bessel function of the first kind.) :

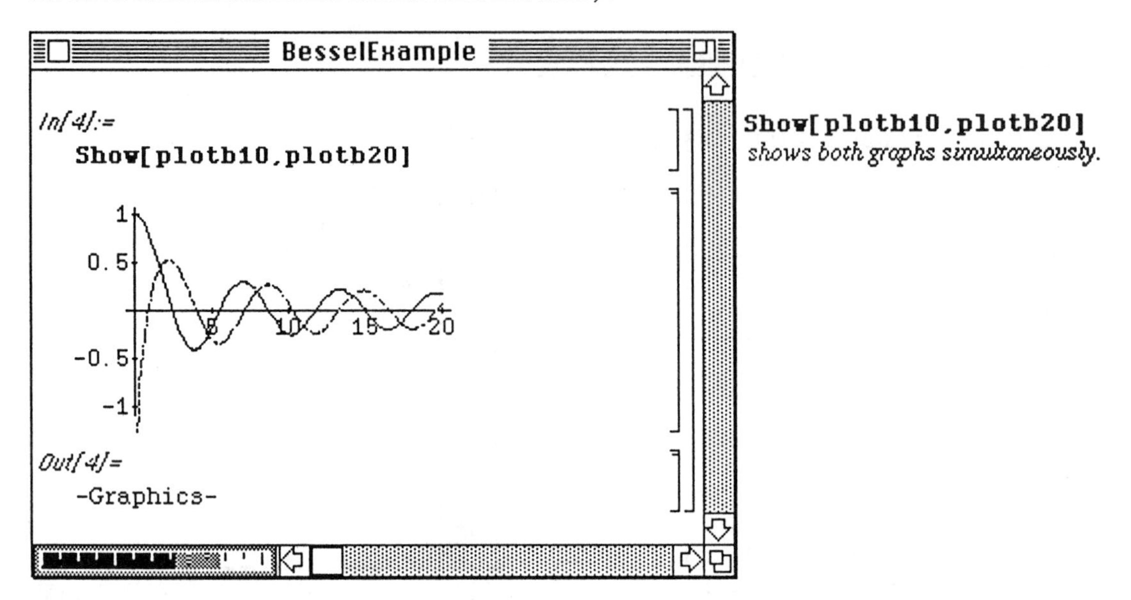

By using the graphs above, the coefficient of the Bessel function of the second kind must be zero in the equation $W(r) = C_1 J_0(k\,r) + C_2 Y_0(k\,r)$. Otherwise, the solution is unbounded at the origin which contradicts the boundary condition. Applying the other boundary condition, $u(R, t) = 0$, leads to the equation $J_0(kR) = 0$. Hence, $k_m = \frac{\alpha_m}{R}$ where $\alpha_m = m^{th}$ zero of $J_0(k\,r)$ and m = 1, 2, 3, ... These zeros of the

Bessel function can be located using *Mathematica* as opposed to simply looking them up in a table. Hopefully, some of the artificial nature of these values is alleviated by using this approach to the problem. In order to find the zeros of the Bessel function of the first kind, the *Mathematica* command **FindRoot[equation,{variable,firstguess}]** is used. **FindRoot** depends on an initial guess to the root of the equation. This initial guess is obtained from the graph **plotb10.**

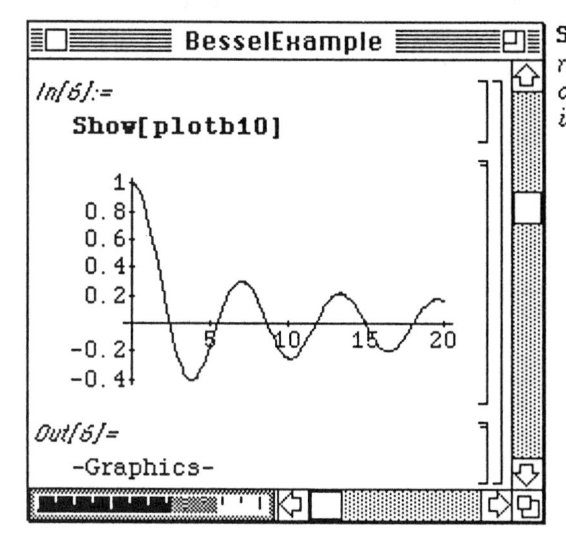

Show[plotb10] *redisplays the graph of the Bessel function of the first kind of order zero on the interval [0,20].*

Using the plot shown above, the Bessel function of the first kind appears to have its first four roots near r = 2.5, 5.5, 8.5, and 11.5. Therefore, the following command numerically determines approximations of the first four roots by using the command **FindRoot** with the initial guesses **start**. Note that start obtains values on the interval 2.5 to 11.5 using a stepsize of 3 units. Hence, the four initial guesses are the same as thos given earlier, 2.5, 5.5, 8.5, and 11.5. (Notice the double equals sign which must always be used with an equation) :

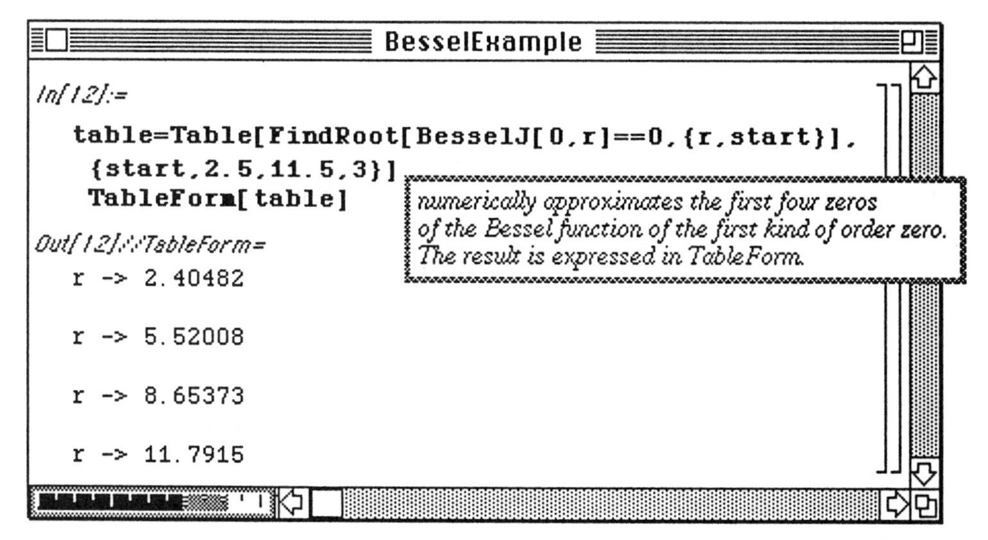

These values will be used to determine the coefficients in the solution.

The solution to $\dfrac{\partial^2 G}{\partial t^2} + \lambda^2 G = 0$ is clearly $G_m(t) = A_m \cos(\lambda_m t) + B_m \sin(\lambda_m t)$.

Thus, the functions $u_m(r,t) = W_m(t)\, G_m(t) = (A_m \cos\lambda_m t + B_m \sin\lambda_m t)\, J_0(k_m r)$ are solutions of (i) satisfying the boundary conditions (ii)-(iii). To obtain a solution which also satisfies the initial conditions

(iv)−(v), the series $u(r,t) = \sum_{m=1}^{\infty} u_m(r,t) = \sum_{m=1}^{\infty} (A_m \cos\lambda_m t + B_m \sin\lambda_m t)\, J_0(k_m r)$

must be considered. In applying the initial condition (iv), the equation

$u(r,0) = \sum_{m=1}^{\infty} A_m J_0(k_m r) = f(r)$ is obtained Using the orthogonality properties of the Bessel

functions, the coefficients A_m and B_m

in $u(r,t) = \sum_{m=1}^{\infty} (A_m \cos\lambda_m t + B_m \sin\lambda_m t)\, J_0(k_m r)$ are found with the integral formula

$A_m = \dfrac{2}{R^2 J_1^{\,2}(\alpha_m)} \displaystyle\int_0^R r\, f(r)\, J_0\!\left(\frac{\alpha_m}{R} r\right) dr$. The coefficient B_m is found with a similar formula:

$B_m = \dfrac{2}{c\, \alpha_m R\, J_1^{\,2}(\alpha_m)} \displaystyle\int_0^R r\, f(r)\, J_0\!\left(\frac{\alpha_m}{R} r\right) dr$. In most cases, these two formulas are difficult

to evaluate For a limited number of functions, integration by parts using $\dfrac{d}{dr}\left[r^{\nu} J_\nu(r)\right] = r^{\nu+1} J_\nu(r)$

is possible. However, even when possible, this calculation is quite tedious and lengthy. Fortunately, *Mathematica* can ease the difficulty of the computation of the coefficients through the use of **NIntegrate**.

For example, consider equations (i)-(v) with $R = 1$, $c = 2$, initial position function

$f(r) = 1 - r^2$ and initial velocity function $g(r) = 0$. Clearly, $B_m = 0$ for all m and

$A_m = \dfrac{2}{J_1^{\,2}(\alpha_m)} \displaystyle\int_0^1 r\,(1-r^2)\, J_0(\alpha_m r)\, dr$. This particular integral can be evaluated exactly with integration by parts, but the determination of the approximate values of the coefficients will be demonstrated. The following examples illustrate how the zeros of the Bessel function are extracted from the table. Similar commands will be included in the calculations which follow:

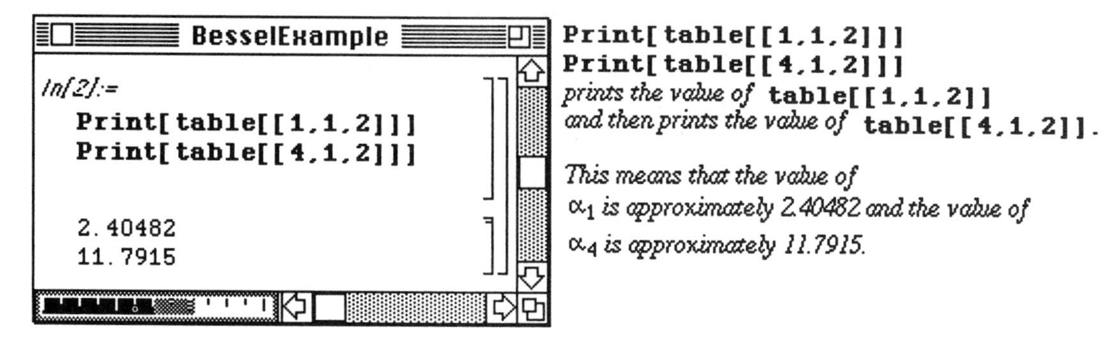

To approximate the first coefficient with `NIntegrate`, the following command is entered :

BesselExample

In[21]:=

```
a[1]=(2/(BesselJ[1,table[[1,1,2]]])^2)*
     NIntegrate[r (1-r^2)
     BesselJ[0,table[[1,1,2]] r],{r,0,1}]
```

Out[21]=

```
1.10802
```

computes an approximation of $\frac{2}{J_1^2(\alpha_1)}\int_0^1 r\,(1-r^2)\,J_0(\alpha_1 r)\,dr$ *and names the result* **a[1]**.

Be particularly careful that brackets and parentheses are nested correctly.

The other coefficients are easily determined as well :

BesselExample

In[22]:=

```
a[2]=(2/(BesselJ[1,table[[2,1,2]]])^2)*
     NIntegrate[ r (1-r^2)
     BesselJ[0,table[[2,1,2]] r],{r,0,1}]
```

Out[22]=

```
-0.139778
```

computes an approximation of

$$\frac{2}{J_1^2(\alpha_2)}\int_0^1 r\,(1-r^2)\,J_0(\alpha_2 r)\,dr$$

and names the result **a[2]**.

In[23]:=

```
a[3]=(2/(BesselJ[1,table[[3,1,2]]])^2)*
     NIntegrate[ r (1-r^2)
     BesselJ[0,table[[3,1,2]] r],{r,0,1}]
```

Out[23]=

```
0.0454765
```

computes an approximation of

$$\frac{2}{J_1^2(\alpha_3)}\int_0^1 r\,(1-r^2)\,J_0(\alpha_3 r)\,dr$$

and names the result **a[3]**.

In[24]:=

```
a[4]=(2/(BesselJ[1,table[[4,1,2]]])^2)*
     NIntegrate[ r (1-r^2)
     BesselJ[0,table[[4,1,2]] r],{r,0,1}]
```

Out[24]=

```
-0.0209908
```

computes an approximation of

$$\frac{2}{J_1^2(\alpha_4)}\int_0^1 r\,(1-r^2)\,J_0(\alpha_4 r)\,dr$$

and names the result **a[4]**.

Hence, the first four terms of the solution are found with the function `r[r,t,j]` defined below. This function is then used to compute the approximate solution by adding the first terms of the series. (Note that in this case only the first four coefficients have been calculated. Therefore, if more terms are desired, similar steps may be followed to compute more zeros of `BesselJ[0,x]` and more series coefficients.)

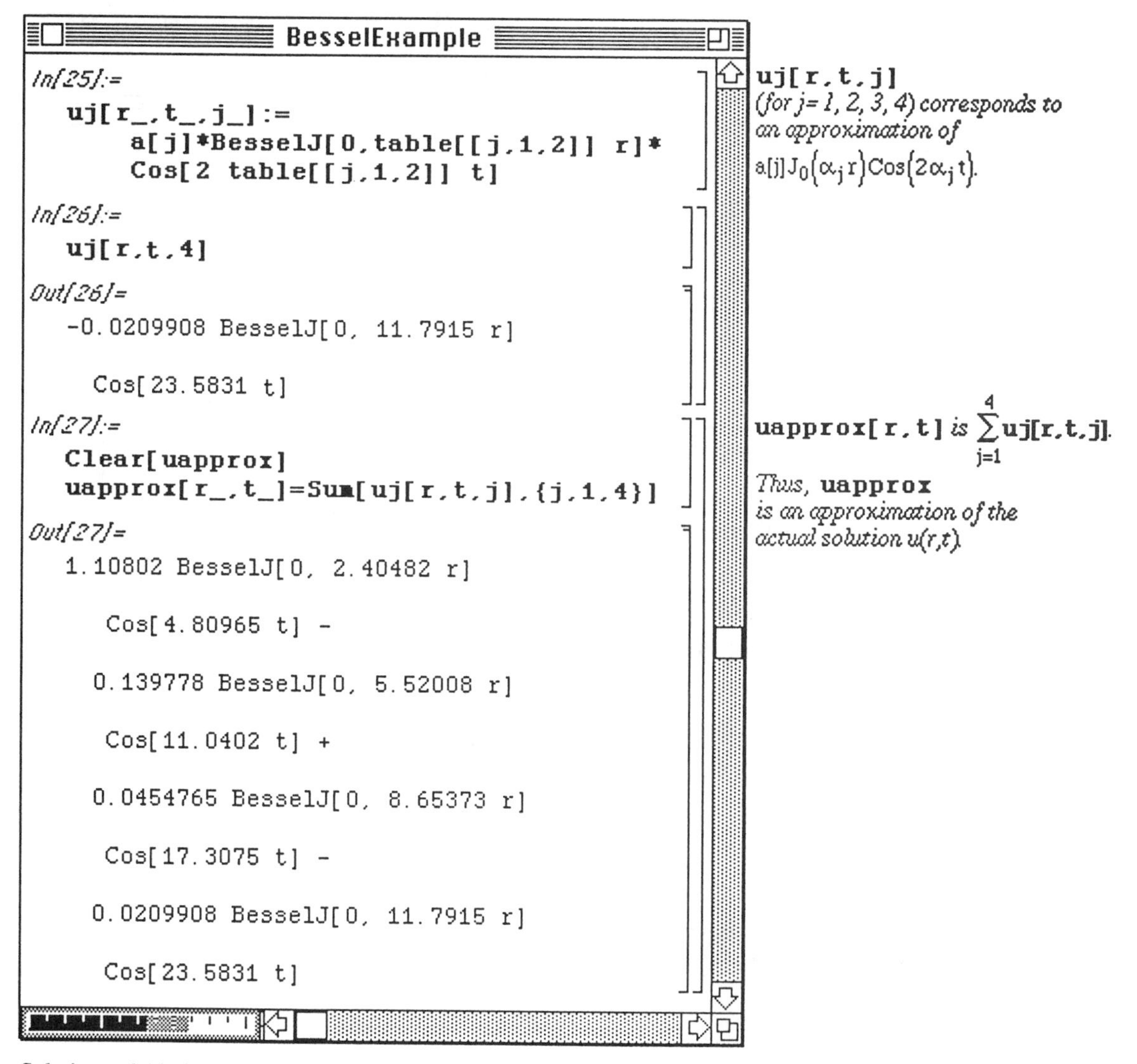

```
BesselExample

In[25]:=
  uj[r_,t_,j_]:=
       a[j]*BesselJ[0,table[[j,1,2]] r]*
       Cos[2 table[[j,1,2]] t]

In[26]:=
  uj[r,t,4]

Out[26]=
  -0.0209908 BesselJ[0, 11.7915 r]

    Cos[23.5831 t]

In[27]:=
  Clear[uapprox]
  uapprox[r_,t_]=Sum[uj[r,t,j],{j,1,4}]

Out[27]=
  1.10802 BesselJ[0, 2.40482 r]

     Cos[4.80965 t] -

    0.139778 BesselJ[0, 5.52008 r]

     Cos[11.0402 t] +

    0.0454765 BesselJ[0, 8.65373 r]

     Cos[17.3075 t] -

    0.0209908 BesselJ[0, 11.7915 r]

     Cos[23.5831 t]
```

`uj[r,t,j]` *(for j= 1, 2, 3, 4) corresponds to an approximation of* $a[j]J_0(\alpha_j r)\text{Cos}(2\alpha_j t)$.

`uapprox[r,t]` *is* $\sum_{j=1}^{4}$`uj[r,t,j]`.

Thus, `uapprox` *is an approximation of the actual solution u(r,t).*

Solutions of this form are hard to visualize. In an attempt to bring true meaning to the "circular drumhead" problem, *Mathematica* can be used to actually see the drumhead either from a side view in two-dimensions or a full view in three.

Plotting in two-dimensions is simple. The question which needs to be answered is : " What shape does the drumhead assume at a particular time t ?" This can be answered using several different methods. The easiest approach involves plotting the solution individually for various values of t. However, this involves much more time than a second method which takes advantage of a `Do` loop. Once plotted, these graphs can be animated to see the actual movement of the drumhead. The following command plots the solution for values of t between 0 and 1 using increments of 0.1. (Notice that r assumes values from r = -1 to r = 1. This seems to contradict the idea that r represents a nonnegative distance from the origin. However, u(r,t) is symmetric about the y-axis.)

Several of these graphs are shown below. These graphs can then be animated.

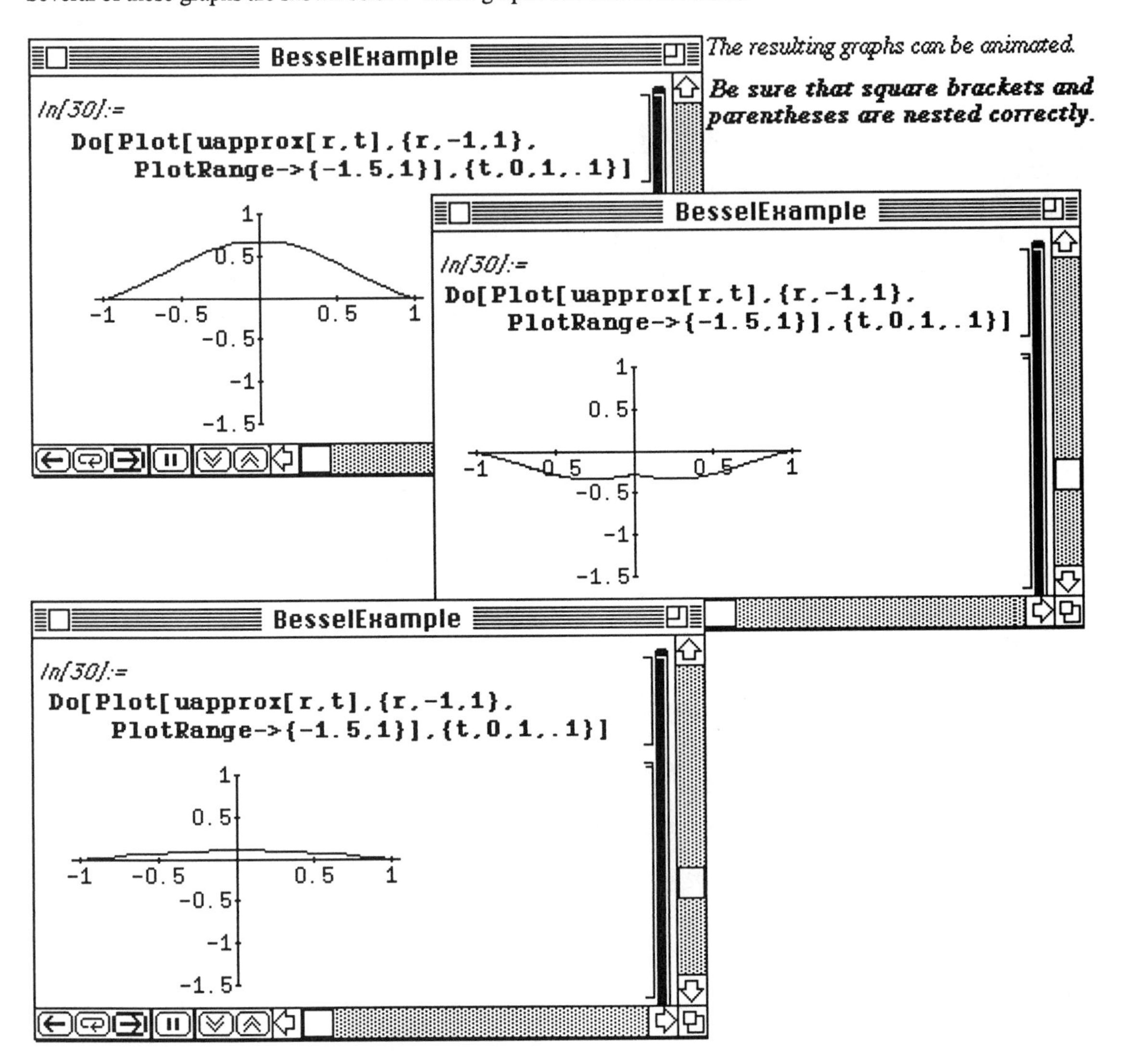

Several of the three-dimensional plots are shown below. These plots were obtained using the command **`solidrev`**. The command **`solidrev`** is discussed in the **Appendix**.

This is the graph of **`uapprox[r,0]`** *on the unit circle:*

This is the graph of **`uapprox[r,.2]`** *on the unit circle:*

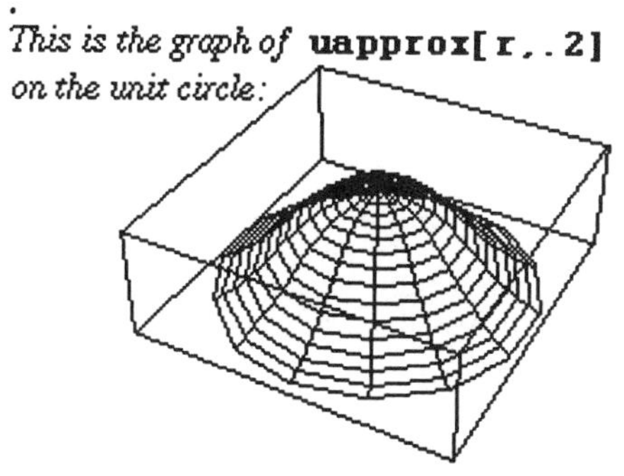

This is the graph of **`uapprox[r,.4]`** *on the unit circle:*

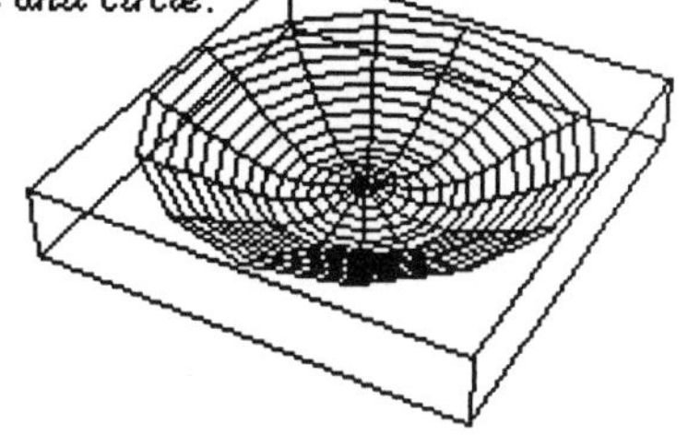

This is the graph of **`uapprox[r,1]`** *on the unit circle:*

6.11 Numerical Solutions of Differential Equations

Version 2.0 of *Mathematica* contains the command **NDSolve** which numerically solves ordinary differential equations with initial conditions. This command is particularly useful when working with nonlinear equations which **DSolve** is unable to solve. As was the case with **DSolve**, **NDSolve** can be used with single equations as well as systems. (Note that enough initial conditions must accompany the differential equation to completely solve the problem in order for **NDSolve** to be successful.)

• Application: The Damped Pendulum Equation

In order to illustrate **NDSolve**, consider the nonlinear pendulum equation , $x'' + .25\ x' + \sin(x) = 0$ with initial conditions $x(0) = 1$, $x'(0) = 0$. The differential equation is defined below as **equation**. (Note the square brackets which must be used with the dependent variable **x[t]** in the definition.)
The syntax for **NDSolve** (to solve a second order initial value problem with dependent variable **x[t]** as is the case here) is as follows :
NDSolve[{eqn,x[t0]==c0,x'[t0]==c1},x[t],{t,t0,t1}].
This finds a numerical solution to **eqn** which is valid over the interval **{t0,t1}** and satisfies the given initial conditions. Since the results are numerical, they are given in terms of the list
`{x[t]->InterpolatingFunction[{t0,t1},<>][t]}}`. The pendulum equation given earlier is solved and plotted below. The list which results from **NDSolve** is called **sol1** while the graph is assigned the name **plot1**. (Note the manner in which the **Plot** command involving the interpolating function is stated.)

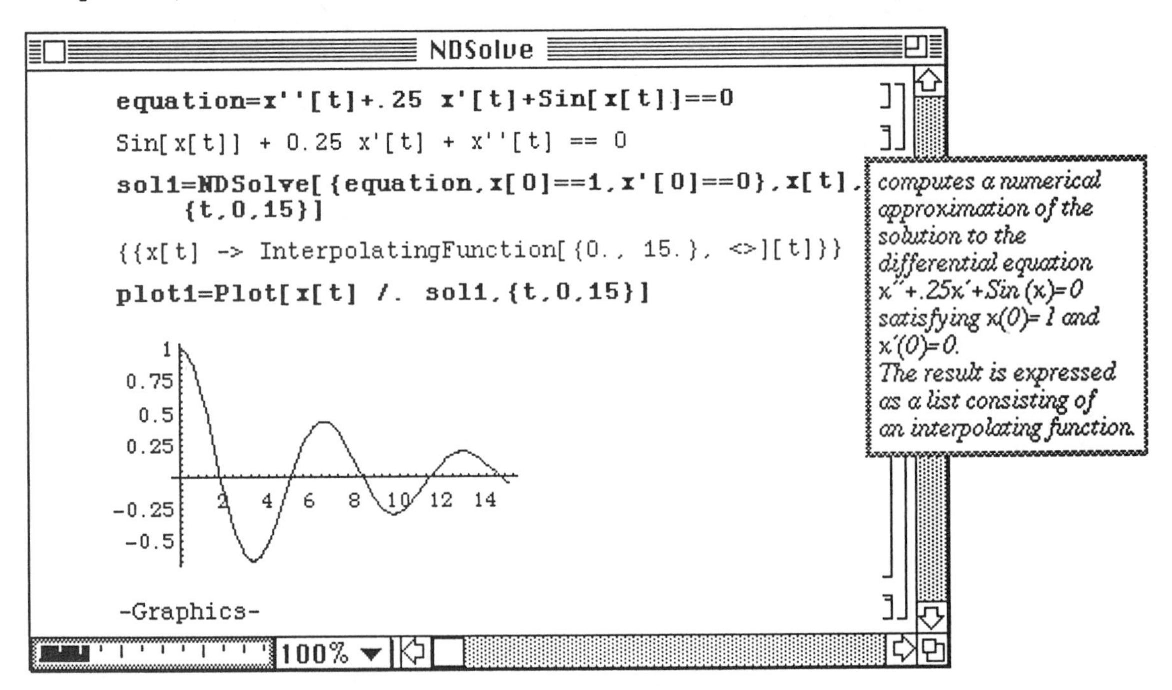

Next, consider the same equation defined in **equation** with the initial conditions $x(0) = 3$, $x'(0) = 0$. This initial value problem is solved and plotted in the same manner as the previous example. These results are named **sol6** and **plot6**, respectively.

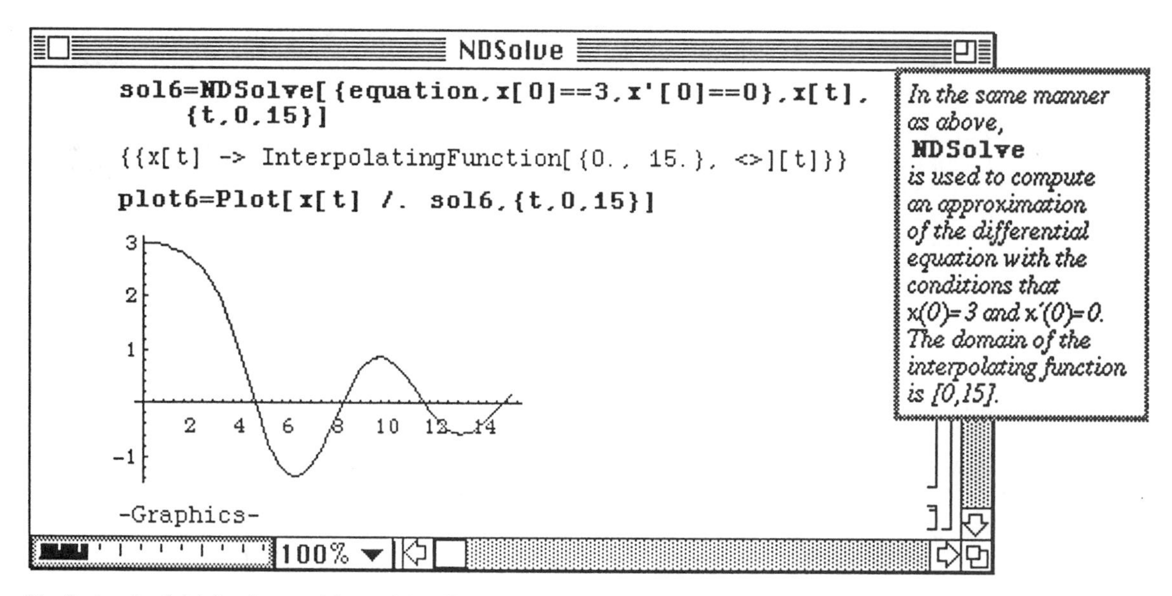

Similarly, the initial value problem with x(0) = 4 and x'(0) = 0 is solved and plotted below in **sol7** and **plot7**. Note the effect that the nonlinear term in **equation** has on the behavior of the solution. The solutions to this equation differ considerably from solutions to the second-order linear differential equations with constant coefficients discussed earlier in *Mathematica* By Example.

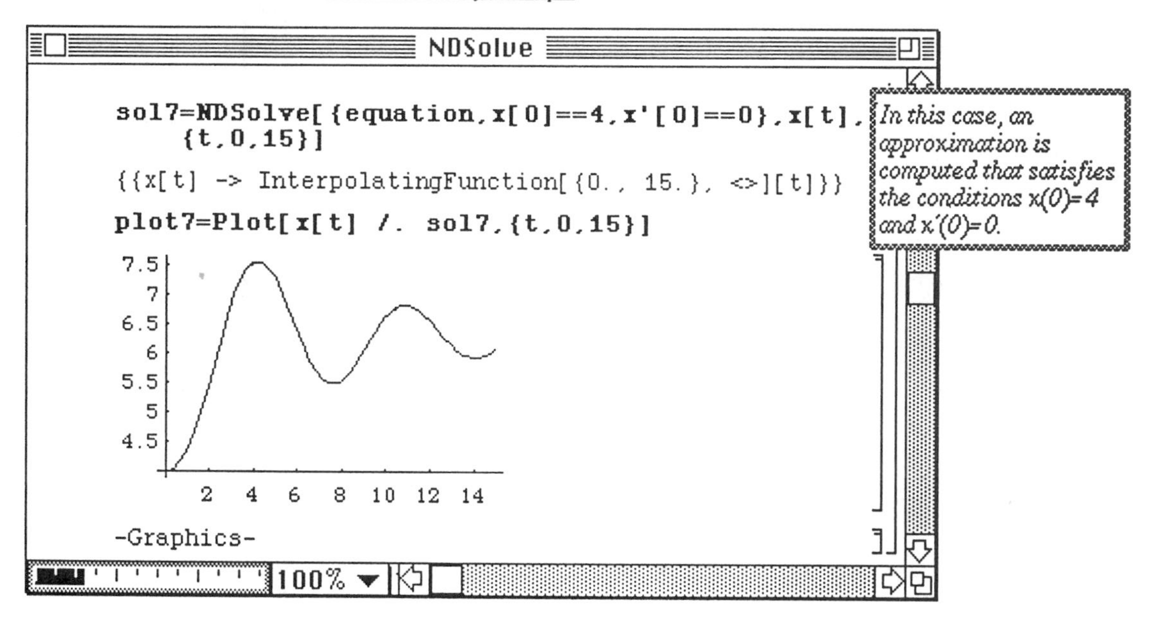

Solutions to the pendulum equation for varying initial conditions are computed with **NDSolve**. The plots for eight other solutions are given in **plot5**, **plot6**, ..., **plot13**, and **plot14**. These solutions are viewed simultaneously with **Show**. First, the graphs are shown in a single graphics cell in groups of four in **multone**, **multtwo**, **multthree**, and **multfour**. In the final command, however, all fourteen graphs are shown simultaneously.

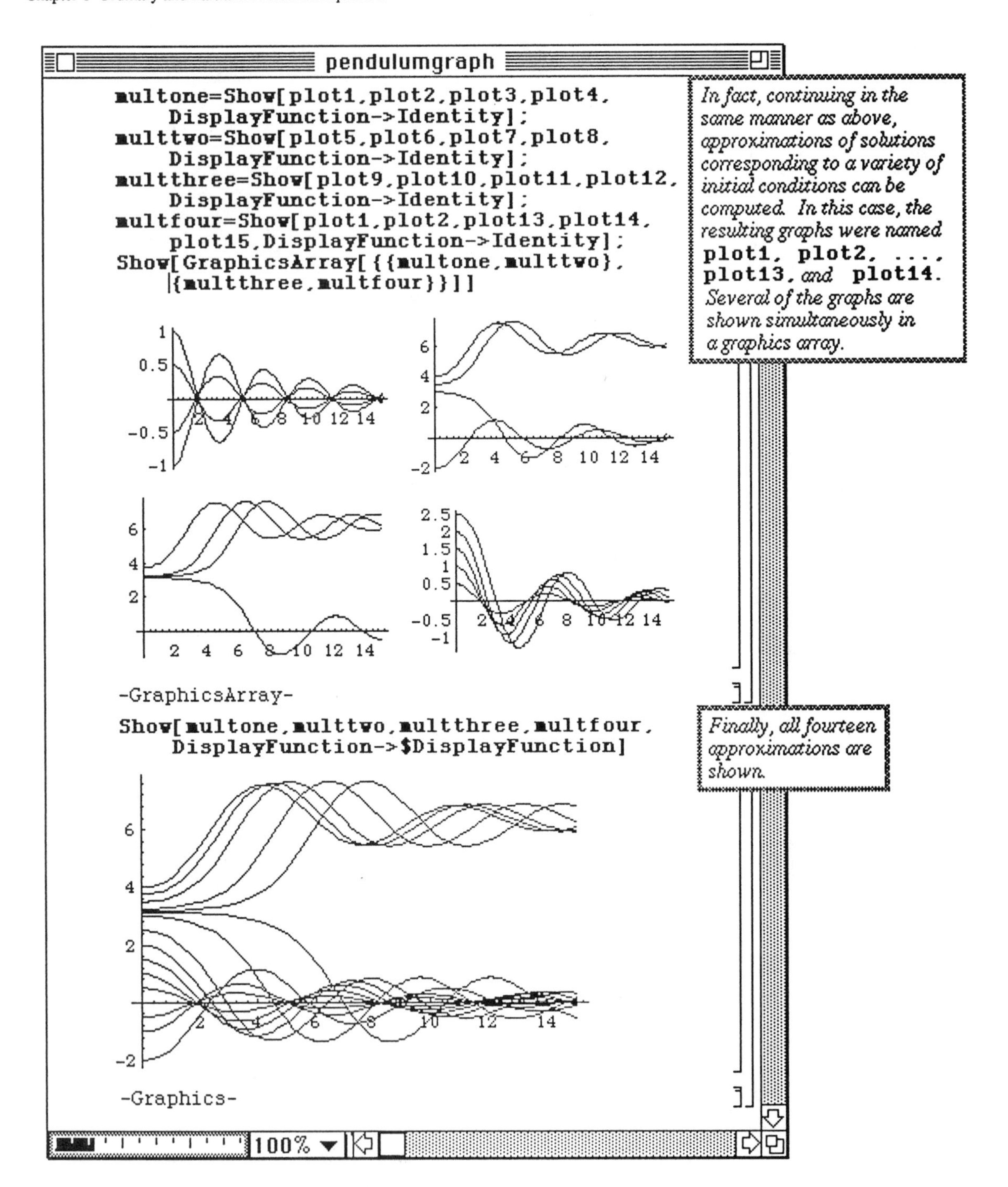
pendulumgraph
multone=Show[plot1,plot2,plot3,plot4,
DisplayFunction->Identity];
multtwo=Show[plot5,plot6,plot7,plot8,
DisplayFunction->Identity];
multthree=Show[plot9,plot10,plot11,plot12,
DisplayFunction->Identity];
multfour=Show[plot1,plot2,plot13,plot14,
plot15,DisplayFunction->Identity];
Show[GraphicsArray[{{multone,multtwo},
{multthree,multfour}}]]
-GraphicsArray-
Show[multone,multtwo,multthree,multfour,
DisplayFunction->$DisplayFunction]
-Graphics-
In fact, continuing in the same manner as above, approximations of solutions corresponding to a variety of initial conditions can be computed. In this case, the resulting graphs were named plot1, plot2, ..., plot13, and plot14. Several of the graphs are shown simultaneously in a graphics array.
Finally, all fourteen approximations are shown.
100%

6.12 Numerical Solutions of Systems of Differential Equations

• Application: Van der Pol's Equation

As indicated earlier **NDSolve** can be used to solve systems of ordinary differential equations. Of course, there is a slight difference in the syntax from the previous case. For a first-order system of two equations with dependent variables **x[t]** and **y[t]**, the correct command is as follows: **NDSolve[{eq1,eq2,x[t0]==c0,y[t0]==c1},{x[t],y[t]},{t,t0,t1}]**. Again, the results are given as an interpolating function and are only valid over the interval **{t0,t1}**. Solving a system of differential equations with **NDSolve** is illustrated below with Van der Pol's equation, x" + e (x^2 - 1) x' + x = 0, x(0) = c0, x'(0) = c1. This second-order equation can be transformed into a system of first-order equations with the substitution, x'= y. Hence, the following first-order system is obtained : x' = y , y' = e (1 - x^2) y - x, x(0) = c0, y(0) = c1.

In the steps which follow, the parameter "e" is assumed to equal 1 (**e=1**), and the equations are named **eq1** and **eq2**, respectively. (Note the square brackets which must accompany the dependent variables, **x[t]** and **y[t]**.) Van der Pol's equation with initial conditions x(0) = .25, y(0) = 0 is solved in **sol** below. In this case, the solution is made up of the ordered pair **{x[t],y[t]}**. Hence, **ParametricPlot** is used to graph the numerical solutions in **plotone**.

Next, the same equation is solved using the initial conditions x(0) = 0, y(0) = -.2. This solution is determined in the same manner as above and is plotted in **plottwo**.

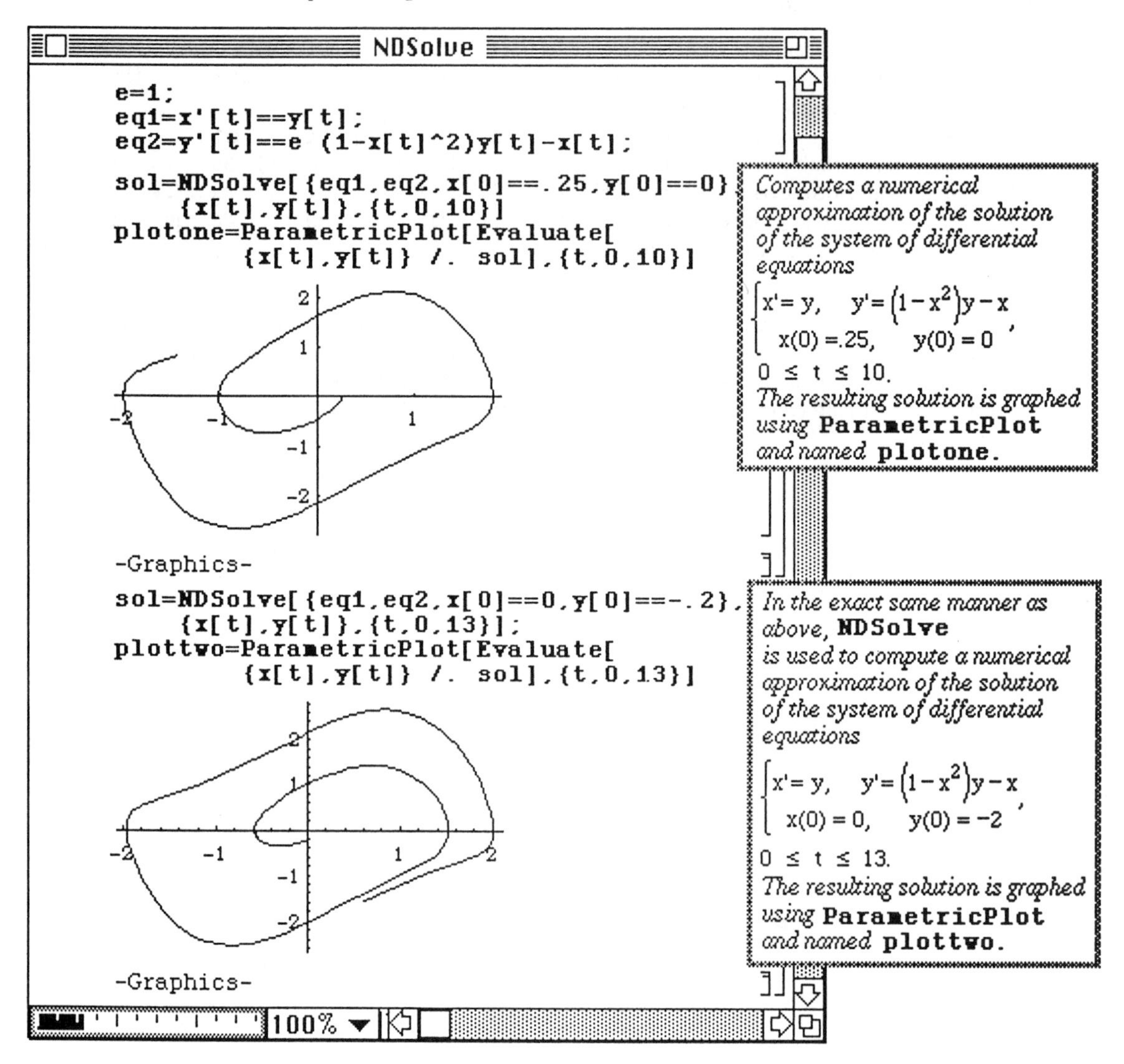

plotthree and **plotfour** are given below. These correspond to solutions to Van der Pol's equation with initial conditions x(0) = -.15, y(0) = 0 and x(0) = 0, y(0) = .1, respectively.

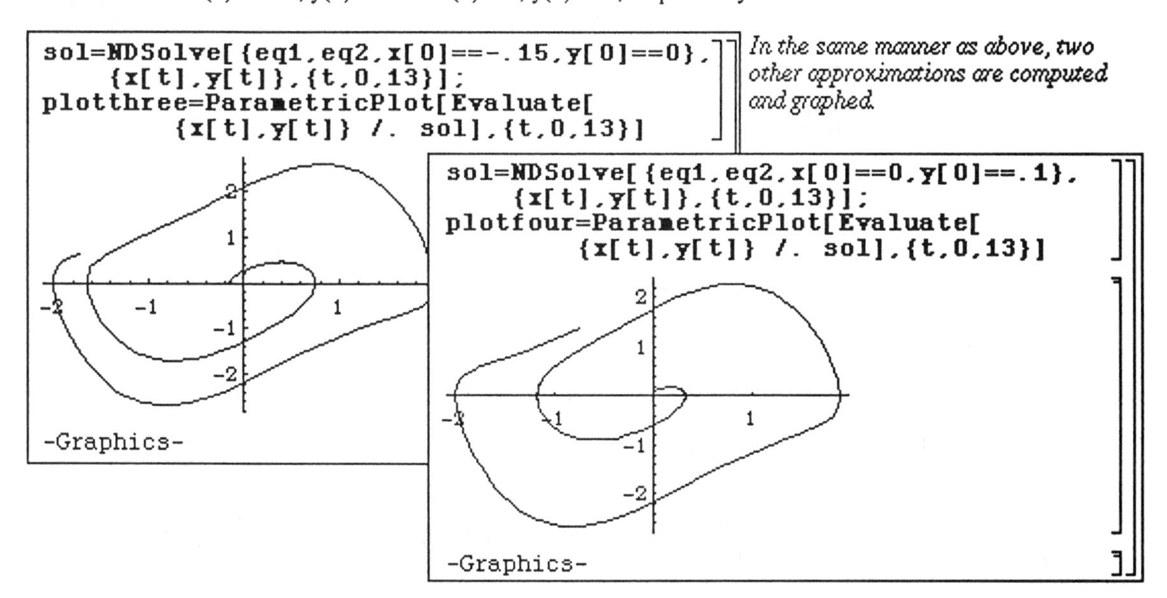

```
sol=NDSolve[{eq1,eq2,x[0]==-.15,y[0]==0},
    {x[t],y[t]},{t,0,13}];
plotthree=ParametricPlot[Evaluate[
        {x[t],y[t]} /. sol],{t,0,13}]
```

-Graphics-

In the same manner as above, two other approximations are computed and graphed.

```
sol=NDSolve[{eq1,eq2,x[0]==0,y[0]==.1},
    {x[t],y[t]},{t,0,13}];
plotfour=ParametricPlot[Evaluate[
        {x[t],y[t]} /. sol],{t,0,13}]
```

-Graphics-

The four approximate solutions to Van der Pol's equation computed to this point with **NDSolve** are displayed in a single cell below using **GraphicsArray** within the **Show** command. This is named **setone**. The graphs are also shown simultaneously in **partone** by using the **Show** command without **GraphicsArray**.

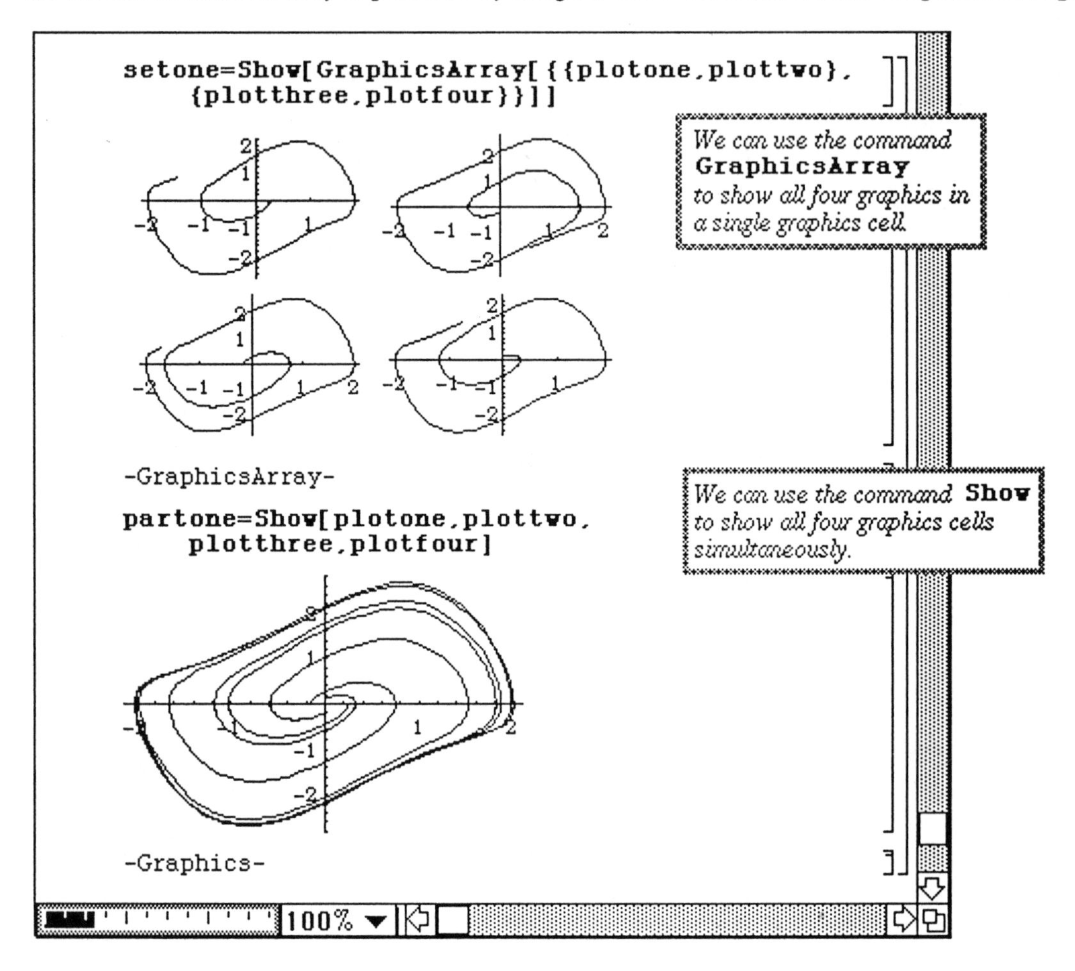

By choosing other initial conditions, several solutions to Van der Pol's equation are computed in the same manner as before. The graphs of these approximate solutions are named **cycle1**, **cycle2**, **cycle3**, and **cycle4**. They are first shown below in the single graphics cell, called **settwo**, and are then displayed simultaneously in **parttwo**. In the final command, the eight approximate solutions to Van der Pol's equation which have been computed with **NDSolve** are displayed simultaneously in **partthree**.

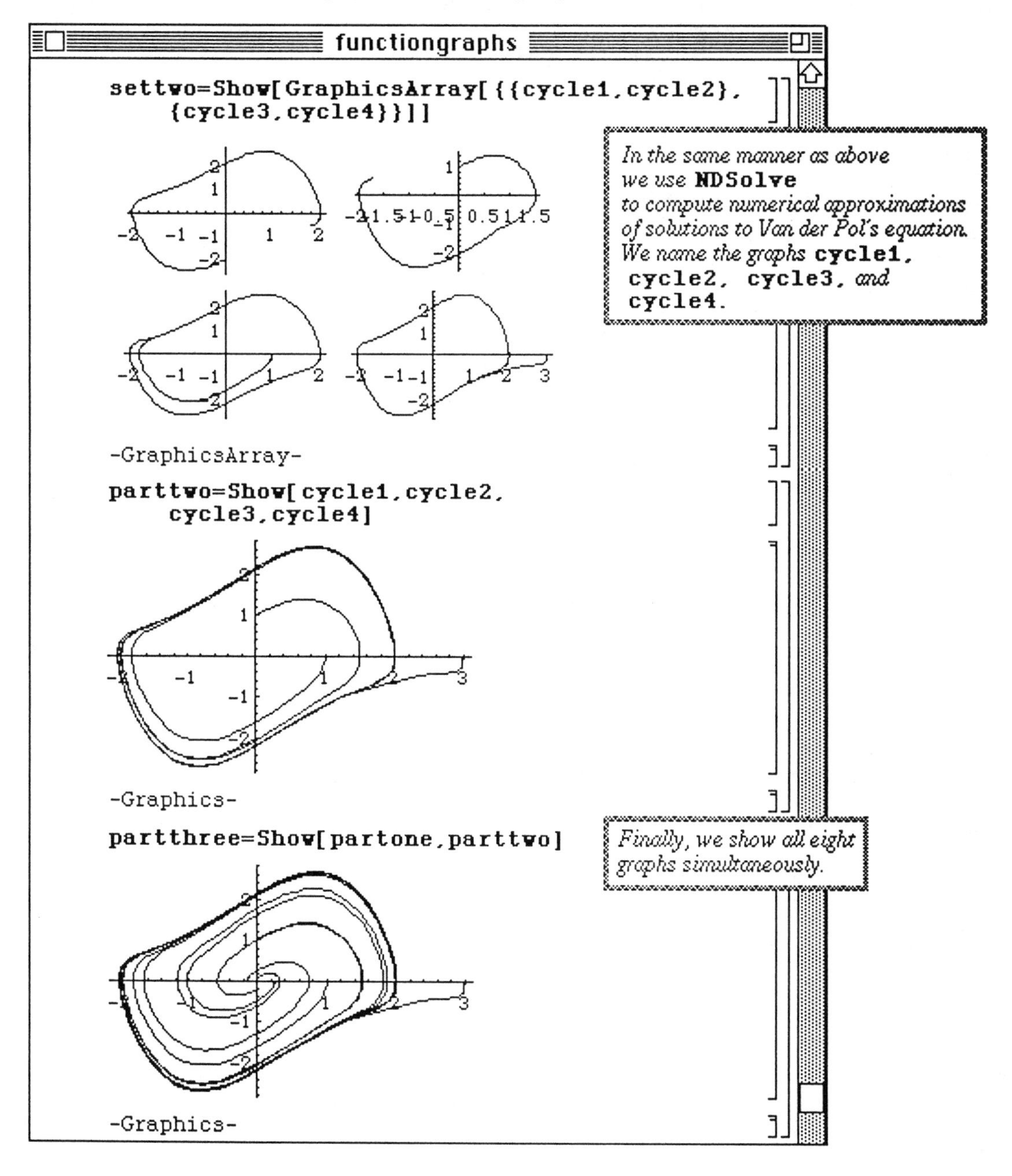

Chapter 7
Introduction to *Mathematica* Packages

Many useful commands and functions which are not automatically performed by *Mathematica* can be found in **Packages**. The following window is obtained by opening the **Packages** folder as shown below. There are thirteen folders located in **Packages** in Version 1.2 (fourteen folders are located in Version 2.0 **Packages**); each folder contains several packages. The contents of each folder can be seen by double-clicking on the appropriate icon.

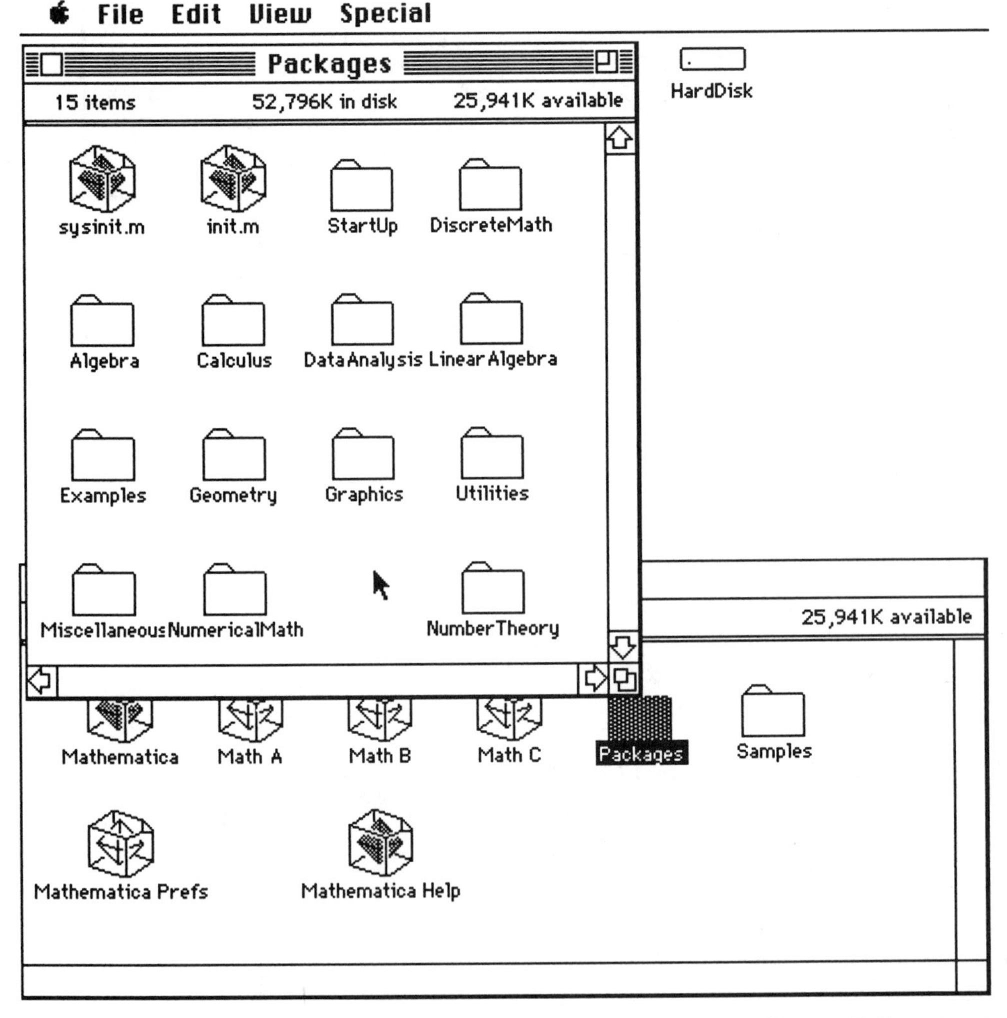

Chapters 7, 8, and **9** contain discussions of some *Mathematica* packages. **Chapter 7** discusses elementary packages from the **Algebra**, **Linear Algebra**, **Calculus**, and **Discrete Math** folders; **Chapter 8** discusses some of the packages contained in the **Graphics** folder; and **Chapter 9** discusses some of the more specialized packages contained in the **Numerical Math** and **Data Analysis** folders.

■ A Note Regarding Packages:

When directly opening a package, notice that the functions defined within the package are listed along with their definitions. Consequently, users can usually determine the purpose of a package by reading the beginning statements and experimenting.

▩ 7.1 Algebra

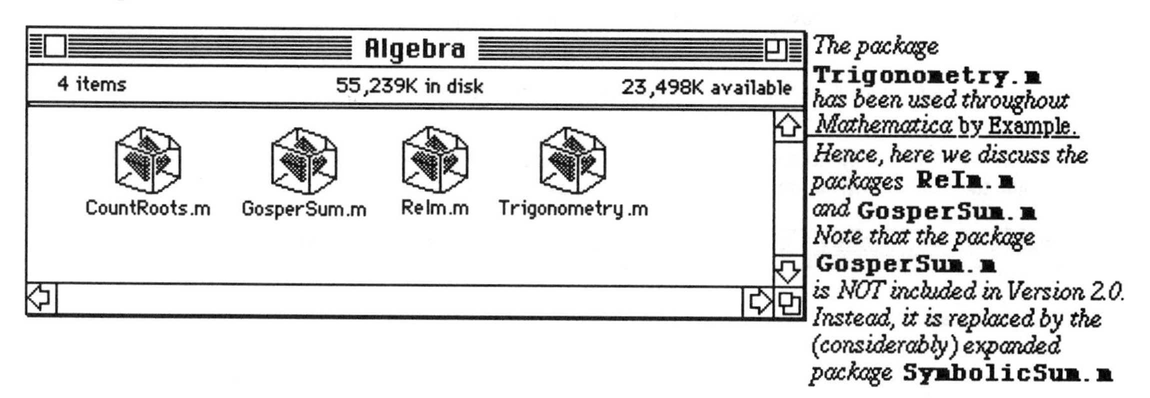

The package **Trigonometry.m** *has been used throughout* *Mathematica* by Example. *Hence, here we discuss the packages* **ReIm.m** *and* **GosperSum.m** *Note that the package* **GosperSum.m** *is NOT included in Version 2.0. Instead, it is replaced by the (considerably) expanded package* **SymbolicSum.m**

■ GosperSum.m

o The **GosperSum.m** package is not included with Version 2.0; instead it is superseded by the package **SymbolicSum.m**.

The symbolic summation of some series of the form $\sum_{k=1}^{n} a_k$, which are useful in many areas of mathematics can be determined with the use of the **GosperSum .m** package.

These calculations are not possible without this package. The built-in *Mathematica* command **Sum** can be used for finite sums as illustrated in the first example below. However, this command cannot find a closed form of the summation in the second example. Therefore, the **GosperSum.m** package must be loaded.

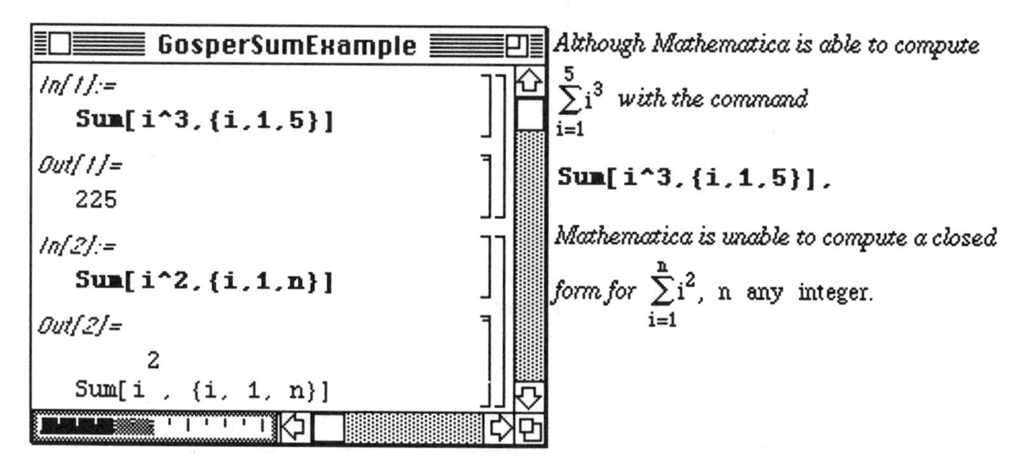

Although Mathematica is able to compute $\sum_{i=1}^{5} i^3$ *with the command*

Sum[i^3,{i,1,5}],

Mathematica is unable to compute a closed form for $\sum_{i=1}^{n} i^2$, n any integer.

□ Example:

The command which can be used to determine a closed form expression of some sums of the form $\sum_{k=kmin}^{kmax} a[k]$ is **GosperSum[a[k],{k, kmin , kmax }]**. After the package **GosperSum .m** is loaded, a table is constructed below which consists of the summation formula for each of the series $\sum_{k=1}^{n} k^i$ for $i = 2, 4, 6,$ and 8. Since **Print** is used in the **Table** command, the results appear in a print cell and can be accessed for later use. Although quite useful in many cases, **GosperSum** cannot compute the closed form summation for many series as illustrated in the second example.

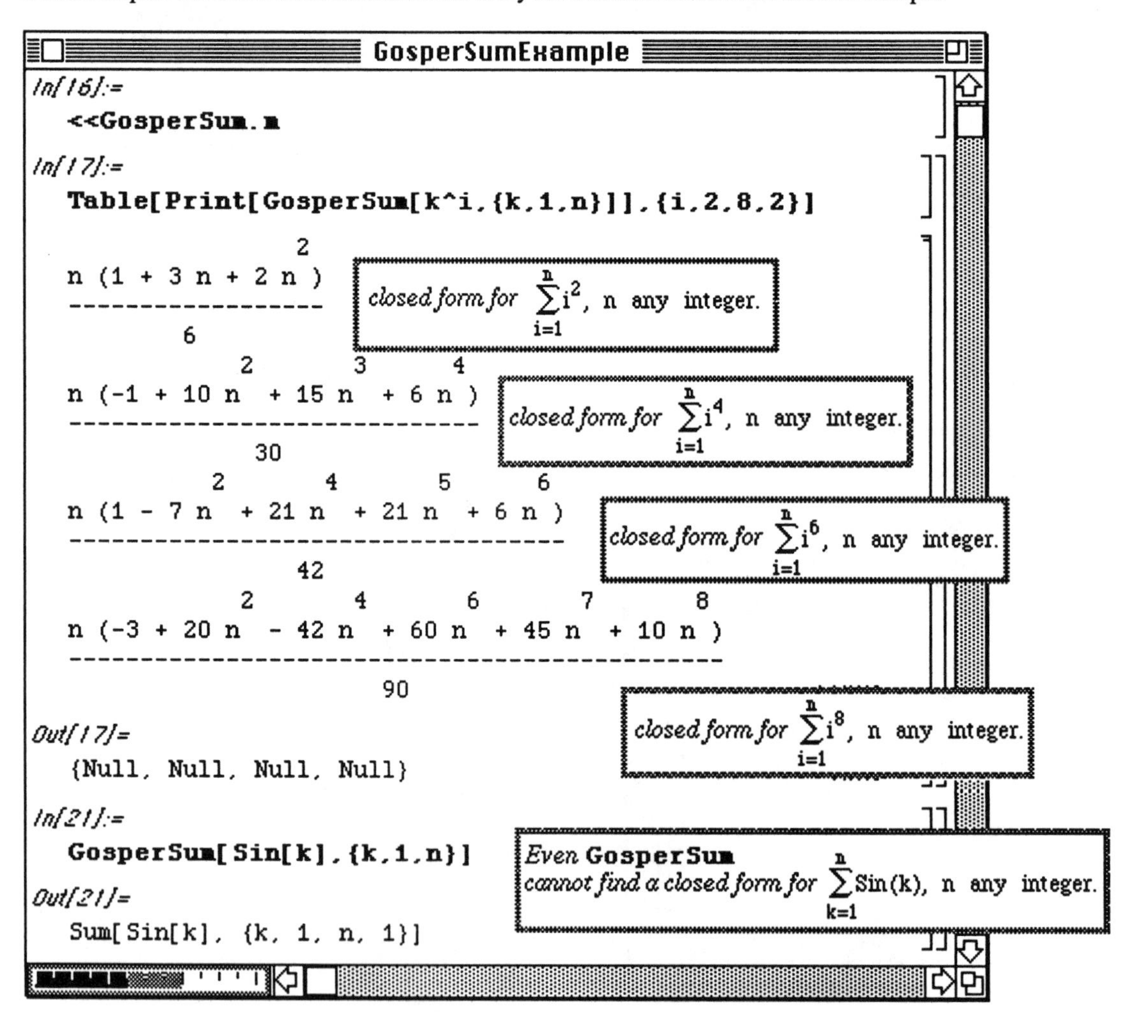

• SymbolicSum.m

o In Version 2.0 the package **SymbolicSum.m** replaces the package **GosperSum.m** from previous versions of *Mathematica*. In general, *Mathematica*'s standard built-in commands cannot compute symbolic sums of the form $\sum_{k=n_1}^{n_2} f(k)$ when n_1 and n_2 are not specific numbers.

The command **SymbolicSum[f[k],{k,n1,n2}]** attempts to write the symbolic sum $\sum_{k=n1}^{n2} \mathbf{f[k]}$ in a closed form when **f[k]** is a rational function.

o Example:

Find closed forms for (i) $\sum_{n=1}^{k}(n+2)(n-3)$ and (ii) $\sum_{n=1}^{k}\frac{1}{n+3}$. For (ii), evaluate when $k=100$.

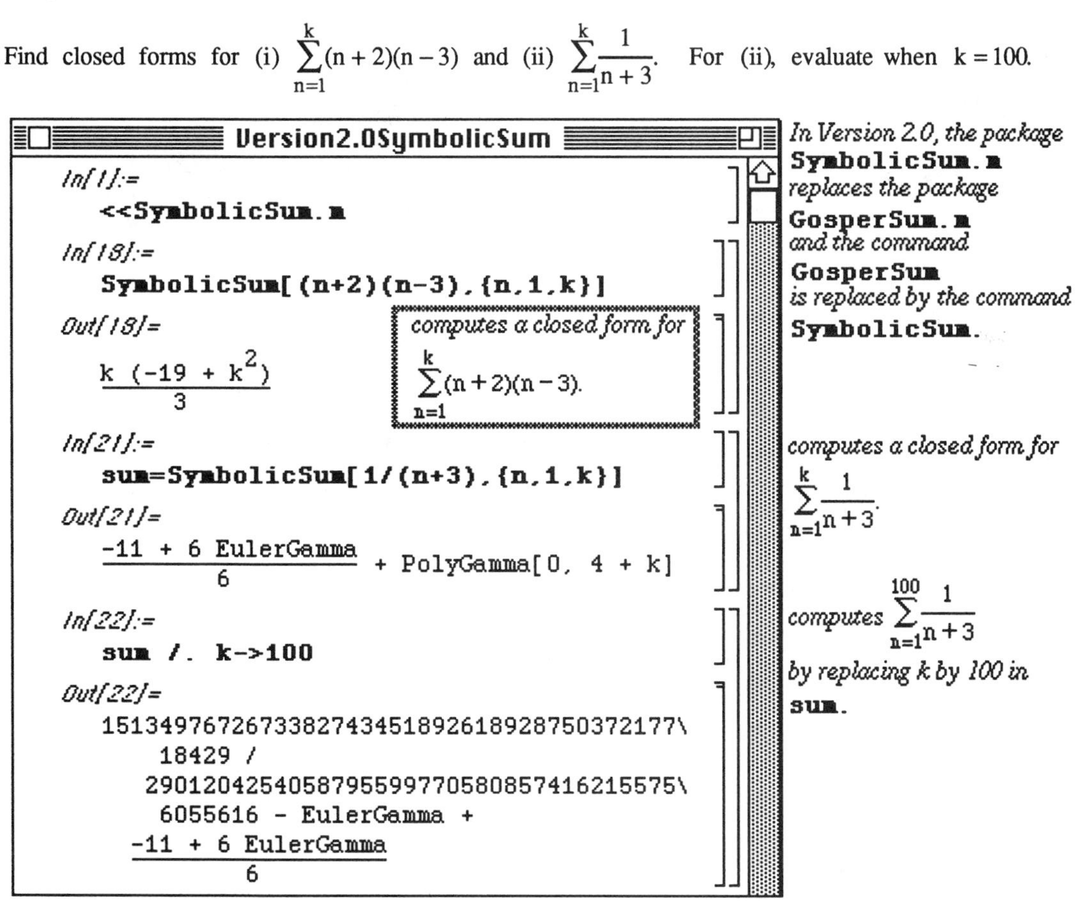

Since $\sum_{n=1}^{100}\frac{1}{n+3}$ is not a symbolic sum, the commands **Sum** and **NSum** can be used to calculate $\sum_{n=1}^{100}\frac{1}{n+3}$:

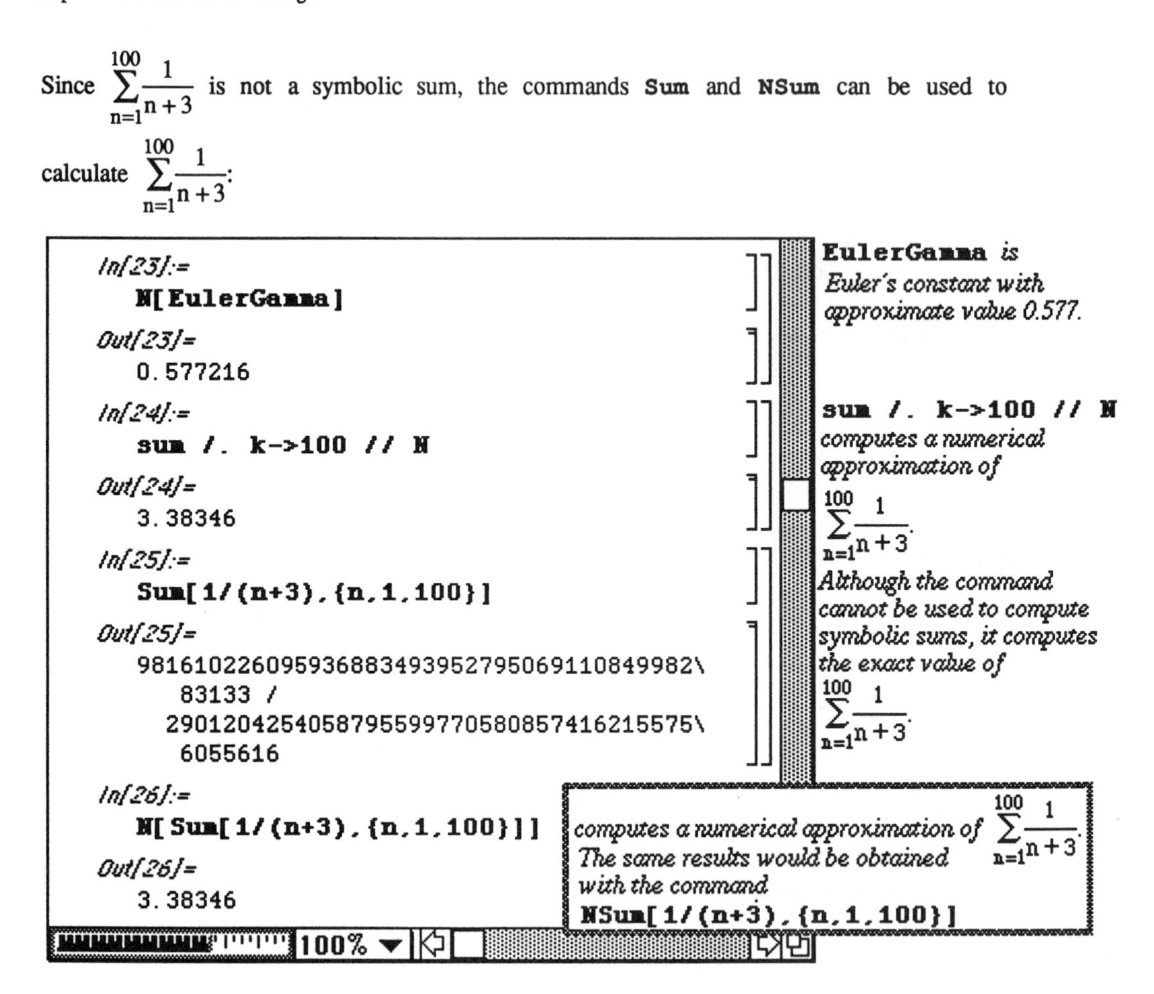

■ ReIm.m

The built-in *Mathematica* functions **Re[z]** and **Im[z]** compute the real and imaginary parts of the complex number **z**. However, these commands are not helpful when working with complex-valued functions as shown below. Since determining the real and imaginary parts of functions is important in many problems such as the solution of differential equations, a technique to find the quantities is necessary.

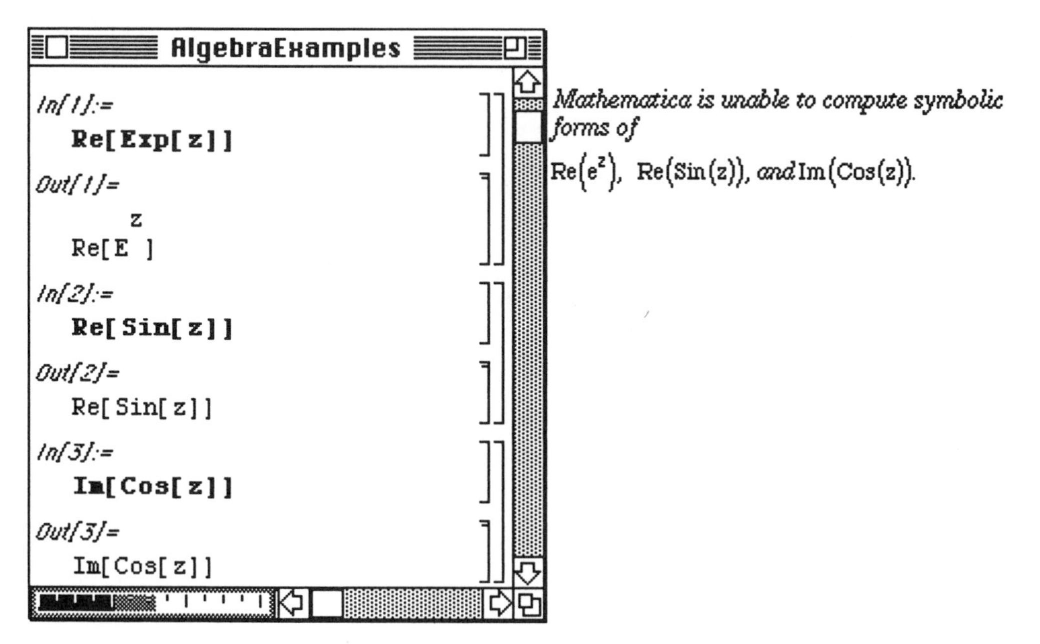

The package **ReIm.m** is loaded so that many of the identities from complex analysis can be used. Hence, simplification can be accomplished in the following way:

$$e^z = e^{x+iy} = e^x e^{iy} = e^x(\cos y + i \sin y).$$

Therefore, the real and imaginary parts are determined where **x=Re[z]** and **y=Im[z]** below. Similar calculations yield **Re[Sin[z]]** and **Im[Cos[z]]**.

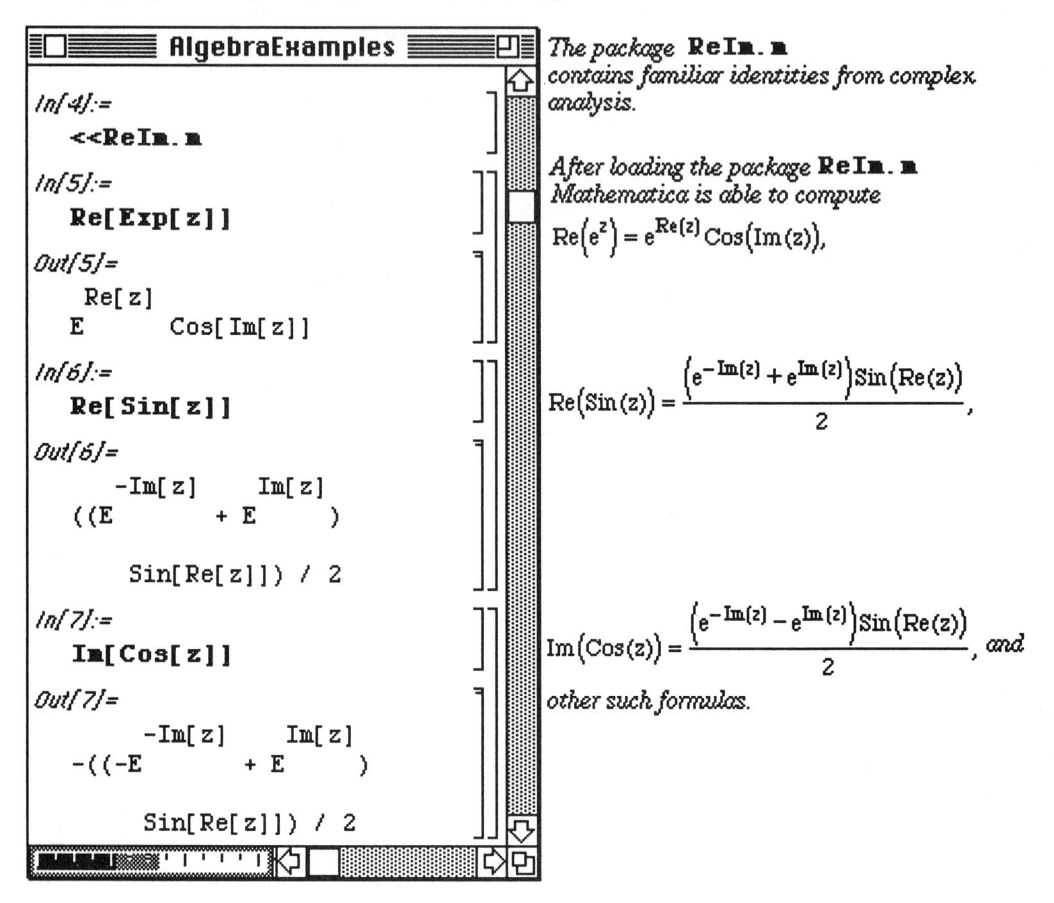

Example:

These techniques are useful in determining the solutions to differential equations with complex eigenvalues. Consider the second order linear equation with constant coefficients, y''+y'+y=0. Problems of this form arise when modeling spring problems. In this case, a spring of mass, m, and spring constant, k , is displaced in a medium with damping coefficient, a, where m = k = a. The solution of the equation is known to be

$y(x) = e^{-x/2}\left(c_1 \operatorname{Cos}\left(\frac{\sqrt{3}x}{2}\right) + c_2 \operatorname{Sin}\left(\frac{\sqrt{3}\,x}{2}\right)\right)$. Unfortunately, since this equation has complex eigenvalues, **DSolve** yields a solution involving complex terms. The **Trigonometry.m** package is useful in the application of DeMoivre's formula. Yet, the results obtained with the command **ComplexToTrig** are less than desirable . Hence, an alternate approach is demonstrated.

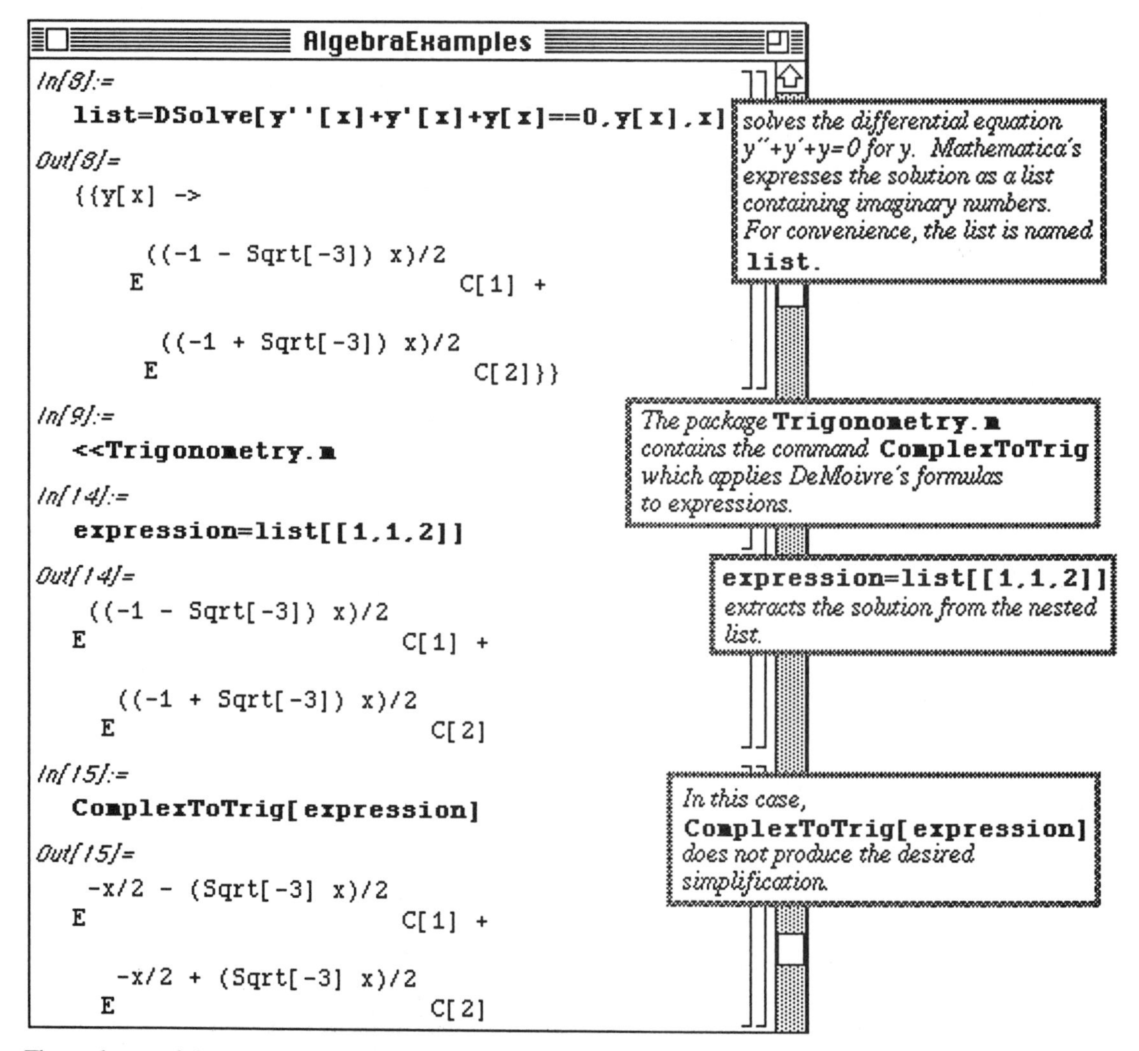

The real part of the solution, **rexpression**, is obtained with **Re[expression]**. Note that in the calculation of **rexpression**, the variable x is considered a complex number.

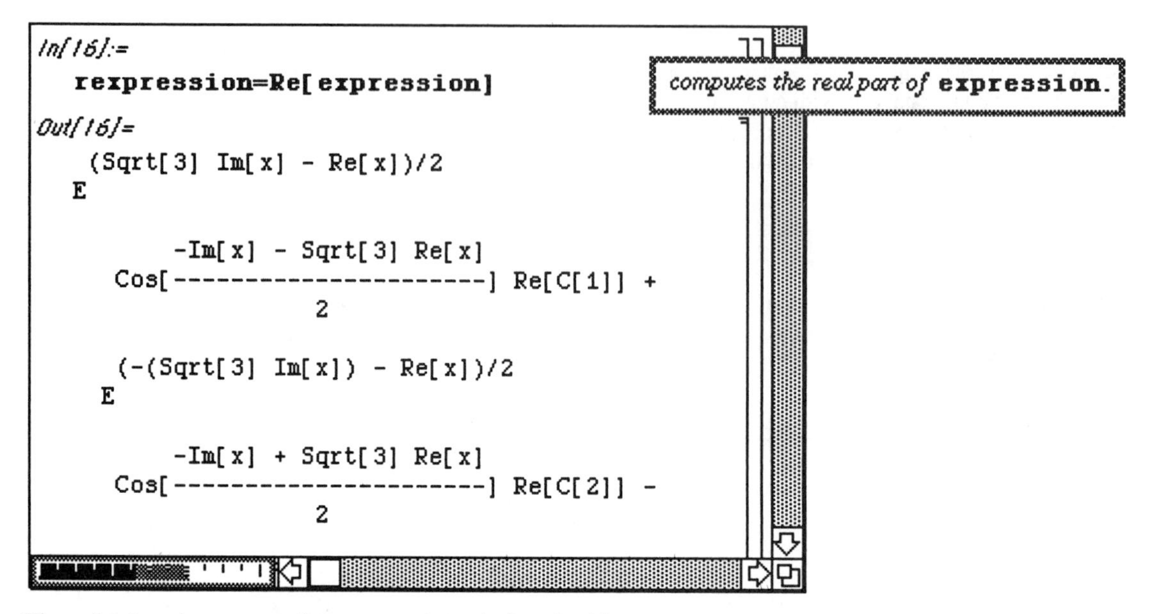

Then, the imaginary part of **expression** is found with **Im[expression]** and called **iexpression**.

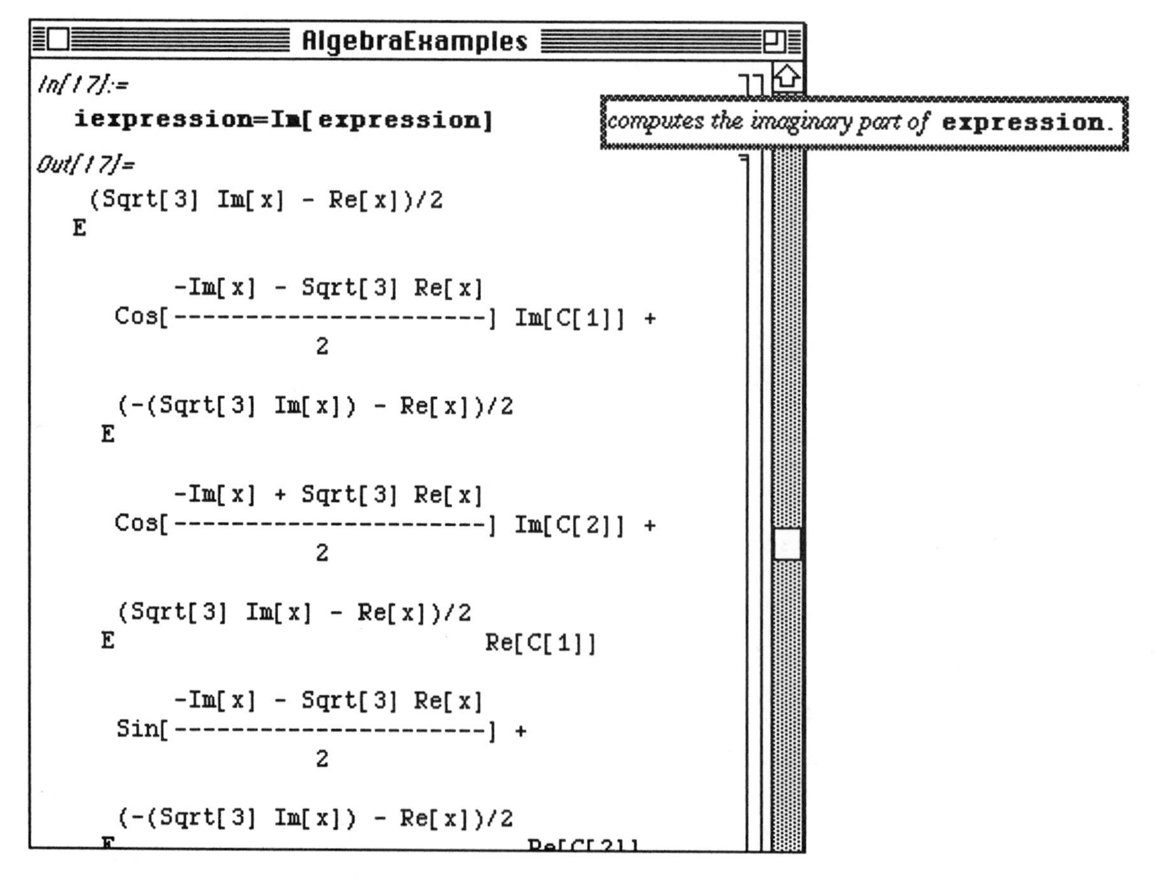

The expressions which result from the commands **Re** and **Im** can be further simplified by establishing several rules. This is done below with **rule**. Since **x** is a real number, **rule** replaces **Im[x]** with **0** and **Re[x]** with **x**. Also, the constants **C[1]** and **C[2]** are real, so they should have zero imaginary parts. This is done in **rule** with **Im[C[1]]->0, Im[C[2]]->0, Re[C[1]]->C[1]**, and **Re[C[2]]->C[2]**. Therefore, the command **rexpression/.rule** applies **rule** to **rexpression**. The result is called **solone**.

```
AlgebraExamples
In[18]:=
  rule={Im[x]->0,Re[x]->x,Im[C[1]]->0,
      Re[C[1]]->C[1],Im[C[2]]->0,
      Re[C[2]]->C[2]}
Out[18]=
  {Im[x] -> 0, Re[x] -> x, Im[C[1]] -> 0,

    Re[C[1]] -> C[1], Im[C[2]] -> 0,

    Re[C[2]] -> C[2]}
In[19]:=
  solone=rexpression /. rule          applies rule
                                      to expression.
Out[19]=
             -(Sqrt[3] x)
  C[1] Cos[------------]
                  2
  ---------------------- +
              x/2
             E

               Sqrt[3] x
    C[2] Cos[---------]
                   2
    -------------------
              x/2
             E
```

Next, **solone** is simplified and factored with **TrigReduce[solone]**. The resulting expression involves the constant, C[1]+C[2] which can be replaced with a single constant C[0]. This is done with **%/.(C[1]+C[2])->C[0]**. This gives the real part of the solution to the differential equation and is named **solutionone**.

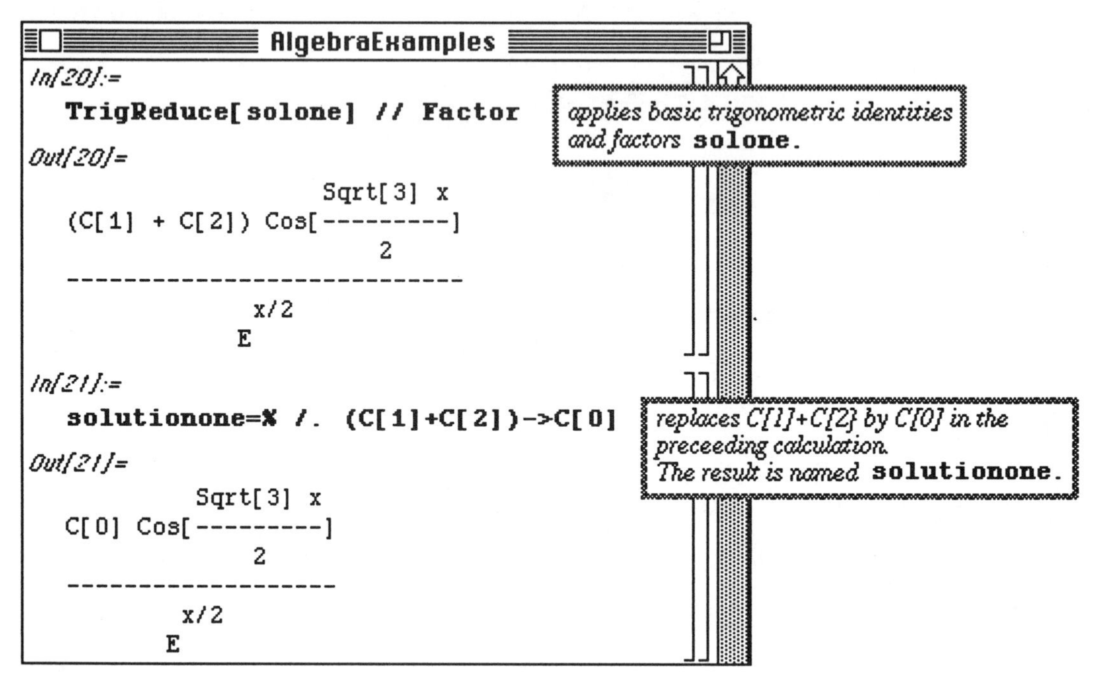

A similar approach leads to the imaginary part of the solution. First, **soltwo** is obtained by applying **rule** to **iexpression**. Then, **soltwo** is simplified and factored. This yields an expression involving the constant, `C[1]-C[2]`. This constant is replaced with the imaginary constant, `I C[1]`, in order to obtain the desired results. The expression which results after substitution of this constant is called **solutiontwo**. Notice that **solutiontwo** involves `I`.

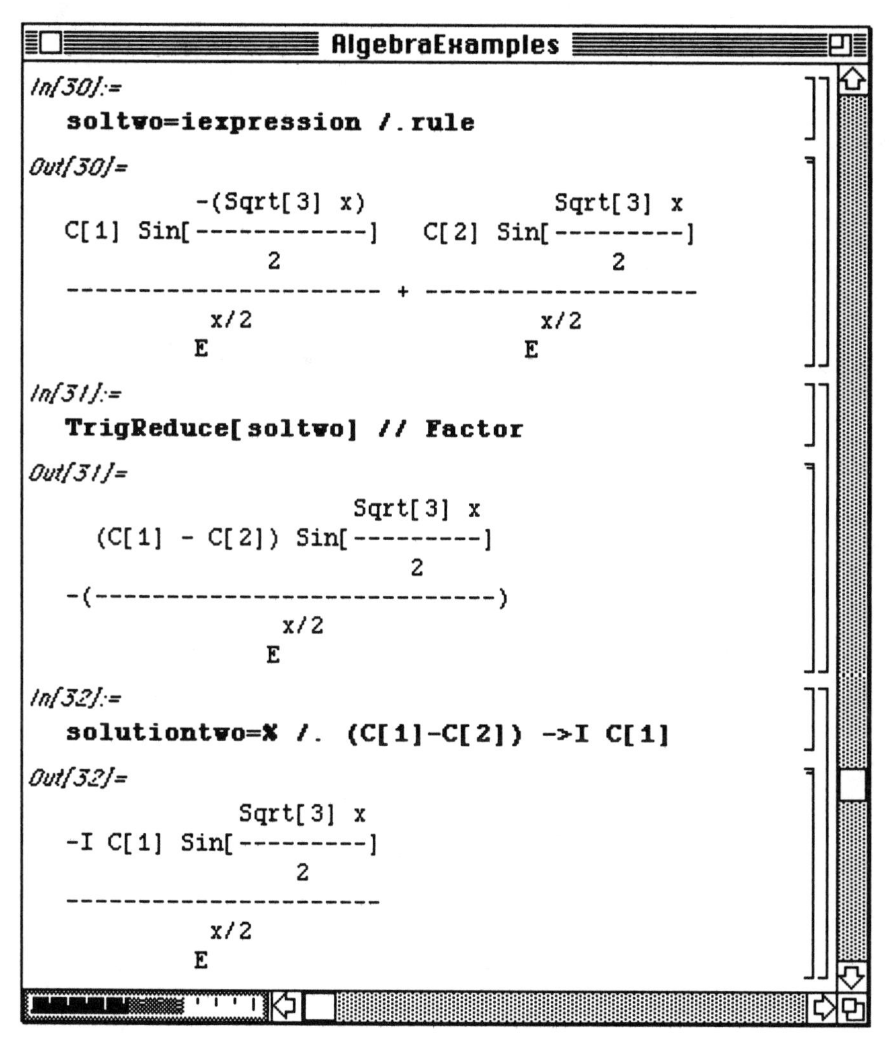

The general solution to the differential equation, **solution**, is therefore obtained with the linear combination of **solutionone** and **solutiontwo**, **solutionone+I solutiontwo**. The **solution** is then verified by the substitution of **solution** into the original differential equation.

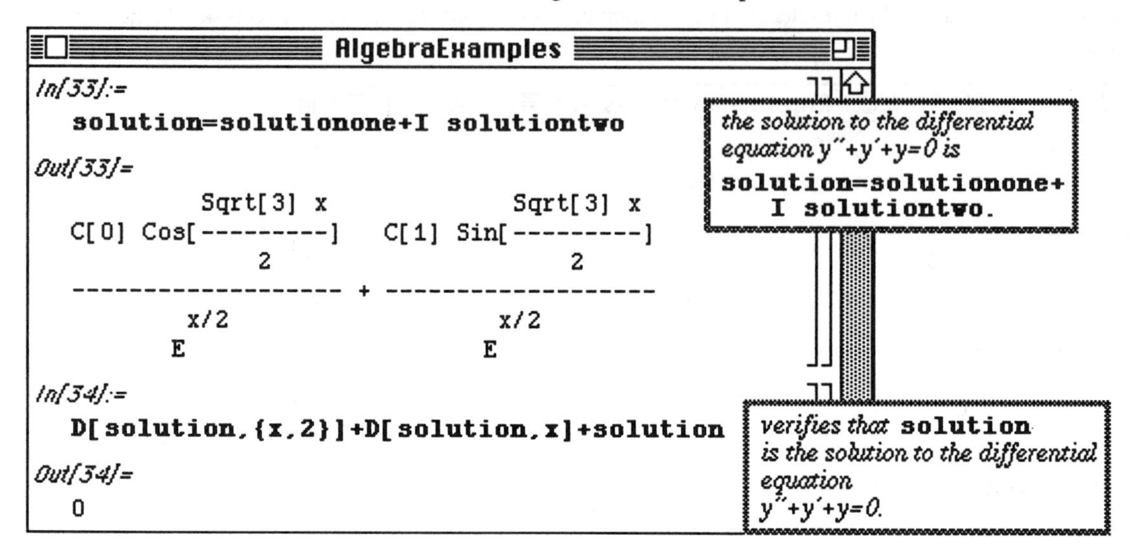

7.2 Linear Algebra

• Cholesky.m

○ **Cholesky.m** is contained within the **Linear Algebra** folder in Version 2.0; **Cholesky.m** is not available with Version 1.2.

The **complex conjugate transpose** of an m x n matrix A is the n x m matrix

$\overline{A}'$ obtained from A by taking the complex conjugate of each element of A and transposing the result.

An $m \times n$ matrix A is **Hermitian** means $A = \overline{A}'$. An $n \times n$ Hermitian matrix A is **positive definite** means the eigenvalues of A are positive. This is equivalent to saying

$\langle Ax, x \rangle = Ax \bullet \overline{x}$ is positive for every nonzero vector x.

The concept of positive definite matrices is of importance in many areas such as physics and geometry. An n x n matrix is **symmetric** if A is equal to its transpose.

The **Cholesky.m** package contains the command **CholeskyDecomposition[a]** which yields a matrix **u** such that

$u^T u = a$ where u^T = the transpose of **u**.

In order to determine this decomposition, however, **a** must be a symmetric, positive definite (and, hence, real) matrix. Therefore, the command **CholeskyDecomposition[a]** can serve as a test to determine if the matrix **a** is positive definite.

o Example:

This command is illustrated below with the 2x2 matrix, **{{2,1},{1,2}}**. The matrix which results is called **u** so that the property stated above can be verified with **a==Transpose[u].u**. Since a value of `True` is obtained, the result is correct.

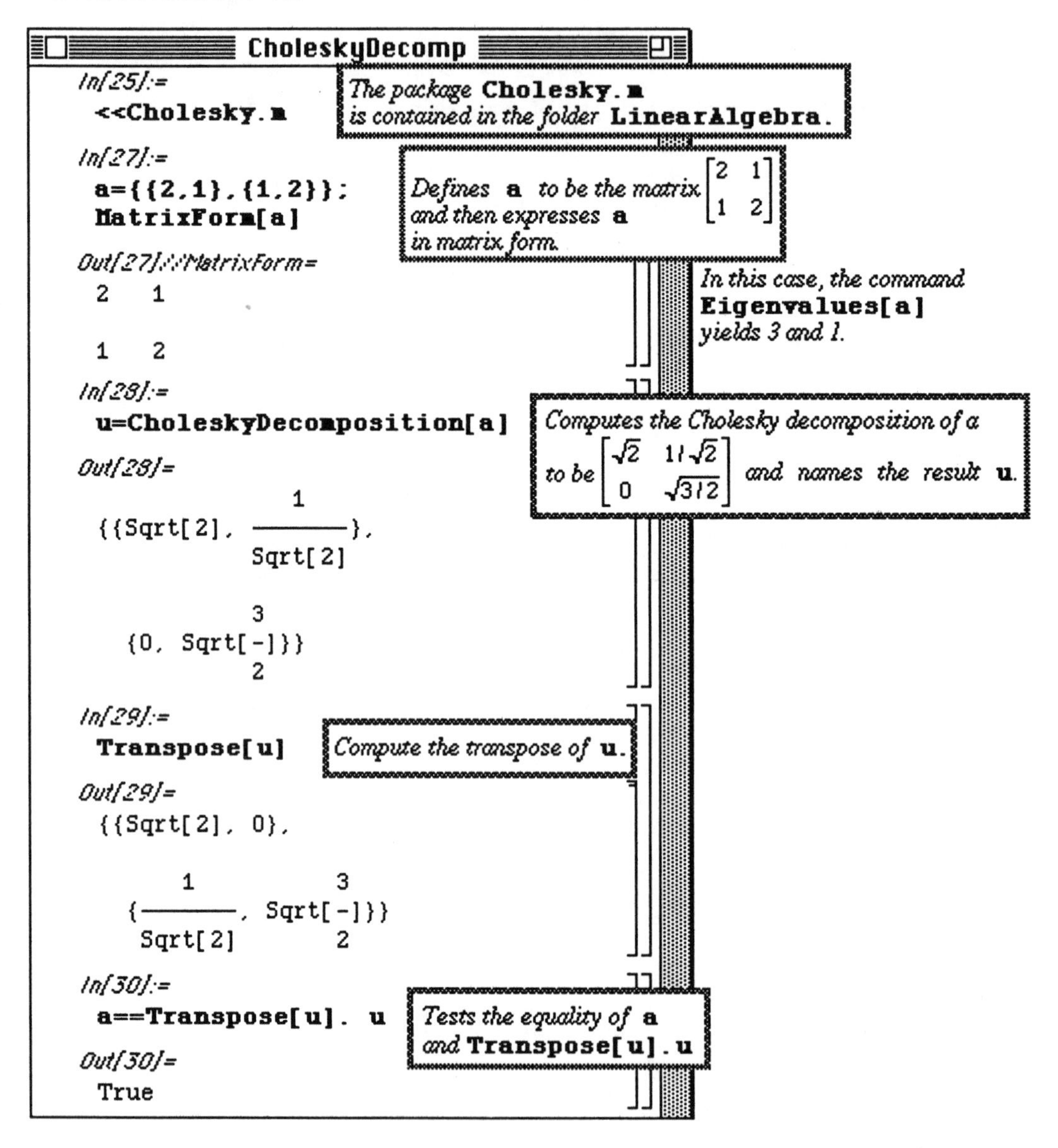

As mentioned above, the **CholeskyDecomposition** command only applies to symmetric, positive definite matrices. However, the command does not check for these properties before trying to perform the decomposition. This is demonstrated in the two examples below. In the following example, a non-symmetric matrix is considered. The resulting matrix is incorrect since a `False` response to **a==Transpose[u].u** is given:

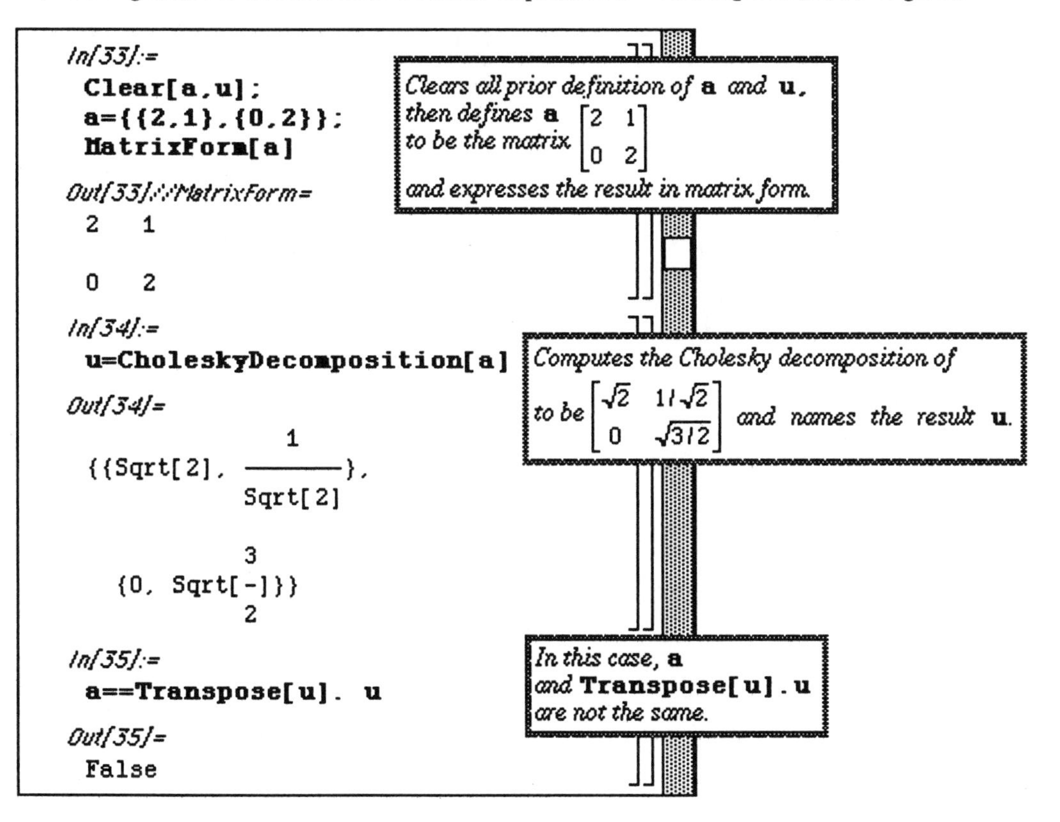

In the next case, the matrix is not positive definite, so several error messages result:

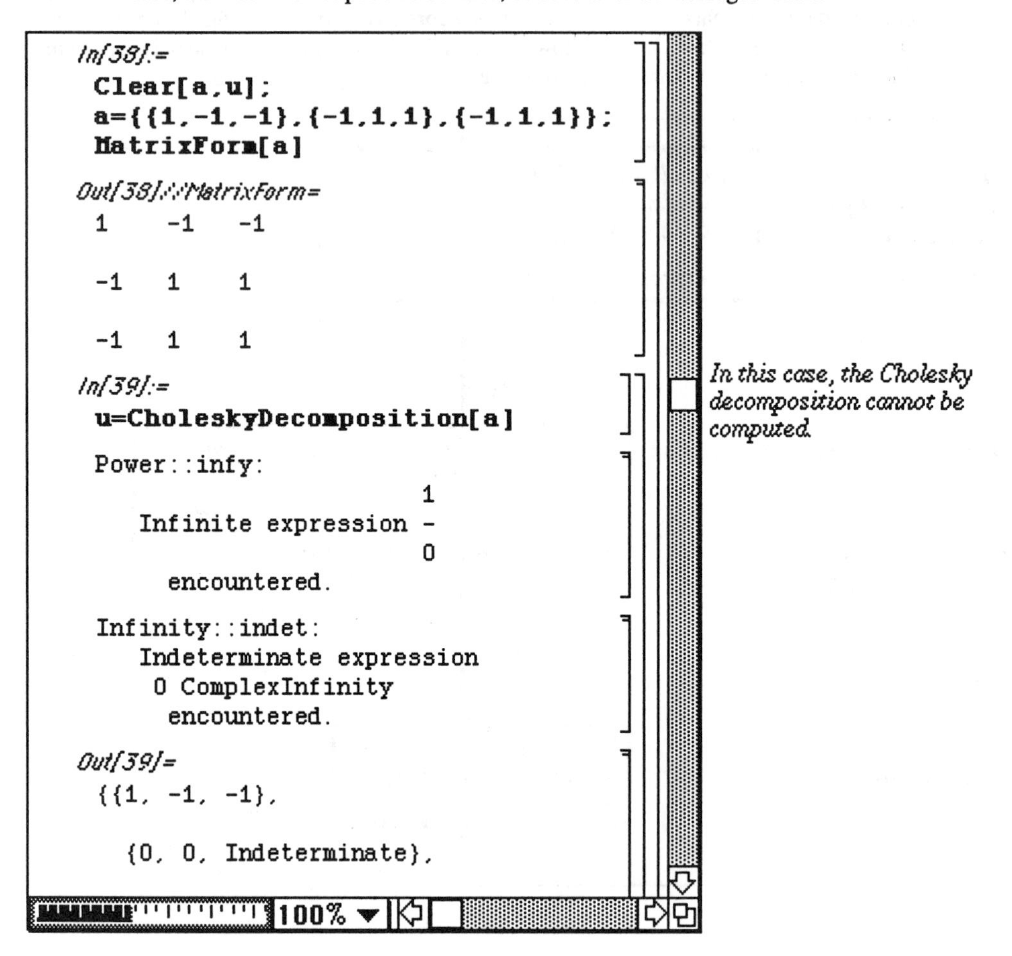

In this case, the Cholesky decomposition cannot be computed.

The matrix which results from the decomposition may contain imaginary numbers. This is shown below with the matrix **a**. However, the resulting matrix is verified since **Transpose[r].r==a**.

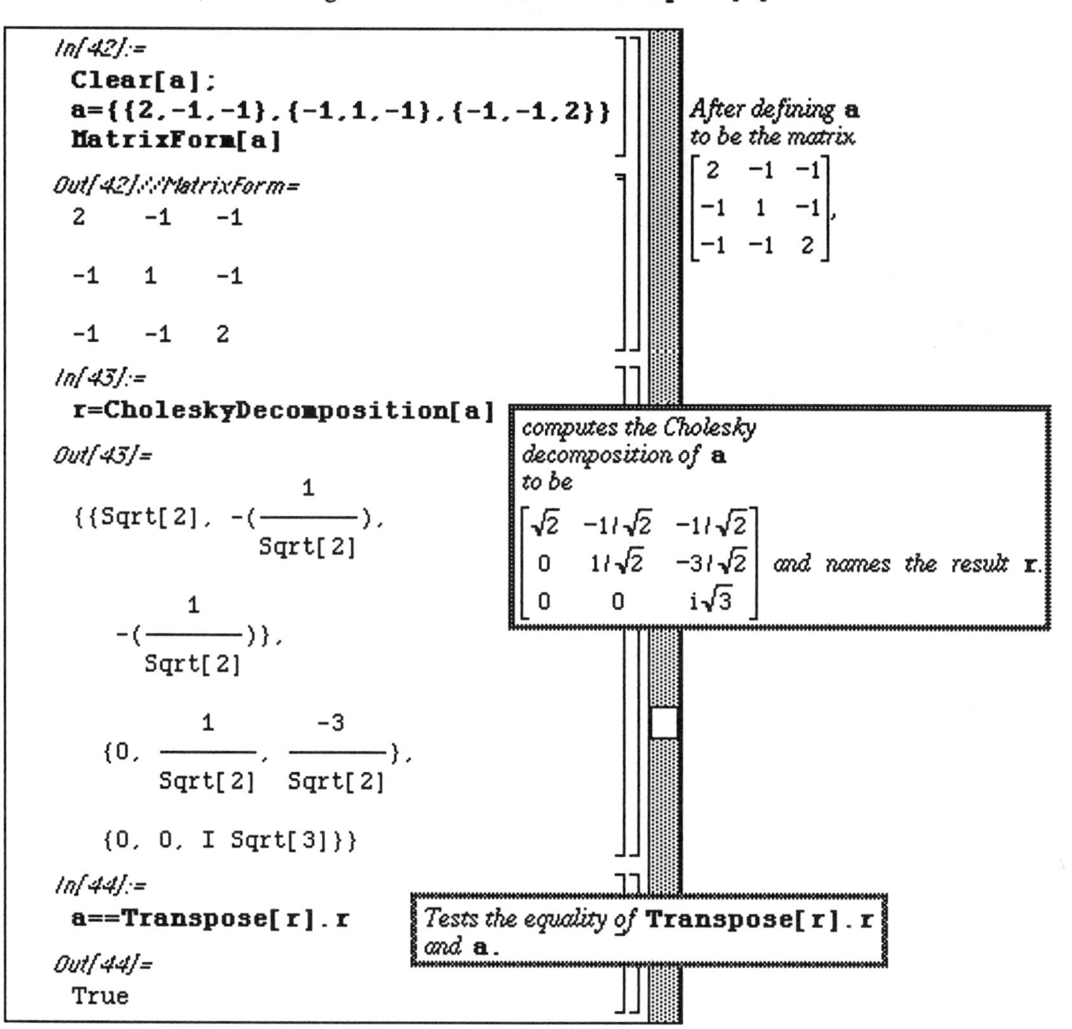

o Application: Quadratic Equations

An application of positive definite matrices is the analysis of quadratic equations.

For any symmetric matrix A, the product $f = x^T Ax$ is a pure quadratic form :

$$x^T Ax = [x_1, x_2, \ldots, x_n] \begin{bmatrix} a_{11} & a_{12} & \cdots & a_{1n} \\ a_{21} & a_{22} & \cdots & a_{2n} \\ \vdots & \vdots & \vdots & \vdots \\ a_{n1} & a_{n2} & \cdots & a_{nn} \end{bmatrix} \begin{bmatrix} x_1 \\ x_2 \\ \vdots \\ x_n \end{bmatrix}$$

$$= a_{11}x_1^2 + a_{12}x_1x_2 + a_{21}x_2x_1 + \cdots + a_{nn}x_n^2$$

$$= \sum_{i=1}^{n}\sum_{j=1}^{n} a_{ij}\, x_i\, x_j \, .$$

If the matrix A is positive definite, then $f > 0$ for all values of x. Consider the following quadratic equation:

$$2x^2 - 2xy + 2y^2 - 2xz + 2yz + 2z^2 = -1.$$

We attempt to determine if this equation has any real roots by considering the matrix A. If the symmetric matrix

$$A = \begin{bmatrix} 2 & -1 & -1 \\ -1 & 2 & 1 \\ -1 & 1 & 2 \end{bmatrix}$$

which results when this equation is represented as $x^T\, Ax$ is positive definite, then $x^T\, Ax > 0$ for all x. Hence, $x^T\, Ax \neq -1$ for any values of x.

The matrix A is defined below as **a**. Note that this matrix yields the appropriate quadratic form as shown with **Expand[Transpose[vec].a.vec]** where **vec={x,y,z}**. The matrix A is then shown to be positive definite by using **CholeskyDecomposition**. Therefore, this equation has no real solutions.

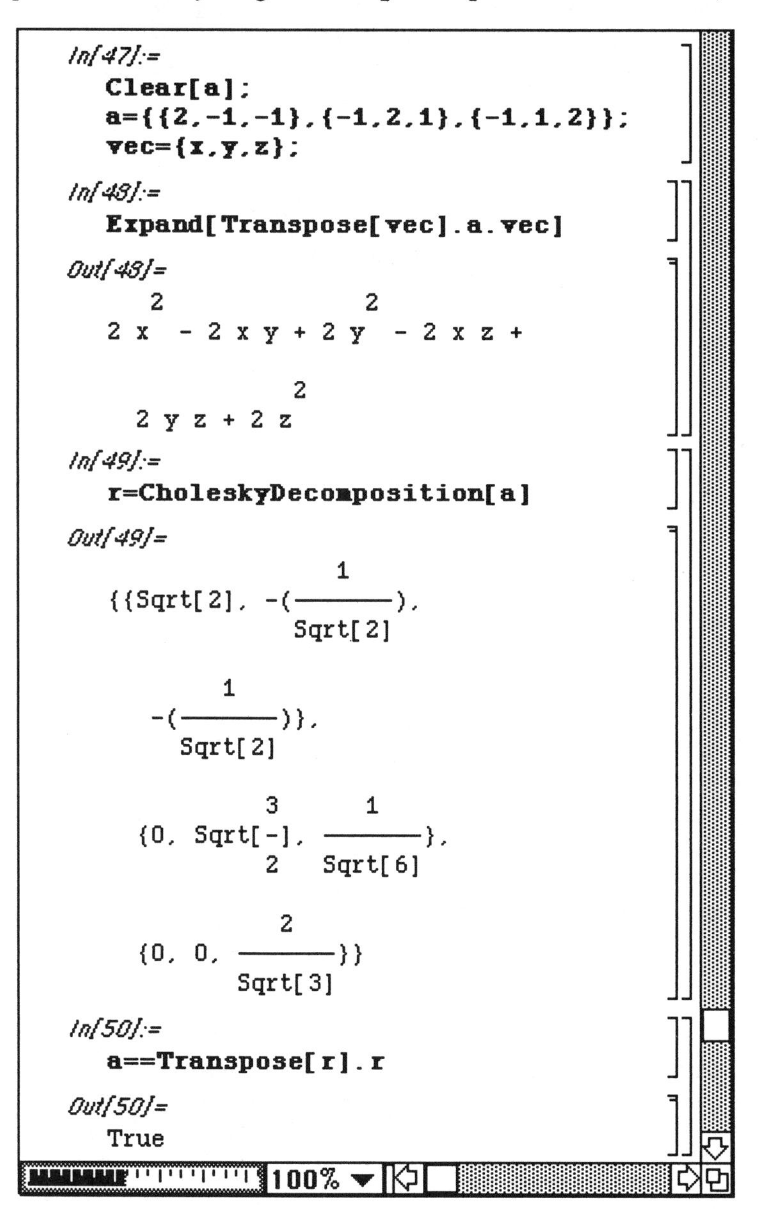
```
In[47]:=
  Clear[a];
  a={{2,-1,-1},{-1,2,1},{-1,1,2}};
  vec={x,y,z};

In[48]:=
  Expand[Transpose[vec].a.vec]

Out[48]=
     2             2
  2 x  - 2 x y + 2 y  - 2 x z +

                 2
    2 y z + 2 z

In[49]:=
  r=CholeskyDecomposition[a]

Out[49]=
                          1
  {{Sqrt[2], -(---------),
                       Sqrt[2]

         1
    -(---------)},
       Sqrt[2]

           3        1
    {0, Sqrt[-], ---------},
           2     Sqrt[6]

              2
    {0, 0, ---------}}
           Sqrt[3]

In[50]:=
  a==Transpose[r].r

Out[50]=
  True
```

■ CrossProduct.m

Neither Version 1.2 nor Version 2.0 contain a built-in command for computing the cross product of two three-dimensional vectors. In order to compute cross products of vectors, Version 1.2 users must load the package **Cross.m** in the **Linear Algebra** folder to use the command **Cross[vec1,vec2]** which computes the cross product of vectors **vec1** and **vec2**; Version 2.0 users must use the package **CrossProduct.m** in the **Linear Algebra** folder to use **CrossProduct[vec1,vec2]**. Of course, the vector which results from the cross product is orthogonal to each of the original vectors. This is verified below as is the property that the cross product of two parallel vectors is the zero vector.

□ **Example:**

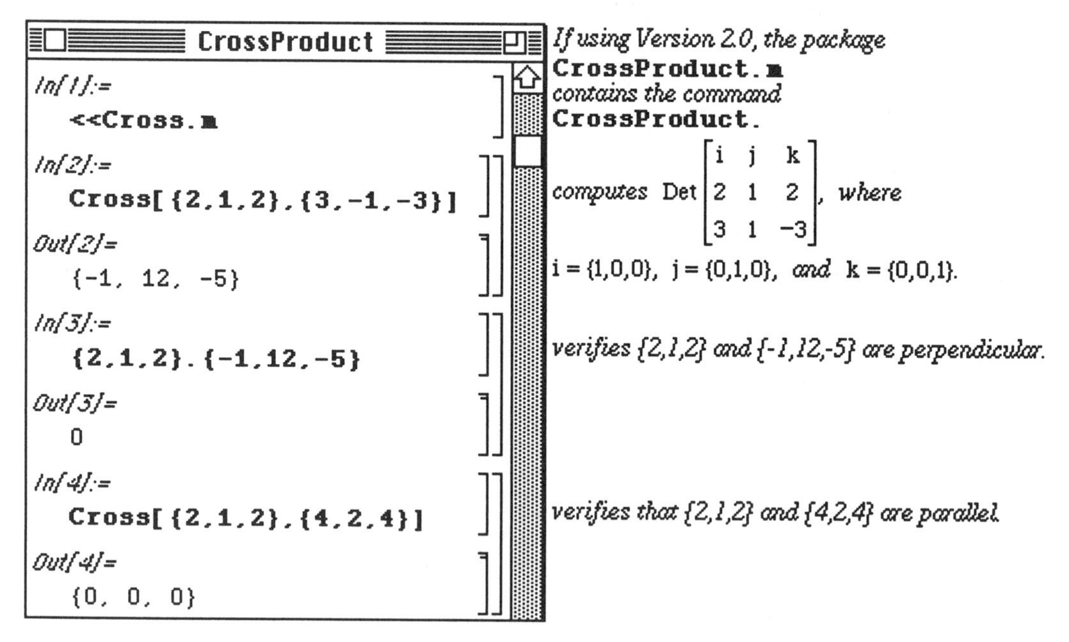

□ **Example:**

One application of the cross product is in computing the area of a triangle the vertices of which are the points P, Q, and R. This area is given by the formula

$$\text{Area} = \frac{1}{2}\left|\overline{PQ} \times \overline{PR}\right|.$$

Mathematica does not contain a built-in function to calculate the length of vectors. Hence, the function **length[v]** which yields this value is defined as the square root of the dot product of **v** with itself. This function is illustrated with the vector **{1,2,-1}** to show that the exact value results. There is no restriction on the dimension of vectors to be used with **length** as seen below with **{1,3,-2,4,0,8}**.

The area of the triangle is computed by, first, defining the points **p**, **q**, and **r**; and then determining the vectors **pq** and **pr**. The numerical approximation of the area is easily found with
`N[.5 length[Cross[pq,pr]]]`.

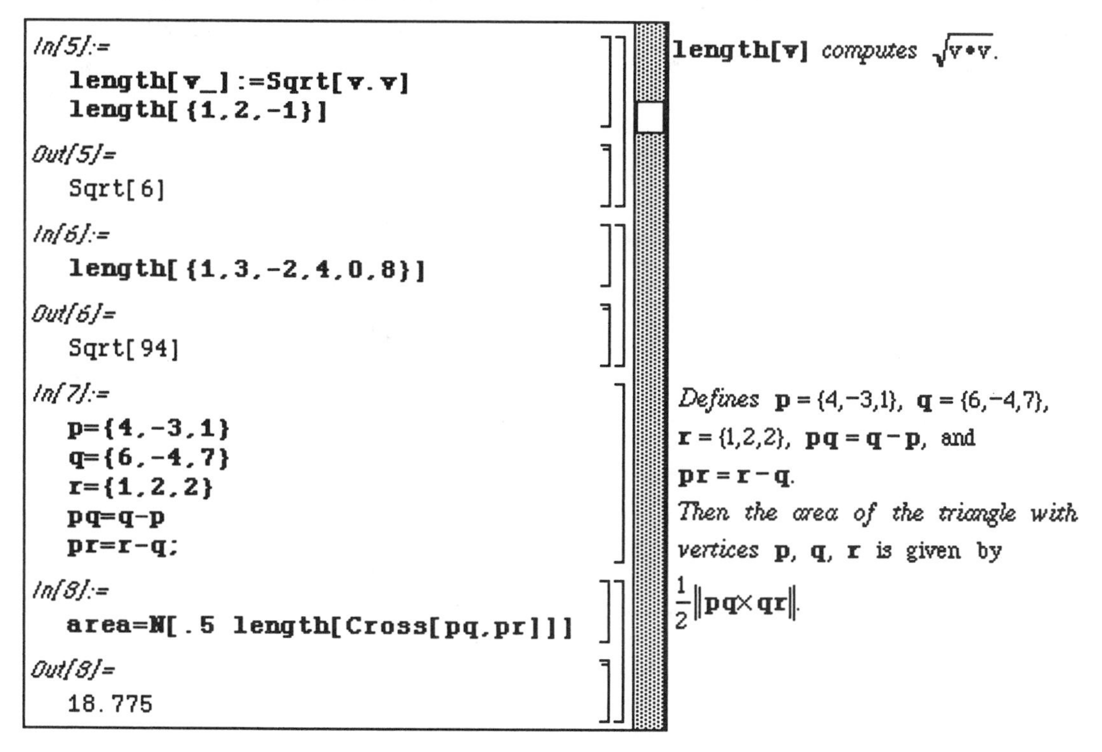

This triangle is shown below:

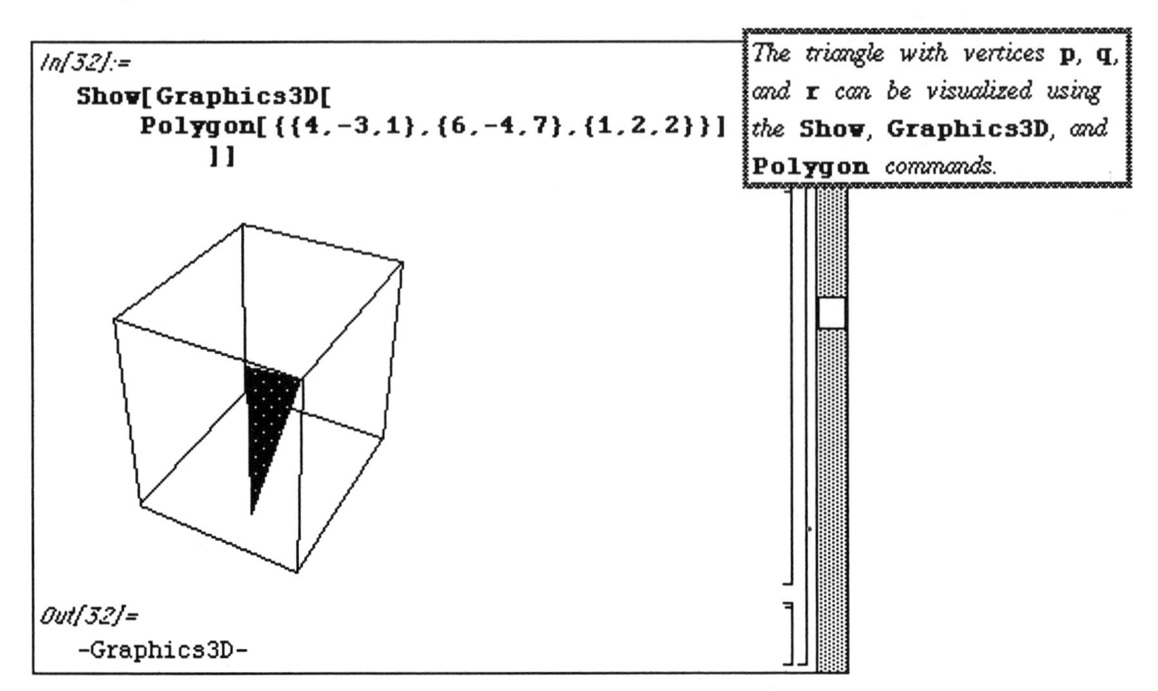

Another similar application of the cross product is in finding the distance from a line containing the points P and Q to a point R not on the line. This distance is known to be

$$d = \frac{1}{\left|\overline{PQ}\right|}\left|\overline{PQ} \times \overline{PR}\right|.$$

A function which computes this distance is defined as **distance** below and is illustrated by finding the distance from the point (2,1,-2) to the line through the points (3,-4,1) and (-1,2,5). Since **distance** is defined in terms of the points P, Q, and R, this distance is found by simply entering the appropriate points.

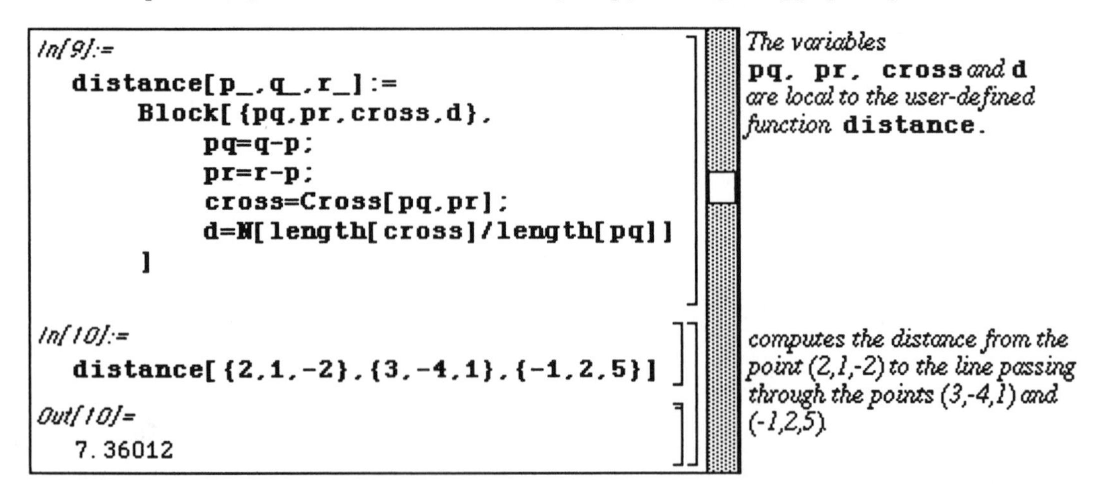

A useful feature of the **Cross[v1,v2]** command is that when numerical vectors are not given, *Mathematica* gives the formula used to compute the cross product. This can be used as a tool in finding vectors with certain properties. For example, the function **cp[p,q,r]** defined below computes the cross product of the vectors **pq** and **pr**. The arbitrary points **{p1,p2,p3}**, **{q1,q2,q3}**, and **{r1,r2,r3}** are used below to show that the cross product formula is obtained. Then, this formula is used to determine the value(s) of **p3** such that the triangle formed by the points **{1,-1,p3}**, **{1,2,0}**, and **{-3,5,-1}** has area **a**. (Hence, the cross product has magnitude 2**a**.) This is accomplished by substituting the known points into **cp** and solving for **p3**.

```
In[11]:=
  Clear[p,q,r];
  cp[p_,q_,r_]:=Cross[q-p,r-p]
  vec=cp[{1,-1,p3},{1,2,0},{-3,5,-1}]
Out[11]=
  {-3 + 3 p3, 4 p3, 12}
In[12]:=
  l=length[vec]
Out[12]=
                       2                  2
  Sqrt[144 + 16 p3  + (-3 + 3 p3) ]
In[14]:=
  sol=Simplify[Solve[l==2a,p3]]
Out[14]=
                                      2
            18 + Sqrt[-14976 + 400 a ]
  {{p3 -> --------------------------},
                       50

                                      2
            18 - Sqrt[-14976 + 400 a ]
    {p3 -> --------------------------}}
                       50
```

The following calculations reveal two possibilities for **p3**. However, these roots are imaginary if **a** < 6.11882 (Negative values of **a** are disregarded.) Therefore, only values of **a** greater than 12.2376 can be considered. An example is worked and verified for **a**=7.

```
In[15]:=
  Solve[-14976+400a^2==0,a]//N
Out[15]=
  {{a -> 6.11882}, {a -> -6.11882}}
In[16]:=
  sol=Solve[l==14,p3]//N
Out[16]=
  {{p3 -> -1.}, {p3 -> 1.72}}
In[17]:=
  p=sol[[1,1,2]]
Out[17]=
  -1.
In[18]:=
  test=cp[{1,-1,p},{1,2,0},{-3,5,-1}]
Out[18]=
  {-6., -4., 12.}
In[19]:=
  length[test]//N
Out[19]=
  14.
```

• MatrixManipulation.m

o Version 2.0 contains the package **MatrixManipulation.m** which contains several commands useful for manipulating matrices. **MatrixManipulation.m** is not contained with Version 1.2. The command **AppendColumns[m1, m2, m3, ...]** yields a new matrix composed of the submatrices **m1, m2**, ..., by joining the columns of **m1, m2**, ... while **AppendRows[m1, m2, ...]** performs a similar operation by joining the rows of the matrices. In each command, the submatrices must have the same number of columns or rows, respectively. Several examples of these commands are given below and viewed in **MatrixForm** to better understand the results. After defining the 2x2 matrices **a** and **b**, the command **AppendColumns[a,b]** appends the columns of **b** to the columns of **a**, and **AppendRows[a,b]** adds the rows of **b** to those of **a**.

o **Example:**

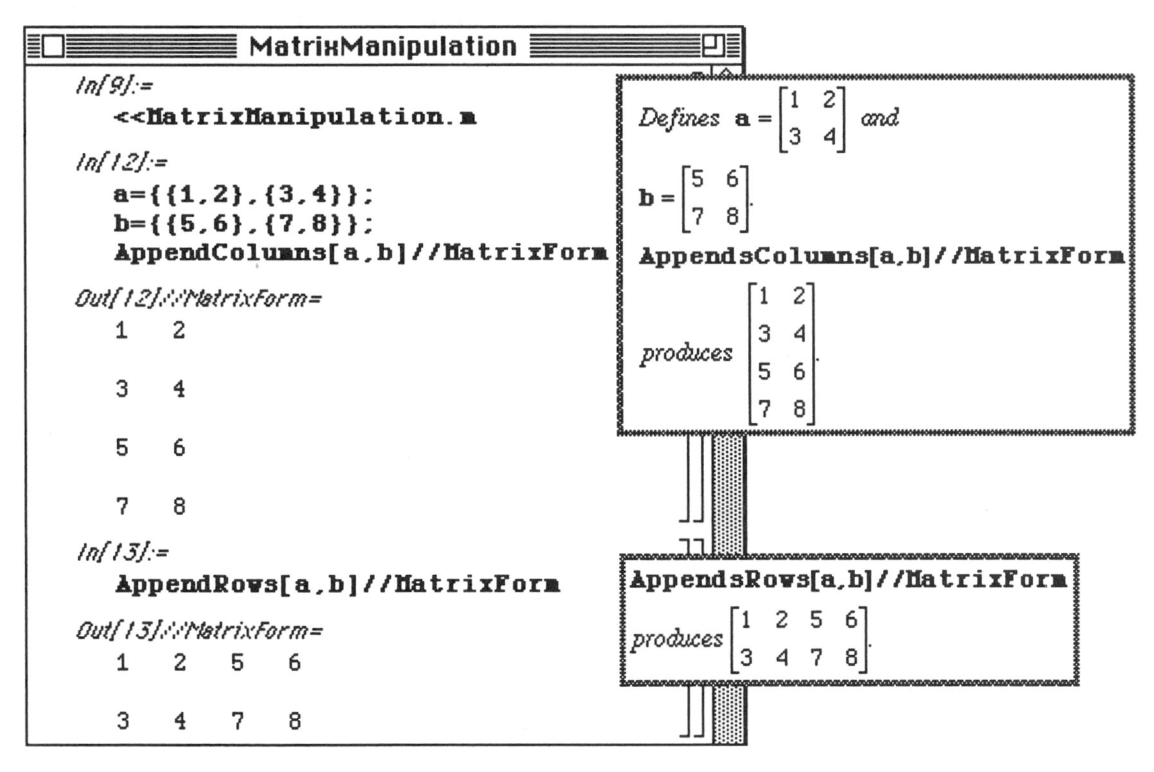

The matrix **c** is then defined to illustrate that more than two matrices can be used as arguments as well as the fact that correct dimensions must be used. The command **AppendColumns[a,c]** is not evaluated since **c** has more columns than **a**.

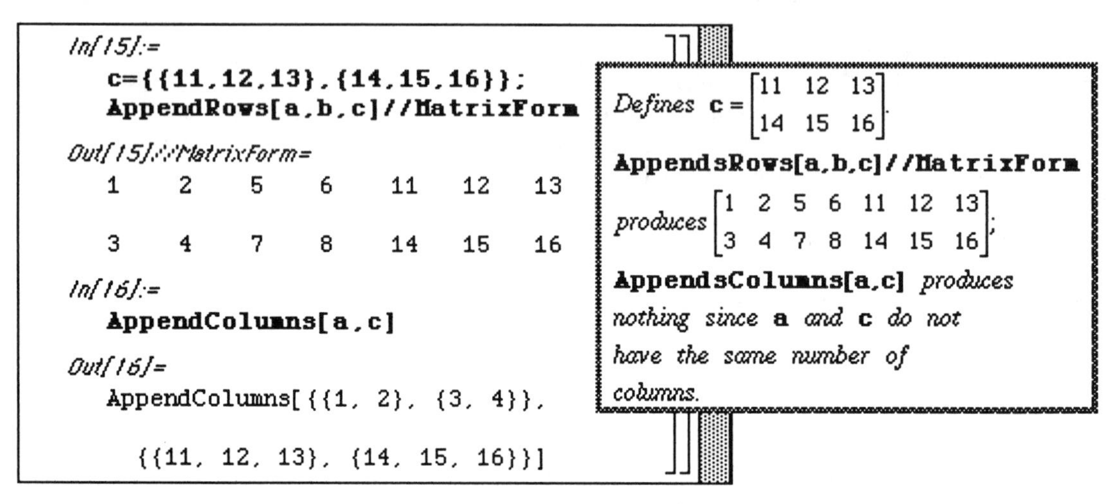

o <u>Application:</u> **Computing the Adjacency Matrix of a Graph**

An application of these commands is the manipulation of the adjacency matrix of a graph. Recall that two vertices of a graph are said to be **<u>adjacent</u>** if there is at least one edge joining them. Consider the graph G with no loops and n vertices labeled 1, 2, ..., n. The **<u>adjacency matrix of G</u>** is the n x n matrix in which the entry in row i and column j is the number of edges joining the vertices i and j. For example, suppose that a graph has the

$$\text{adjacency matrix } A=\begin{bmatrix}0&1&0&1\\1&0&1&2\\0&1&0&1\\1&2&1&0\end{bmatrix}.$$

This matrix is represented as **adj** below. Then, suppose that two more vertices are added to the graph with vertex 5 adjacent to vertices 2 and 3; and vertex 6 adjacent to vertices 1 and 4. Instead of defining a new adjacency matrix for the revised graph (which can be quite cumbersome in many cases), these additions can be made with **AppendColumns** and **AppendRows**.

This is done below in **addtocols** and **addtorows**. The rows added in **addtocols** represent the edges from vertices 5 and 6 to vertices 1, 2, 3, and 4 while the columns added in **addtorows** give these same edges from the original set of vertices to vertices 5 and 6.

```
In[18]:=
  adj={{0,1,0,1},{1,0,1,2},{0,1,0,1},{1,2,1,0}};
  addtocols=AppendColumns[adj,{{0,1,1,0},{1,0,0,1}}]

Out[18]=
  {{0, 1, 0, 1}, {1, 0, 1, 2}, {0, 1, 0, 1}, {1, 2, 1, 0},

    {0, 1, 1, 0}, {1, 0, 0, 1}}

In[20]:=
  addtorows=AppendRows[addtocols,
                  {{0,1},{1,0},{1,0},{0,1},{0,1},{1,0}}];
  addtorows//MatrixForm

Out[20]//MatrixForm=
  0   1   0   1   0   1

  1   0   1   2   1   0

  0   1   0   1   1   0

  1   2   1   0   0   1

  0   1   1   0   0   1

  1   0   0   1   1   0
```

An interesting fact concerning an adjacency matrix M is that the (i,j) th-element of the kth power of M represents the number of walks of length k from vertex i to vertex j. A **walk of length k** in a graph is a succession of k edges. This is important in problems in which the number of ways to travel between two locations must be determined.

Using the matrix given in **addtorows**, the number of walks of length 2 between every vertex pair is determined from **twowalks**. For example, there are 4 walks of length 2 from vertex 4 to vertex 5 as seen with **twowalks[[4,5]]**. The number of walks of length 3 are found in **threewalks**.

```
In[21]:=
  twowalks=addtorows.addtorows
Out[21]=
  {{3, 2, 2, 3, 2, 1}, {2, 7, 3, 2, 1, 4},
    {2, 3, 3, 2, 1, 2}, {3, 2, 2, 7, 4, 1},
    {2, 1, 1, 4, 3, 0}, {1, 4, 2, 1, 0, 3}}
In[22]:=
  twowalks[[4,5]]
Out[22]=
  4
In[23]:=
  threewalks=twowalks.addtorows
Out[23]=
  {{6, 13, 7, 10, 5, 8},
    {13, 10, 10, 23, 14, 5},
    {7, 10, 6, 13, 8, 5},
    {10, 23, 13, 10, 5, 14},
    {5, 14, 8, 5, 2, 9}, {8, 5, 5, 14, 9, 2}}
```

As the power increases, the built-in command **MatrixPower[matrix,k]** is useful. This is used below to find the number of walks of length 10.

```
In[24]:=
  tenwalks=MatrixPower[adj,10]
Out[24]=
  {{17408, 28160, 17408, 28160},
    {28160, 46080, 28160, 45056},
    {17408, 28160, 17408, 28160},
    {28160, 45056, 28160, 46080}}
```

MatrixManipulation.m also contains several well-known matrices such as the Hilbert matrix. The

Hilbert matrix H is given by $H = \begin{bmatrix} 1 & \frac{1}{2} & \frac{1}{3} & \cdots \\ \frac{1}{2} & \frac{1}{3} & \frac{1}{4} & \cdots \\ \frac{1}{3} & \frac{1}{4} & \frac{1}{5} & \cdots \\ \vdots & & & \end{bmatrix}$.

The n x n Hilbert matrix is found with **HilbertMatrix[n]** and is illustrated with **HilbertMatrix[3]** below. This matrix is named **hm3** and its inverse is called **hm3inv**. This is verified with **hm3.hm3inv** which yields the 3 x 3 identity matrix. Next, the 10 x 10 Hilbert matrix, **hm10**, is computed so that the system of equations **Ax** = **b** can be solved where **A** = **hm10** and **b={1,0,0,0,0,0,0,0,0,0}**.

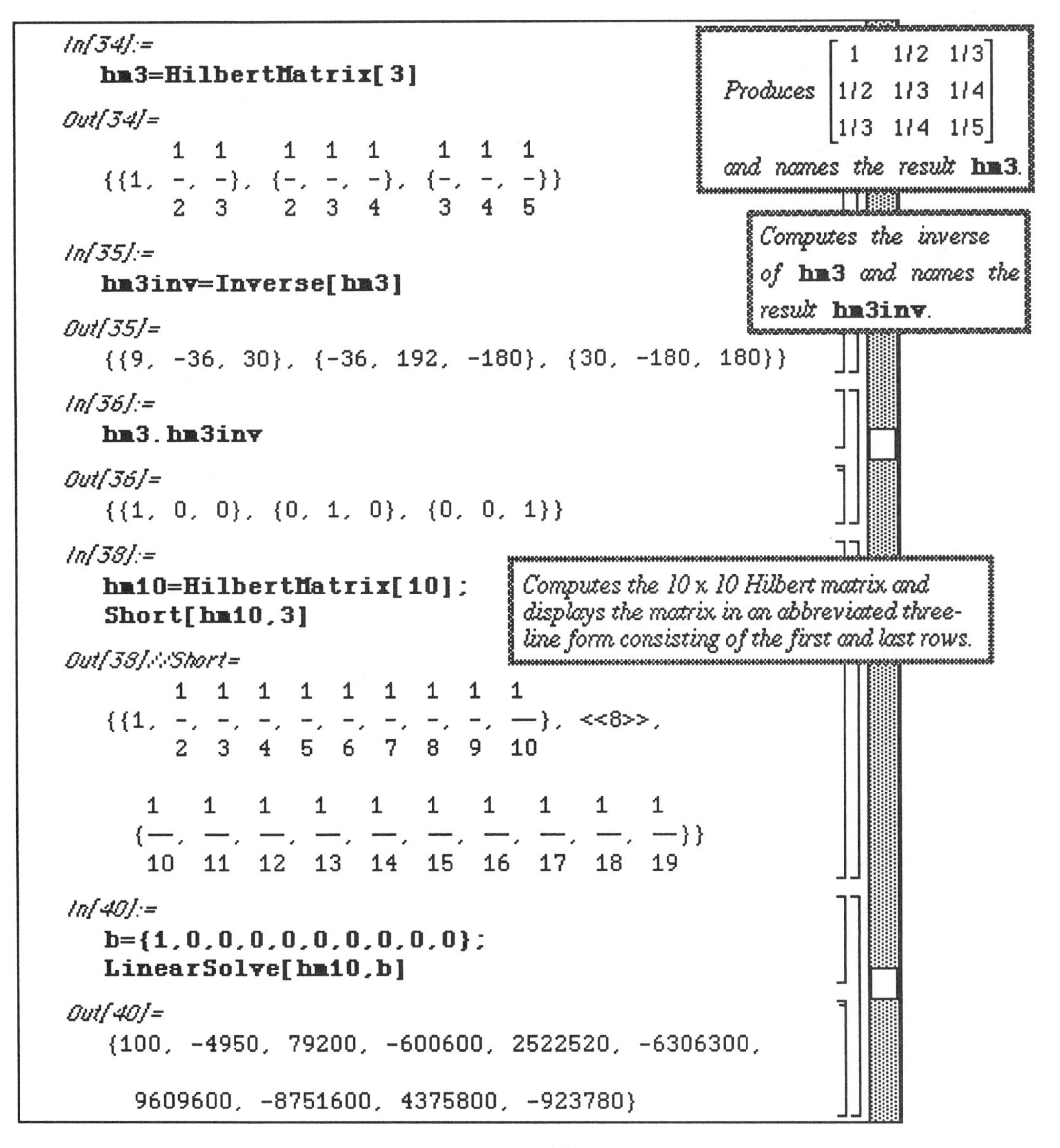

This system is also solved with **b1={.75,0,0,0,0,0,0,0,0,0}** to illustrate that this system is sensitive to small changes.

```
In[40]:=
    b={1,0,0,0,0,0,0,0,0,0};
    LinearSolve[hm10,b]

Out[40]=
    {100, -4950, 79200, -600600, 2522520, -6306300,

      9609600, -8751600, 4375800, -923780}

In[42]:=
    b1={.75,0,0,0,0,0,0,0,0,0};
    LinearSolve[hm10,b1]

Out[42]=
                                                    6
    {75., -3712.5, 59400., -450450., 1.89189 10 ,

                   6             6             6
      -4.72972 10 , 7.2072 10 , -6.5637 10 ,

                  6
      3.28185 10 , -692835.}
```

The sensitivity of the system **Ax=b** can be measured with the **condition number** of **A**. This number is defined in several ways, one of which is based on the 1-norm of **A**:

For a nonsingular m × m matrix A, the **condition number of A** is denoted c(A) and is defined

by $c(A) = \|A\| \left\|A^{-1}\right\|$ where $\|A\| = \text{Max}\left\{\sum_{i=1}^{m} |a_{ij}|\right\}$ for j = 1, 2, ..., m.

If c(A) is small, then A is said to be **well-conditioned**; A is **ill-conditioned** otherwise.

The function **norm[matrix]** is defined below to compute the 1-norm of **matrix**. This definition is given in terms of a general m x n matrix. Note that this definition involves the built-in *Mathematica* command **Dimensions[matrix]** which gives the dimensions of an mxn matrix in the form {m,n} as demonstrated below with **Dimensions[hm3]**.

The 1-norm is calculated for **hm3** and **hm3inv** using the function **norm**. The condition number of **matrix** is then defined as the product of the norm of **matrix** and the norm of its inverse. This function is called **cnum** and is illustrated with **hm3** to reveal a very large number. This is expected based on the results from the previous problem in which small changes in the original system involving **hm3** led to large changes in the solution.

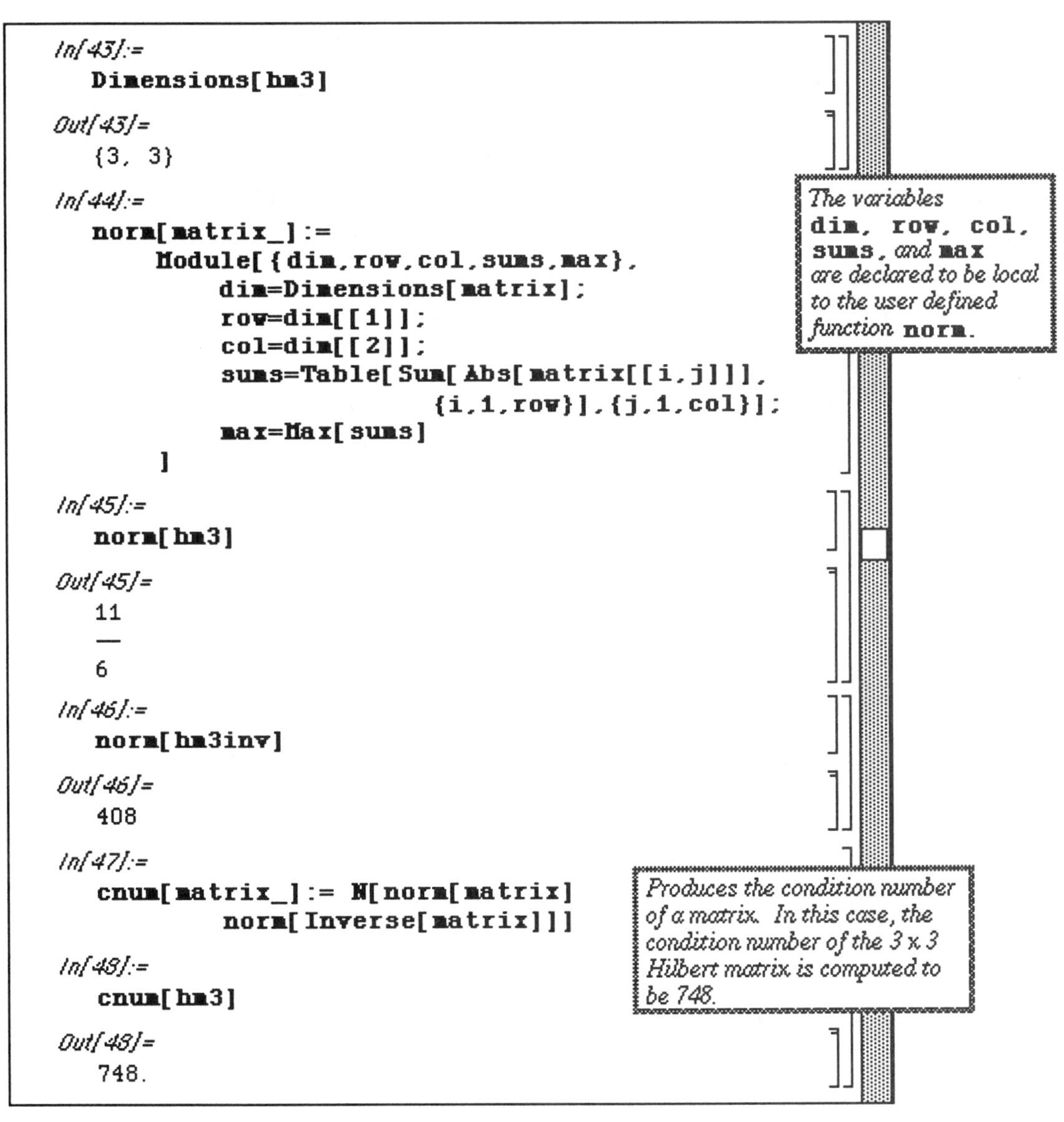

```
In[43]:=
  Dimensions[hm3]

Out[43]=
  {3, 3}

In[44]:=
  norm[matrix_]:=
      Module[{dim,row,col,sums,max},
           dim=Dimensions[matrix];
           row=dim[[1]];
           col=dim[[2]];
           sums=Table[Sum[Abs[matrix[[i,j]]],
                          {i,1,row}],{j,1,col}];
           max=Max[sums]
      ]

In[45]:=
  norm[hm3]

Out[45]=
  11
  --
  6

In[46]:=
  norm[hm3inv]

Out[46]=
  408

In[47]:=
  cnum[matrix_]:= N[norm[matrix]
          norm[Inverse[matrix]]]

In[48]:=
  cnum[hm3]

Out[48]=
  748.
```

The variables **dim, row, col, sums,** *and* **max** *are declared to be local to the user defined function* **norm.**

Produces the condition number of a matrix. In this case, the condition number of the 3 x 3 Hilbert matrix is computed to be 748.

Mathematica has a built-in command which can be used in determining the condition number. **SingularValues[m]** returns the list `{u,w,v}` where `w` is a list of non-zero singular values. The condition number is the ratio of the largest singular value of the matrix to the smallest one. The other information in the output of **SingularValues[m]** can be used to represent the matrix **m** as **Transpose[u].DiagonalMatrix[w].v**. This is known as the singular value decomposition of **m**.

Hence, the condition number can be found by determining the ratio of the values in the list `w`. This is shown below with matrix **a**. The definition of condition number used in the built-in function differs from the one stated earlier since the value obtained with **SingularValues** and that found with **cnum[a]** differ. At any rate, each method yields a very large condition number. Hence, **a** is ill-conditioned, so numerical methods used to solve systems involving **a** are unreliable.

Unfortunately , the built-in command does not work with all matrices. An error message is given with **SingularValues[hm3]**. Therefore, the earlier procedure for calculating the condition number may prove to be more useful.

```
In[50]:=
   a={{34.9,23.6},{22.9,15.6}};
   s=SingularValues[a]

Out[50]=
   {{{-0.835497, -0.549495}, {-0.549495, 0.835497}},

     {50.4255, 0.0793249},

     {{-0.827801, -0.561022}, {-0.561022, 0.827801}}}

In[51]:=
   s[[2]]

Out[51]=
   {50.4255, 0.0793249}

In[52]:=
   cdn=Max[s[[2]]]/Min[s[[2]]]

Out[52]=
   635.683

In[53]:=
   cnum[a]

Out[53]=
   845.325

In[54]:=
   sing=SingularValues[hm3]

   SingularValues::svdf:
      SingularValues has received a matrix with
        infinite precision.

   SingularValues::svdf:
      SingularValues has received a matrix with
        infinite precision.

Out[54]=
                            1  1    1  1  1    1  1  1
   SingularValues[{{1, -, -}, {-, -, -}, {-, -, -}}]
                            2  3    2  3  4    3  4  5
```

100%

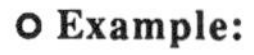

• Orthogonalization.m

Version 2.0's **Orthogonalization.m** package contains useful commands for working with vector spaces. The main command in this package is **GramSchmidt[veclist,options]** which produces an orthonormal basis for the vectors in **veclist**. Recall that two properties of an orthonormal basis are that the inner product of any two basis vectors is 0, and the norm of each basis vector is 1. In the first example below, a basis for the list of vectors in **vecs** is found with **GramSchmidt[vecs]**.

o Example:

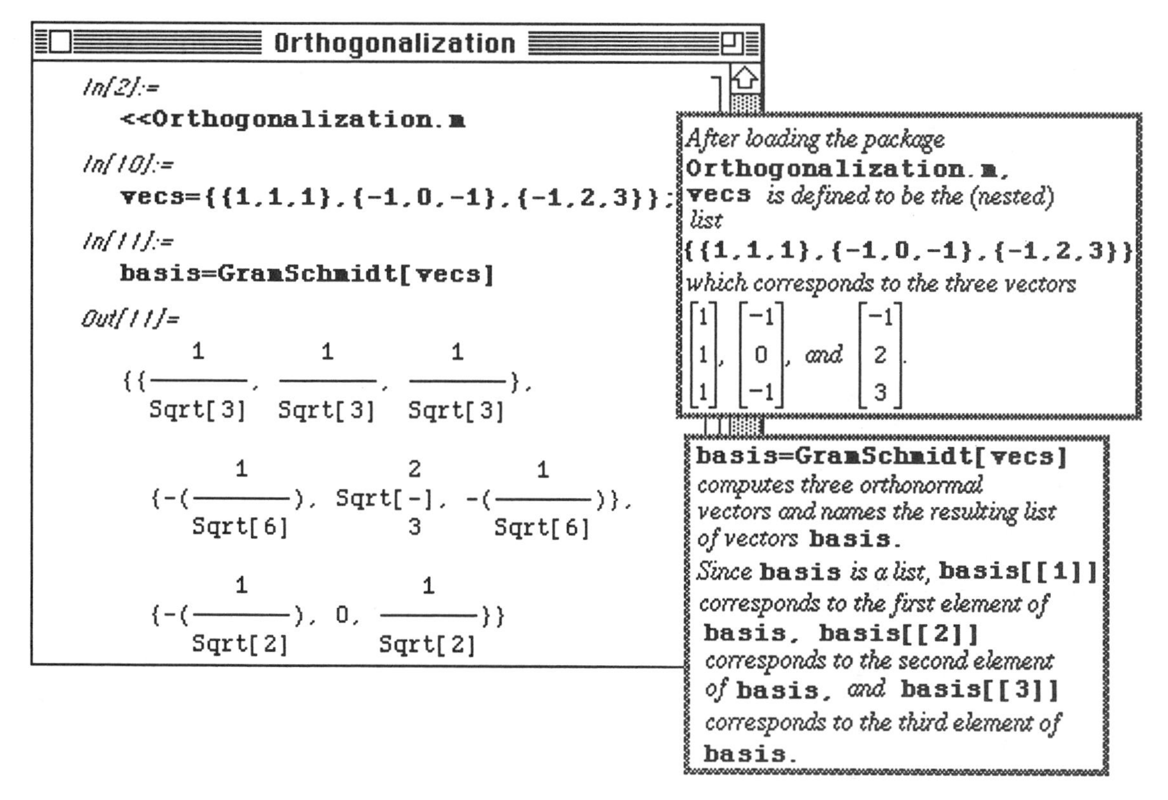

Some of the previously mentioned properties are verified using the orthonormal basis found in **basis**.

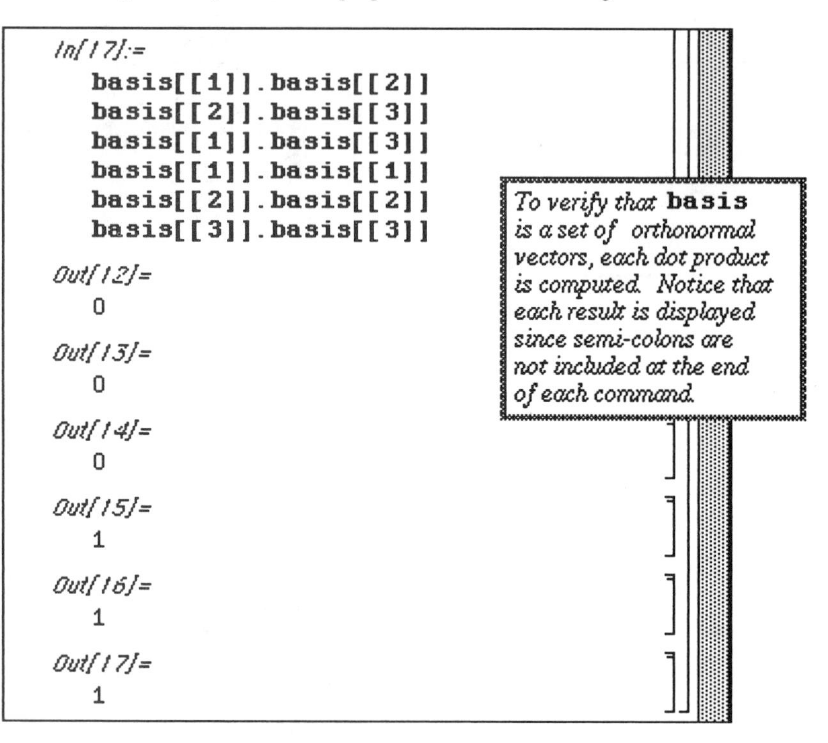

o Application: **Distance**

One application of the determination of an orthonormal set of basis vectors is in the calculation of the distance of a point **x** from a subspace W. If **u** is an orthonormal basis for W, then the distance from **x** to S is defined as the magnitude of the component of **x** which is orthogonal to each vector in W. This vector is found by projecting **x** onto W using the basis vectors in **u**. More formally, the distance is given by the formula :

$\|\mathbf{x} - \mathrm{proj}_W \mathbf{x}\|$ where $\mathrm{proj}_W \mathbf{x}$ is the projection of **x** onto W.

In the example which follows, the distance of the point **x** = (4,1,-7) to the subspace W which consists of all vectors of the form (a,b,b) is found. Every vector in W can be written as the linear combination of the vectors `{1,0,0}` and `{0,1,1}`. Hence, the orthogonal basis for W is found in **onbasis** below using `{1,0,0}` and `{0,1,1}`.

The elements of **onbasis** are extracted in the usual manner as illustrated with **onbasis[[1]]**. The vector **x** is defined and the projection of **x** onto W is determined in **proj**. The orthogonal component is given in **diff**, and finally, the length of this vector is found in the standard way by using the dot product.

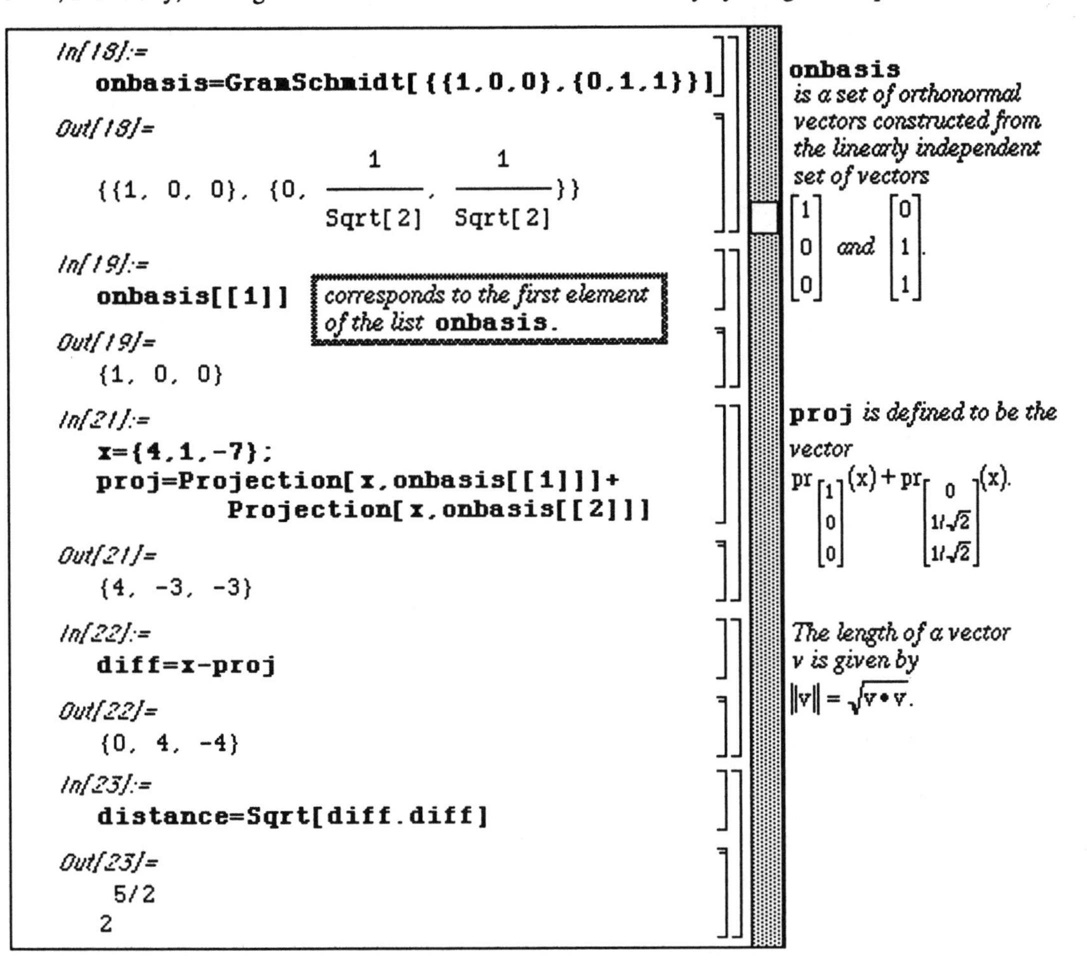

onbasis *is a set of orthonormal vectors constructed from the linearly independent set of vectors* $\begin{bmatrix}1\\0\\0\end{bmatrix}$ *and* $\begin{bmatrix}0\\1\\1\end{bmatrix}$.

proj *is defined to be the vector* $\mathrm{pr}_{\begin{bmatrix}1\\0\\0\end{bmatrix}}(x)+\mathrm{pr}_{\begin{bmatrix}0\\1/\sqrt{2}\\1/\sqrt{2}\end{bmatrix}}(x)$.

The length of a vector v is given by $\|v\| = \sqrt{v \cdot v}$.

The above calculations to determine the distance from a point to a subspace can be generalized in the function **distance** below. The arguments of distance are the point **v** and a list of basis vectors, **vecs**, for W. The function performs all of the necessary calculations to find the desired distance. To verify that **distance** does yield the correct output, it is demonstrated with the same problem as before and gives the same result. First, all previously used definitions are cleared.

```
In[24]:=
  Clear[v,proj,onbasis,vecs,x,diff,distance]

In[25]:=
  distance[v_,vecs_]:=Module[{d,onbasis,proj,diff},
                    onbasis=GramSchmidt[Evaluate[vecs]];
                    proj=Sum[Projection[v,onbasis[[i]]],
                         {i,1,Length[onbasis]}];
                    diff=v-proj;
                    d=Sqrt[diff.diff]
                    ]

In[26]:=
  distance[{4,1,-7},{{1,0,0},{0,1,1}}]

Out[26]=
   5/2
  2
```

One of the options available to **GramSchmidt** is that of **InnerProduct**. In the earlier examples the dot product (the default value of **InnerProduct**) was used as the the inner product. However, with this option, other vector spaces such as function spaces can be considered.

The **inner product of two continuous functions f and g** is given by

$$(f,g) = \int_a^b w(x)\, f(x)\, g(x)\, dx$$

where $f, g \in C[a,b]$ and $w(x)$ is a nonnegative weight function on (a,b).

The **InnerProduct** option must be given in terms of a pure function. In order to do this, the syntax (option->(function&) must be used. Hence, the inner product with $w(x) = 1$ is defined with **InnerProduct->(Integrate[#1 #2,{x,-1,1}]&)** where **#1** and **#2** represent the first and second variables of the inner product, respectively.

The first command below yields the orthonormal basis from **{1,x,x^2,x^3,x^4}**, the basis for the space of fourth-order polynomials. Another **GramSchmidt** option is **Normalized->False**. When this is used, the basis vectors which result are orthogonal but not of length 1. The first command is repeated with this command and called **legendre5**.

```
In[2]:=
  GramSchmidt[{1,x,x^2,x^3,x^4},
              InnerProduct->(Integrate[#1 #2,{x,-1,1}]&)]

Out[2]=
       1             3               5     1      2
  {--------, Sqrt[-] x, 3 Sqrt[-] (-(-) + x ),
   Sqrt[2]           2               8     3

                                                    1      2
                                               6 (-(-) + x )
                                       1     4       3
                                105 (-(-) + x  - -------------)
         7    -3 x    3                5               7
    5 Sqrt[-] (---- + x ), ---------------------------------}
         8     5                             7/2
                                            2

In[3]:=
  legendre5=GramSchmidt[{1,x,x^2,x^3,x^4},
            InnerProduct->(Integrate[#1 #2,{x,-1,1}]&),
            Normalized->False]

Out[3]=
                                                     1      2
                                                6 (-(-) + x )
          1      2   -3 x     3     1      4          3
  {1, x, -(-) + x , ---- + x , -(-) + x  - -------------}
          3           5             5              7
```

On the other hand, when the option **Normalized->True** is used, the basis vectors which result are orthogonal with length 1. In the following example, several integrals are computed to verify that the resulting basis vectors are orthogonal and have length 1:

```
In[2]:=
  gs=GramSchmidt[{1,x,x^2,x^3,x^4},
                 InnerProduct->(Integrate[#1 #2,{x,-1,1}]&),
                 Normalized->True]
Out[2]=
       1              3            5       1     2
  {---------, Sqrt[-] x, 3 Sqrt[-] (-(-) + x ),
   Sqrt[2]            2            8       3
                                                  1     2
                                             6 (-(-) + x )
                                  1     4         3
                           105 (-(-) + x  -  -------------)
            7    -3 x    3        5                7
    5 Sqrt[-] (---- + x ), ---------------------------------}
            8     5                     7/2
                                       2
In[3]:=
  gs[[1]]
Out[3]=
      1
  -------
  Sqrt[2]
In[4]:=
  Integrate[gs[[1]] gs[[2]],{x,-1,1}]
Out[4]=
  0
In[5]:=
  Sqrt[Integrate[ gs[[1]]^2,{x,-1,1}]]
Out[5]=
  1
In[6]:=
  Integrate[gs[[1]] gs[[4]],{x,-1,1}]
Out[6]=
  0
In[7]:=
  Sqrt[Integrate[ gs[[4]]^2,{x,-1,1}]]
Out[7]=
  1
```

computes $\int_{-1}^{1}\frac{1}{\sqrt{2}}\sqrt{\frac{3}{2}}x\,dx$.

computes $\sqrt{\int_{-1}^{1}\left(\frac{1}{\sqrt{2}}\right)^2 dx}$.

computes $\int_{-1}^{1}\frac{1}{\sqrt{2}}5\sqrt{\frac{7}{8}}\left(x^3-\frac{3}{5}x\right)dx$.

computes $\sqrt{\int_{-1}^{1}\left(5\sqrt{\frac{7}{8}}\left(x^3-\frac{3}{5}x\right)\right)^2 dx}$.

The polynomials which result are normalized in **lpoly** so that each member equals 1 at **x**=1. This yields a list of the first five Legendre polynomials and is accomplished by dividing each entry in **legendre5** with its value at **x**=1. In this case, the Legendre polynomials were the result of an example illustrating the **InnerProduct** option with **GramSchmidt**. Nevertheless, *Mathematica* contains the built-in command **LegendreP[n,x]** which gives these polynomial.

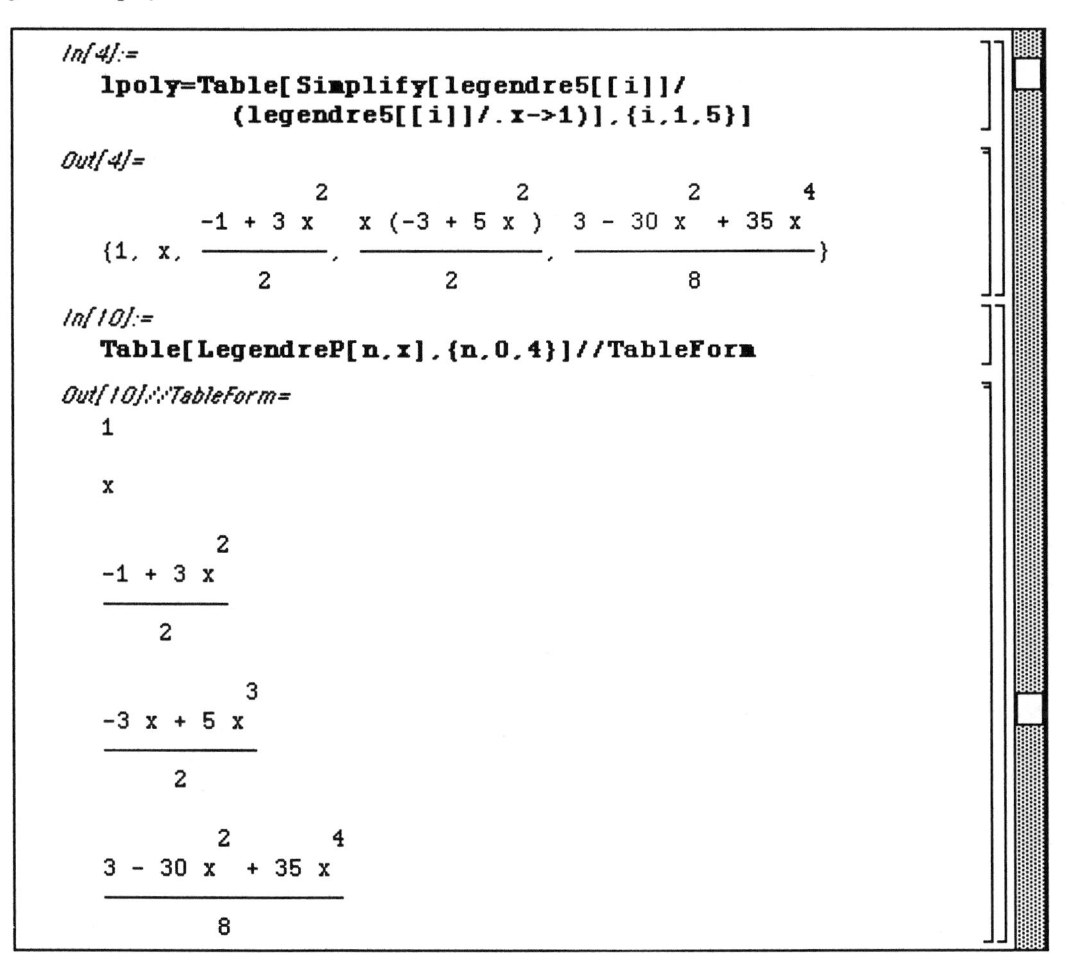

A useful purpose for obtaining an orthogonal set of basis vectors is in approximating functions with polynomials. If a function f(x) is to be approximated with a polynomial of degree n, then the closest polynomial is computed by projecting f(x) onto each of the first n (or n+1) basis members of the space of n-th order polynomials. Hence, approximation can be conducted by using a basis made up of the Legendre polynomials created earlier. In order to illustrate this technique, the command **Projection[vec1,vec2,options]** which also appears in **Orthogonalization.m** must be discussed. This command projects **vec1** onto **vec2** and can employ the same option of **InnerProduct** as was shown earlier with **GramSchmidt**. The function **proj** is defined below to project a vector (the function) **v** onto the basis vectors in **basis** using the appropriate inner product for function spaces. This function is then demonstrated by projecting **Exp[x]** onto the orthogonal basis obtained earlier in **lpoly**. Hence, this calculation results in the approximation of **Exp[x]** using the first 5 Legendre polynomials. The approximating polynomial is then simplified and expressed with numerical coefficients in **app**. Both functions are plotted simultaneously to show the closeness of the approximation, and then the difference is plotted to better illustrate the accuracy of the approximation with Legendre polynomials.

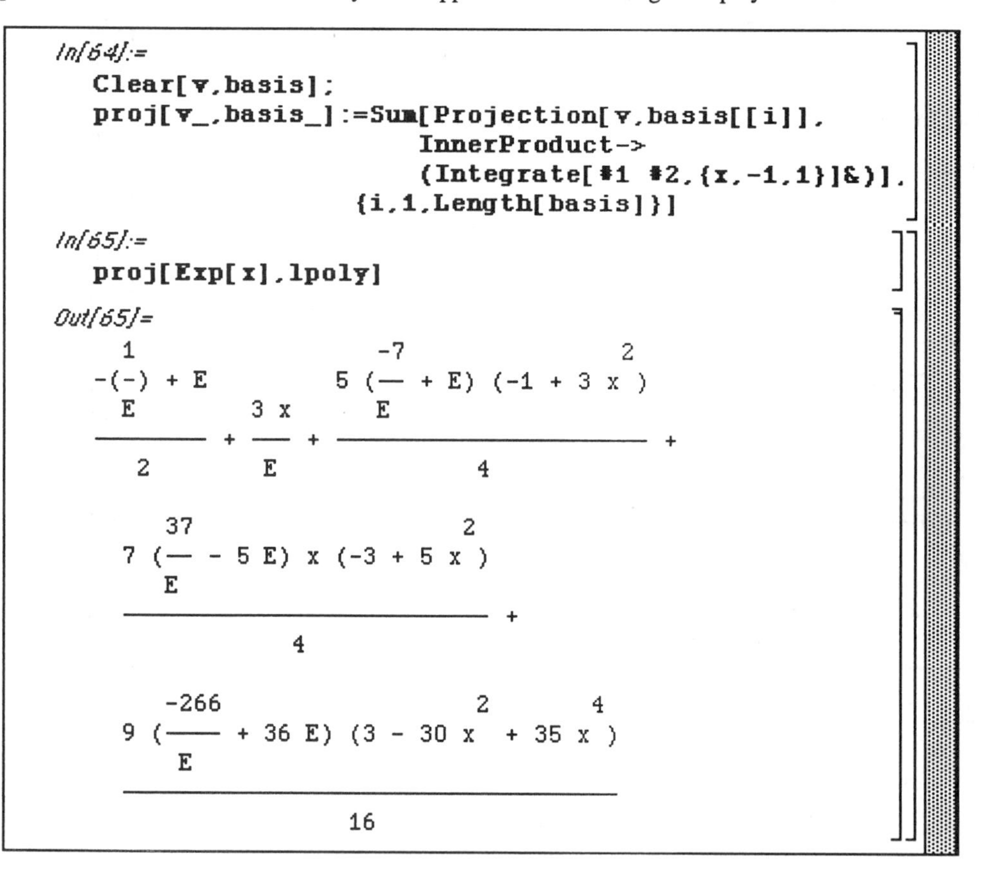

```
In[64]:=
  Clear[v,basis];
  proj[v_,basis_]:=Sum[Projection[v,basis[[i]],
                        InnerProduct->
                        (Integrate[#1 #2,{x,-1,1}]&)],
                  {i,1,Length[basis]}]

In[65]:=
  proj[Exp[x],lpoly]

Out[65]=
  1                        -7               2
  -(-) + E               5 (-- + E) (-1 + 3 x )
  E           3 x           E
  --------- + --- + ------------------------------- +
     2         E                 4

     37                    2
   7 (-- - 5 E) x (-3 + 5 x )
      E
   ---------------------------- +
               4

     -266                      2        4
   9 (---- + 36 E) (3 - 30 x  + 35 x )
       E
   -----------------------------------
                   16
```

The approximating polynomial is then simplified and expressed with numerical coefficients in **app**.

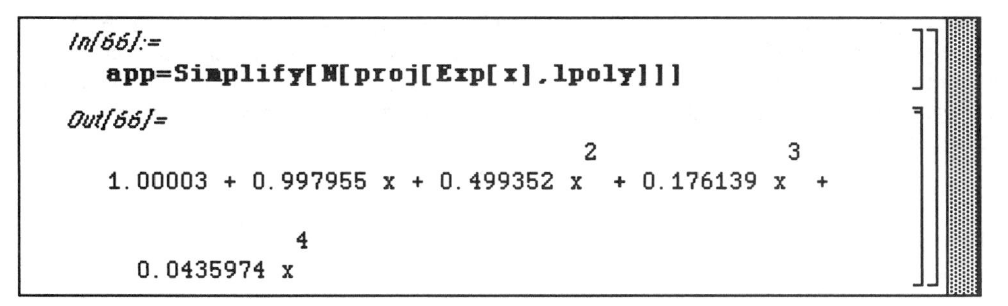

Both functions are plotted simultaneously to show the closeness of the approximation, and then the difference is plotted to better illustrate the accuracy of the approximation with Legendre polynomials.

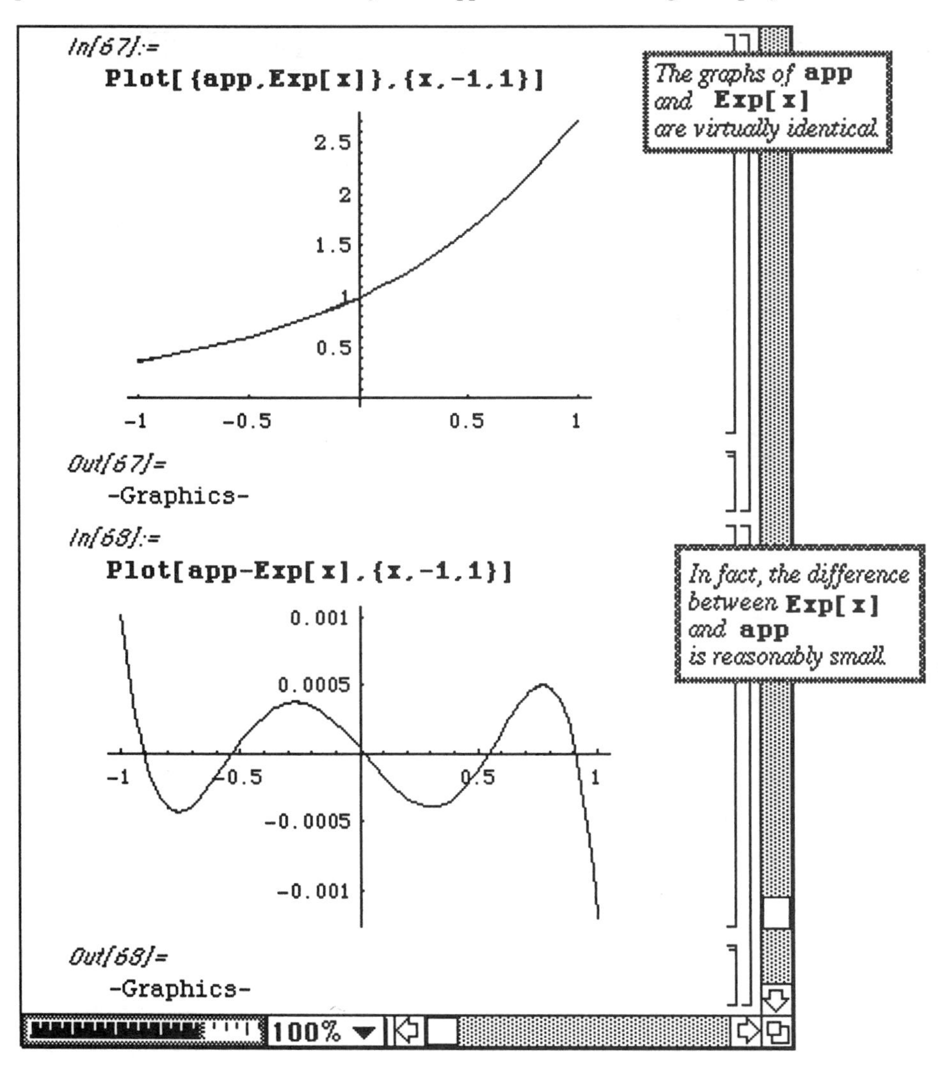

The graphs of **app** and **Exp[x]** are virtually identical.

In fact, the difference between **Exp[x]** and **app** is reasonably small.

• Tridiagonal.m

Version 2.0's **Tridiagonal.m** package offers the command **TridiagonalSolve[a,b,c,r]** to solve the system of equations **Ax=r** where **A** is a tridiagonal matrix with diagonal **b**, upper diagonal **c**, and lower diagonal **a**. Matrices of this type arise in many areas of applied mathematics and can be rather large in dimensions. Therefore, this command may be more useful than other commands available for solving systems of linear equations. In the first example below, the following system is solved:

o Example:

$$\begin{bmatrix} 2 & -1 & 0 \\ -1 & 2 & -1 \\ 0 & -1 & 2 \end{bmatrix} \begin{bmatrix} x_1 \\ x_2 \\ x_3 \end{bmatrix} = \frac{\pi^2}{4} \begin{bmatrix} 1 \\ 0 \\ -1 \end{bmatrix}.$$

Notice that the solution of the system is given as the list {x1,x2,x3}.

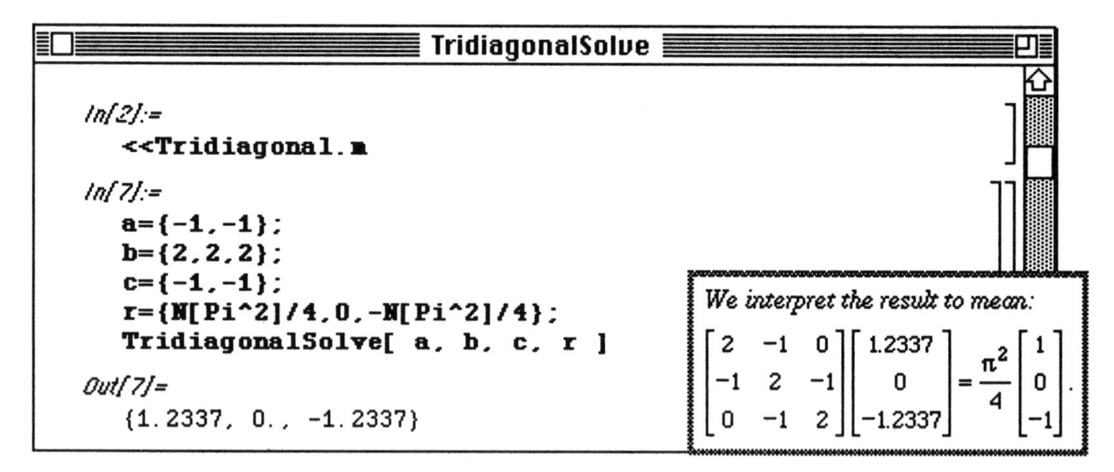

$$\begin{bmatrix} 2 & -1 & 0 \\ -1 & 2 & -1 \\ 0 & -1 & 2 \end{bmatrix} \begin{bmatrix} 1.2337 \\ 0 \\ -1.2337 \end{bmatrix} = \frac{\pi^2}{4} \begin{bmatrix} 1 \\ 0 \\ -1 \end{bmatrix}.$$

□ Example:

As previously indicated, tridiagonal systems are common in many fields. One example of this is the numerical solution to differential equations. Consider the boundary value problem

$$-\frac{d^2u}{dx^2} = f(x), \quad 0 \le x \le 1$$
$$u(0) = 0\,,\ u(1) = 0.$$

(This is the problem which describes the steady-state case of the heat equation with fixed end temperatures of zero degrees and heat source f(x).) In order to solve this problem numerically, it must be changed from a continuous problem to a discrete one. This is accomplished first by providing a finite amount of information about f at the equally spaced points x = h, x = 2h, ..., x = nh. An approximate solution is computed at these values of x. At the endpoints, x = 0 and x = 1 = (n+1)h, the solution must be zero from the boundary conditions. Hence, the approximate solutions at the endpoints are known to be

$$u_0 = 0 \text{ and } u_{n+1} = 0.$$

The derivatives in the differential equation are replaced by the difference quotients

$$\frac{du}{dx} = \frac{u(x+h) - u(x-h)}{2h}$$

and

$$\frac{d^2u}{dx^2} = \frac{u(x+h) - 2u(x) + u(x-h)}{h^2}.$$

Hence, at a typical mesh point x = jh, the differential equation is replaced by the discrete analog

$$-u_{j+1} + 2u_j - u_{j-1} = h^2 f(jh).$$

There are n equations of the form, one for each value of j = 1,..., n. (Note that h = 1/(n+1)). Therefore, a numerical solution to this differential equation is obtained by solving the tridiagonal system which results from the discrete problem. The function **soln** which sets up and solves the system for any value of **n** and any function **f** is defined below. The output is given as a list of ordered pairs in which the first coordinate represents the meshpoint while the second coordinate gives the approximate solution at the corresponding meshpoint. In the example which follows, the boundary-value problem

$$-\frac{d^2u}{dx^2} = 4\pi^2 \sin(2\pi x), \quad 0 \le x \le 1$$
$$u(0) = 0\,,\ u(1) = 0.$$

is solved. Of course, the exact solution to this problem is known to be $u(x) = \sin 2\pi x$. Hence, the approximate solutions found with this finite-difference method can be compared to the true solution.

The solution is approximated, first, for **n** = 6. The points given by **soln** are then plotted with **ListPlot** and shown simultaneously with the graph of the exact solution.

```
In[9]:=
   Clear[a,b,c,r]
   soln[f_,n_]:=
           Module[{h,a,b,c,r,vec,table},
                h=1/(n+1);
                a=Table[-1,{i,1,n-1}];
                b=Table[2,{i,1,n}];
                c=Table[-1,{i,1,n-1}];
                r=Table[ h^2 N[f[i h]],{i,1,n}]//N;
                vec=TridiagonalSolve[a,b,c,r]//N;
                table=Table[{i h,vec[[i]]},{i,1,n}]
           ]
```

For the function **soln, h a,b,c,r,vec** *and* **table** *are defined to be local variables.*

```
In[10]:=
   f[x_]=4 Pi^2 Sin[2 Pi x]
Out[10]=
       2
   4 Pi  Sin[2 Pi x]
```

For this example, define $f(x) = 4\pi^2 \sin(2\pi x)$.

```
In[11]:=
   soln[f,6]
Out[11]=
     1              2             3              4
   {{-, 0.836508}, {-, 1.04311}, {-, 0.464227}, {-, -0.464227},
     7              7             7              7

     5              6
    {-, -1.04311}, {-, -0.836508}}
     7              7
```

The points given by **soln** are then plotted with **ListPlot** and shown simultaneously with the graph of the exact solution.

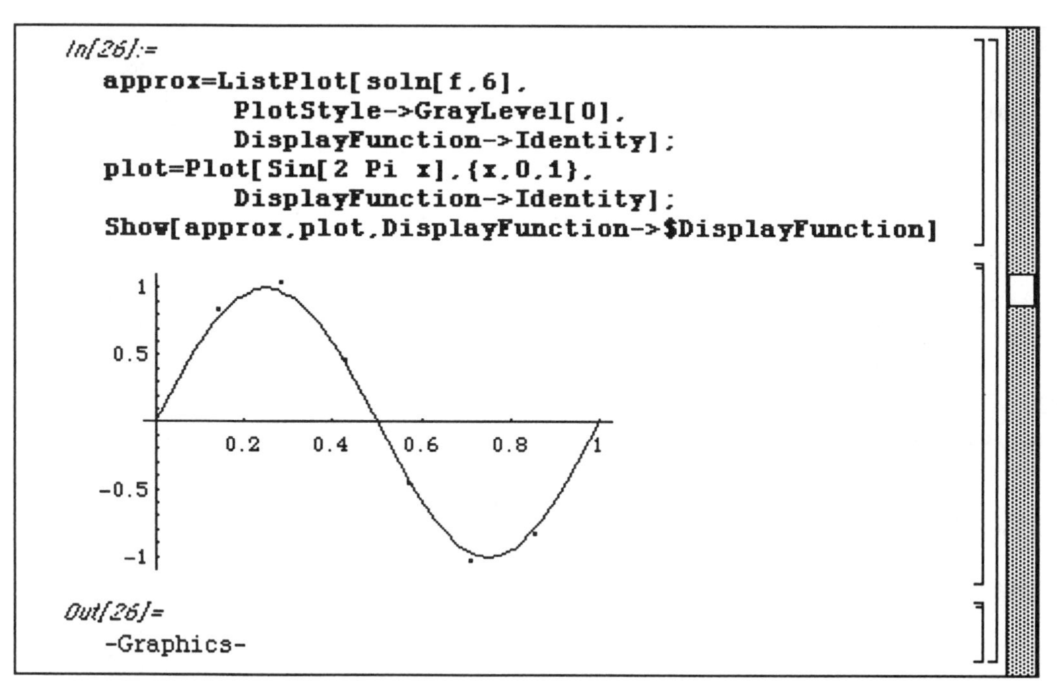

Similar steps are followed below with **n** = 20 and **n** = 50 to demonstrate the improved approximation as **n** increases.

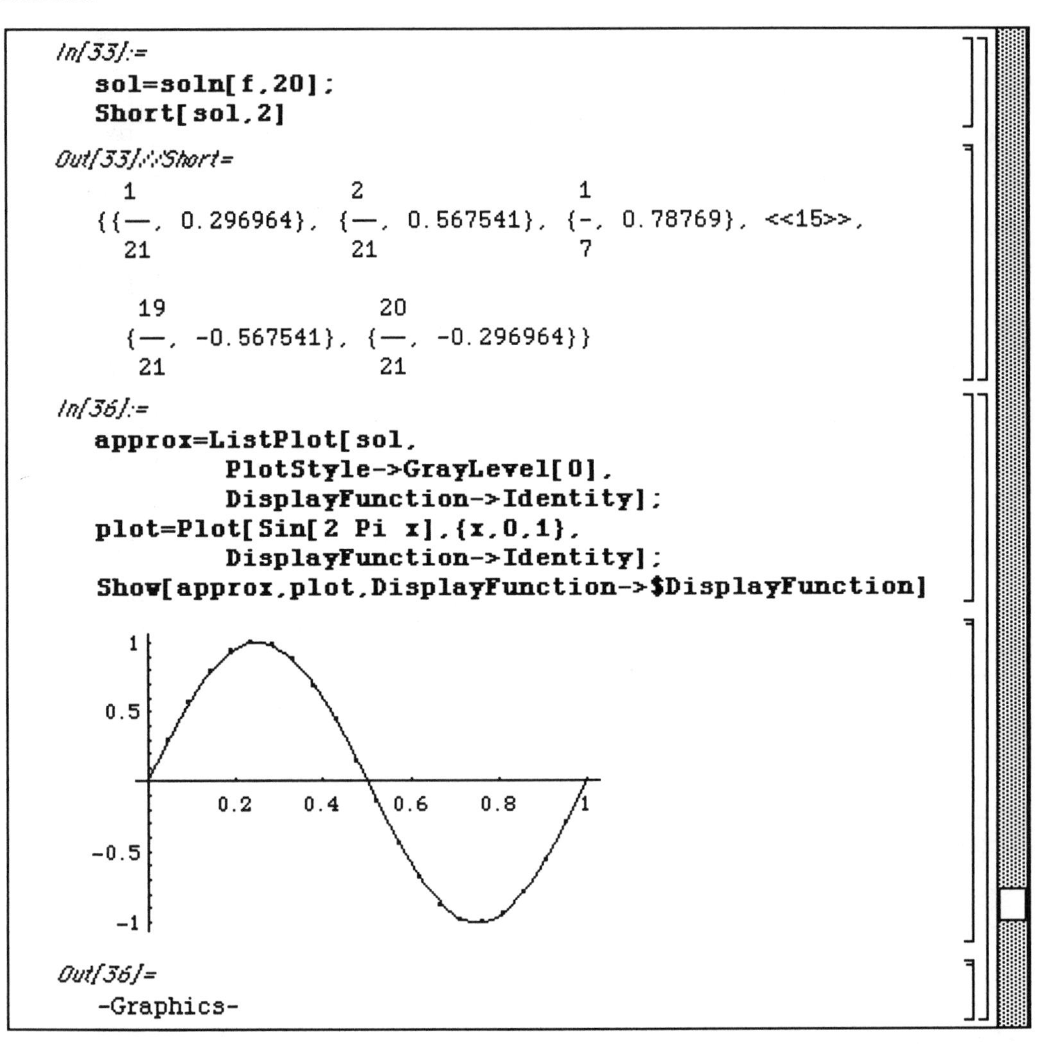

The calculations for **n**=50 are given below. Note the improvement of the approximation which results:

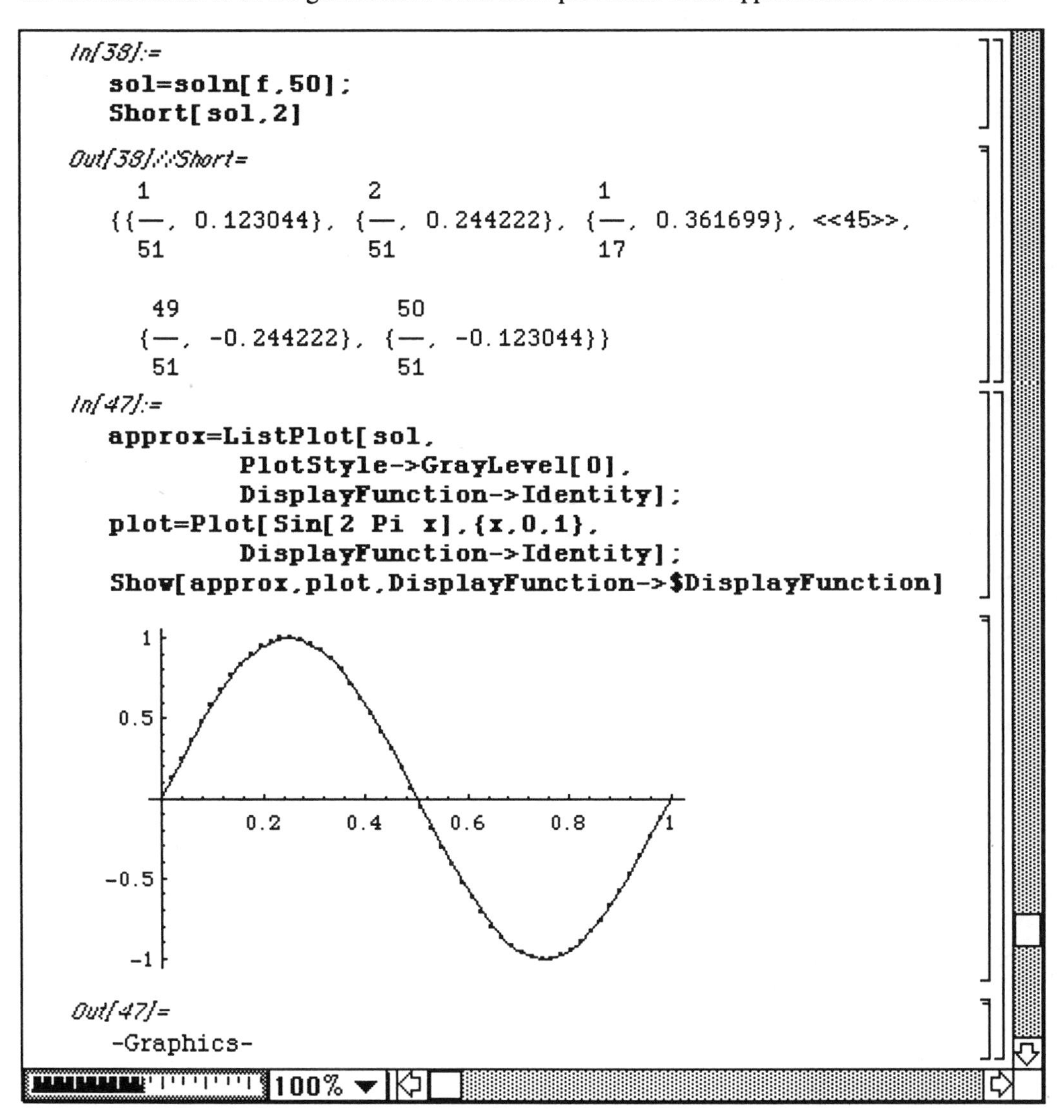

```
In[38]:=
   sol=soln[f,50];
   Short[sol,2]

Out[38]//Short=
     1              2              1
  {{—, 0.123044}, {—, 0.244222}, {—, 0.361699}, <<45>>,
    51             51             17

      49                 50
    {—, -0.244222}, {—, -0.123044}}
     51                 51

In[47]:=
   approx=ListPlot[sol,
            PlotStyle->GrayLevel[0],
            DisplayFunction->Identity];
   plot=Plot[Sin[2 Pi x],{x,0,1},
            DisplayFunction->Identity];
   Show[approx,plot,DisplayFunction->$DisplayFunction]

Out[47]=
   -Graphics-
```

7.3 Version 2.0 Calculus

The packages contained in the Version 2.0 **Calculus** folder appear as follows:

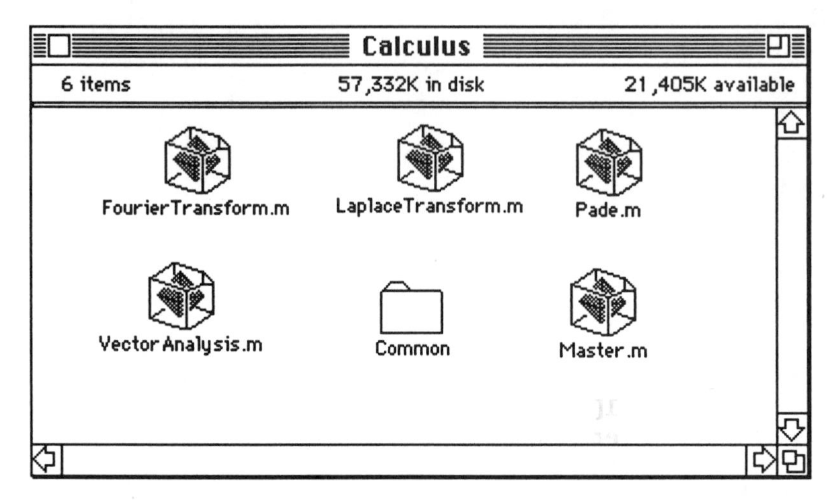

The packages **FourierTransform.m** and **LaplaceTransform.m** are discussed here; **VectorAnalysis.m** was discussed in **Chapter 5**.

• LaplaceTransform.m

Commands which can be used to compute Laplace transforms and inverse Laplace transforms are located in the **LaplaceTransform.m** package. The command **InverseLaplaceTransform[f[s],s,t]** computes the inverse Laplace transform of **f[s]** and the result is a function of **t** while **LaplaceTransform[g[t],t,s]** yields the Laplace transform of **g[t]** as a function of **s**. Several examples are given below.

o Example:

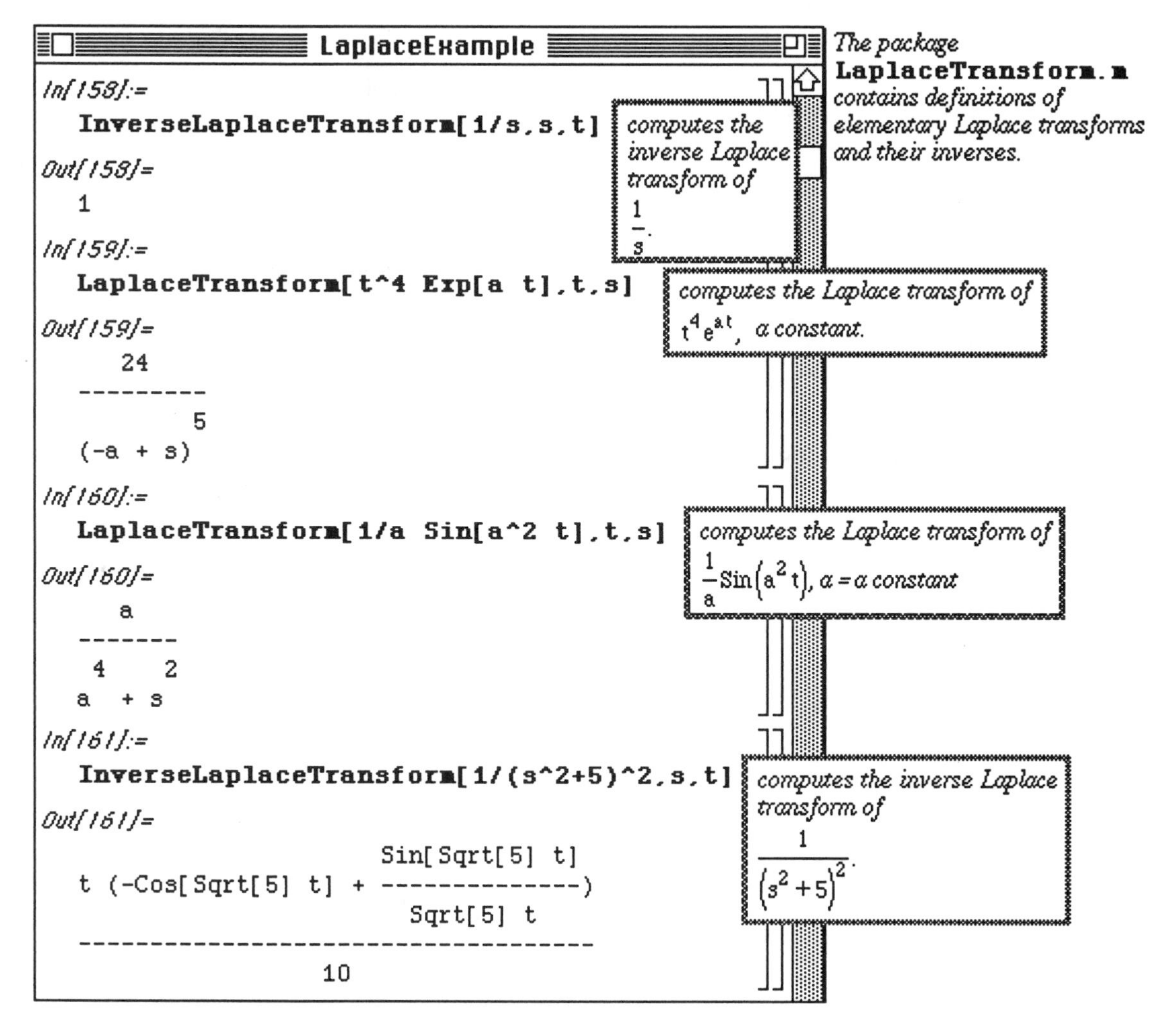

The package **LaplaceTransform.m** *contains definitions of elementary Laplace transforms and their inverses.*

Of course, an application of Laplace transforms is the solution of differential equations. In order to use these techniques, however, several important properties of Laplace transforms must be recalled. Two of these are investigated below with *Mathematica* and will be used later. Note how the derivative of the Laplace transform is represented in the second problem.

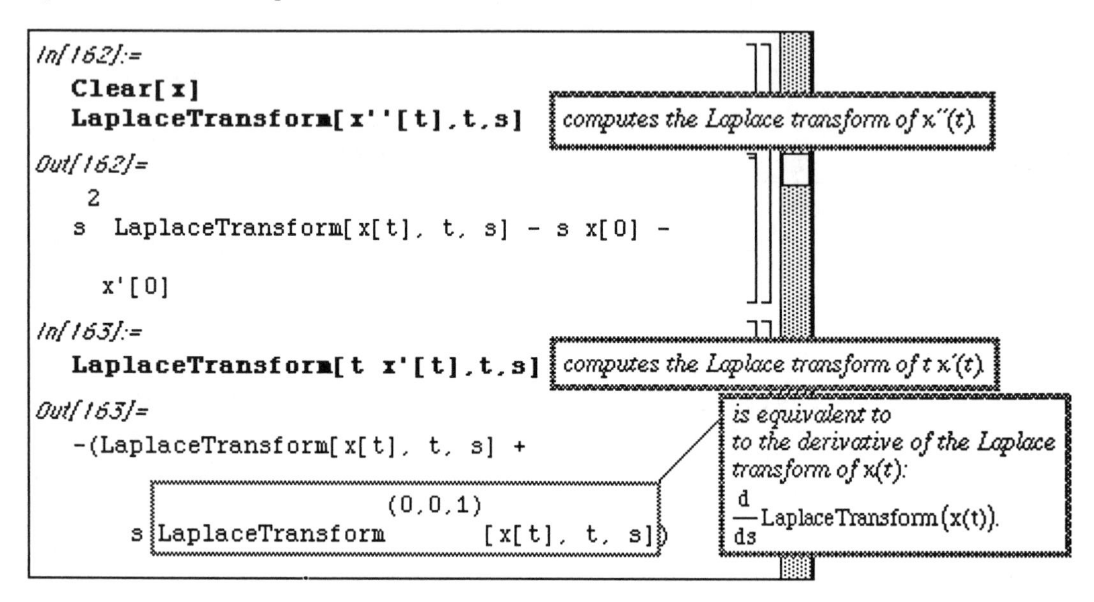

o Application: Solutions of Ordinary Differential Equations

Consider the following second-order system of ordinary differential equations with initial conditions:

$$\begin{cases} x'' + 10\,x - 4\,y = 0,\ x(0) = 0,\ x'(0) = 3/2 \\ -4\,x + y'' + 4\,y = 0,\ y(0) = 0,\ y'(0) = -3/2 \end{cases}$$

Laplace transforms can be used to solve this system. After clearing all definitions, the Laplace transform of the first equation is found and called **leq1**. Similarly, the Laplace transform of the second equation is found and called **leq2** .

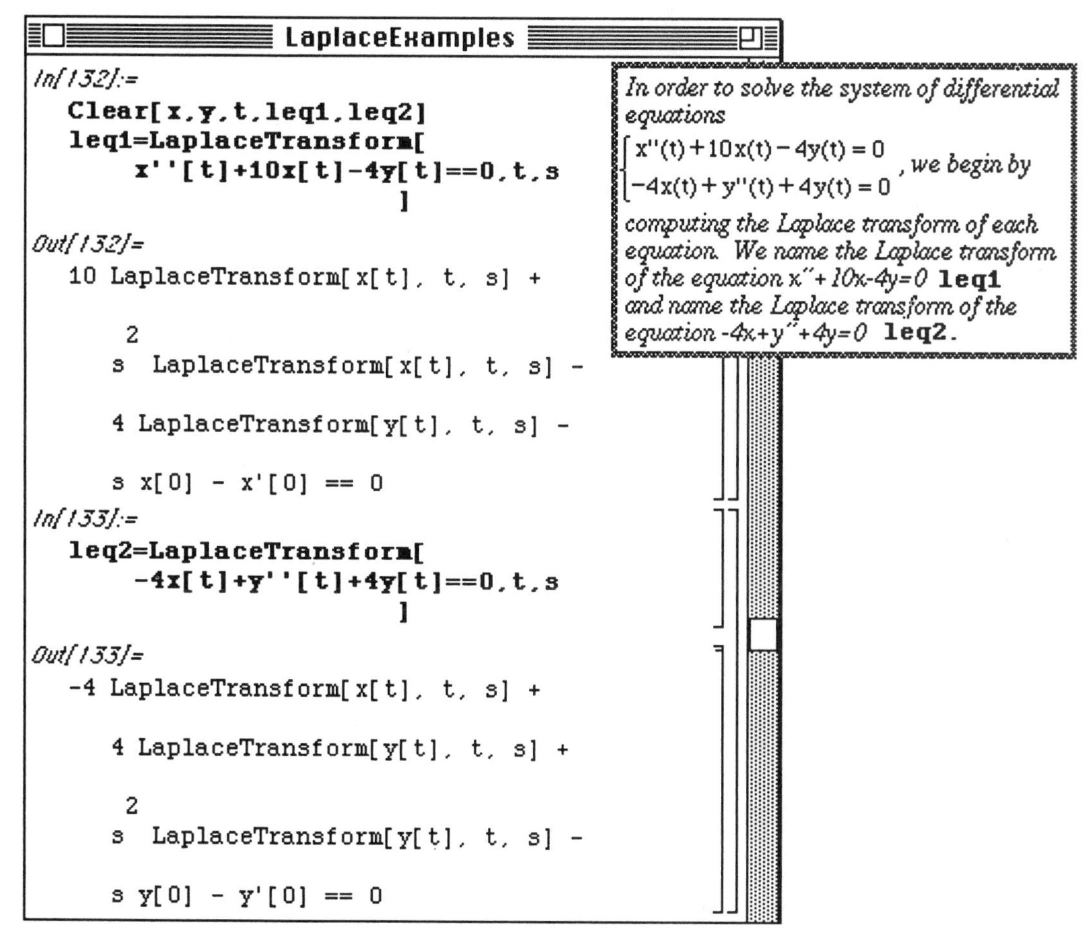
LaplaceExamples

```
In[132]:=
   Clear[x,y,t,leq1,leq2]
   leq1=LaplaceTransform[
        x''[t]+10x[t]-4y[t]==0,t,s
                              ]
Out[132]=
   10 LaplaceTransform[x[t], t, s] +

      2
     s  LaplaceTransform[x[t], t, s] -

     4 LaplaceTransform[y[t], t, s] -

     s x[0] - x'[0] == 0
In[133]:=
   leq2=LaplaceTransform[
        -4x[t]+y''[t]+4y[t]==0,t,s
                              ]
Out[133]=
   -4 LaplaceTransform[x[t], t, s] +

     4 LaplaceTransform[y[t], t, s] +

      2
     s  LaplaceTransform[y[t], t, s] -

     s y[0] - y'[0] == 0
```

In order to solve the system of differential equations $\begin{cases} x''(t)+10x(t)-4y(t)=0 \\ -4x(t)+y''(t)+4y(t)=0 \end{cases}$*, we begin by computing the Laplace transform of each equation. We name the Laplace transform of the equation x''+10x-4y=0* **leq1** *and name the Laplace transform of the equation -4x+y''+4y=0* **leq2.**

The usual convention when working with Laplace transforms is to use the capital letter X when referring to the Laplace transform of x. Hence, in the above calculations, **LaplaceTransform[x[t],t,s]** is replaced with **capx** and **LaplaceTransform[y[t],t,s]** is replaced with **capy**. Also the initial values **x[0]=0,x'[0]=3/2,y[0]=0,** and **y'[0]=-3/2** can be substituted into the previous calculations.

These substitutions are made after defining the list of replacements, called **rule**, below. Then, **leq1/.rule** and **leq2/.rule** make all of the replacements given in **rule**. (**leq1/.rule** is an equation involving **capx** while **leq2/.rule** involves **capy**). Hence, the two equations which result are solved for **capx** and **capy**, respectively, in the **Solve** command. The result is named **solution**.

In[134]:=

```
rule={LaplaceTransform[x[t],t,s]->capx,
     x[0]->0,x'[0]->3/2,
     y[0]->0,y'[0]->-3/2,
     LaplaceTransform[y[t],t,s]->capy};
```

In[135]:=

```
solution=Solve[
     {leq1 /. rule,leq2 /.rule},
          {capx,capy}]
```

Out[135]=

```
{{capx ->

                                        2
          -24              3 (8 + 2 s )
   ---------------- + -----------------\
            2      4           2      4
   96 + 56 s  + 4 s   96 + 56 s  + 4 s

   , capy ->

                                        2
           24              3 (20 + 2 s )
   ---------------- - -----------------}
            2      4           2      4
   96 + 56 s  + 4 s   96 + 56 s  + 4 s

     }
```

We must solve **leq1** *and* **leq2** *for* **LaplaceTransform[x[t],t,s]** *and* **LaplaceTransform[y[t],t,s]** *with the conditions that* $x(0)=0$, $x'(0)=3/2$, $y(0)=0$, *and* $y'(0)=-3/2$. *Hence, for convenience, define* **rule** *to substitute the appropriate values into* **leq1** *and* **leq2**.

solution=Solve[{leq1 /. rule, leq2 /. rule}, {capx,capy}] *first replaces the expressions* **LaplaceTransform[x[t],t,s]**, **LaplaceTransform[y[t],t,s]**, **x[0]**, **x'[0]**, **y[0]**, *and* **y'[0]** *by* **capx**, **capy**, **0**, **3/2**, **0**, *and* **-3/2**, *then solves the resulting system for* **capx** *and* **capy**.
The resulting list is named **solution**.

Note that **solution** is a list. Therefore, the formulas for **capx** and **capy** must be extracted. This is accomplished with **solution[[1,1,2]]** and **solution[[1,2,2]]**, respectively. This is shown below:

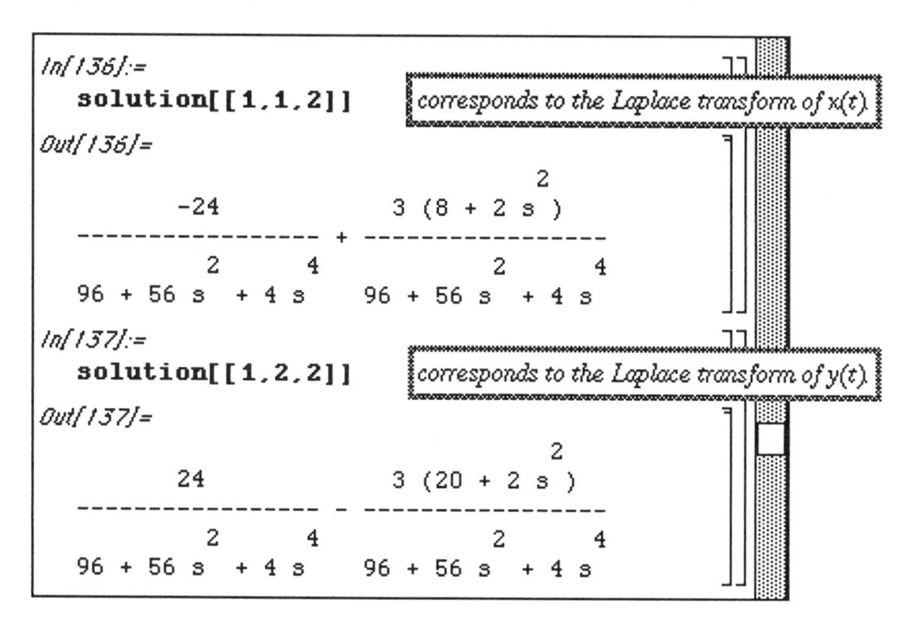

The solution **(x[t],y[t])** is now obtained through the inverse Laplace transform of the formulas of the Laplace transform of x, **capx**, and that of y, **capy**. These are determined below with the **InverseLaplaceTransform** command. Finally, the solution is graphed with **ParametricPlot** for values of **t** from **t** = **0** to **t** = **2 Pi**.

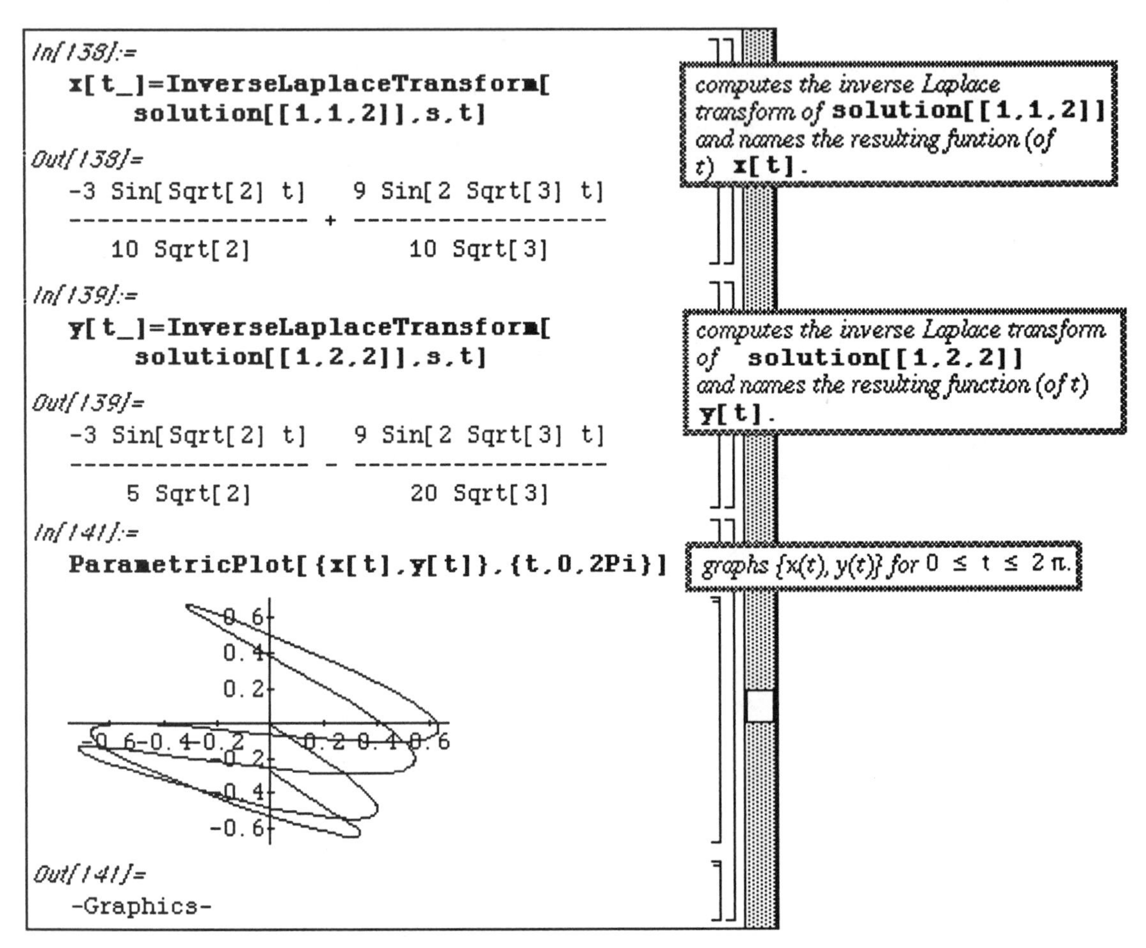

• FourierTransform.m

The **FourierTransform.m** package contains two commands useful for the computation of Fourier series. The first of these is **FourierTrigSeries[f[x],{x,-L,L},n]** which computes the first **n** terms (the **n**th partial sum) of the Fourier series of the periodic extension of **f[x]** on the interval **{-L,L}**. This series is given by the formula

$$a_0 + \sum_{n=1}^{\infty}\left(a_n \cos\frac{n\pi x}{L} + b_n \sin\frac{n\pi x}{L}\right)$$

where

$$a_0 = \frac{1}{2L}\int_{-L}^{L} f(x)\,dx$$

$$a_n = \frac{1}{L}\int_{-L}^{L} f(x)\cos\frac{n\pi x}{L}\,dx \quad , n = 1,2,...$$

$$b_n = \frac{1}{L}\int_{-L}^{L} f(x)\sin\frac{n\pi x}{L}\,dx \quad , n = 1,2,... .$$

Since these integrals are rather complicated in most cases, the **FourierTrigSeries** command is quite useful. This package also includes the command **NFourierTrigSeries** which gives the Fourier series in terms of numerically approximated Fourier series coefficients. These two commands are illustrated below.

□ Example:

First, the fourth partial sum of the Fourier series for the periodic extension of **f[x] = x** is computed with **FourierTrigSeries** and named **serexact**. Next, the same is given numerically for the function **f[x] = Abs[x]**. This result is called **serapprox**.

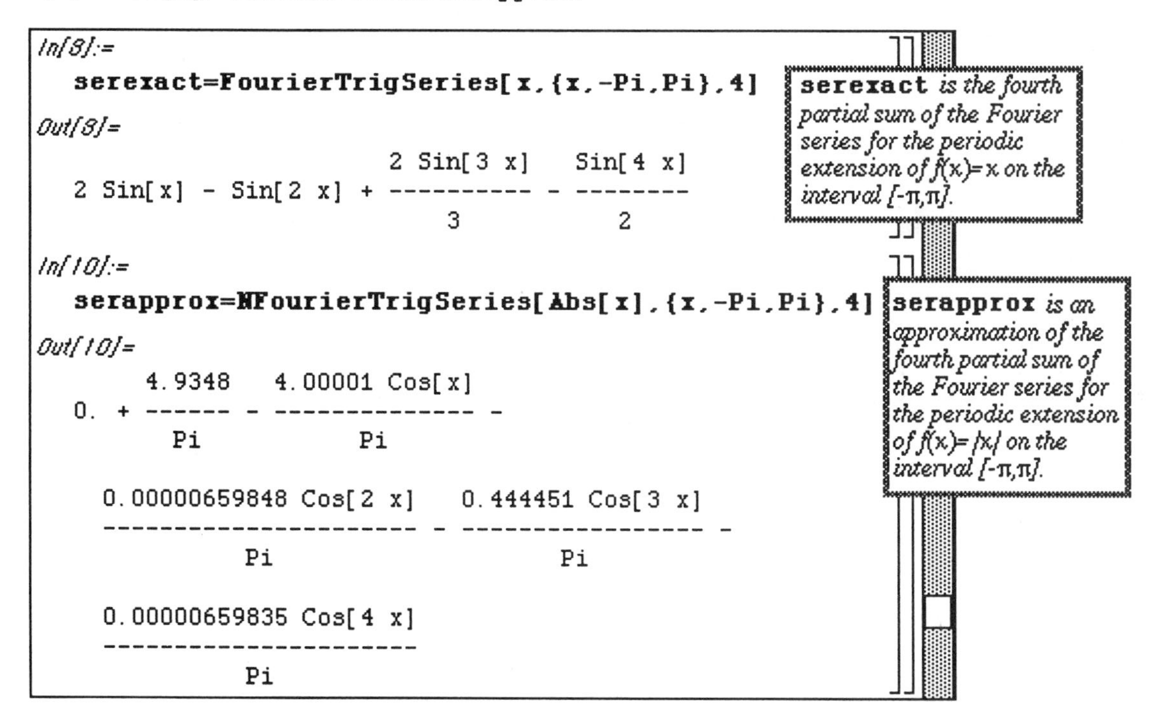

The graphs of **f[x] = x** and **f[x] = Abs[x]** along with the fourth partial sums of the corresponding Fourier series, **serexact** and **serapprox**, can be displayed in the same graphics cell. This is accomplished in the single command below. Recall that the **Plot** option **DisplayFunction->Identity** causes the graph to be suppressed initially. Then, the command **Show[GraphicsArray[{{plotx,plotserexact},{plotabsx,plotserapprox}}]]** displays all four graphs in one cell. Also, because of the grouping within **GraphicsArray**, the plots are displayed in the appropriate order. The Fourier series approximation of each function can easily be viewed in this manner.

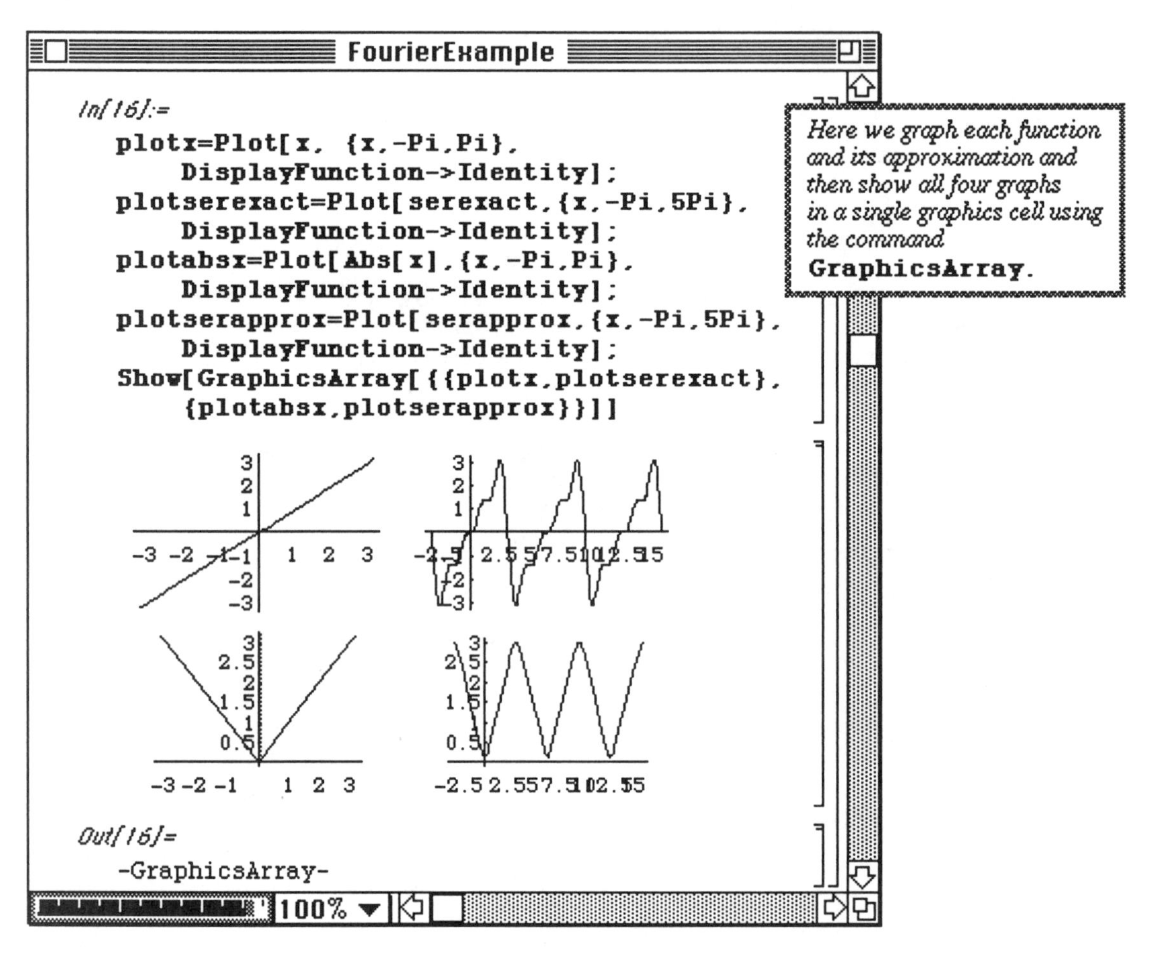

□ **Example:**

The Fourier series of piecewise-defined functions can also be computed. This is illustrated with the function **f[x]** below. After making the correct definition of this function, a table called **approxs** which consists of the second, sixth, and tenth partial sums of the Fourier series for the periodic extension of **f** is constructed with **NFourierSeries**. (A shortened output for **approxs** is given below.) Since **approxs** is a list of three elements, the second partial sum is given by **approxs[[1]]**, the sixth by **approxs[[2]]**, and the tenth by **approxs[[3]]**.

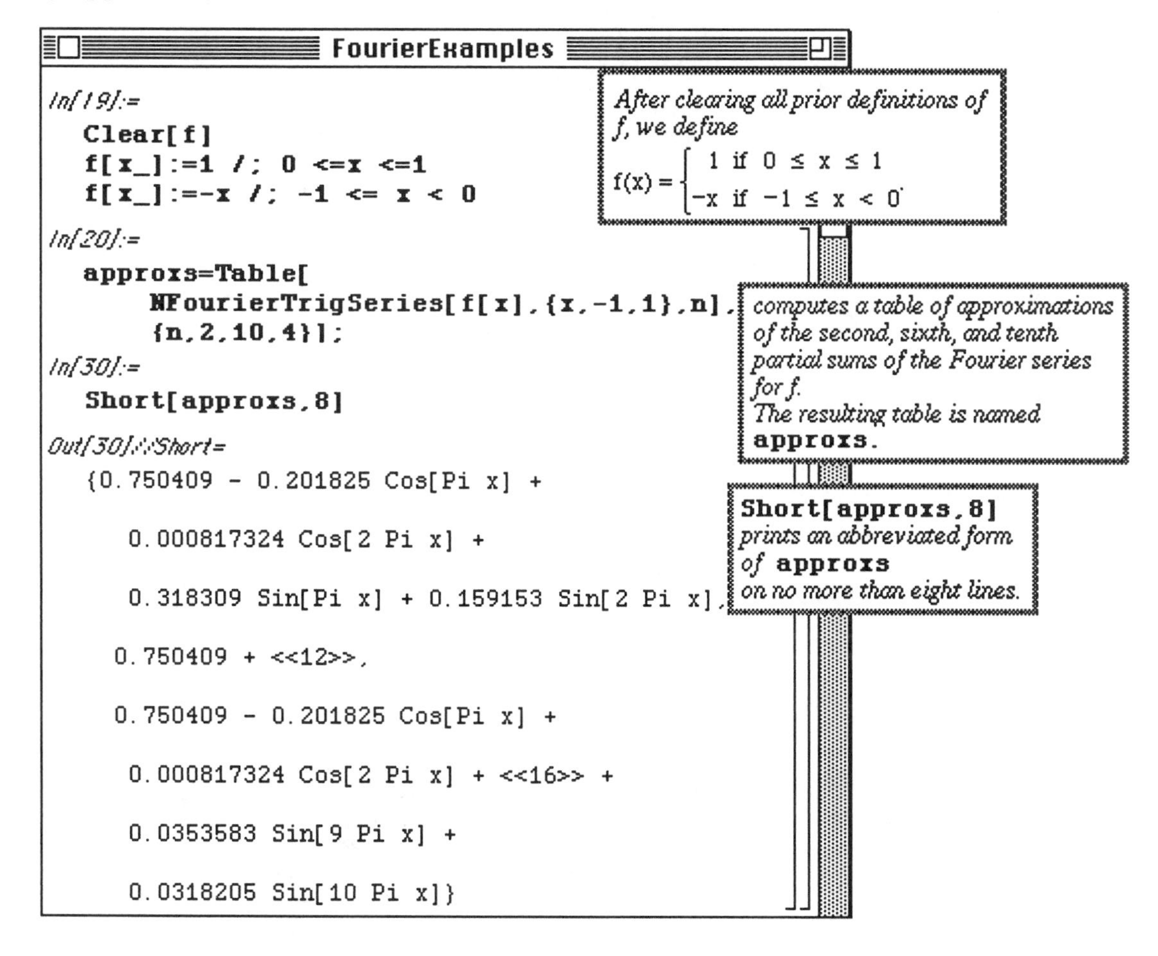

The accuracy of the approximation of f[x] by the partial sums of the Fourier series can be seen by plotting **approxs[[1]]**, **approxs[[2]]**, and **approxs[[3]]**. The graphs below clearly show that the approximation becomes more accurate (the corresponding curves move closer to the function **f**) as more terms of the Fourier series are used.

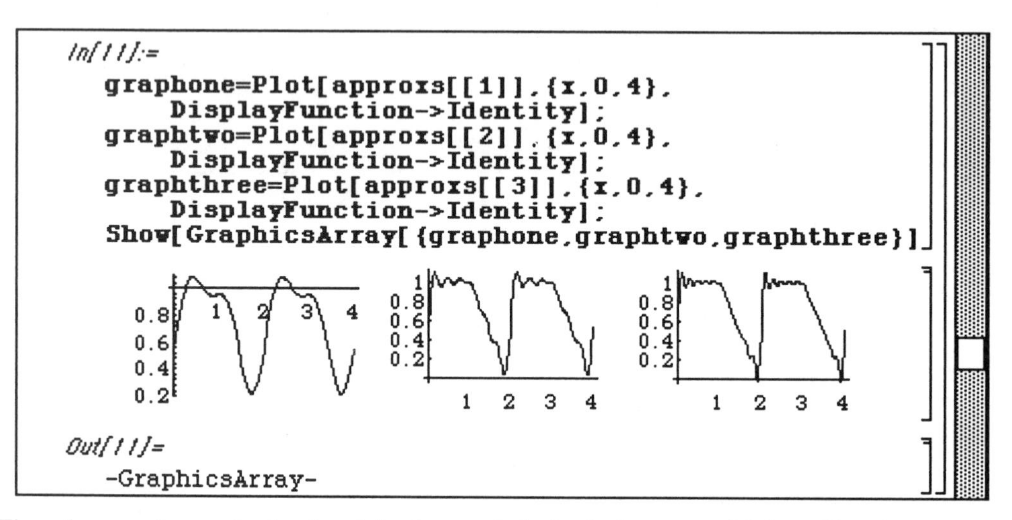

o These three graphs can also be plotted simultaneously in the single command below. Note that **Evaluate[approxs]** must be used in the **Plot** command just as **Release** was used in Version 1.2. The three are given in varying **GrayLevel** with the tenth partial sum represented by the darkest curve.

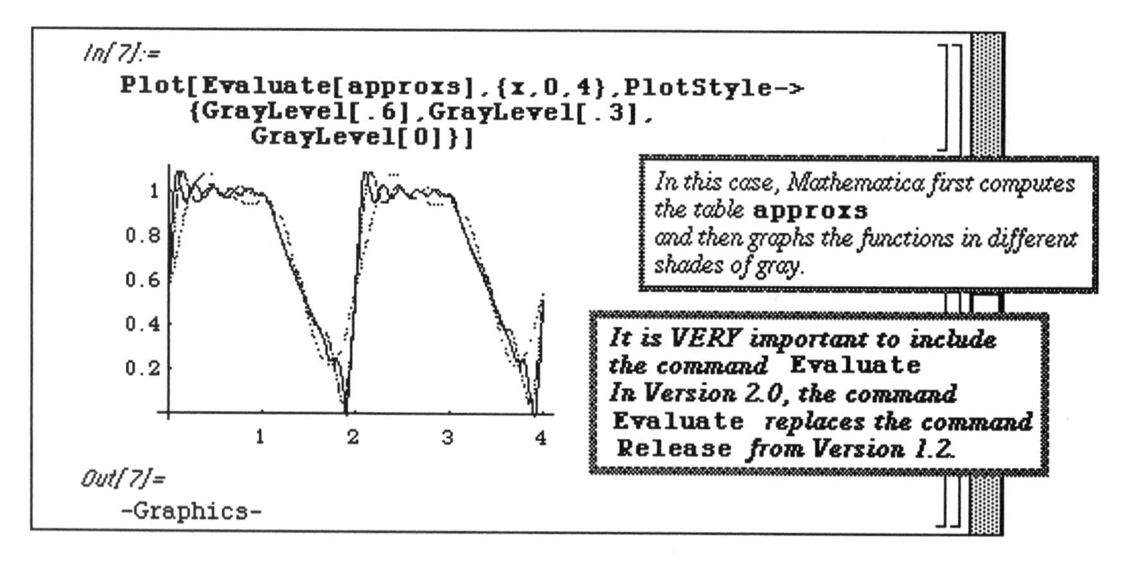

7.4 Discrete Math

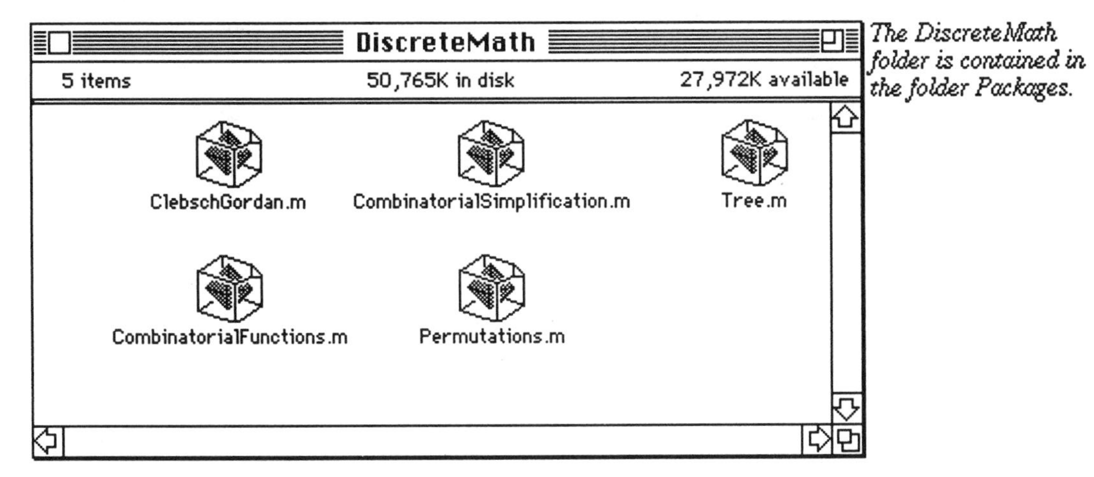

The DiscreteMath folder is contained in the folder Packages.

CombinatorialFunctions.m

In addition to the built-in *Mathematica* functions **Binomial[m,n]** and **Multinomial[n1,n2,n3,...,nm]**, the package **CombinatorialFunctions.m**, included with both Version 1.2 and Version 2.0, provides commands which deal with subfactorials, Catalan and Fibonacci numbers as well as the Hoftstadter functions. We first discuss the built-in functions.

Binomial[n,m] gives the binomial coefficient $\binom{n}{m} = \frac{n!}{m!\,(n-m)!}$. (The number of ways to choose **m** objects from a collection of **n** objects, without consideration to order.)

Multinomial[n_1,n_2,...,n_m] gives the multinomial coefficient $\frac{N!}{n_1!\,n_2!\cdots n_m!}$, where $\sum_{i=1}^{m} n_i = N$. Hence, the multinomial coefficient gives the number of ways to partition N distinct objects into m sets, each having size n_i .

Several examples of these functions are given below prior to loading the package :

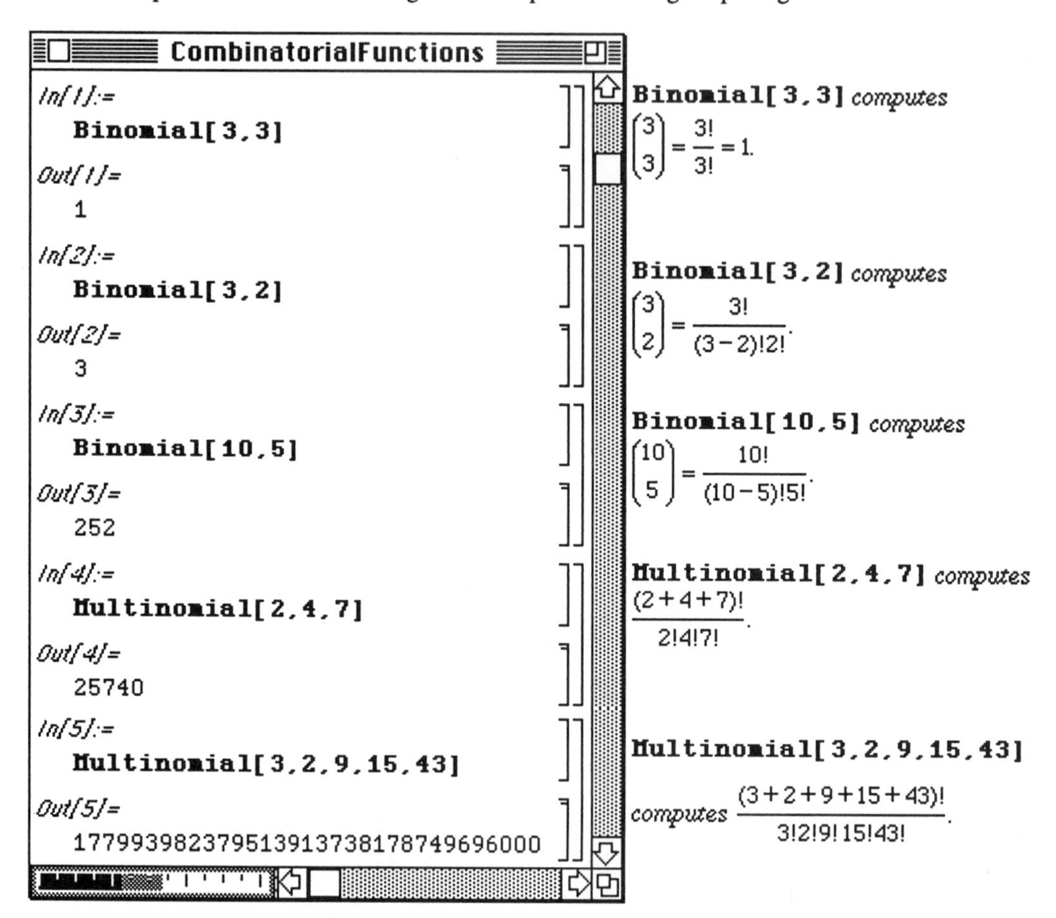

Typical problems which involve determining the binomial and multinomial coefficients are as follows :

□ **Example:**

In how many ways can an unordered 5-card poker hand be selected from a deck of 52 cards ?

Solution :

The answer is $\binom{52}{5}$ and is easily computed by entering **`Binomial[52,5]`**.

□ **Example:**

How many strings can be formed from the letters in *Mathematica* ?

Solution : The solution is determined using the multinomial coefficient. In the word *Mathematica*, the letter A appears 3 times while the letters M and T each appear twice. All others appear only once and there are 11 total letters. Therefore, the correct value is found with **`Multinomial[3,2,2,1,1,1,1]`** . (The 1's must not be omitted because the command uses them to determine the numerator in the calculation.)

Answers to both examples are computed below:

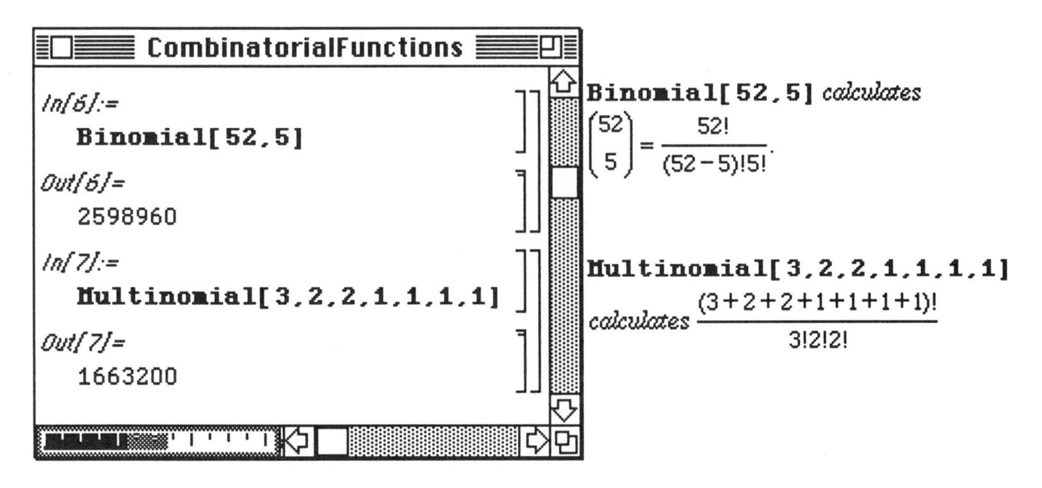

□ **Example:**

A useful combinatorial function which can be used only after loading **CombinatorialFunctions.m**, is **Subfactorial[n]**. This gives the number of permutations of n objects which leave no object fixed and is determined using the formula **n! Sum[(-1)^k/k!,{k,0,n}]**.
After loading **CombinatorialFunctions.m**, several examples are given. **Subfactorial[4]** is also found using the formula **4! Sum[(-1)^k/k!,{k,0,4}]** to verify that the same value results from each command.

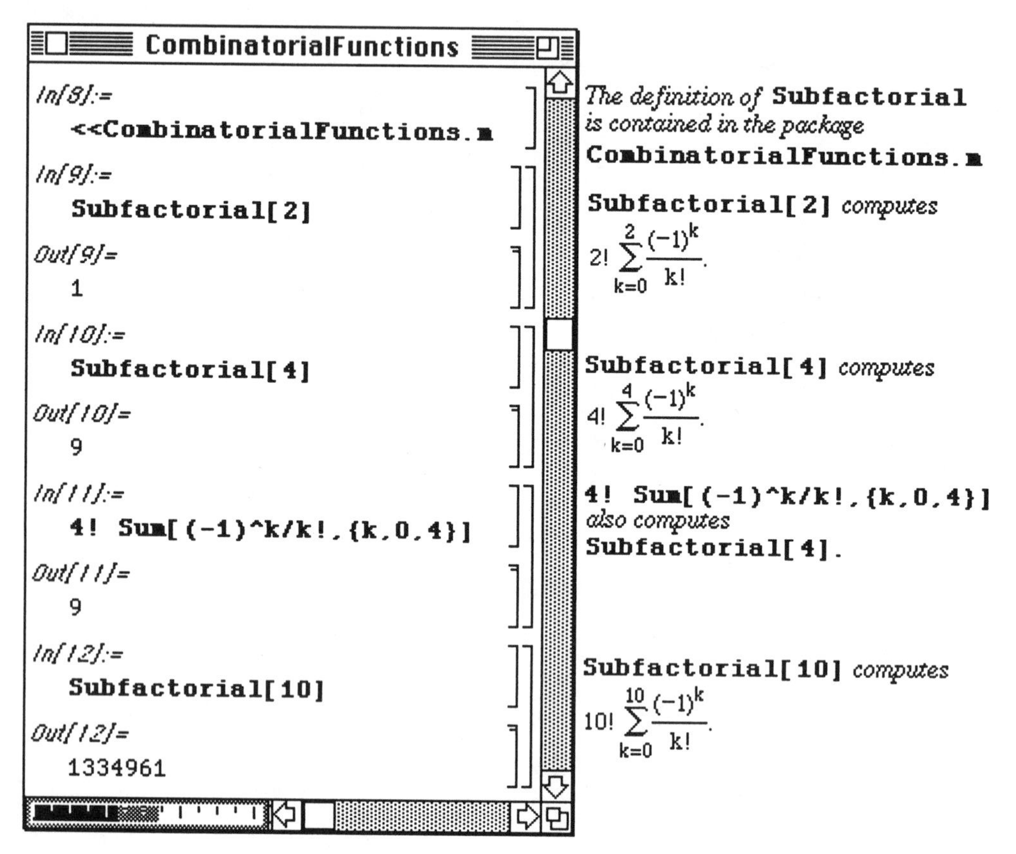

Another function found in **CombinatorialFunctions.m** determines the Catalan numbers. These numbers are found with the formula **Binomial[2n,n]/(n+1)** and are named after the Belgian mathematician Eugene-Charles Catalan (1814-1894) who discovered an elementary derivation of the formula. The problem which led to this discovery was that of determining the number of ways to divide a convex (n+2)-sided polygon, n greater than or equal to 1, into triangles by drawing (n-1) lines through the corners that do not intersect in the interior of the polygon. This number is found with **CatalanNumber[n]** which gives the nth Catalan number.

□ Example:

For example, there are five ways to divide a convex pentagon (n=3) into triangles by drawing two nonintersecting lined through the corners. This is given with **CatalanNumber[3]**. A table of Catalan numbers can also be created as illustrated below.

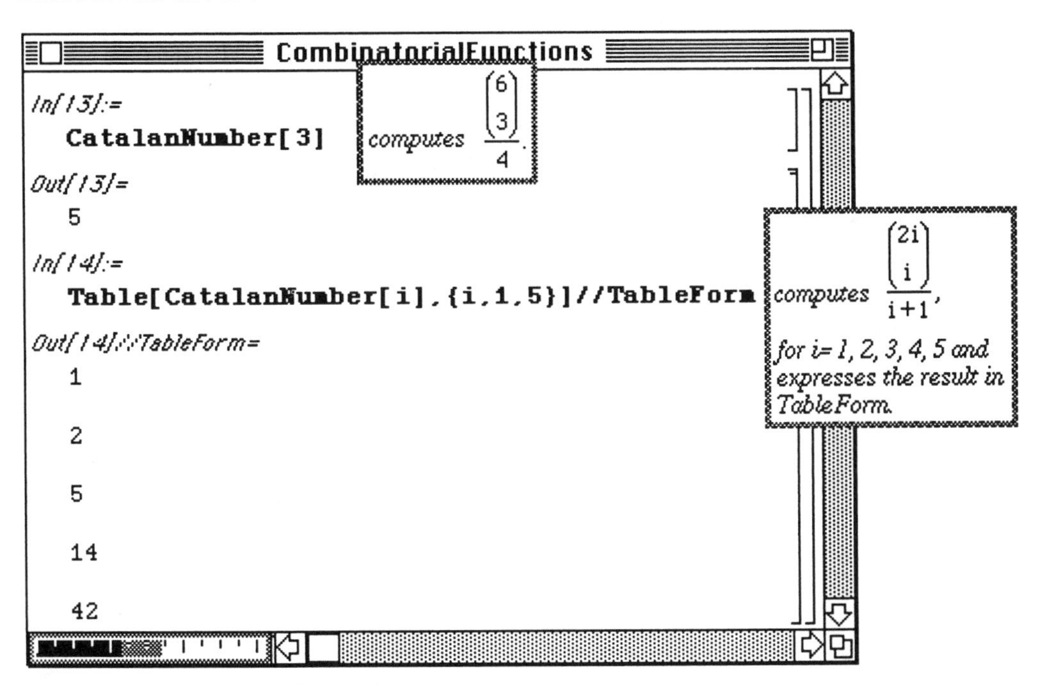

Another useful sequence is that of the Fibonacci numbers named for the Italian merchant and mathematician, Leonardo Fibonacci (1170-1250). These numbers are encountered in many areas as well and first arose in a puzzle about rabbits. One interesting fact about this sequence involves the ratio of successive terms. The limiting value of this ratio is:

GoldenRatio $-1=\frac{1+\sqrt{5}}{2}-1=\frac{\sqrt{5}-1}{2}\approx 0.61803.$

□ **Example:**

This is demonstrated below. First, a table of the first 16 Fibonacci numbers is created with **`Table`** and **`Fibonacci[n]`** which yields the nth Fibonacci number. This is called **`table1`**. Then, a table of the ratios of successive Fibonacci numbers is computed using **`table1`** to illustrate that the limit is the Golden Mean-1. A numerical approximation of the Golden Mean is also computed for comparison. Note that the first Fibonaaci number is found with index, **`n`** = 0. Thus, the nth Fibonacci number is found with **`Fibonacci[n-1]`**.

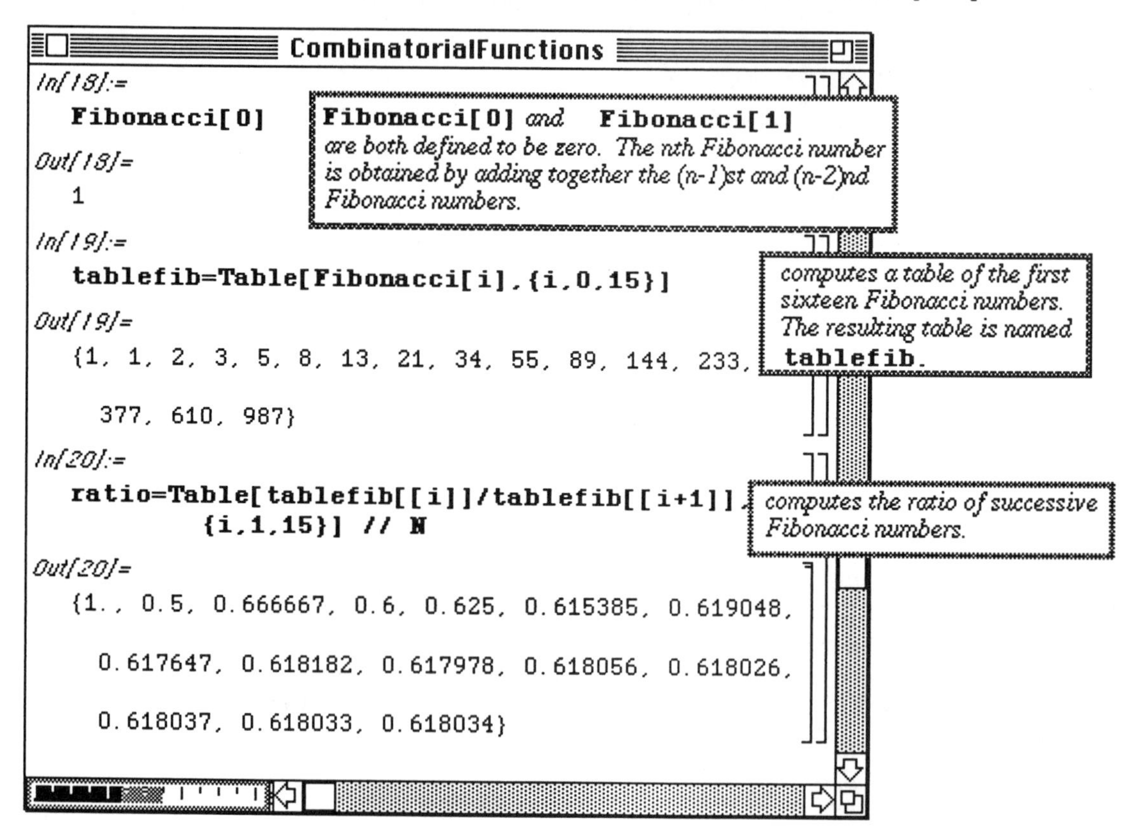

□ **Example:**

It should be noted that calculation of **`Fibonacci[n]`** for **`n`** > 10 is quite slow. Hence, an alternate approach is suggested. A table of the first 250 Fibonacci numbers can easily be found as follows (a shortened table of five lines is requested to save space).

Mathematica is designed to allow it to "remember" computed function values. In general, functions defined in the form

`f[x_]:=f[x]=`*`function definition`* will retain all values computed for later use.

The calculation of the first 250 Fibonacci numbers required approximately 1200 seconds in computation time with **Fibonacci[n]**, but the output was practically instantaneous when using the commands below.

CombinatorialFunctions

```
In[23]:=
  fib[0]=1
  fib[1]=1
  fib[n_]:=fib[n]=fib[n-1]+fib[n-2]
  table=Table[fib[i],{i,0,250}]
  Short[table,5]

Out[23]//Short=
  {1, 1, 2, 3, 5, <<242>>, 30161<<43>>0501,

    48801977467930020767542949510206990049732\

     287771475874, 7896325826131730509282738\

     943634332893686268675876375}

Time: 12.77 seconds
```

fib[n_]:=fib[n]= *tells Mathematica to remember the values of* **fib[n]** *that it computes.*

A table containing the values **fib[50]**, **fib[100]**, **fib[150]**, and **fib[200]** is given below.

Since **fib[50]**, **fib[100]**, **fib[150]**, and **fib[200]** were computed above, we note that recalling them is nearly instantaneous. Using the function **Fibonacci[n]** to compute **Fibonacci[50]**, **Fibonacci[100]**, **Fibonacci[150]**, and **Fibonacci[200]** takes considerable time.

CombinatorialFunctions

```
In[25]:=
  Table[fib[50 n],{n,1,4}] // TableForm

Out[25]//TableForm=
  12586269025

  354224848179261915075

  9969216677189303386214405760200

  280571172992510140037611932413038677189525

Time: 0.42 seconds
```

Since Mathematica has already computed **fib[50]**, **fib[100]**, **fib[150]**, *and* **fib[200]**, *representing them is almost immediate.*

Note that for Mathematica to compute **Fibonacci[200]** *takes considerable time.*

□ **Application:**

Some interesting facts concerning Fibonacci numbers can now be investigated.

Consider the sequence $\{T_n\}_{n=1}^{\infty}$ where $T_n = \dfrac{F_{5n}}{5F_n}$, F_n being the nth Fibonacci number.

Theorem: If T_n is prime, then n is either prime or a power of 5.

(Another interesting fact about this sequence is that all numbers end in 1.) By computing the following table with the sequence of Fibonacci numbers in **fib[n]** found above, these properties can be investigated.

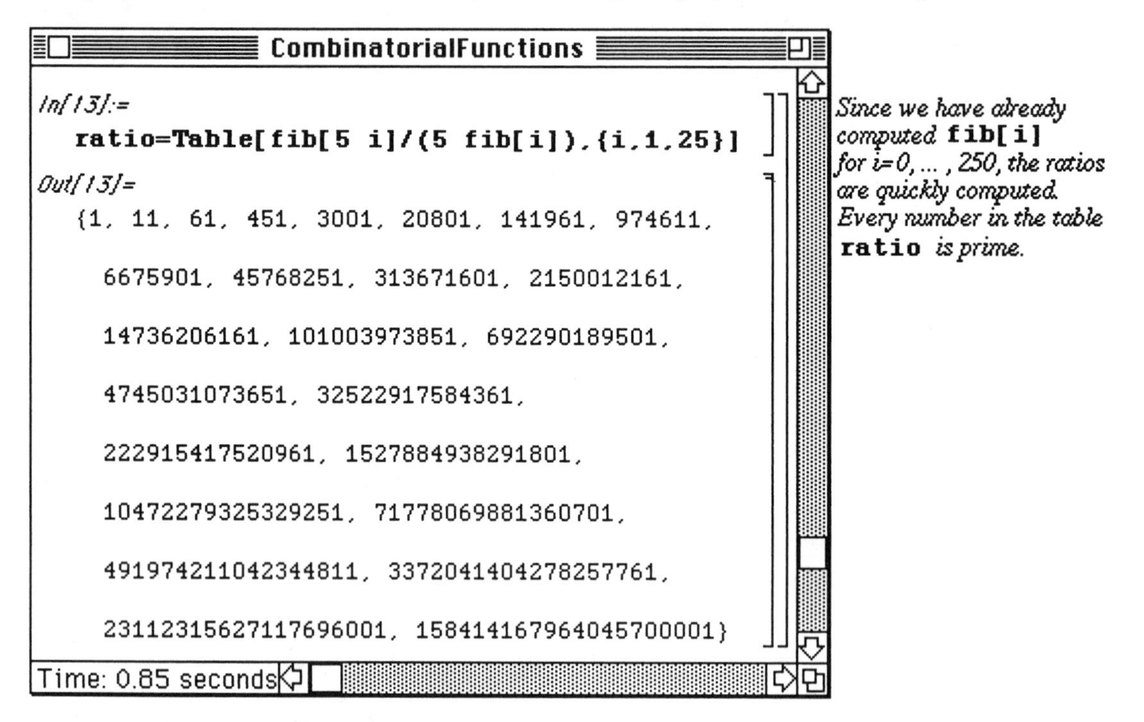

The **Hofstadter function**, Hof, is recursively defined by
(i) Hof(1)=Hof(2)=1; and
(ii) Hof(n)=Hof[n-Hof(n-1))+Hof(n-Hof(n-2))

Mathematica's definition of the Hofstadter function is contained in the package **CombinatorialFunctions.m** and is denoted by **Hofstadter[n]**.

□ **Example:**

A table of the first thirty values of the Hofstadter function is given below :

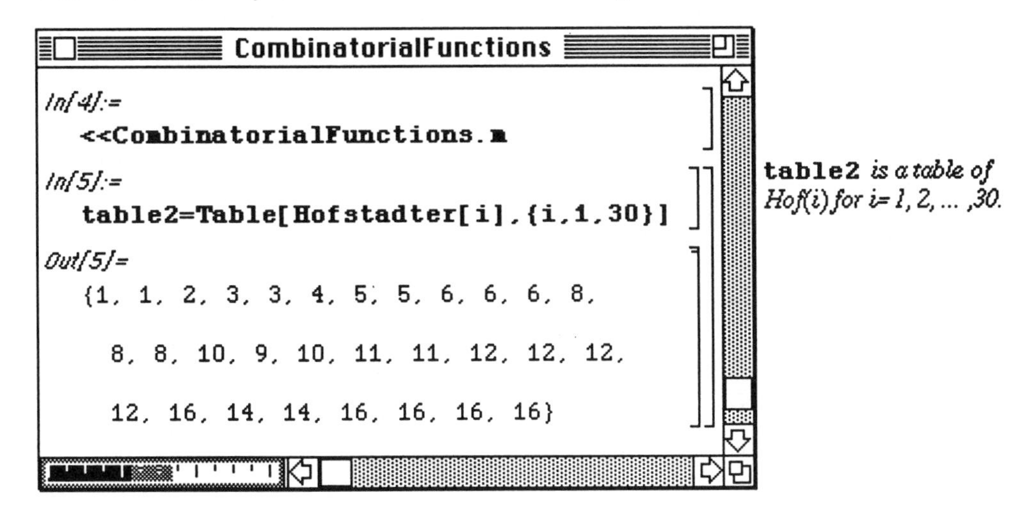

`table2` *is a table of Hof(i) for i= 1, 2, ... ,30.*

■ CombinatorialSimplification.m

The package **CombinatorialSimplification.m** includes several rules which can be used to simplify certain expressions which *Mathematica* does not automatically simplify. For specific integers m and n, **n!** computes the factorial of n; **Binomial[m,n]** computes $\binom{m}{n} = \frac{m!}{(m-n)!n!}$.

Unfortunately, if m and n are not specific integers, *Mathematica* does not automatically simplify the following expressions involving factorials and binomial coefficients:

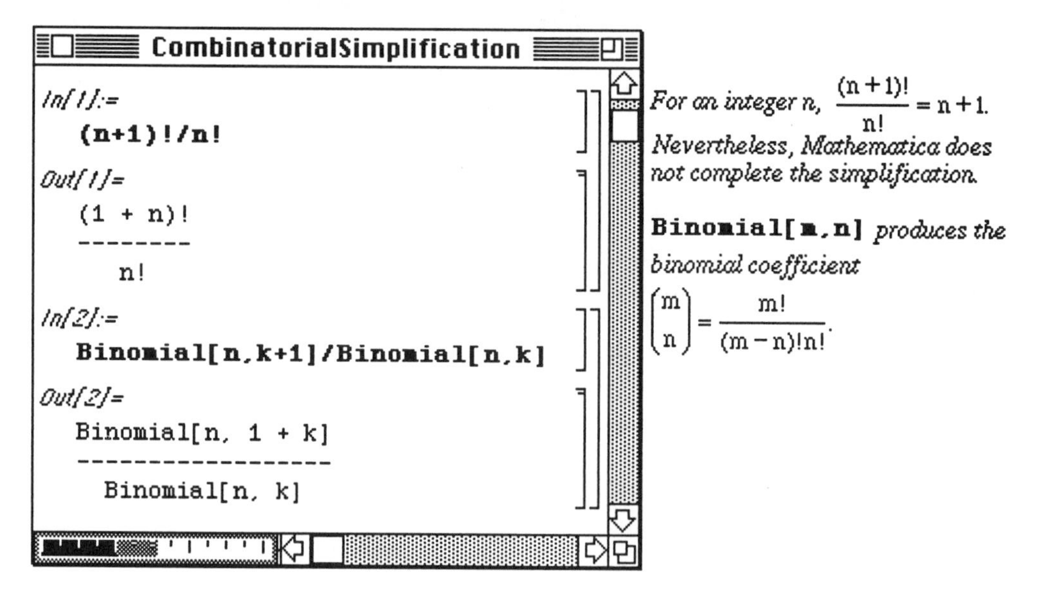

For an integer n, $\frac{(n+1)!}{n!} = n+1$. *Nevertheless, Mathematica does not complete the simplification.*

Binomial[m,n] *produces the binomial coefficient* $\binom{m}{n} = \frac{m!}{(m-n)!n!}$.

☐ Example:

Fortunately, these expressions can be simplified with after loading the package **CombinatorialSimplification.m**. since **CombinatorialSimplification.m** redefines the built-in combinatorial functions so that *Mathematica* symbolically reduces combinatorial functions:

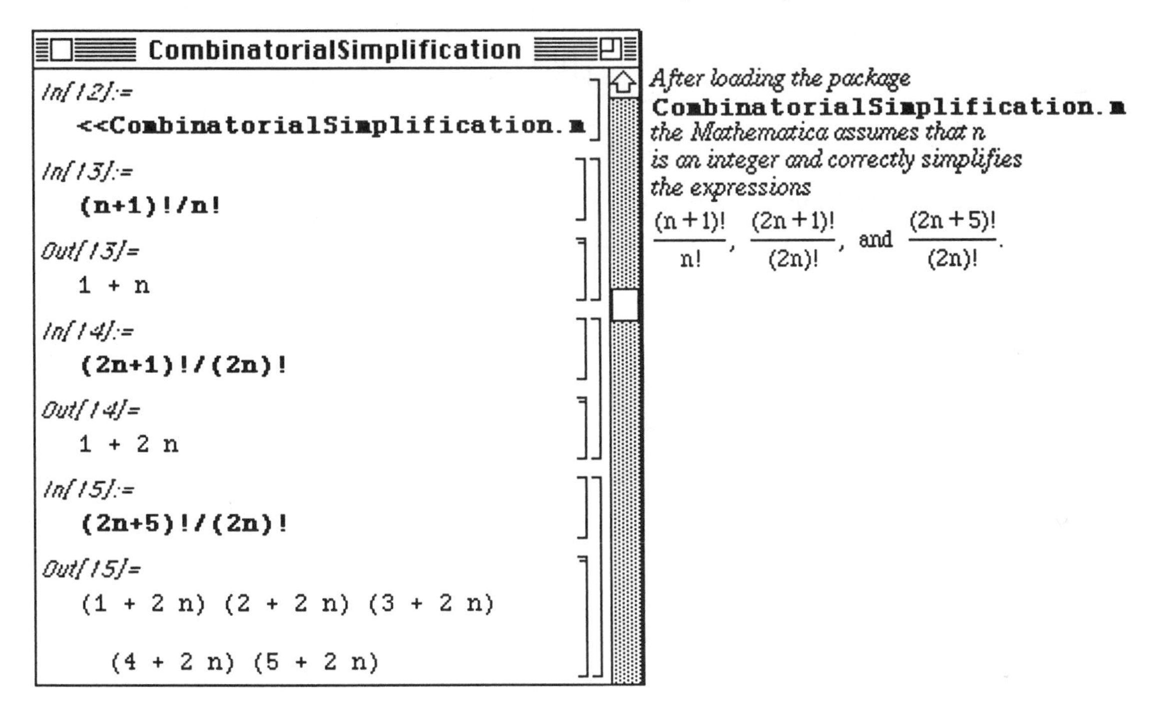

`Binomial[n,m]` is also redefined for the quotients `Binomial[n,k]/Binomial[n,k-1]`:

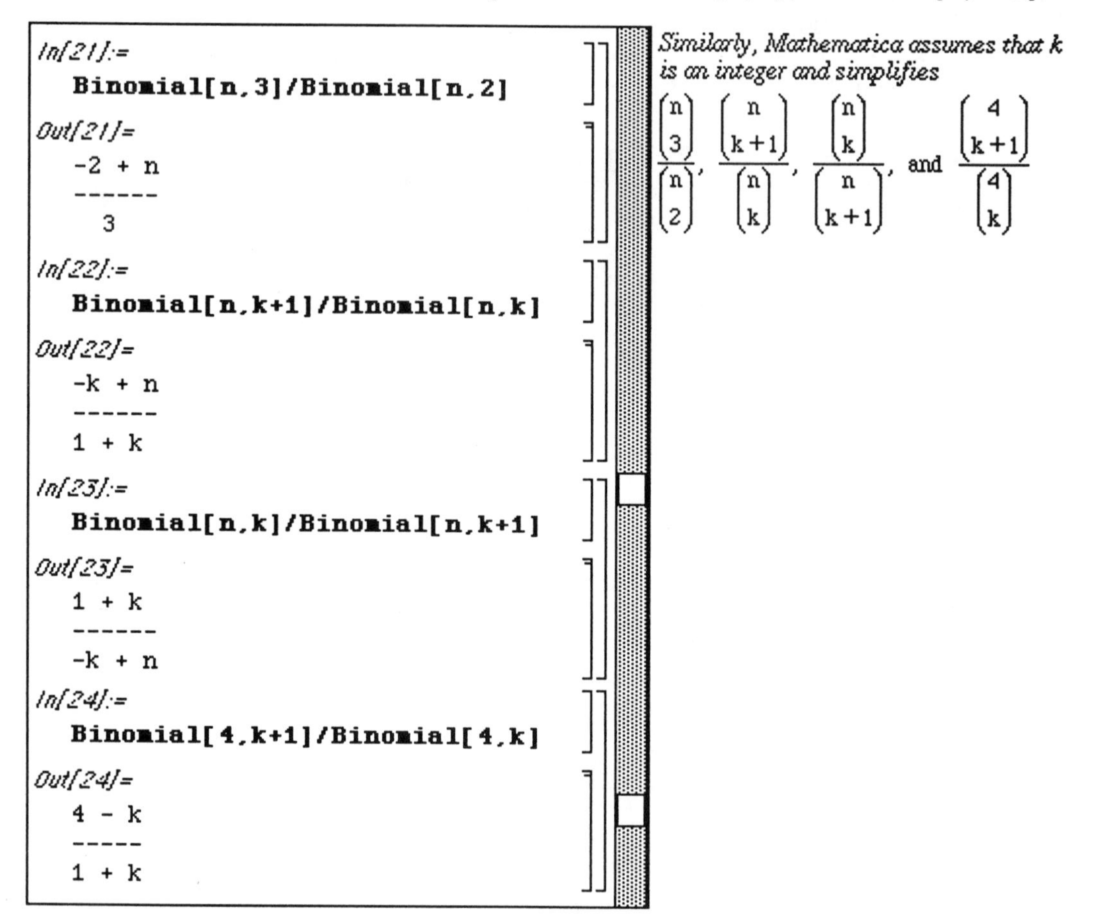

```
In[21]:=
  Binomial[n,3]/Binomial[n,2]
Out[21]=
  -2 + n
  ------
    3
In[22]:=
  Binomial[n,k+1]/Binomial[n,k]
Out[22]=
  -k + n
  ------
  1 + k
In[23]:=
  Binomial[n,k]/Binomial[n,k+1]
Out[23]=
  1 + k
  ------
  -k + n
In[24]:=
  Binomial[4,k+1]/Binomial[4,k]
Out[24]=
  4 - k
  -----
  1 + k
```

Similarly, Mathematica assumes that k is an integer and simplifies $\frac{\binom{n}{3}}{\binom{n}{2}}$, $\frac{\binom{n}{k+1}}{\binom{n}{k}}$, $\frac{\binom{n}{k}}{\binom{n}{k+1}}$, and $\frac{\binom{4}{k+1}}{\binom{4}{k}}$

((Binomial[n,k+1]/Binomial[n,k])^2)/
((Binomial[n,k+1]/Binomial[n,k])^p denotes the expression

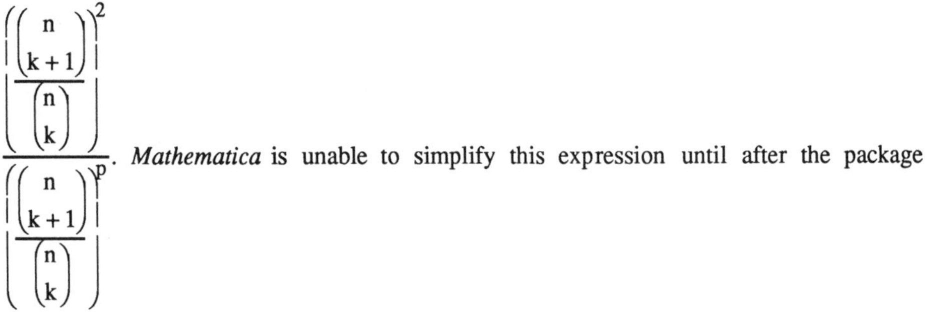

$$\frac{\left(\frac{\binom{n}{k+1}}{\binom{n}{k}}\right)^2}{\left(\frac{\binom{n}{k+1}}{\binom{n}{k}}\right)^p}.$$

Mathematica is unable to simplify this expression until after the package has been loaded:

```
CombinatorialSimplification

In[3]:=
  ((Binomial[n,k+1]/Binomial[n,k])^2)/
  ((Binomial[n,k+1]/Binomial[n,k])^p)

Out[3]=
                     -2 + p                      2 - p
  Binomial[n, k]          Binomial[n, 1 + k]

In[4]:=
  <<CombinatorialSimplification.m

In[5]:=
  ((Binomial[n,k+1]/Binomial[n,k])^2)/
  ((Binomial[n,k+1]/Binomial[n,k])^p)

Out[5]=
          -2 + p         2 - p
  (1 + k)        (-k + n)
```

Upon closing the **CombinatorialSimplification.m** package, all new definitions are cleared. Hence, any built-in function which was redefined is unaltered by the changes made in the package.

■ Permutations.m

A **permutation** of n distinct elements $x_1, x_2, \ldots, x_n$ is an ordering of the n elements.

`Permutations.m` gives several commands in addition to the built-in commands which are helpful in working with permutations. *Mathematica* already includes the function **`Permutation[list]`** which gives all possible permutations of the list of n elements **`list`**. This package includes **`PermutationQ[list]`** which gives a value of `True` if **`list`** is a possible permutation of n distinct elements and `False` otherwise. Therefore, if any of the n elements is omitted or repeated in **`list`**, a value of `False` is given. Several examples that illustrate these ideas are given below. These calculations can be made after loading the package :

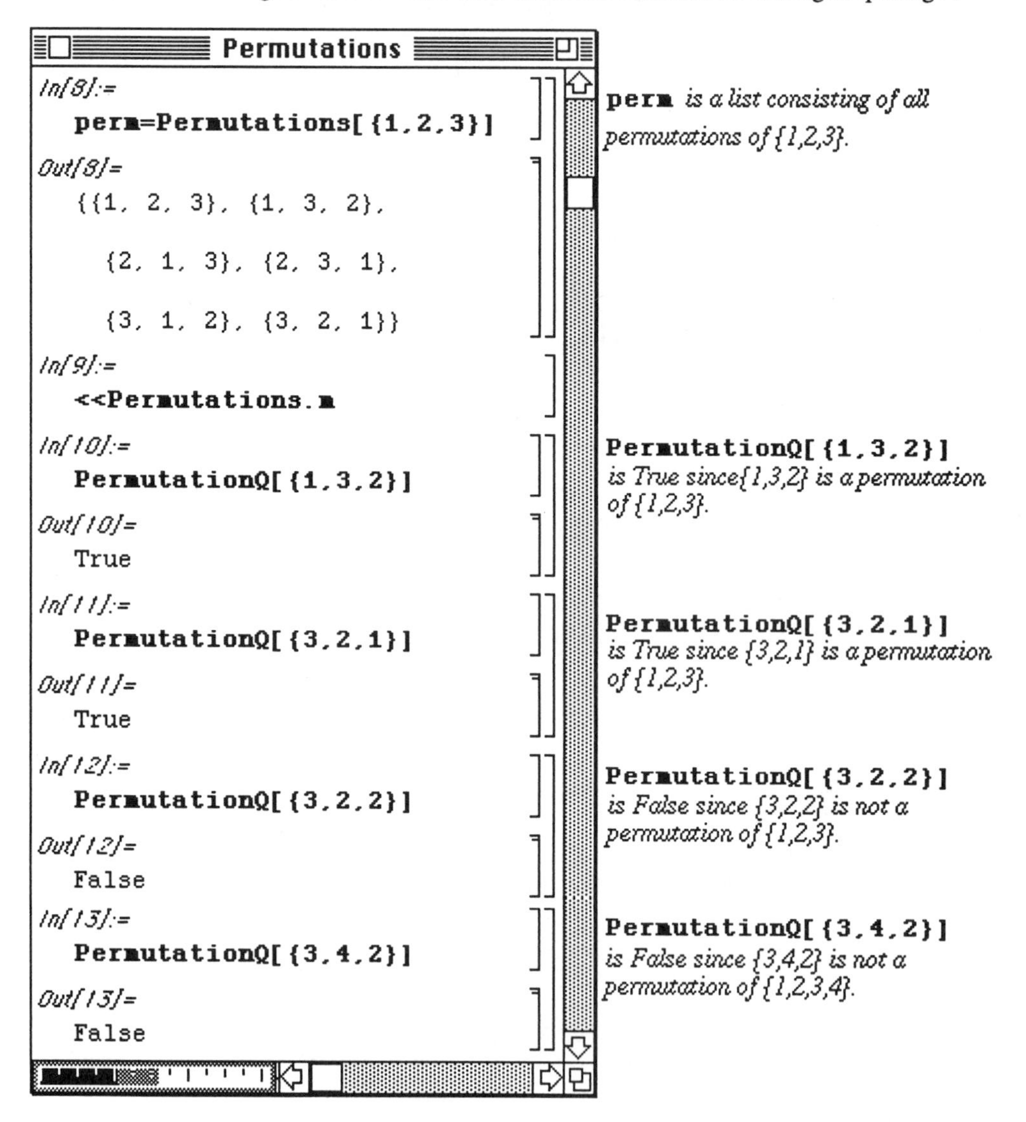

`perm` *is a list consisting of all permutations of {1,2,3}.*

`PermutationQ[{1,3,2}]` *is True since {1,3,2} is a permutation of {1,2,3}.*

`PermutationQ[{3,2,1}]` *is True since {3,2,1} is a permutation of {1,2,3}.*

`PermutationQ[{3,2,2}]` *is False since {3,2,2} is not a permutation of {1,2,3}.*

`PermutationQ[{3,4,2}]` *is False since {3,4,2} is not a permutation of {1,2,3,4}.*

The package also includes the command **RandomPermutation[n]** which yields one possible permutation of the **n** elements in the list **{1,2,3,...n}**. This permutation is selected at random. Notice in the example which follows that two different permutations are given with **RandomPermutation[15]**. Two other commands found in the package are **ToCycles[permutation]** which gives **permutation** as a list of cyclic permutations and **FromCycles[cycles]** which returns **permutation** to its original form.

Notice in the example which follows that two different permutations are given with **RandomPermutation[15]**.

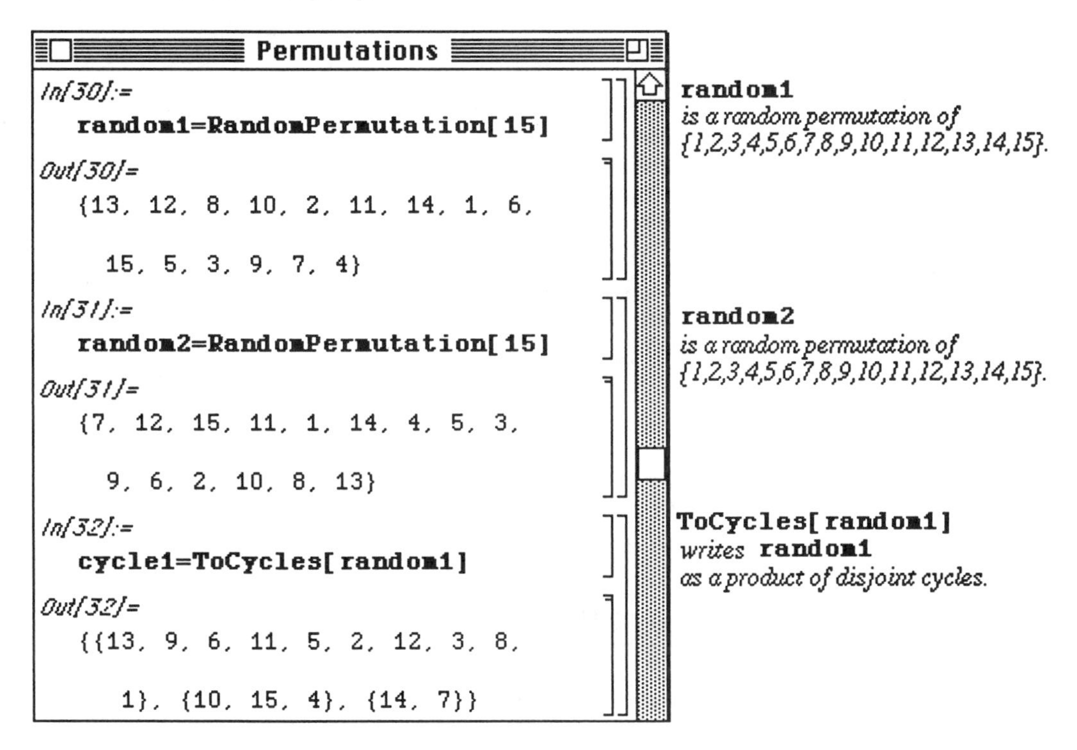

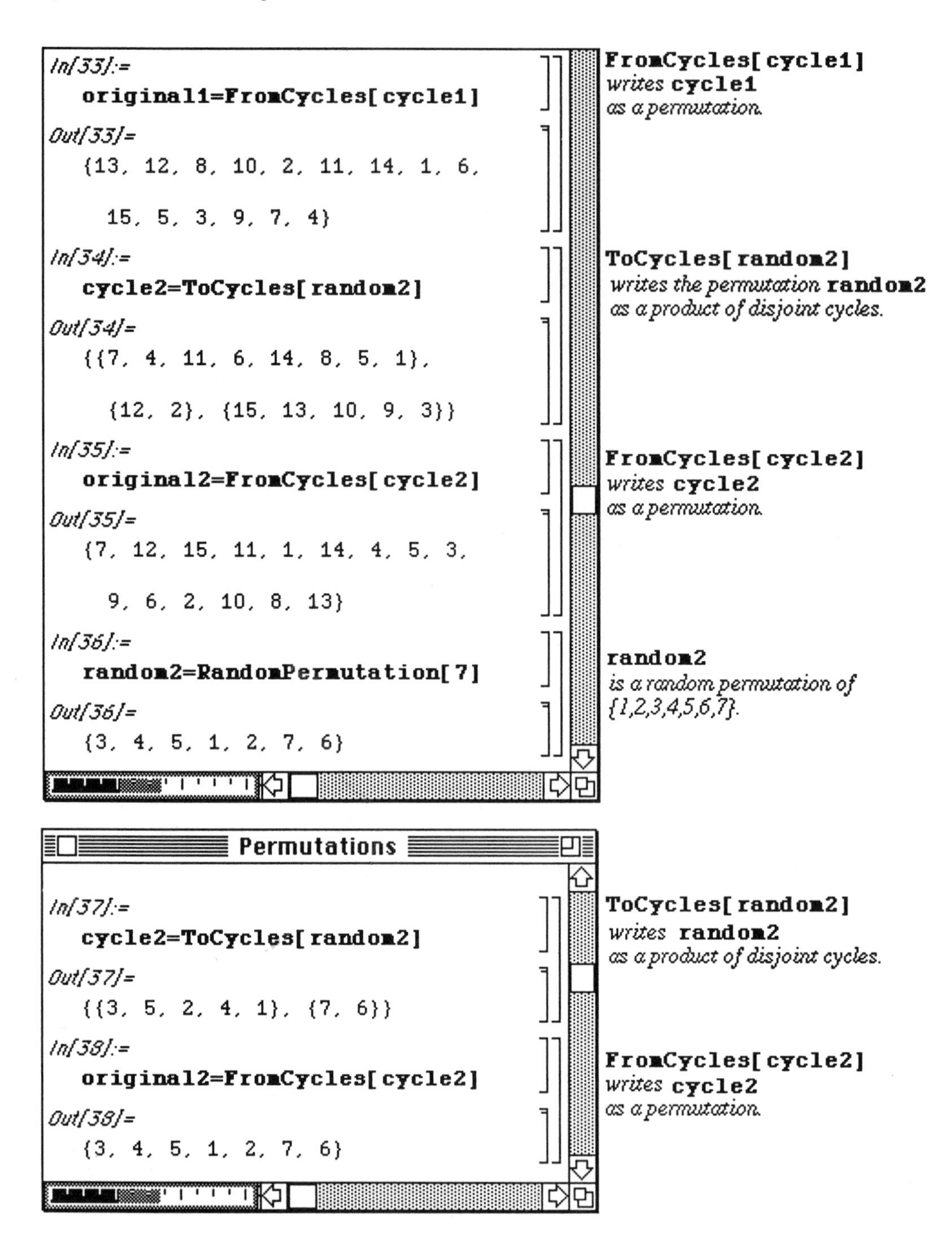

FromCycles[cycle1] *writes* **cycle1** *as a permutation.*

ToCycles[random2] *writes the permutation* **random2** *as a product of disjoint cycles.*

FromCycles[cycle2] *writes* **cycle2** *as a permutation.*

random2 *is a random permutation of {1,2,3,4,5,6,7}.*

ToCycles[random2] *writes* **random2** *as a product of disjoint cycles.*

FromCycles[cycle2] *writes* **cycle2** *as a permutation.*

Chapter 8
Some Graphics Packages

■ **Chapter 8** introduces several of the graphics packages available with *Mathematica.* Differences between Version 1.2 and Version 2.0 are discussed where appropriate.

■ Graphics

Opening the Version 1.2 **Graphics** folder yields the following window. Each package shown below contains one or more *Mathematica* commands or functions which cannot be used without loading the appropriate package.

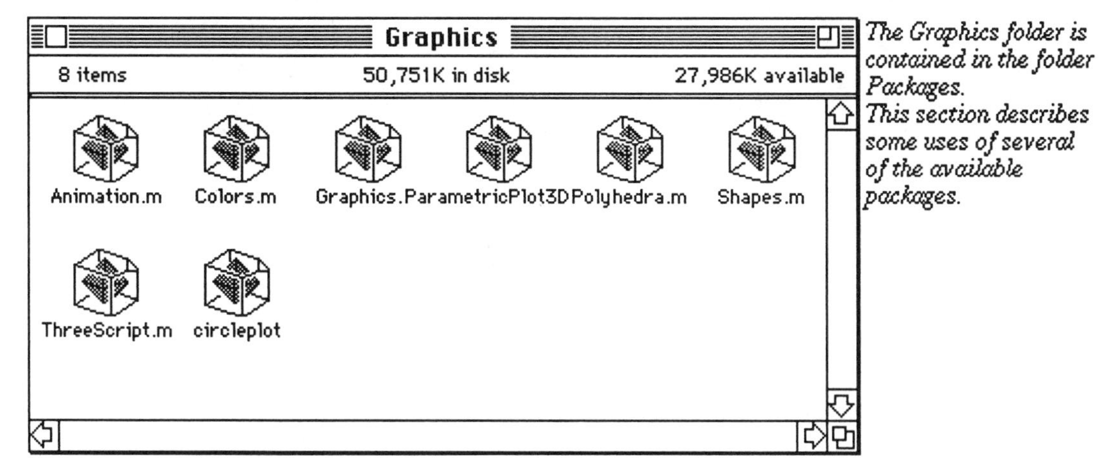

The Graphics folder is contained in the folder Packages. This section describes some uses of several of the available packages.

■ 8.1 Graphics.m

Loading **Graphics.m** enables the user to take advantage of several commands which will improve the graphing capabilities previously available. The first command discussed below, **PolarPlot**, allows for the graphing of functions given in polar coordinates (r,θ). This command should be entered in the following manner: **PolarPlot[function[var],{var,var1,var2},options]** where **var** represents the angular coordinate θ and **var** varies from **var1** to **var2**. This command produces the graph of the function **r=function[var]**.

□ **Example:**

In the first example below, the graph of r = 1-2 sinθ is given. It is followed by the graph of r = 1+2 cosθ which is plotted with the **GrayLevel** option. Notice that the graphs are named **polarone** and **polartwo**, respectively, for later use.

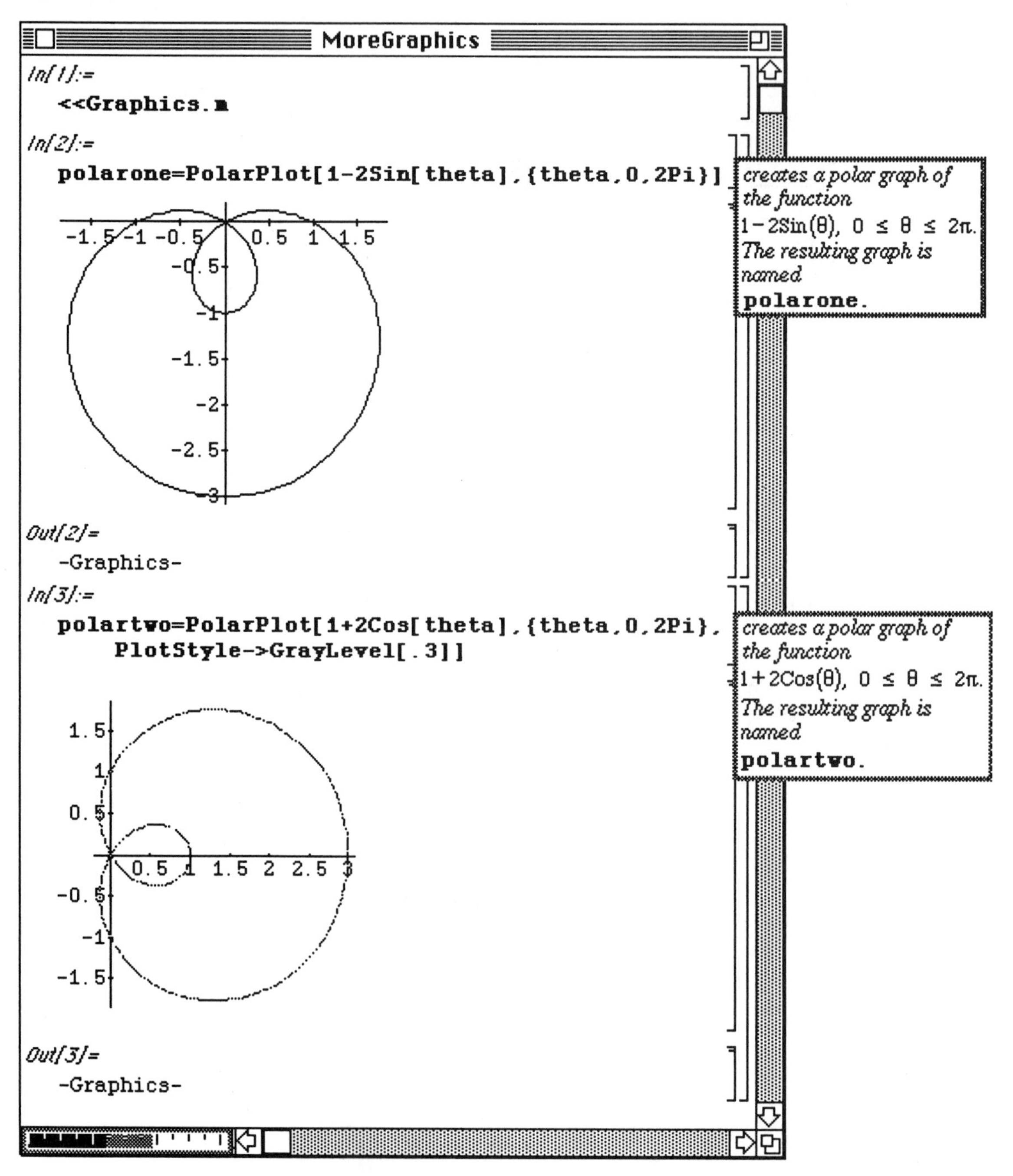

Since the two previous graphs were assigned names, they can easily be recalled and graphed together with **Show[polarone,polartwo]**.

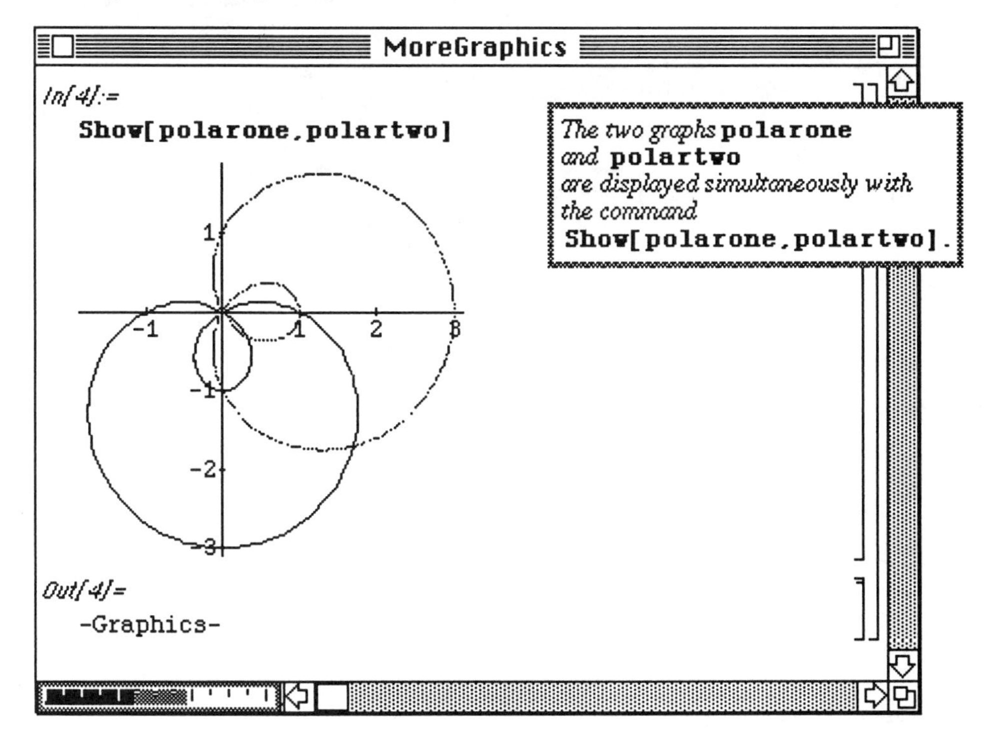

□ **Example:**

The following command produces the "spiral-shaped" graph of **Log[1+r]**. Note that **r** represents the angular coordinate.

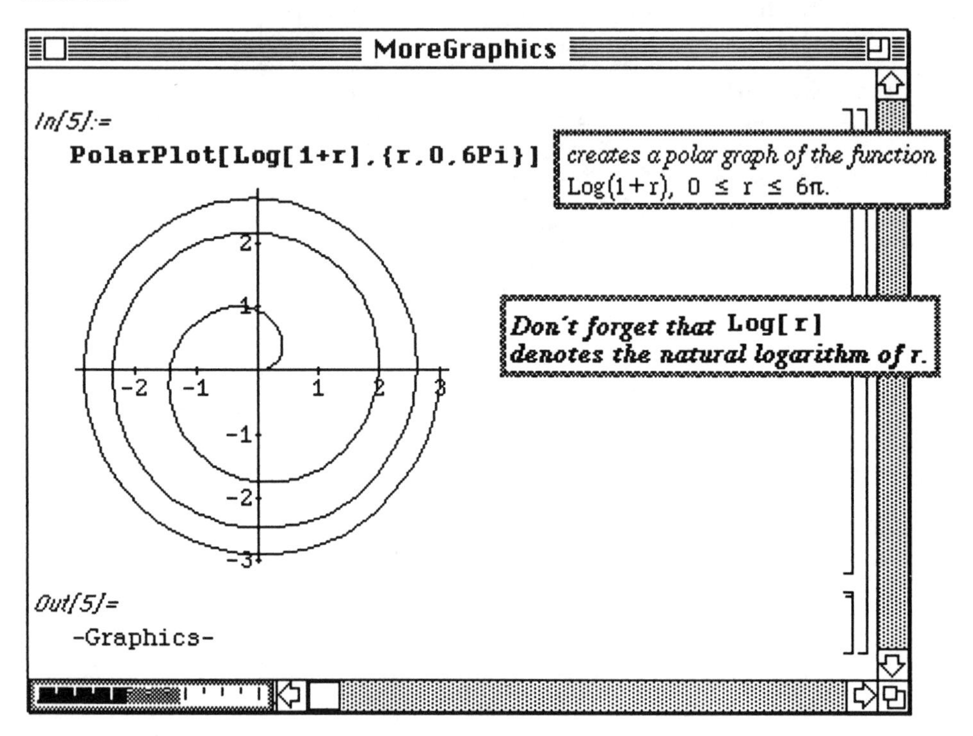

Another useful command in **Graphics.m** is **PolarListPlot**. This enables the user to be able to plot a list of points given in polar coordinates. This command is stated in the following manner:

PolarListPlot[list,options] where **list** a list of points. A convenient way to enter **list** is in the form of a table.

□ Example:

In the example which follows, a table of values of sin(2t) is created where t varies from 0 to 2π, using increments of $2\pi/90$. In this case, the variable t represents the angular coordinate. The graph is given below and named **`listplotone`**.

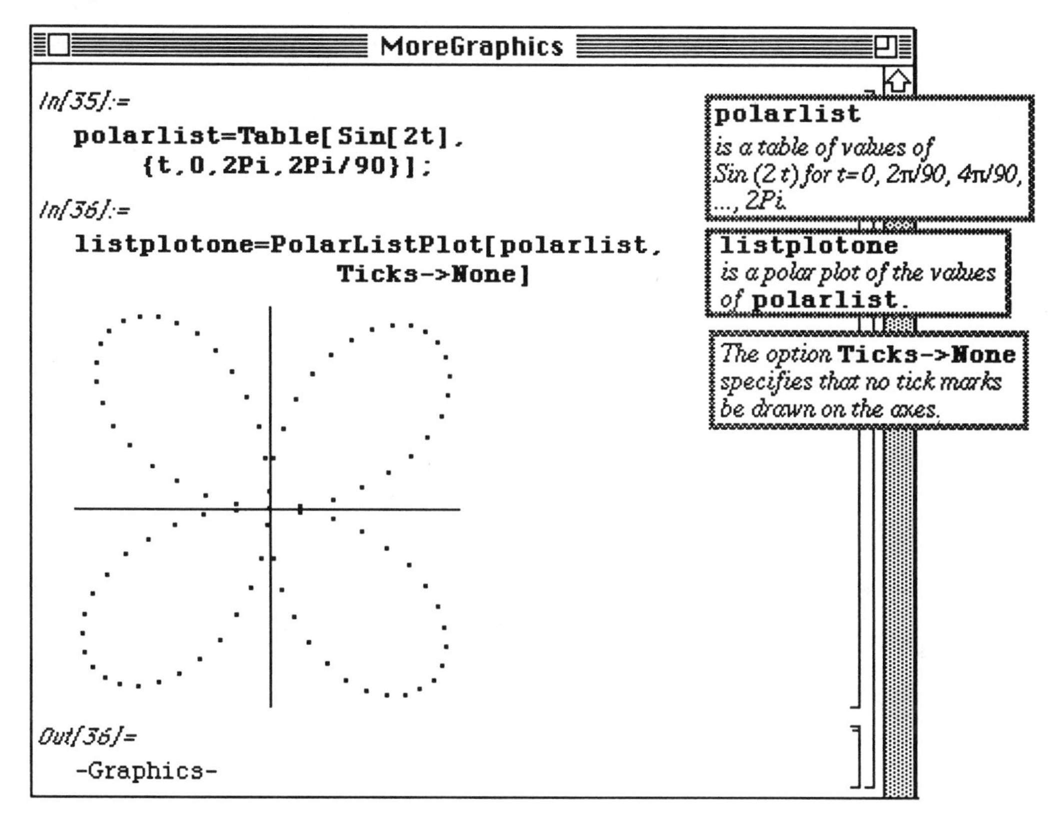

Next, the graph of the list of values of 3/2 cos(3t) is produced. In the **PolarListPlot** command, the **PlotJoined->True** option is used, so the points are connected. This plot is called **listplottwo**. Notice the overlap in the graph; the three-petal rose is produced for t between 0 and π. Hence, the curve is retraced for t between π and 2π.

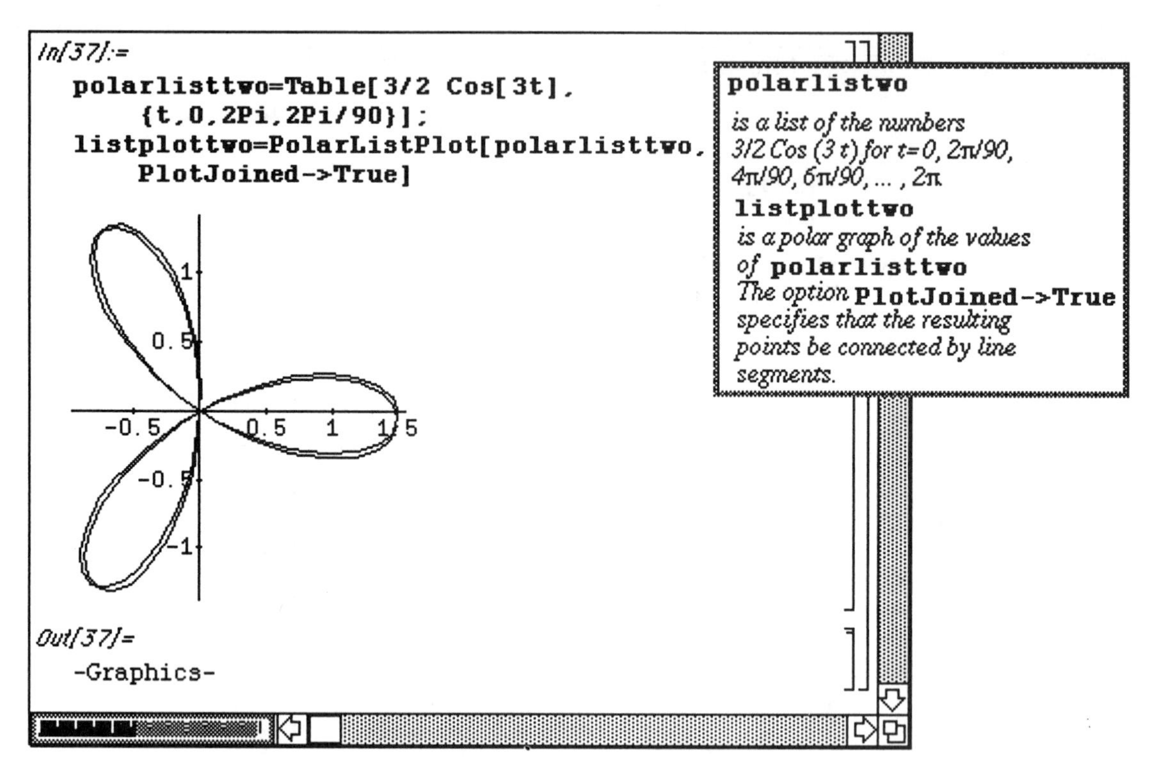

The previous plots can be displayed simultaneously with **Show**.

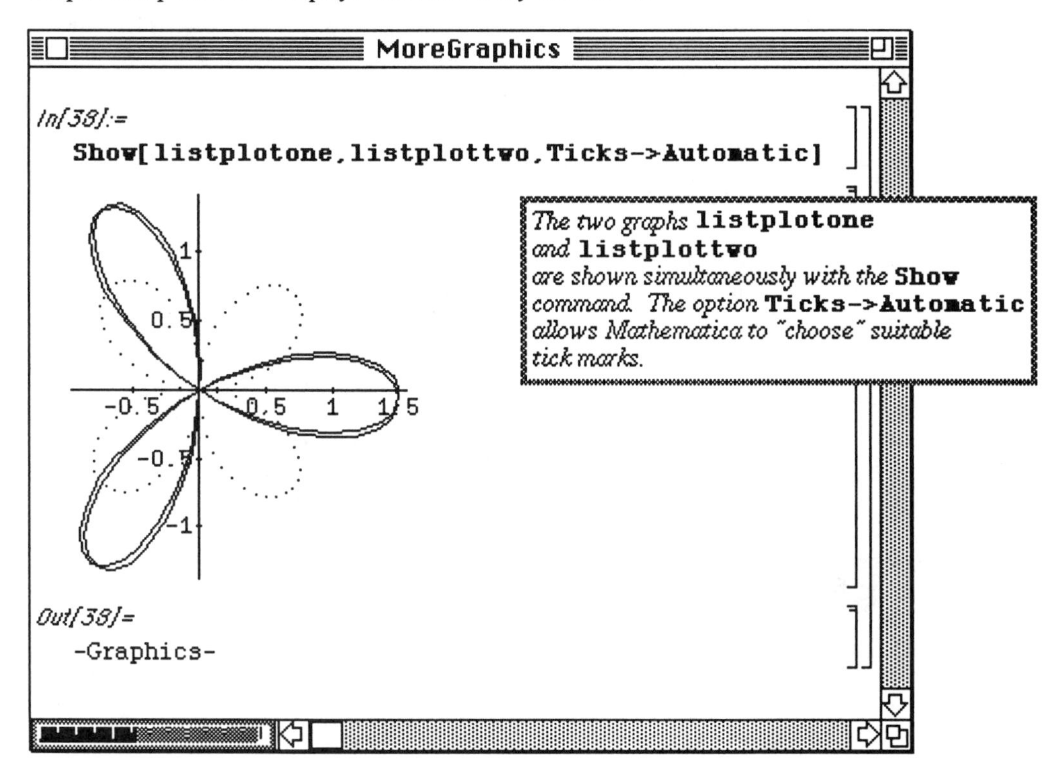

Bar graphs can be drawn using **Graphics.m** with **BarChart[list]**. For each number in **list**, *Mathematica* draws a rectangle of that height. These rectangles are drawn in order from left to right. The position of the element is given beneath each rectangle. These numbers are quite small in the following graph, but resizing the graph, which will be described next, is possible.

□ **Example:**

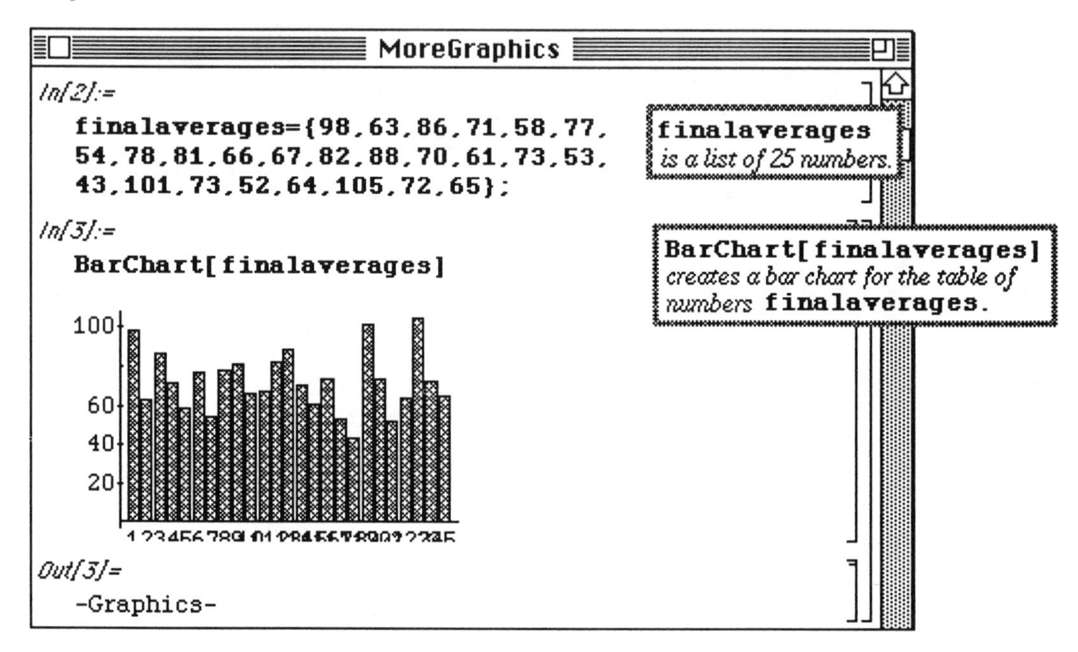

Resizing is accomplished by clicking once anywhere inside the graphics cell. This encloses the graph in a box as shown below. If the cursor is moved to the lower right-hand corner and dragged, the graph can be enlarged to the desired size. Notice below that the numbers beneath the graph are now readable.

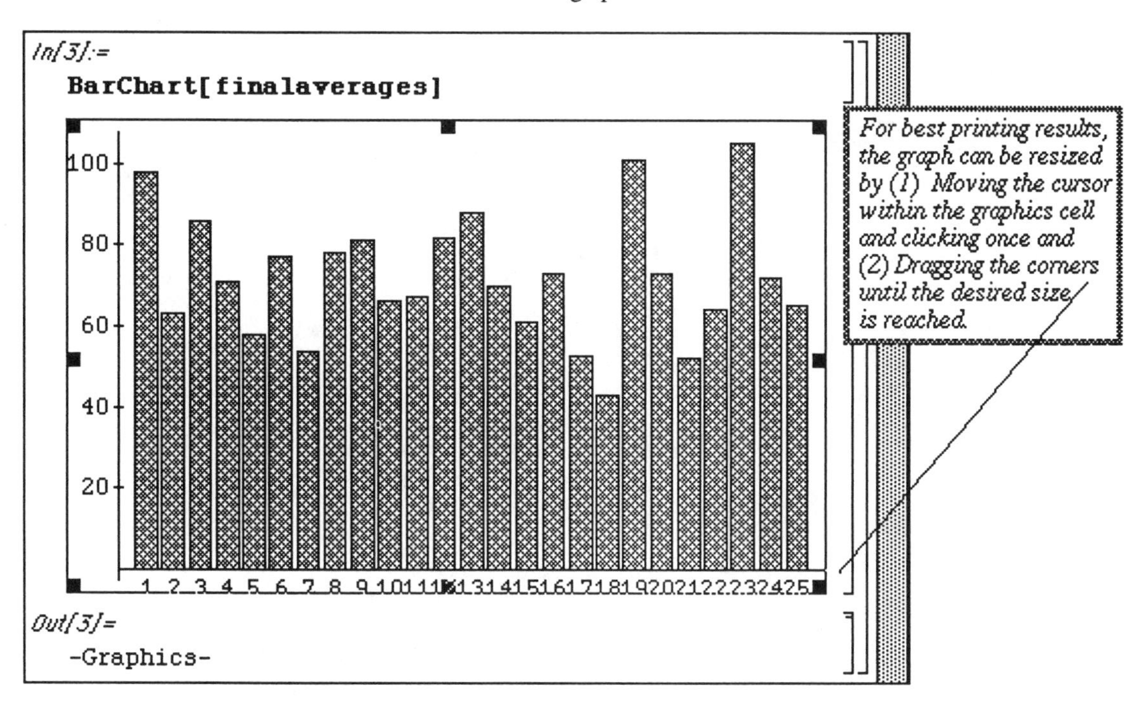

Before producing **BarChart[list]**, the elements of **list** can first be sorted. **Sort[list]** sorts the elements of **list** from smallest to largest. In the following example, the grades found in **finalaverages** are sorted and named **sort**. A shortened 2-line output of **sort** is given with **Short[sort,2]**,and the bar graph corresponding to the sorted list is produced with **BarChart[sort]**. The options of **BarChart** are listed below along with the default values. To use an option with **BarChart** simply enter the command as **BarChart[list,options]**.

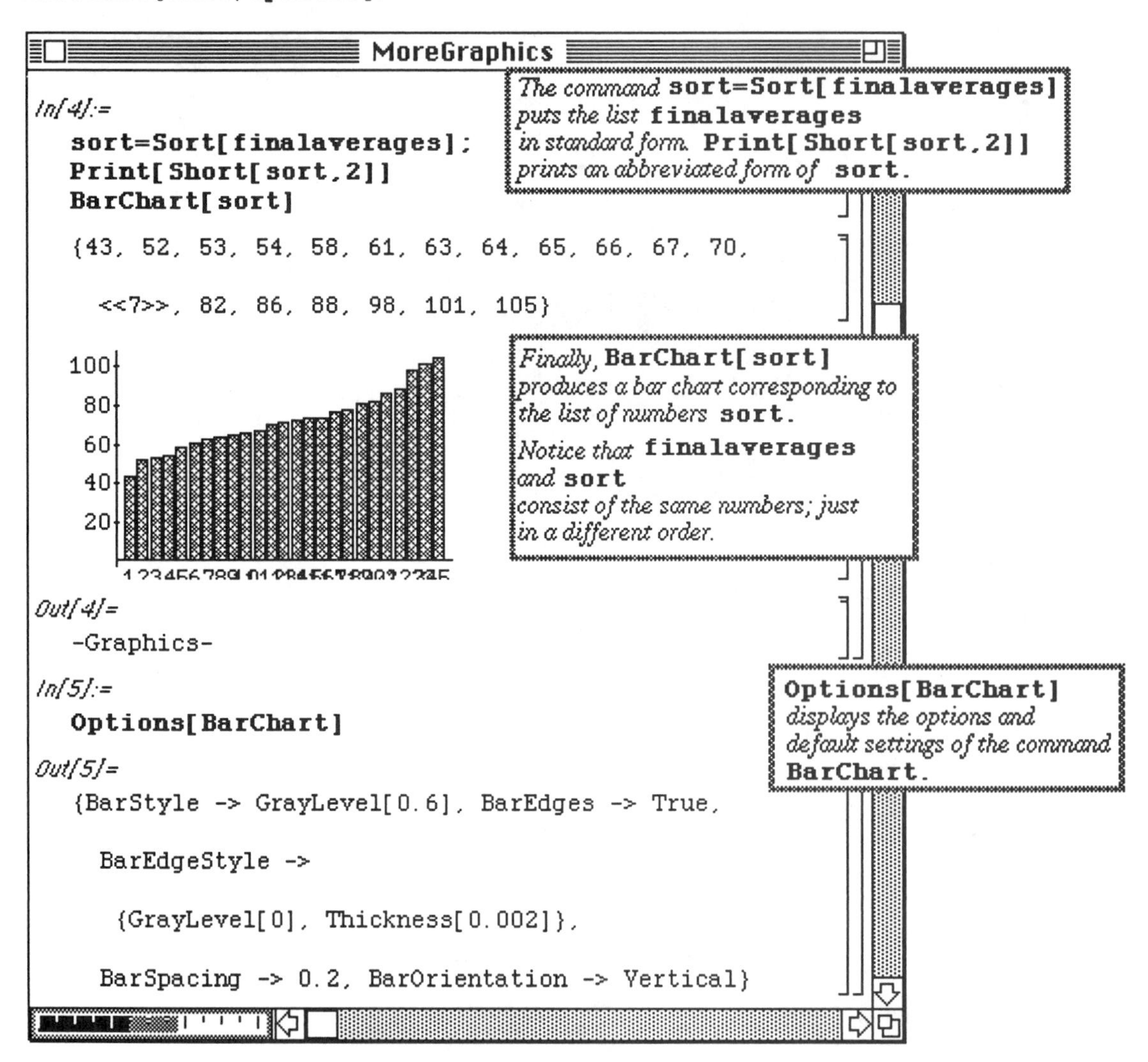

Pie charts are also possible with *Mathematica* by making use of the **PieChart[list]** command found in **Graphics.m**. This can be done in two ways. In the first example below, a pie chart is created from the list of numbers, **percents**. **PieChart[percents]** produces a pie chart in which each segment of the pie represents a number in **percents**. The segments are numbered to correspond to the position in **percents**. Notice that the sum of the numbers in **percents** is 1. However, a pie chart can be created which depicts both a quantity and a description. The list **description** is given below. Note that each element of **description** contains a number and a description. These represent portions of a governmental budget.

□ **Example:**

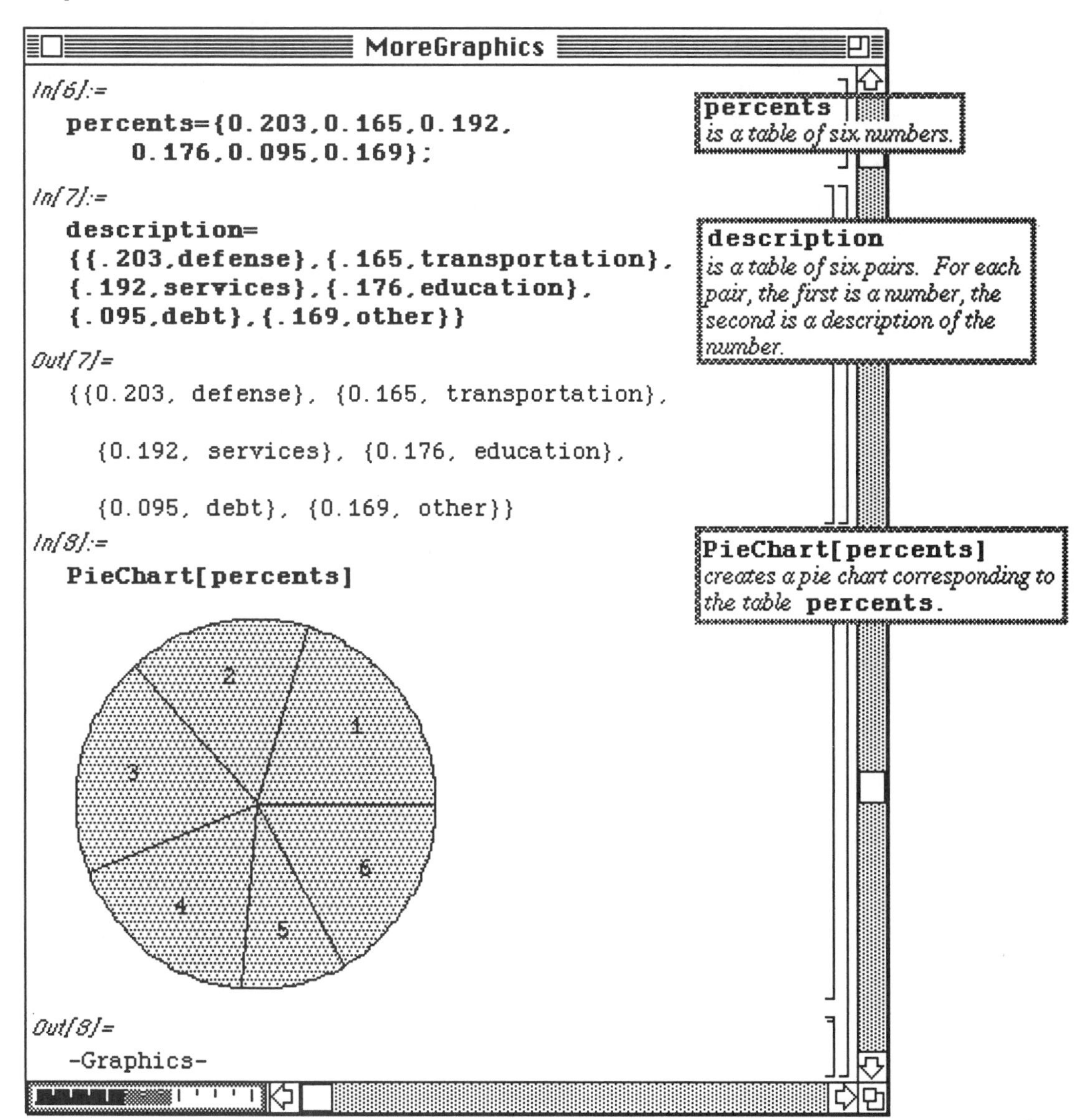

Hence, a labeled pie chart is given with **PieChart[description]**. This chart demonstrates the percentage of the budget that is allotted to each area.

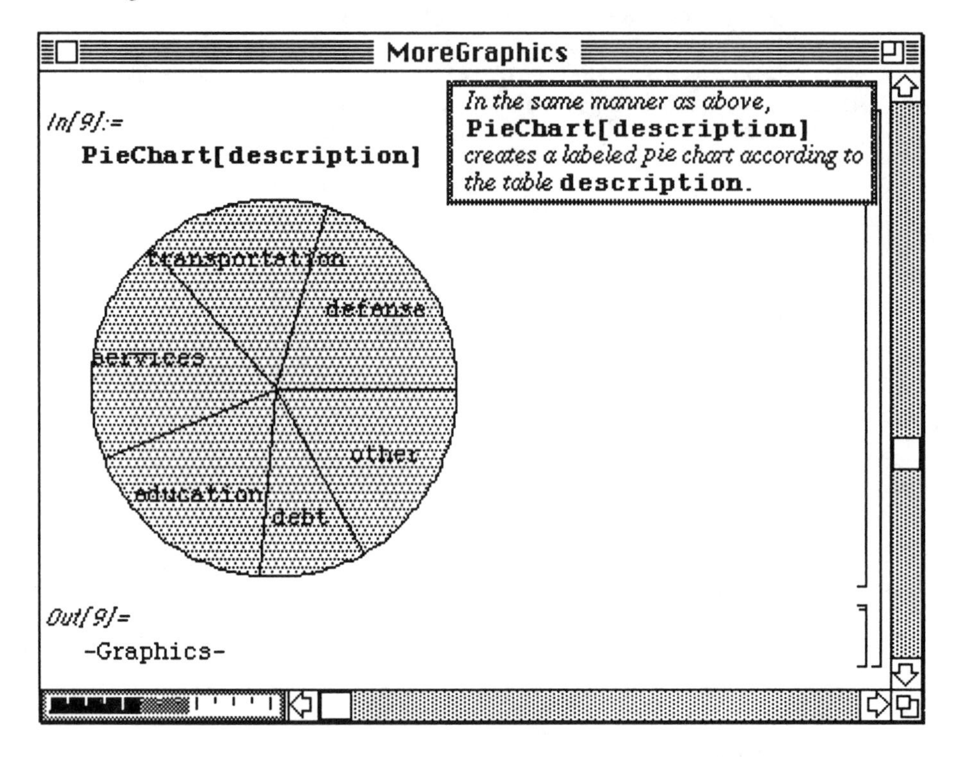

■ 8.2 Polyhedra.m

Pictures of polyhedra can be produced with **Polyhedra.m**. Many geometrical properties of polyhedra are stored in this package, so some pictures can be obtained by specifying a desired polyhedra with **Show[Polyhedron[Shape]]**. Since stored polyhedra include the icosahedron, dodecahedron, octahedron, cube, and tetrahedron. **Shape** is one of the following: **Icosahedron**, **Dodecahedron**, **Octahedron**, **Cube**, or **Tetrahedron** . If unspecified, the center is taken to be (0,0,0). A cube centered at the origin in produced in the first example below.

Several polyhedra can be shown simultaneously and, thus, complicated three-dimensional objects can be constructed. However, another command which involves more options must first be introduced. Three-dimensional graphics objects are created but not displayed with **Graphics3D[Polyhedron[{x0,y0,z0},scale]]** where **Polyhedron** is the desired polyhedron from the list of stored polyhedra: **Icosahedron**, **Dodecahedron**, **Octahedron**, **Cube**, and **Tetrahedron**. **{x0,y0,z0}** represents the center, and **scale** adjusts the size. The default value of **scale** is 1, so **scale** >1 produces a larger polyhedron and **scale** <1, a smaller one. The second command below creates and stores (as **figure1**) the graphics of a dodecahedron centered at the origin using **scale** = 1/2. Since **Show** is not used, the picture is not shown.

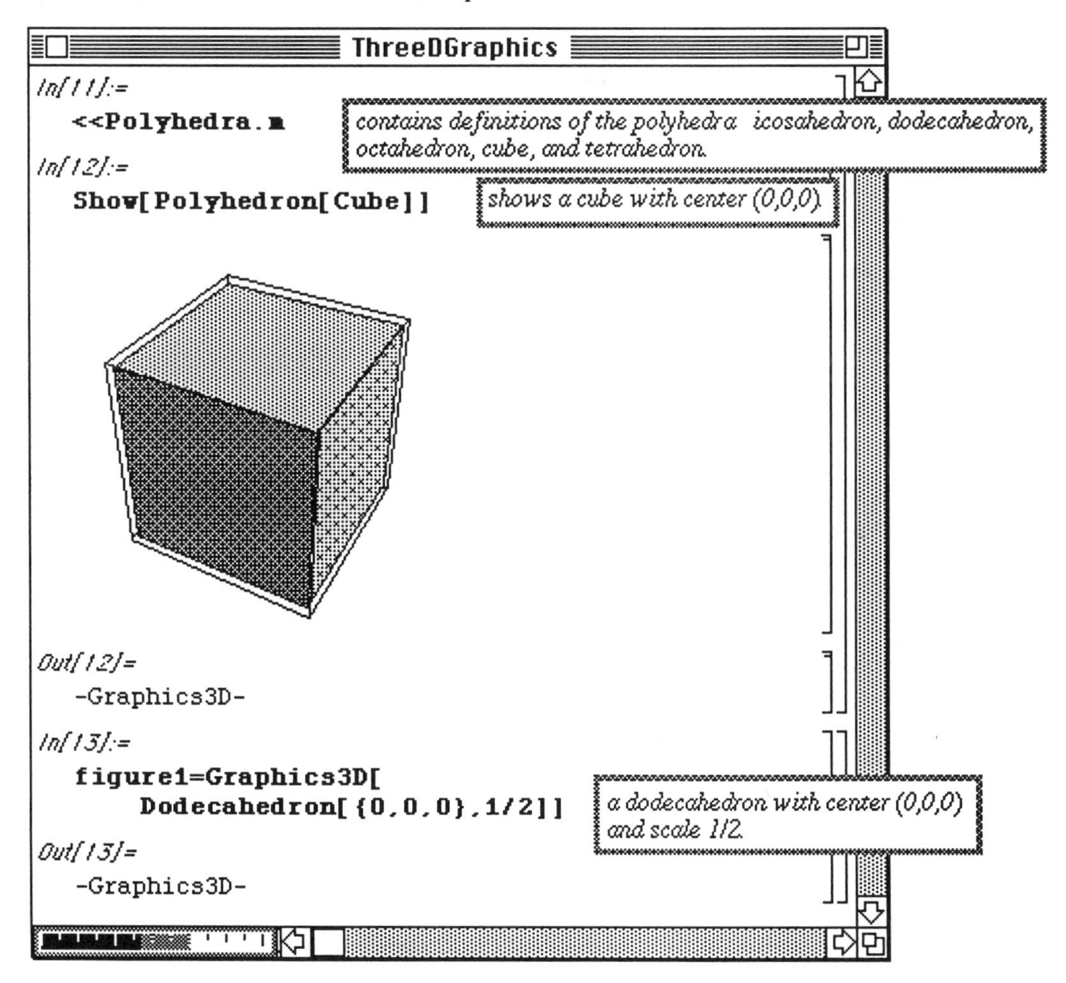

□ Example:

Next, the graphics of an octahedron centered at **{Cos[Pi/3],Sin[Pi/3],0}** and **scale** = **1/3** is created and stored as **figure2**. Also, a tetrahedron with center **{Cos[2Pi/3],Sin[2Pi/3],1/3}** and **scale** = **1/4** is stored as **figure3**. Since the graphics of each polyhedra was named, they can be shown simultaneously with **Show[figure1,figure2,figure3]**.

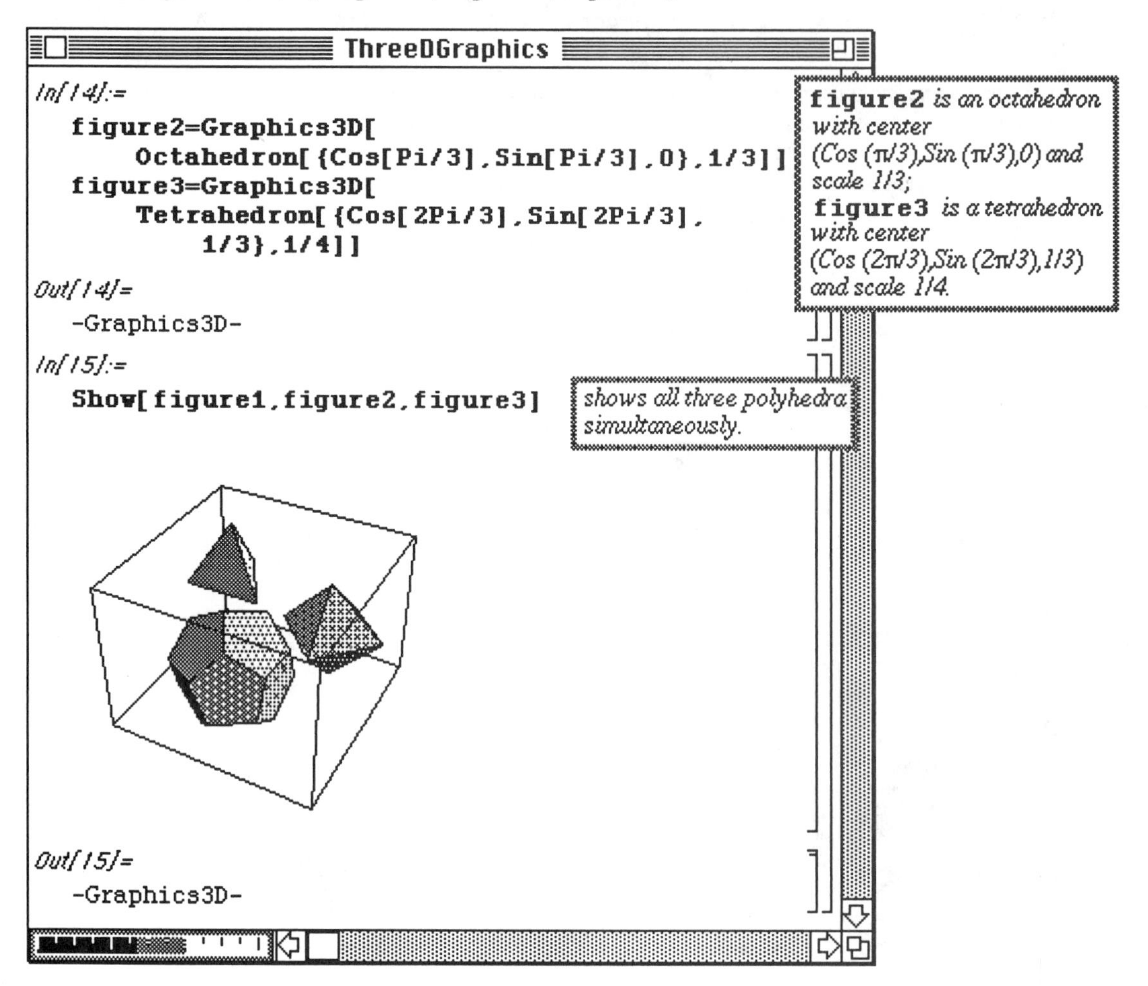

Another command available in **Polyhedra.m** is **Stellate[Polyhedron[Shape],ratio]** where **Shape** is again one of the following: **Icosahedron**, **Dodecahedron**, **Octahedron**, **Cube**, or **Tetrahedron**. This takes the symbolic representation of the polyhedron and represents it as a stellated polyhedron. (Each face is replaced by a stellate.) A function **a[i]** is defined below as **Stellate[Polyhedron[Dodecadron],i]**. This function is then used in a **Do** loop to produce the graphics of stellated dodecahedra for values of **i** from **i=.25** to **i=2** using increments of **.25**. The first graph with **i=.25** is shown below.

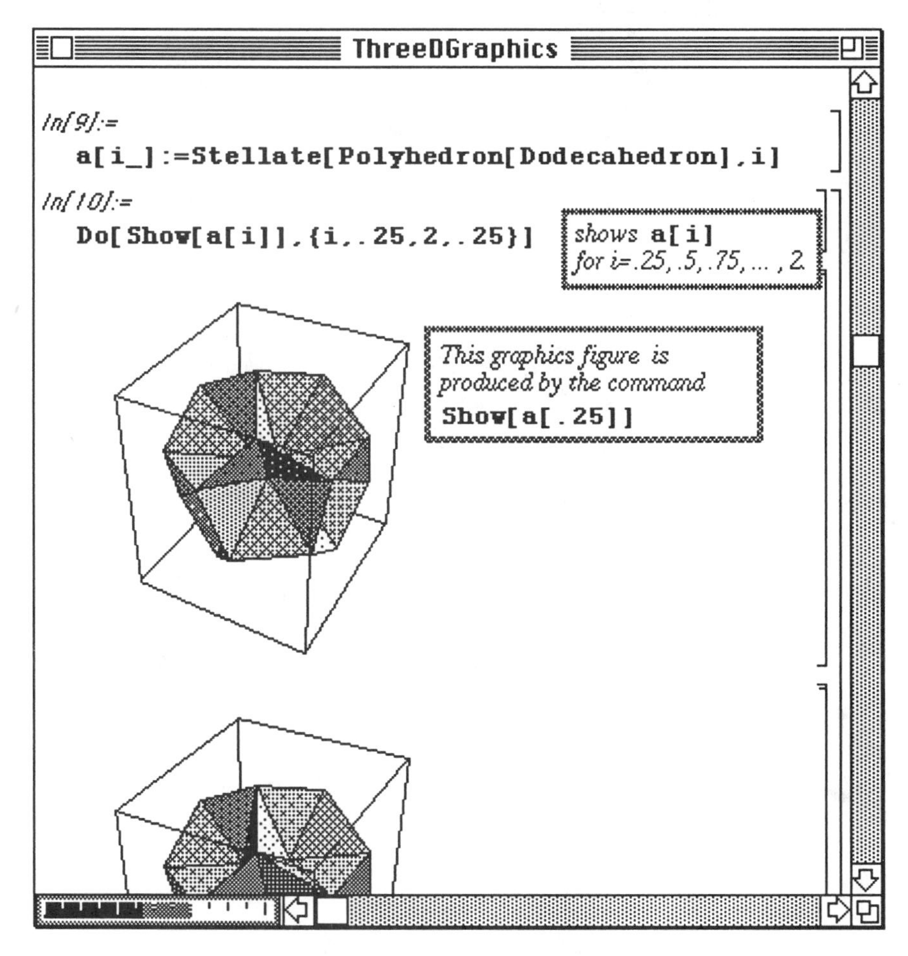

Notice how the pictures change with **ratio**. If **ratio** <1, the object is concave. If **ratio** >1, the object is convex. Shown below are three pictures. In the first, **ratio** = **.25** <1 , so the object produced is concave. In the third, **ratio** = **2** >1, so a convex object is given. Both can be compared to the middle picture which is simply a dodecahedron (**ratio** = 1). The graphics obtained with this **Do** loop can be animated to observe the changes which take place as **ratio** changes.

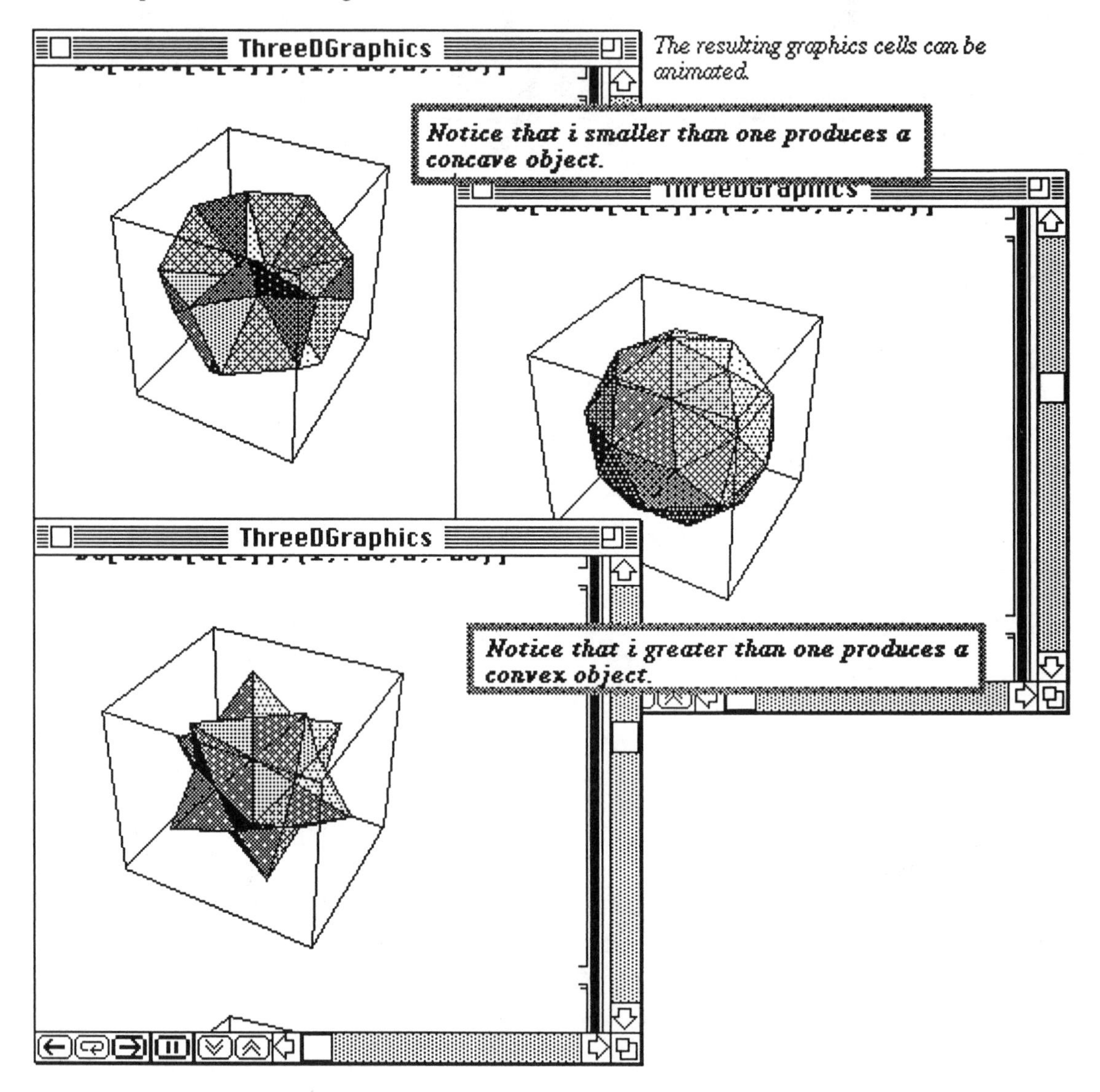

■ 8.3 Shapes.m

Shapes.m contains commands which produce the graphics of many shapes commonly used in mathematics. As with the all graphics objects, different shapes may be combined and shown simultaneuously to create more complicated objects.

□ **Example:**

Illustrated first below is **MoebiusStrip[outerradius,innerradius,n]** where **innerradius** and **outerradius** are the inner and outer radii, respectively, and the Moebius strip is approximated using 2**n** polygons. (**MoebiusStrip** actually produces a list of polygons which are displayed with **Show**.) In the example below, the graphics are produced for a Moebius Strip with inner radius 2 and outer radius 4. The graph uses 60 polygons and is called **msone**. The list of polygons created with **MoebiusStrip** is visualized with **Show[msone]**.

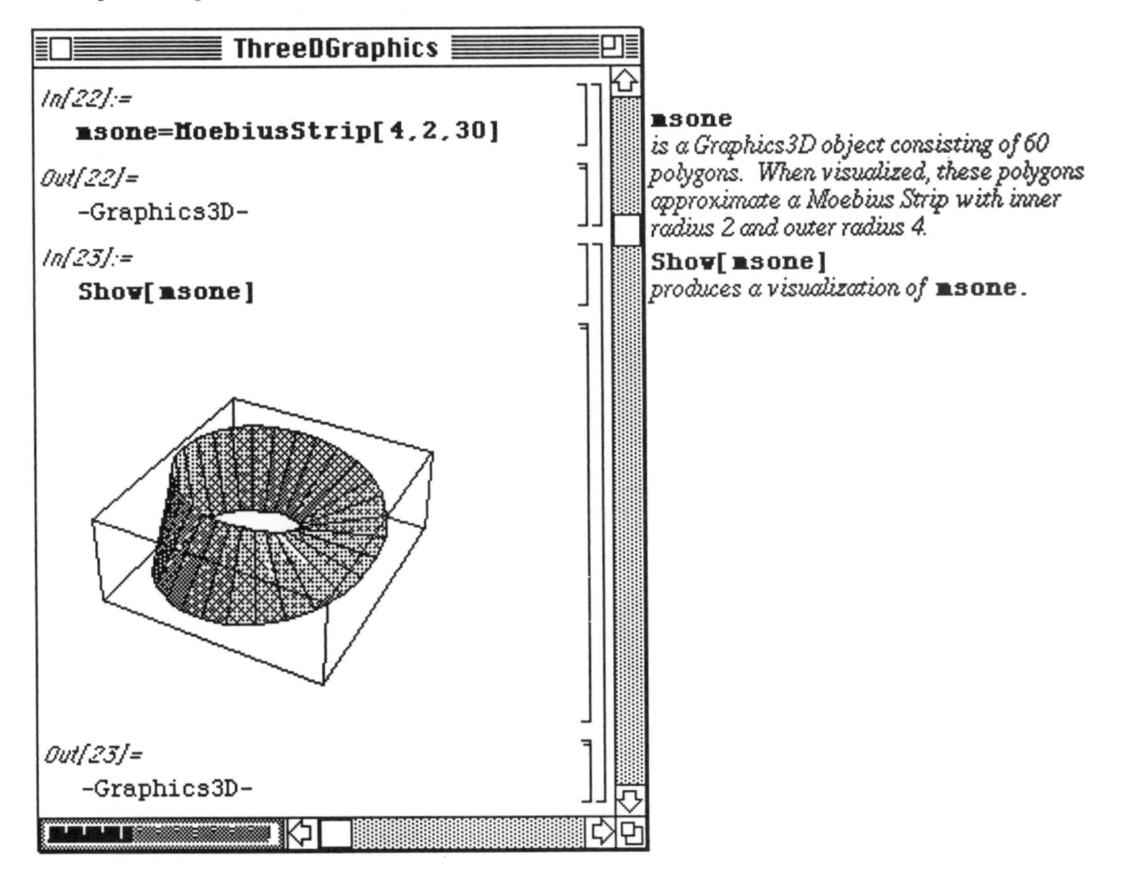

□ **Example:**

Another shape which can be approximated in this package is that of a torus. This is accomplished with **`Torus[outerradius,innerradius,m,n]`** where **`m`** times **`n`** polygons are used to approximate the shape of the torus. A torus of inner radius .5 and outer radius 1 is approximated with 300 polygons and called **`torusone`**. The approximation is then shown with **`Show[torusone]`**.

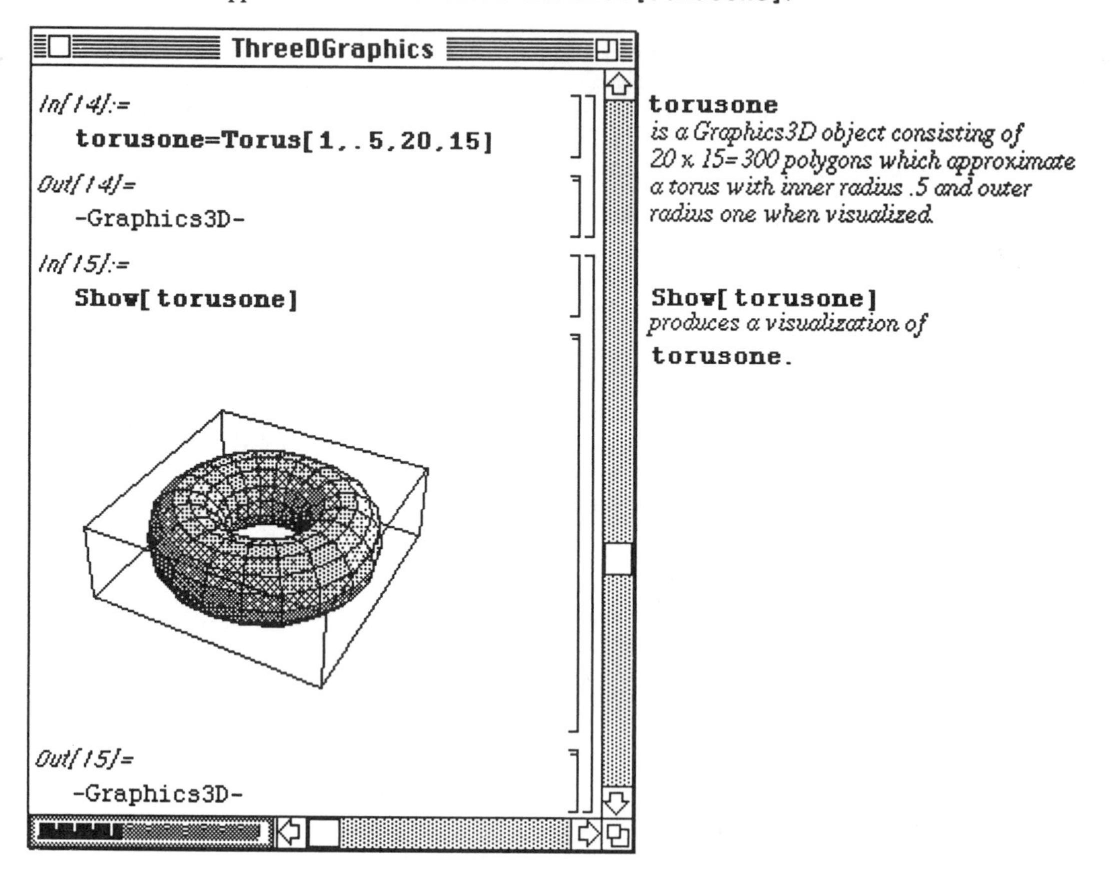

torusone
is a Graphics3D object consisting of 20 x 15=300 polygons which approximate a torus with inner radius .5 and outer radius one when visualized.

Show[torusone]
produces a visualization of **torusone.**

□ **Example:**

The command **Sphere[r,m,n]** produces an approximation of a sphere of radius = **r** using **m** times **n** polygons. The approximation of a sphere of radius 1 is obtained below using 225 polygons.

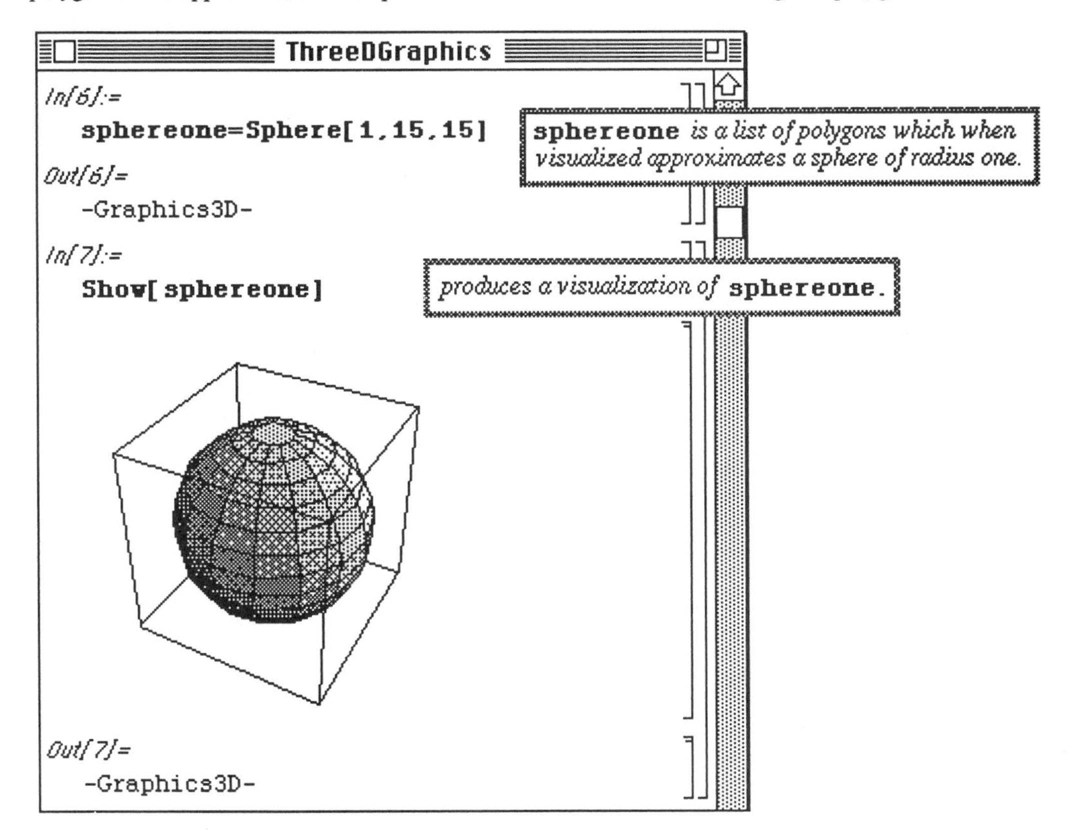

Several other commands are available for visualizing the lists of polygons produced by the commands found in **Shapes.m**. The command **WireFrame[polygonlist]** replaces each polygon in **polygonlist** by closed lines, so the shape resembles that of a wire frame when visualized.

□ **Example:**

In the example below, a list of 144 polygons to approximate a sphere of radius 2 is obtained with **Sphere[2,12,12]**. This list is called **spheretwo**. **WireFrame** is then applied to this list of polygons, and the list of closed lines **wiretwo** which results is visualized with **Show[wiretwo,Boxed->False]**. (The **Show** option, **Boxed->False**, causes no box to be drawn around the sphere.)

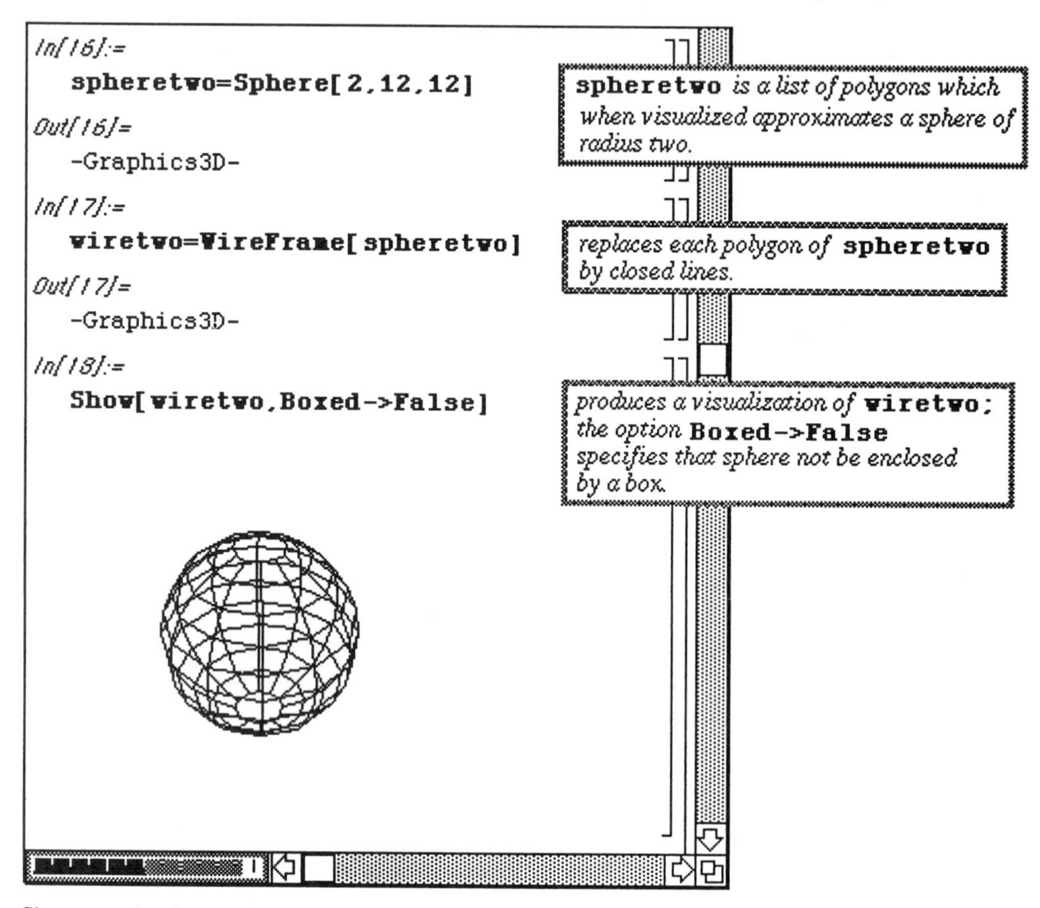

Shapes can be viewed simultaneously using the commands previously discussed.

□ **Example:**

The following example shows how the approximations of the sphere of radius 1, **sphereone**, and the wire frame form of the sphere of radius 2,**wiretwo**, are shown together.

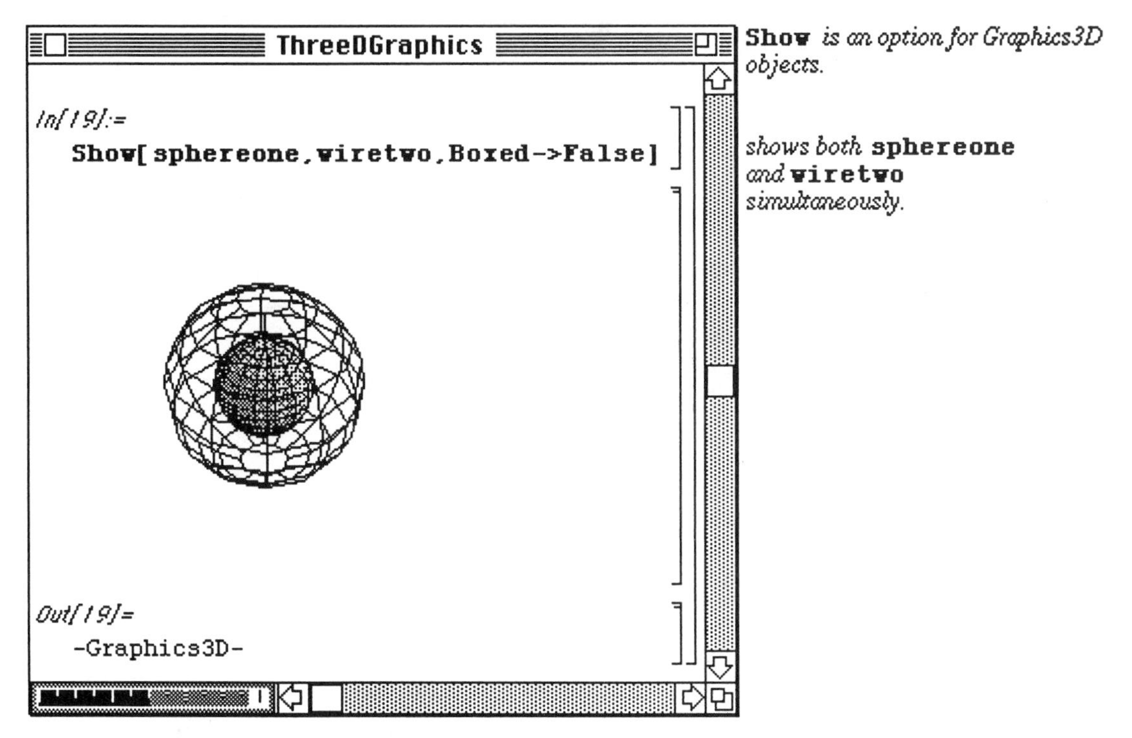

Show *is an option for Graphics3D objects.*

shows both **sphereone** *and* **wiretwo** *simultaneously.*

Below, the graphs of two Moebius strips are shown simultaneously by first computing the lists of approximating polygons in **msone** and **mstwo**. **Show** is then used to visualize these objects. Two other useful commands in **Shapes.m** are illustrated in the second example.

RotateShape[shape,xrotate,yrotate,zrotate] causes **shape** to be rotated **xrotate** units about the x-axis, **yrotate** units about the y-axis, and **zrotate** units about the z-axis.

In the example below, **msone** is rotated $\pi/2$ units about the y-axis.

The other command introduced is **TranslateShape[shape,{x0,y0,z0}]** which translates **shape** **x0** units along the x-axis, **y0** units along the y-axis, and **z0** units along the z-axis.

□ **Example:**

The Moebius strip obtained through rotation in **msone** is then translated 2 units along the x-axis and called **mstwo**. The two are then shown simultaneously with **mstwo** being the Moebius strip to the right of **msone**.

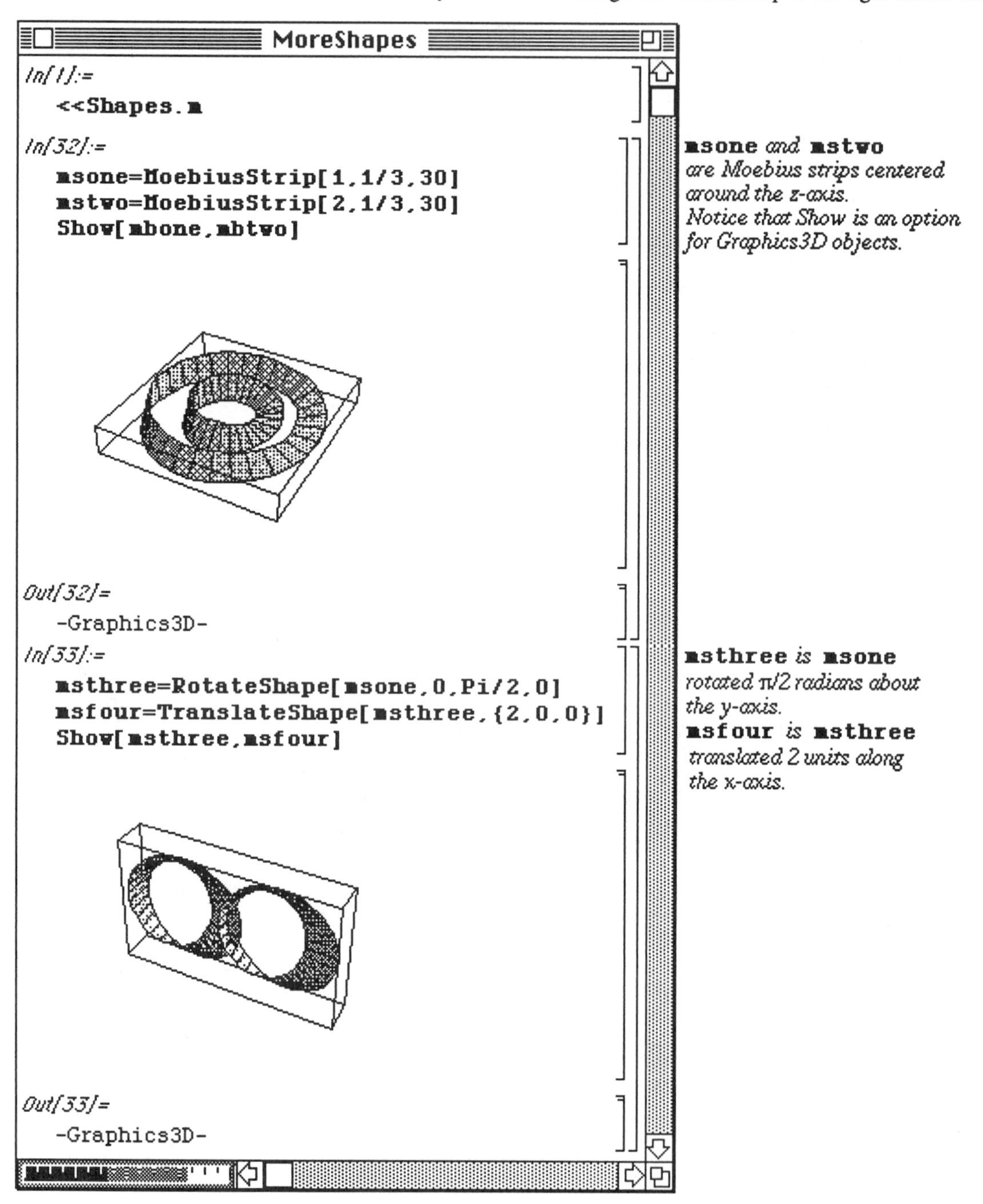

The Moebius strip **mstwo** given previously is then rotated π/2 radians about both the x- and z-axes. This is named **msfive**. Then, **msfive** is translated (-2) units along the y-axis to obtain **mssix**. **msfive** and **mssix** are then visualized below with **Show** where **mssix** is the Moebius strip to the rear of **msfive**. Several **Graphics3D** objects can be viewed simultaneously as shown below.

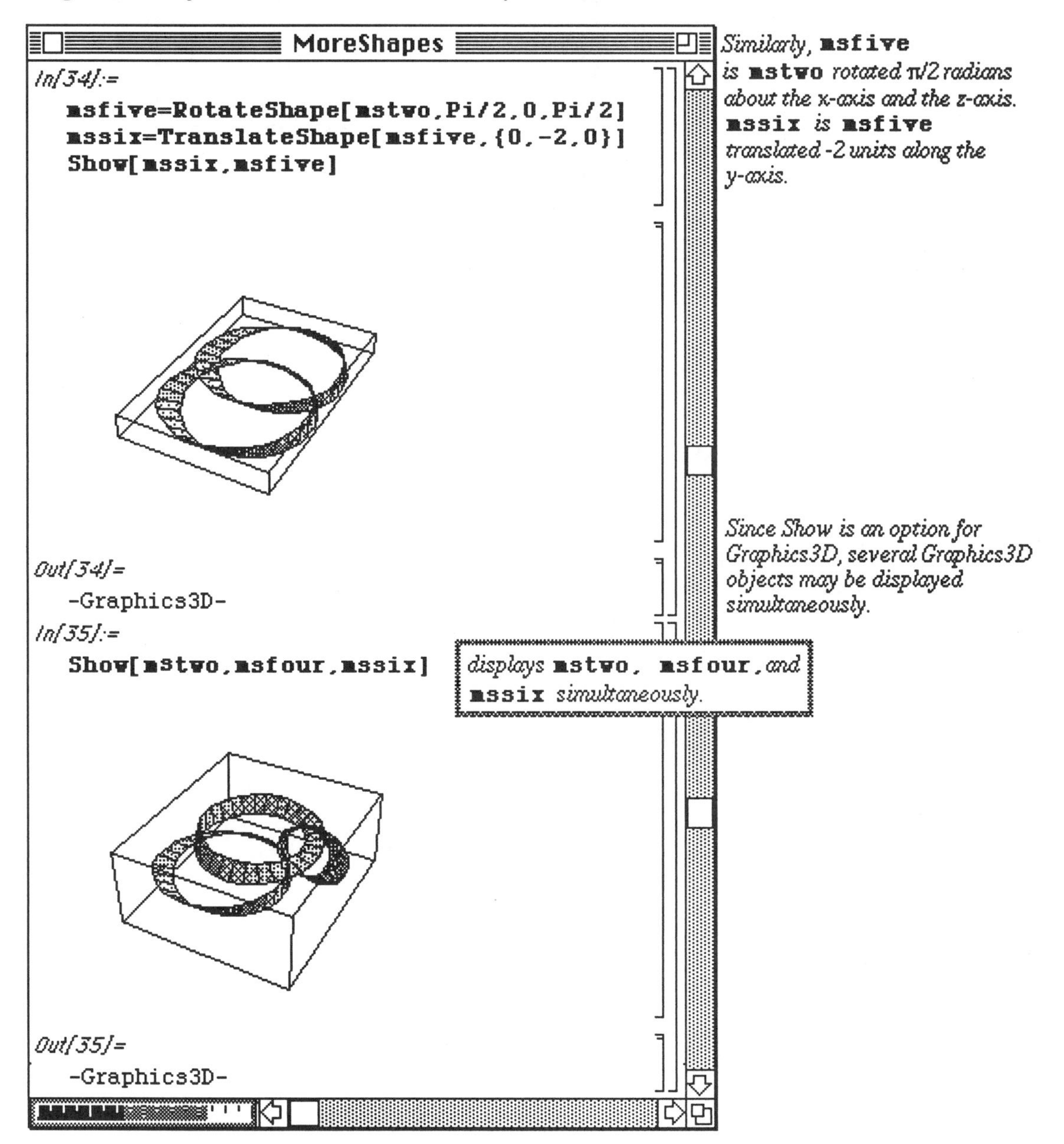

Yet another shape that can be graphed in **Shapes.m** is **Helix[r,h,m,n]** which approximates a helix with half height **h** and **m** turns using **m** * **n** (where **n** =20**r**) polygons.

□ **Example:**

Shown below is a helix of half height 3 with 5 turns. The list of polygons which approximate the helix is found in **helixtwo**. Hence, **Show[helixtwo]** displays the helix.

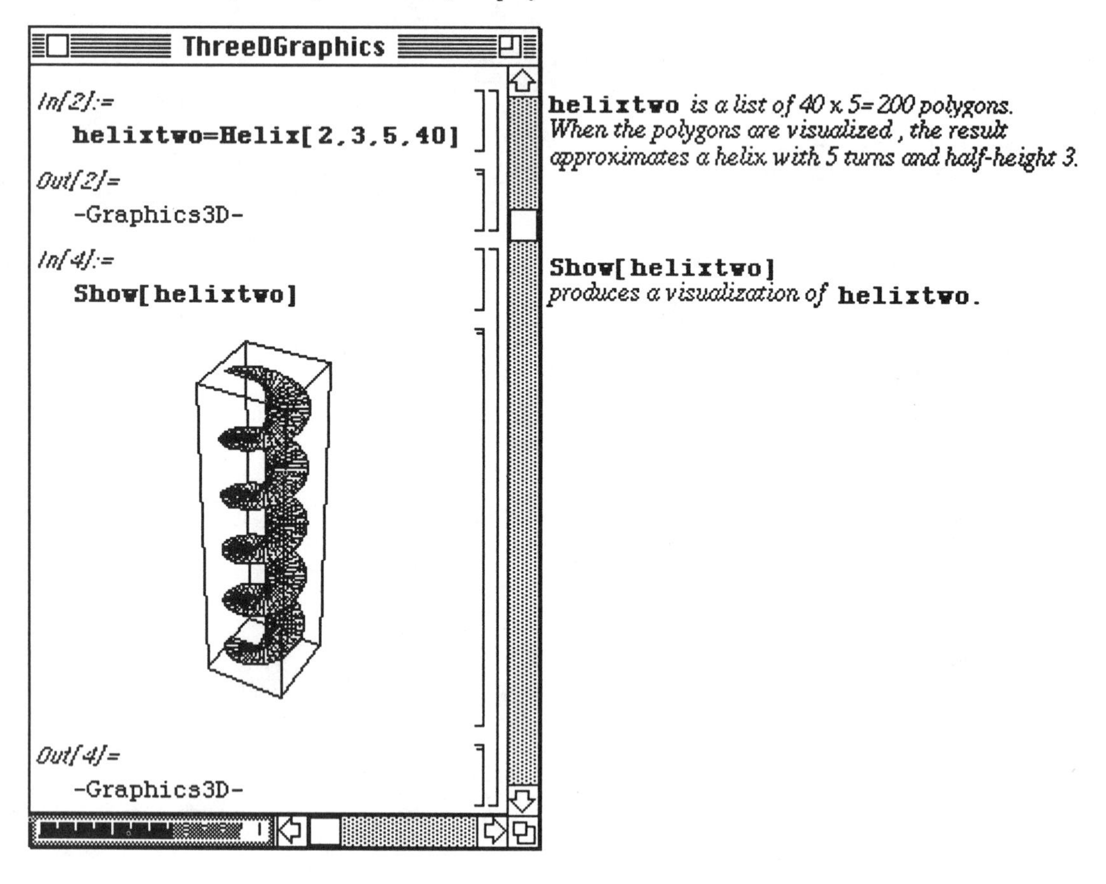

A **Do** loop which shows the rotation of helixtwo **n*Pi/2** radians (where **n** varies from **n** = 0 to **n** = 12) about the z-axis is defined below. This loop produces 13 graphics cells which can be animated to view the rotation of the helix about the z-axis. Several of these cells are shown below.

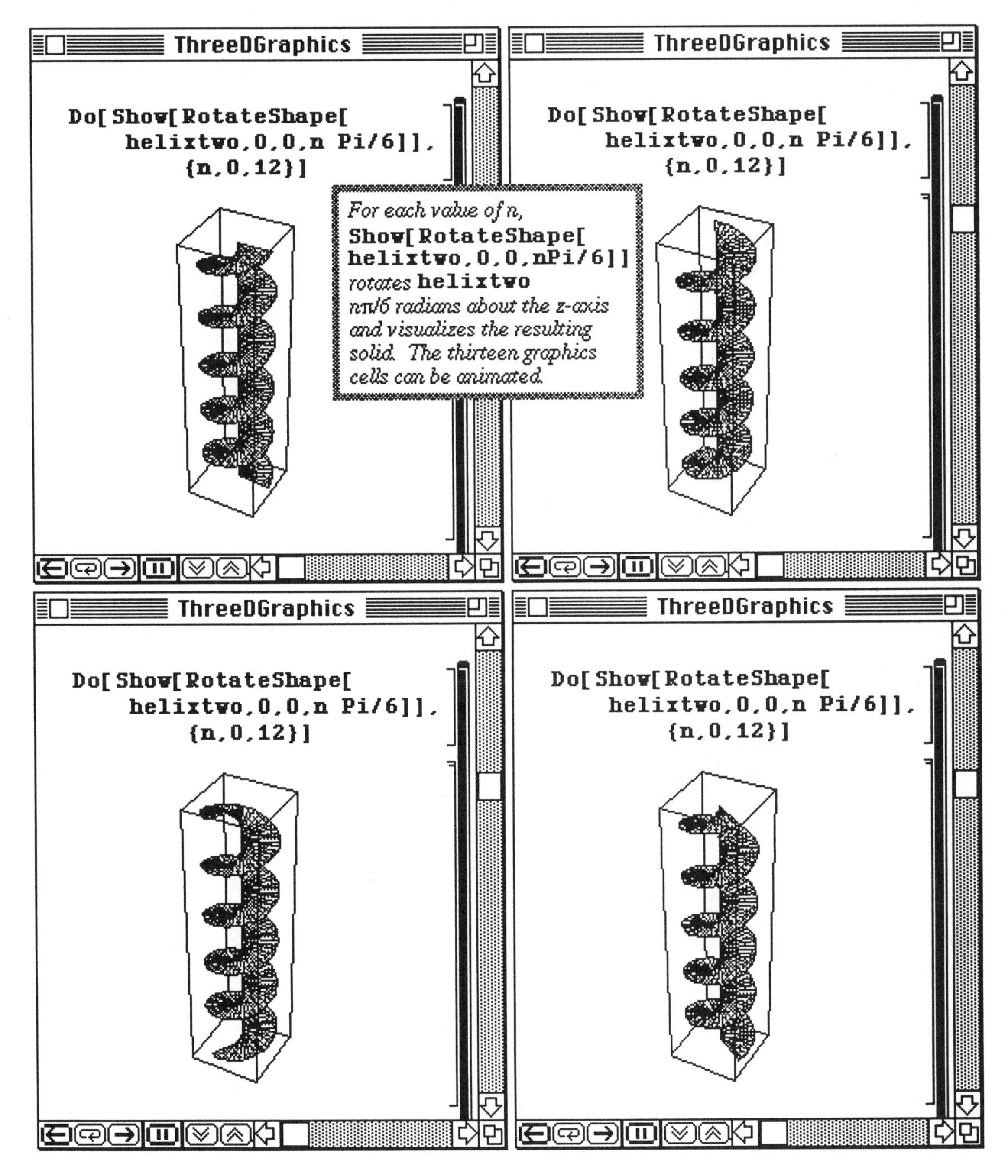

□ **Example:**

In the following example, we define a function **surface[n]** which simultaneously graphs the tori obtained with **torusj[1,1/2,{0,9/5(j-1),0}]** for **j =1** to **j =n**. The result allows us to visualize a surface of genus n.

The command **torusj**, given in the first line below, is defined in terms of **Torus** and may seem redundant. However, defining it in this manner enables *Mathematica* to remember the previous torus as it proceeds through the loop contained within the **Block**. Hence, **torusj[j-1]** does not have to be recomputed in order to find **torusj[j]**. After the loop is completed for **j =n**, the table of tori found in **s** is displayed with **Show**. The surface obtained with **n =4** is shown below.

o In Version 2.0, **Block**, although still recognized, has been replaced by the command **Module**.

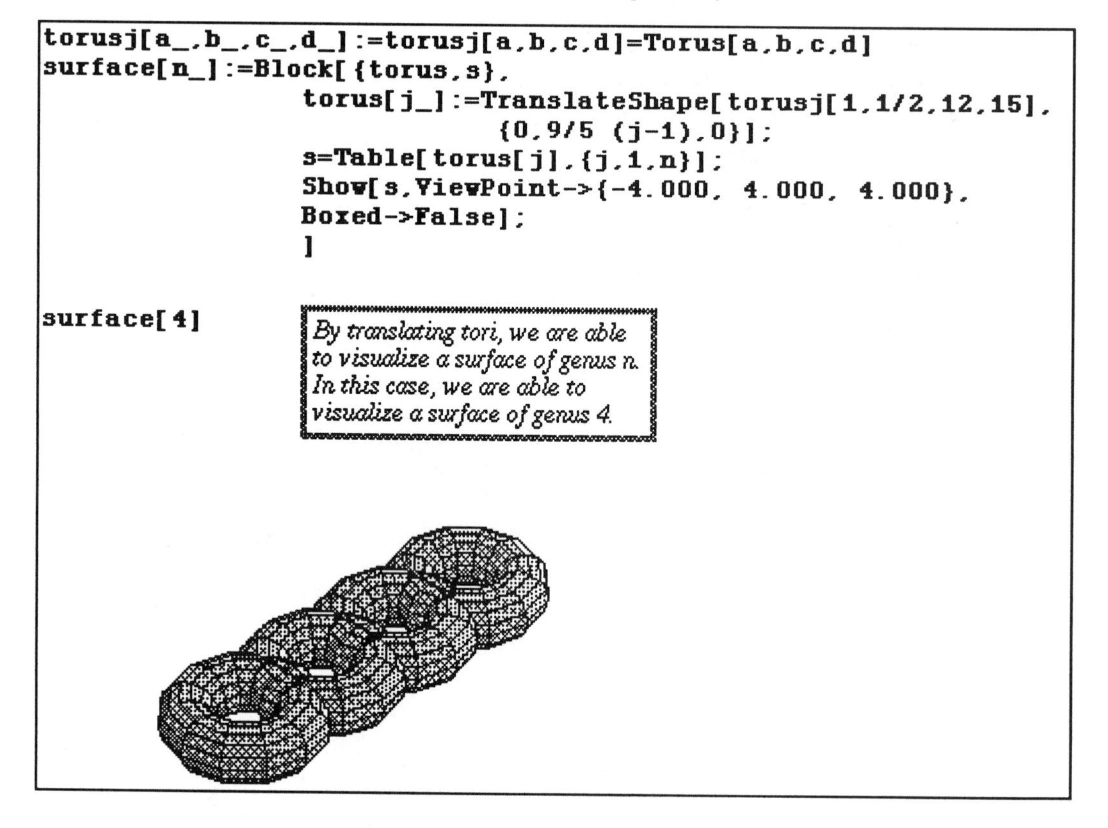

```
torusj[a_,b_,c_,d_]:=torusj[a,b,c,d]=Torus[a,b,c,d]
surface[n_]:=Block[{torus,s},
                     torus[j_]:=TranslateShape[torusj[1,1/2,12,15],
                                    {0,9/5 (j-1),0}];
                     s=Table[torus[j],{j,1,n}];
                     Show[s,ViewPoint->{-4.000, 4.000, 4.000},
                     Boxed->False];
                     ]

surface[4]
```

By translating tori, we are able to visualize a surface of genus n. In this case, we are able to visualize a surface of genus 4.

■ 8.4 ParametricPlot3D.m

ParametricPlot3D.m provides the capabilities to produce three-dimensional graphs of functions which depend on two parameters. These equations exist in many different forms. The example below illustrates how the surface of revolution generated by a function of one variable revolved about the y-axis can be generated with **ParametricPlot3D**. The command is entered in the following way:
ParametricPlot3D[{x[u,v],y[u,v],z[u,v]},{u,u0,u1},{v,v0,v1},options] where **x**, **y**, and **z** are defined in terms of the parameters **u** and **v**. The limits on the parameters are **{u0,u1}** and **{v0,v1}**.

o In Version 2.0, **ParametricPlot3D[{x[t],y[t],z[t]},{t,t0,t1},options]** where **x**, **y**, and **z** are defined in terms of the parameter **t** graphs the vector-valued function **{x[t],y[t],z[t]}** for **t0** $\leq$ **t** $\leq$ **t1**.

In addition to the command **ParametricPlot3D**, Versions 1.2 and 2.0 of the package **ParametricPlot3D.m** contain the command **SphericalPlot[r[theta,phi], {theta,theta0,theta1},{phi,phi0,phi1},options]** which is used to create a graph of **r[theta,pi]** using spherical coordinates; and

o the Version 2.0 edition of **ParametricPlot3D.m** contains the command **CylindricalPlot[z[r,theta],{z,z0,z1},{theta,theta0,theta1},options]** which is used to graph **z[r,theta]** using cylindrical coordinates.

Since the surface of revolution generated by revolving $y = f(x) = \dfrac{\text{Cos}(\pi x)}{x^3+1}$ about the y-axis is radially symmetric (**z** does not depend on the angular coordinate), the surface is visualized by representing **x** and **y** in polar coordinates and replacing **x** in the function **Cos[2 Pi x]/(x+1)** with **u** to form the equation for **z** in terms of the parameters.

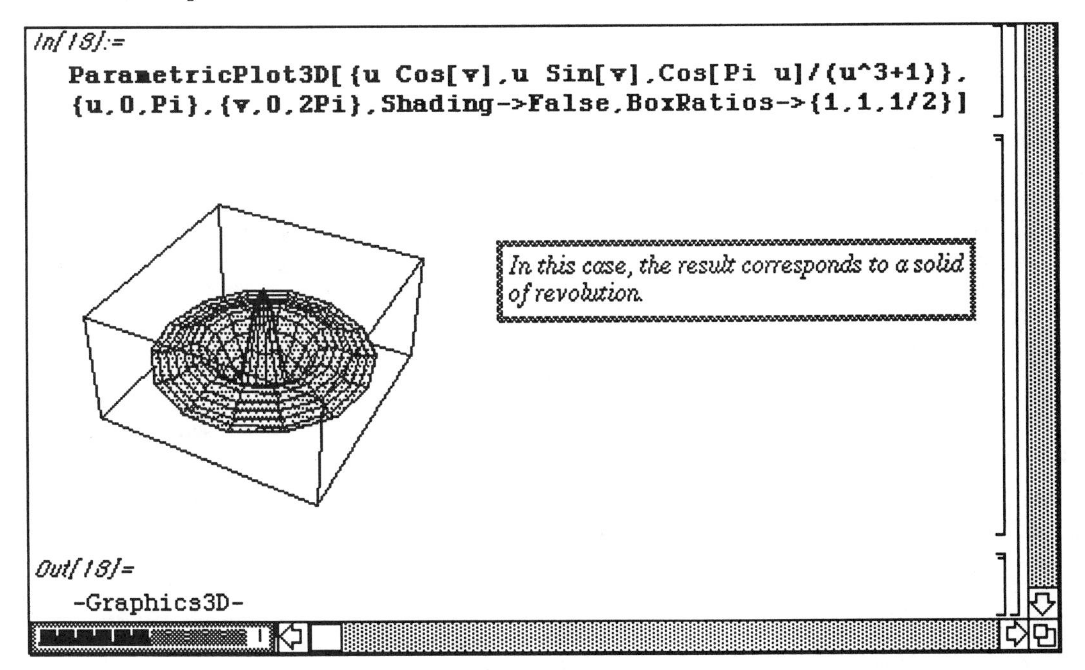

In order to change the perspective from which hree-dimensional graphics are viewed, the **3D ViewPoint Selector** which is shown below can be used. The window seen below is obtained by selecting **Prepare Input** under **Action** in the *Mathematica* Menu. The sub-menu found under **Prepare Input** contains this **3D ViewPoint Selector**. The numbers in the three boxes on the right represent the coordinates of the viewpoint. The x and y coordinates are changed by dragging the boxes beneath the graphics window. Similarly, the z coordinate is increased and decreased by dragging the box which is along the right side of the graphics window. Once a desirable viewpoint is located, it can be pasted into the **ParametricPlot3D** command by clicking the **Paste** button.

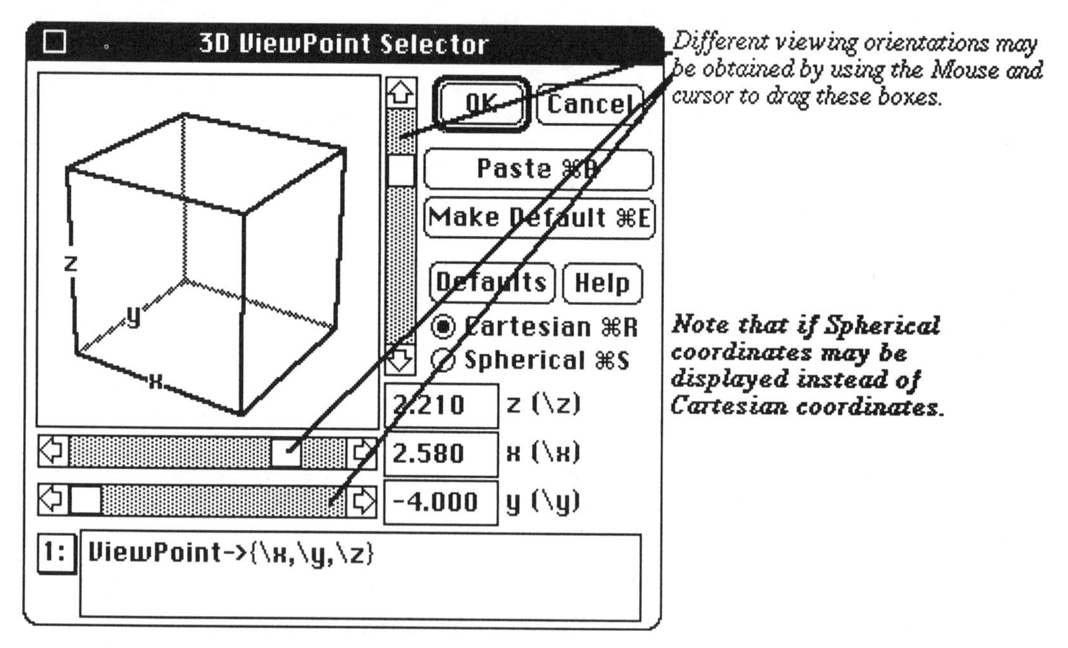

In the next example, all three components **{x,y,z}** depend on two parameters, **r** and **t**. Again, the **x** and **y** coordinates are defined as functions of the polar coordinates, **r** and **t**, where **t** represents the angular coordinate. The third coordinate is defined as the function **f[r,t]**. The graph of this function of two parameters is plotted for values of **r** from **r=0** to **r=2** using increments of **2/25** and for **t=0** to **t=2Pi** with **2Pi/25** increments. Several options are illustrated in this command. The graph is not enclosed in a box, the axes are given automatically, the ratios of the **x,y**, and **z** coordinates are 1-1-1, and the viewpoint is selected to be the point **{0.390,-4.000,3.360}**.

Another command included in the Version 1.2 edition of **Parameter icPlot3D.m** is **SpaceCurve** which allows curves that depend on one variable to be plotted in three dimensions. This command is entered as **SpaceCurve[{x,y,z},{t,t0,t1},options]** where the coordinates **x**, **y**, and **z** depend on the parameter **t** and **t** varies from **t=t0** to **t=t1**. The options for **SpaceCurve** are the same as those for **ParametericPlot3D**.

In the second example below, a spiral shaped curve is plotted. All three coordinates increase as t increases, so the points along the curve move away from the origin. The spiraling is due to the sine and cosine terms in the **x** and **y** coordinates.

o In Version 2.0, the command **SpaceCurve** has been replaced by the command **ParametricPlot3D**. If using Version 2.0, **ParametricPlot3D[{x[t],y[t],z[t]},{t,t0,t1}]** yields the same result as the command **SpaceCurve[{x[t],y[t],z[t]},{t,t0,t1}]** from earlier versions.

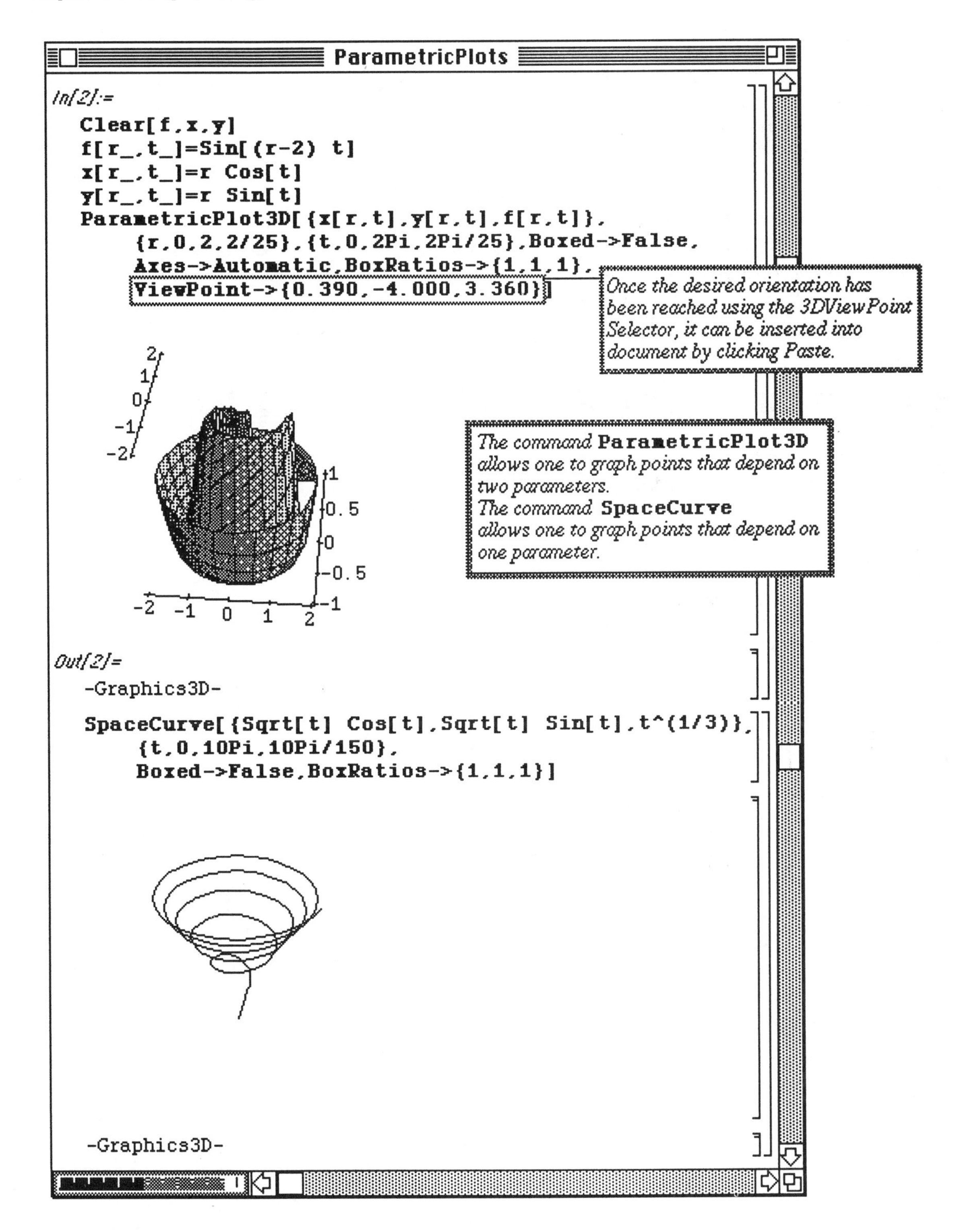
ParametricPlots
In[2]:=
Clear[f,x,y]
f[r_,t_]=Sin[(r-2) t]
x[r_,t_]=r Cos[t]
y[r_,t_]=r Sin[t]
ParametricPlot3D[{x[r,t],y[r,t],f[r,t]},
{r,0,2,2/25},{t,0,2Pi,2Pi/25},Boxed->False,
Axes->Automatic,BoxRatios->{1,1,1},
ViewPoint->{0.390,-4.000,3.360}]
Once the desired orientation has been reached using the 3DViewPoint Selector, it can be inserted into document by clicking Paste.
The command ParametricPlot3D allows one to graph points that depend on two parameters.
The command SpaceCurve allows one to graph points that depend on one parameter.
Out[2]=
-Graphics3D-
SpaceCurve[{Sqrt[t] Cos[t],Sqrt[t] Sin[t],t^(1/3)},
{t,0,10Pi,10Pi/150},
Boxed->False,BoxRatios->{1,1,1}]
-Graphics3D-

□ **Example:**

The following example shows how functions which have been previously defined can be graphed with **SpaceCurve**. A function **v[t]** is defined in terms of the parameter **t**. Then, a function **norm[t]** is given. This function is simply the square root of the sum of the squares of the components of **v[t]**. Since **v** is a list, **Length[v]** represents the number of components of **v** which in this case is 3.

In this case, defining **norm[v_]:=Sqrt[v.v]** produces the same result.

Finally, a function **w[t]** is defined as **v[t]/norm[t]**.
The graph of **v** is plotted with **SpaceCurve** and called **scone**. Notice that **Release[v[t]]** must be entered in this command in order for the components of **v[t]** to be evaluated at the various values of **t**. The curve is plotted for **t=0** to **t=2Pi** using small increment sizes of **2Pi/300**.

o If using Version 2.0, use **ParametricPlot3D** instead of **SpaceCurve**; use **Evaluate** instead of **Release**.

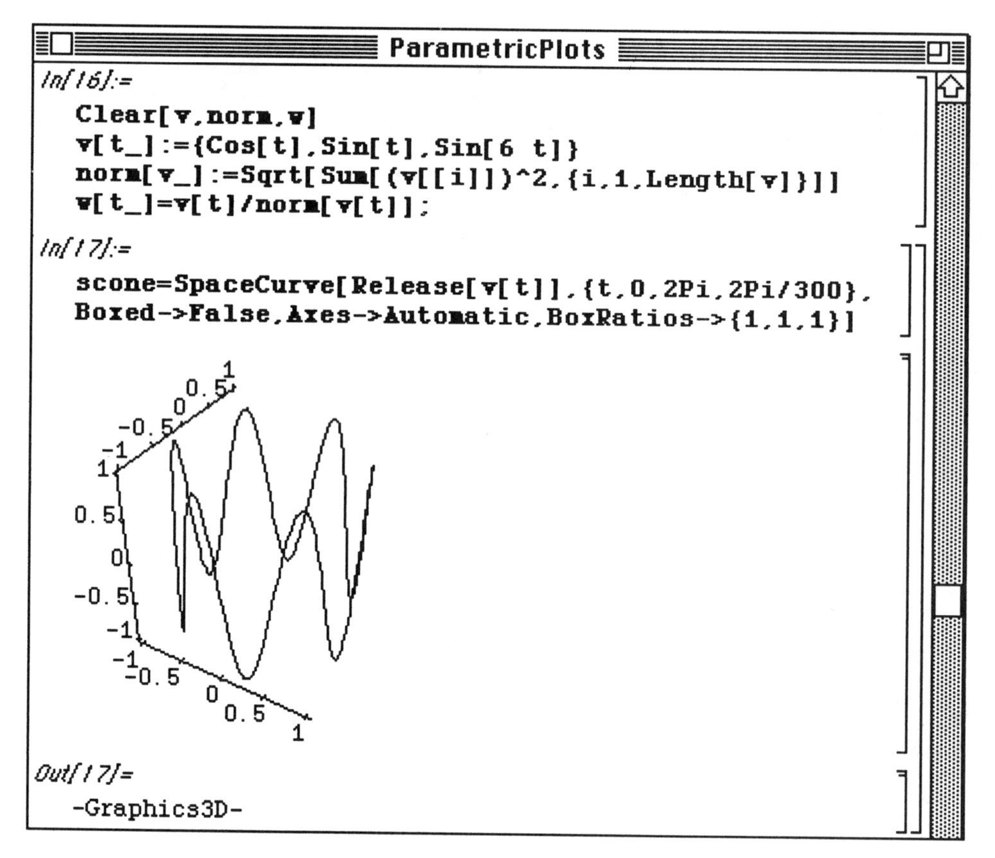

The graph of **w[t]** is then plotted with a similar command. Again, **Release[w[t]]** must be used in order to plot this function of one parameter.

o If using Version 2.0, be sure to use **Evaluate** instead of **Release**.

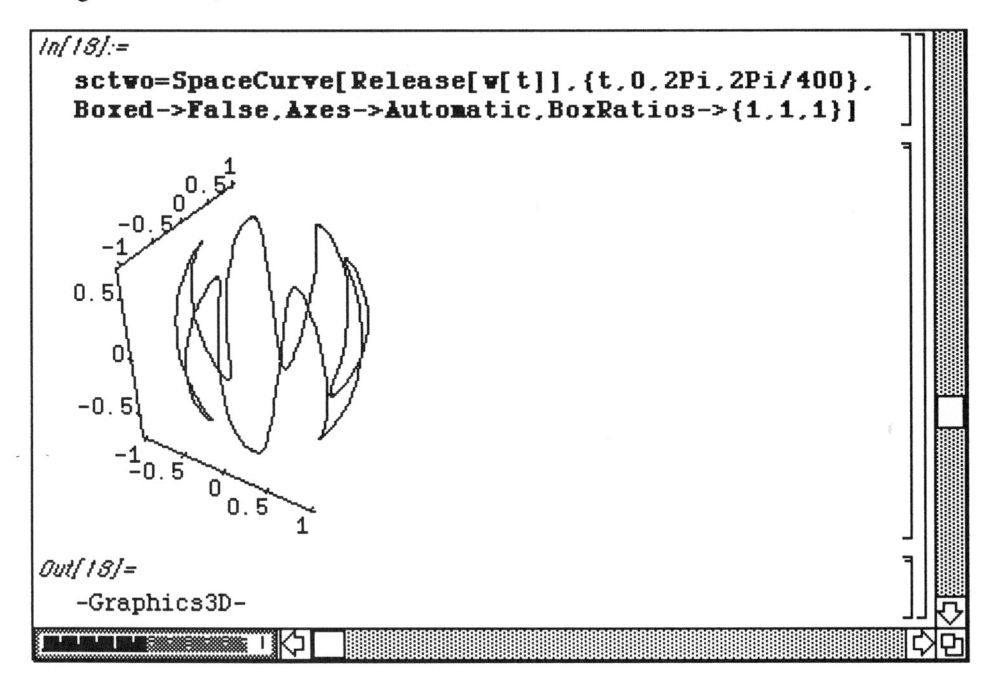

Version 2.0 Graphics

The following window shows the contents of the Graphics folder in Version 2.0 of *Mathematica*. Notice that several new packages are included. In this section, we discuss ImplicitPlot.m,

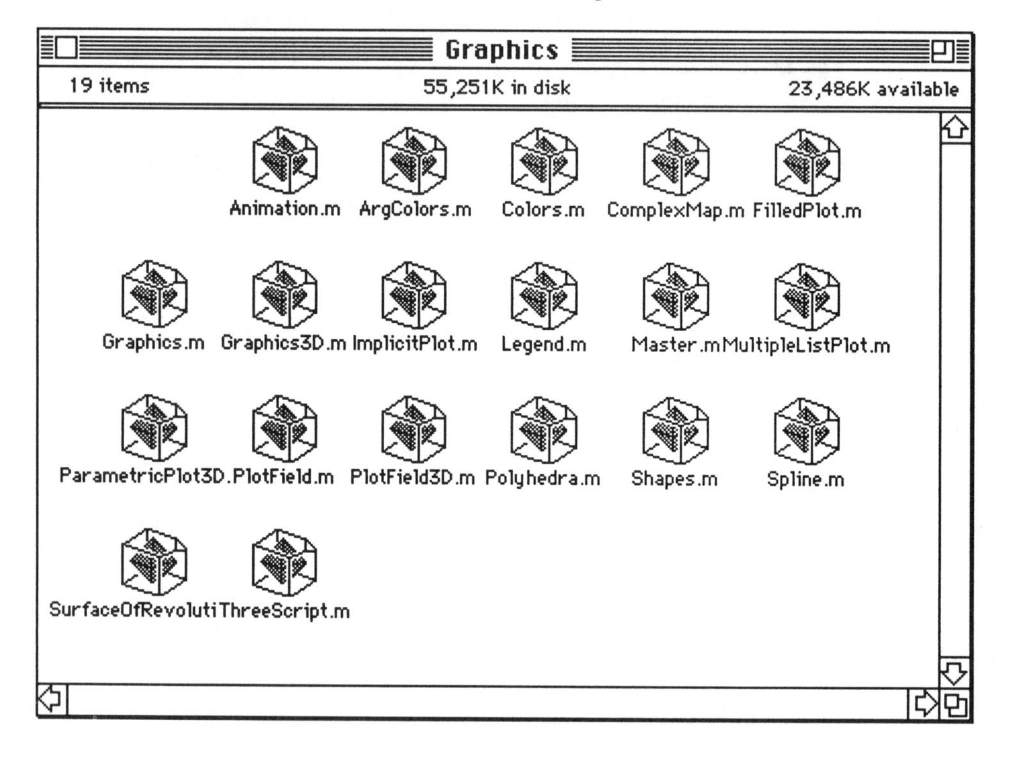

• 8.5 ImplicitPlot.m

Plotting an implicit equation was demonstrated earlier using the commands available in Version 1.2. This involved a rather lengthy procedure in which the equation was graphed in pieces. Fortunately, this task is not required in Version 2.0 which contains the **Graphics** package **ImplicitPlot.m**. This package includes the command **ImplicitPlot[equation,{x,xmin,xmax}]** which graphs the implicit equation, **equation**, from **x** = **xmin** to **x** = **xmax**. The set of y-values displayed may be specified by entering the command **ImplicitPlot[equation,{x,xmin,xmax},{y,ymin,ymax}]**.

o Example:

After loading **ImplicitPlot.m**, this command is demonstrated with the same equation that was plotted earlier with Version 1.2 in **Chapter 4** on page 237:

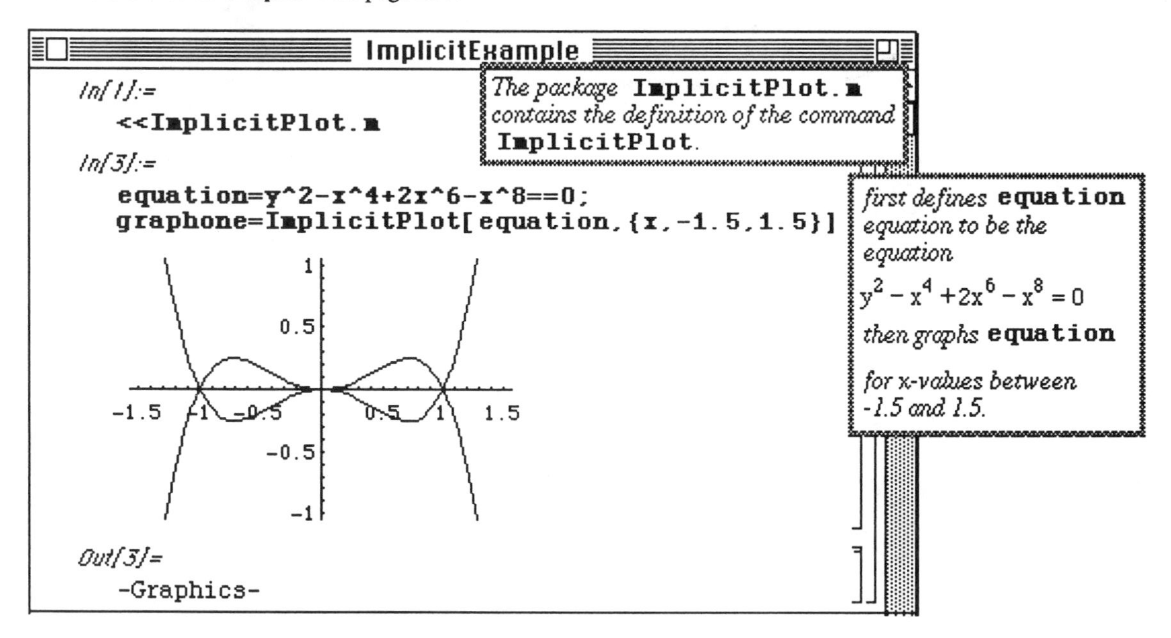

○ Example:

Implicit equations can be plotted simultaneously, as with the command **Plot**, with **ImplicitPlot[{eq1,eq2,...,eqn},{x,xmin,xmax}]** and **ImplicitPlot[{eq1,eq2,...,eqn},{x,xmin,xmax},{ymin,ymax}]**. This is shown below. Recall that a double equals sign must be used with each equation.

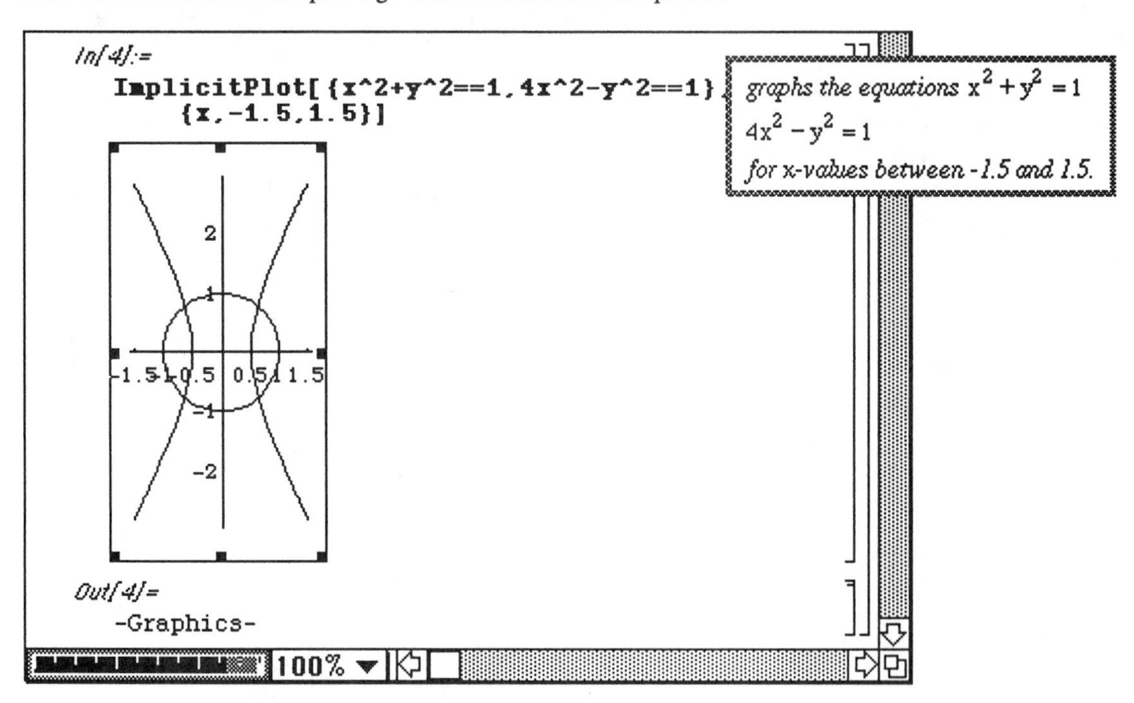

□ Example:

Conic sections can also be plotted with **ImplicitPlot**. A table of conic equations of the form **conic=a x^2+b y^2==1** is produced below.

The values of **a** are found in **alist={-2,-1,1,2}** while those for **b** are in **blist={1,2}**. The eight equations found in **conicequations** result when **conic** is evaluated at the values in **alist** and **blist**. Note that **conicequations** is a list of pairs of elements.
Hence, the conic equations can be extracted from this list with **conicequations[[i,j]]**, where **i**=1,2,3,4 and **j**=1,2.

A function which yields the graphics of each equation is defined as
graph[i,j]=ImplicitPlot[conicequations[[i,j]],{x,-2,2}, DisplayFunction->Identity].
The option **DisplayFunction->Identity** causes the plot to be suppressed. Then, the table of graphics for each equation in **conicequations** is produced with
graphics=Table[graph[i,j],{i,1,4},{j,1,2}].

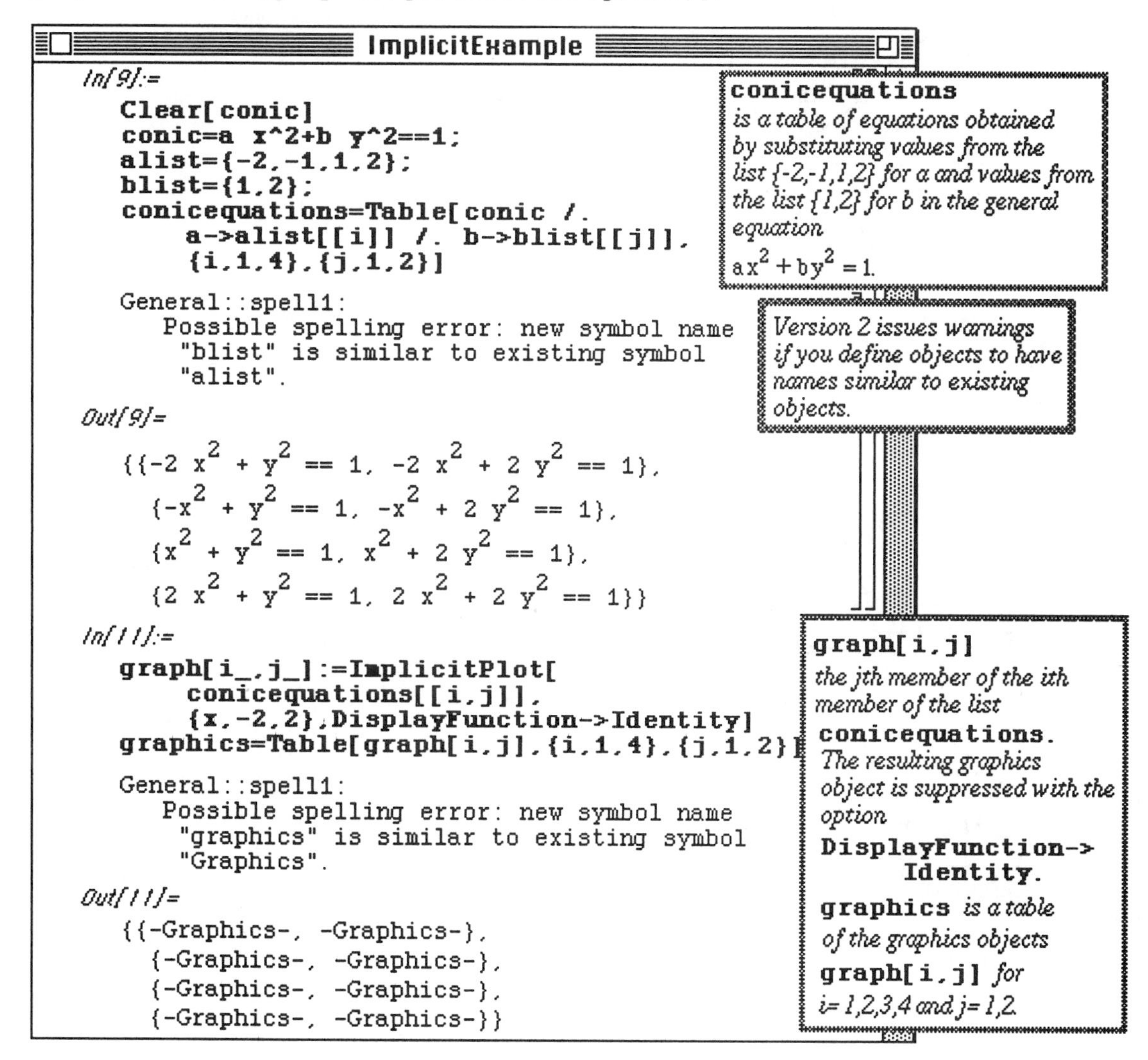

The graphics contained in **graphics** can be visualized with **Show[GraphicsArray[graphics]]**. **GraphicsArray[graphics]** sets up similar rectangular areas to display each graph in **graphics**.

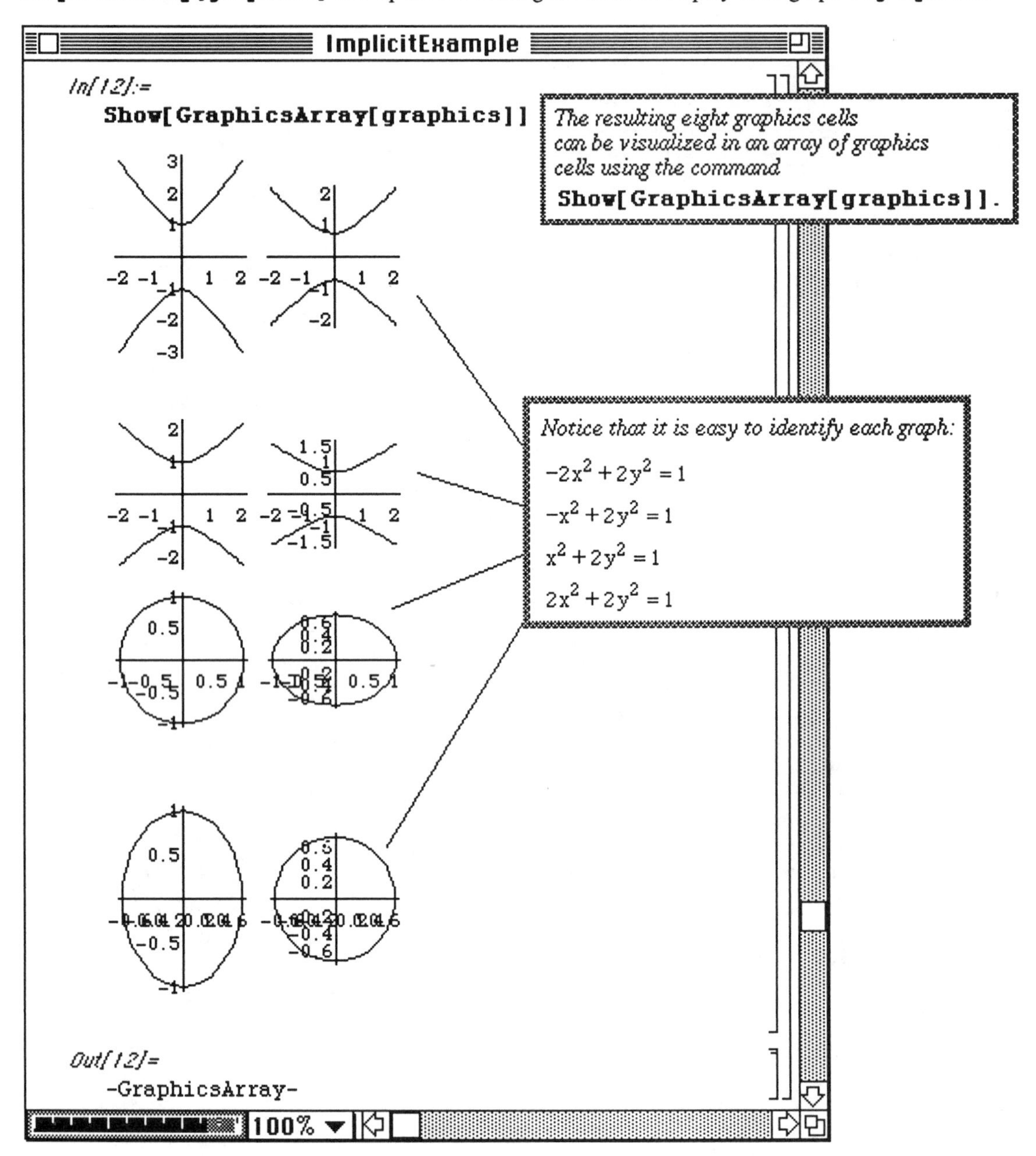

• 8.6 PlotField.m

The package **`PlotField.m`** contains the commands **`PlotVectorField`** and **`PlotGradientField`** which are useful in many areas of physics and engineering.

`PlotVectorField[vector[x,y],{x,xmin,xmax},{y,ymin,ymax}]` graphs the vector field given by the vector-valued function, **`vector[x,y]`**. This is illustrated below with the vector field **`f[x,y]`**.

o Example:

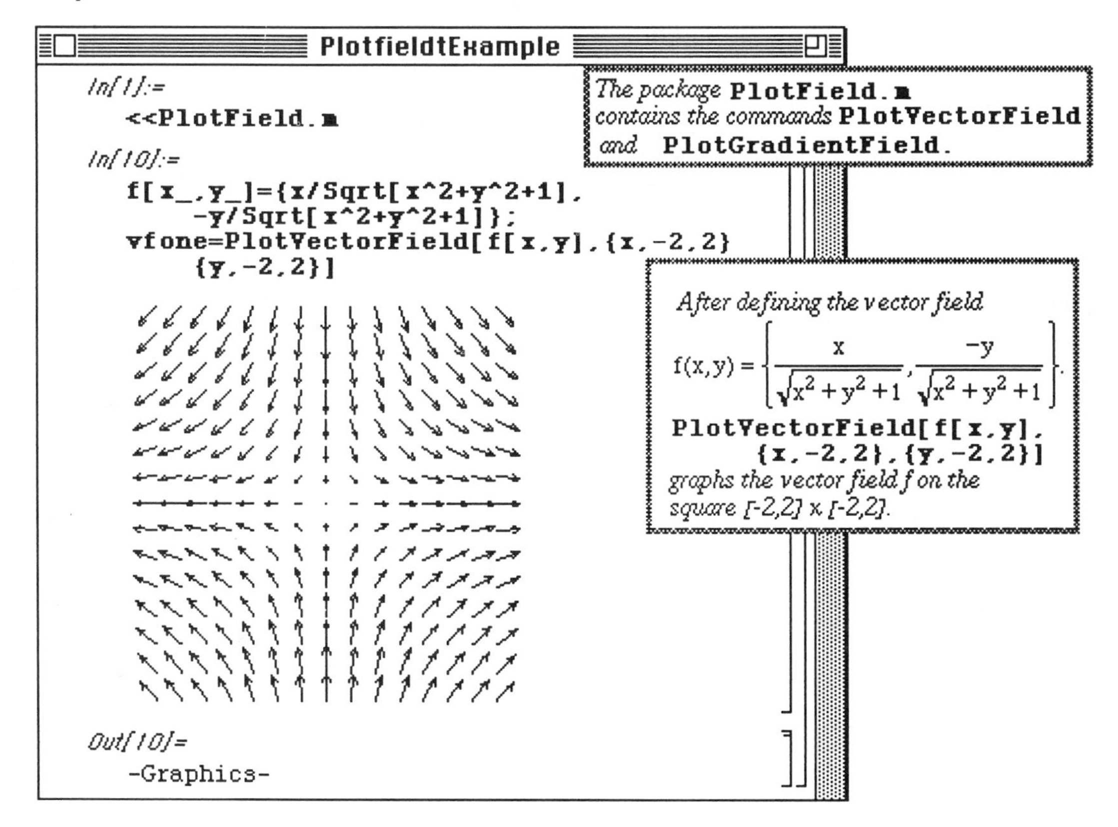

The command **PlotGradientField[function[x,y],{x,xmin,xmax},{y,ymin,ymax}]** graphs the gradient field of the function, **function[x,y]**. This is done by first computing the gradient of **function[x,y]** (which yields a vector field) and then plotting the gradient. This is shown below with the function **w[x,y]=Cos[4x^2+8y^2]**.

o **Example:**

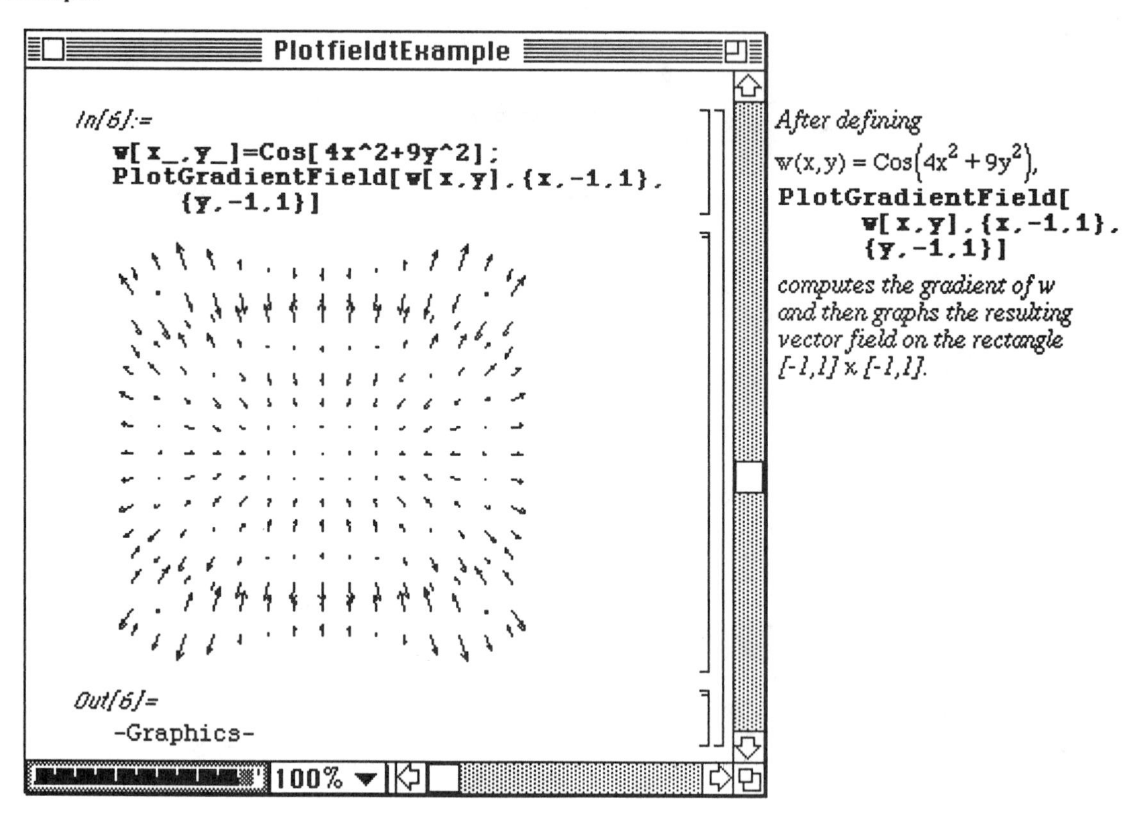

• 8.7 PlotField3D.m

Vector fields can be plotted in three dimensions as well. The commands needed to plot these fields are found in the **PlotField3D.m** package. The syntax for the **PlotGradientField3D** and **PlotVectorField3D** commands are similar to those used in the two-dimensional case in the previous section with the additional z-component.

o Example:

PlotGradientField3D is shown below with the three dimensional function, **Cos[x y z]**.

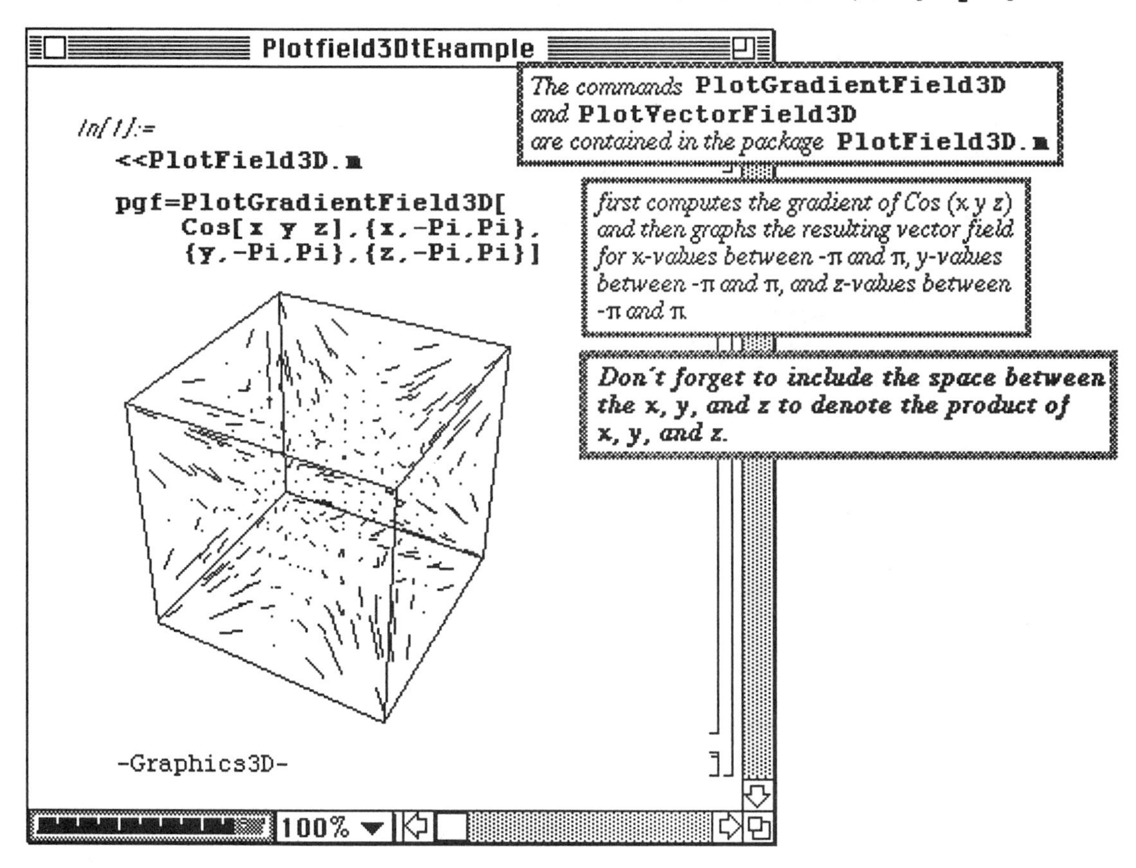

□ Example:

The vector field, **vf[x,y,z]={y^2-z,z^2+x,x^2-y}**, is plotted below using the command **PlotVectorField3D**. This graph is named **vfone** for later use.

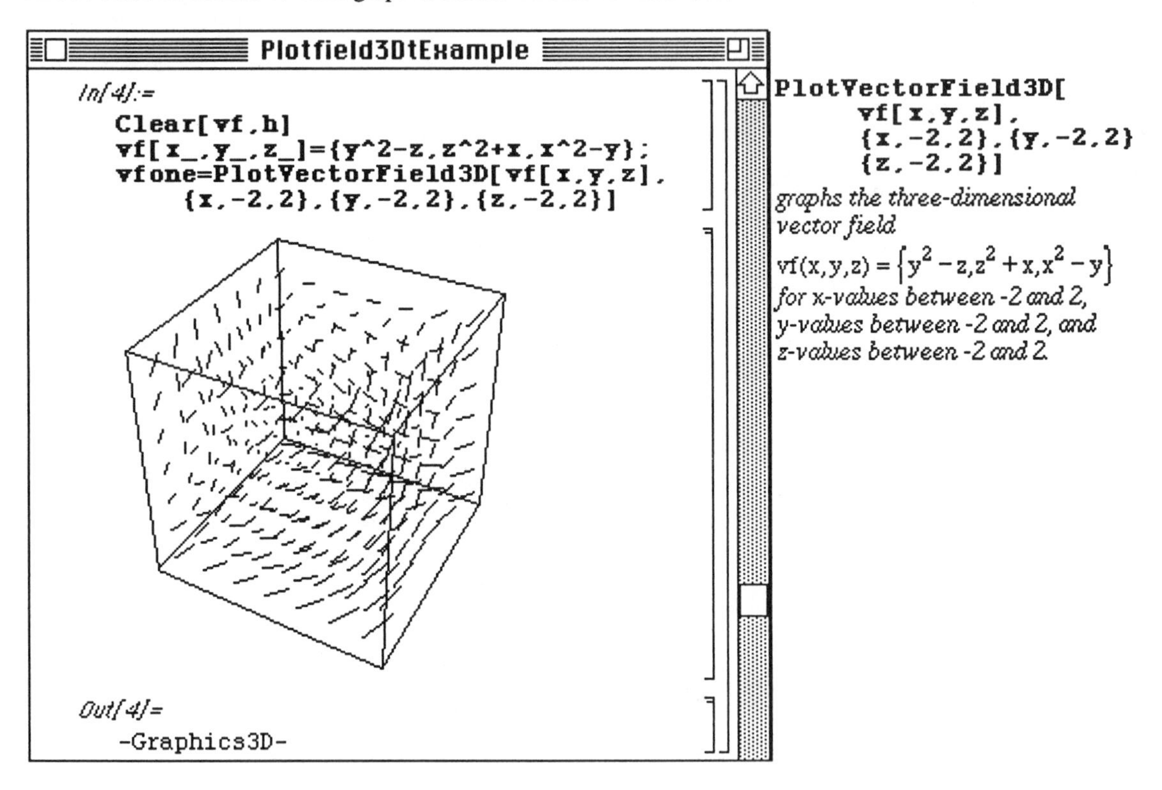

□ Example:

The function **h[x,y,z]=x^2+y^2+z+4** is defined below and the gradient field for **h[x,y,z]**, called **vgone**, is graphed with **PlotGradientField3D**. Notice that since the definition of **h** and the **PlotGradientField3D** command are enclosed in the same input cell and no semi-colon follows the definition of **h**, the formula for **h** is given as part of the output.

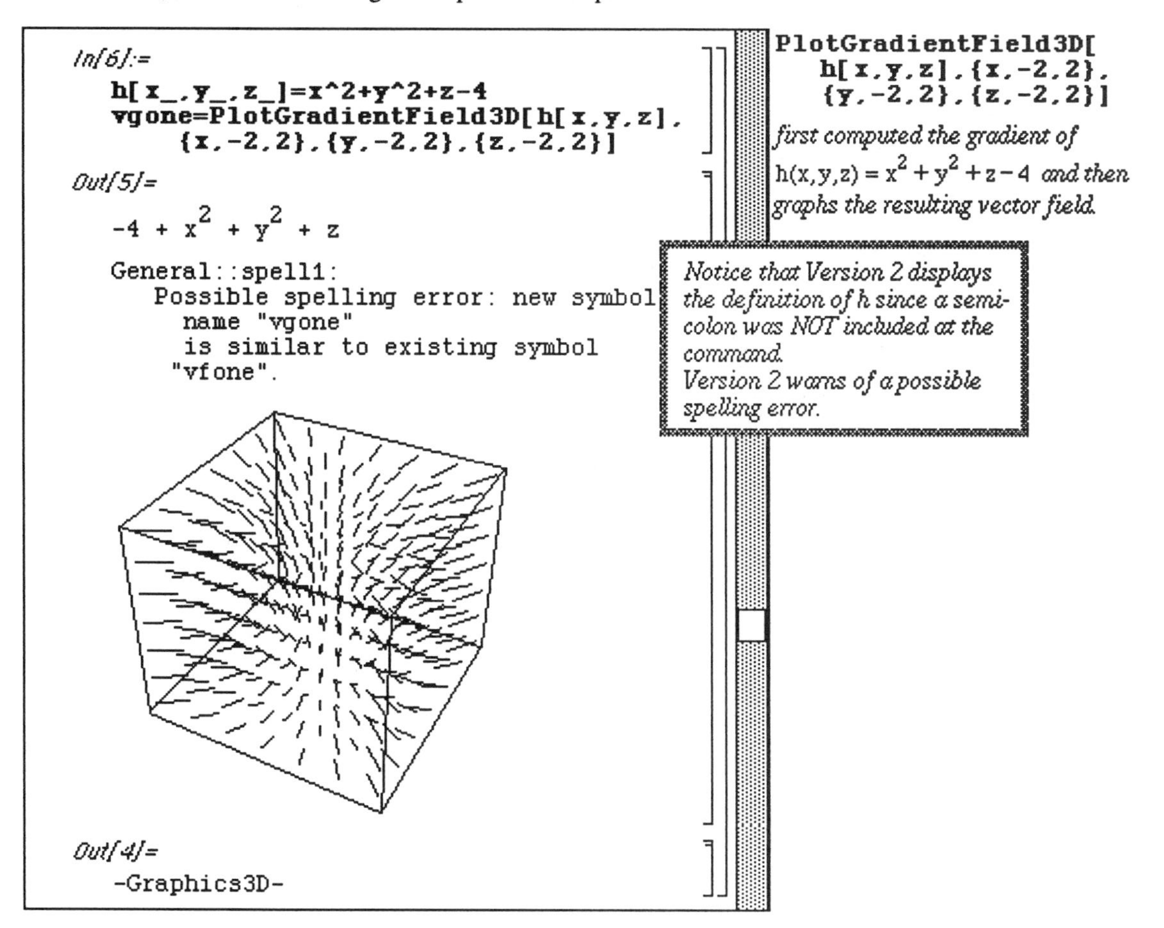

The three-dimensional graphs, **vfone** and **vgone**, obtained earlier can be viewed together with **Show[GraphicsArray[{vfone,vgone}]**.

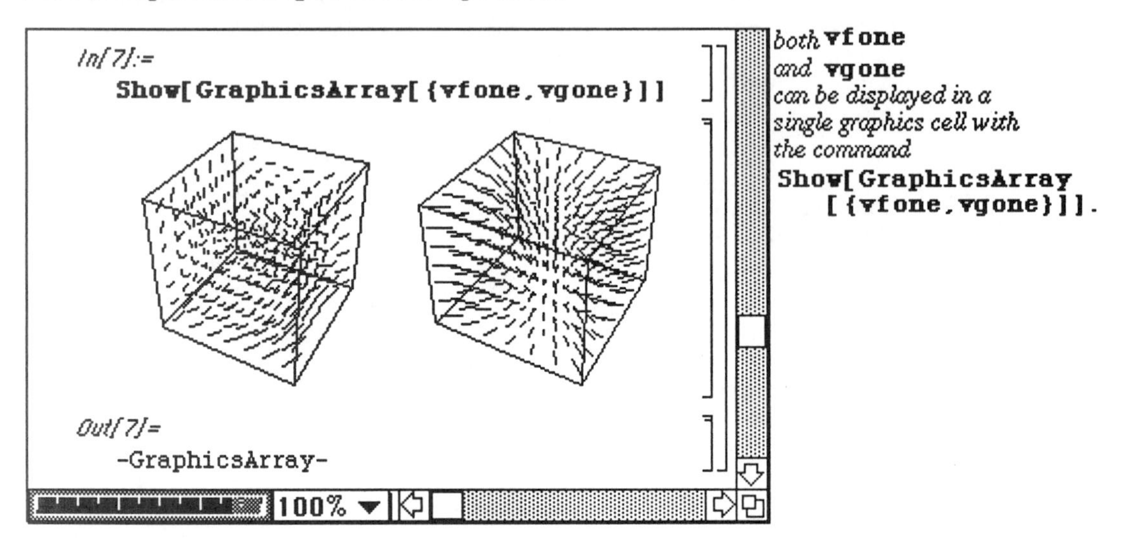

• 8.8 ComplexMap.m

A problem of interest in complex analysis is finding the image of a complex-valued function f(z). The package **`ComplexMap.m`** provides several commands which are useful in solving problems of this type. **`CartesianMap[f[z],{{x0,x1},{y0,y1}]`** gives the image of **`f[z]`** using Cartesian coordinate grid lines over the rectangular region **`{x0,x1}x{y0,y1}`**.

This is illustrated below with the functions **`id[z]=z`** and **`f[z]=(z-1)/(z+1)`**. Since **`id[z]`** is the identity map, each point in the domain is mapped to itself. Hence, the Cartesian grid, called **`cmid`**, is unchanged upon application of **`id[z]`**. (This region can therefore be viewed as the domain of **`f[z]`**.) The second graph, **`cmf`**, illustrates the effects that **`f[z]`** has on the points in **`cmid`**.

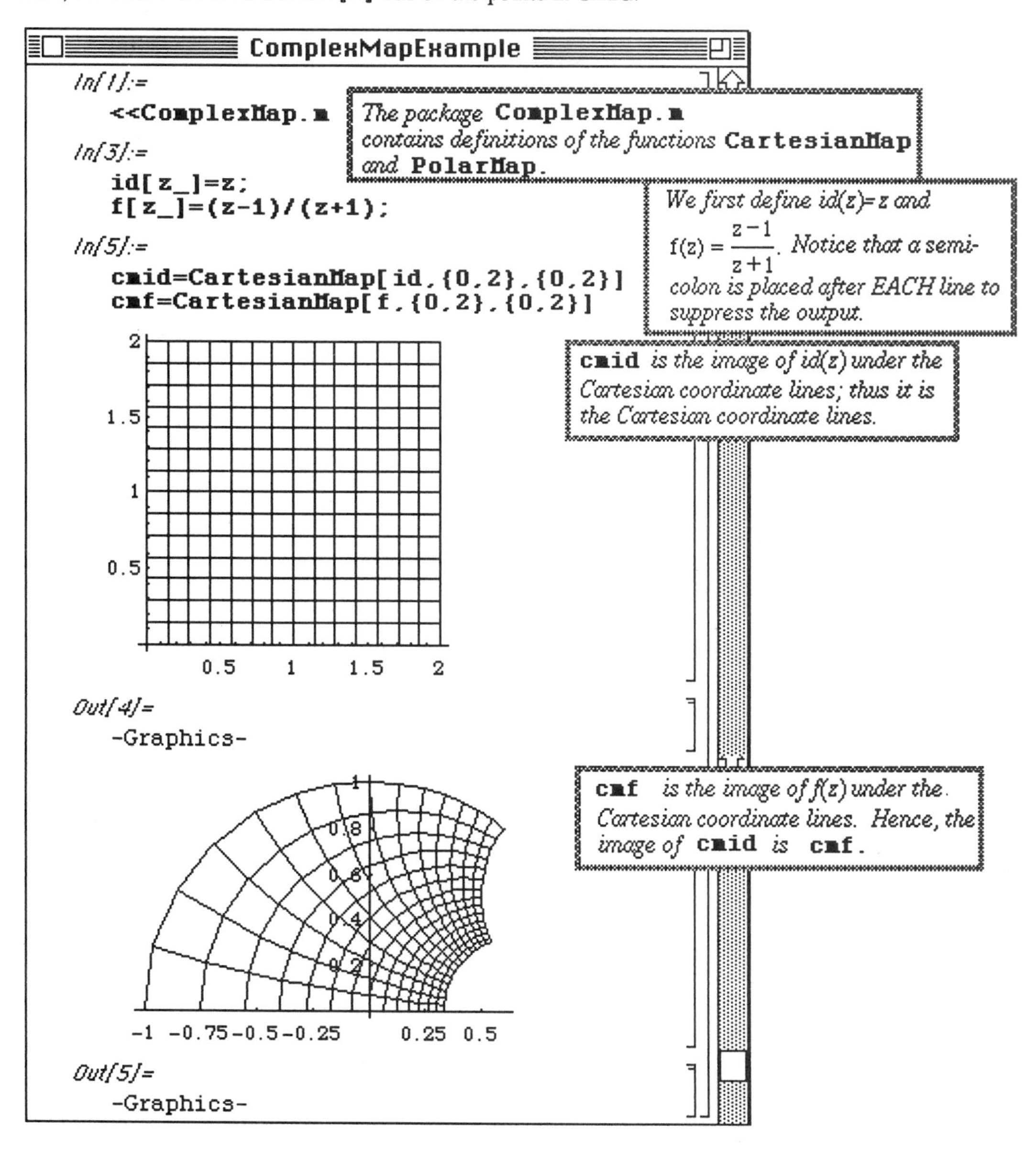

The two graphics objects, **cmid** and **cmf**, can be viewed in a single graphics cell with **Show[GraphicsArray[{cmid,cmf}]**. This gives the usual manner in which the domain and image of a function are illustrated.

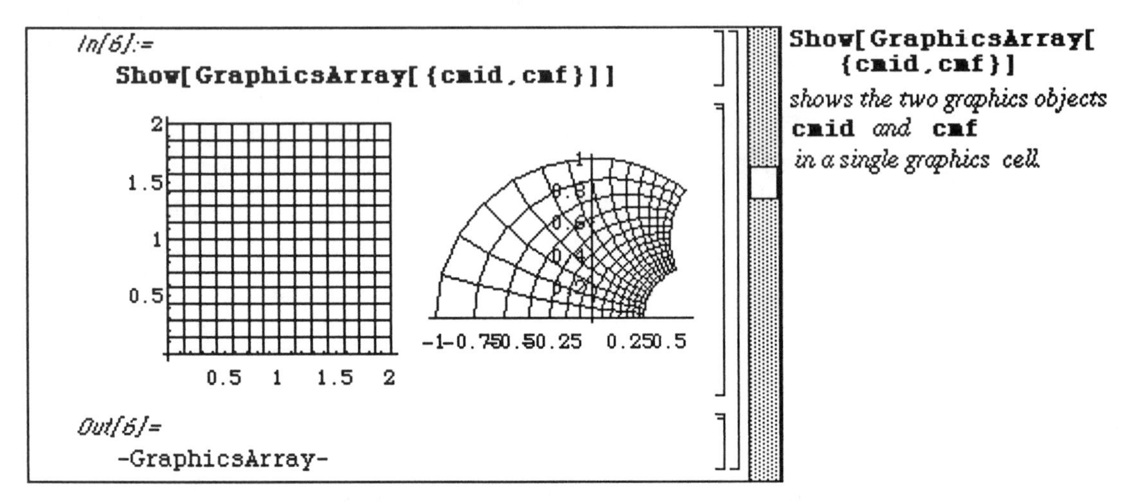

In addition to Cartesian coordinates, polar coordinates can also be used. This is done with **PolarMap[f[z],{r0,r1},{theta0,theta1}]** which produces the image of **f[z]** over the circular region **R** bounded by limits placed on the polar coordinates r and θ:

R : **r0** ≤ r ≤ **r1** , **theta0** ≤ θ ≤ **theta1** .

The following problem is worked in a method similar to that of the previous problem involving Cartesian coordinates. However, many of the graphs are suppressed by using the **`DisplayFunction->Identity`** option. Again, the identity map, **`id[z]=z`**, is used to produce the polar grid, called **`pmid`**, to be viewed as the domain of the function **`f[z]`**. The image of **`f`**, named **`pmf`**, is then determined with **`PolarMap`** and the two are displayed with **`Show[GraphicsArray[{pmid,pmf}]]`**.

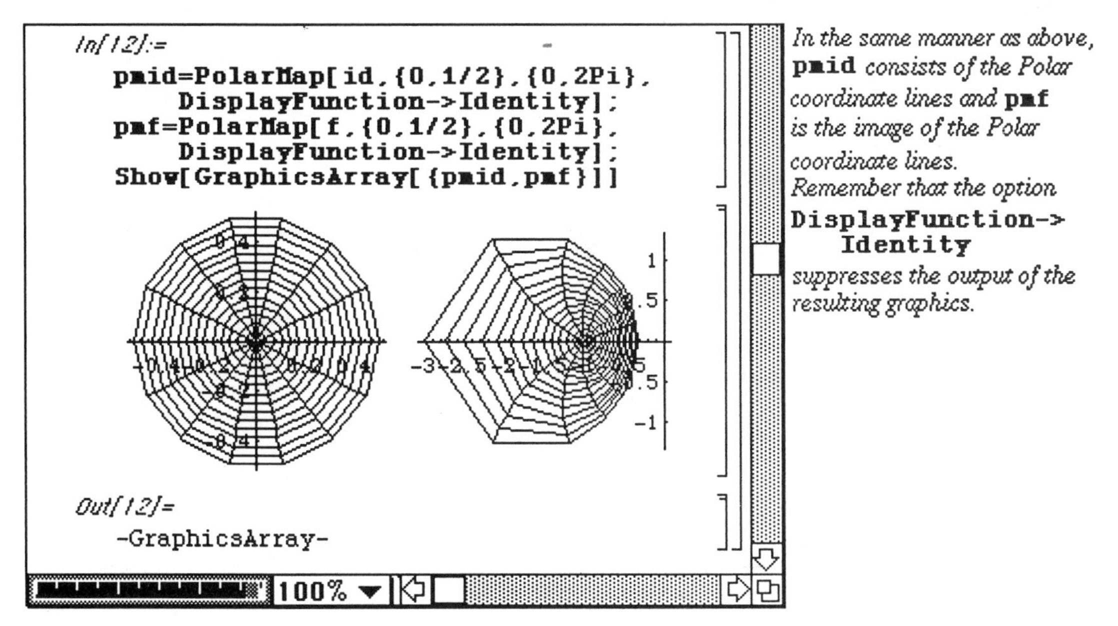

The following example illustrates several ideas. First, the built-in *Mathematica* function **`Identity`** can be used to produce the domain grid for a function as opposed to defining **`id[z]=z`** which was done in the previous examples. Also, **`GraphicsArray`** can be used to plot graphics cells in a desired order. For example, the domain and image of a function can be displayed consecutively.

This is done below for the functions **`w[z]=(1-Cos[z])/(1+Cos[z])`** and **`m[z]=(z-2)/(2z-1)`**. The domain and image of **`w`** are called **`cmid`** and **`cmw`**, respectively, while those of **`m`** are named **`pmid`** and **`pmm`**. These graphics objects are shown in the appropriate order with the command
`Show[GraphicsArray[{{cmid,cmw},{pmid,pmm}},AspectRatio->1]`. (Notice the grouping of {domain, image} within **`GraphicsArray`**.)

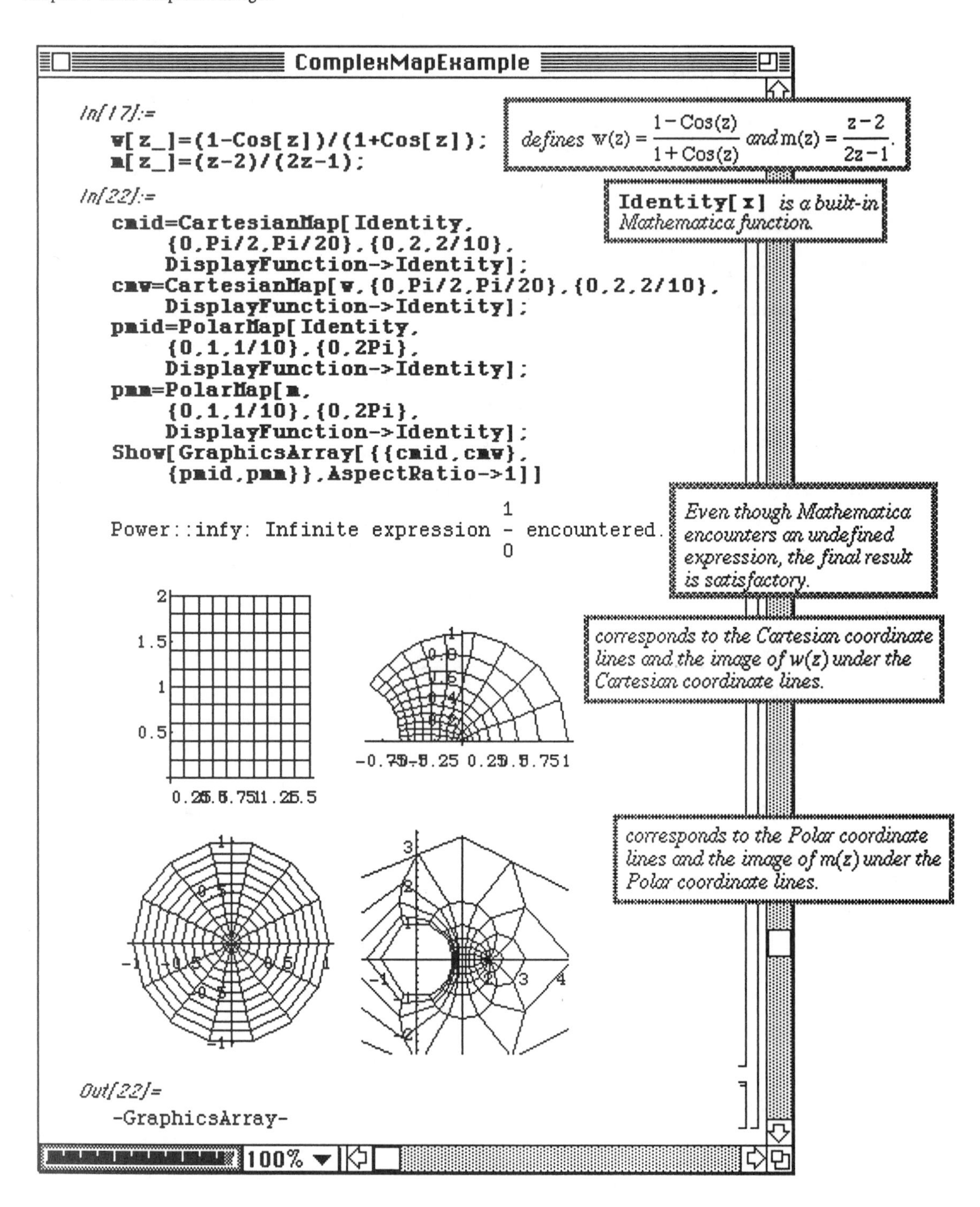

ComplexMapExample
In[17]:=
w[z_]=(1-Cos[z])/(1+Cos[z]);
m[z_]=(z-2)/(2z-1);
defines w(z) = (1-Cos(z))/(1+Cos(z)) and m(z) = (z-2)/(2z-1).
In[22]:=
cmid=CartesianMap[Identity,
{0,Pi/2,Pi/20},{0,2,2/10},
DisplayFunction->Identity];
cmw=CartesianMap[w,{0,Pi/2,Pi/20},{0,2,2/10},
DisplayFunction->Identity];
pmid=PolarMap[Identity,
{0,1,1/10},{0,2Pi},
DisplayFunction->Identity];
pmm=PolarMap[m,
{0,1,1/10},{0,2Pi},
DisplayFunction->Identity];
Show[GraphicsArray[{{cmid,cmw},
{pmid,pmm}},AspectRatio->1]]
Identity[x] is a built-in Mathematica function.
Power::infy: Infinite expression 1/0 encountered.
Even though Mathematica encounters an undefined expression, the final result is satisfactory.
corresponds to the Cartesian coordinate lines and the image of w(z) under the Cartesian coordinate lines.
corresponds to the Polar coordinate lines and the image of m(z) under the Polar coordinate lines.
Out[22]=
-GraphicsArray-
100%

Chapter 9
Some Special Packages

Chapter 9 discusses some of the more specialized packages available with *Mathematica*.

● Numerical Math

The packages within the **`Numerical Math`** folder in Version 2.0 are shown below:

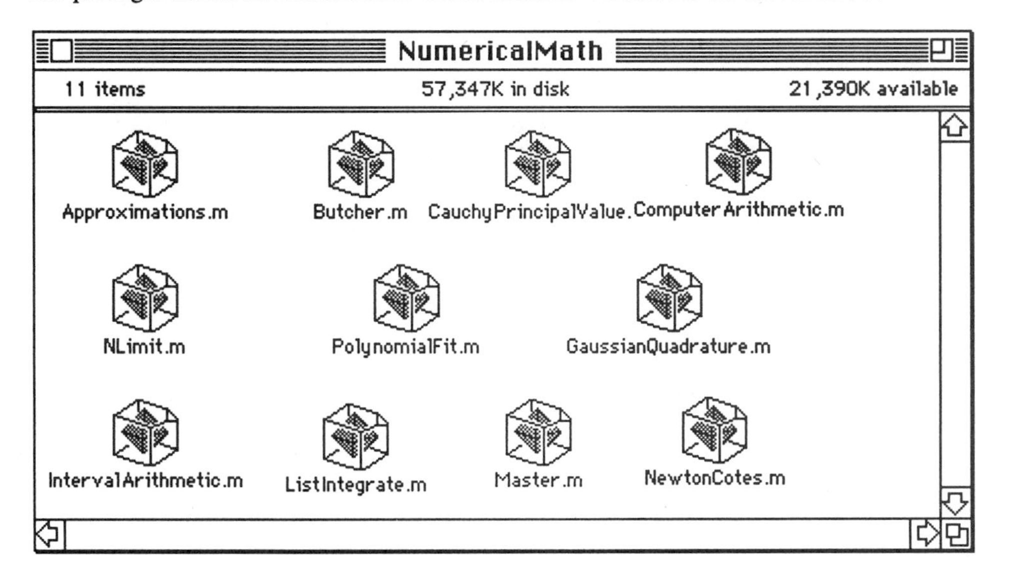

• 9.1 Approximations.m

o Although the examples done here were completed with Version 2.0, Version 1.2 also contains the package **Approximations.m** in the **Numerical Math** folder.

The package **Approximations.m** contains useful commands for the approximation of functions with rational functions. The first command discussed is that of **RationalInterpolation[function,{x,m,n},{x,x0,x1},options]** which gives the interpolating rational function P(x,y)/Q(x,y) on the interval from **x0** to **x1** where the the degree of P(x,y) is **m** and that of Q(x,y) is **n**.

o **Example:**

This command is illustrated below with **Sqrt[1-4x^2]**. The interpolating rational function **rint1** is found and then this approximation is compared to the original function by investigating the error function, **Abs[rint1 -Sqrt[1-4x^2]]**.

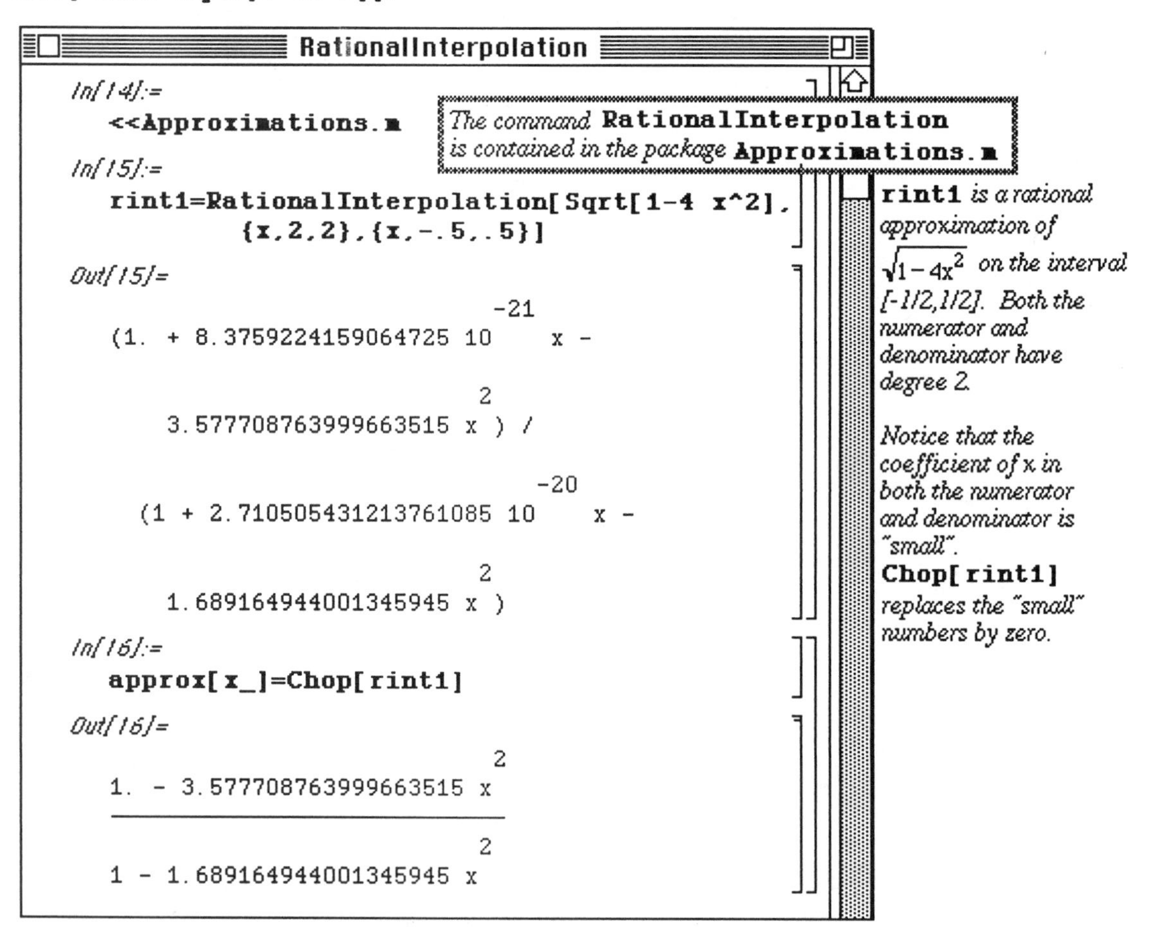

This error, **eplot1**, is graphed below.

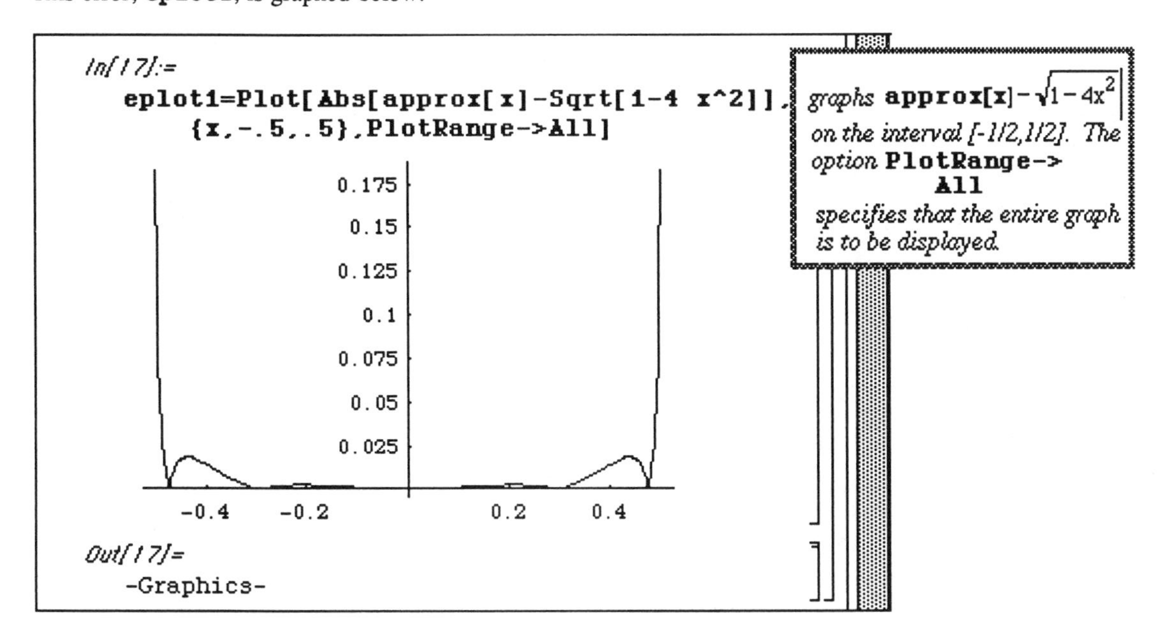

Several other interpolating rational functions are computed below. **rint2** is a function of order 3 in both the numerator and denominator. Similarly, **rint3** is of order 4 in both the numerator and denominator. The output of these functions is suppressed. Finally, the error for each rational interpolating function is graphed in order to compare accuracy. Clearly, **rint3** yields the best approximation of the three.

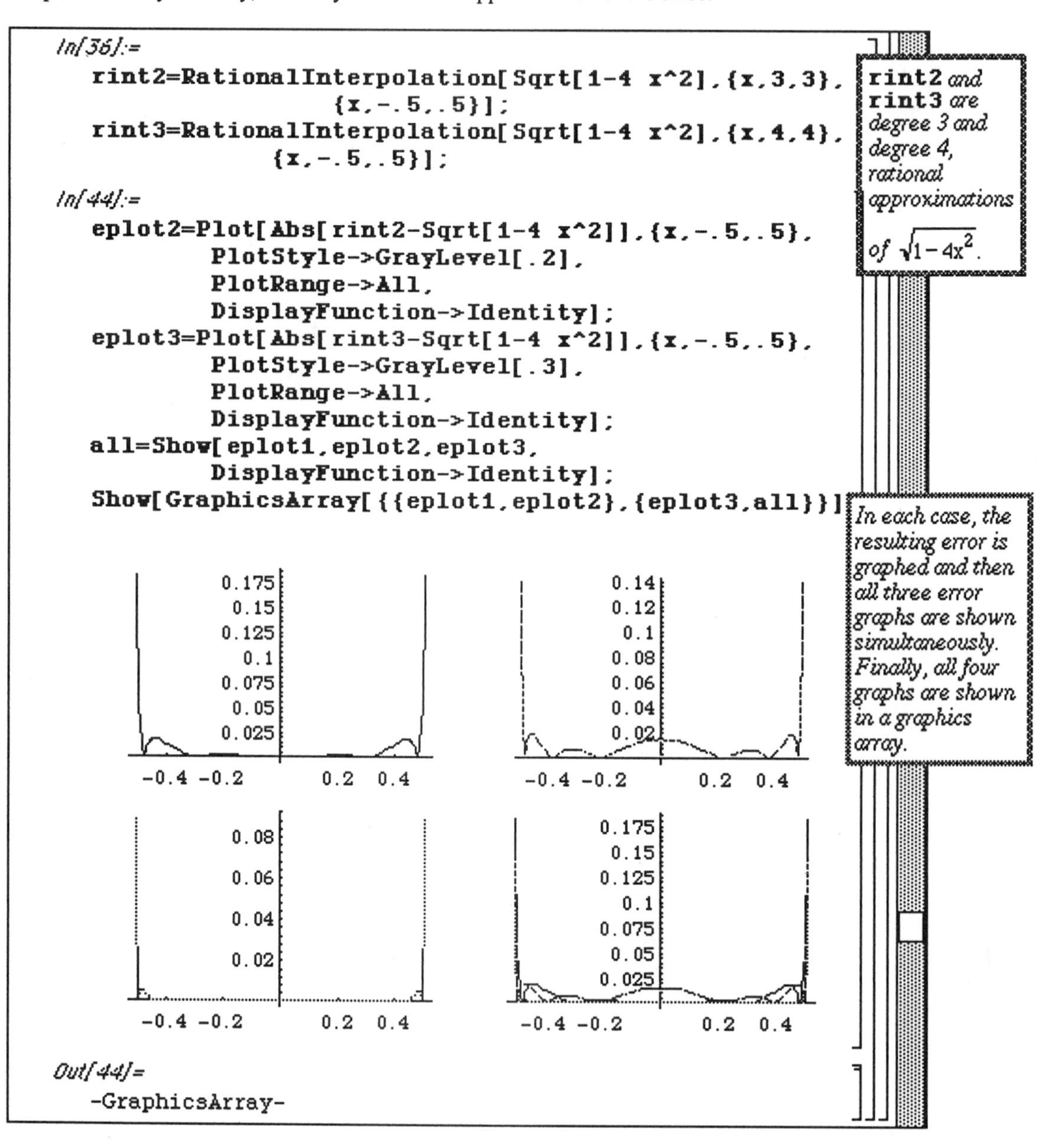

Also located in the **Approximations.m** package is the command **MiniMaxApproximation[f[x],{x,{x0,x1},m,k}]** which improves on the approximation found with **RationalInterpolation[f[x],{x,m,k},{x,x0,x1}]**. Although the syntax differs, both of these commands yield a rational function of order **m** in the numerator and **k** in the denominator which approximates **f[x]** over the interval **{x0,x1}**. Another difference in the commands occurs in the output. While **RationalInterpolation** gives only the approximating rational function, **MiniMaxApproximation** gives some additional information of the form **{abscissalist,{approx,maxerror}}** where **abscissalist** is a list of abscissa at which the maximum error occurs, **approx** is the desired rational approximation, and **maxerror** is the value of the minimax error.

o **Example:**

These two commands are investigated below with **f[x]=Exp[x]** on the interval **{-1,1}** using a rational function of order **2** in the numerator and **1** in the denominator. First, the approximating rational function is found with **RationalInterpolation** and named **rint1**. Note that this command yields only the approximating function as output. Next, an approximation is found with **MiniMaxApproximation**. These results (which are given in the form mentioned earlier) are named **mmax1**.

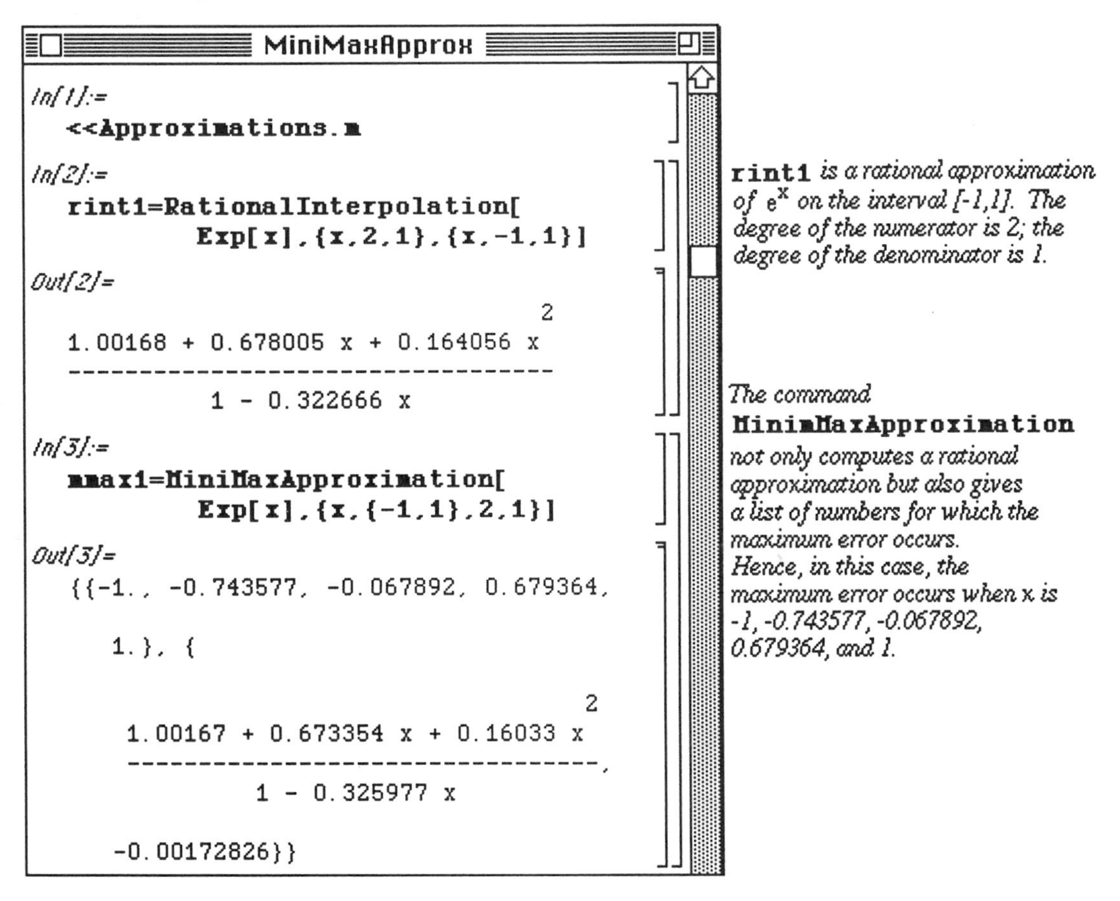

In order to work with the approximating function obtained with **MiniMaxApproximation**, the technique of extracting an element from a list must be used. Since **mmax1** is a list of two parts, the second of which contains the approximating function, **mmax1[[2,1]]** yields the desired rational function from the list. This function is called **apx1**. After extracting the approximating function **apx1**, the error for both approximations is investigated. This is done by observing the error function for each. The function **rint1error** which represents the error of the **RationalInterpolation** approximation is given by **Abs[rint1-Exp[x]]**. Likewise, the error of the **MiniMaxApproximation** , called **mmax1error**, is given by **apx1-Exp[x]** . These error functions are plotted simultaneously below. Notice how the critical points of the **mmax1error** curve (graphed in the lighter print) correspond to the values in **abscissalist**.

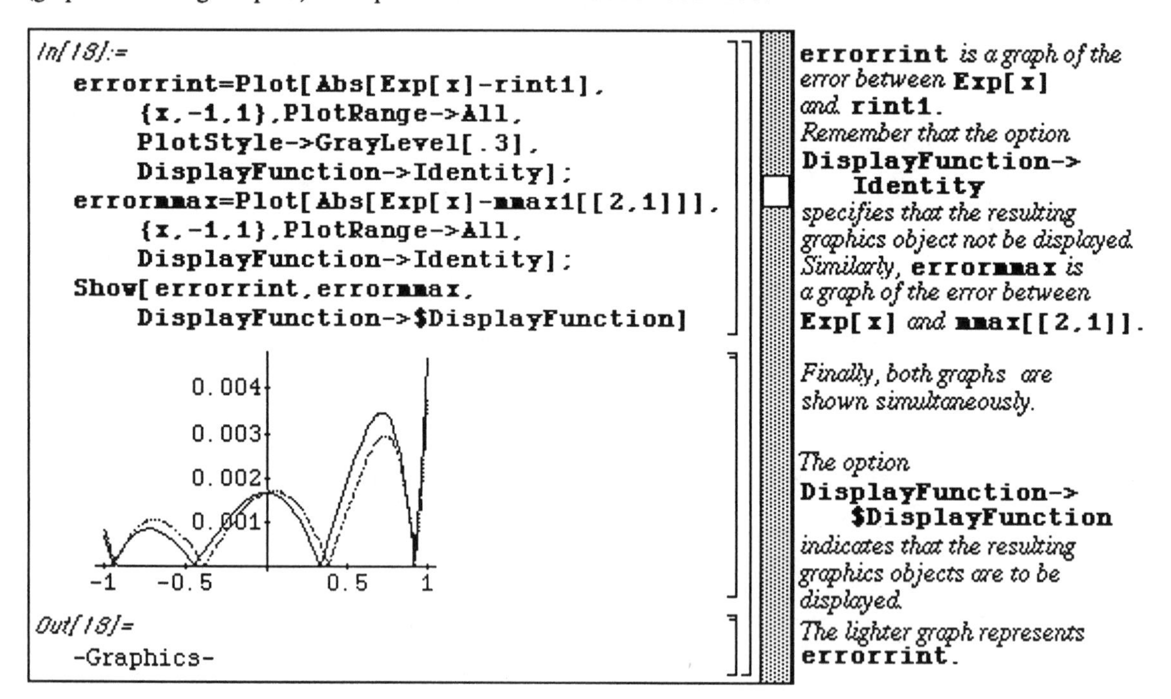

• 9.2 GaussianQuadrature.m

o The package **GaussianQuadrature.m** is contained only in the Version 2.0 **Numerical Math** and is not included with Version 1.2.

Numerical integration by Gaussian quadrature is based on the Lagrange interpolation formula

$$p(x) = \sum_{i=1}^{n} f(x_i)\, \ell_i(x) \quad \text{where} \quad \ell_i(x) = \prod_{\substack{j=1 \\ j \neq i}}^{n} \left(\frac{x - x_j}{x_i - x_j} \right).$$

If this formula provides a good approximation of f, then the integral of p yields a good approximation to the integral of f. Therefore,

$$\int_a^b f(x)\,dx \approx \int_a^b p(x)\,dx = \sum_{i=1}^{n} f(x_i) \int_a^b \ell_i(x)\,dx = \sum_{i=1}^{n} A_i f(x_i)$$

where A_i represents the weights and x_i the nodes for $i = 1,2,\ldots,n$.

The *Mathematica* command **GaussianQuadratureWeights[n,a,b]** which is located in the **GaussianQuadrature.m** package determines the values of these weights and nodes. The output is given in the form of ordered pairs where the first entry in each pair gives the node while the second entry represents the corresponding weight. Several examples of this command are given below for different values of **n** using the same interval from -2 to 2. Note that the calculation of the weights and nodes is independent of the function f.

o **Example:**

```
GaussianQuadrature(NumMath)

In[1]:=
   <<GaussianQuadrature.m

In[2]:=
   GaussianQuadratureWeights[2,-2,2]

Out[2]=
   {{-1.1547, 2.}, {1.1547, 2.}}

In[3]:=
   GaussianQuadratureWeights[3,-2,2]

Out[3]=
   {{-1.54919, 1.11111}, {0., 1.77778}, {1.54919, 1.11111}}

In[4]:=
   GaussianQuadratureWeights[4,-2,2]

Out[4]=
   {{-1.72227, 0.69571}, {-0.679962, 1.30429}, {0.679962, 1.30429},

     {1.72227, 0.69571}}
```

Since the output appears in the form of a list, the weights and nodes can be extracted from the output. This is illustrated below by assigning the name of **gqw** to the expression which results from the command **GaussianQuadratureWeights[3,-1,1]**. Hence, **gqw[[1,1]]** gives the first node on the interval [-1,1], and **gqw[[1,2]]** gives the weight which corresponds to this node. Therefore, the integral of a function f(x) from -1 to 1 can be approximated with the Gaussian quadrature formula given earlier using the values obtained with **gqw**. This is done in **gqint** below. Since no function is specified, the general integral formula results.

o Example:

In the next command, however, the function **f[x]=Exp[-(Cos[x])^2]** is defined. Thus, a numerical approximation of the integral of **f** from x = -1 to x = 1 is given.

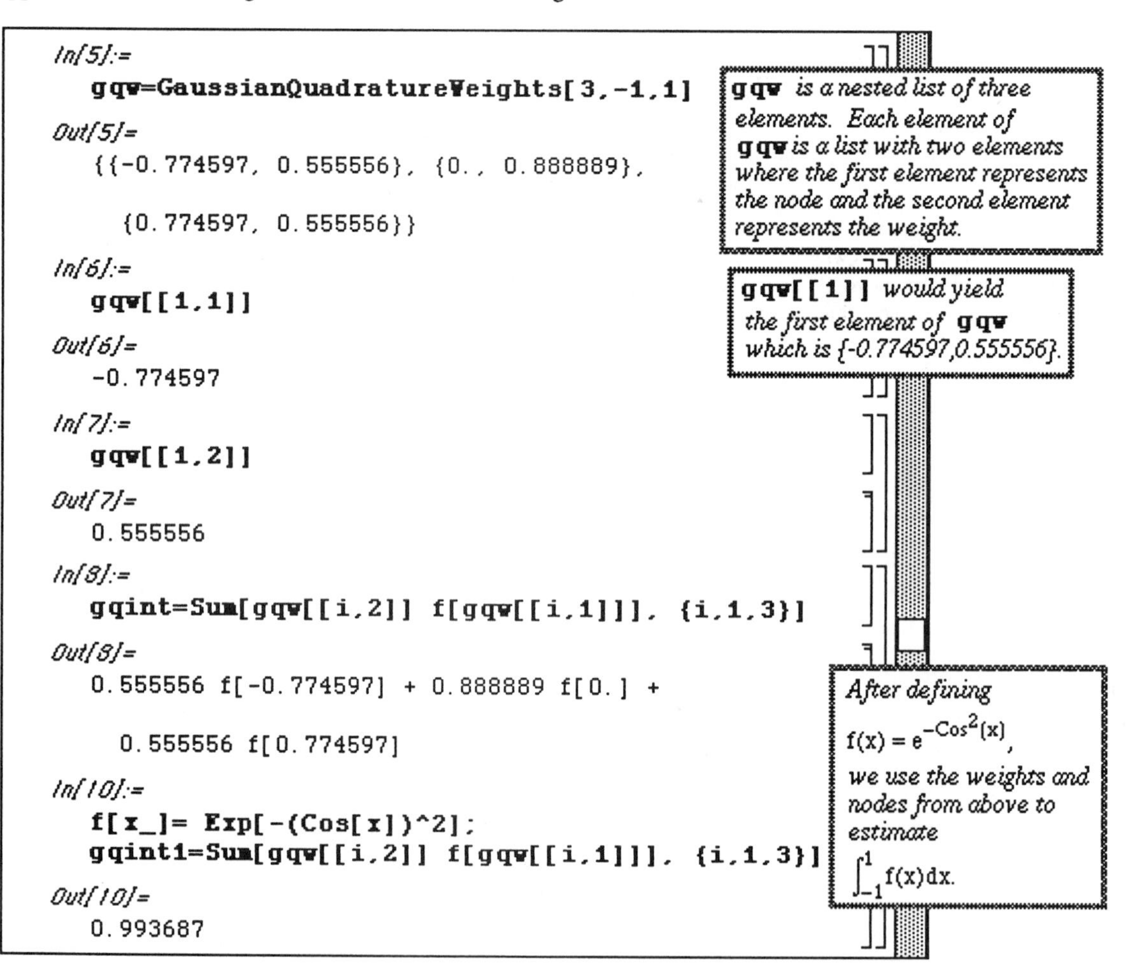

A function which evaluates the Gaussian quadrature for any value of **n** is defined below in **gauss[n]**. A table of approximations of the integral of **f** given above is then created for **n** = 2 to **n** = 10 and placed in **TableForm**. This procedure is useful in comparing the approximations obtained with Gaussian quadrature and can be repeated for other integrals.

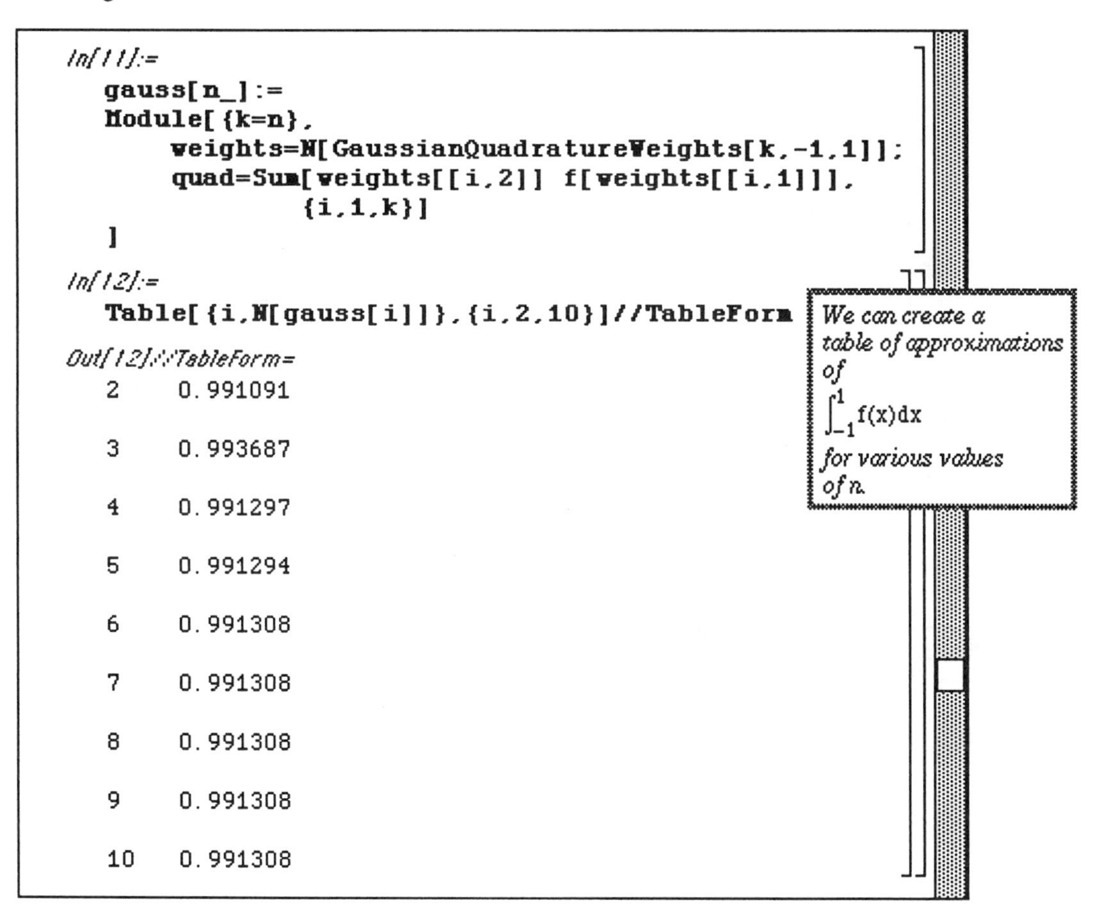

Gaussian Quadrature procedures are useful when the exact value of an integral cannot be computed. This is the case with $\int_{-1}^{1} e^{-\mathrm{Cos}^2(x)}$ as shown below. However, in this case, our results can be verified with **NIntegrate**:

```
In[2]:=
   f[x_]=Exp[-(Cos[x])^2];
   Integrate[f[x],{x,-1,1}]

   Syntax::bktwrn:
      Warning: "f (b+a x^dg)" should probably be "f [b+a x^dg]".
        (line 169 of "Integrate`mainalgorithm`")

   Syntax::bktwrn:
      Warning: "f (Denominator[a]^Abs[b])" should probably be
       "f [Denominator[a]^Abs[b]]".
        (line 1227 of "Integrate`mainalgorithm`")
Out[2]=
                     2
             -Cos[x]
   Integrate[E        , {x, -1, 1}]
In[3]:=
   NIntegrate[f[x],{x,-1,1}]
Out[3]=
   0.991308
```

Mathematica cannot compute the exact value of $\int_{-1}^{1} f(x)dx$ *although a numerical approximation which agrees with the above can be calculated using* **NIntegrate.**

• 9.3 NLimit.m

○ **`NLimit.m`** is included with Version 2.0 but not earlier versions.

The **`NLimit.m`** package contains which are useful in the calculation of limits and derivatives. These are **`NLimit[f[x], x->x0]`** and **`ND[f[x],x,x0]`**. **`NLimit[f[x], x->x0]`** computes the numerical limit of $\lim_{x \to x0} f[x]$.

The value of **`x0`** can be either **`Infinity`** or **`-Infinity`**. However, a limit is not given with **`NLimit`** if the limit is **`Infinity`** or **`-Infinity`**. This command may be used when the built-in function **`Limit[f[x], x->x0]`** fails.

○ **Example:**

Compute $\lim_{x \to \infty} \frac{e^x}{x!}$.

Illustrated below is the calculation of the limit as **`x`** approaches **`Infinity`** of the function **`Exp[x]/(x!)`**. First, the limit is attempted using **`Limit`**. Since this is unsuccessful, a second attempt is made with **`NLimit`** to yield a limit of 0.

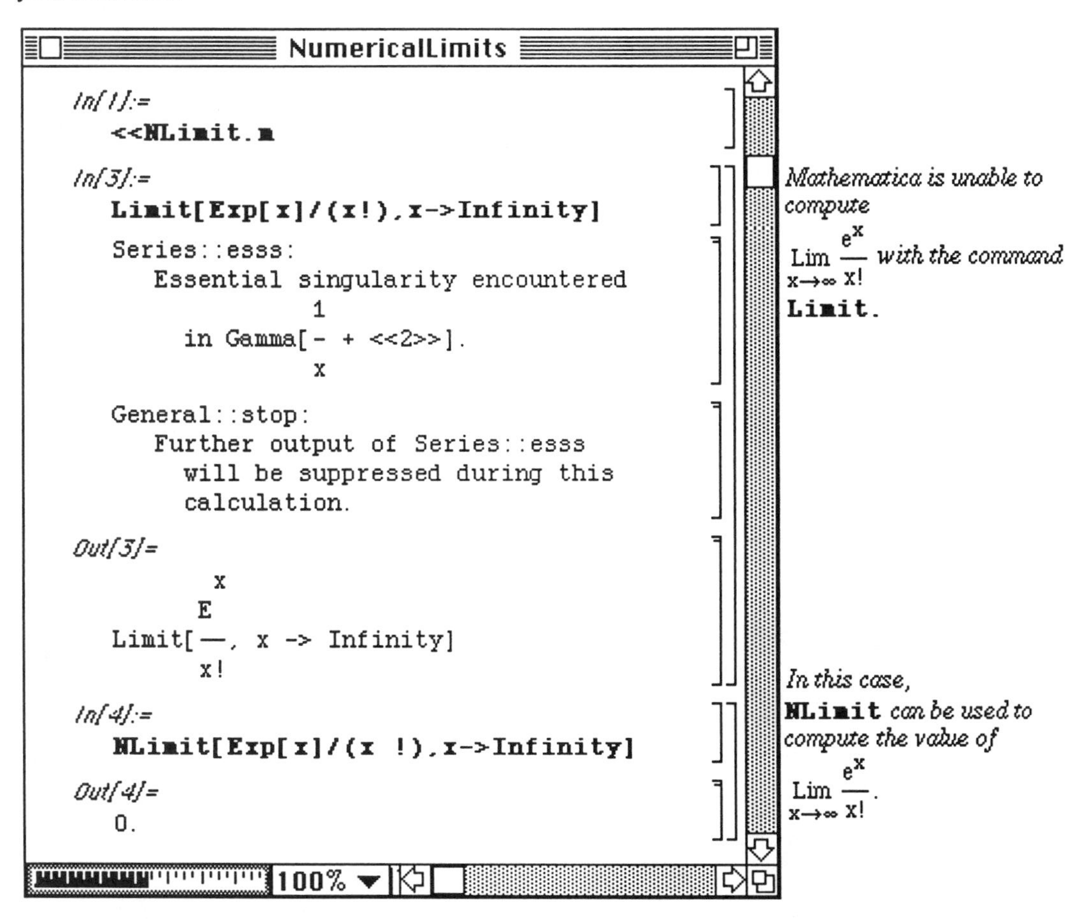

Similarly, **Limit** is unsuccessful in computing the limit as **x** approaches **Infinity** of the function **(x^5)/Exp[x]**. However, **NLimit** is used to obtain a numerical approximation of this limit. Note that the result is quite close to 0. The built-in *Mathematica* command **Chop[expression]** replaces all approximate real numbers in **expression** which are less than 10^(-10) in magnitude with the number 0. Hence, **Chop[%]** yields the correct limit.

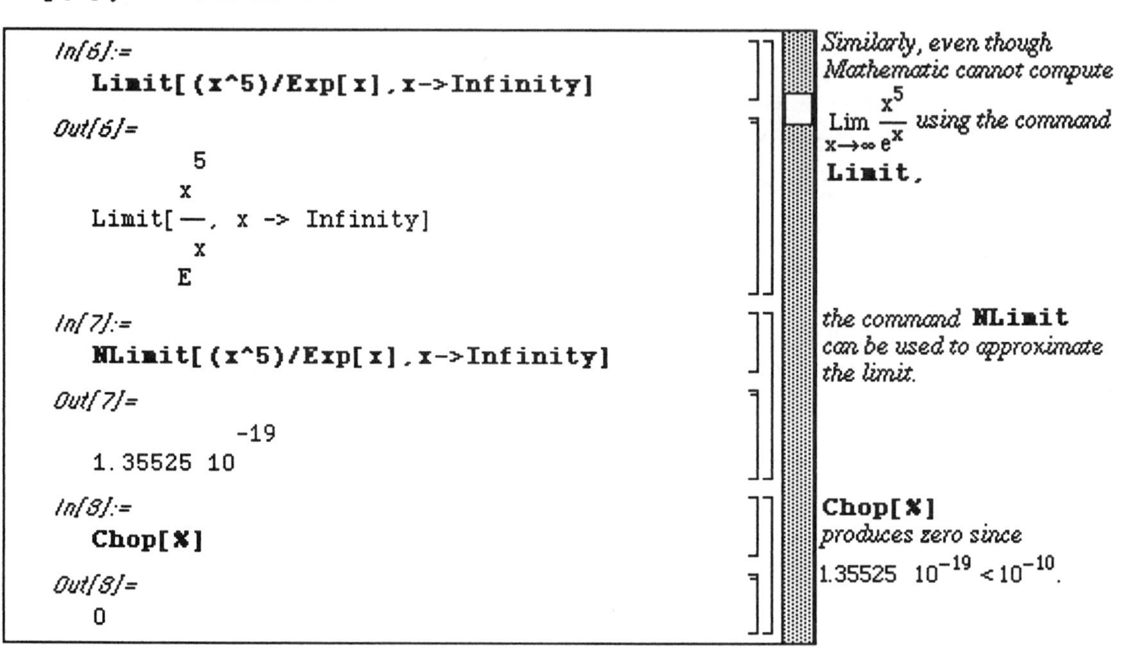

o Example:

Special care must be taken when dealing with limits which achieve the value of **Infinity** or **-Infinity**. **NLimit** cannot calculate limits of this type ! In these cases, **NLimit** may yield an incorrect limit or no limit at all. In the first example which follows, **NLimit** gives a value which obviously is not the limit of the given function. The limit of this function is **Infinity** as substantiated with the graph which follows as well as with well-known properties of functions of this type. In the second example, **NLimit** does not compute the limit as **x** approaches **Infinity** of the function x^2. This limit is clearly **Infinity**.

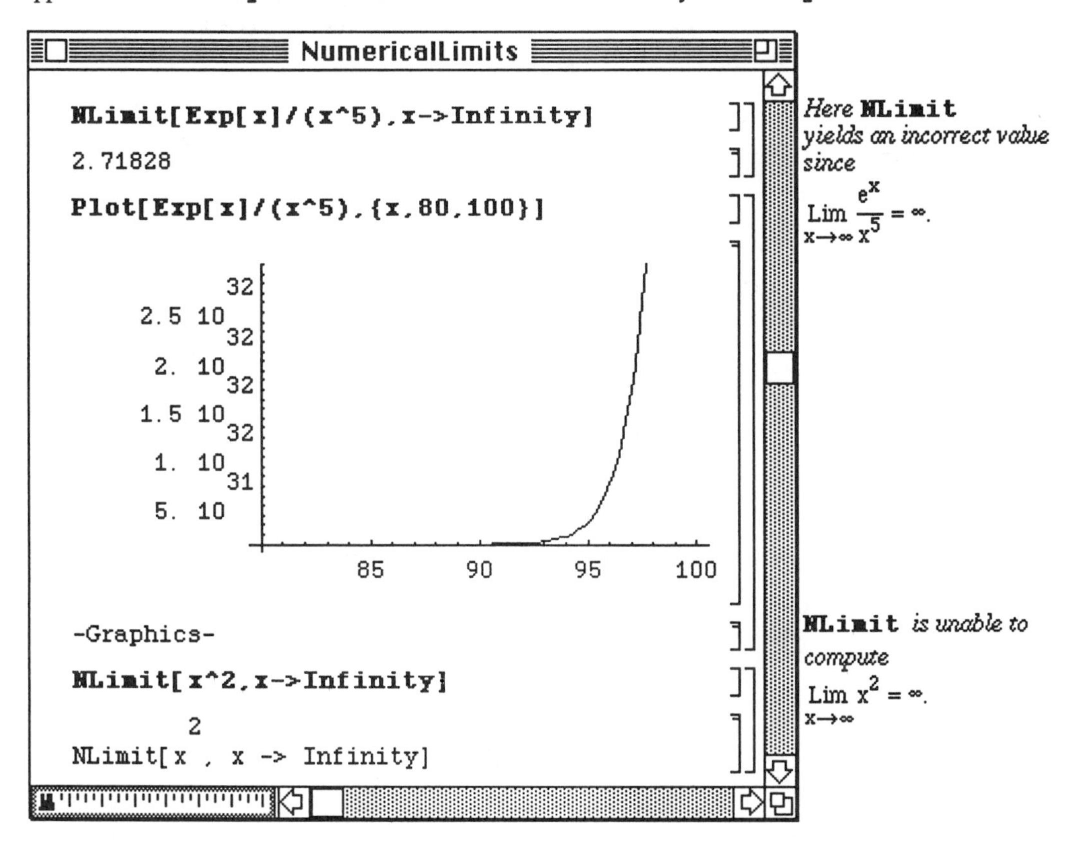

Here **NLimit** *yields an incorrect value since* $\lim_{x\to\infty} \frac{e^x}{x^5} = \infty$.

NLimit *is unable to compute* $\lim_{x\to\infty} x^2 = \infty$.

When dealing with certain functions, the calculation of a numerical derivative may be necessary. The command **ND[f[x],x,x0]** gives the numerical approximation of **f'[x0]**. This method is based on Richardson extrapolation and does not yield a formula for the derivative. It simply gives an approximation of the derivative at the value **x = x0**.

o Example:

The derivative of **(x!)^2** at **x = 1** is approximated below. Next, an attempt to illustrate the accuracy of **ND[f[x],x,x0]** is performed with the function **Tan[x]**.

The function $(x!)^2$ is the same as the function $(\Gamma(x))^2$ where $\Gamma(x) = \int_0^\infty e^{-t}t^{x-1}dt$

$$= \lim_{n\to\infty} \frac{n!n^x}{x(x+1)\ldots(x+n)}$$

is the Gamma function which is given by the command **Gamma[a]**.

$\Psi(x) = \Gamma'(x)$ is called the Digamma Function and is given by the command **Polygamma[x]**.

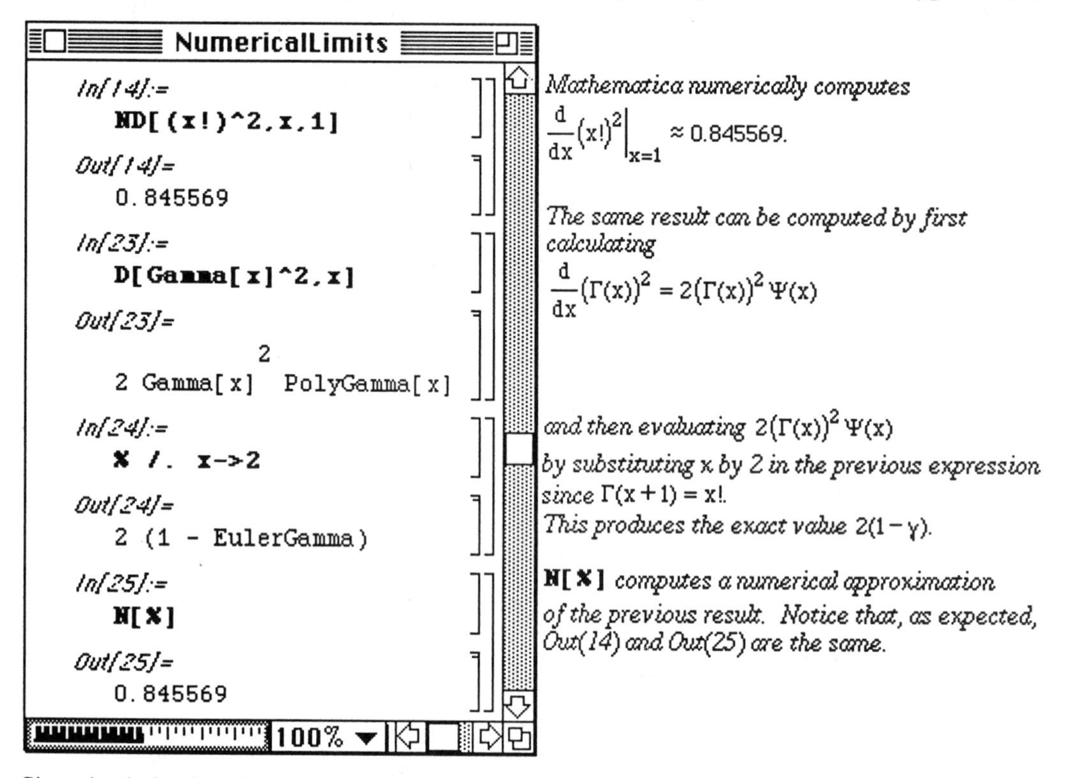

Since the derivative of **Tan[x]** is known to be **(Sec[x])^2**, the numerical values obtained using **ND** are compared to the values of **(Sec[x])^2** for **x = -1.5** to **x = 1.5**. A table of numerical approximations of the derivative of Tan(x) at values of x in the interval [-1.5,1.5] using **ND** are given in **dtable**.

This table of ordered pairs is created in order to plot the approximated derivative. These points are plotted simultaneously with **(Sec[x])^2** to show the accuracy of this approximation.

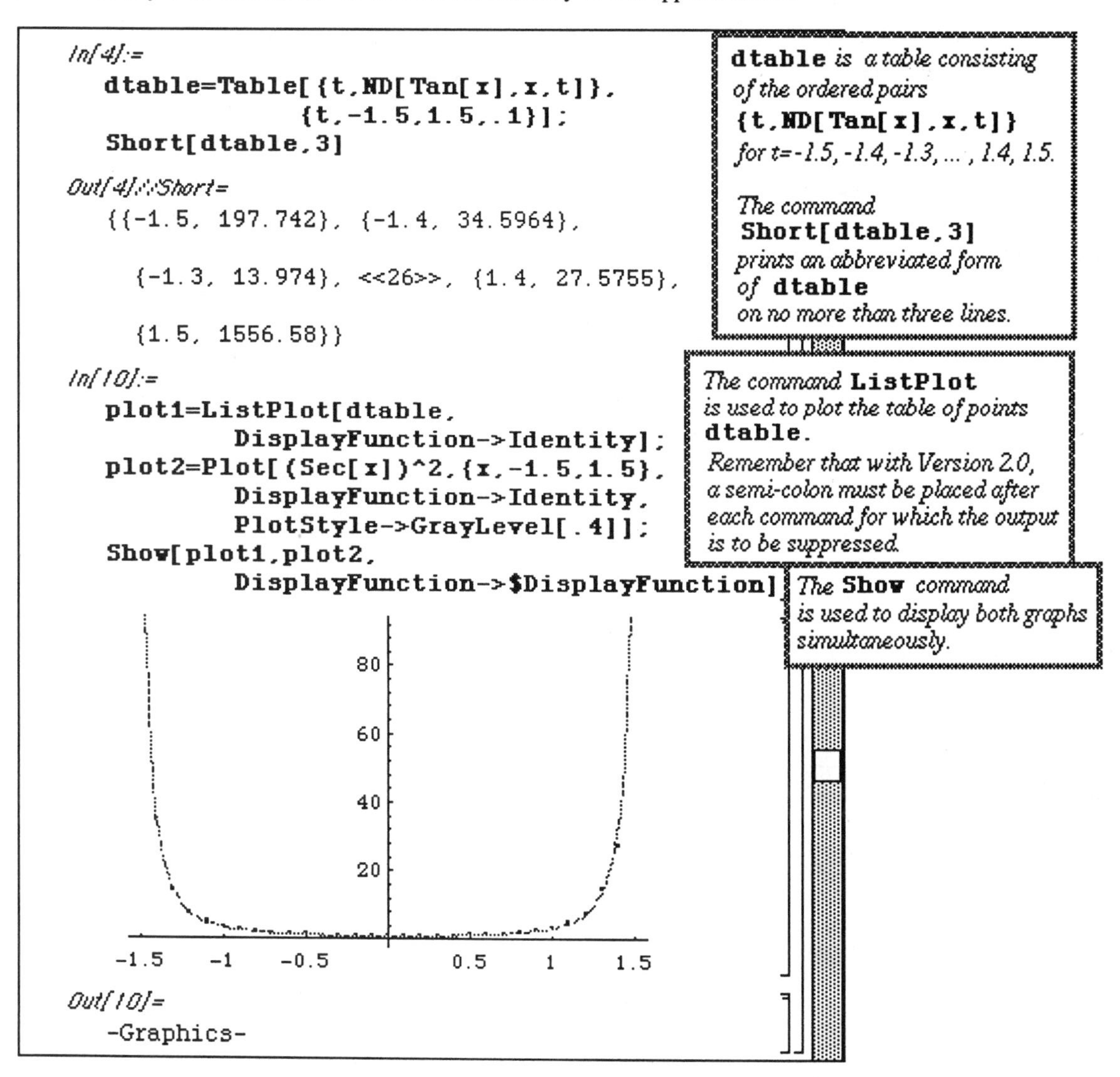

□ **Displaying Points with Versions 1.2 and 2.0:**

Since the points are difficult to see in the previous plot, the command **PointSize[n]** is used to increase the size of the points. In order to increase their size, the ordered-pairs must be specified as **points**. Given an ordered pair **{a,b}**, **Point[{a,b}]** specifies the graphics object which represents a point with coordinates **{a,b}**. The graphics object may then be displayed with **Show[Point[{a,b}]]**. Therefore, the command **Map[Point,dtable]** produces the list of points

```
{Point[{-1.5,197.742}],Point[{-1.4,34.5964}],...,
        Point[{1.5,1556.58}]}.
```

Given a list of points orderedpairs, **Show[Graphics[{PointSize[n],orderedpairs]]** displays orderedpairs according to the sixe given by **PointSize[n]**. Consequently, **plot3** represents the points created by **Map[Point,dtable]** displayed in size **.015**. The resulting graph is not displayed because the option **DisplayFunction->Identity** is included. Instead both **plot2** and **plot3** are displayed together by including the option **DisplayFunction->$DisplayFunction**.

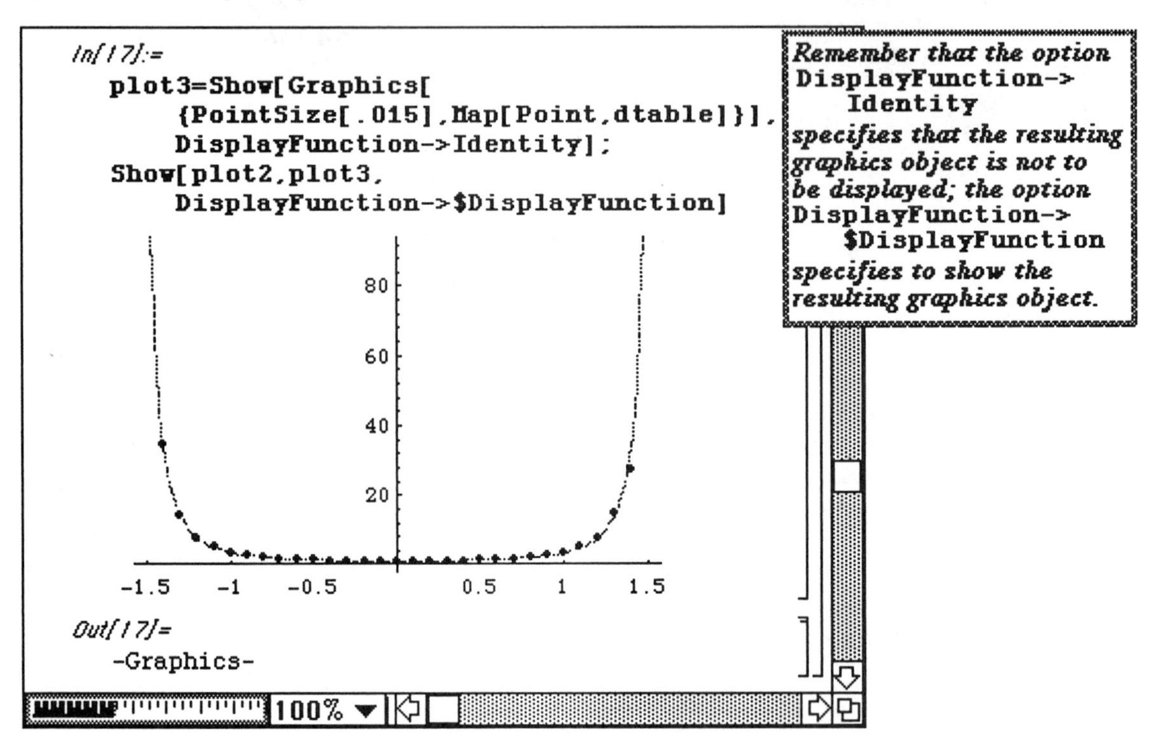

• 9.4 PolynomialFit.m

Version 2.0 includes the package **PolynomialFit.m** which offers the command **PolynomialFit[datalist,n]** that can be used to approximate a list of data in **datalist** with a polynomial of degree **n**. The evaluation of the approximating polynomial is based on Chebyshev polynomials and the formula for this polynomial is not given as output as it was with the built-in **Fit** command. Instead, the command **PolynomialFit[datalist,n]** yields the true function, FittingPolynomial[<>,n] which can be used to investigate the accuracy of the approximation. (Note that **datalist** is not a list of ordered pairs.)

A list of data is given below in **values**. After loading this package, this data is approximated with a polynomial of degree 2. This polynomial is named **approx1** and is found with **PolynomialFit[values,2]**. The true function **approx1** can be plotted as shown in **plot1**. (Note that square brackets must be used with **approx1** in the **Plot** command.) It can also be evaluated for any value of x as illustrated below with **approx1[4.5]**.

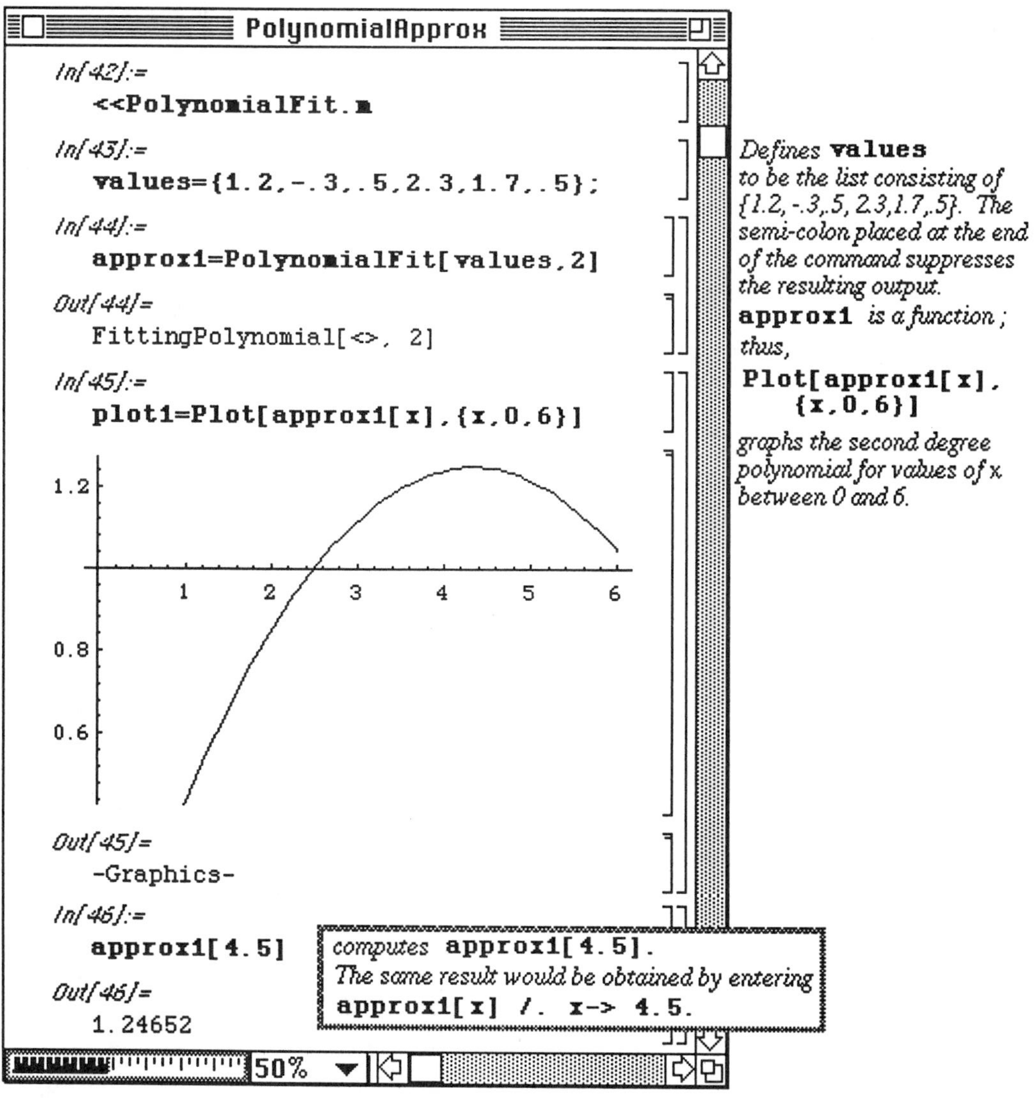

Defines **values** *to be the list consisting of {1.2,-.3,.5,2.3,1.7,.5}. The semi-colon placed at the end of the command suppresses the resulting output.*

approx1 *is a function; thus,*

Plot[approx1[x], {x,0,6}]

graphs the second degree polynomial for values of x between 0 and 6.

computes **approx1[4.5]**. *The same result would be obtained by entering* **approx1[x] /. x-> 4.5**.

In order to investigate how well the approximating polynomial **approx1** fits the data in **values**, the two are plotted simultaneously. Recall that the **ListPlot** option **DisplayFunction->Identity** causes the graph of the data points in **points** to be suppressed initially. Then, **DisplayFunction->$DisplayFunction** is used in the **Show** command to display **plot1** and **points**.

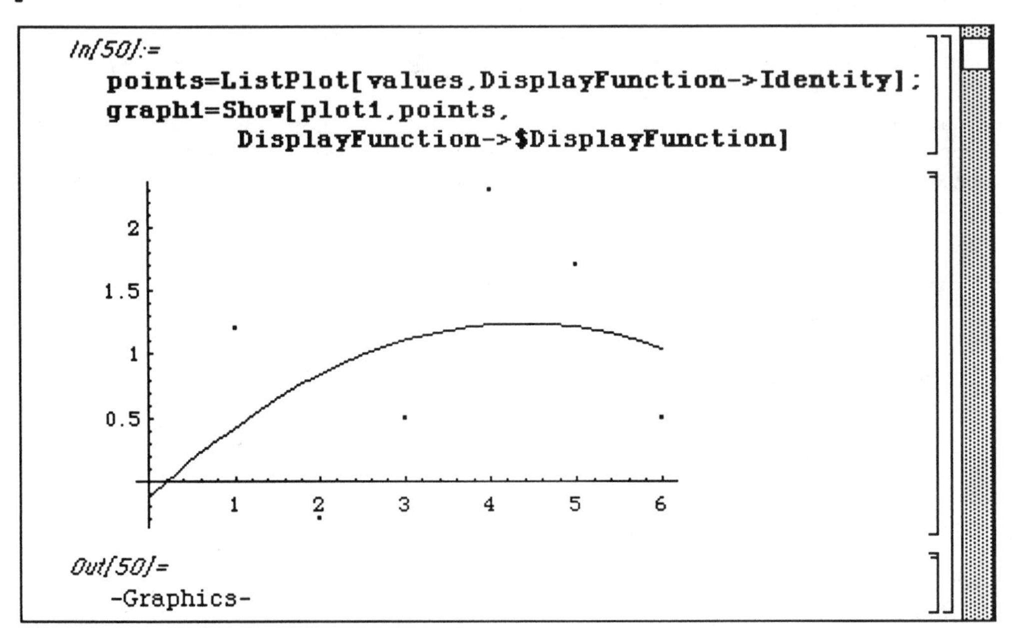

Since the second order polynomial determined above does not yield an accurate approximation, the polynomial of order 4 is computed below. This polynomial is called **approx2** and is found with **PolynomialFit[values,4]**. In a manner similar to that used above, the data and the approximating polynomial are plotted together.

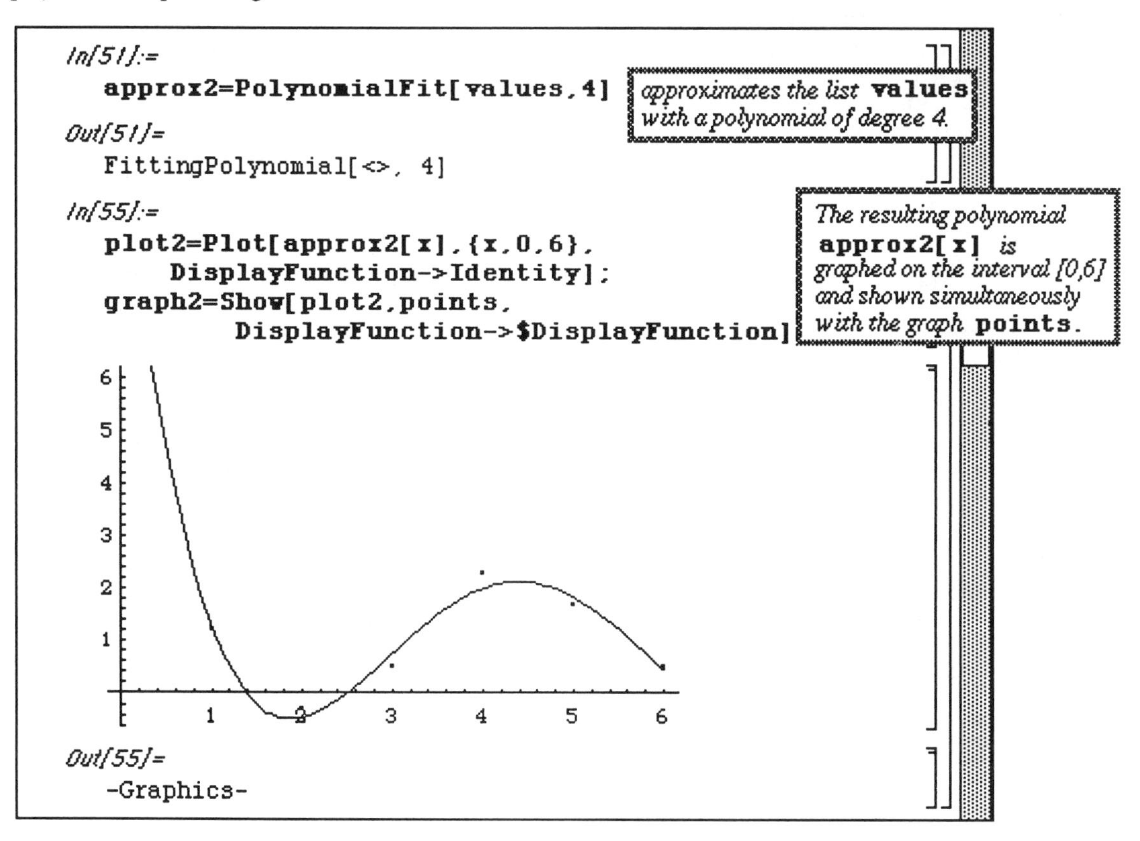

In an attempt to improve on the approximation, the sixth order polynomial, **approx3**, is found and plotted below. This appears to be an accurate approximation of the data after plotting all three simultaneously.

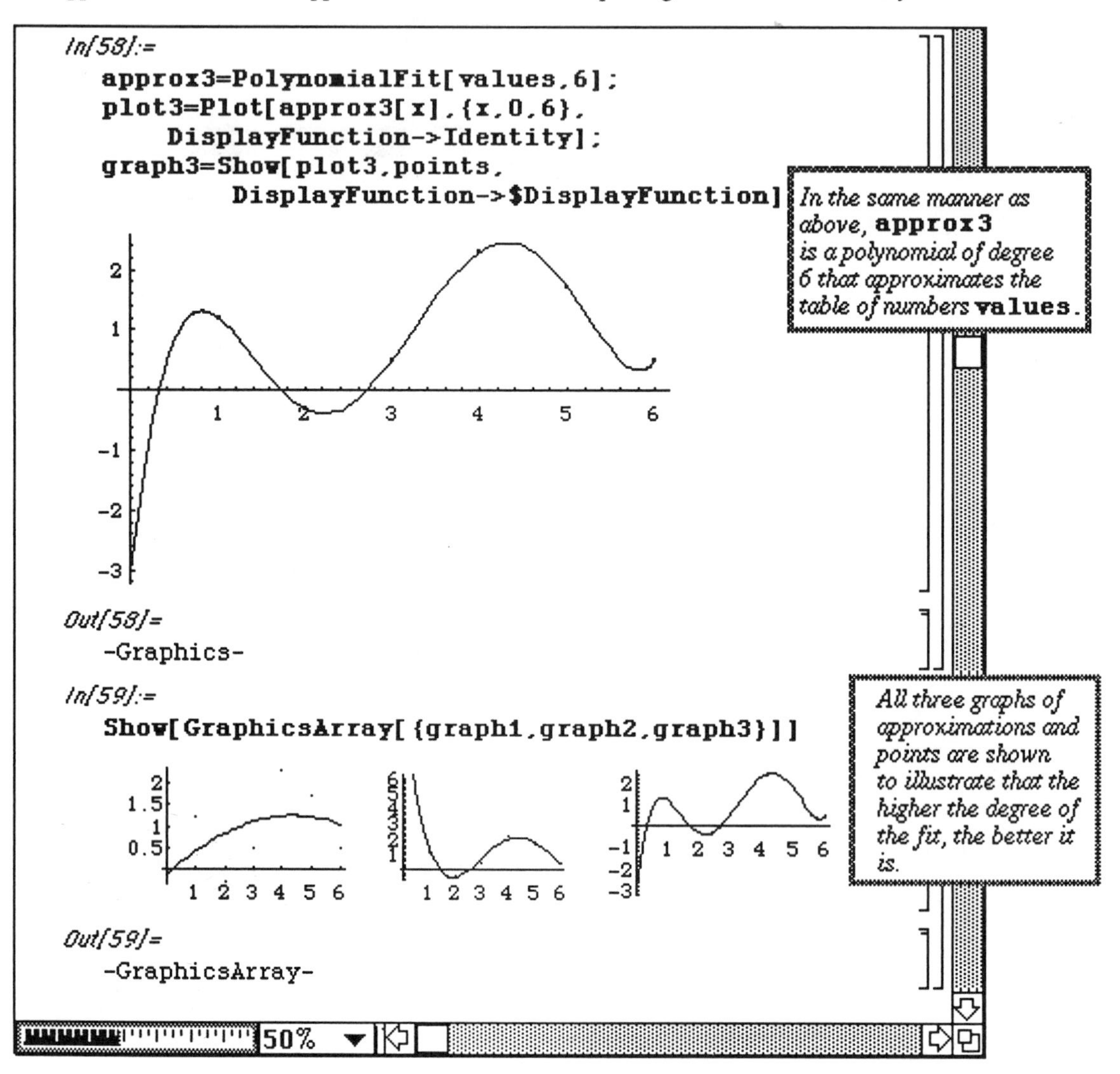

o Example:

The determination of an approximating polynomial has many applications. For instance, this method can be used on existing data to estimate future values such as population. Consider the following census data:

Year	Population (million)
1950	150.7
1960	179.3
1970	203.3
1980	226.5
1990	252.7

This data are entered as **data** below. Polynomials of order 2 and 3 are calculated and plotted below along with the data points. Both polynomials, **poly2** and **poly3**, appear to be accurate approximations. Since **ListPlot** is used to plot the data points, one unit along the x-axis represents 10 years.

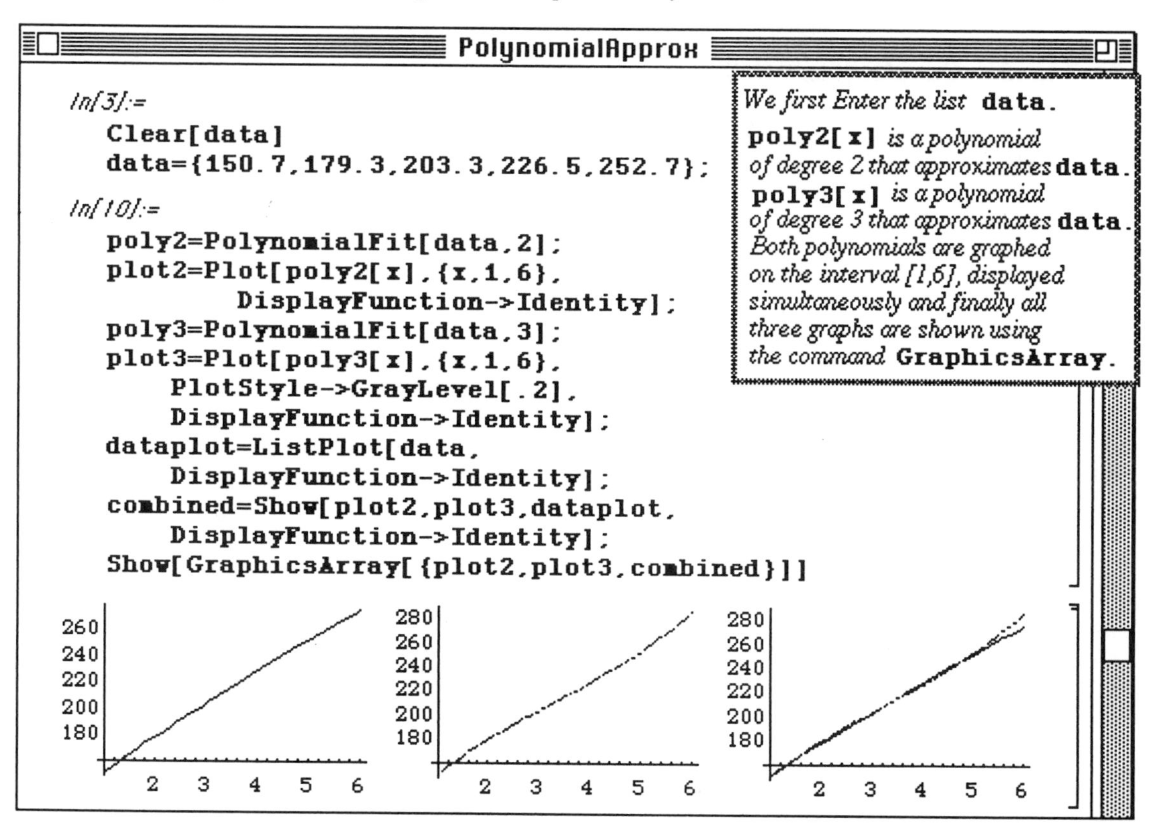

poly2[6] and **poly3[6]** yield the approximate population in the year 2000 based on each polynomial. These values are given below.

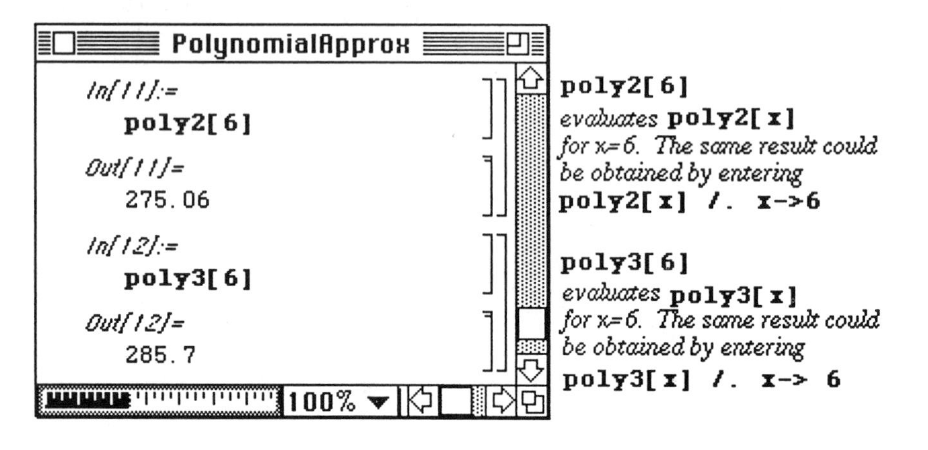

■ 9.5 RungeKutta.m

o In Version 1.2, **RungeKutta.m** is contained in the **Numerical Math** folder; in Version 2.0 **RungeKutta.m** is contained in the **Programming Examples** folder.

The Runge-Kutta method is used to numerically approximate the solution of the initial value problem

$y' = f(x,y)$

$y(x_0) = y_0$,

at $x_1 = x_0 + h$, $x_2 = x_1 + h$, ...

This method can be used with systems of equations as well as a single equation. The steps towards solving ordinary differential equations of this form by the Runge-Kutta method are given in the package **Runge-Kutta.m** in the Numerical Methods folder located in **Packages**. It should be noted that the package be used to solve those differential equations which are classified as nonstiff. Note the following definition of a stiff differential equation.

The differential equation $y' = f(x,y)$ is stiff if the ratio of the largest eigenvalue to the smallest eigenvalue of the matrix $\frac{\partial f}{\partial y}(x,y)$ is very large.

The syntax for the command is given as follows:

```
RungeKutta[ListofODEexpressions,ListOfVariables,
             ListOfInitialValues,
            Finalx,ErrorTolerance,(options...)]
```

where **ListofODEexpressions** is made up of the components of f(x,y), **ListOfVariables** is the list of both independent and dependent variables, **ListOfInitialValues** is the initial vector (including the initial value of the independent variable x), **Finalx** is the value of x at which a solution is sought, and **ErrorTolerance** is the local error tolerance which must be met at each integration. In giving **ListofODEexpressions** , **ListOfVariables** , and **ListOfInitialValues**, the variables must be entered in the same order in which the system of differential equations is stated with the independent variable listed first.

The **RungeKutta** options are given below along with their default values :

```
WorkingPrecision->Precision[N[1]],
MaximumStepSize->Finalx-Initialx,
InitialStepSize->MaximumStepSize,
ProgressTrace->False.
```

ProgressTrace->True causes the value of x, the step-size, and the local error to be given as the calculation is performed.

□ Example:

The following example illustrates how the package is used to solve the following initial value problem:

$$\begin{cases} y' = 2x + y \\ y(0) = 1 \end{cases}$$

Notice that the differential equation <u>must</u> be entered in the form of a list even though there is only one equation. **ErrorTolerance** is selected to be **10^(-5)** and the solution, y, is found at values of x between 0 and 1. Since a value of **False** is assumed by **ProgressTrace**, no progress report is given. The elements given in the output are of the form {x, y} where y is the approximate solution to the differential equation at x.

```
In[20]:=
  <<RungeKutta.m

In[21]:=
  sol1=RungeKutta[{2 x +y},{x,y},{0,1},1,10^(-5),
          WorkingPrecision->20, InitialStepSize->.1,
          MaximumStepSize->0.2, ProgressTrace->False]

Out[21]=
  {{0., 1.}, {0.1, 1.115512755}, {0.3, 1.4495764739790405334},

    {0.5, 1.9461639361434221276}, {0.7, 2.6412583489668909842},

    {0.9, 3.5788097018513151164}, {1., 4.154845894401927461}}
```

If **True** is used with **ProgressTrace**, the following output is given:

```
In[23]:=
   RungeKutta[{2 x +y},{x,y},{0,1},1,10^(-5),
       WorkingPrecision->20, InitialStepSize->.1,
       MaximumStepSize->0.2, ProgressTrace->True]

   RungeKutta::progress  x = 0., step = 0.1, local error =

                        -8
     2.08760499561 10
   RungeKutta::progress  x = 0.1, step = 0.2, local error =

                         -7
     5.452752959092 10
   RungeKutta::progress  x = 0.3, step = 0.2, local error =

                         -7
     4.960625430711 10
   RungeKutta::progress  x = 0.5, step = 0.2, local error =

                         -7
     4.464407969928 10
   RungeKutta::progress  x = 0.7, step = 0.2, local error =

                         -7
     4.024343595717 10
   RungeKutta::progress  x = 0.9, step = 0.1, local error =

                        -8
     1.37858326798 10
Out[23]=
   {{0., 1.}, {0.1, 1.115512755}, {0.3, 1.4495764739790405334},

     {0.5, 1.946163936143421276}, {0.7, 2.6412583489668909842},

     {0.9, 3.5788097018513151164}, {1., 4.154845894401927461}}
```

□ **Example:**

As stated earlier, the **RungeKutta** command can also be used to solve systems of first-order ordinary differential equations. Consider the following initial value problem :

$$\begin{cases} y' = z \\ z' = 2xz - 4y \\ y(0) = 1, z(0) = 0 \end{cases}$$

Here there are two equations which are entered as the list **{z,2xz-4y}** for **ListofODEexpressions** in the **RungeKutta** command. The variables must, therefore, be given as the list **{x,y,z}**. The initial conditions are given for y and z at x = 0, so **ListOfInitialValues** is the list **{0,1,0}** which represents the initial values for the variables **{x,y,z}**, respectively. As in the previous example, the local tolerance is taken to be **10^(-5)** and the solution is approximated for values of x between 0 and 1. Each element in the output consists of the value of x and the approximate values of y and z found with this procedure.

```
In[24]:=
  sol2=RungeKutta[ {z, 2 x z-4 y}, {x,y,z}, {0,1,0}, 1, 10^-5,
                   WorkingPrecision -> 20, InitialStepSize -> 0.1,
                   MaximumStepSize -> 0.2, ProgressTrace -> False ]

Out[24]=
  {{0., 1., 0.}, {0.1, 0.9799999817919399506, -0.4000000190929186766},

    {0.2652985946997730703, 0.8592329761222134417,

     -1.0611951756237893048}, {0.4353952520609771713,

     0.6208612134185668668, -1.7415828078592385648},

    {0.6258389127874684977, 0.2166499887699848187,

     -2.5033597842530430796}, {0.8258389127874684977,

     -0.3640218792002295708, -3.3033631282921385819},

    {1., -1.0000032370017912819, -4.0000094146543892189}}
```

If less precision is desired, a smaller value can be used for **WorkingPrecision**. The following command shows how this change affects the results obtained with the previous command. (A value of **10** is used below as compared to **20** in the previous example.)

```
In[25]:=
  RungeKutta[ {z, 2 x z-4 y}, {x,y,z}, {0,1,0}, 1, 10^-5,
              WorkingPrecision -> 10, InitialStepSize -> 0.1,
              MaximumStepSize -> 0.2, ProgressTrace -> False ]

Out[25]=
  {{0., 1., 0.}, {0.1, 0.98, -0.4}, {0.265299, 0.859233, -1.0612},

    {0.435395, 0.620861, -1.74158}, {0.625839, 0.21665, -2.50336},

    {0.825839, -0.364022, -3.30336}, {1., -1., -4.00001}}
```

Solutions found using **RungaKutta** can easily be plotted in two dimensions with the command **PlotODESolution[solutionlist,m,n,options]** where **m** and **n** are the components of the solution vector to be plotted.

In the previous problem, the solution was represented as the list {x,y,z}. Hence, a graph of the y and z values at each x can be generated with **PlotODESolution[%,2,3]**. The values of **2** and **3** in this command represent y and z, respectively. The solution list to the previous problem was conveniently named **sol2** for use with **PlotODESolution**.

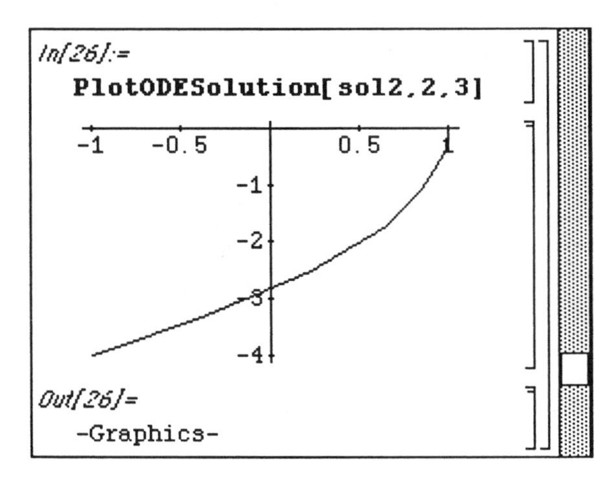

Solutions to a single ordinary differential equation can also be plotted with **PlotODESolution**. Since each element in the solution list has only two components, x and y, the command is entered in the following way to produce a graph of the approximate solution y.

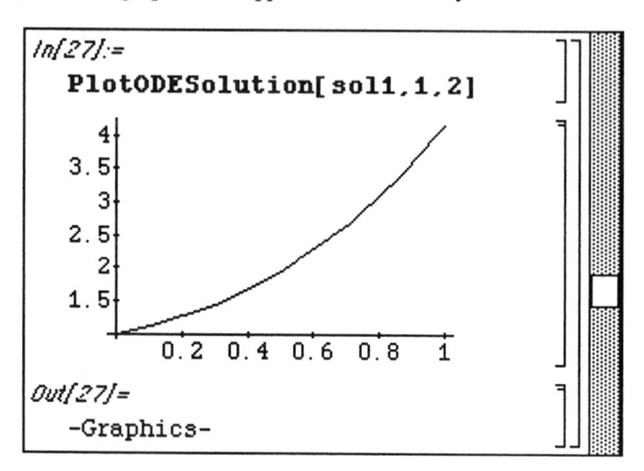

As indicated earlier, **ListPlot** options can be included in the **PlotODESolution** command. Several options are illustrated in the following command.

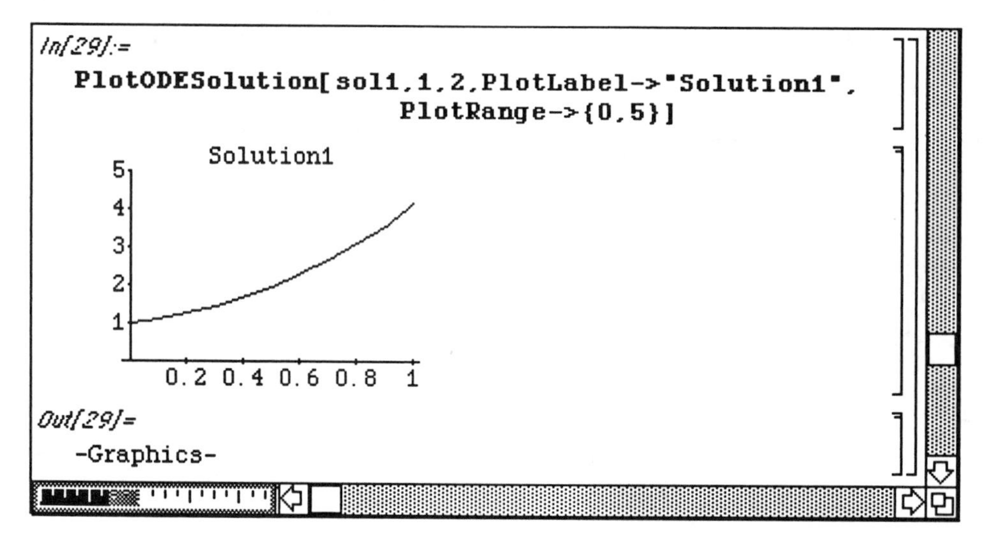

▩ Version 1.2 Data Analysis

o The Version 1.2 **Data Analysis** folder depicted below contains several packages useful in the area of statistics In Version 2.0, the **Data Analysis** folder is replaced by the **Statistics** folder.

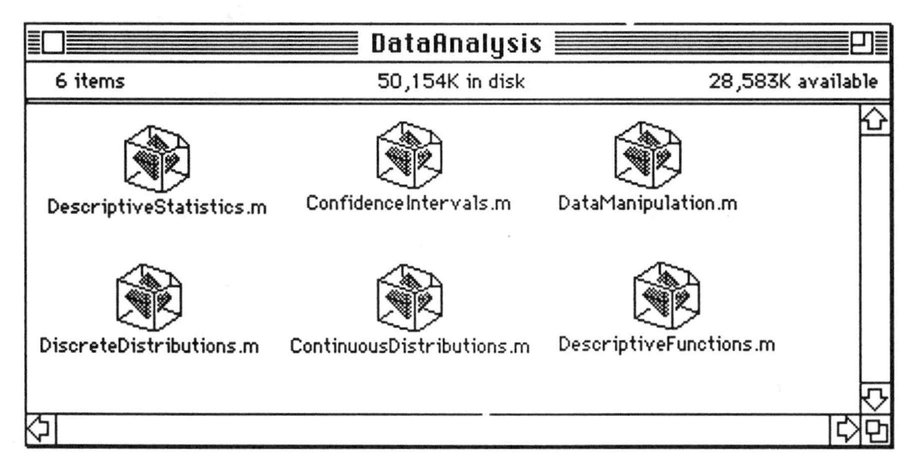

■ 9.6 Continuous Distributions.m and DescriptiveStatistics.m

The *Mathematica* packages **ContinuousDistributions.m** and **DescriptiveStatistics.m** contain many useful commands which can be used to solve problems in statistics.

o In Version 2.0, **ContinuousDistributions.m** and **DescriptiveStatistics.m** are contained in the Statistics folder.

These packages are first loaded below in order that several examples can be shown. **ContinuousDistributions.m** includes the following distributions: **BetaDistribution, CauchyDistribution, ChiDistribution, ChiSquareDistribution, ExponentialDistribution, ExtremeValueDistribution, FRatioDistribution, GammaDistribution, NormalDistribution, HalfNormalDistribution, LaplaceDistribution, LogNormalDistribution, LogisticDistribution, RayleighDistribution, StudentTDistribution, UniformDistribution,** and **WeibullDistribution.**

The **mean value** of a continuous distribution with probability density f is given by $\mu = \int_{-\infty}^{+\infty} x f(x) dx$.

The **variance** of a distribution is given by $\sigma^2 = \int_{-\infty}^{+\infty} (x-\mu)^2 f(x) dx$; the **standard deviation** is given by $\sigma = +\sqrt{\sigma^2} = +\sqrt{\int_{-\infty}^{+\infty} (x-\mu)^2 f(x) dx}$.

The **DescriptiveStatistics.m** package includes several functions which can be applied to these distributions. For example, **Density[Distribution,t]** gives the density function which corresponds to **Distribution**. **Mean[Distribution]** gives the mean of **Distribution** and **Variance[Distribution]** gives the variance of **Distribution.** These values are commonly known for the Normal distribution and are given below using these newly introduced commands .

□ **Example:**

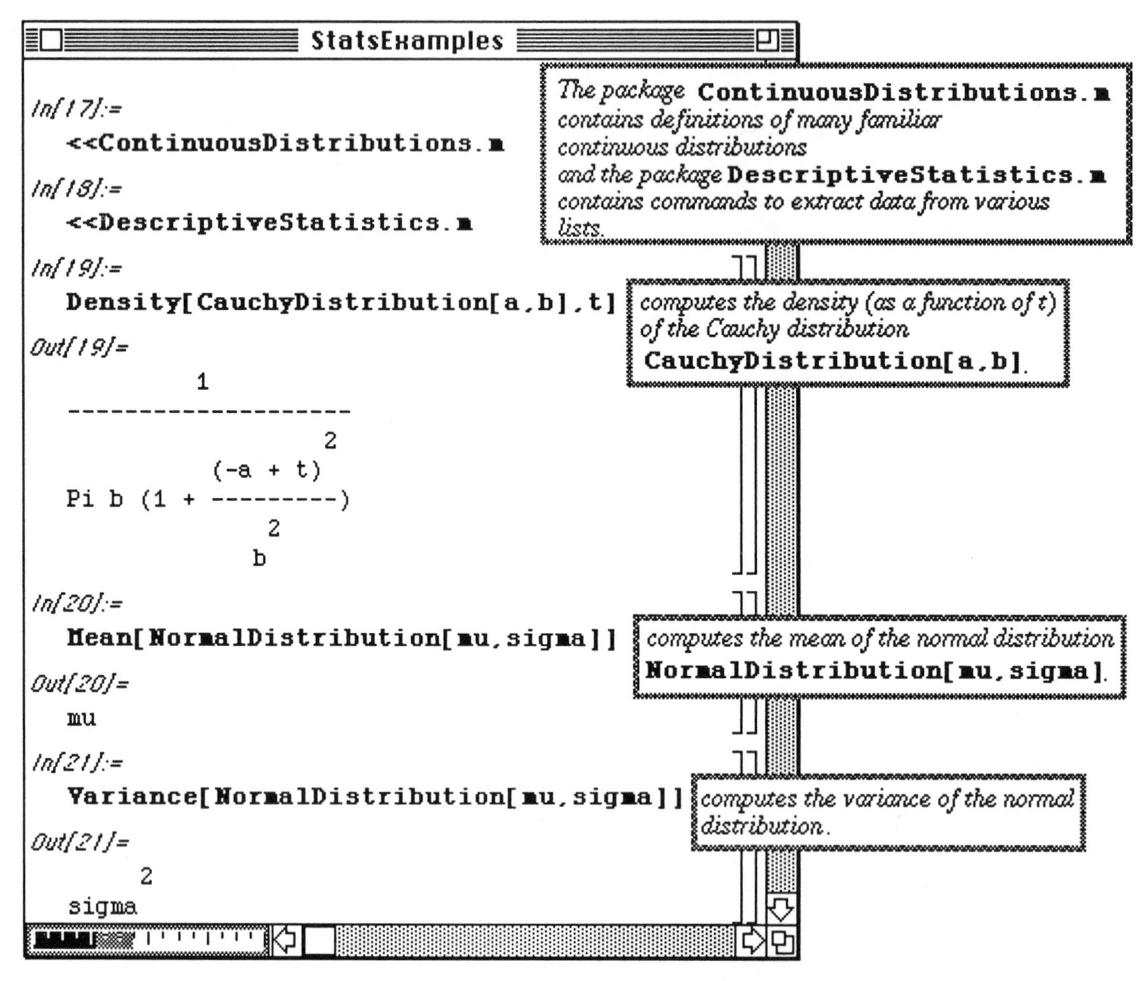

The following example illustrates how a set of data is analyzed with *Mathematica*. The command **`Table[Random[NormalDistribution[mu,sigma]],{n}]`** generates a random list of **`n`** numbers which approximates a normal distribution of mean **`mu`** and standard deviation **`sigma`**.

□ **Example:**

In this example, a list of **`175`** numbers which approximate the normal distribution of mean **`75`** and standard deviation **`10`** is produced and stored in **`table1`**. The values in **`table1`** are rounded to integers with **`Map[Floor,table1]`** where **`Floor[x]`** gives the greatest integer not larger than **`x`**. **`Map`** applies **`Floor`** to each value in **`table1`**. This new table is called **`table2`**. The integer values in **`table2`** are then sorted with **`Sort[table2]`** and named **`table3`**. A shortened list of these sorted integers is given with **`Short[table3,3]`**.

Since the command **`Random`** is used, each time these commands are executed, different results are obtained.

The mean of **table3** is computed by adding the integers in **table3** and dividing by the number of integers found in **table3**. This is accomplished in the single command **Apply[Plus,table3]/Length[table3]//N.**
Of course, the mean can also be computed with **Mean[table3]** which is illustrated below.

StatsExamples

In[22]:=
```
table1=Table[Random[
    NormalDistribution[75,10]],{175}];
table2=Map[Floor,table1]
table3=Sort[table2]
Short[table3,3]
```
Out[22]//Short=
```
{49, 49, 53, 54, 54, 56, 57, 58, 59, 60,

  60, 60, 61, 61, 61, <<153>>, 88, 89, 90,

  91, 93, 93, 97}
```
In[23]:=
```
Apply[Plus,table3]/
    Length[table3] // N
```
computes a numerical value of the mean of **table3.**

Out[23]=
```
73.6629
```
In[24]:=
```
Mean[table3]
```
computes an exact value of the mean of **table3.**

Out[24]=
```
12891
-----
 175
```

table1 *is a table of 175 "random" numbers that approximate a normal distribution of mean 75 and variance 100.*
Map[Floor,table1] *computes* **Floor[table1[[i]]]** *for each i= 1, 2, ... , 175 and names the resulting list* **table2.**
Sort[table2] *sorts* **table2** *according to the standard increasing order and names the resulting list* **table3.**
Short[table3,3] *displays an abbreviated form of* **table3** *on no more than three lines.*

In order to compare the mean of **table3** obtained by the two methods, a numerical approximation of **Mean[table3]** is requested to show that the same result is achieved by each approach. **Median[table3]** determines the median of **table3** and **Quartiles[table3]//N** determines the numerical approximation of the quartiles of **table3**. **MeanDeviation[table3]//N** gives a numerical value of the mean deviation and **N[MedianDeviation[table3]]** gives a numerical value for the median deviation of **table3**.

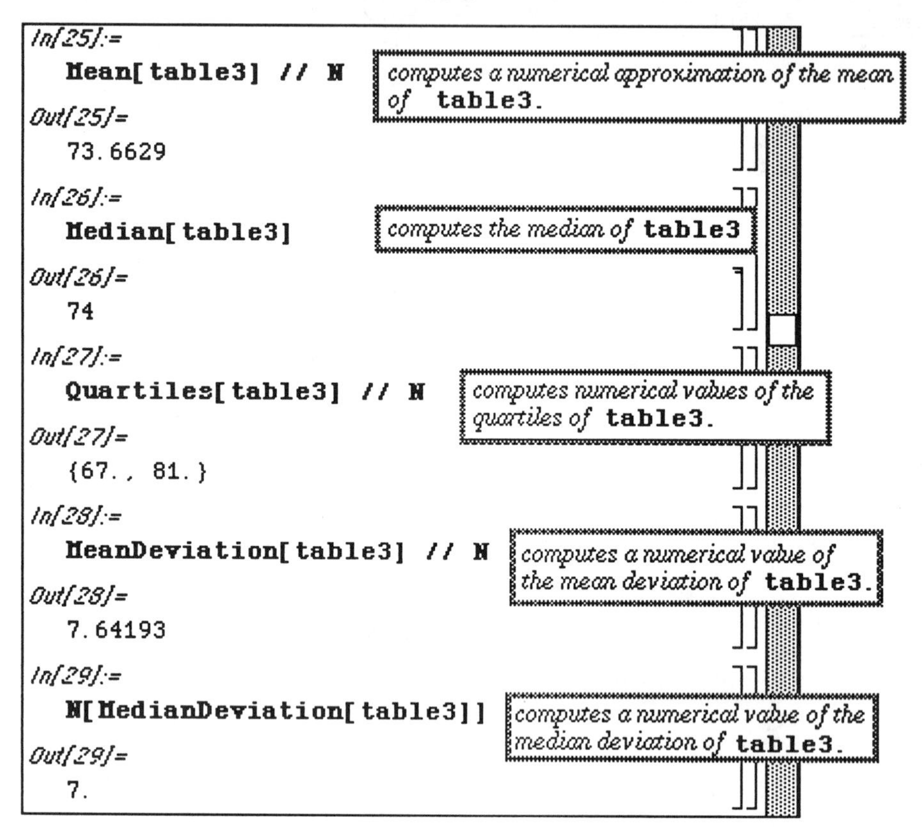

A summary of some of the information concerning **table3** is obtained with the command **DispersionReport[table3]//N**. These numerical values are shown below:

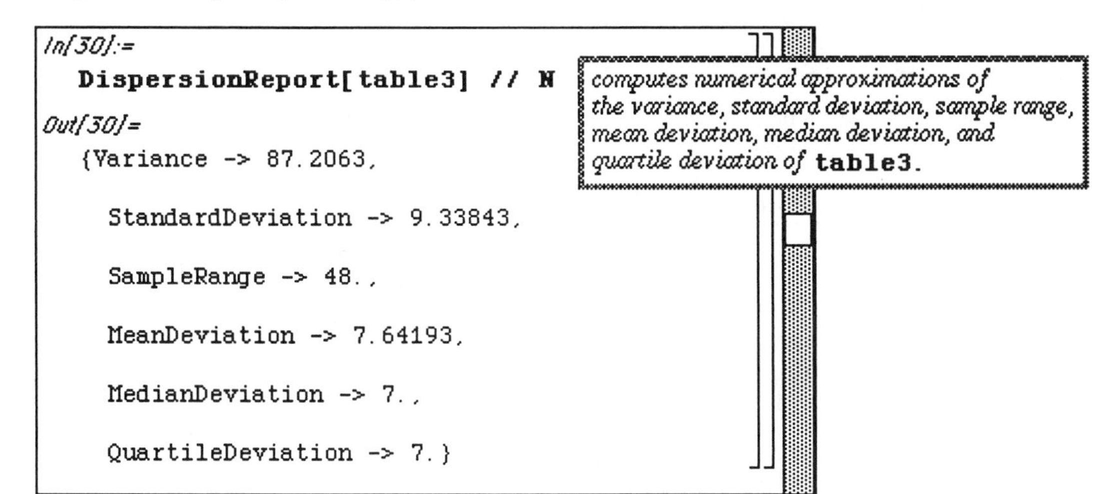

```
In[30]:=
  DispersionReport[table3] // N
Out[30]=
  {Variance -> 87.2063,
    StandardDeviation -> 9.33843,
    SampleRange -> 48.,
    MeanDeviation -> 7.64193,
    MedianDeviation -> 7.,
    QuartileDeviation -> 7.}
```

computes numerical approximations of the variance, standard deviation, sample range, mean deviation, median deviation, and quartile deviation of **table3**.

□ Example:

In the following example, the function **countlist[table]** is defined. This function counts the number of times each distinct member of the list **table** appears in **table**. First the variables **j** and **list** are defined to be local to the function **countlist**. Second, **table** is sorted with **Sort[table]** and this sorted list is named **list**. Then, **Union[list]** removes all duplicates from **list** to obtain a new list, **list2**. Finally, **Count[list,list2[[j]]]** gives the number of elements in **list** that match each element in the list of distinct elements, **list2[[j]]**. In other words, the number of times each element in **list** appears is given. This number is divided by **Length[list]** (the total number of elements in **list**) to yield the portion of list that each element comprises. This function is illustrated below with **table3** given earlier. The output lists the distinct elements of **table3** along with a numerical value which indicates the portion of **table3** that each element constitutes. Hence, if {number,p} is an element of **table3**, then the probability of choosing number from the list of numbers in **table3** is p.

Also defined below is the function **between[list, {a,b}]** which counts the number of elements of **list** which fall on the interval from **a** to **b**, including the endpoints. This is done with the command **Sum[Count[list,j],{j,a,b}]** where **Count[list,j]** gives the number of elements of **list** which match **j** (i.e., the number of times **j** appears in **list**). This is done for each value of **j** from **j** = **a** to **j** = **b** and then the sum of these numbers is taken to yield the total number of values between **a** and **b**.

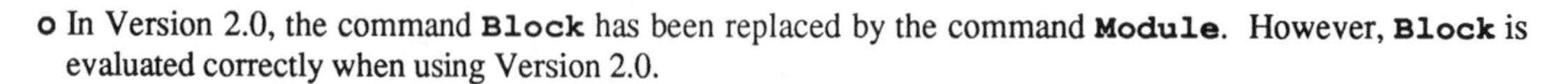

o In Version 2.0, the command **Block** has been replaced by the command **Module**. However, **Block** is evaluated correctly when using Version 2.0.

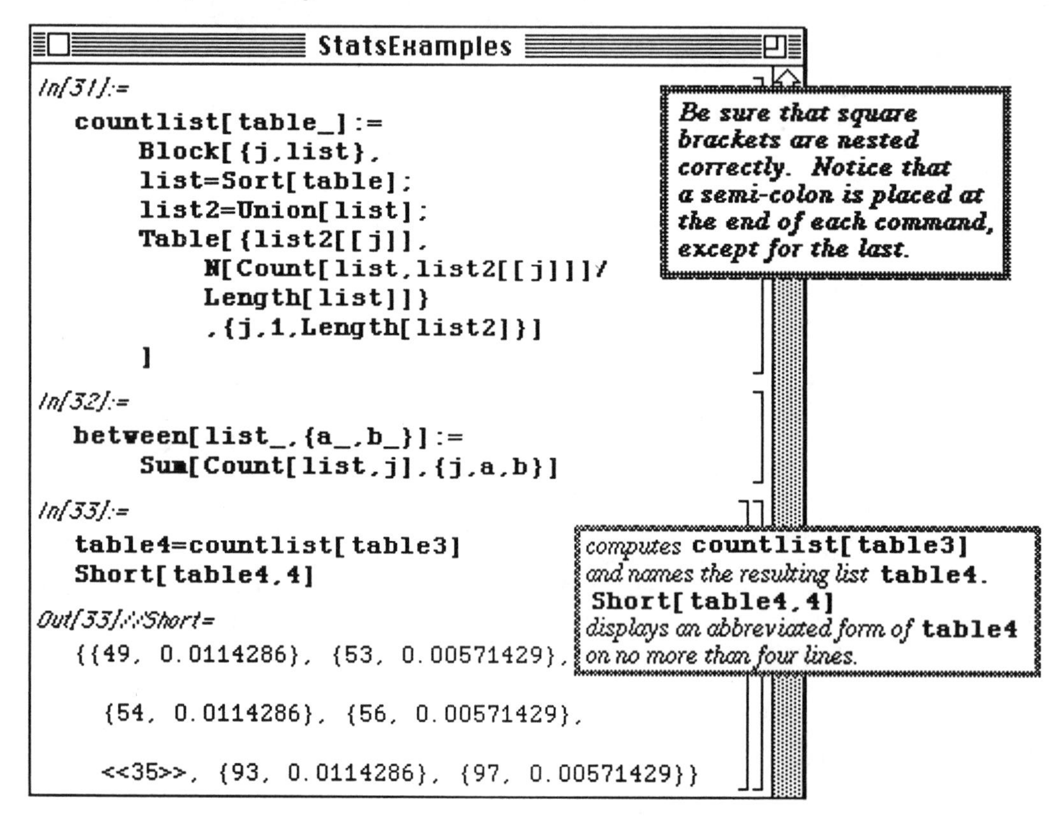

The sum of the numerical values given in **table4** should equal 1. This is verified below with **Apply[Plus,table4][[2]]**. The command **Apply[Plus,table4]** adds the corresponding components of the members of **table4** and is, therefore, of the form of each element of **table4**. Hence, the sum of the second components is given with **Apply[Plus,table4][[2]]**.

A table is then compiled which gives an interval breakdown of the values found in **table3**. This is accomplished by choosing several intervals, {45,55}, {55,65},...,{95,105}, and using the function **between**. The output is given in the form of a table as shown below.

In[34]:=

```
Apply[Plus,table4][[2]]
```

Out[34]=

```
1.
```

In[35]:=

```
Table[{45+10 i,55+10i,
    between[table3,{45+10 i,55+10i}]},
    {i,0,5}]//TableForm
```

Out[35]//TableForm=

```
45   55   5
55   65   33
65   75   70
75   85   61
85   95   23
95   105  1
```

Hence, there are five elements of **table3** *between 45 and 55; 33 elements between 55 and 65; 70 elements between 65 and 75; 61 elements between 75 and 85; 23 elements between 85 and 95; and 1 element between 95 and 105.*

The elements of **table4** can be graphed along with the normal distribution with mean 75 and variance 10 to show how well these values approximate the density function of this distribution. Since each member of **table4** is an ordered pair, the elements in **table4** are plotted with **ListPlot** below. However, the graph is not shown initially, because the option **DisplayFunction->Identity** is used. The density function of the normal distribution is plotted for values of t between 49 and 97. This plot is named **plot1**.
The second graph below shows **plot** and **plot1** simultaneously. The option **DisplayFunction->$DisplayFunction** allows **plot** to be displayed. The **Show** command also specifies that the axes meet at the point {47,0}.

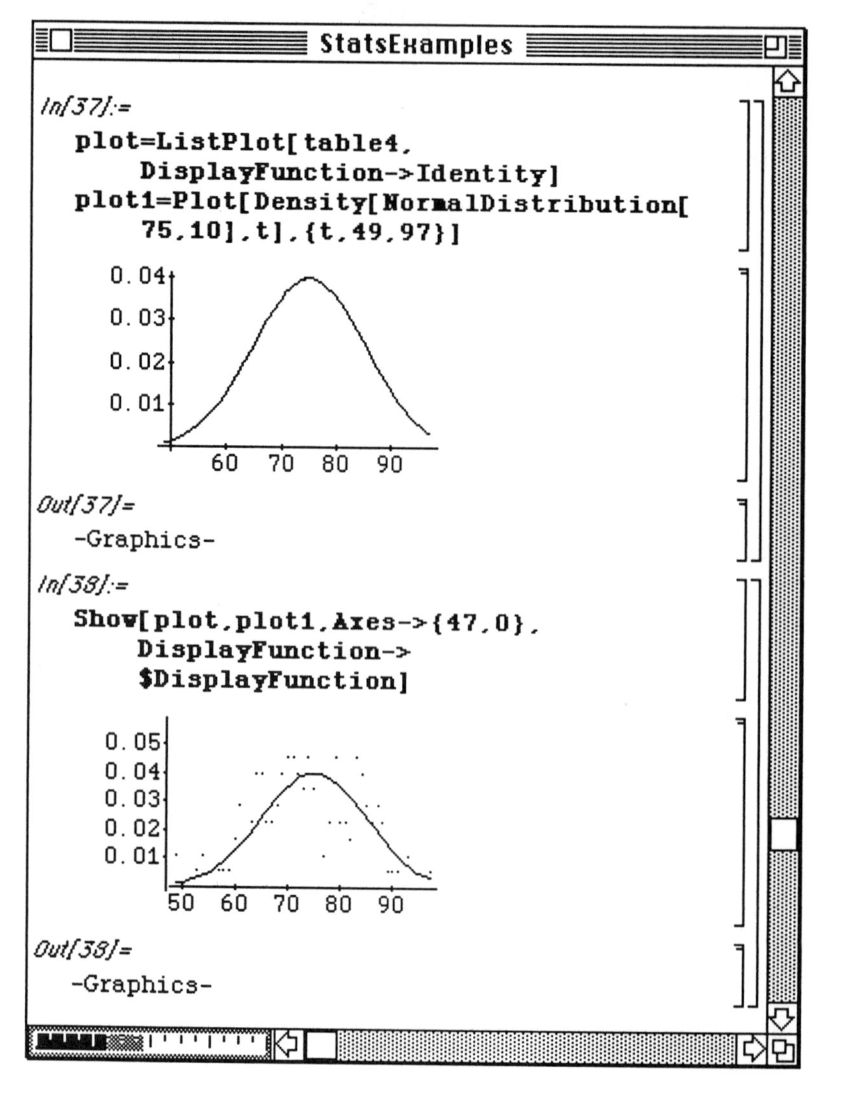

● Version 2.0 Statistics

In Version 2.0, the **`Data Analysis`** folder is replaced by the following **`Statistics`** folder. Some of the added packages are discussed below.

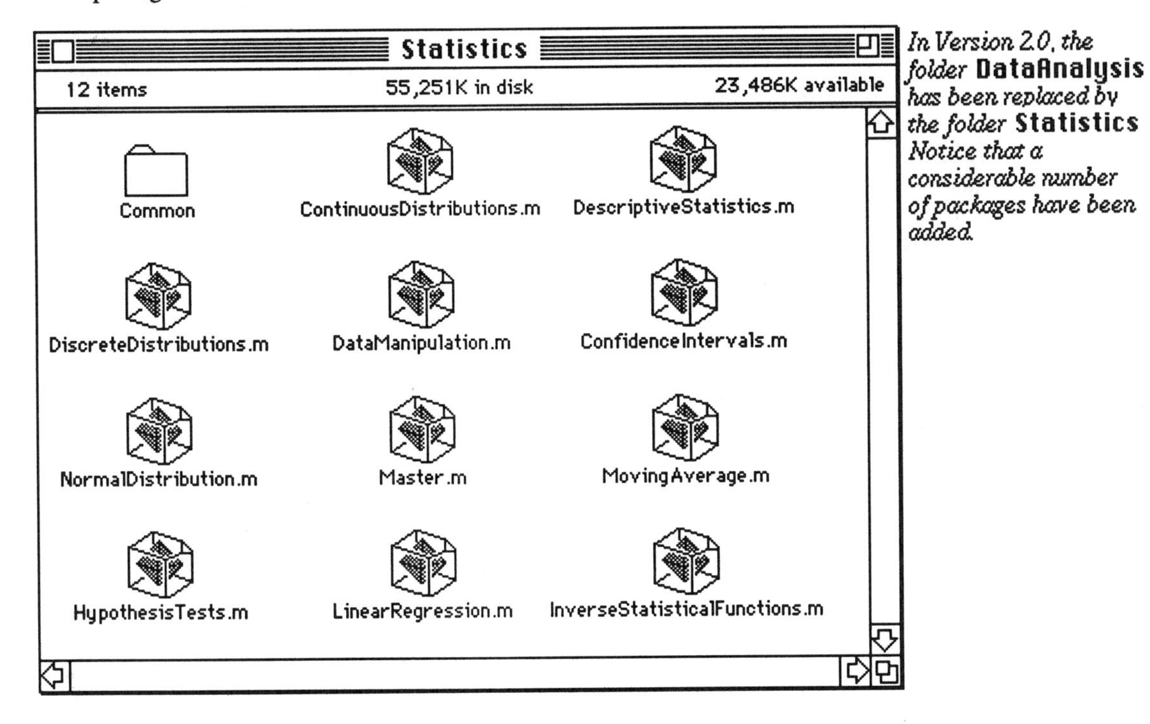

In Version 2.0, the folder **DataAnalysis** *has been replaced by the folder* **Statistics** *Notice that a considerable number of packages have been added.*

• 9.7 HypothesisTests.m

The Version 2.0 **HypothesisTests.m** package contains useful hypothesis test commands for solving problems in statistics. The first command shown below is **MeanTest[list,mu0,options]** which is used to test the null hypothesis that the population mean equals **mu0**. Each command in this package yields the appropriate P-value. Recall that the P-value of a hypothesis test is equal to the smallest significance level at which the null hypothesis can be rejected. Hence, if the P-value is less than or equal to the specified significance level, then the null hypothesis is rejected. Otherwise, the null hypothesis is not rejected.

o Example:

Consider the data collected by the U.S. Energy Information Administration on residential energy expenditures. According to this agency, the mean residential energy expenditure of all American families was $1123 in 1985. The expenditures of 15 upper-level families is given in **energy**. The hypothesis that the population mean equals $1123 is tested with this data (with **mu0** = 1123). The command **MeanTest[energy, 1123]** gives the one-sided P-value. Assuming a significance level of .05, the P-value obtained is compared to .5(.05) = .025. Since .0014959 < .025, the null hypothesis is rejected.

A helpful option of all of the tests to be discussed in this section is that of **FullReport->True**. This option is illustrated with the same problem to reveal the sample mean, the test statistic, the number of degrees of freedom (when applicable), and type of test statistic used. These commands use the test statistics based on the normal distribution, the Student's t-distribution, the chi-square distribution, or the F-distribution. The `FullReport` shows that in this case, the Student's t-distribution was employed.

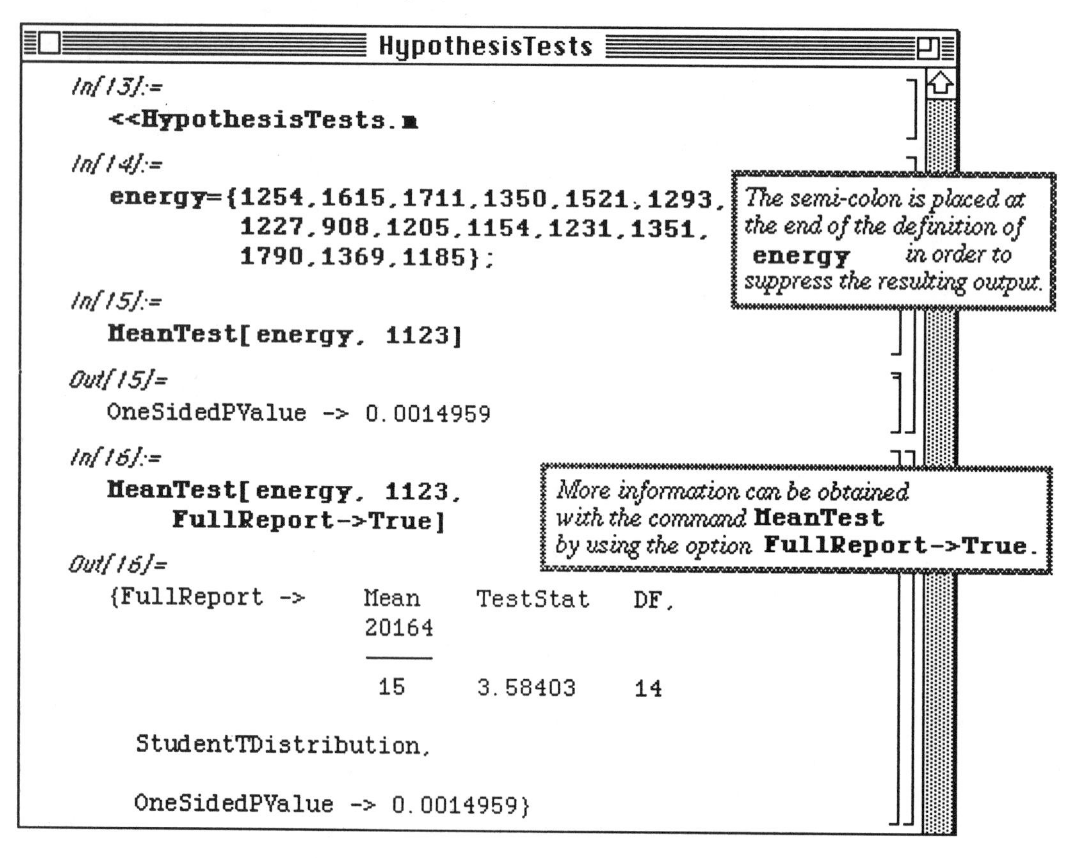

Another option is **KnownVariance->var0**. When this option is used, the normal test statistic is used to obtain the P-value as expected. This is also shown below.

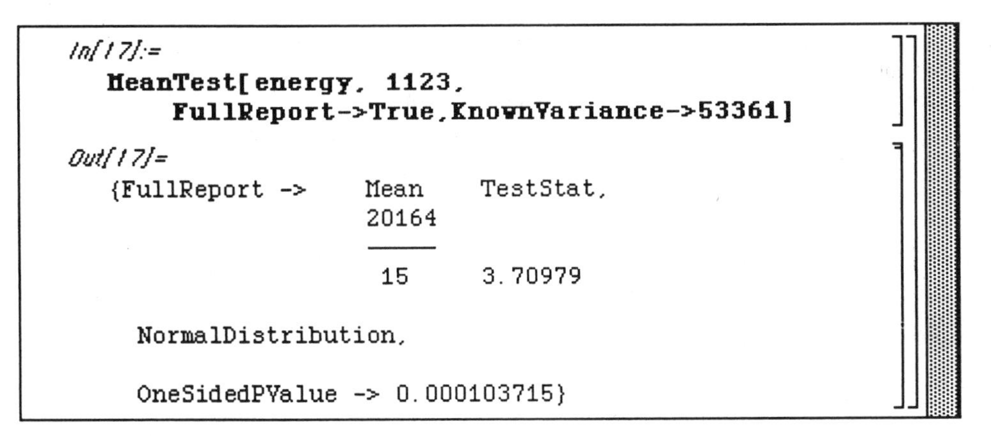

```
In[17]:=
  MeanTest[energy, 1123,
      FullReport->True,KnownVariance->53361]

Out[17]=
  {FullReport ->    Mean     TestStat,
                    20164
                    -----
                     15      3.70979

    NormalDistribution,

    OneSidedPValue -> 0.000103715}
```

o Example:

Consider another set of data which gives the daily intake of calcium (in milligrams) for 35 people with an income below the poverty level. This is given in **calcium** below.

```
calcium={879,1096,701,986,828,1077,703,
         555,422,997,473,702,508,530,
         513,720,944,673,574,707,864,
         1199,743,1325,655,1043,599,1008,
         705,180,287,542,893,1052,473};
```

A sample list indicating the daily intake of calcium for 35 people with income below the poverty level.

The recommended daily allowance (RDA) is 800 milligrams. A nutritionist states that the average person with an income below the poverty level gets less than the RDA of 800 mg. This claim is tested below with **MeanTest**. This is a one-sided test. Suppose that the significance level is again .05. Since the P-value which results (.124765) is greater than .05, the null hypothesis (that the calcium intake of people of poverty level incomes is equal to 800 mg) is not rejected. Another option available to **MeanTest** is **KnownStandardDeviation->sigma**. Suppose that the standard deviation is assumed to be 262. When the sample standard deviation equals the known standard deviation, *Mathematica* uses the normal distribution with **MeanTest**. This is indicated in the `FullReport`.

```
In[19]:=
  MeanTest[calcium,800,
              FullReport->True]
Out[19]=
  {FullReport ->    Mean      TestStat   DF,
                    26156
                    -----
                     35       -1.17153   34

    StudentTDistribution, OneSidedPValue -> 0.124765}
In[20]:=
  MeanTest[calcium,800,
              FullReport->True,
              KnownStandardDeviation -> 262]
Out[20]=
  {FullReport ->    Mean      TestStat,
                    26156
                    -----
                     35       -1.18967

    NormalDistribution, OneSidedPValue -> 0.117089}
```

Mathematica can also be used to conduct hypothesis tests for the means of two normal populations with equal standard deviations using independent samples. This is accomplished with the command **MeanDifferenceTest[list1,list2,diff0,options]** where the two populations are given in **list1** and **list2**. The value **diff0** is compared to **Mean[list1]-Mean[list2]**.

o Example:

The data given in **below** represents the daily protein intake in grams for 10 people with incomes below the poverty level while **above** represents that of 15 people with incomes above the poverty level. The null hypothesis is that the below-poverty mean is not less than the above-poverty mean. Hence, the alternative hypothesis is that the below-poverty mean is less than the above-poverty mean. If the means of these two populations are equal, the difference of the respective means equals zero. Hence, **diff0 = 0** in the command below. The P-value given is .00934066 which is less than a significance level of .05. Thus, the null hypothesis is rejected. In the `FullReport` which follows, the test statistic used in the determination of this P-value is given.

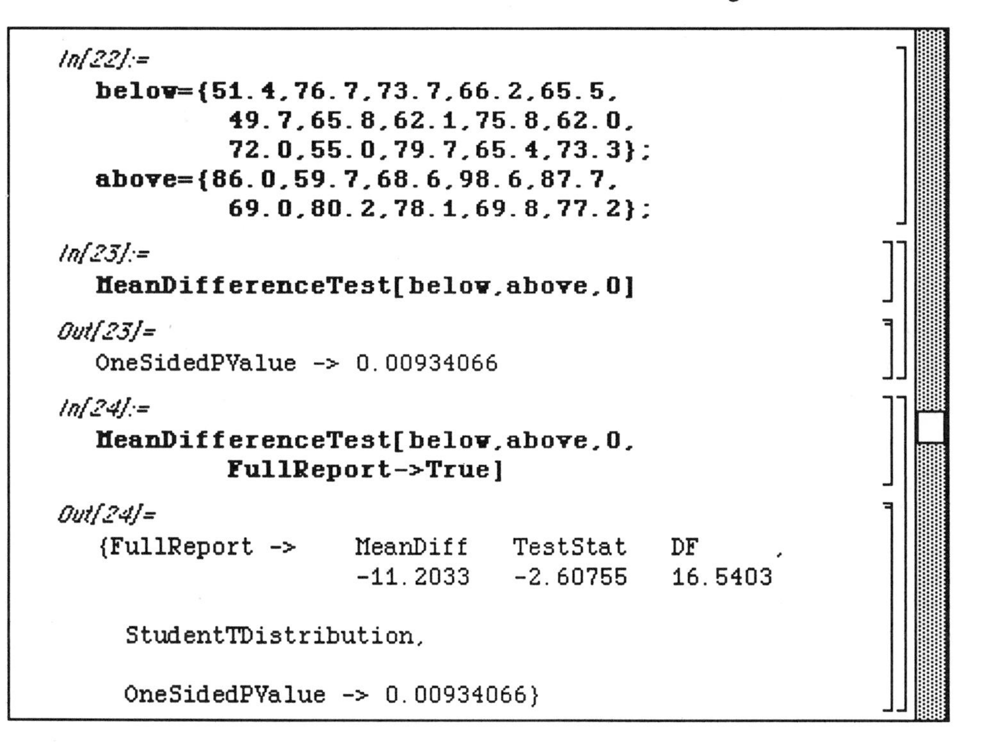

```
In[22]:=
  below={51.4,76.7,73.7,66.2,65.5,
           49.7,65.8,62.1,75.8,62.0,
           72.0,55.0,79.7,65.4,73.3};
  above={86.0,59.7,68.6,98.6,87.7,
           69.0,80.2,78.1,69.8,77.2};

In[23]:=
  MeanDifferenceTest[below,above,0]

Out[23]=
  OneSidedPValue -> 0.00934066

In[24]:=
  MeanDifferenceTest[below,above,0,
           FullReport->True]

Out[24]=
  {FullReport ->    MeanDiff   TestStat   DF       ,
                    -11.2033   -2.60755   16.5403

    StudentTDistribution,

    OneSidedPValue -> 0.00934066}
```

o Example:

Consider the data below which is used to compare the lifetimes of two brands of water heaters. The lifetimes (in years) of one brand is given in the list **heater1** while that of the other brand is given in **heater2**. In this problem, the null hypothesis is that the mean of the lifetimes of **heater1** equals that of **heater2**. Hence, this is a two-sided test. Assume a significance level of .05. The P-value for this test is determined from the Student's t-distribution. Since this value is greater than .025, the null hypothesis is not rejected. Hence, the data do not provide sufficient information to conclude that the two brands of water heaters have different mean lifetimes.

```
In[26]:=
  heater1={6.9,7.2,7.6,7.3,6.6,5.7,
           7.8,6.2,5.5,7.4,8.2,6.9};
  heater2={8.7,8.6,11.2,7.0,6.1,6.1,
           8.7,7.5,6.3,6.7,7.7,7.0,
           7.8,7.5,10.7};

In[27]:=
  MeanDifferenceTest[heater1,heater2,0,
                FullReport->True]

Out[27]=
  {FullReport ->    MeanDiff     TestStat   DF        ,
                    -0.898333    -1.94567   22.2285

    StudentTDistribution,

    OneSidedPValue -> 0.0322241}
```

In addition to hypothesis tests on the mean of a population as was illustrated with **MeanTest**, *Mathematica* is able to consider hypothesis tests for a population standard deviation (or variance). This is done with **VarianceTest[list,var0,options]**. In problems of this type, the null hypothesis is that the variation of the population in **list** equals **var0**. This command uses the same options as the previously used commands.

o Example:

An analysis is conducted on the data in **diameter** which gives the diameter in millimeters of 20 bolts produced by a hardware manufacturer. It has been determined that an acceptable standard deviation for bolt diameters is .09 millimeters. Therefore, **var0 = (.09)^2** in the **VarianceTest** command which follows. The `FullReport` indicates that the chi-square distribution is used as is expected. If a significance level of .05 is assumed, the null hypothesis is rejected. Hence, the bolts produced by the manufacturer have diameter less than .09 millimeters.

```
In[17]:=
  diameter={10.03,10.08,10.05,10.03,
            9.89,9.95,9.97,9.99,
            9.99,10.00,10.03,10.08,
            9.96,9.94,9.98,10.02,
            10.10,10.01,10.05,9.98};

In[18]:=
  VarianceTest[diameter, (.09)^2,
                   FullReport->True]

Out[18]=
  {FullReport ->    Variance      TestStat   DF,
                    0.00272921    6.73879    19

    ChiSquare Distribution, OneSidedPValue -> 0.00451858}
```

• 9.8 ConfidenceIntervals.m

o In Version 2.0, **ConfidenceIntervals.m** is contained in **Statistics**; in Version 1.2 **ConfidenceIntervals.m** is contained in **Data Analysis**.

An important concept in statistics is that of confidence intervals. For the two-sided hypothesis test,

$$H_0 : \mu = \mu_0$$
$$H_a : \mu \neq \mu_0$$

at a significance level α, the null hypothesis is not rejected if μ_0 lies in the $(1 - \alpha)$ - level confidence interval for μ and, the null hypothesis is rejected if μ_0 does not lie in the $(1 - \alpha)$ - level confidence interval for μ.

These confidence intervals can be determined for the normal, Student's t, chi-square, and F distributions for several types of hypothesis tests using commands found in **ConfidenceIntervals.m**.

□ **Example:**

In the first example below, the data used in the **HypothesisTests.m** section concerning the residential energy expenditures of 15 upper level families located in **energy** is considered. In the previous section, the null hypothesis that the mean expenditure equals $1123 was tested. Since this is a test concerning the mean, the command **MeanCI[list,options]** is used. Therefore, the confidence interval for this data is determined with **MeanCI[energy]//N**. (**//N** requests that numerical values be given as opposed to exact.) Since 1123 does not lie in this interval, the null hypothesis is rejected as it was in the previous section. One of the options available is that of **ConfidenceLevel->alpha** with default value **.95**. The (.10)-level confidence interval is determined with the **ConfidenceLevel->.90** option. Of course, a smaller interval is the result.

ConfidenceIntervals

```
In[1]:=
<<ConfidenceIntervals.m

In[2]:=
energy={1254,1615,1711,1350,1521,1293,
        1227,908,1205,1154,1231,1351,
        1790,1369,1185};

In[3]:=
MeanCI[energy]//N

Out[3]=
{1211.85, 1476.68}

In[4]:=
MeanCI[energy,ConfidenceLevel -> .90]//N

Out[4]=
{1235.53, 1453.}
```

In this case, it is 95% certain that the population mean lies between 1211.85 and 1476.68.

In this case, it is 90% certain that the population mean lies between 1235.53 and 1453.

□ **Example:**

In the case of the daily calcium intake in which the standard deviation was known, the null hypothesis that the daily intake of people with incomes below the poverty level equals 800 milligrams was not rejected. This test is investigated below with the data given in **calcium**. This illustrates another option, **KnownStandardDeviation->sd**. For this problem, **sd** = 262. The confidence interval found with **MeanCI** shows that 800 is in the interval. Thus, the null hypothesis is not rejected.

```
In[6]:=
   calcium={879,1096,701,986,828,1077,703,
                555,422,997,473,702,508,530,
                513,720,944,673,574,707,864,
                1199,743,1325,655,1043,599,1008,
                705,180,287,542,893,1052,473};
   MeanCI[calcium,KnownStandardDeviation -> 262]//N
Out[6]=
   {660.515, 834.113}
```

In this case, the confidence interval is (660.515,834.113).

Hypothesis tests for the comparison of two population means can also be considered with confidence intervals. Again, referring to data from the previous section, the command **MeanDifferenceCI[list1,list2,options]** is explored.

□ **Example:**

Recall that **below** represents the daily protein intake in grams for 10 people with incomes below the poverty level while **above** represents that of 15 people with incomes above the poverty level. The null hypothesis is that the below-poverty mean is not less than the above-poverty mean. The confidence interval obtained below does not contain zero (the hypothesized mean difference). Hence, the null hypothesis is rejected. Therefore, the data indicate that the average person with an income below the poverty level gets less protein than the average person with an income above the poverty level.

```
In[9]:=
   below={51.4,76.7,73.7,66.2,65.5,
            49.7,65.8,62.1,75.8,62.0,
            72.0,55.0,79.7,65.4,73.3};
   above={86.0,59.7,68.6,98.6,87.7,
            69.0,80.2,78.1,69.8,77.2};
   MeanDifferenceCI[below, above]

   General::spell1:
      Possible spelling error: new symbol name
       "below" is similar to existing symbol "Below"

   General::spell1:
      Possible spelling error: new symbol name
       "above" is similar to existing symbol "Above"

Out[9]=
   {-20.2874, -2.11927}
```

Version 2.0 warns of possible spelling errors. In this case, the warning message can be ignored since no error was made.

Mathematica can also determine the confidence interval of the variance of a population. (Note that the variance is used instead of the standard deviation.)

□ **Example:**

Recall the information collected by a hardware manufacturer giving the diameter of 20 bolts produced by the company. It was determined that an acceptable standard deviation for bolt diameters is .09 millimeters. Hence, the null hypothesis is that the variance of the diameters equals (.09)^2 = .0081. The command which determines the confidence interval for variance is `VarianceCI[list,options]`. When this command is used with **`diameter`**, the interval obtained does not contain .0081. Thus, the null hypothesis is rejected. However, the manufacturer can be 95% confident that the variance of the diameters of all 10-millimeters bolts produced is somewhere between .00157843 and .00582214.

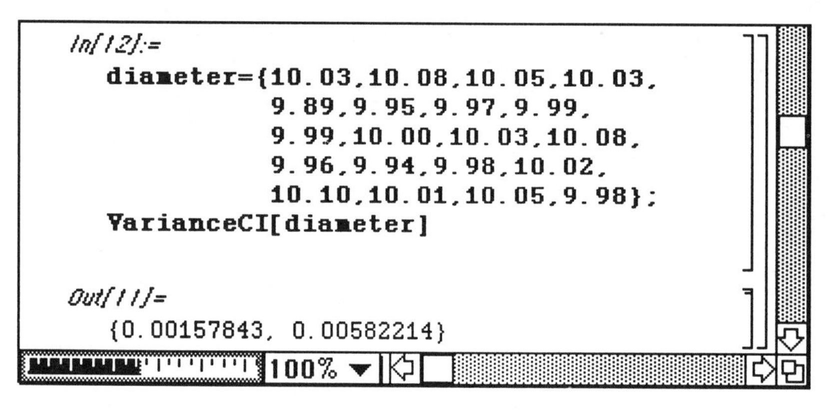

• 9.9 LinearRegression.m

The Version 2.0 **LinearRegression.m** package contains the command **Regress[data,functions,vars]** which leads to the determination of the linear regression equation for the data points in **data** as well as other useful information concerning **data**. In the **Regress** command, the regression equation is formed from the linear combination of functions given in the list **functions**. Also, **vars** represents the variables in the regression equation. Since *Mathematica* can be used for multiple regression, the number of functions and variables will depend on the type of problem to be solved.

o Example:

Consider the following list of data in **data4**. The regression equation for **data4** is found with **Regress**. The coefficients of the regression equation appear in the column labeled `Estimate` under `ParameterTable` with 1.5 as the coefficient of x and -.25 the coefficient of 1. Hence, the regression equation is y = 1.5 x -.25. The other information given is discussed in later examples.

```
Regression

In[1]:=
   <<LinearRegression.m

In[5]:=
   data4={{1,1},{1,2},{2,2},{4,6}};
   Regress[data4,{x,1},x]

Out[5]=
   {ParameterTable ->

           Estimate   SE          TStat

              PValue
           1.5        0.322749    4.64758      0.04\

              33108

       x   -0.25      0.756913    -0.330289

       1      0.772571

       , RSquared -> 0.915254,

     AdjustedRSquared -> 0.872881,
     EstimatedVariance -> 0.625,
```

```
ANOVATable ->

            DoF    SoS       MeanSS    FRatio

                   PValue                      }
            1      13.5      13.5      21.6

                   0.0433108

            2      1.25      0.625

   Model

   Error    3      14.75

   Total
```

o Example:

Next, consider the age versus price data collected on a particular type of sports car. This data is as follows and is defined below in **`price`**.

Age(years)	Price($100s)
5	85
4	103
6	70
5	82
5	89
5	98
6	66
6	95
2	169
7	70
7	48

The regression equation is found to be y = 195.468 - 20.2613 x. Notice in the output that there is not enough room to include all of the information in `ANOVATable` on one line. Therefore, the remaining information concerning `Model`, `Error`, and `Total` is given beneath these headings. Most of the information provided is self-explanatory. Since the value of `RSquared` (the coefficient of determination) is near one (.853373), a good deal of the variation in the sampled prices is explained by the regression line. This implies that age is useful in predicting price.
Another way to analyze the provided information is through hypothesis tests for the slope of the regression line. The variable x is a useful predictor of y if they are linearly related. This can be tested by forming the hypothesis test :

$H_0 : \beta_1 = 0$ (x is not useful for predicting y)

$H_a : \beta_1 \neq 0$ (x is useful for predicting y)

where the regression line is of the form $y = \beta_0 + \beta_1 x$.

The final entry in the row corresponding to x is the P-value for the hypothesis test. If a significance level of .05 is used, the null hypothesis is rejected since .00004 < .025.

```
In[9]:=
  price={{5,85},{4,103},{6,70},{5,82},
         {5,89},{5,98},{6,66},{6,95},
         {2,169},{7,70},{7,48}};
  Regress[price,{1,x},x]

Out[9]=
  {ParameterTable ->

       Estimate    SE         TStat       PValue
     1  195.468    15.2403    12.8257     0

     x  -20.2613   2.79951    -7.23743    0.0000488191

    RSquared -> 0.853373, AdjustedRSquared -> 0.837081,

    EstimatedVariance -> 158.17,

    ANOVATable ->

            DoF   SoS        MeanSS     FRatio

               PValue
            1     8285.01    8285.01    52.3804    0.0000\

               488191

            9     1423.53    158.17

     Model

     Error  10    9708.55

     Total

     }
```

The regression equation is plotted below with the **ListPlot** of the data in **price**. As expected, as the age of the car increases, the price decreases.

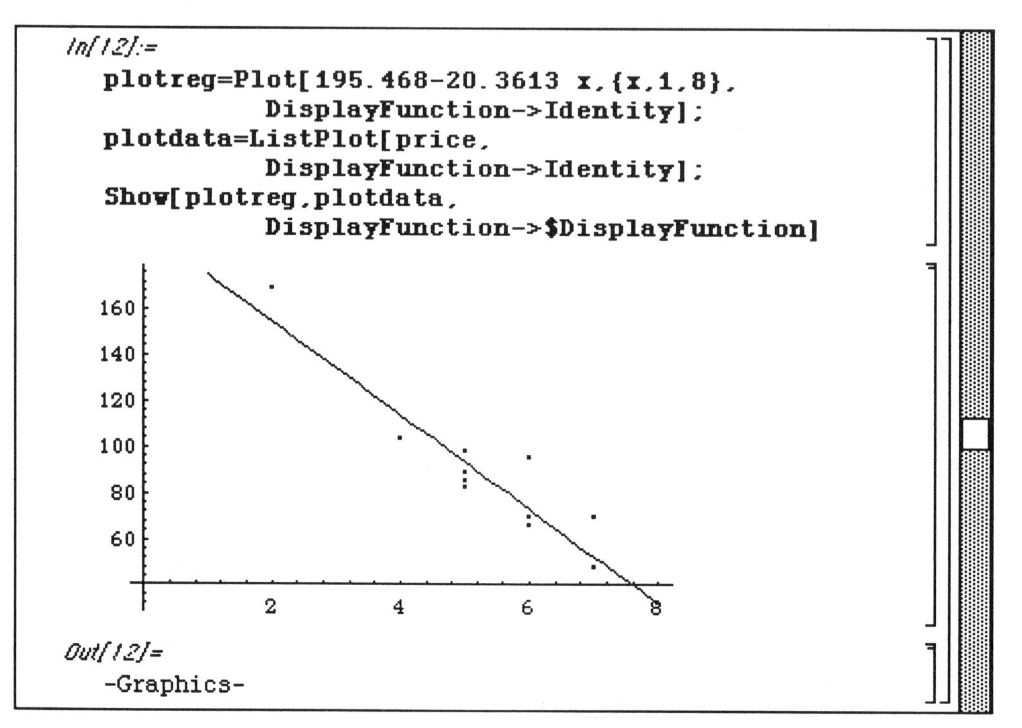

As indicated before, **Regress** can be used for multiple regression. This is demonstrated below with the addition of the number of miles driven (in thousands) to the data in the previous example.

o Example:

It was determined above that 85.3% (`RSquared*100`) of the variation in the price data is explained by age. Then, by using this additional information (mileage), perhaps this variation is better explained. The revised data is given in **extra** below. The variables in the new regression equation are **x1** and **x2** which represent age and miles, respectively. Using the **Regress** command below, the equation is found to be y = 183.035 - 9.50427 **x1** - .821483 **x2**. Again, the coefficients are located under the column labeled `Estimate`. To determine if the variation is better explained with the additional data, the value of `RSquared` is considered. In this case, `RSquared` = .936115. Hence, 93.6% of the price variation is explained by age and miles driven. Since age only explained 85.3% of the variation, the multiple regression equation provides a much better explanation of the variation in the price data than the simple linear equation found in the previous example.
As was the case above, hypothesis tests can be used with multiple regression.

Assuming a regression equation of the form $y = \beta_0 + \beta_1 x_1 + \beta_2 x_2$,the following hypothesis test is formed :

$H_0 : \beta_1 = \beta_2 = 0$

H_a: At least one of β_1 and β_2 is not zero.

This test depends on the the F-statistic. The P-value for this test appears under `PValue` in the `ANOVATable` in the output below. Since this P-value (.0000166571) is much smaller than a reasonable significance level, the null hypothesis is rejected. Hence, the previous findings that age and mileage are good predictors of price is verified.

```
In[14]:=
  extra={{5,57,85},{4,40,103},{6,77,70},{5,60,82},
         {5,49,89},{5,47,98},{6,58,66},{6,39,95},
         {2,8,169},{7,69,70},{7,89,48}};
  Regress[extra,{1,x1,x2},{x1,x2}]

Out[14]=
  {ParameterTable ->

            Estimate    SE         TStat      PValue   ,
       1    183.035     11.3476    16.1298    0

       x1   -9.50427    3.87419    -2.45323   0.0397362

       x2   -0.821483   0.255207   -3.21889   0.0122595

    RSquared -> 0.936115, AdjustedRSquared -> 0.920144,

    EstimatedVariance -> 77.529,
    ANOVATable ->


     Model
                                                                }
     Error

     Total

        DoF   SoS       MeanSS    FRatio    PValue
        2     9088.31   4544.16   58.6124   0.0000166571

        8     620.232   77.529

        10    9708.55
```

Chapter 10
Getting Help from *Mathematica* and Making *Mathematica* Do What You Want

10.1 Getting Help from *Mathematica*

Help Commands

Becoming competent with *Mathematica* can take a serious investment of time. Hopefully, messages that result from syntax errors are viewed lightheartedly. Ideally, instead of becoming frustrated, beginning *Mathematica* users will find it challenging and fun to locate the source of errors. In this process, it is natural that one will become more proficient with *Mathematica* .

One way to obtain information about commands and functions is the command **?**. **?Name** gives information on the *Mathematica* function **Name**

Example:

The following window shows how information is obtained on the command **Solve** as well as the form in which this information is given. Notice how the description includes the particular forms in which the command **Solve** should be entered. This can be quite helpful in attempting to use the command.
The command **?** can be used in several different ways. For example, **?letter*** gives a list of all *Mathematica* commands which begin with **letter** . This is illustrated below with the letter **N**.

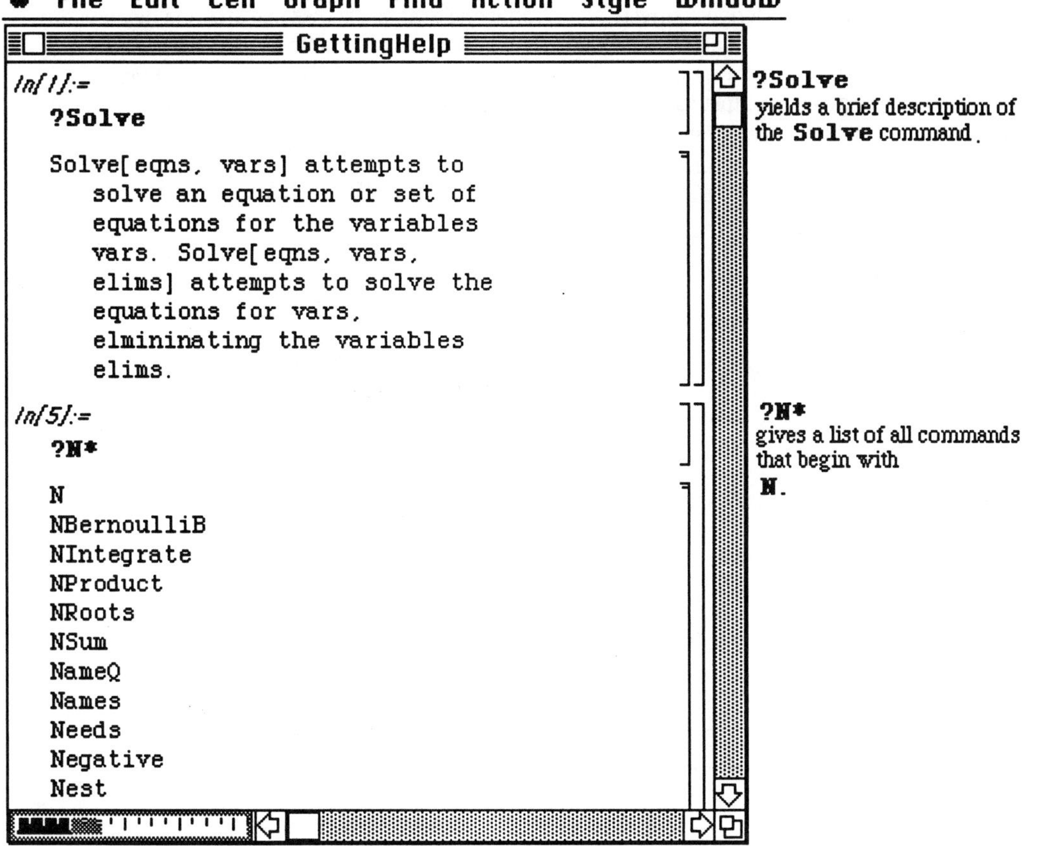

?Solve yields a brief description of the **Solve** command.

?N* gives a list of all commands that begin with **N**.

Another useful application of **?** is in determining the definition of functions. This is especially helpful in verifying that a user-defined function has been defined correctly. The following example shows that after the function **f** is defined, the command **f** gives the definition of **f**. Notice that after **f** is cleared, the formula for **f** is no longer known. Hence, **?f** yields nothing. If the symbol " := " is used in the definition of a function, the function's formula is not automatically given as output as it is below when the equals sign is used in the definition . Therefore, **?** is of particular help in obtaining the definition.

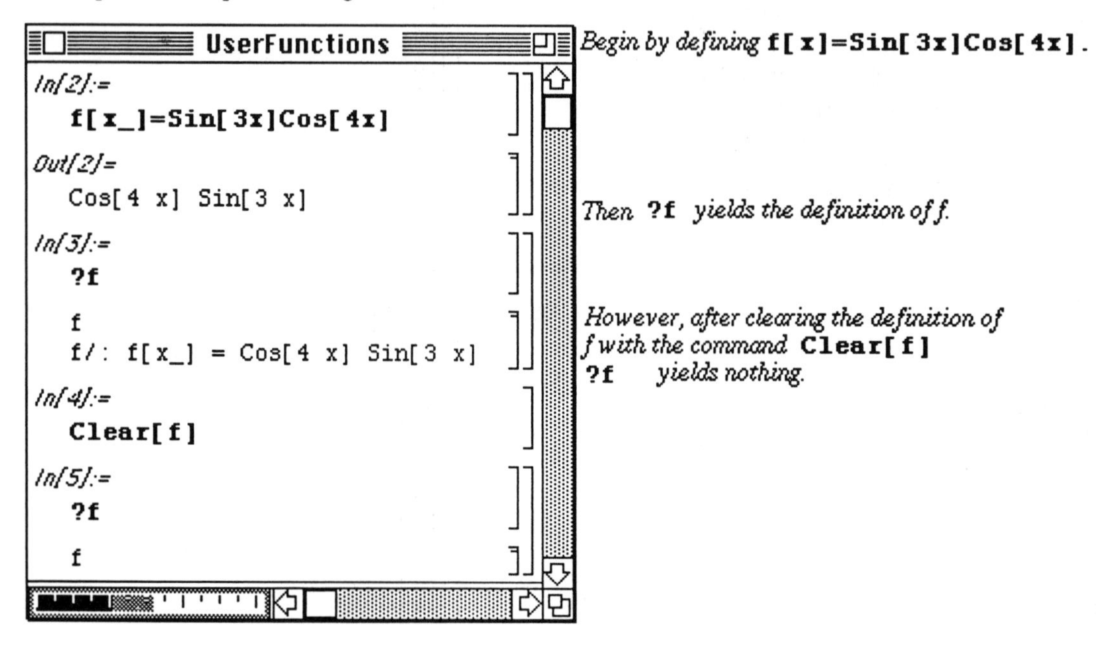

Begin by defining **f[x]=Sin[3x]Cos[4x]**.

Then **?f** *yields the definition of f.*

However, after clearing the definition of f with the command **Clear[f]** **?f** *yields nothing.*

□ **Example:**

Several other forms of the **?** command are shown below :

?*letters gives all *Mathematica* commands which end in **letters**.
?letters* gives all *Mathematica* commands which begin with **letters**.
?function gives a description of the built-in *Mathematica* function, **function**.
Examples which illustrate these commands follow :

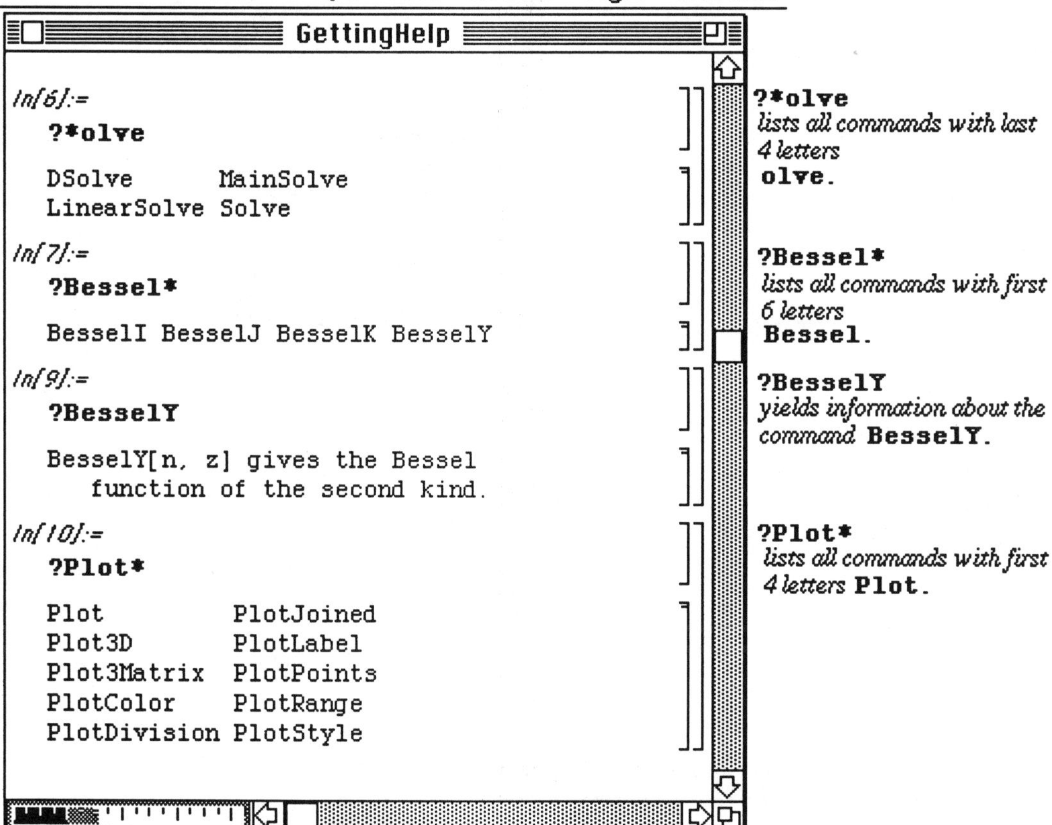

Yet another form of the command `?` is `?*letters*`. This command gives a list of all *Mathematica* commands that contain `letters`. Several examples which illustrate this are given below.

 File Edit Cell Graph Find Action Style Window

GettingHelp

```
In[10]:=
  ?*olve*

  DSolve      Solve
  LinearSolve SolveAlways
  MainSolve

In[11]:=
  ?*grate*

  Integrate  NIntegrate

In[12]:=
  ?*lot*

  ContourPlot
  Cyclotomic
  DensityPlot
  ListContourPlot
  ListDensityPlot
  ListPlot
  ListPlot3D
  ParametricPlot
  Plot
  Plot3D
  Plot3Matrix
  PlotColor
  PlotDivision
  PlotJoined
  PlotLabel
```

?*olve* *gives a list of all the commands containing the letters olve.*

?*grate* *gives a list of all the commands containing the letters grate.*

?*lot* *gives a list of all the commands containing the letters lot.*

- Version 2.0 users will notice that some commands from earlier versions of *Mathematica* have been made obsolete with the release of Version 2.0. In these cases, *Mathematica* is able to tell what command replaces the outdated command.

- **Example:**

For example, in Version 2.0 the functions **TrigExpand** and **TrigCanonical** from previous versions of *Mathematica* are obsolete. Nevertheless, the Version 2.0 command **Expand[expression,Trig->True]** performs the same function as the command **TrigExpand[expression]** from earlier versions.

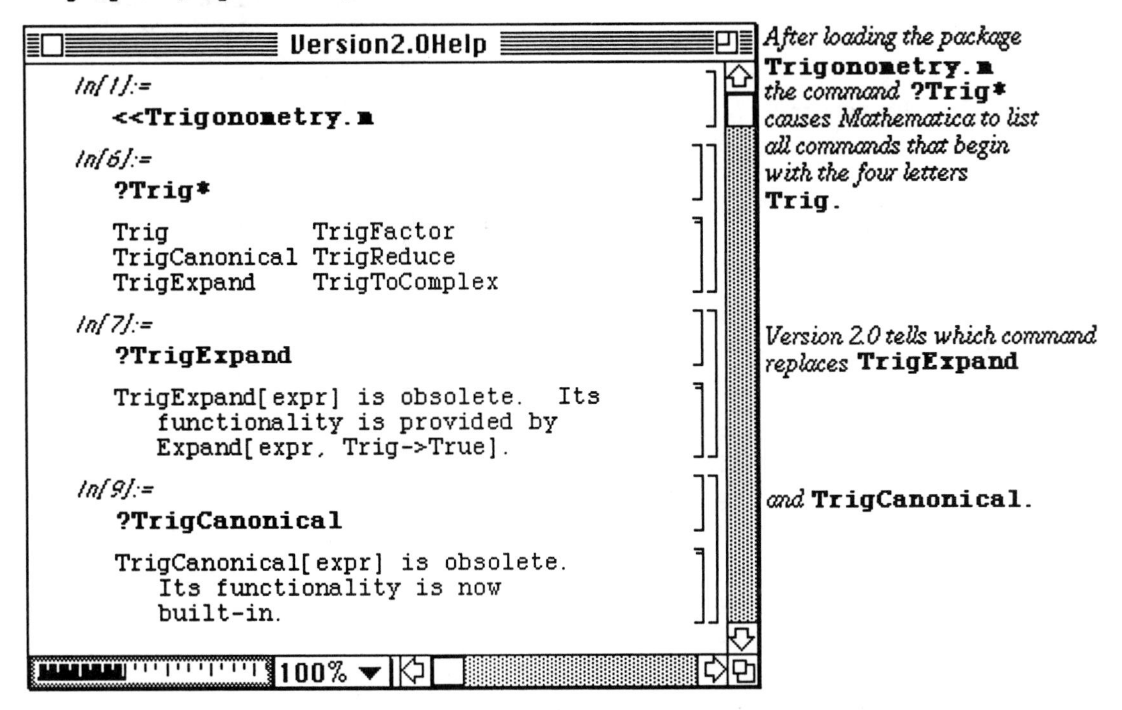

After loading the package **Trigonometry.m** *the command* **?Trig*** *causes Mathematica to list all commands that begin with the four letters* **Trig.**

Version 2.0 tells which command replaces **TrigExpand**

and **TrigCanonical.**

Another way to obtain information on *Mathematica* commands is the command **Options**. **Options[Command]** gives all of the available options associated with **Command**. This is quite useful when working with a *Mathematica* command such as **Plot** which has many options. Notice that the default value (the value automatically assumed by *Mathematica*) for each option is given in the output.

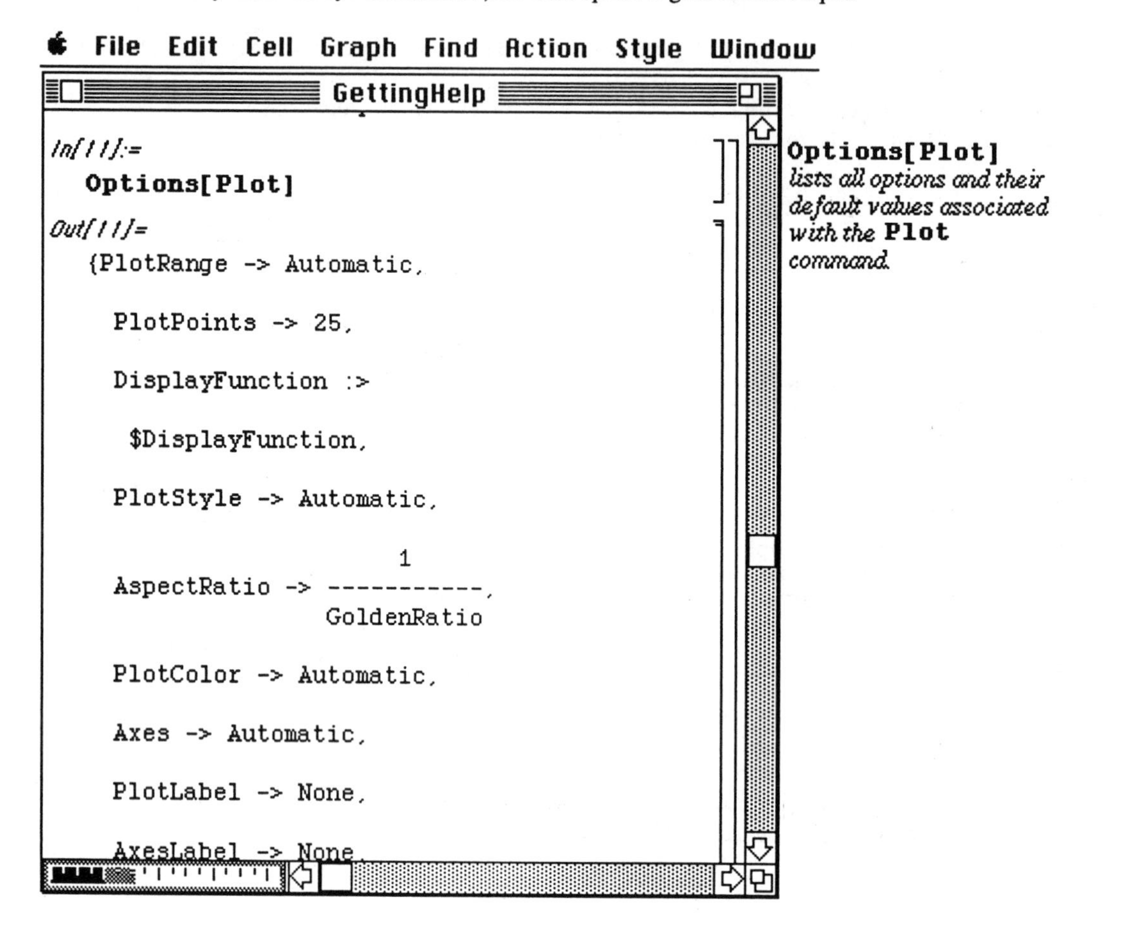

Options[Plot] *lists all options and their default values associated with the* **Plot** *command.*

The command **??Command** gives a brief description of **Command** as well as the list of **Options** available to **Command**.

□ **Example:**

This is illustrated below with the *Mathematica* command **ListPlot**.

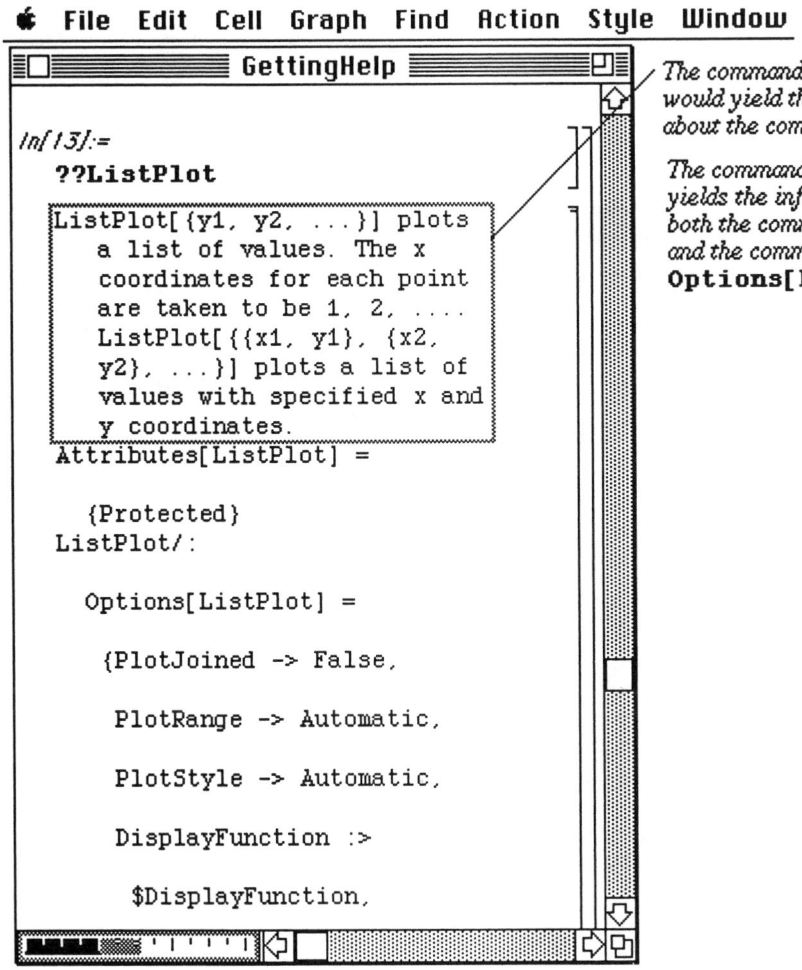

 File Edit Cell Graph Find Action Style Window

GettingHelp

In[13]:=

```
??ListPlot
ListPlot[{y1, y2, ...}] plots
   a list of values. The x
   coordinates for each point
   are taken to be 1, 2, ....
   ListPlot[{{x1, y1}, {x2,
   y2}, ...}] plots a list of
   values with specified x and
   y coordinates.
Attributes[ListPlot] =

  {Protected}
ListPlot/:

  Options[ListPlot] =

   {PlotJoined -> False,

    PlotRange -> Automatic,

    PlotStyle -> Automatic,

    DisplayFunction :>

     $DisplayFunction,
```

The command **?ListPlot** *would yield this information about the command* **ListPlot**.

The command **??ListPlot** *yields the information given by both the command* **?ListPlot** *and the command* **Options[ListPlot]**.

Yet another method for acquiring information on *Mathematica* commands is through the use of **Complete Selection**. This is located under **Prepare Input** in the **Action** submenu and is useful when attempting to complete a command.

◻ Example:

For example, if the user wishes to use a command which begins with **Polynomial**, but does not remember the rest of the command, help can be obtained in the following manner : (1) Type the word **Polynomial**, (2) Move the cursor to the **Action** heading and use to the mouse to obtain the **Action** submenu, (3) Choose **Complete Selection** from the submenu (This causes a list of commands which begin with **Polynomial** to be displayed), (4) Move the cursor to the desired command in the list and click. The correct command is then completed on the screen.

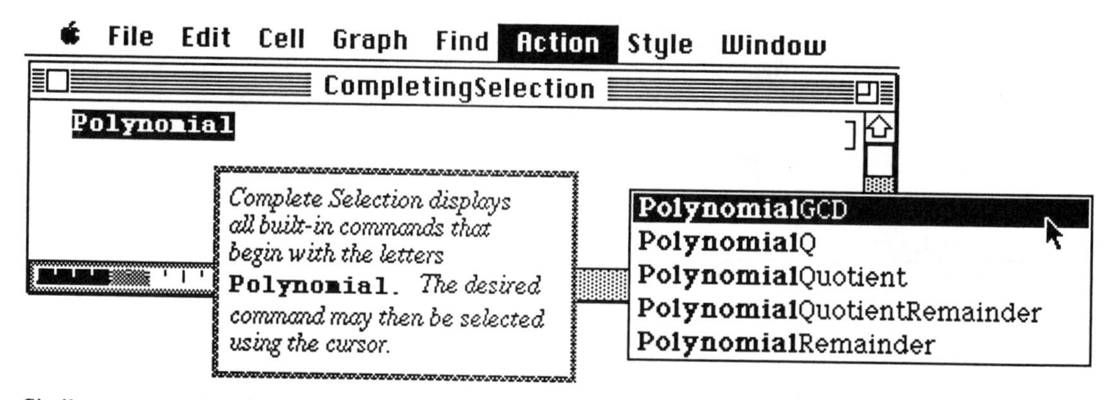

Similar steps may be taken to obtain the proper syntax of a command. The following windows illustrate how **Prepare Input** can be chosen from the **Action** submenu to yield the arguments of **NRoots**.

File Edit Cell Graph Find Action Style Window

CompletingSelection

NRoots

After the command has been written, go to **Action** *and select* **Prepare Input** *and then select* **Make Template.**

The same steps which were described above for using **Complete Selection** are followed in this case with the exception that **Make Template** is chosen instead of **Complete Selection**. Note that if other *Mathematica* commands begin with the word **NRoots**, then they would all be displayed as they were in the previous example with **Polynomial**.

File Edit Cell Graph Find Action Style Window

CompletingSelection

```
NRoots[lhs==rhs, var]
```

Make Template automatically fills in the command with "typical syntax." Notice that more information is obtained using the command **?NRoots** *or* **??NRoots**.

■ *Mathematica* Help

The ***Mathematica* Help** file is located in the *Mathematica* folder. The information in the window below is given when this file is opened. This describes what is contained in the file and points out that it should be rarely opened. No changes should be made to ***Mathematica* Help**, since this would only cause confusion. Information can be printed from this file, however, by using **Print...** or **Print Selection...** from **File.**

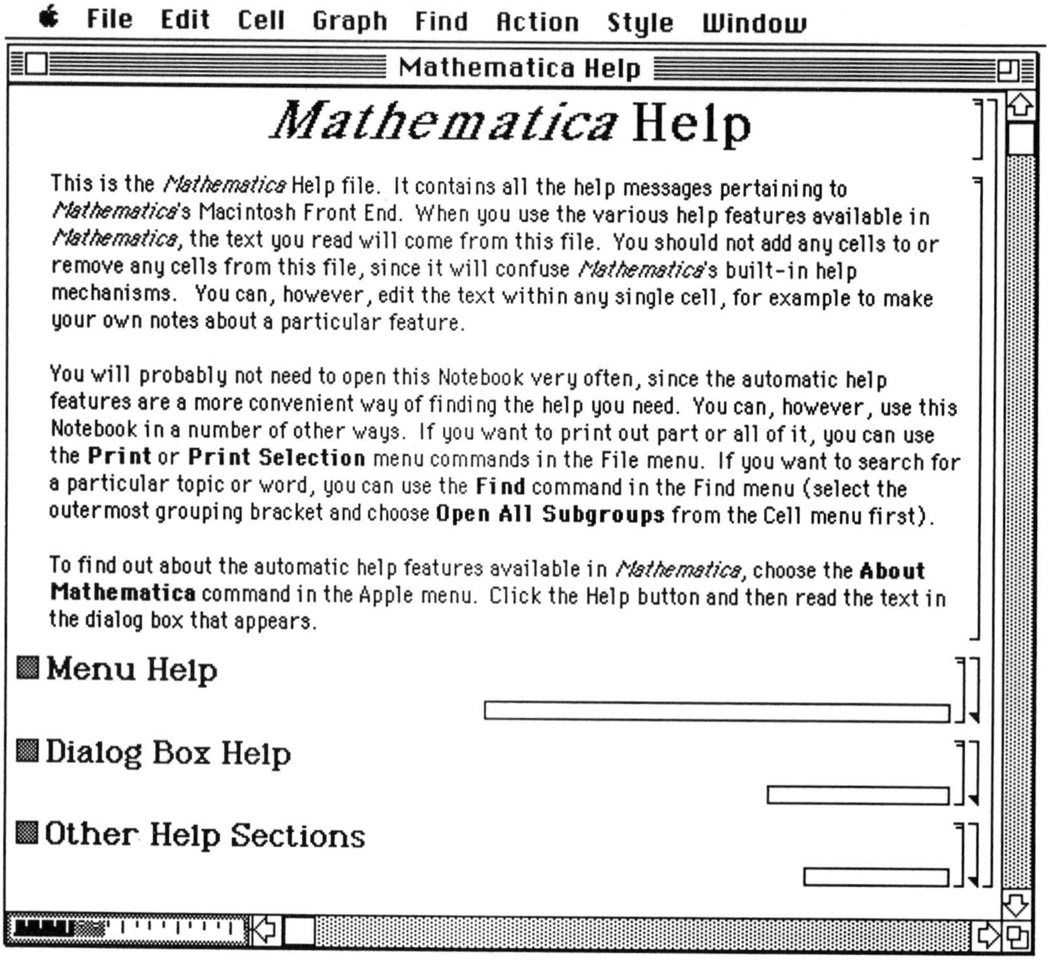

File Edit Cell Graph Find Action Style Window

Mathematica Help

Mathematica Help

This is the *Mathematica* Help file. It contains all the help messages pertaining to *Mathematica*'s Macintosh Front End. When you use the various help features available in *Mathematica*, the text you read will come from this file. You should not add any cells to or remove any cells from this file, since it will confuse *Mathematica*'s built-in help mechanisms. You can, however, edit the text within any single cell, for example to make your own notes about a particular feature.

You will probably not need to open this Notebook very often, since the automatic help features are a more convenient way of finding the help you need. You can, however, use this Notebook in a number of other ways. If you want to print out part or all of it, you can use the **Print** or **Print Selection** menu commands in the File menu. If you want to search for a particular topic or word, you can use the **Find** command in the Find menu (select the outermost grouping bracket and choose **Open All Subgroups** from the Cell menu first).

To find out about the automatic help features available in *Mathematica*, choose the **About Mathematica** command in the Apple menu. Click the Help button and then read the text in the dialog box that appears.

Menu Help

Dialog Box Help

Other Help Sections

□ **Example:**

The following window appears when the command `Plot[Sin[x],{x,-2Pi,2Pi}]` is selected and **Explain Selection...** is chosen from under the apple icon on the *Mathematica* Menu. The window gives a description of the cell and explains how the cell can be altered.

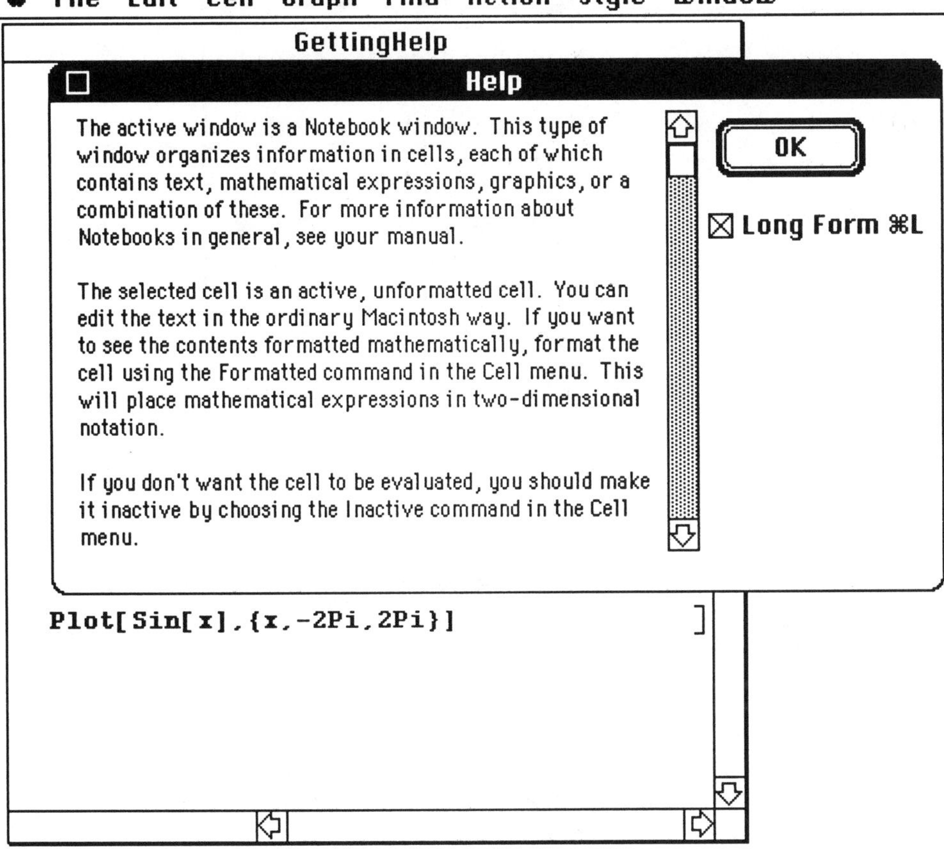

• Version 2.0 Help and Kernel Help

The folder Kernel Help is installed on the desk top when Version 2.0 of *Mathematica* is installed on the computer.

The folder **Kernel Help** contains sixteen notebooks that provide examples of nearly every built-in *Mathematica* command. Since these notebooks can be opened during a *Mathematica* session, they can be particularly helpful sources of documentation.

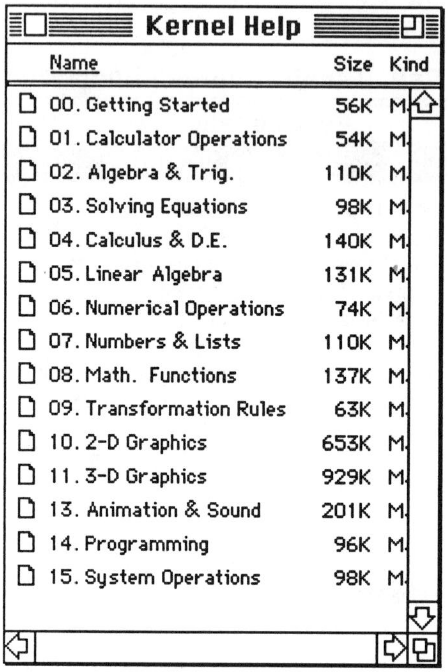

Kernel Help

Name	Size	Kind
00. Getting Started	56K	M
01. Calculator Operations	54K	M
02. Algebra & Trig.	110K	M
03. Solving Equations	98K	M
04. Calculus & D.E.	140K	M
05. Linear Algebra	131K	M
06. Numerical Operations	74K	M
07. Numbers & Lists	110K	M
08. Math. Functions	137K	M
09. Transformation Rules	63K	M
10. 2-D Graphics	653K	M
11. 3-D Graphics	929K	M
13. Animation & Sound	201K	M
14. Programming	96K	M
15. System Operations	98K	M

The first two notebooks, **Getting Started** and **Calculator Operations**, look as follows when they are opened:

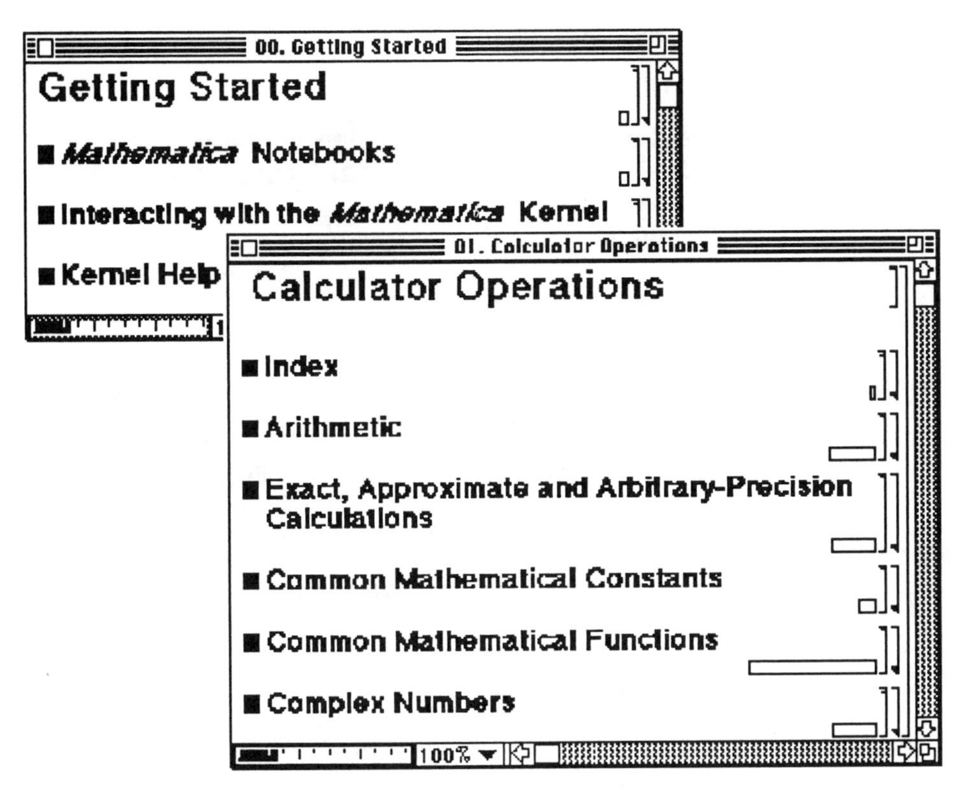

Similarly, **Calculus and Differential Equations**, **Built-in Mathematical Functions**, and **Programming** look the same:

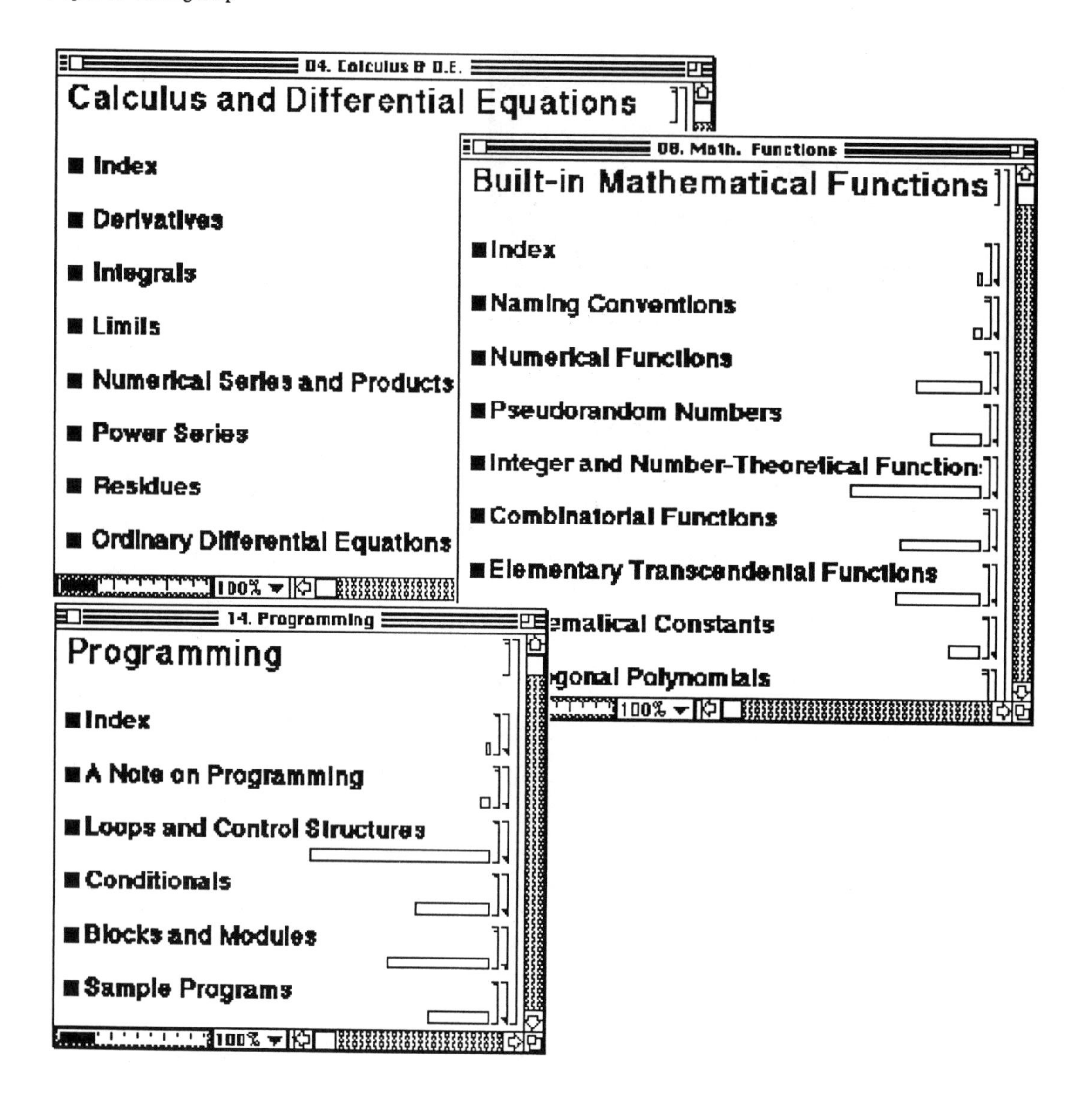
04. Calculus & D.E.
Calculus and Differential Equations
Index
Derivatives
Integrals
Limits
Numerical Series and Products
Power Series
Residues
Ordinary Differential Equations
100%
08. Math. Functions
Built-in Mathematical Functions
Index
Naming Conventions
Numerical Functions
Pseudorandom Numbers
Integer and Number-Theoretical Function
Combinatorial Functions
Elementary Transcendental Functions
ematical Constants
gonal Polynomials
100%
14. Programming
Programming
Index
A Note on Programming
Loops and Control Structures
Conditionals
Blocks and Modules
Sample Programs
100%

■ 10.2 The `init.m` file

The **`init.m`** file gives the user the opportunity to supply information to be read in each time the *Mathematica* kernel is started. Since this file is read each time the kernel is started, any information included in the file is automatically loaded. The **`init.m`** file is found inside of the **`Packages`** folder. The window obtained by double-clicking on the **`Packages`** folder is shown below:

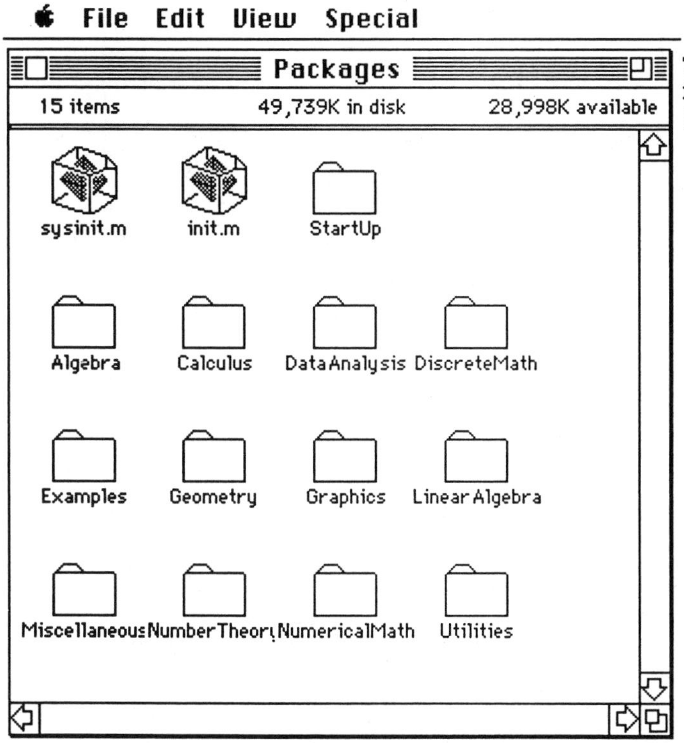

File Edit View Special
Packages
15 items 49,739K in disk 28,998K available
sysinit.m init.m StartUp
Algebra Calculus DataAnalysis DiscreteMath
Examples Geometry Graphics LinearAlgebra
Miscellaneous NumberTheory NumericalMath Utilities

The file **`init.m`** is contained in the folder **`Packages`**.

Double-clicking on the **init.m** icon yields the following window which gives a brief description of the file and how new information can be entered.

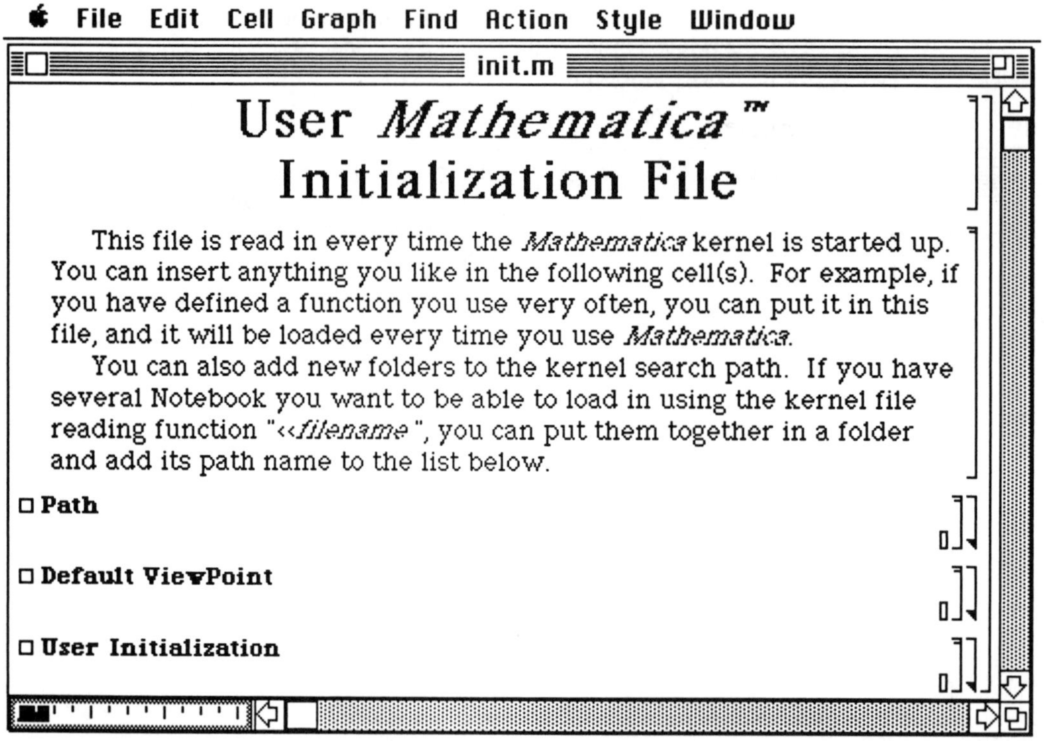

File Edit Cell Graph Find Action Style Window

init.m

User *Mathematica*™ Initialization File

This file is read in every time the *Mathematica* kernel is started up. You can insert anything you like in the following cell(s). For example, if you have defined a function you use very often, you can put it in this file, and it will be loaded every time you use *Mathematica*.

You can also add new folders to the kernel search path. If you have several Notebook you want to be able to load in using the kernel file reading function "<<*filename*", you can put them together in a folder and add its path name to the list below.

Path

Default ViewPoint

User Initialization

A useful property of the **init.m** file is user initialization of functions and commands. This gives the user the opportunity to "customize" his or her own commands or define functions which the user often employs. The following is an illustration of how this is done. If the user prefers not to capitalize the trigonometric functions, they can be redefined with small letters in terms of the associated built-in *Mathematica* commands. The same can be done with the constant Pi. Below, the user chooses to use a five-place decimal approximation of the well-known constant and name it **pi**. Hence, whenever the *Mathematica* kernel is opened, these new definitions will be read in and, thus, will be recognized when used.

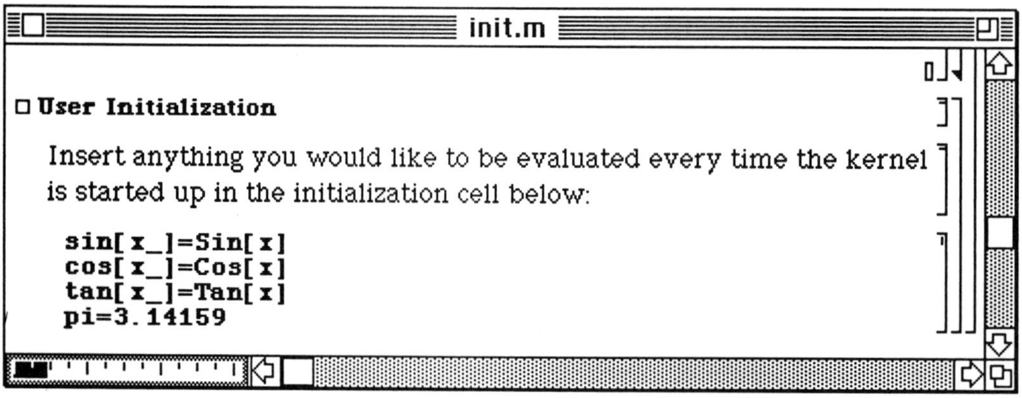

init.m

User Initialization

Insert anything you would like to be evaluated every time the kernel is started up in the initialization cell below:

```
sin[x_]=Sin[x]
cos[x_]=Cos[x]
tan[x_]=Tan[x]
pi=3.14159
```

After initializing, *Mathematica* uses the user-defined functions to give the correct results. Of course, the built-in commands would be properly evaluated as well, so these newly-defined commands have not replaced the original ones.

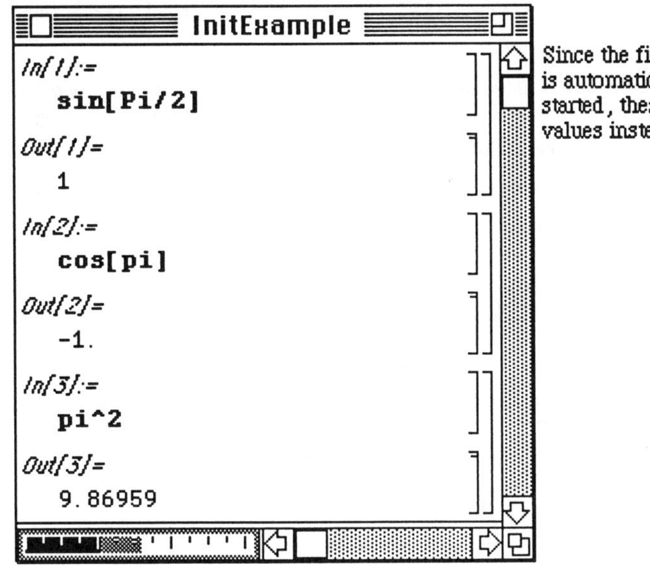

Since the file **init.m** is automatically loaded when the kernel is started, these commands yield the desired values instead of error messages.

10.3 Explanation of the *Mathematica* Menu

The Version 1.2 Menu

For a complete discussion of the menu as it appears in either Version 1.2 or Version 2.0, see the User's Guide for the Macintosh in the *Mathematica* software package.

The menu below which indicates the editing capabilities of *Mathematica* appears under **Edit** in the *Mathematica* Menu. The usual Macintosh editing features are included in this list along with the addition of several commands for working with *Mathematica* cells. The main item to notice in the list is **Settings**. These **Settings** options are discussed in greater detail below.

Edit

Edit
Can't Undo
Cut
Copy
Paste
Clear
Paste and Discard
Convert Clipboard
Select All Cells
Nesting
Divide Cell
Merge Cells
Settings

In Version 2.0, **Settings** has been changed to **Preferences**; **Divide Cell** and **Merge Cells** are found under **Cell**.

The **Settings** options which are available under **Edit** are accessed by moving the cursor to the **Edit** heading on the *Mathematica* Menu and dragging the cursor to the last entry which is **Settings**. A list of six **Settings** options is then displayed: **Display**, **Graphics**, **Color**, **Animation**, **Action**, and **Startup**. To obtain a particular **Settings** window, move the cursor to the desired **Settings** option and release the mouse button.

The first of these options discussed are those found under **Startup Settings**. Instead of loading certain files or tables during a *Mathematica* session, the user has the option of requesting that they be loaded automatically. This is done on the window below by placing an X in the box next to the package which is to be loaded each time a *Mathematica* session begins. (To place the X in a box, move the cursor inside the box and click once with the mouse; to remove the X, place the cursor on the X which is to be deleted and click once with the mouse.)

Startup Settings

Stack size (number of KBytes):

Current: 256 Requested: 256

At startup load these packages:

- ☒ Messages (msg.m) ⌘M
- ☒ Function information (info.m) ⌘F
- ☐ Integration rules (IntegralTables.m) ⌘I
- ☐ Elliptic functions (Elliptic.m) ⌘E
- ☐ Series functions (Series.m) ⌘S

☐ Automatically start local kernel ⌘A

OK | Apply | Defaults | Help | Cancel

The checked boxes indicate the packages that Mathematica automatically loads when the Mathematica kernel is loaded. Many users like to have both **msg.m** and **info.m** loaded each time. In addition, if you are frequently computing definite integrals, you may also want Mathematica to load the package **IntegralTables.m**

Several settings can be made under **Action Settings**. These are listed on the **Action Settings** window which is shown below. The user has selected those options which are checked. Some of the more useful of these options include, displaying the computation time of calculations and having the output displayed so that it fits in a window of specified width.

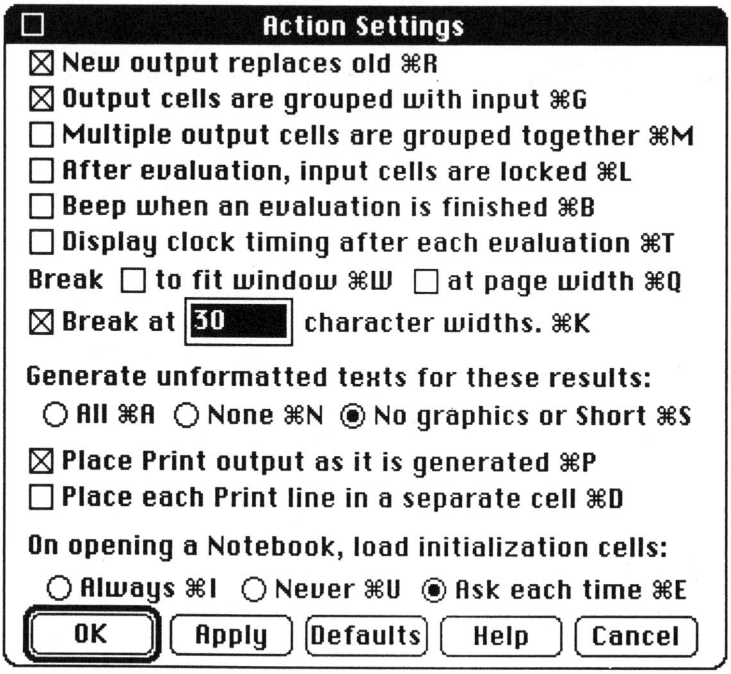

Action Settings

- ☒ New output replaces old ⌘R
- ☒ Output cells are grouped with input ⌘G
- ☐ Multiple output cells are grouped together ⌘M
- ☐ After evaluation, input cells are locked ⌘L
- ☐ Beep when an evaluation is finished ⌘B
- ☐ Display clock timing after each evaluation ⌘T

Break ☐ to fit window ⌘W ☐ at page width ⌘Q

☒ Break at 30 character widths. ⌘K

Generate unformatted texts for these results:

○ All ⌘A ○ None ⌘N ◉ No graphics or Short ⌘S

- ☒ Place Print output as it is generated ⌘P
- ☐ Place each Print line in a separate cell ⌘D

On opening a Notebook, load initialization cells:

○ Always ⌘I ○ Never ⌘U ◉ Ask each time ⌘E

OK | Apply | Defaults | Help | Cancel

Users can customize Mathematica by modifying various options.

A checked box means that Mathematica will perform the option.

In the following window, the user has requested that 78 characters be included on each line of output. This may lead to the results being in a form which are more easily read, since longer expressions can be printed on one line.

Action settings:

Action Settings

☒ New output replaces old ⌘R
☒ Output cells are grouped with input ⌘G
☐ Multiple output cells are grouped together ⌘M
☐ After evaluation, input cells are locked ⌘L
☐ Beep when an evaluation is finished ⌘B
☒ Display clock timing after each evaluation ⌘T
Break ☒ to fit window ⌘W ☐ at page width ⌘Q
☐ Break at 78 character widths. ⌘K

Generate unformatted texts for these results:
○ All ⌘A ○ None ⌘N ◉ No graphics or Short ⌘S

☒ Place Print output as it is generated ⌘P
☐ Place each Print line in a separate cell ⌘D

On opening a Notebook, load initialization cells:
○ Always ⌘I ◉ Never ⌘U ○ Ask each time ⌘E

OK | Apply | Defaults | Help | Cancel

When working with the animation of graphics, the user may find the options located under **Animation Settings** helpful. Most importantly, the user can set the speed of animation by increasing or decreasing the number of frames viewed per second. This number can be changed by simply typing the desired value for the speed (when the box is darkened) or by moving the cursor to the box and clicking the mouse button once to obtain a vertical flashing cursor. Then, any changes can be made in the usual manner. Other options include the order in which the animation is viewed. The user selects the desired order by placing a dot in the circle corresponding to the appropriate order (**Forward**, **Backward**, or **Cyclic**). Dots are placed and removed in the same manner as boxes are checked.

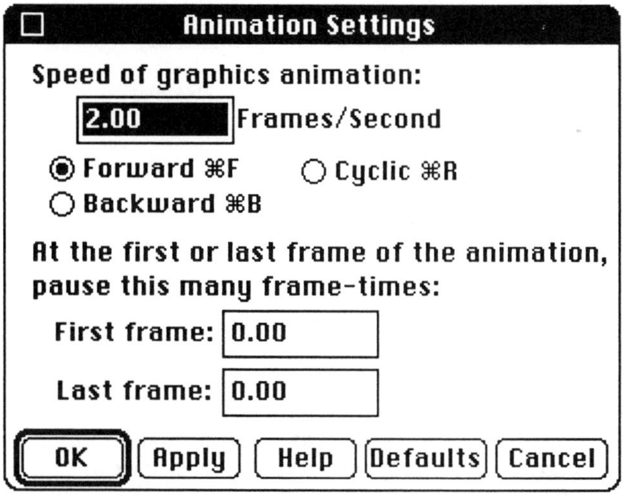

Animation Settings

Speed of graphics animation:

2.00 Frames/Second

Forward ⌘F   Cyclic ⌘R

Backward ⌘B

At the first or last frame of the animation, pause this many frame-times:

First frame: 0.00

Last frame: 0.00

OK | Apply | Help | Defaults | Cancel

Animation options can also be modified through the different settings.

Cell options are found under the **Cell** heading in the *Mathematica* Menu. These options allow the user to create a *Mathematica* notebook in any form desired. These options are used by selecting a cell or cells in a *Mathematica* notebook and then choosing the appropriate **Cell** option. One of the more useful **Cell** options is that of grouping cells. This **Group Cells** option is illustrated in greater detail in a subsequent window. Another option is that of locking cells. If the **Locked** option is chosen, then no changes can be made in that cell. (Output cells are automatically locked.)

o In Version 2.0, the options **Page Break** are found under **Style** and the options **Formatted**, **Inactive**, **Initialization**, **Locked**, **Closed**, and **Fixed Height** are contained under **Attributes** which is also found under **Style**.

Cell

Formatted	*The Cell Options allow notebooks to be customized.*
Inactive	
PostScript	
Locked	
Closed	*Initialization cells are cells within a notebook that are automatically evaluated when the Mathematica kernel is started.*
Initialization	
Fixed Height	
Page Break	
Group Cells	*Cells may be grouped, ungrouped, opened or closed.*
Ungroup Cells	
Open All Subgroups	
Close All Subgroups	
Closed Group	
Evaluation Group	

The following sequence of windows illustrates how **Group Cells** is implemented. Consecutive cells can be grouped by first selecting the cells and then choosing the **Group Cells** option. When cells are grouped, they can be closed simultaneously by double-clicking on the outermost cell which encloses the group. In the first window below, all of the cells are selected by first selecting the uppermost cell and then dragging down through all of the cells. After these cells are selected (as shown in the second window), choosing **Group Cells** from the list of **Cell** options causes the selected cells to be enclosed in a single (outermost) cell. This is displayed in the third window below.

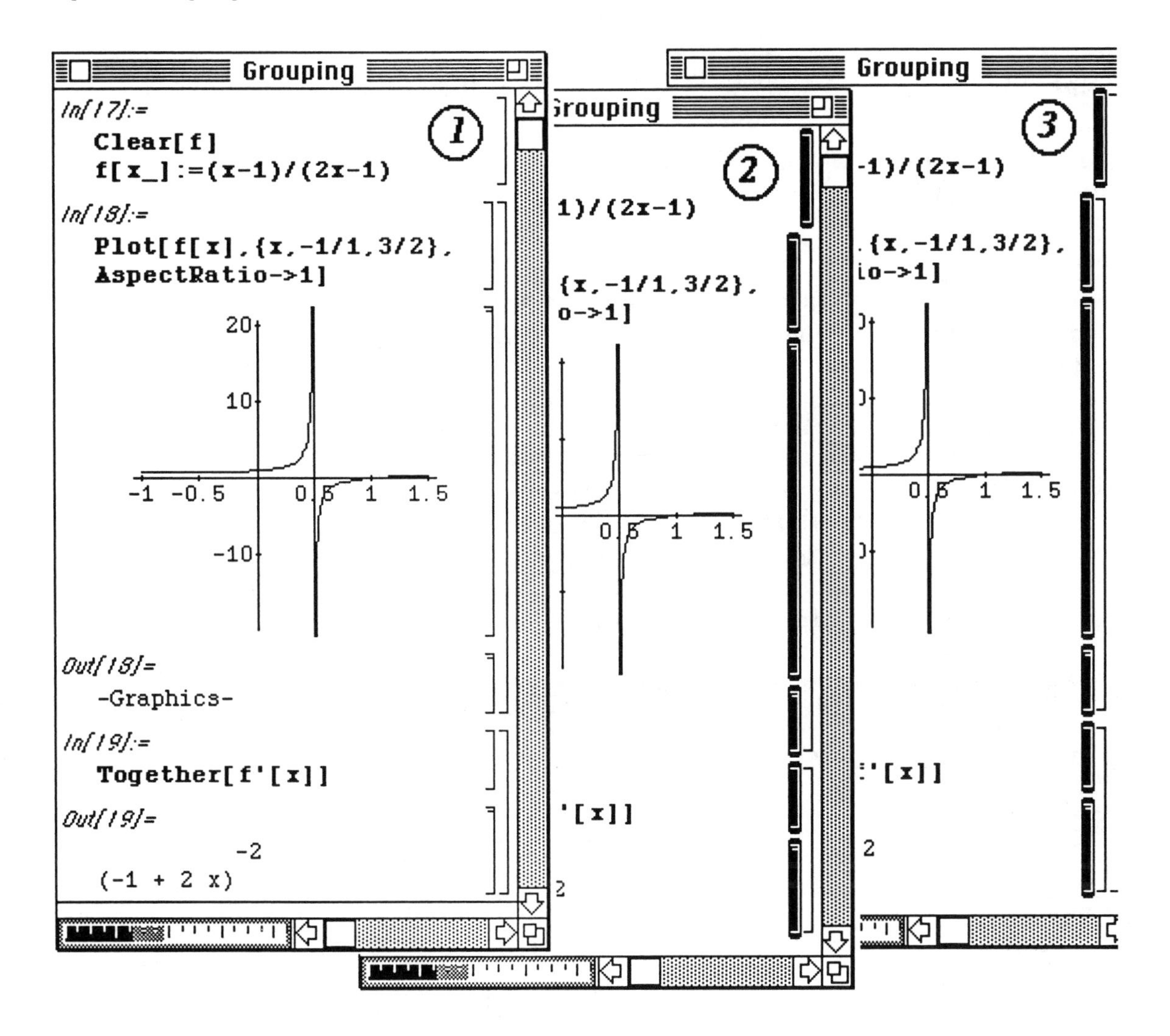

Grouping
In[17]:=
Clear[f]
f[x_]:=(x-1)/(2x-1)
In[18]:=
Plot[f[x],{x,-1/1,3/2},
AspectRatio->1]
20
10
-1 -0.5 0.5 1 1.5
-10
Out[18]=
-Graphics-
In[19]:=
Together[f'[x]]
Out[19]=
-2
(-1 + 2 x)
1
2
3

Action

- Prepare Input
- Evaluate Selection
- Evaluate Next Input
- Evaluate Notebook
- Evaluate Initialization
- Interrupt
- Kernels
- Clear Kernel History
- Quit Kernel

When **Prepare Input** is selected, this menu appears:

- Copy Input from Above
- Copy Output from Above
- Complete Selection
- Make Template
- 3D ViewPoint Selector...
- Color Selector...

Interrupt can be used to stop *Mathematica* calculations. The **Interrupt** menu is discussed below.

An important option located under **Action** is **Interrupt**. This can be used to interrupt a calculation if the calculation is taking longer than it should or if the user notices a mistake in the command which would lead to undesirable results. The following window appears when **Interrupt** is selected from the menu. This window includes several options. **Step** and **Trace** display the calculations that *Mathematica* is performing. These calculations appear within the window to the right of the four options. **Abort** causes the calculation to be stopped. In the case of **Abort**, the current *Mathematica* session may be continued. On the other hand, **Quit Kernel** causes the calculation to cease, and the user must exit and restart *Mathematica* if more calculations are desired.

Local Kernel Interrupt

- Continue
- Step ⌘S
- Trace ⌘T
- Abort ⌘A
- Quit Kernel ⌘Q
- Help

If a calculation is taking longer than expected and one wishes to abort the calculation, click abort. To quit the kernel completely (in which case, one must completely exit the Mathematica session and restart to continue using Mathematica) click quit kernel.
In addition, one may view the calculations Mathematica is performing by clicking on either step or trace.

- In Version 2.0, users can abort a calculation directly from the **Action** menu. However, to **Trace** a calculation, Version 2.0 users must select **Enter Dialog** from the Version 2.0 **Action** menu and then use the commands **Stack** or **Trace**.

Mathematica notebooks can be customized by taking advantage of the features under **Style** and **Window**.

Style

Font
Face
Size
Color
Format
Cell Style
Uniform Style
Default Style
All Default Styles...

In Version 1.2, text **Font, Face, Size, Color,** and **Format,** in addition to **Cell Style** can be changed from the **Style** menu.

Format contains alignment, scrolling, and word-wrapping options as well as cell variations or "dingbats"

Cell styles are modified by going to **Window** and selecting **Styles Window**. Once the **Styles Window** appears on the screen, fonts, sizes, and faces of the various cell types are modified using **Font**, **Face**, and **Size** found under **Style**. Selecting **All Default Styles...** resets all cell styles to their defaults.

- In Version 2.0, **Style** is considerably expanded, containing many of the options found under **Cell** and **Window** in Version 1.2.

Window

Stack Windows **Tile Windows Wide** **Tile Windows Tall**	*Stack Windows, Tile Windows Wide, and Tile Windows Tall are various ways of viewing several open Mathematica notebooks simultaneously.*
Network Window **Defaults Window** **Styles Window** **Clipboard Window**	*The Defaults Window shows Mathematica's default values which may be modified. Similarly, the Styles Window shows the font and size for each type of cell which may also be changed.*
(open notebooks)	*All open notebooks are shown; when a notebook is selected, it is brought to the front of the screen and becomes the active window.*

Several options are found under the **Window** heading on the *Mathematica* Menu. The first group of options deals with viewing several *Mathematica* notebooks at once. These options, **Stack Windows**, **Tile Windows Wide**, and **Tile Windows Tall**, are illustrated individually after the menu below.

Two of the windows which may be viewed are the **Defaults Window** and the **Styles Window**. The **Defaults Window** is displayed and explained below. The **Styles Window** displays all of the styles (font, face, and size) used for each particular type of cell in the *Mathematica* notebook. These styles can be changed by selecting a cell (or cells) and choosing another font, face, or size. This window, therefore, allows the user to customize the notebook.

The last entry in the list of **Window** options is a list of all open *Mathematica* notebooks. Hence, an opened notebook can be brought to the front of the screen by selecting it from the list with the cursor.

If **Stack Windows** is chosen, then the notebook windows are stacked one behind the other so that only the notebook in front can be fully viewed. The other open notebooks can only be partially viewed. However, a notebook can be brought to the front by simply moving the cursor to that notebook and clicking the mouse button once.

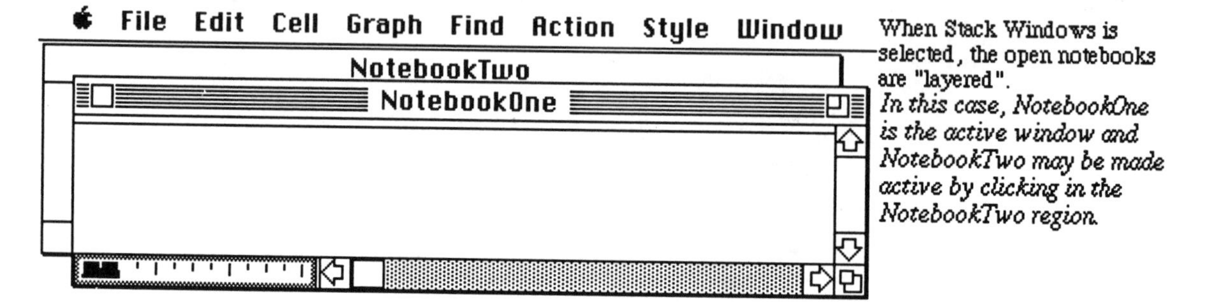

When Stack Windows is selected, the open notebooks are "layered".
In this case, NotebookOne is the active window and NotebookTwo may be made active by clicking in the NotebookTwo region.

Tile Windows Tall, illustrated below with **NotebookOne** and **NotebookTwo**, changes the width of each notebook window so that they fit side-by-side on the screen. (Note that each window has a complete boundary.)

File Edit Cell Graph Find Action Style Window

NotebookTwo

NotebookOne

In this case, the two open notebooks, NotebookOne and NotebookTwo, are viewed simultaneously after selecting Tile Windows Tall.

NotebookOne is the active window. NotebookTwo may be made active by moving the cursor within the NotebookTwo region and clicking once.

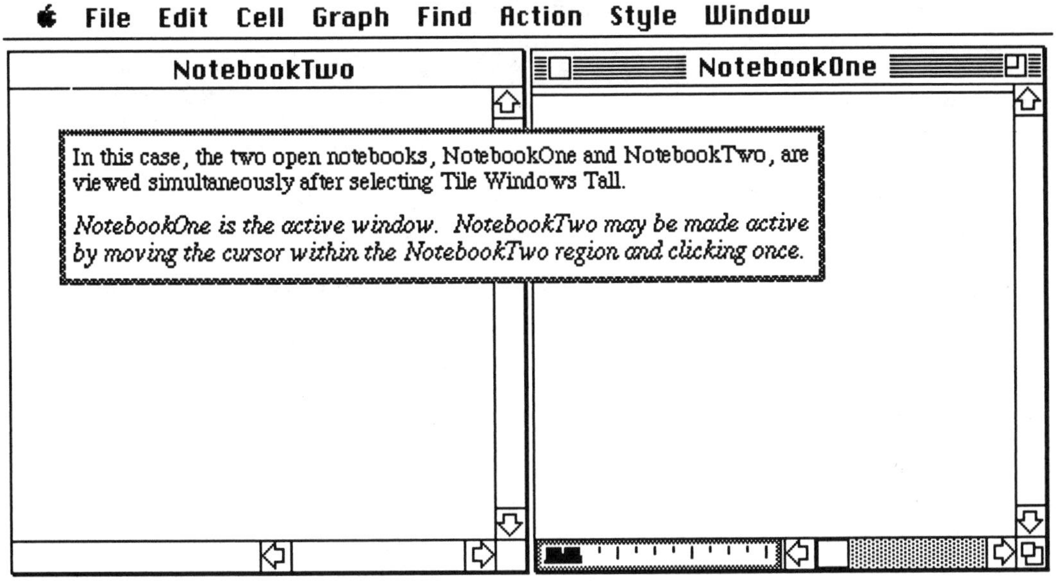

Tile Windows Wide alters the height of each notebook window so that the reduced windows fit on the computer screen simultaneously. (Again, each window has a complete boundary.) **NotebookOne** and **NotebookTwo** are displayed in the following manner with **Tile Windows Wide**.

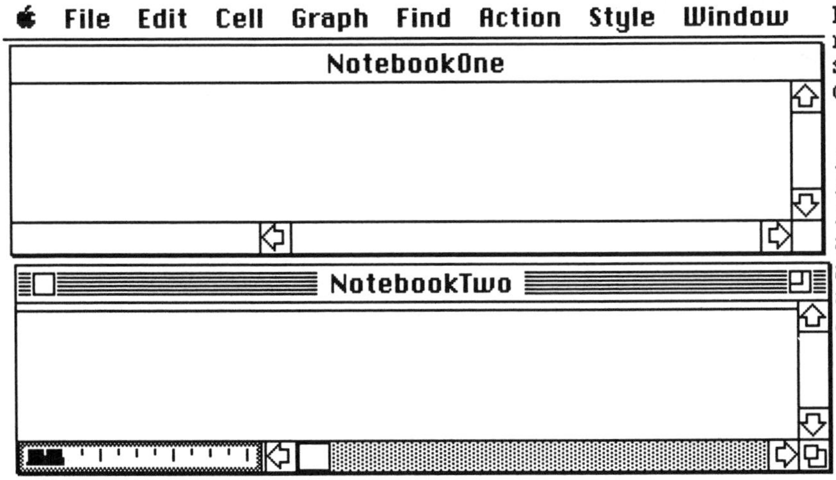

File Edit Cell Graph Find Action Style Window

NotebookOne

NotebookTwo

In this case, two open notebooks are viewed simultaneously using the option Tile Windows Wide.

In this case, NotebookTwo is the active window. NotebookOne can be made the active window by moving the cursor within the NotebookOne region and clicking once.

The defaults of the fonts, faces, and sizes used in each of the cell types can be viewed in the **Defaults Window**. This window also includes the graphics size. The window obtained when **Default Window** is selected is shown below with a description of the steps necessary for changing the default size of graphics cells. Changes in the cell styles are accomplished through opening the **Styles Window**.

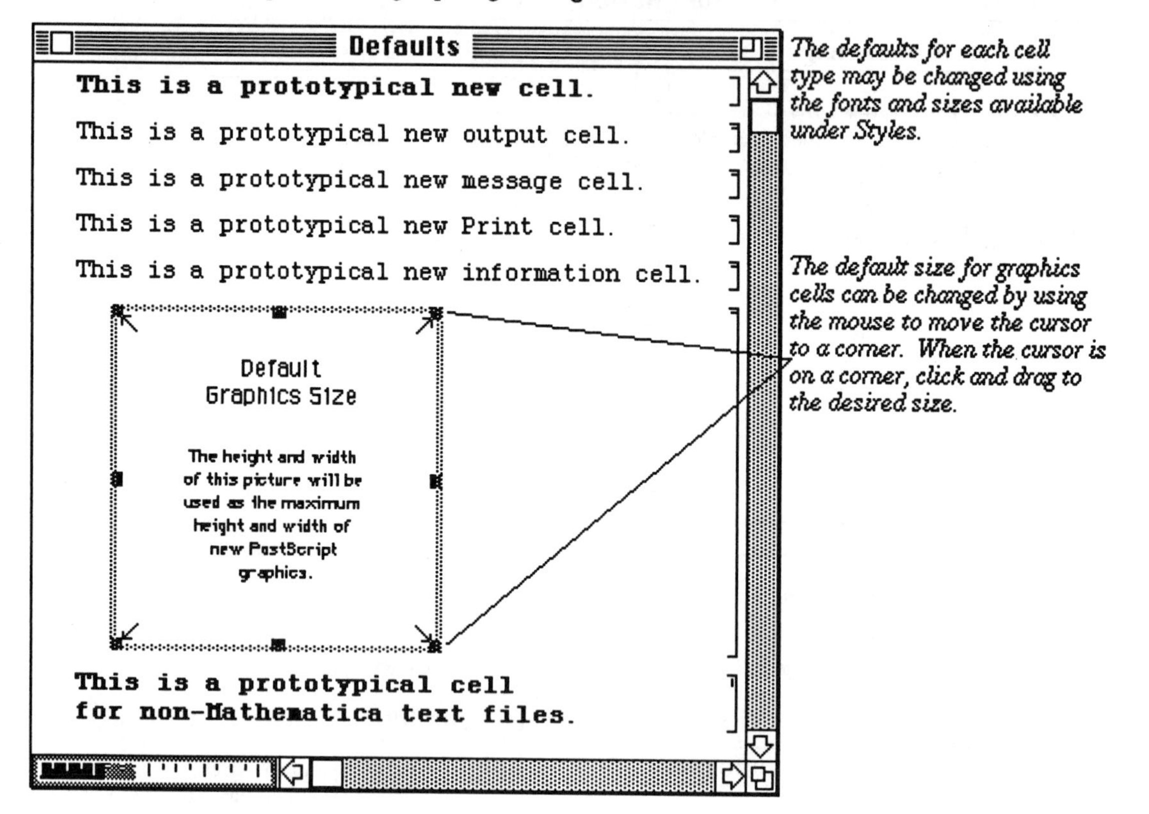

• The Version 2.0 Menu

The *Mathematica* Menu in Version 2.0 appears at first glance to be identical to that of Version 1.2. Upon further inspection, however, the user easily notices that many of the features in Version 1.2 have been rearranged under the menu headings. Version 2.0 also has a **Short Menu** option which is located under **Edit**. The **Long Menu** is displayed if a check mark is placed beside **Long Menu**. This can be changed to **Short Menu** by clicking once on this check mark. Both menus are shown below.

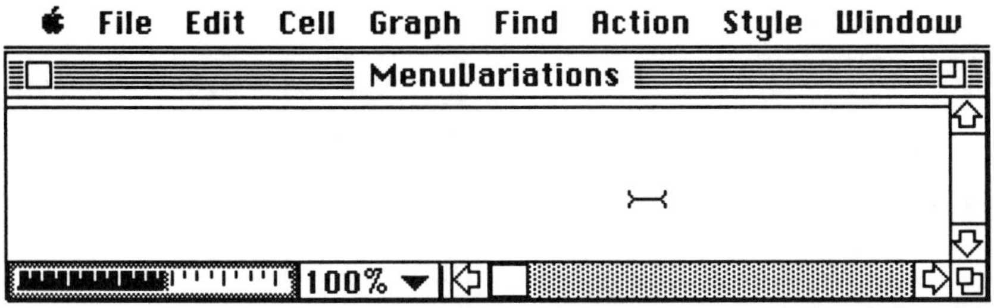

File Edit Cell Graph Find Action Style Window *Long Menu*

MenuVariations

100%

Short Menu File Edit Action Style

MenuVariations

100%

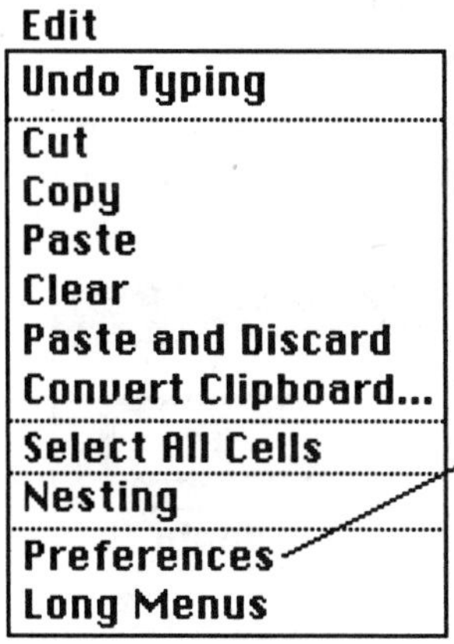

Edit

Undo Typing
Cut
Copy
Paste
Clear
Paste and Discard
Convert Clipboard...
Select All Cells
Nesting
Preferences
Long Menus

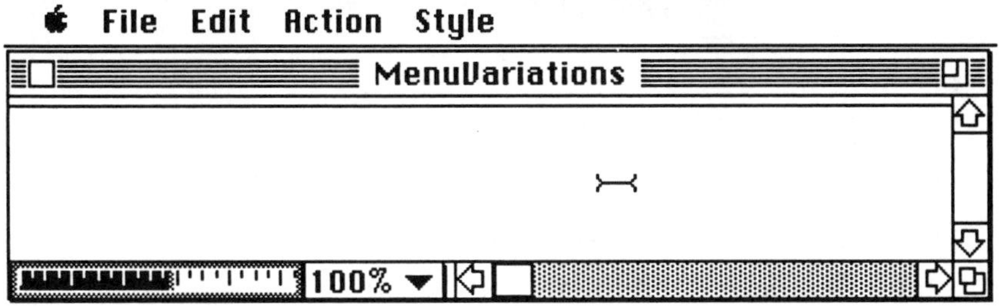

Version 2.0 **Edit** Settings differ somewhat from earlier versions. **Action** and other settings are contained under **Preferences**.

■ Action Settings for Version 2.0:

The **Action Preferences** shown below are the same as those in **Action Settings** in Version 1.2, however.

Action Preferences

☒ Output cells are grouped with input
☐ Multiple output cells are grouped together
☐ After evaluation, input cells are locked
☐ Beep when an evaluation is finished
☐ Display clock timing after each evaluation
☐ Append kernel name to In/Out names

Break ☒ to fit window ☐ at page width
☐ Break at 78 character widths

Generate unformatted text for these results:
○ All ○ None ◉ No graphics or Short

☐ Place Print output as it is generated
☐ Place each Print line in a separate cell

On opening a Notebook, load initialization cells:
○ Always ○ Never ◉ Ask each time

OK Apply Defaults Help Cancel

The Action Settings available in Version 2.0.

The changes in **Startup Preferences** are clear. The option of automatically loading **msg.m**, **info.m**, **IntegralTables.m**, **Elliptic.m**, and **Series.m** at the beginning of each *Mathematica* session is not offered in Version 2.0 as it was in Version 1.2.

The Startup Preferences in Version 2.0 are substantially different from prior versions. In particular, packages such as **IntegralTables.m** *are automatically loaded when the kernel is started.*

Startup Preferences

Stack size (number of KBytes):
Current: 512 Requested: 512

☐ Automatically start a kernel on launch

When automatically starting a kernel,
start the following kernel:
Local Kernel ▾

OK Apply Defaults Help Cancel

Display Preferences are basically the same as **Display Settings** in Version 1.2 with the exception of the cell sizing options which were included in Version 1.2. Version 2.0 includes several options such as Gray Areas jump-scroll and Measurement units, however, which were not found in Version 1.2.

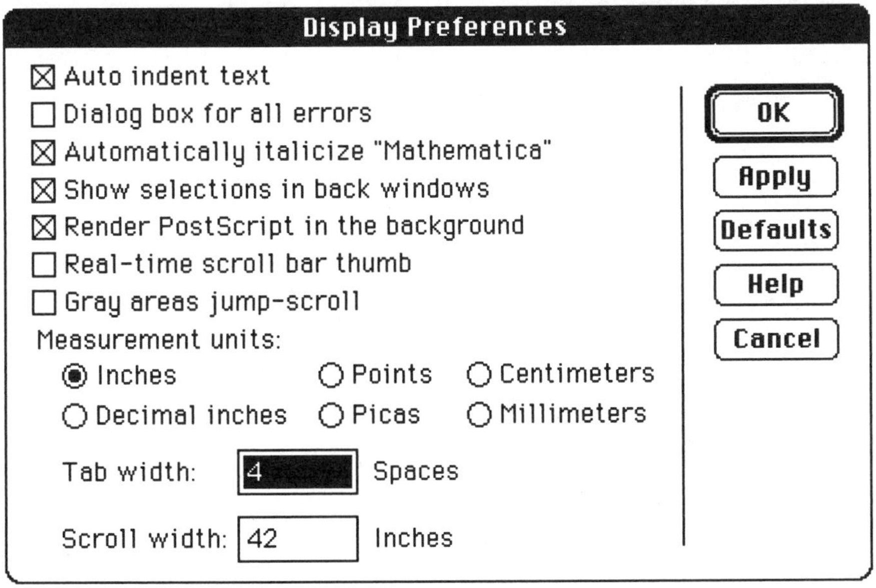

Display Preferences

- ☒ Auto indent text
- ☐ Dialog box for all errors
- ☒ Automatically italicize "Mathematica"
- ☒ Show selections in back windows
- ☒ Render PostScript in the background
- ☐ Real-time scroll bar thumb
- ☐ Gray areas jump-scroll

Measurement units:

◉ Inches	○ Points	○ Centimeters
○ Decimal inches	○ Picas	○ Millimeters

Tab width: 4 Spaces

Scroll width: 42 Inches

OK | Apply | Defaults | Help | Cancel

Display Preferences available in Version 2.0.

Version 2.0 **Cell** settings differ substantially from earlier versions. Notice that **Divide Cell** and **Merge Cells**, which were contained under **Edit** in prior versions, are now contained under **Cell**.

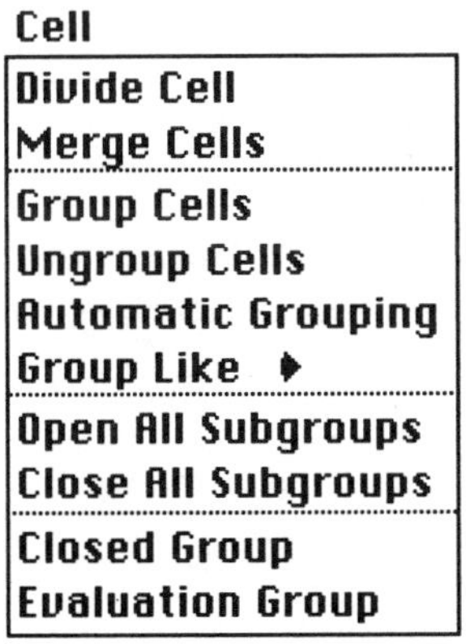

Cell

- Divide Cell
- Merge Cells
- Group Cells
- Ungroup Cells
- Automatic Grouping
- Group Like ▶
- Open All Subgroups
- Close All Subgroups
- Closed Group
- Evaluation Group

The menu which accompanies the **Windows** heading no longer contains the windows for **Network**, **Defaults**, **Styles**, and **Clipboard**. However, the remaining menu members perform the same tasks as those in Version 1.2.

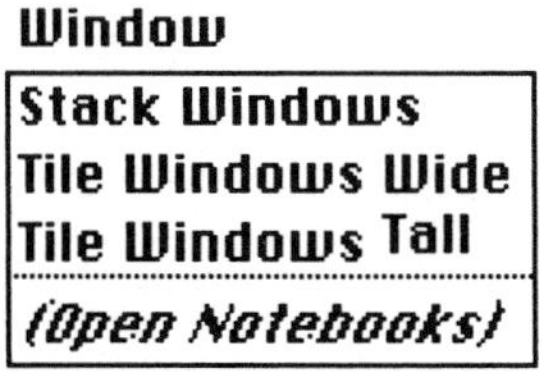

Window

- Stack Windows
- Tile Windows Wide
- Tile Windows Tall
- *(Open Notebooks)*

Stack Windows, **Tile Windows wide**, and **Tile Windows Tall** perform the same task as in Version 1.2.

The main difference in **Action** is found under **Interrupt Calculation**. The window which was displayed with **Interrupt** in Version 1.2 no longer exists. Instead, the options found in this window are listed when **Interrupt Calculation** is selected. These are shown under **Interrupt Calculation** in the window below even though this list is actually hidden until **Interrupt Calculation** is selected.

Action

Action menu
Prepare Input ▶
Edit Connections...
Terminals ▶
Current Kernel ▶
Notebook's Kernel ▶
Connect Remote Kernel...
Quit/Disconnect Kernel
Evaluate Selection
Evaluate Next Input
Evaluate in Dialog
Don't Evaluate
Evaluate Notebook
Evaluate Initialization
Interrupt Calculation
Abort Calculation
Abort to Top
Enter Dialog
Exit Dialog
Auto Save after Each Result

In Version 2.0, a calculation may be aborted by selecting **Action** and then **Abort Calculation**. Version 2.0 is able to stop calculations much faster than prior version. Also note that if the option **Auto Save After Each Result** is checked, then *Mathematica* will save the file after each calculation. When lengthy calculations are being performed and there is fear that the computer may crash, this option can often help avoid heartache.

- While *Mathematica* is performing calculations, Version 2.0 users can select **Enter Dialog** which pauses the current calculation and allows the user to perform other calculations. **Exit Dialog** causes the suspended calculation to resume.

Style

Cell Style ▸
Attributes ▸
Font ▸
FaceSize ▸
Leading ▸
AlignmentText Color ▸
Backgroung Color ▸
Page Breaks ▸
Formatter ▸
Evaluator ▸
Show Ruler
Edit Styles...
Uniform Style
All Default Styles...

In earlier versions, **Page Break** was contained under **Cell**; now it is contained under **Style**.

Similarly, **Edit Styles** was contained under **Windows** as **Styles Window**. In Version 2.0, the styles may be changed by selecting **Styles** and then **Edit Styles**.

The menu under **Style** has several changes although many of these changes are only in appearance. **Attributes** contains the option **Dingbats...** which allows for the use of symbols in creating a notebook. These symbols include the circle and block which were found in the section and subsection styles in Version 1.2. Also under **Attributes** are many of the options such as **Formatted**, **Locked**, **Closed**, and **Fixed Height** found under **Cell** in Version 1.2. Useful additions located under **Face** include **Superscript**, **Subscript**, and **Overstrike**. Another obvious change is **Show Ruler**. If this is chosen, then a ruler which includes three alignment options (left, right, and center) is displayed in the notebook. The **Styles Window** which no longer appears under **Windows** is opened by selecting **Edit Styles...** in Version 2.0.

10.4 Some Common Errors and Their Remedies

Learning to recognize and correct errors will alleviate many of the frustrations that some first-time users encounter when working with *Mathematica* and will enable the user to make the most of *Mathematica*'s vast capabilities. Some of the more common errors and their remedies are illustrated below.

o A list of all *Mathematica* warning messages is contained in Technical Report: *Mathematica* Warning Messages, by David Withoff which is included in the *Mathematica* packaging box.

One of the most commonly made mistakes occurs when using built -in *Mathematica* commands, functions, or constants. The user must always remember to use square brackets and/or capital letters. Several examples are shown below which demonstrate these types of errors.
In the first example, *Mathematica* interprets the command `sin(pi/2)` as `sin*(pi/2)` which certainly was not intended. The user also failed to capitalize the Sine function and the constant `Pi`.
The second example demonstrates that even if capital letters are used correctly, the absence of square brackets yields almost the same output as the first command. In this case, however, the expressions are capitalized.
Finally, the third example shows that when square brackets are used, *Mathematica* interprets the command correctly and gives the exact value of `Sin[Pi/2]`which, of course, is 1.

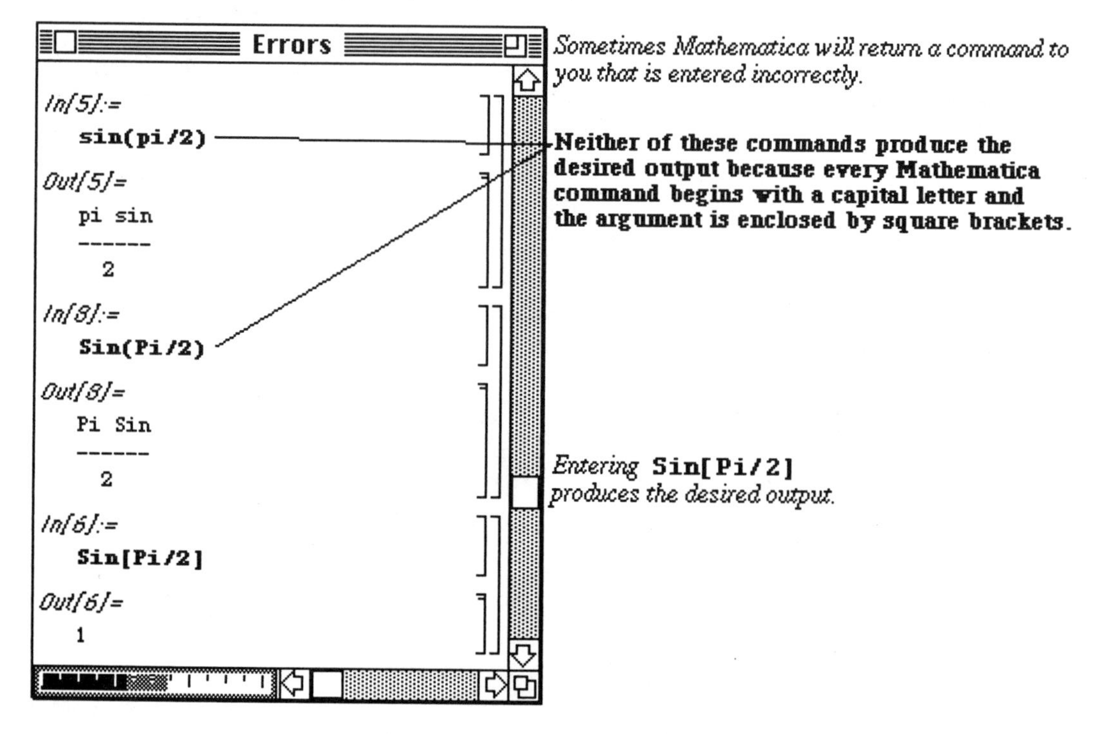

Error messages can be disturbing to receive. In some cases, however, they can be ignored. For example, in the following window, the user attempts to load a package, **`IntegralTables.m`**, which has already been loaded. Hence, the error messages are given, but they may be deleted.

o Version 2.0 users need never load the package **`IntegralTables.m`**. Neverthess, if one reads in a package and then reads it in again, messages of this sort often appear.

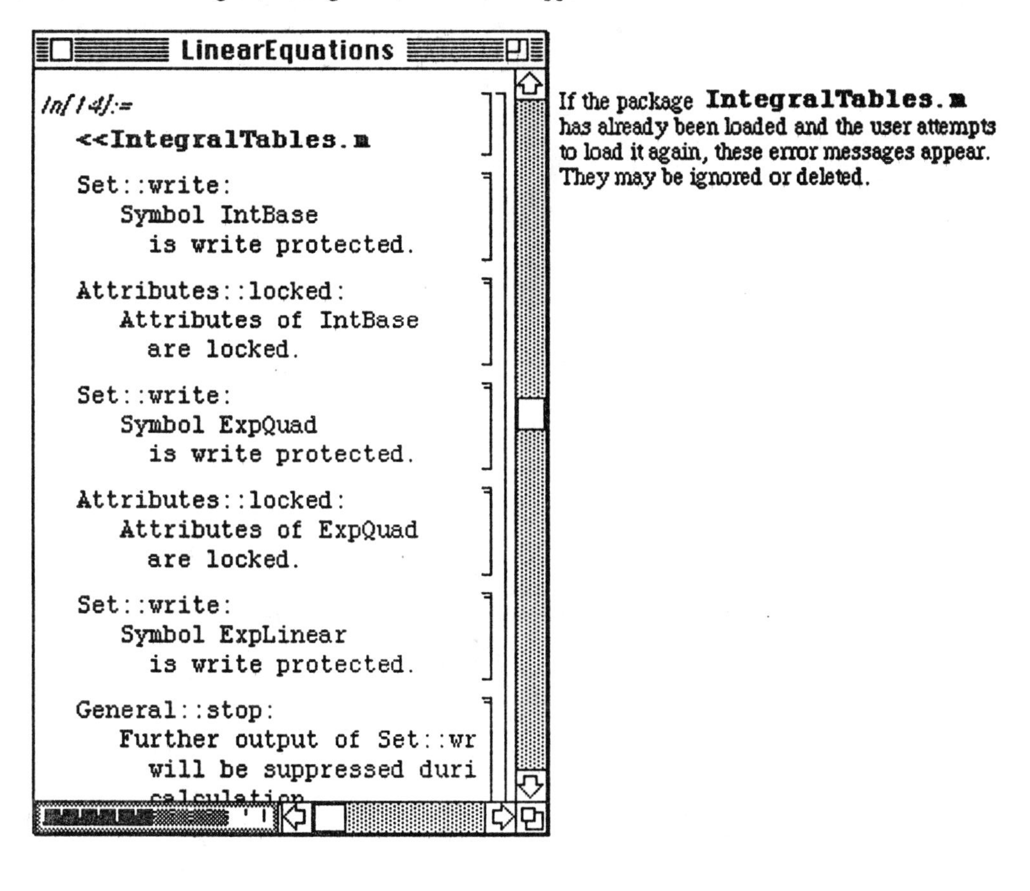

If the package **`IntegralTables.m`** has already been loaded and the user attempts to load it again, these error messages appear. They may be ignored or deleted.

The following error is quite common to new *Mathematica* users. The user attempts to plot a function **g** without properly defining the function beforehand. Hence, there is no function to graph. After receiving the error message, an easy way to check that this is the problem is to use **?g** . If the function is undefined, the output is simply g as shown below. Otherwise, the formula for **g** would be displayed.

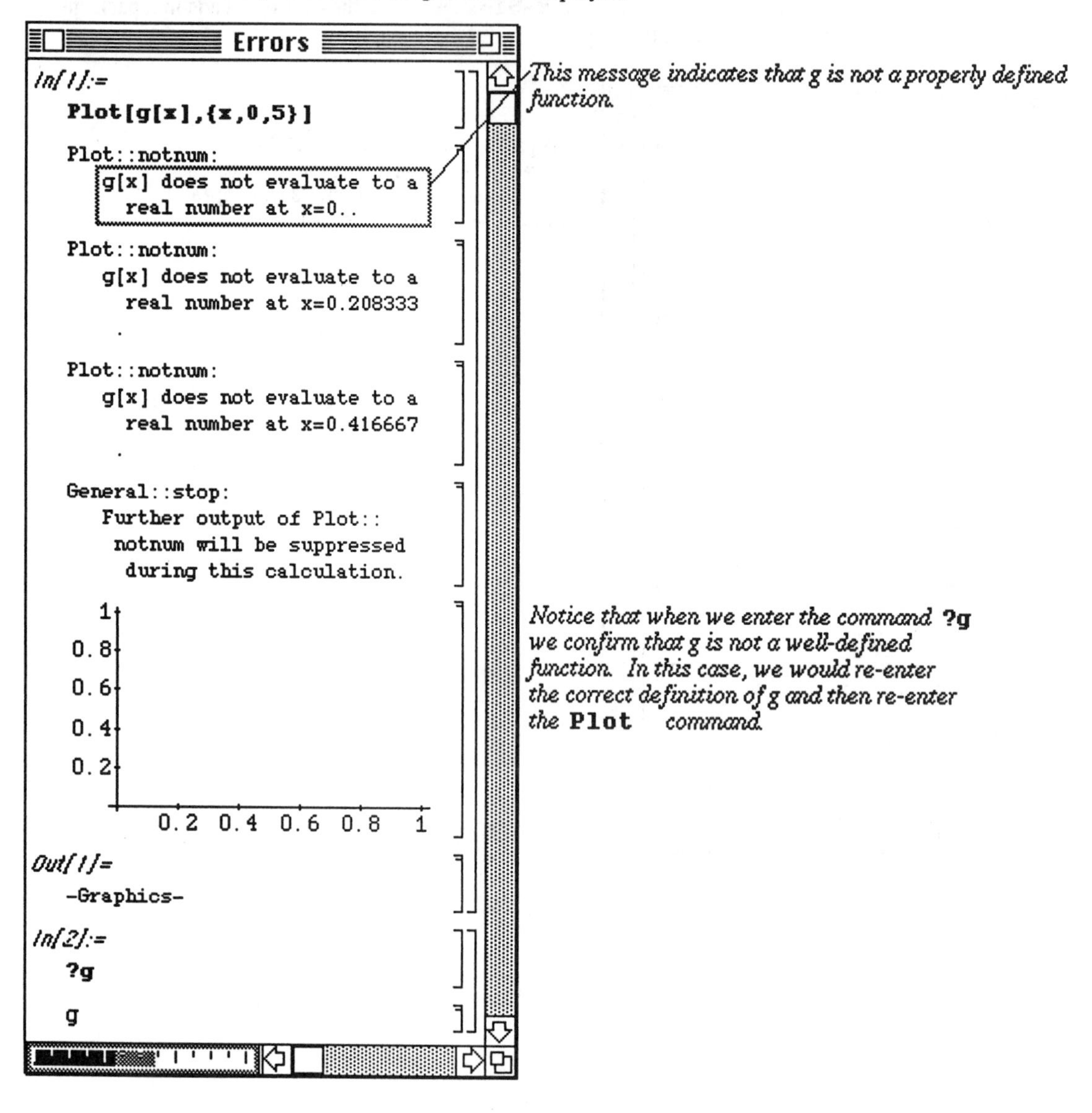

Another common mistake occurs when trying to work with the elements of a table produced with two indices. In the example below, a table of Legendre polynomials is formed in **lps**. Note that the output is in the form of a list in which every element is itself a list of two elements. (*Mathematica* computes the polynomials in pairs, one pair for each value of **n**.) Therefore, the **Plot** command as it is stated below cannot plot the members of **lps**.

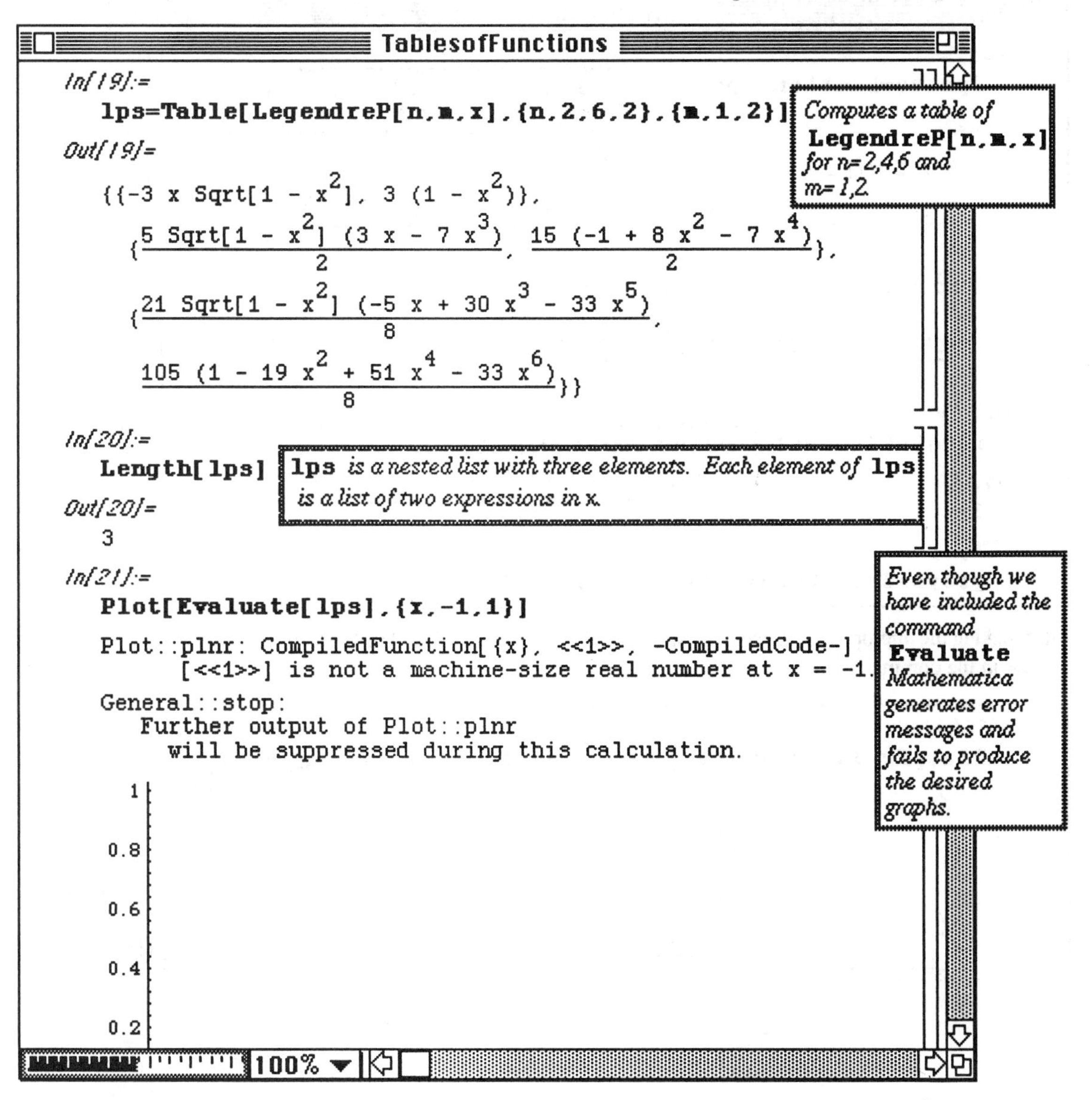

To remedy the problem, **Flatten[lps]** must be used before trying to plot the Legendre polynomials in **lps**. Notice that **Flatten[lps]** removes the inner brackets contained in **lps** and converts it to a list of length 6 called **lpstwo**. A table of **GreyLevel** values is created in **grays**.

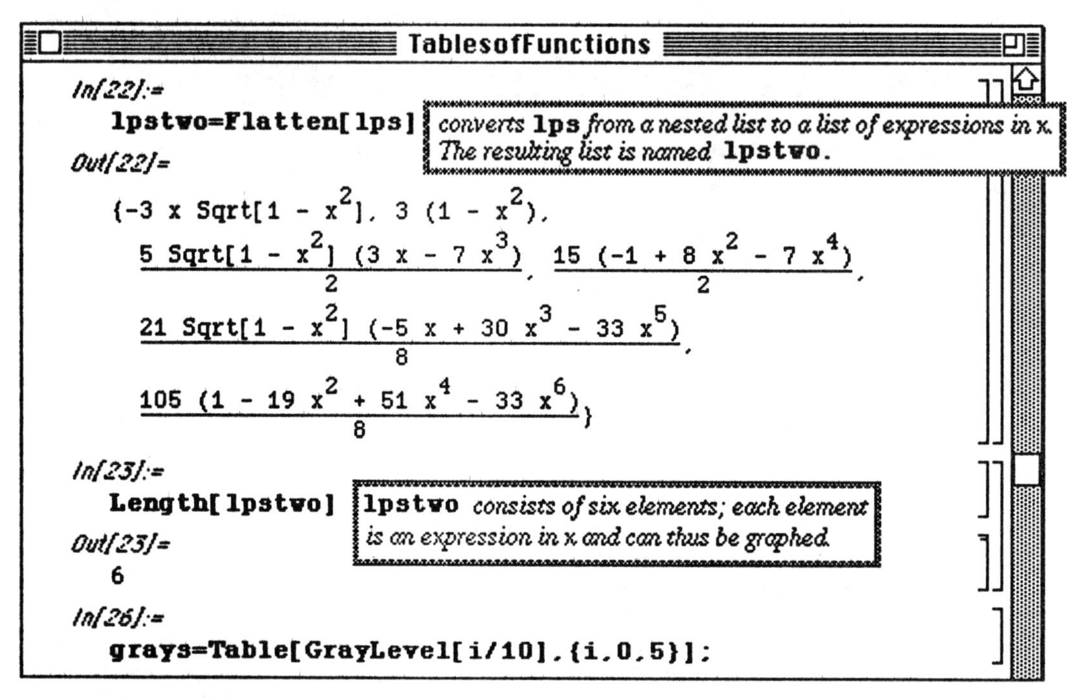

After making the appropriate changes, the six Legendre polynomials found in **lpstwo** are correctly plotted using **grays** in the command below.

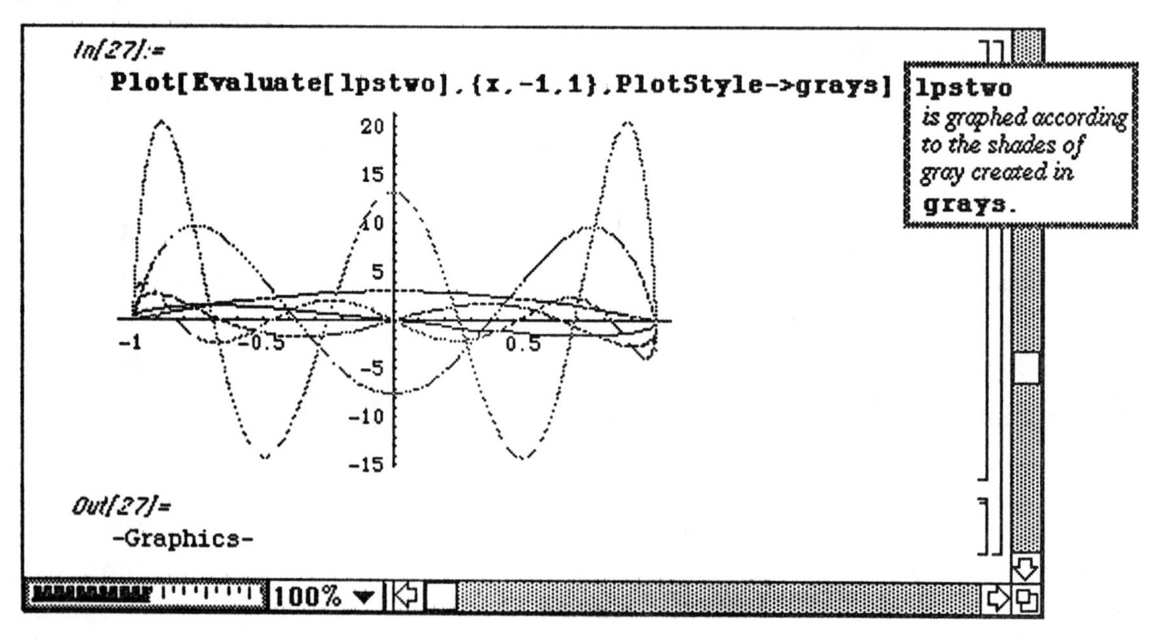

Throughout Mathematica by Example, user-defined functions have always been defined using lower-case letters. Since every built-in command begins with a capital letter, we have been sure to avoid any ambiguity with built-in functions. Nevertheless, if one does attempt to define a function that conflicts with a built-in function, errors like the following result:

In this case, we have attempted to define Sin(x) = x^2 Cos(x). *The definition conflicts with the built-in function* **Sin[x]** *Consequently, the user-definition is refused and* **Sin[Pi/2]** *returns the "correct" value.*

10.5 Additional References

Additional References regarding Macintosh Computers:

- Getting Started With Your Macintosh, Apple Computer, Inc.
- Macintosh Reference, AppleComputer, Inc.

Additional references regarding *Mathematica*:

- Crandall, Richard E., *Mathematica* for the Sciences, Addison-Wesley Publishing Co. (1991);
- Gray, Theodore and Glynn, Jerry, Exploring Mathematics with *Mathematica*, Addison-Wesley Publishing Co. (1991);
- Maeder, Roman, Programming in *Mathematica*, Addison-Wesley Publishing Co. (1990);
- Wagon, Stan, *Mathematica* in Action, W. H. Freeman and Co. (1991);
- Wolfram Research, Inc., *Mathematica*: A System for Doing Mathematics by Computer, User's Guide for the Macintosh (1991);
- Wolfram, Stephen, *Mathematica*: A System for Doing Mathematics by Computer, Addison-Wesley Publishing Co. (1988);
- Wolfram, Stephen, *Mathematica*: A System for Doing Mathematics by Computer, Second Edition, Addison-Wesley Publishing Co, (1991); and
- The *Mathematica* Journal, published quarterly by the Advanced Book Program, Addison-Wesley Publishing Co.

Additional references regarding the mathematical topics that appeared in *Mathematica* by Example:

- Arnold, Steven F., Mathematical Statistics, Prentice-Hall (1990);
- Cheney, Ward and Kincaid, David, Numerical Mathematics and Computing, Second Edition, Brooks/Cole Publishing Co. (1985);
- Hillier, Frederick S. and Lieberman, Gerald L., Introduction to Operations Research, Fifth Edition, McGraw-Hill Publishing Co. (1990);
- Jordan, D. W. and Smith, P., Nonlinear Ordinary Differential Equations, Second Edition, Oxford University Press (1988);
- Kreyszig, Erwin, Advanced Engineering Mathematics, Sixth Edition, John Wiley & Sons (1988);
- Powers, David L., Boundary Value Problems, Second Edition, Academic Press (1979);
- Strang, Gilbert, Linear Algebra and its Applications, Third Edition, Harcout Brace Jovanovich, Publishers (1988);
- Weiss, Neil A. and Hassett, Matthew J., Introductory Statistics, Second Edition, Addison Wesley Publishing Co. (1991); and
- Wilson, R. J., and Watkins, J. J., Graphs: An Introductory Approach, John Wiley & Sons (1990).

Appendix
Introduction to Programming in *Mathematica*

The **Appendix** provides a brief introduction to programming in *Mathematica*. Examples include some of the programs that were used to create some of the graphics objects in Mathematica By Example. However, users that intend to become proficient *Mathematica* programmers should refer to Maeder's book Programming in Mathematica.

In Version 1.2, local variables are declared using the command **Block**. The following example illustrates the use of local variables within a **Block**. Notice that changes in the local variable **j** in the function **value** do not affect the value of the previously defined global variable **j** = 0. This value remains zero although the local variable **j** has value 4.

- In Version 2.0, the command **Module** replaces the command **Block**; although **Block** is still supported under Version 2.0.

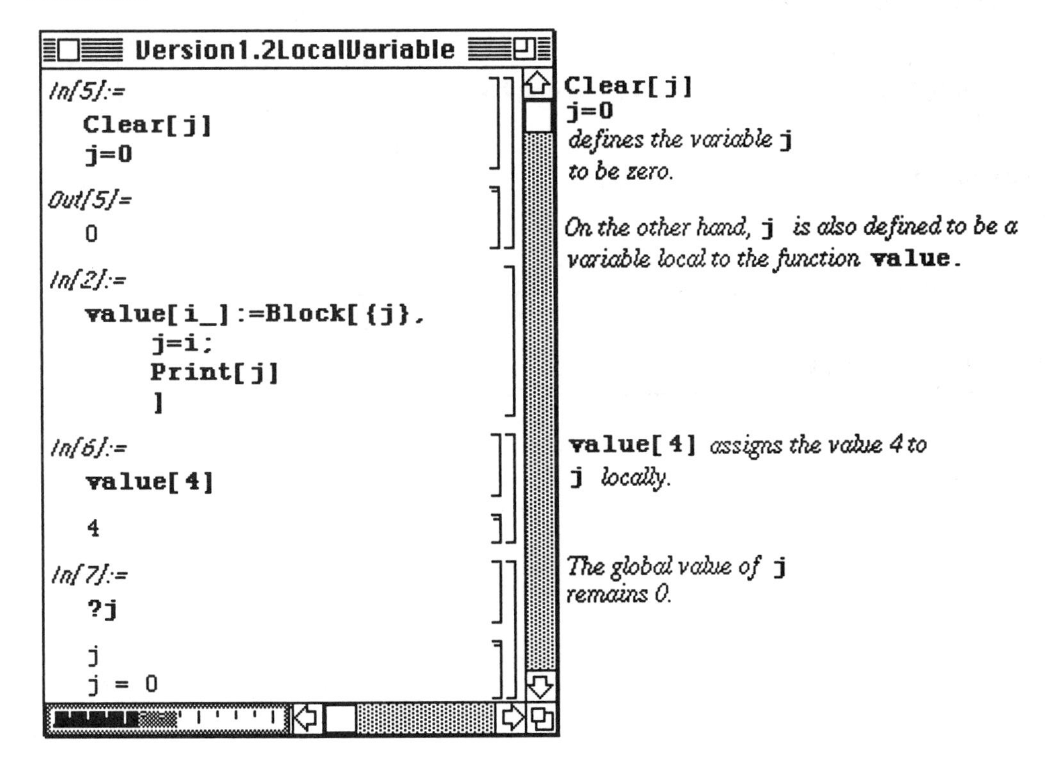

Clear[j]
j=0
defines the variable **j** *to be zero.*

On the other hand, **j** *is also defined to be a variable local to the function* **value**.

value[4] *assigns the value 4 to* **j** *locally.*

The global value of **j** *remains 0.*

Functions can be defined to perform various tasks using *Mathematica* programming skills. This is shown below with the definition of the function **arclength** which calculates the length of the curve given by **r** over the interval from t = **a** to t = **b**. This function depends on the local variables **rprime**, **length**, and **integrand** which give the derivative of **r**, the number of components of **r**, and the integrand given in the integral formula to determine arc length, respectively. A particular function **r[t]** is then defined to illustrate the use of **arclength**.

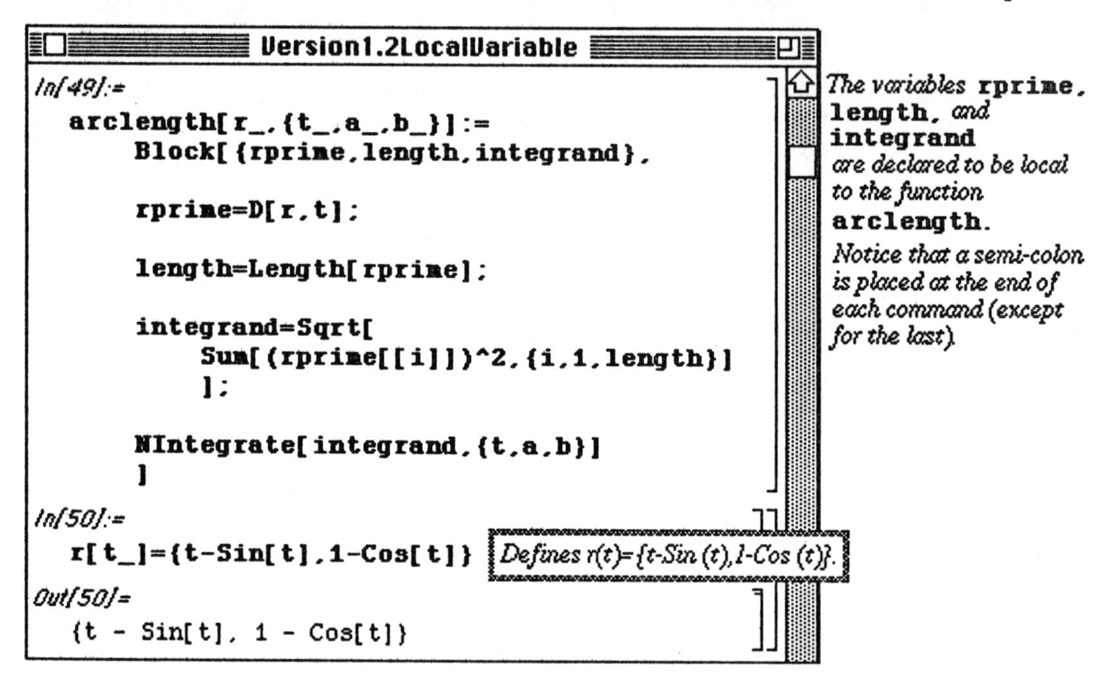

To better understand the use of **arclength**, the function **r[t]** is plotted below from **t = 0** to **t = 4Pi**. Then, the length of this curve is determined with **arclength**. Note that the dependent variable is not of importance in the use of this function as **s** is used in this command instead of the variable **t**.

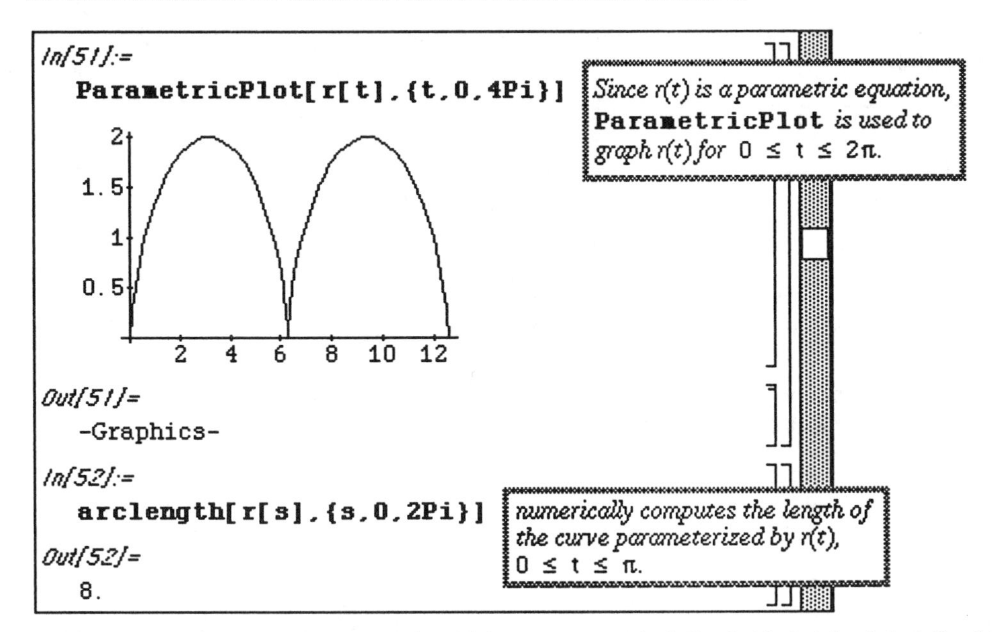

Another point of interest is the manner in which **arclength** is defined. Notice that it is defined in terms of the number of components in the function **r**. Hence, **arclength** can be used with functions of more than two components as illustrated below with the function **v[t]**.

After plotting this curve with **SpaceCurve** (located in the package **ParametricPlot3D.m**), the length of this curve in three-dimensions is found with **arclength**. Note that **Release** must be used with **v[t]** in order for it to be evaluated at various values of **t** in the **SpaceCurve** command.

o In Version 2.0, the command **Release** is replaced by the command **Evaluate**; **SpaceCurve** is replaced by **ParametricPlot3D**.

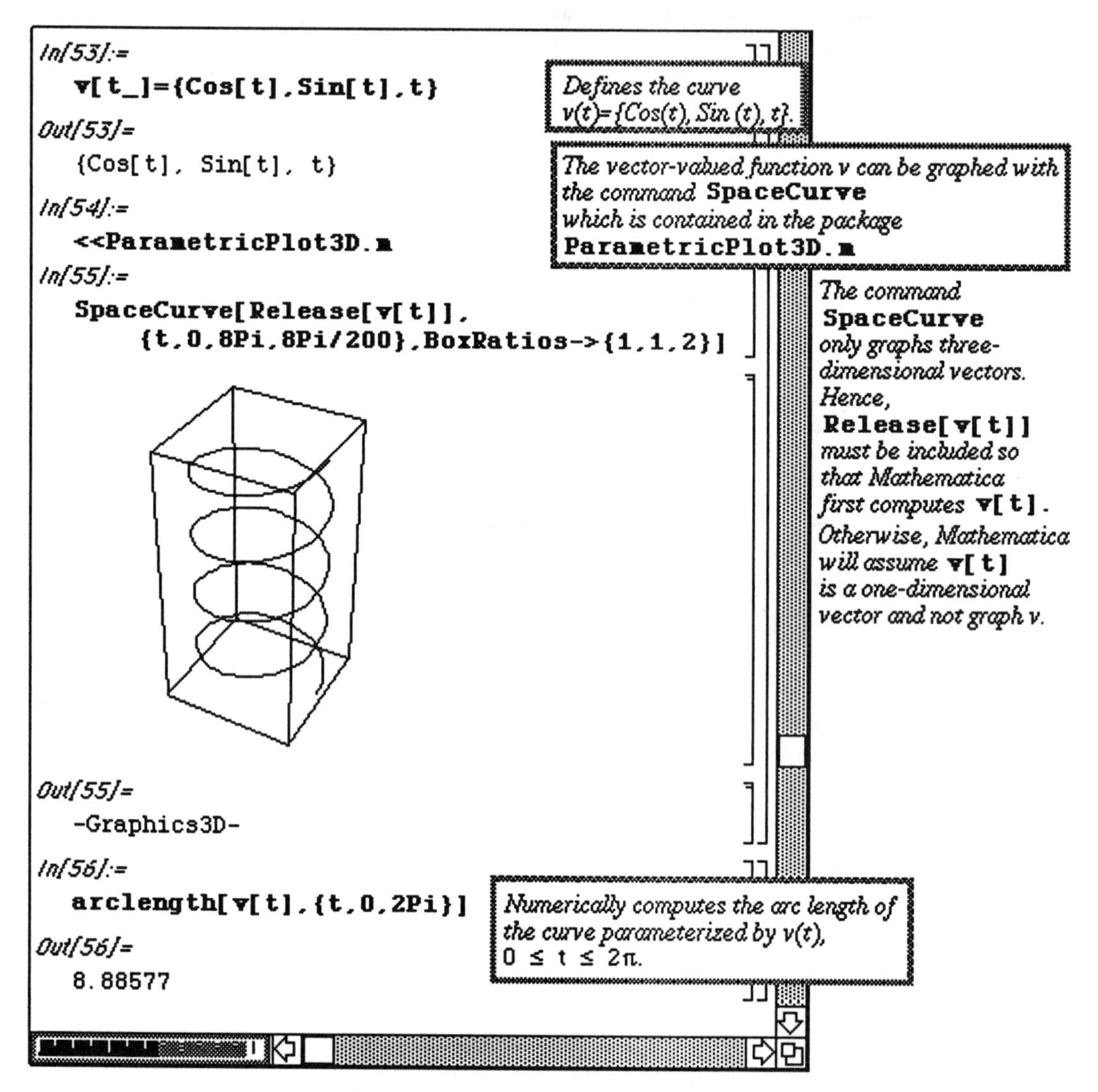

Mathematica also includes several typical programming techniques. These include the **If** statement and the **Do** loop. Before illustrating these ideas, however, several built-in *Mathematica* commands must be introduced. These include **Divisors[n]** which lists all divisors of the integer **n**, including **n**, and **Drop[list,-1]** which deletes the last element of **list** and returns the resulting list. In the example below, the divisors of 6 are computed with **Divisors[6]**. This list is called **div6**. Next, the last term in **div6** is removed with **Drop[div6,-1]** and the resulting list named **divs**. Finally, the sum of the elements of **divs** is found with **Apply[Plus,divs]**. Similar steps will be used in the example which follows.

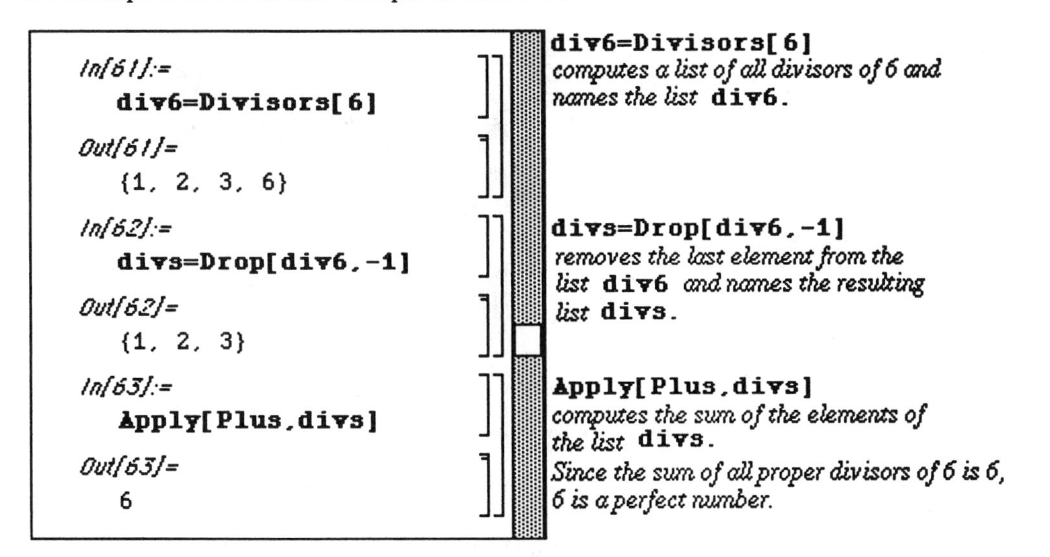

The calculations previously discussed can be used to find perfect numbers. Recall that a number **n** is perfect if the sum of its divisors (not including **n**) equals the number **n** itself. A function **perfectq[n]** is defined below using the steps illustrated above. Note that there is an **If** statement within this function. Syntax for an **If** statement is **If[condition,then,else]**. In the case of the function below, if the sum of the divisors is **n**, then a value of **yes** is assumed while a **no** is assumed otherwise. Next, the function **printp[j]** is defined to print a number if it is perfect. Finally, a **Do** loop is used to find all of the perfect numbers between 1 and 10,000. Note that the loop **Do[expression,{i,imin,imax}]** evaluates **expression** from **i = imin** to **i = imax**.

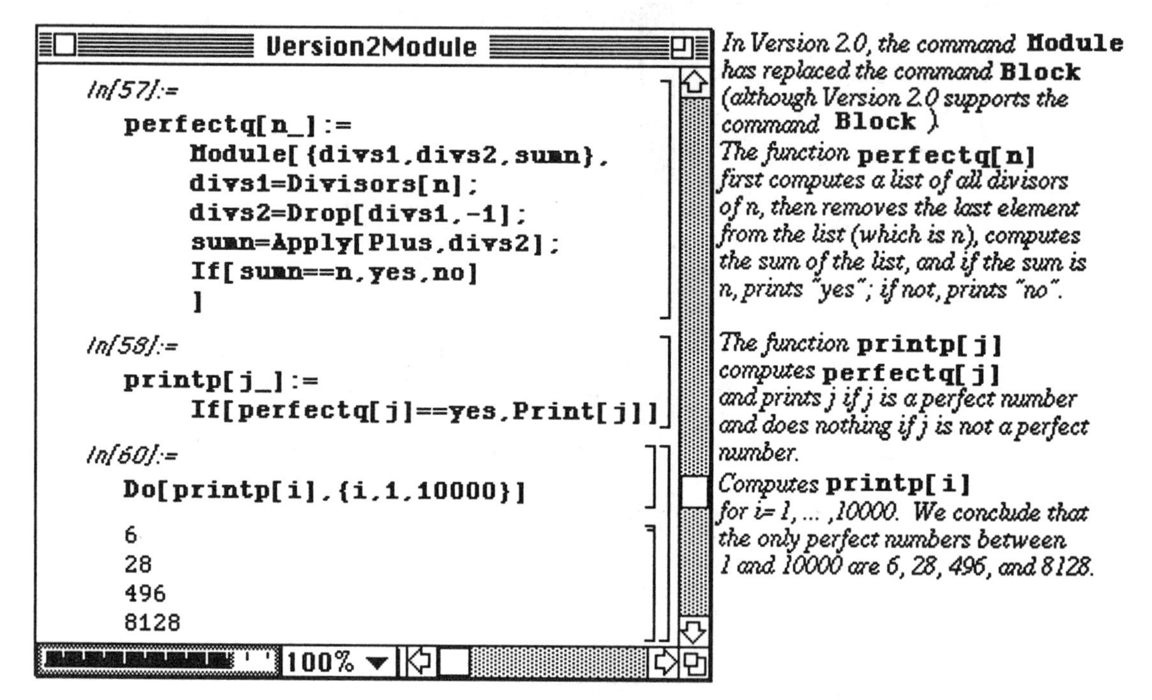

□ **Example:**

The command **solidrev** was used to create the solids of revolution found in **Chapters 3** and **6**. A brief description of **solidrev** is given below. Notice that the arguments of this command include the function **f**, the domain **{a,b}**, the axis about which **f** is to be revolved (either **xaxis** or **yaxis**), and the **solid** option which graphs the resulting solid of revolution.

```
solidrev::usage="solidrev[f,{a,b},axis]
yields a three-dimensional meshed
image of the function f[x] defined
on the domain [a,b] revolved about
the xaxis or yaxis.solidrev[f,{a,b},axis,solid]
yields a solid surface.
The interval [a,b] is automatically
divided into 10 subintervals.  This
may be changed by substituting {a,b,n} for {a,b}
where n is the desired number of subintervals."
```

```
solidrev[f_,{a_,b_,m_:Automatic},axis_,ll_:Automatic]:=
Block[

     {n,ll,xaxis,yaxis,un,list1,s,t,list2,q,list4,poly},
     xaxis=0;
     yaxis=1;
     uu=axis;

     If[m===Automatic,n=10,n=m];

     list1=Table[{x,f[x]},{x,a,b,(b-a)/n}] // N;

     list2=If[uu==0,
             s[{x_,y_}]:=Table[{x ,y sincostab[[i,2]],
                  y sincostab[[i,1]]},{i,1,Length[sincostab]}];
             Map[s, list1],

             t[{x_,y_}]:=Table[{x sincostab[[i,1]],
                  x sincostab[[i,2]],y},{i,1,Length[sincostab]}];
             Map[t, list1]
             ];

     un[k_]:=Partition[k,2,1];

     list3=Map[un,list2];

     q[i_,j_]:=Join[list3[[i,j]],Reverse[list3[[i+1,j]]]];

     list4=Flatten[Table[q[i,j],{j,1,Length[list3[[1]]]},
          {i,1,Length[list3]-1}],1];

     poly=If[ll===Automatic,Map[Line,list4],Map[Polygon,list4]];

     Show[Graphics3D[poly]]
];
```

□ Example of solidrev:

An illustration of the use of **solidrev** is given below. A function **f** is first defined and plotted.

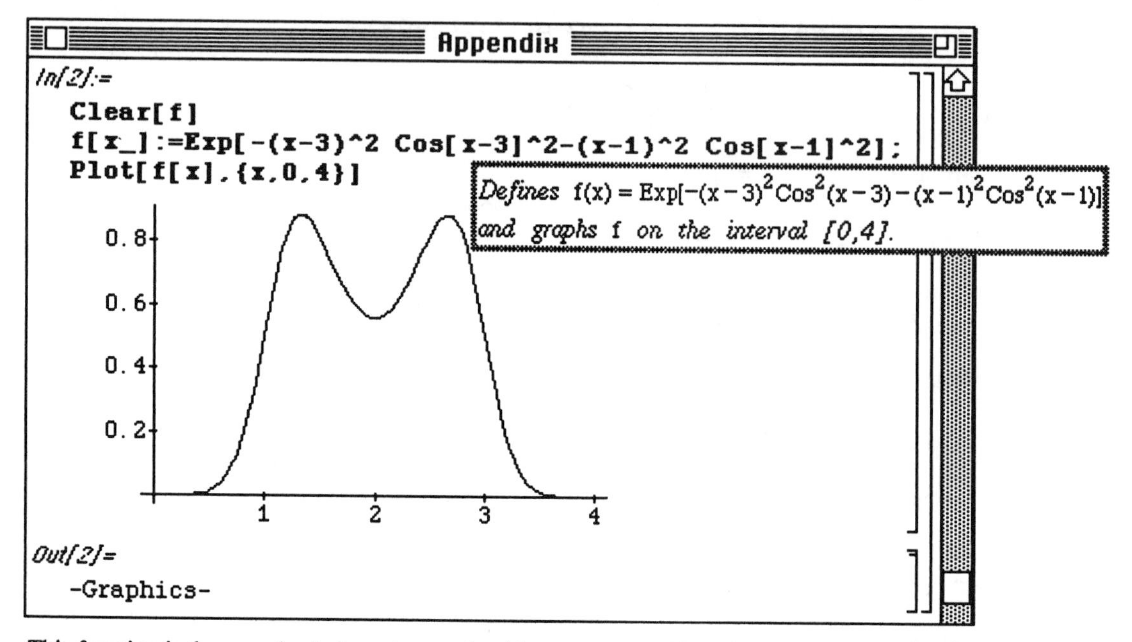

This function is then revolved about the x-axis with **solidrev**. Note the manner in which the arguments are entered in **solidrev**.

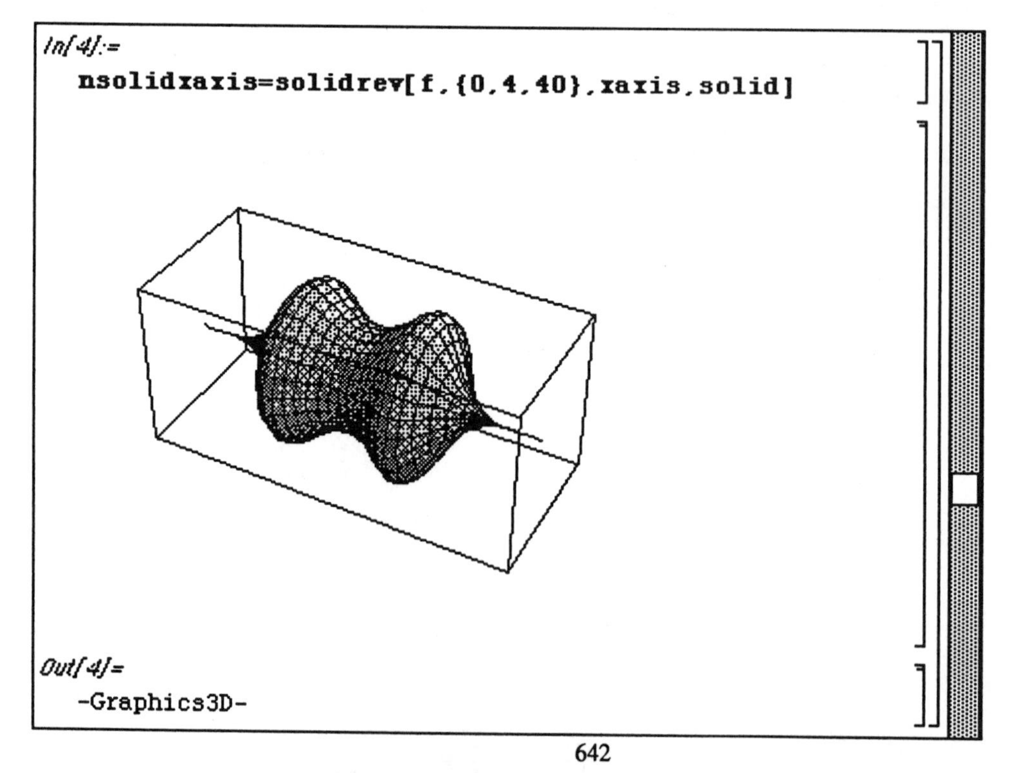

The same function is then revolved about the y-axis.

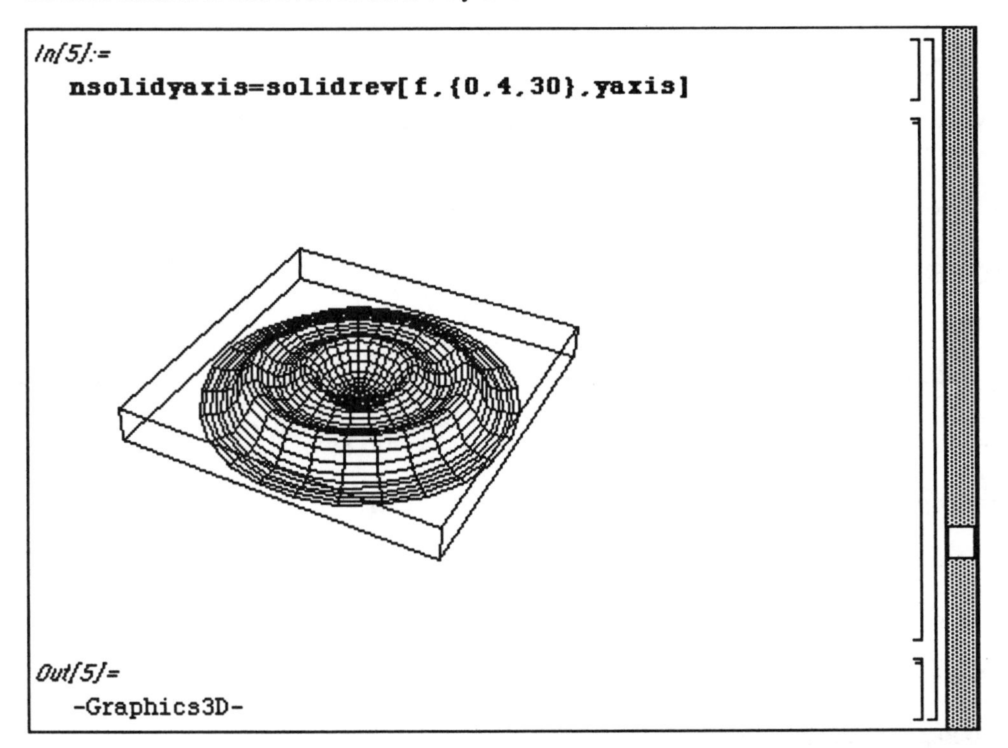

Some other graphics which were illustrated earlier in *Mathematica* By Example without explanation occurred in the section on Lagrange multipliers. These graphics were produced with the function **lagrangem** below. This function graphs the curves **f** and **g** for values of x over the interval **xmin** to **xmax** by evaluating each function at **n** points and joining the points obtained with line segments. Graphs of this type are useful in giving a geometrical impression of where the optimum values of **f** occur subject to the constraint **g**.

□ **Example:**

```
lagrangem[{f_,g_,cc_:0},
     {xmin_,xmax_,n_:15},u_:{1,1,.4}]:=

     Block[{values,graphf,graphg,y,ycoords,t1,t2,coords,ff,
          fpoints},

     values=Table[N[xmin +i(xmax-xmin)/n],{i,0,n}];

     y[k_]:=NRoots[g[k,y]==0,y];

     ycoords=Map[y,values];

     t1=Table[{values[[i]],ycoords[[i,1,2]],cc},{i,1,n+1}];

     t2=Table[{values[[i]],ycoords[[i,2,2]],cc},{i,1,n+1}];

     graphg1=Graphics3D[{GrayLevel[.3],Line[t1]}];

     graphg2=Graphics3D[{GrayLevel[.3],Line[t2]}];

     ff[{a_,b_,c_}]:={a,b,f[a,b]};

     fpoints1=Map[ff,t1];

     fpoints2=Map[ff,t2];

     graphf1=Graphics3D[Line[fpoints1]];

     graphf2=Graphics3D[Line[fpoints2]];

     Show[graphf1,graphf2,graphg1,graphg2,Axes->Automatic,

          Boxed->False,BoxRatios->u,ViewPoint->{3.880,0.950,2.220}]
     ]
```

□ Example of lagrangem:

The function **lagrangem** is illustrated below with the two functions, **f** and **g**. This is done over the interval from -2 to 2 using 100 points. Notice where the maximum and minimum values of the function **f** occur.

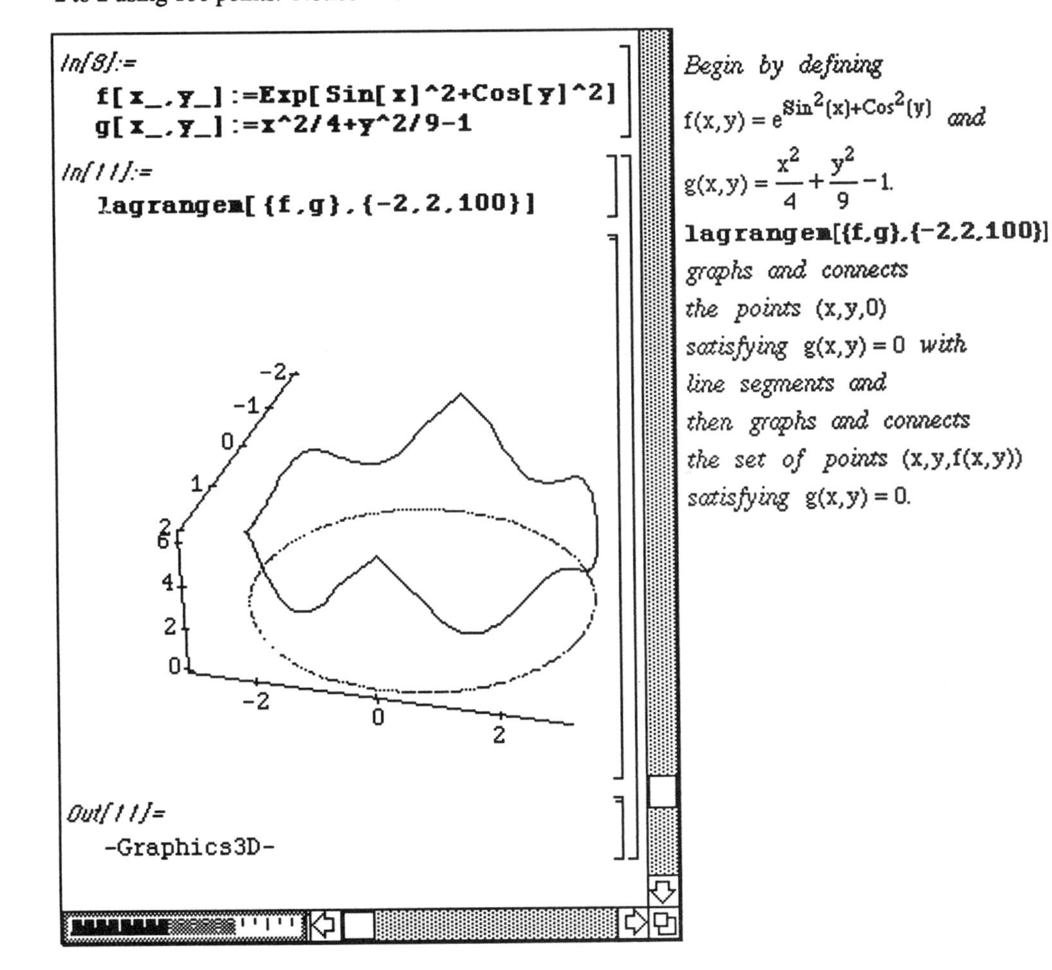

Begin by defining $f(x,y) = e^{Sin^2(x)+Cos^2(y)}$ *and* $g(x,y) = \frac{x^2}{4}+\frac{y^2}{9}-1.$ **lagrangem[{f,g},{-2,2,100}]** *graphs and connects the points* (x,y,0) *satisfying* g(x,y) = 0 *with line segments and then graphs and connects the set of points* (x,y,f(x,y)) *satisfying* g(x,y) = 0.

Index

(o) implies obsolete in Version 2.0;
(2) implies applicable only to Version 2.0.

Index

(o) implies obsolete in Version 2.0;
(2) implies applicable only to Version 2.0.

Index

(o) implies obsolete in Version 2.0;
(2) implies applicable only to Version 2.0.

Index

(o) implies obsolete in Version 2.0;
(2) implies applicable only to Version 2.0.

Index

(o) implies obsolete in Version 2.0;
(2) implies applicable only to Version 2.0.

Index

(o) implies obsolete in Version 2.0;
(2) implies applicable only to Version 2.0.

Q

R

S

Index

(o) implies obsolete in Version 2.0;
(2) implies applicable only to Version 2.0.

(o) implies obsolete in Version 2.0;
(2) implies applicable only to Version 2.0.

W